Linear and Nonlinear Optimization

Linear and Nonlinear Optimization

SECOND EDITION

Igor Griva
Stephen G. Nash
Ariela Sofer

George Mason University
Fairfax, Virginia

siam ® Society for Industrial and Applied Mathematics • Philadelphia

Library of Congress Cataloging-in-Publication Data

Griva, Igor.
  Linear and nonlinear optimization / Igor Griva, Stephen G. Nash, Ariela Sofer. -- 2nd ed.
      p. cm.
  Includes bibliographical references and index.
  ISBN 978-0-898716-61-0
  1. Linear programming. 2. Nonlinear programming. I. Nash, Stephen (Stephen G.) II. Sofer, Ariela. III. Title.
  T57.74.G75 2008
  519.7'2--dc22

                                                                    2008032477

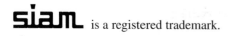

Contents

Preface

This book provides an introduction to the applications, theory, and algorithms of linear and nonlinear optimization. The emphasis is on practical aspects—modern algorithms, as well as the influence of theory on the interpretation of solutions or on the design of software. Two important goals of this book are to present linear and nonlinear optimization in an integrated setting, and to incorporate up-to-date interior-point methods in linear and nonlinear optimization.

As an illustration of this unified approach, almost every algorithm in this book is presented in the form of a General Optimization Algorithm. This algorithm has two major steps: an optimality test, and a step that improves the estimate of the solution. This framework is general enough to encompass the simplex method and various interior-point methods for linear programming, as well as Newton's method and active-set methods for nonlinear optimization. The optimality test in this algorithm motivates the discussion of optimality conditions for a variety of problems. The step procedure motivates the discussion of feasible directions (for constrained problems) and Newton's method and its variants (for nonlinear problems).

In general, there is an attempt to develop the material from a small number of basic concepts, emphasizing the interrelationships among the many topics. Our hope is that, by emphasizing a few fundamental principles, it will be easier to understand and assimilate the vast panorama of linear and nonlinear optimization.

We have attempted to make accessible a number of topics that are not often found in textbooks. Within linear programming, we have emphasized the importance of sparse matrices on the design of algorithms, described computational techniques used in sophisticated software packages, and derived the primal-dual interior-point method together with the predictor-corrector technique. Within nonlinear optimization, we have included discussions of truncated-Newton methods for large problems, convergence theory for trust-region methods, filter methods, and techniques for alleviating the ill-conditioning in barrier methods. We hope that the book serves as a useful introduction to research papers in these areas.

The book was designed for use in courses and course sequences that discuss both linear and nonlinear optimization. We have used consistent approaches when discussing the two topics, often using the same terminology and notation in order to emphasize the similarities between the two topics. However, it can also be used in traditional (and separate) courses in Linear Programming and Nonlinear Optimization—in fact, that is the way we use it in the courses that we teach. At the end of this preface are chapter descriptions and course outlines indicating these possibilities.

We have also used the book for more advanced courses. The later chapters (and the later sections within chapters) contain a great deal of material that would be difficult to cover in an introductory course. The Notes at the ends of many sections contain pointers to research papers and other references, and it would be straightforward to use such materials to supplement the book.

The book is divided into four parts plus appendices. Part I (Basics) contains material that might be used in a number of different topics. It is not intended that all of this material be presented in the classroom. Some of it might be irrelevant (as the sample course outlines illustrate). In other cases, material might be familiar to the students from other courses, or simple enough to be assigned as a reading exercise. The material in Part I could also be taught in stages, as it is needed. In a course on Nonlinear Optimization, for example, Chapter 4 (Representation of Linear Constraints) could be delayed until after Part III (Unconstrained Optimization). Our intention in designing Part I was to make the book as flexible as possible, and instructors should feel free to exploit this flexibility.

Part II (Linear Programming) and Part III (Unconstrained Optimization) are independent of each other. Either one could be taught or read before the other. In addition, it is not necessary to cover Part II before going on to Part IV (Nonlinear Optimization), although the material in Part IV will benefit from an understanding of Linear Programming. The material in the appendices may already be familiar. If not, it could either be presented in class or left for students to read independently.

Many sections in the book can be omitted without interrupting the flow of the discussions (detailed information on this is given below). Proofs of theorems and lemmas can similarly be omitted. Roughly speaking, it is possible to skip later sections within a chapter and later chapters within a part and move on to later chapters in the book. The book was organized in this way so that it would be accessible to a wider audience, as well as to increase its flexibility.

Many of the exercises are computational. In some cases, pencil-and-paper techniques would suffice, but the use of a computer is recommended. We have not specified how the computer might be used, and we leave this up to the instructor. In courses with an emphasis on modeling, a specialized linear or nonlinear optimization package might be appropriate. In other courses, the students might be asked to program algorithms themselves. We leave these decisions up to the instructor. Some information about software packages can be found in Appendix C. In addition, some exercises depend on auxiliary data sets that can be found on the web site for the book:

http://www.siam.org/books/ot108

In our own classes, we use the MATLAB® software package for class demonstrations and homework assignments. It allows us to demonstrate a great many techniques easily, and it allows students to program individual algorithms without much difficulty. It also includes (in its toolboxes) prepared algorithms for many of the optimization problems that we discuss.

We have gone to considerable effort to ensure the accuracy of the material in this book. Even so, we expect that some errors remain. For this reason, we have set up an online page for errata. It can be obtained at the book Web site.

Using This Book

This book is designed to be flexible. It can be read and taught in many different ways.

The material in the appendices can be taught as needed, or left to the students to read independently. Also, all formally identified proofs can be omitted.

Part II (Linear Programming) and Part III (Unconstrained Optimization) are independent of each other. Part II does not assume any knowledge of Calculus. Part IV (Nonlinear Optimization) does not assume that Part II has been read (with the exception of Section 14.4.1).

The only "essential" chapters in Part II are Chapters 4 (Geometry of Linear Programming), 5 (The Simplex Method), and 6 (Duality). The only "essential" chapter in Part III is Chapter 11 (Basics of Unconstrained Optimization). The other chapters can be skipped.

We now describe the chapters individually, pointing out various ways they can be used. The sample course outlines that follow indicate how chapters might be selected to construct individual courses (based on a 15-week semester).[1]

Part I: Basics

- *Chapter* 1: *Optimization Models.* This chapter is self-contained and describes a variety of optimization models. Sections 1.3–1.5 are independent of one another. Section 1.6 includes more realistic models and assumes that the reader is familiar with the basic models described in the earlier sections. The subsections of Section 1.6 are independent of one another.

- *Chapter* 2: *Fundamentals of Optimization.* For Part II, only Sections 2.1–2.4 are needed (and Section 2.3.1 can be omitted). For Parts III and IV the whole chapter is relevant.

- *Chapter* 3: *Representation of Linear Constraints.* Sections 3.3.2–3.3.4 can be omitted (although Section 3.3.2 is needed for Part IV). This chapter is only relevant to Parts II and IV; it is not needed for Part III.

Part II: Linear Programming

- *Chapter* 4: *Geometry of Linear Programming.* All sections of this chapter are needed in Part II.

- *Chapter* 5: *The Simplex Method.* Sections 5.1 and 5.2 are the most important. How the rest of the chapter is used depends on the goals of the instructor, in particular with regard to tableaus. In a number of examples, we use the full simplex tableau to display data for linear programs. Thus, it is necessary to be able to read these tableaus to extract information. This is the only use we make of the tableaus elsewhere in the book. It is not necessary to be able to manipulate these tableaus.

---

[1]Throughout the book, the number of a section or subsection begins with the chapter number. That is, Section 10.3 refers to the third section in Chapter 10, and Section 16.7.2 refers to the second subsection in the seventh section of Chapter 16. Also, a reference to Appendix A.9 refers to the ninth section of Appendix A. A similar system is used for tables, examples, theorems, etc.; Figure 8.10 refers to the tenth figure in Chapter 8, for example. For exercises, however, the chapter number is omitted, e.g., Exercise 4.7 is the seventh exercise in Section 4 of the current chapter (unless another chapter is specified).

- *Chapter* 6: *Duality and Sensitivity.* Sections 6.1 and 6.2 are the most important. The remaining sections can be skipped, if desired. If taught, we recommend that Sections 6.3–6.5 be taught in order, although Section 6.3 is only used in a minor way in the remaining two sections. It would be possible to stop after any section. Note: The remaining chapters in Part II are independent of each other.

- *Chapter* 7: *Enhancements of the Simplex Method.* The sections in this chapter are independent of each other. The instructor is free to pick and choose material, with one partial exception: the discussion of the decomposition principle is easier to understand if column generation has already been read.

- *Chapter* 8: *Network Problems.* In this chapter, the sections must be taught in order. It would be possible to stop after any section.

- *Chapter* 9: *Computational Complexity of Linear Programming.* The first two sections contain basic material used in Sections 9.3–9.5. Ideally, the remaining sections should be taught in order, although Sections 9.4 and 9.5 are independent of each other. Even if some topics are not of interest, at least the introductory paragraphs of each section should be read. (Section 9.5 requires some knowledge of statistics.)

- *Chapter* 10: *Interior-Point Methods for Linear Programming.* Sections 10.1 and 10.2 are the most important. The later sections could be skipped but, if taught, Sections 10.4–10.6 should be taught in order. Section 10.4 reviews some fundamental concepts from nonlinear optimization needed in Sections 10.5–10.6.

Part III: Unconstrained Optimization

- *Chapter* 11: *Basics of Unconstrained Optimization.* We recommend reading all of this chapter (with the exception of the proofs). If desired, either Section 11.5 or Section 11.6 could be omitted, but not both. Chapters 12 and 13 could be omitted. Chapter 13 makes more sense if taught after Chapter 12, but in fact, only Section 13.5 makes explicit use of the material in Chapter 12.

- *Chapter* 12: *Methods for Unconstrained Optimization.* Sections 12.1–12.3 are the most important. All the remaining sections and subsections can be taught independently of each other.

- *Chapter* 13: *Low-Storage Methods for Unconstrained Problems.* Once Sections 13.1 and 13.2 have been taught, the remaining sections are independent of each other.

Part IV: Nonlinear Optimization

- *Chapter* 14: *Optimality Conditions for Constrained Problems.* We recommend reading Sections 14.1–14.6. The rest of the chapter may be omitted. Within Section 14.8, Sections 14.8.3 and 14.8.5 can be taught without teaching the remaining subsections, although Section 14.8.5 depends on Section 14.8.3. (The discussion of nonlinear duality in Section 14.8 is only needed in Sections 16.6–16.8 of Chapter 16.)

- *Chapter* 15: *Feasible-Point Methods.* We recommend reading Sections 15.1–15.4 (although Section 15.4.1 could be omitted). These sections explain how to solve problems with linear constraints. Sections 15.5–15.7 discuss methods for problems with nonlinear constraints. Sections 15.5 and 15.6 are independent of each other, but Section 15.7 depends on Section 15.5.

- *Chapter* 16: *Penalty and Barrier Methods.* We recommend reading Sections 16.1 and 16.2 (although Section 16.2.3 could be omitted). If more of the chapter is covered, then Section 16.3 should be read. Sections 16.4–16.8 are independent of each other. Sections 16.6–16.8 use Section 14.8.3 of Chapter 14.

Changes in the Second Edition

The overall structure of the book has not changed in the new addition, and the major topic areas are the same. However, we have updated certain topics to reflect developments since the first edition appeared. We list the major changes here.

Chapter 1 has been expanded to include examples of more realistic optimization models (Section 1.6). The description of interior-point methods for linear programming has been thoroughly revised and restructured (Chapter 10). The discussion of derivative-free methods has been extensively revised to reflect advances in theory and algorithms (Section 12.5). In Part IV we have added material on filter methods (Section 15.7), nonlinear primal-dual methods (Section 16.7), and semidefinite programming (Section 16.8). In addition, numerous smaller changes have been made throughout the book.

Some material from the first edition has been omitted here. The most notable examples are the chapter on nonlinear least-squares data fitting, and the sections on interior-point methods for convex programming. These topics from the first edition are available at the book Web site (see above for the URL).

Sample Course Outlines

We provide below some sample outlines for courses that might use this book. If a section is listed without mention of subsections, then it is assumed that all the subsections will be taught. If a subsection is specified, then the unmentioned subsections may be omitted.

Proposed Course Outline: Linear Programming

I: Foundations

Chapter 1. Optimization Models
    1. Introduction
    3. Linear Equations
    4. Linear Optimization
    7. Optimization Applications
        1. Crew Scheduling and Fleet Scheduling

Chapter 2. Fundamentals of Optimization
    1. Introduction
    2. Feasibility and Optimality
    3. Convexity
    4. The General Optimization Algorithm

---

[2]Not all the applications need be taught.
[3]The material in Chapter 3 is not needed until Part IV.

Proposed Course Outline: Introduction to Optimization

I: Foundations

---

[4]Not all the applications need be taught.

Acknowledgments

We owe a great deal of thanks to the people who have assisted us in preparing this second edition of this book. In particular, we would like to thank the following individuals for reviewing various portions of the manuscript and providing helpful advice and guidance: Erling Andersen, Bob Bixby, Sanjay Mehrotra, Hans Mittelmann, Michael Overton, Virginia Torczon, and Bob Vanderbei. We are especially grateful to Sara Murphy at SIAM for guiding us through the preparation of the manuscript.

Special thanks also to Galina Spivak, whose design for the front cover skillfully conveys, in our minds, the spirit of the book.

We continue to be grateful to those individuals who contributed to the preparation of the first edition. These include: Kurt Anstreicher, John Anzalone, Todd Beltracchi, Dimitri Bertsekas, Bob Bixby, Paul Boggs, Dennis Bricker, Tony Chan, Jessie Cohen, Andrew Conn, Blaine Crowthers, John Dennis, Peter Foellbach, John Forrest, Bob Fourer, Christoph Luitpold Frommel, Saul Gass, David Gay, James Ho, Sharon Holland, Jeffrey Horn, Soonam Kahng, Przemyslaw Kowalik, Michael Lewis, Lorin Lund, Irvin Lustig, Maureen Mackin, Eric Munson, Arkadii Nemirovsky, Florian Potra, Michael Rothkopf, Michael Saunders, David Shanno, Eric Smith, Martin Smith, Pete Stewart, André Tits, Michael Todd, Virginia Torczon, Luis Vicente, Don Wagner, Bing Wang, and Tjalling Ypma.

While preparing the first edition, we received valuable support from the National Science Foundation. We also benefited from the facilities of the National Institute of Standards and Technology and Rice University.

Igor Griva
Stephen G. Nash
Ariela Sofer

Part I

Basics

Chapter 1

Optimization Models

1.1 Introduction

Optimization models attempt to express, in mathematical terms, the goal of solving a problem in the "best" way. That might mean running a business to maximize profit, minimize loss, maximize efficiency, or minimize risk. It might mean designing a bridge to minimize weight or maximize strength. It might mean selecting a flight plan for an aircraft to minimize time or fuel use. The desire to solve a problem in an optimal way is so common that optimization models arise in almost every area of application. They have even been used to explain the laws of nature, as in Fermat's derivation of the law of refraction for light.

Optimization models have been used for centuries, since their purpose is so appealing. In recent times they have come to be essential, as businesses become larger and more complicated, and as engineering designs become more ambitious. In many circumstances it is no longer possible, or economically feasible, for decisions to be made without the aid of such models. In a large, multinational corporation, for example, a minor percentage improvement in operations might lead to a multimillion dollar increase in profit, but achieving this improvement might require analyzing all divisions of the corporation, a gargantuan task. Likewise, it would be virtually impossible to design a new computer chip involving millions of transistors without the aid of such models.

Such large models, with all the complexity and subtlety that they can represent, would be of little value if they could not be solved. The last few decades have witnessed astonishing improvements in computer hardware and software, and these advances have made optimization models a practical tool in business, science, and engineering. It is now possible to solve problems with thousands or even millions of variables. The theory and algorithms that make this possible form a large portion of this book.

In the first part of this chapter we give some simple examples of optimization models. They are grouped in categories, where the divisions reflect the properties of the models as well as the differences in the techniques used to solve them. We include also a discussion of systems of linear equations, which are not normally considered to be optimization models. However, linear equations are often included as constraints in optimization models, and their solution is an important step in the solution of many optimization problems.

3

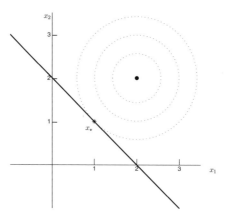

Figure 1.1. *Nonlinear optimization problem. The feasible set is the dark line.*

In the last section of this chapter we give some examples of applications of optimization. These examples reflect families of problems that are either in wide use, or—at the time of writing of this edition of the book—are subject of intense research. The examples reflect the tastes of the authors; by no means do they constitute a broad or representative sample of the myriad applications where optimization is in use today.

1.2   Optimization: An Informal Introduction

Consider the problem of finding the point on the line $x_1 + x_2 = 2$ that is closest to the point $(2, 2)^T$ (see Figure 1.1) . The problem can be written as

$$\begin{array}{ll} \text{minimize} & f(x) = (x_1 - 2)^2 + (x_2 - 2)^2 \\ \text{subject to} & x_1 + x_2 = 2. \end{array}$$

It is easy, of course, to see that the problem has an optimum at $x_\star = (1, 1)^T$.

This problem is an example of an optimization problem. Optimization problems typically minimize or maximize a function f (called the *objective function*) in a set of points S (called the *feasible set*). Commonly, the feasible set is defined by some constraints on the variables. In this example our objective function is the nonlinear function $f(x) = (x_1 - 2)^2 + (x_2 - 2)^2$, and the feasible set S is defined by a single linear constraint $x_1 + x_2 = 2$. The feasible set could also be defined by multiple constraints. An example is the problem

$$\begin{array}{ll} \text{minimize} & f(x) = x_1 \\ \text{subject to} & x_1^2 \leq x_2 \\ & x_1^2 + x_2^2 \leq 2. \end{array}$$

The feasible set S for this problem is shown in Figure 1.2; it is easy to see that the optimal point is $x_\star = (-1, 1)^T$. It is possible to have an *unconstrained optimization* problem where

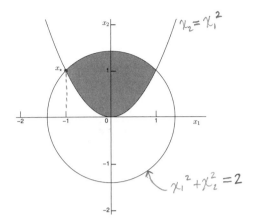

Figure 1.2. *Nonlinear optimization problem with inequality constraints.*

there are no constraints, as in the example

$$\text{minimize } f(x) = (e^{x_1} - 1)^2 + (x_2 - 1)^2.$$

The feasible set S here is the entire two-dimensional space. The minimizer is $x_\star = (0, 1)^T$, since the function value is zero at this point and positive elsewhere.

We see from these examples that the feasible set can be defined by equality constraints or inequality constraints or no constraints at all. The functions defining the objective function and the constraints may be linear or nonlinear. The examples above are *nonlinear optimization* problems since at least some of the functions involved are nonlinear. If the objective function and the constraints are all linear, the problem is a *linear optimization problem* or *linear program*. An example is the problem

$$\begin{aligned}
\text{maximize} \quad & f(x) = 2x_1 + x_2 \\
\text{subject to} \quad & x_1 + x_2 \le 1 \\
& x_1 \ge 0, \ x_2 \ge 0.
\end{aligned}$$

Figure 1.3 shows the feasible set. The optimal solution is clearly $x_\star = (1, 0)^T$.

Consider now the nonlinear optimization problem

$$\begin{aligned}
\text{maximize} \quad & f(x) = (x_1 + x_2)^2 \\
\text{subject to} \quad & x_1 x_2 \ge 0 \\
& -2 \le x_1 \le 1 \\
& -2 \le x_2 \le 1.
\end{aligned}$$

The feasible set is shown in Figure 1.4. The point $x_c = (1, 1)^T$ has an objective value of $f(x_c) = 4$, which is a higher objective value than any of its "nearby" feasible points. It is therefore called a *local optimizer*. In contrast the point $x_\star = (-2, -2)^T$ has an objective value $f(x_\star) = 16$ which is the best among all feasible points. It is called a *global optimizer*.

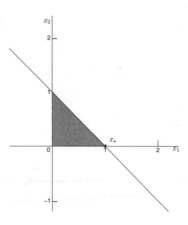

Figure 1.3. *Linear optimization problem. The feasible region is shaded.*

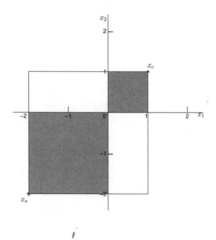

Figure 1.4. *Local and global solutions. The feasible region is shaded.*

The methods we consider in this book focus on finding local optima. We will usually assume that the problem functions and their first and second derivatives are continuous. We can then use derivative information at a given point to anticipate the behavior of the problem functions at "nearby" points and use this to determine whether the point is a local solution and if not, to find a better point. The derivative information cannot usually anticipate the behavior of the functions at points "farther away," and hence cannot determine whether a local solution is also the global solution. One exception is when the problem solved is a *convex optimization problem*, in which any local optimizer is also a global optimizer (see Section 2.3). Luckily, linear programs are convex so that for this important family of problems, local solutions are also global.

* Linear Problems are convex!
  e.g., local optimum ⟹ global optimum

It may seem odd to give so much attention to finding local optimizers when they are not always guaranteed to be global optimizers. However, most global optimization algorithms seek the global optimum by finding local solutions to a sequence of subproblems generated by some methodical approximation to the original problem; the techniques described in the book are suitable for these subproblems. In addition, for some applications a local solution may be sufficient, or the user might be satisfied with an improvement on the objective value. Of course, some applications require finding a global solution. The drawback is that for a problem that is not convex (or not known to be convex), finding a global solution can require substantially more computational effort than finding a local solution.

Our book will also assume that the variables of the problems are continuous, that is, they can take a continuous range of real values. For this reason the problems we consider are also referred to as *continuous optimization problems*. Many variables such as length, volume, weight, and time are by nature continuous, and even though we cannot compute or measure them to infinite precision, it is plausible in the optimization to assume that they are continuous. On the other hand, variables such as the number of people to be hired, the number of flights to dispatch per day, or the number of new plants to be opened can assume only integer values. Problems where the variables can only take on integer values are called *discrete optimization problems* or, in the case where all problem functions are linear, *integer programming problems*. In a few applications it is sufficient to solve the problem ignoring the integrality restriction, and once a solution is obtained, to round off the variables to their nearest integer. Unfortunately rounding off of a solution does not guarantee that it is optimal, or even that it is feasible, so this approach is often inadequate.

While a discussion of discrete optimization is beyond the scope of this book, we will mention that such problems are much harder than their continuous counterparts for much the same reason global optimization is harder than local optimization. Since at a given point we only have information of the behavior of the function at "nearby points," there are no straightforward conditions that can determine whether a given feasible solution is optimal. Hence the solution process must rule out either explicitly or implicitly every other feasible solution. Thus the search for an integer solution requires the solution of a potentially large sequence of continuous optimization subproblems. Typically the first of these subproblems is a *relaxed problem*, in which the integrality requirement on each variable is relaxed (omitted) and replaced by a (continuous) constraint on the range of the variable. If, for example, a variable x_j is restricted to be either 0, 1, or 2, the relaxed constraint would be $0 \le x_j \le 2$. Subsequent subproblems would typically include additional continuous constraints. The subproblems would be solved by continuous optimization methods such as those described in the book.

Continuous optimization is the basis for the solution of many applied problems, both discrete and continuous, convex or nonconvex. The examples in this chapter reflect just a small fraction of such applications.

1.3 Linear Equations

Systems of linear equations are central to almost all optimization algorithms and form a part of a great many optimization models. They are used in this section to represent a data-fitting example. A slight generalization of this example will lead to the important problem of least-squares data fitting. Linear equations are also used to represent constraints in a model.

Finally, solving systems of linear equations is an important step in the simplex method for linear programming and Newton's method for nonlinear optimization, and is a technique used to determine dual variables (Lagrange multipliers) in both settings. In this chapter we only give examples of linear equations. Techniques for their solution are discussed in Appendix A.

Our example is based on Figure 1.5. The points marked by ● are assumed to lie on the graph of a quadratic function. These points, denoted by $(t_i, b_i)^T$, have the coordinates $(2, 1)^T$, $(3, 6)^T$, and $(5, 4)^T$. The quadratic function can be written as

$$b(t) = x_1 + x_2 t + x_3 t^2,$$

where x_1, x_2, and x_3 are three unknown parameters that determine the quadratic. The three data points define three equations of the form $b(t_i) = b_i$:

$$x_1 + x_2(2) + x_3(2)^2 = 1$$
$$x_1 + x_2(3) + x_3(3)^2 = 6$$
$$x_1 + x_2(5) + x_3(5)^2 = 4$$

or

$$x_1 + 2x_2 + 4x_3 = 1$$
$$x_1 + 3x_2 + 9x_3 = 6$$
$$x_1 + 5x_2 + 25x_3 = 4.$$

The solution is $(x_1, x_2, x_3)^T = (-21, 15, -2)^T$, or

$$b(t) = -21 + 15t - 2t^2,$$

and is graphed in Figure 1.5.

This approach to data fitting has many applications. It is not unique to fitting data by a quadratic function. If the data were thought to have some sort of periodic component (perhaps a daily fluctuation), then a more appropriate model might be

$$b(t) = x_1 + x_2 t + x_3 \sin t,$$

and the system of equations would have the form

$$x_1 + x_2(2) + x_3(\sin 2) = 1$$
$$x_1 + x_2(3) + x_3(\sin 3) = 6$$
$$x_1 + x_2(5) + x_3(\sin 5) = 4.$$

Also, there is nothing special about having three data points and three terms in the model. If we wish to associate the data-fitting problem with a system of linear equations, then the number of data points and the number of model terms must be the same. However, through the use of least-squares models (see Section 1.5), it would be possible to have more data points than model terms. In fact, this is often the case. Least-squares techniques are also appropriate if there are measurement errors in the data (also a common occurrence).

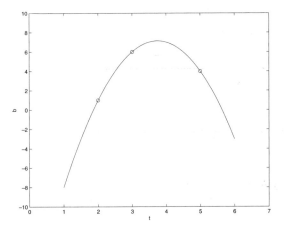

Figure 1.5. *Fitting a quadratic function to data.*

$$b(t) = x_1 + x_2 t + x_3 t^2$$

Let us return to the example of the quadratic model. We can write the system of equations in matrix form as

$$\begin{pmatrix} 1 & 2 & 4 \\ 1 & 3 & 9 \\ 1 & 5 & 25 \end{pmatrix} \begin{pmatrix} x_1 \\ x_2 \\ x_3 \end{pmatrix} = \begin{pmatrix} 1 \\ 6 \\ 4 \end{pmatrix},$$

or more generally,

$$\begin{pmatrix} 1 & t_1 & t_1^2 \\ 1 & t_2 & t_2^2 \\ 1 & t_3 & t_3^2 \end{pmatrix} \begin{pmatrix} x_1 \\ x_2 \\ x_3 \end{pmatrix} = \begin{pmatrix} b_1 \\ b_2 \\ b_3 \end{pmatrix}.$$

If there were n data points and the model were of the form

$$b(t) = x_1 + x_2 t + \cdots + x_n t^{n-1},$$

then the system would have the form

$$\begin{pmatrix} 1 & t_1 & \cdots & t_1^{n-1} \\ 1 & t_2 & \cdots & t_2^{n-1} \\ & & \vdots & \\ 1 & t_n & \cdots & t_n^{n-1} \end{pmatrix} \begin{pmatrix} x_1 \\ x_2 \\ \vdots \\ x_n \end{pmatrix} = \begin{pmatrix} b_1 \\ b_2 \\ \vdots \\ b_n \end{pmatrix}.$$

We will often denote such a system of linear equations as $Ax = b$.

For these examples the number of data points is equal to the number of variables. Equivalently the matrix A has the same number of rows and columns. We refer to this as a "square" system because of the shape of the matrix A. It is also possible to consider problems with unequal numbers of data points and variables. Such examples, called "rectangular," are discussed in Section 1.5.

Table 1.1. *Cabinet data.*

Cabinet	Wood	Labor	Revenue
Bookshelf	10	2	100
With Doors	12	4	150
With Drawers	25	8	200
Custom	20	12	400

1.4 Linear Optimization

A linear optimization model (also knows as a "linear program") involves the optimization of a linear function subject to linear constraints on the variables. Although linear functions are simple functions, they arise frequently in economics, production planning, networks, scheduling, and other applications. We will consider several examples. Further examples are included in Section 1.7 and in Chapters 5–8. In particular, examples of network models are discussed in Section 8.2.

Suppose that a manufacturer of kitchen cabinets is trying to maximize the weekly revenue of a factory. Various orders have come in that the company could accept. They include bookcases with open shelves, cabinets with doors, cabinets with drawers, and custom-designed cabinets. Table 1.1 indicates the quantities of materials and labor required to assemble the four types of cabinets, as well as the revenue earned.

Suppose that 5000 units of wood and 1500 units of labor are available. Let $x_1, \ldots, x_4$ represent the number of cabinets of each type made (x_1 for bookshelves, x_2 for cabinets with doors, etc.). Then the corresponding linear programming model might be

$$\text{maximize} \quad z = 100x_1 + 150x_2 + 200x_3 + 400x_4$$
$$\text{subject to} \quad 10x_1 + 12x_2 + 25x_3 + 20x_4 \leq 5000$$
$$2x_1 + 4x_2 + 8x_3 + 12x_4 \leq 1500$$
$$x_1, x_2, x_3, x_4 \geq 0.$$

This problem can easily be expanded from four products (bookshelves, cabinets with doors, cabinets without doors, etc.) to any number of products n, and from two resources (wood and labor) to any number of resources m. Denoting the unit profit from product j by c_j, the amount available of resource i by b_i, and the amount of resource i used by a unit of product j by a_{ij}, the problem can be written in the form

$$\text{maximize} \quad z = \sum_{j=1}^{n} c_j x_j$$
$$\text{subject to} \quad \sum_{j=1}^{n} a_{ij} \leq b_i, \quad i = 1, \ldots, m$$
$$x_j \geq 0, \qquad j = 1, \ldots, n.$$

The problem can be written in a more compact manner by introducing matrix-vector notation. Letting $x = (x_1, \ldots, x_n)^T$, $c = (c_1, \ldots, c_n)^T$, $b = (b_1, \ldots, b_m)^T$, and denoting the matrix

Table 1.2. *Work times (in minutes).*

Worker	Information	Policy	Claim
1	10	28	31
2	15	22	42
3	13	18	35
4	19	25	29
5	17	23	33

of coefficients a_{ij} by A, the problem becomes

$$\text{maximize} \quad z = c^T x = \sum_{j=1}^{n} c_j x_j$$
$$\text{subject to} \quad Ax \leq b$$
$$x \geq 0.$$

This is a typical example of a linear program. Here a linear objective function is to be maximized subject to linear inequality constraints and nonnegativity constraints on the variables. In the general case, the objective of a linear program may be either maximized or minimized, the constraints may involve a combination of inequalities and equalities, and the variables may be either restricted in sign or unrestricted. Although these may appear as different forms, it is easy to convert from one form to another.

As another example, consider the assignment of jobs to workers. Suppose that an insurance office handles three types of work: requests for information, new policies, and claims. There are five workers. Based on a study of office operations, the average work times (in minutes) for the workers are known; see Table 1.2.

The company would like to minimize the overall elapsed time for handling a (long) sequence of tasks, by appropriately assigning a fraction of each type of task to each worker. Let p_i be the fraction of information calls assigned to worker i, q_i the fraction of new policy calls, and r_i the fraction of claims; t will represent the elapsed time. Then a linear programming model for this situation would be

$$\text{minimize} \quad z = t$$
$$\text{subject to} \quad p_1 + p_2 + p_3 + p_4 + p_5 = 1$$
$$q_1 + q_2 + q_3 + q_4 + q_5 = 1$$
$$r_1 + r_2 + r_3 + r_4 + r_5 = 1$$
$$10p_1 + 28q_1 + 31r_1 \leq t$$
$$15p_2 + 22q_2 + 42r_2 \leq t$$
$$13p_3 + 18q_3 + 35r_3 \leq t$$
$$19p_4 + 25q_4 + 29r_4 \leq t$$
$$17p_5 + 23q_5 + 33r_5 \leq t$$
$$p_i, q_i, r_i \geq 0, \ i = 1, \ldots, 5.$$

The constraints in this model assure that t is no less than the overall elapsed time. Since the objective is to minimize t, at the optimal solution t will be equal to the elapsed time.

The problems we have introduced so far are small, involving only a handful of variables and constraints. Many real-life applications involve much larger problems, with possibly hundreds of thousands of variables and constraints. Section 1.7 discusses some of these applications.

Exercise[5]

4.1. Consider the production scheduling problem of the perfume Polly named after a famous celebrity. The manufacturer of the perfume must plan production for the first four months of the year and anticipates a demand of 4000, 5000, 6000, and 4500 gallons in January, February, March, and April, respectively. At the beginning of the year the company has an inventory of 2000 gallons. The company is planning on issuing a new and improved perfume called Pollygone in May, so that all Polly produced must be sold by the end of April. Assume that the production cost for January and February is $5 per gallon and this will rise to $5.5 per gallon in March and April. The company can hold any amount produced in a certain month over to the next month at an inventory cost of $1 per unit. Formulate a linear optimization model that will minimize the costs incurred in meeting the demand for Polly in the period January through April. Assume for simplicity that any amount produced in a given month may be used to fulfill demand for that month.

1.5   Least-Squares Data Fitting

Let us re-examine the quadratic model from Section 1.3:

$$b(t) = x_1 + x_2 t + x_3 t^2.$$

For the data points $(2, 1)$, $(3, 6)$, and $(5, 4)$ we obtained the linear system

$$\begin{pmatrix} 1 & 2 & 4 \\ 1 & 3 & 9 \\ 1 & 5 & 25 \end{pmatrix} \begin{pmatrix} x_1 \\ x_2 \\ x_3 \end{pmatrix} = \begin{pmatrix} 1 \\ 6 \\ 4 \end{pmatrix}$$

with solution $x = (-21, 15, -2)^T$ so that

$$b(t) = -21 + 15t - 2t^2.$$

It is easy to check that the three data points satisfy this equation.

Suppose that the data points had been obtained from an experiment, with an observation made at times $t_1 = 2$, $t_2 = 3$, and $t_3 = 5$. If another observation were made at $t_4 = 7$, then (assuming that the quadratic model is correct) it should satisfy

$$b(7) = -21 + 15 \times 7 - 2 \times 7^2 = -14.$$

If the observed value at $t_4 = 7$ were not equal to -14, then the observation would not be consistent with the model.

---

[5] See Footnote 1 in the Preface for an explanation of the Exercise numbering within chapters.

It is common when collecting data to gather more data points than there are variables in the model. This is true in political polls where hundreds or thousands of people will be asked which candidate they plan to vote for (so that there is only one variable). It is also true in scientific experiments where repeated measurements will be made of a desired quantity. It is expected that each of the measurements will be in error, and that the observations will be used collectively in the hope of obtaining a better result than any individual measurement provides. (The collective result may only be better in the sense that the bound on its error will be smaller. Since the true value is often unknown, the actual errors cannot be measured.)

Since each of the measurements is considered to be in error, it is no longer sensible to ask that the model equation (in our case $b(t) = x_1 + x_2t + x_3t^2$) be solved exactly. Instead we will try to make components of the "residual vector"

$$r = b - Ax = \begin{pmatrix} b_1 - (x_1 + x_2t_1 + x_3t_1^2) \\ b_2 - (x_1 + x_2t_2 + x_3t_2^2) \\ \vdots \\ b_m - (x_1 + x_2t_m + x_3t_m^2) \end{pmatrix}$$

small in some sense.

The most commonly used approach is called "least squares" data fitting, where we try to minimize the sum of the squares of the components of r:

$$\underset{x}{\text{minimize}} \ r_1^2 + \cdots + r_m^2 = \sum_{i=1}^{m} [b_i - (x_1 + x_2t_i + x_3t_i^2)]^2.$$

Under appropriate assumptions about the errors in the observations, it can be shown that this is an optimal way of selecting the coefficients x.

If the fourth data point was $(7, -14)^T$, then the least-squares approach would give $x = (-21, 15, -2)^T$, since this choice of x would make $r = 0$. In this case the graph of the model would pass through all four data points. However, if the fourth data point was $(7, -15)^T$, then the least-squares solution would be

$$x = \begin{pmatrix} -21.9422 \\ 15.6193 \\ -2.0892 \end{pmatrix}.$$

The corresponding residual vector would be

$$r = b - Ax = \begin{pmatrix} 0.0603 \\ -0.1131 \\ 0.0754 \\ -0.0226 \end{pmatrix}.$$

None of the residuals is zero, and so the graph of the model does not pass through any of the data points. This is typical in least-squares models.

If the residuals can be written as $r = b - Ax$, then the model is "linear." This name is used because each of the coefficients x_j occurs linearly in the model. It does not mean that the model terms are linear in t. In fact, the model above has a quadratic term x_3t^2. Other examples of linear models would be

$$b(t) = x_1 + x_2 \sin t + x_3 \sin 2t + \cdots + x_{k+1} \sin kt$$
$$b(t) = x_1 + \frac{x_2}{1 + t^2}.$$

"Nonlinear" models are also possible. Some examples are

$$b(t) = x_1 + x_2 e^{x_3 t} + x_4 e^{x_5 t}$$
$$b(t) = x_1 + \frac{x_2}{1 + x_3 t^2}.$$

In these models there are nonlinear relationships among the coefficients x_j. A nonlinear least-squares model can be written in the form

$$\text{minimize } f(x) = \sum_{i=1}^{m} r_i(x)^2,$$

where $r_i(x)$ represents the residual at t_i. For example,

$$r_i(x) \equiv b_i - (x_1 + x_2 e^{x_3 t_i} + x_4 e^{x_5 t_i})$$

for the first nonlinear model above. We can also write this as

$$f(x) = r(x)^T r(x).$$

If the model is linear, then $r(x) = b - Ax$ and $f(x)$ can be shown to be a quadratic function. See the Exercises.

Nonlinear least squares models are examples of unconstrained minimization problems, that is, they correspond to the minimization of a nonlinear function without constraints on the variables. In fact, they are one of the most commonly encountered unconstrained minimization problems.

Exercises

5.1. Prove that for the linear least-squares problem with $r(x) = b - Ax$, the objective $f(x) = r(x)^T r(x)$ is a quadratic function.

1.6   Nonlinear Optimization

A nonlinear optimization model (also referred to as a "nonlinear program") consists of the optimization of a function subject to constraints, where any of the functions may be nonlinear. This is the most general type of model that we will consider in this book. It includes all the other types of models as special cases.

Nonlinear optimization models arise often in science and engineering. For example, the volume of a sphere is a nonlinear function of its radius, the energy dissipated in an electric

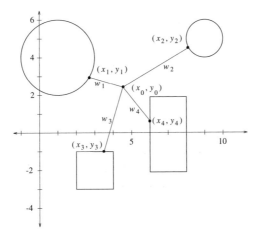

Figure 1.6. *Electrical connections.*

circuit is a nonlinear function of the resistances, the size of an animal population is a nonlinear function of the birth and death rates, etc. We will develop two specific examples here.

Suppose that four buildings are to be connected by electrical wires. The positions of the buildings are illustrated in Figure 1.6. The first two buildings are circular: one at $(1, 4)^T$ with radius 2, the second at $(9, 5)^T$ with radius 1. The third building is square with sides of length 2 centered at $(3, -2)^T$. The fourth building is rectangular with height 4 and width 2 centered at $(7, 0)^T$. The electrical wires will be joined at some central point $(x_0, y_0)^T$ and will connect to building i at position $(x_i, y_i)^T$. The objective is to minimize the amount of wire used. Let w_i be the length of the wire connecting building i to $(x_0, y_0)^T$. A model for this problem is

$$\text{minimize} \quad z = w_1 + w_2 + w_3 + w_4$$
$$\text{subject to} \quad w_i = \sqrt{(x_i - x_0)^2 + (y_i - y_0)^2}, \ i = 1, 2, 3, 4,$$
$$(x_1 - 1)^2 + (y_1 - 4)^2 \leq 4$$
$$(x_2 - 9)^2 + (y_2 - 5)^2 \leq 1$$
$$2 \leq x_3 \leq 4$$
$$-3 \leq y_3 \leq -1$$
$$6 \leq x_4 \leq 8$$
$$-2 \leq y_4 \leq 2.$$

We assume here for simplicity that the wires can be routed through the buildings (if necessary) at no additional cost.

The constraints in nonlinear optimization problems are often written so that the right-hand sides are equal to zero. For the above model this would correspond to using constraints of the form

$$w_i - \sqrt{(x_i - x_0)^2 + (y_i - y_0)^2} = 0, \ i = 1, 2, 3, 4,$$

and so forth. This is just a cosmetic change to the model.

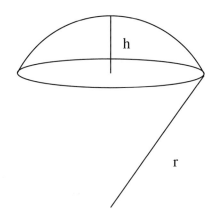

Figure 1.7. *Archimedes' problem.*

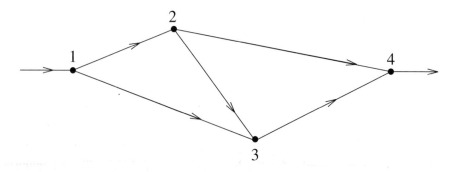

Figure 1.8. *Traffic network.*

As a second example we consider a problem posed by Archimedes. Figure 1.7 illustrates a portion of a sphere with radius r, where the height of the spherical segment is h. The problem is to choose r and h so as to maximize the volume of the segment, but where the surface area A of the segment is fixed. The model is

$$\begin{aligned} \text{maximize} \quad & v(r, h) = \pi h^2(r - \tfrac{h}{3}) \\ \text{subject to} \quad & 2\pi r h = A. \end{aligned}$$

Archimedes was able to prove that the solution was a hemisphere (i.e., $h = r$).

As another illustration of how nonlinear models can arise, consider the network in Figure 1.8. This represents a set of road intersections, and the arrows indicate the direction of traffic. If few cars are on the roads, the travel times between intersections can be considered as constants, but if the traffic is heavy, the travel times can increase dramatically.

Let us focus on the travel time between a pair of intersections i and j. Let $t_{i,j}$ be the (constant) travel time when the traffic is light, let $x_{i,j}$ be the number of cars entering the road per hour, let $c_{i,j}$ be the capacity of the road, that is, the maximum number of cars entering per hour, and let $\alpha_{i,j}$ be a constant reflecting the rate at which travel time increases as the traffic get heavier. (The constant $\alpha_{i,j}$ might be selected using data collected about the road system.)

Then the travel time between intersections i and j could be modeled by

$$T_{i,j}(x_{i,j}) = t_{i,j} + \alpha_{i,j} \frac{x_{i,j}}{1 - x_{i,j}/c_{i,j}}.$$

If there is no traffic on the road ($x_{i,j} = 0$), then the travel time is $t_{i,j}$. If $x_{i,j}$ approaches the capacity of the road $c_{i,j}$, then the travel time tends to $+\infty$. $T_{i,j}$ is a nonlinear function of $x_{i,j}$.

Suppose we wished to minimize the total travel time through the network for a volume of X cars per hour. Then our model would be

$$\text{minimize } f(x) = \sum x_{i,j} T_{i,j}(x_{i,j})$$

subject to the constraints

$$x_{1,2} + x_{1,3} = X$$
$$x_{2,3} + x_{2,4} - x_{1,2} = 0$$
$$x_{3,4} - x_{1,3} - x_{2,3} = 0$$
$$x_{2,4} + x_{3,4} = X$$
$$0 \le x_{i,j} \le c_{ij}.$$

The equations ensure that all cars entering an intersection also leave an intersection. The objective sums up the travel times for all the cars.

A potential snag with this formulation is that if the traffic volume reaches capacity on any arc ($x_{i,j} = c_{i,j}$), the objective function becomes undefined, which will cause optimization software to fail. A number of measures could be invoked to prevent this situation. One alternative is to slightly lower the upper bounds on the variables, so that $x_{i,j} \le c_{i,j} - \epsilon$, where ϵ is a small positive number. Alternatively we could increase each denominator in the objective by a small positive amount ϵ, thus forcing the denominator to have a value of at least ϵ and thereby avoiding division by zero.

Our last example is the problem of finding the minimum distance from a point r to the set $\{x : a^T x = b\}$. In two dimensions the points in the set S define a line, and in three dimensions they define a plane; in the more general case, the set is called a *hyperplane*. The least-distance problem can be written as

$$\begin{aligned} \text{minimize} \quad & f(x) = \tfrac{1}{2}(x - r)^T(x - r) \\ \text{subject to} \quad & a^T x = b. \end{aligned}$$

(The coefficient of one half in the objective is included for convenience; it allows for simpler formulas when analyzing the problem.) Unlike most nonlinear problems this one has a closed-form solution. It is given by

$$x = r + \frac{b - a^T r}{a^T a} a.$$

(See the Exercises for Section 14.2.)

The minimum distance problem is an example of a *quadratic program*. In general, a quadratic program involves the minimization of a quadratic function subject to linear constraints. An example is the problem

$$\begin{array}{ll} \text{minimize} & f(x) = \frac{1}{2}x^T Q x \\ \text{subject to} & Ax \geq b. \end{array}$$

Quadratic programs for which the matrix Q is positive definite are relatively easy to solve, compared to other nonlinear problems.

1.7   Optimization Applications

In this section we present a number of applications that are of current interest to practitioners or researchers. The models we present are but a few of the numerous applications where optimization is making a significant impact.

We start by presenting two problems arising in the optimization of airline operations—the crew scheduling and fleet scheduling problems. Both problems are large linear programs with the added restriction that the variables must take on integer values.

Next we discuss an approach for pattern classification known as support vector machines. Given a set of points that all belong to one of two classes, the idea is to estimate a function that will automatically classify to which of the two classes a new point belongs. In particular we discuss the case where the classifying function is linear. The resulting problem is a quadratic program. This topic is developed further in Chapter 14. Also in this section we discuss a portfolio optimization problem that attempts to balance between the competing goals of maximizing expected returns and minimizing risk in investment planning. This too is a quadratic program.

Next we will discuss two optimization problems arising from medical applications. One problem arises from planning for treatment of cancer by radiation, where the conflicting goals of providing sufficient radiation to the tumor and limiting the dosage to nearby vital organs give rise to a plethora of models which cover the spectrum from linear through quadratic to nonlinear. The other problem arises from positron emission tomography (PET) image reconstruction, where a model of the image that best fits the scan data gives rise to a linearly constrained nonlinear problem. In both applications the optimization problems can be very large and challenging to solve.

Finally we use optimization to find the shape of a hanging cable with minimum potential energy. We present several models of the problem and emphasize the importance of certain modeling issues.

1.7.1   Crew Scheduling and Fleet Scheduling

Consider an airline that operates 2000 flights per day serving 100 cities worldwide, with 400 aircraft of 10 different types, each requiring a flight crew. The airline must design a flight schedule that meets the passenger demand, the maintenance requirements on aircraft, and all other safety regulations and labor contract rules, while trying to be cost effective in order to maximize profit.

This planning problem is extremely complex. For this reason many airlines used a phased planning cycle that breaks the problem into smaller steps. While more manageable, the individual steps themselves can also be complex.

Arguably the most challenging of these is the *crew scheduling problem*, that assigns crews (pilots and flight attendants) to flights. Economically it is a significant problem, since the cost of crews is second only to the cost of fuel in an airline's operating expenses. Saving even 1% of this cost can save the airline hundreds of millions of dollars annually. Computationally it is a difficult problem since it involves a linear model, which is not only very large, but also involves integer variables, which necessitates multiple solutions of linear programs.

In planning the crew activities, the flight schedule is subdivided into "legs," representing a nonstop flight from one city to another. If a plane flew from, say, New York via Chicago to Los Angeles, this would be considered as two legs. A large airline would typically have hundreds of flight legs per day. The planning period might be a day, a week, or a month.

The crews themselves are certified for particular aircraft, and this restricts how personnel can be assigned to legs. In addition, there are union rules and federal laws that constrain the crew assignments.

To set up the model, the airline first specifies a set of possible crew assignments. One of these assignments might correspond to sending a crew from New York (their home city) to a sequence of cities and then back to New York. Each such round trip is called a "pairing." The number of pairings grows exponentially with the number of legs, and for a large airline, the number of pairings may easily run into the billions, even for the shorter planning period of one week.

The variables in the model are $\{x_j\}$, where x_j is 1 if a particular pairing is selected as part of the total schedule, and 0 otherwise. Let the total number of pairings be N. The majority of the constraints correspond to the requirement that each leg in the planning period be covered by exactly one pairing. For the ith leg, the constraint has the form

$$\sum_{j=1}^{N} a_{i,j} x_j = 1,$$

where the constant $a_{i,j} = 1$ if a particular pairing includes leg i, and zero otherwise. There is one such constraint for every leg in the schedule.

The columns of the matrix A correspond to the pairings, and each pairing must represent a round trip that is technically and legally feasible. For example, if a crew flies from New York to Chicago, it cannot then immediately fly out of Denver. The pairing makes sense if it makes sense chronologically, includes minimum rests between flights, satisfies regulations on maximum flying time, and so forth. This places many restrictions on how the pairings are generated, and hence on the coefficients $a_{i,j}$. The resulting columns of A are typically very sparse, with many zeros, and just a few ones, corresponding to the legs of the roundtrip.

The cost c_j of a pairing is a function of the duration of the pairing, the number of flight hours, and "penalties" that may be associated with the pairing. For example, extra wages and expenses must be paid if the crew spends a night away from its home city, or it may be necessary to transport a crew from one city to another for them to complete the pairing.

The basic model has the form

$$\text{minimize} \quad z = c^T x$$
$$\text{subject to} \quad \sum_{j=1}^{N} a_{i,j} x_j = 1$$
$$x_j = 0 \text{ or } 1.$$

The problem is a linear program with the additional requirement that the variables take on integer values (here—zero and one), hence it is an integer programming problem. As mentioned in Section 1.2, such problems are most commonly solved by solving a sequence of linear programs, where the integrality restrictions are relaxed and replaced by a (continuous) constraint on the range of the variable. The range should ideally be as tight as possible, yet should not exclude the optimal solution. For a zero-one problem the relaxed constraints for the first subproblem would typically be $0 \leq x_j \leq 1$ for all j. Subsequent problems are variants of the relaxed problem, usually with additional constraints or an adjusted objective function.

Crew scheduling problems can be very large. A major effort is required just to generate the possible pairings. Commonly, only a partial model is generated, corresponding to a subset of the possible pairings. Even so, problems with millions of variables are typical.

Linear programs of this size (even ignoring the integrality restriction) are difficult to solve. They demand all the resources of the most sophisticated software. The special structure of the matrix A (and in particular its sparsity—the large number of zero entries) and the latest algorithmic techniques must be used. Many of these techniques are discussed in Part II.

The crew scheduling problem is typically the last step in an airline's schedule planning. The first step begins about several months prior to the actual service when the airline selects the optimal set of flight legs to be included in its schedule. The flight schedule lists the schedule of flight legs by departure time, destination, and arrival time.

The next step is fleet assignment, which determines which type of aircraft will fly each leg of the schedule. Airline fleets are made up of many different types of aircraft, which differ in capacity and in operational characteristics such as speed, fuel burn rates, landing weights, range, and maintenance costs. Allocating an aircraft that is too small will result in loss of revenue from passengers turned away, while allocating an aircraft that is too big will result in too many empty seats to cover the high expenses. The airline's problem is to determine the best aircraft to use for each flight leg such that capacity is matched to demand while minimizing the operating cost.

This problem is frequently represented as a *time-line network*. The network includes a line called a "time-line" for each airport, with nodes positioned along the line in chronological order at each arrival and departure time. Each flight is represented by an arc in the network. Thus for example a flight leaving Washington Dulles (IAD) at 6:00 am (Eastern Standard Time) and arriving at Denver (DEN) at 10:00 am (Eastern Standard Time) would be represented by an arc connecting the 6:00 am node on the IAD time-line to the 10:00 am node on the DEN time-line. (In practice, the arrival time is adjusted to account for the time it takes to prepare the aircraft for the next flight, but we will ignore that here.) In addition to the flight arcs we create an arc from each node on a time line to the consecutive node on the

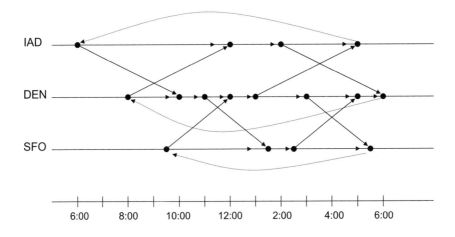

Figure 1.9. *Time-line network.*

time line, and (assuming the schedule is repeated daily) an arc from the last node returning to the first node. The flow on these arcs represents aircraft on the ground that are waiting for their flight.

Figure 1.9 illustrates a time-line network for an airline that has two flights a day each from IAD to DEN, DEN to IAD, DEN to SFO (San Francisco), and SFO to DEN.

Define now x_{ij} to be the number of aircraft of type i on arc j. Any feasible fleet assignment solution must satisfy the following constraints: (i) Covering constraints: each flight leg must be covered by exactly one aircraft; (ii) Flow-balance constraints: for each node of the network the total number of aircraft of type i entering the node must equal the total number of aircraft of type i exiting the node; (iii) Fleet size constraints: the number of aircraft used of each type must not exceed the number of aircraft available. The objective is to minimize the total cost of the assignment. The problem is by nature integer, but it is generally solved by a series of linear programs where the integrality restrictions are relaxed.

Once the fleet is assigned, the individual aircraft of the fleet must be assigned to their flights. This is known as the aircraft routing problem. The planning must take into account the required maintenance for each aircraft. To meet safety regulations, an airline might typically maintain aircraft every 40–45 hours of flying with the maximum time between checks restricted to three to four calendar days. The problem is to determine the most cost effective assignment of aircraft of a single fleet to the scheduled flights, so that all flight legs are covered and aircraft maintenance requirements are satisfied.

The last step of the planning cycle is the task of crew scheduling. Breaking down the full planning cycle into steps helps make the planning more manageable, but ultimately it leads to suboptimal schedules (see Exercise 7.2). Researchers are therefore investigating methods that combine two or more of the planning phases together for more profitable schedules.

Exercises

7.1. Formulate the fleet scheduling problem corresponding to Figure 1.9.

7.2. Consider an airline that has scheduled the flight legs for the next month. It has done so by breaking down the planning cycle into a sequence of steps: first determine the optimal fleet for this schedule; next route the aircraft within the fleet to the flight legs; and finally assign crews for each of the flight legs. Discuss why this makes the planning more manageable but likely leads to suboptimal schedules.

1.7.2   Support Vector Machines

Suppose that you have a set of data points that you have classified in one of two ways: either they have a certain stated property or they do not. These data points might represent the subject titles of email messages, which are classified as either being legitimate email or spam; or they may represent medical data such as age, sex, weight, blood pressure, cholesterol levels, and genetic traits of patients that have been classified either as high risk or as low risk for a heart attack; or they may represent some features of handwritten digits such as ratio of height to width, curvature, that have been classified either as (say) zero or not zero. Suppose now that you obtain a new data point. Your goal is to determine whether this new point does or does not have the stated property. The set of techniques for doing this is broadly referred to as *pattern classification*. The main idea is to identify some rule based on the existing data (referred to as the *training data*) that characterizes the set of points that have the property, which can then be used to determine whether a new point has the property.

In its simplest form classification uses linear functions to provide the characterization. Suppose we have a set of m training data $x_i \in \Re^n$ with classification y_i, where either $y_i = 1$ or $y_i = -1$. A two-dimensional example is shown in the left-hand side of Figure 1.10, where the two classes of points are designated by circles of different shades. Suppose it is possible to find some hyperplane $w^T x + b = 0$ which separates the positive points from the negative. Ideally we would like to have a sharp separation of the positive points from the negative. Thus we will require

$$w^T x_i + b \geq +1 \quad \text{for } y_i = +1,$$
$$w^T x_i + b \leq -1 \quad \text{for } y_i = -1.$$

There is nothing special about the separation coefficients $\pm$ on the right-hand side of the above inequalities. The coefficients w and b of the hyperplane can always be scaled so that the separation will be ± 1.

To obtain the best results we would like the hyperplanes separating the positive points from the negative to be as far apart as possible. From basic geometric principles it can be shown that the distance between the two hyperplanes (that is, the *separation margin*) is $2/\|w\|$. Thus among all separating hyperplanes we should seek the one that maximizes this margin. This is equivalent to minimizing $w^T w$. The resulting problem is to determine the coefficients w and b that solve

$$\begin{aligned} \text{minimize} \quad & f(w, b) = \tfrac{1}{2} w^T w \\ \text{subject to} \quad & y_i(w^T x_i + b) \geq 1, \quad i = 1, \ldots, m. \end{aligned}$$

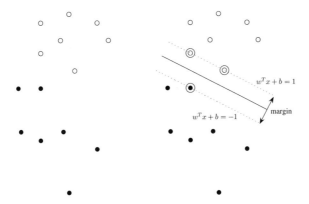

Figure 1.10. *Linear separating hyperplane for the separable case.*

The coefficient $\frac{1}{2}$ in the objective is included for convenience; it results in simpler formulas when analyzing the problem.

The right-hand side of Figure 1.10 shows the solution of our two-dimensional example. The training points that lie on the boundary of either of the hyperplanes are called the *support vectors*; they are highlighted by larger circles. Removal of these points from our training set would change the coefficients of the hyperplanes. Removal of the other training points would leave the coefficients unchanged. The method is called a "support vector machine" because support vectors are used for classifying data as part of a machine (computerized) learning process.

Once the coefficients w and b of the separating hyperplane are found from the training data, we can use the value of the function $f(x) = w^T x + b$ (our "learning machine") to predict whether a new point $\bar{x}$ has the property of interest or not, depending on the sign of $f(\bar{x})$.

So far we have assumed that the data set was separable, that is, a hyperplane separating the positive points from the negative points exists. For the case where the data set is not separable, we can refine the approach to the separable case. We will now allow the points to violate the equations of the separating hyperplane, but we will impose a penalty for the violation. Letting the nonnegative variable ξ_i denote the amount by which the point x_i violates the constraint at the margin, we now require

$$w^T x_i + b \geq +1 - \xi_i \quad \text{for } y_i = +1$$
$$w^T x_i + b \leq -1 + \xi_i \quad \text{for } y_i = -1.$$

A common way to impose the penalty is to add to the objective a term proportional to the sum of the violations. The added penalty term takes the form $C \sum_{i=1}^{m} \xi$ to the objective, where the larger the value of the parameter C, the larger the penalty for violating

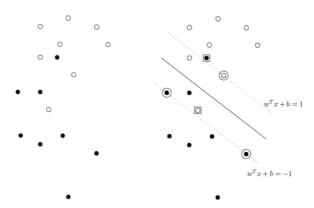

Figure 1.11. *Linear separating hyperplane for the nonseparable case.*

the separation. Our problem is now to find w, b, and ξ that solve

$$\begin{aligned}
\text{minimize} \quad & f(w, b, \xi) = \tfrac{1}{2}w^Tw + C \sum_{i=1}^{m} \xi_i \\
\text{subject to} \quad & y_i(w^Tx_i + b) \geq 1 - \xi_i, \quad i = 1, \ldots, m, \\
& \xi_i \geq 0.
\end{aligned}$$

Figure 1.11 shows an example of the nonseparable case and the resulting separating hyperplane. We see in this example that two of the points (indicated in the figure by the extra squares) are misclassified, since they lie on the incorrect side of the hyperplane $w^Tx+b = 0$.

In later chapters of this book we will see that many problems have a companion problem called the dual problem, that there are important relations between a problem and its dual, and that these relations sometimes lead to insights for solving the problem. In Section 14.8 we will discuss the dual of the problem of finding the hyperplanes with the largest separation margin. We will show that the dual problem directly identifies the support vectors, and that the dual formulation can give rise to a rich family of nonlinear classifications that are often more useful and more accurate than the linear hyperplane classification we presented here.

Exercises

7.1. Consider two classes of data, where the points

$$(13.3), \ (0.31.5), \ (2, 4.2), \ (2.2, 2.9), \ (1.7, 3.6), \ (3, 4), \ (1, 4)$$

possess a certain property and the points

$$(1.8, 1.5), \ (3.4, 3.6), \ (0.2, 2.5), \ (1, 1.3), \ (1, 2.5), \ (3, 1.1), \ (2, 0.1)$$

do not possess this property. Use optimization software to compute the maximum margin hyperplane that separates the two classes of points. Are the classes indeed separable? What are the support vectors? Repeat the problem when the first class includes also the point $(0.2, 2.5)$ and the second class includes the point $(1.7, 3.6)$.

7.2. In this project we create a support vector machine for breast cancer diagnosis. We use the Wisconsin Diagnosis Breast Cancer Database (WDBC) made publicly available by Wolberg, Street, and Mangasarian of the University of Wisconsin. A link to the data base is made available on the Web page for this book, http://www.siam.org/books/ot108. There are two files: wdbc.data and wdbc.names. The file wdbc.names gives more details about the data, and you should read it to understand the context. The file wdbc.data gives $N = 569$ data vectors. Each data vector (in row form) has $n = 32$ components. The first component is the patient number, and the second is either "M" or "B" depending on whether the data is malignant or benign. You may manually change the entries "M" to "+1" and "B" to "−1". These entries are the indicators y_i. Elements 3 through 32 of each row i form a 30-dimensional vector x_i^T of observations.

(i) Use the first 500 data vectors as your training set. Use a modeling language to formulate the problem for the nonseparable case, using $C = 1000$. Solve the problem and display the separating hyperplane. Determine whether the data are indeed separable.

(ii) Use the output of the run to predict whether the remaining 69 patients have cancer. Compare your prediction to the actual patients' medical status. Evaluate the accuracy (proportion of correct predictions), the sensitivity (proportion of positive diagnoses for patients with the disease), and the specificity (the proportion of negative diagnoses for patients without the disease).

1.7.3 Portfolio Optimization

Suppose that an investor wishes to select a set of assets to achieve a good return on the investment while controlling risks of losses. The use of nonlinear models to manage investments began in the 1950s with the pioneering work of Nobel Prize laureate Harry Markowitz, who demonstrated how to reduce the risk of investment by selecting a *portfolio* of stocks rather than picking individual attractive stocks, and established the trade-off between reward and risk in investment portfolios.

An investment portfolio is defined by the vector $x = (x_1, \ldots, x_n)$, where x_j denotes the proportion of the investment to be invested in asset j. Letting μ_j denote the expected rate of return of asset j, the expected rate of return of the portfolio is $\mu^T x$.

Let Σ be the matrix of variances and covariances of the assets' returns. The entry $\Sigma_{j,j}$ is the variance of investment j. A high variance indicates high volatility or high risk; a low variance indicates stability or low risk. The entry $\Sigma_{i,j}$ is the covariance of investments i and j. A positive value of $\Sigma_{i,j}$ indicates assets whose values usually move in the same direction, as often occurs with stocks of companies in the same industry. A negative value indicates assets whose values generally move in opposite directions—a desirable feature in a diversified portfolio. Markowitz defined the risk of the portfolio to be its expected variance $x^T\Sigma x$.

Our optimization problem has two conflicting objectives: to maximize the return $\mu^T x$, and to minimize the risk $x^T\Sigma x$. The relative importance of these objectives will vary depending on the investor's tolerance for risk. We introduce a nonnegative parameter α that reflects the investor's trade-off between risk and return. The objective function in the model will be some combination of the two objectives, parameterized by α, leading to the model

$$\text{maximize } f(x) = \mu^T x - \alpha x^T \Sigma x$$

subject to the constraints

$$\sum_i x_i = 1 \quad \text{and} \quad x \geq 0.$$

The value of α reflects the investor's aversion to risk. A large value indicates a reluctance to take on risk, with an emphasis on the stability of the investment. A low value indicates a high tolerance for risk with an emphasis on the expected return of the investment.

It can be difficult to choose a sensible value for α. For this reason it is common to solve this model for a range of values of this parameter. This can reveal how sensitive the solution is to considerations of risk. The solution of the problem for any value of α is called *efficient* indicating that there is no other portfolio that has a larger expected return and a smaller variance.

There are of course some limitations to our model. First, we do not generally know the theoretical (joint) distribution of the assets' return and will need to estimate the mean and variance from historical data. Denoting the estimate of μ by r and the estimate of Σ by V, the actual problem we solve is

$$\begin{aligned} \text{maximize} \quad & r^T x - \alpha x^T V x \\ \text{subject to} \quad & \sum_i x_i = 1 \\ & x_i \geq 0. \end{aligned}$$

Second, investors should be aware that past performance is no indicator of future returns. Finally, we note that the matrix V is *dense*; that is, it has many nonzero elements. As a result, when the number of assets is large, computations involving V can be expensive thus making the optimization problem computationally difficult.

To illustrate portfolio optimization, consider an investor who is planning a portfolio based on four stocks. Data on the rates of return of the stocks in the last six periods are given in Table 1.3.

Using this information we estimate the mean of the rate of return as

$$r = (\, 0.0667 \quad 0.0900 \quad 0.0717 \quad 0.0733 \,),$$

and the variance as

$$V = \begin{pmatrix} 0.00019 & 0.00065 & 0.00004 & 0.00038 \\ 0.00065 & 0.00883 & 0.00218 & 0.00327 \\ 0.00004 & 0.00218 & 0.00125 & 0.00063 \\ 0.00308 & 0.00327 & 0.00063 & 0.00162 \end{pmatrix}.$$

Table 1.3. *Past rates of return of stocks.*

Period	Stock 1	Stock 2	Stock 3	Stock 4
1	0.08	0.05	0.01	0.08
2	0.06	0.17	0.09	0.12
3	0.07	0.05	0.10	0.07
4	0.04	−0.07	0.04	−0.01
5	0.08	0.12	0.08	0.09
6	0.07	0.22	0.11	0.09

Table 1.4. *Optimal portfolio for selected values of α.*

α	Stock 1	Stock 2	Stock 3	Stock 4	Mean	Variance
1	0	1	0	0	0.090	8.8×10^{-3}
2	0.12	0.65	0.23	0	0.083	4.5×10^{-3}
5	0.57	0.19	0.24	0	0.072	8.0×10^{-4}
10	0.71	0.04	0.25	0	0.069	2.6×10^{-4}
100	0.87	0	0.13	0	0.067	1.7×10^{-4}

The solution of the optimization problem for a selection of values of the parameter α is given in Table 1.4. Figure 1.12 plots the rate of return against the variance of the optimized portfolios for a continuous range of values of α. The curved line is called the *efficient frontier* since it depicts the collection of all efficient points. The figure also shows the rate of return and variance obtained when allocating the entire portfolio to one stock only. In this example, a person who has a high tolerance for risk may choose to invest entirely in Stock 2, whereas a person who is extremely cautious may choose to invest entirely in Stock 1. Investing only in Stock 3, or only in Stock 4, or half in Stock 1 and half in Stock 2 are not recommended strategies for anyone, since they are dominated by strategies that have both higher return and lower risk.

Exercises

7.1. How would the formulation to the problem change if a risk-free asset (such as government treasury bills at a fixed rate of return) is also being considered?

7.2. An investor wants to put together a portfolio consisting of the 30 stocks used to determine the Dow Jones industrial average. Use 25 weekly returns ending on the last Friday of last month to find the optimal portfolio. Experiment with different values of the parameter α and plot the corresponding points on the efficient frontier. You will need access to a nonlinear optimization solver. You may need to use a modeling language to formulate the problem for input to the solver.

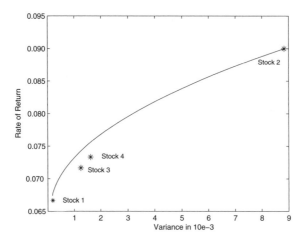

Figure 1.12. *Efficient frontier.*

1.7.4   Intensity Modulated Radiation Treatment Planning

Radiotherapy is the treatment of cancerous tissues with external beams of radiation. As a beam of radiation passes through the body, energy is deposited at points along its path, and as this happens the beam intensity gradually decreases (this is called attenuation). The radiation dosage is the amount of energy deposited locally per unit mass. High doses of radiation can kill cancerous cells, but will also damage nearby healthy cells. If vital organs receive too much radiation, serious complications may arise. Some limited damage to healthy cells may be tolerable however, since normal cells repair themselves more effectively than cancerous cells. If the radiation dosage is limited, the surrounding organs can continue to function and may eventually recover. The goal of the radiation treatment planning is to design a treatment that will kill the cancer in its entirety but limit the damage to surrounding healthy tissue.

To keep the radiation levels of normal healthy tissue low, the treatment typically uses several beams of radiation delivered from different angles. Intensity modulated radiation therapy (IMRT) is an important recent advance that allows each beam to be broken into hundreds (or possibly thousands) of beamlets of varying intensity. This is achieved using a set of metallic leaves (called collimators) that can sequentially move from open to closed position, thus filtering the radiation in a way that not only allows for the modulation of the intensity of the beam, but also enables control of its shape. This enables more accurate radiation treatment. This is particularly important in cases where the tumor has an unusual shape as is the case when it is wrapped around the spinal cord, or when it is close to a vital structure such as the optic nerve.

A simplified example of the desired goals for treatment of a hypothetical prostate cancer patient is given in Table 1.5. Radiation dosage is measured in a unit call Gray (Gy). One Gy is equal to one Joule of energy deposited in one kilogram. The planning target volume (PTV) describes a region large enough to incorporate the diseased organ, the

Table 1.5. *Sample treatment specifications.*

Volume	Requirement	
PTV excluding rectum overlap	Prescription dose	80 Gy
	Maximum dose	82 Gy
	Minimum dose	78 Gy
	95% of volume $\geq$	79 Gy
PTV/rectum overlap	Prescription dose	74 Gy
	Maximum dose	77 Gy
	Minimum dose	74 Gy
Rectum	Maximum dose	76 Gy
	70% of volume $\leq$	32 Gy
Bladder	Maximum dose	78 Gy
	70% of volume $\leq$	32 Gy

cancerous cells, as well as a margin to account for patient movement during the treatment. Organs at risk are the rectum and the bladder. Since the PTV may overlap with the rectum, different treatment specifications are given for the primary region where the PTV is distinct from the rectum, and for the region where they overlap. The specifications for the primary region, for example, include a desired "prescription" dose of 80 Gy at every cell, a minimum dose value of 78 Gy, a maximum dose of 82 Gy, and finally, a "dose-volume" requirement that specifies that 95% of the cells in this region must receive at least 79 Gy. The treatment specification for the bladder includes an upper limit of 72 Gy for the entire organ and a dose-volume requirement that 70% of the organ must receive 32 Gy or less.

To determine the treatment plan we will need to define a volume of interest that includes the PTV and any nearby tissue that may be adversely affected by the treatment. We will divide this volume into a three-dimensional grid of small boxes called *voxels*. We will denote the dose deposited in voxel i by d_i. A key decision in the treatment planning is the *fluence map*—the radiation intensity of the beamlets in each beam. Let x_j denote the intensity of beamlet j. Then the total radiation dosage deposited in the volume of interest is given approximately by the equation

$$d = Ax.$$

The matrix A is called the *fluence matrix* and is assumed to be known. Its components $a_{i,j}$ represent the amount of dose absorbed by voxel i per unit intensity emission from beamlet j.

The problem is therefore to find a fluence map x that yields a radiation dose d that meets the requirements specified by the physician, as in Table 1.5. As such, this seems to be a *feasibility problem*, namely one of finding a feasible solution, rather than an optimization problem. Unfortunately the treatment requirements are usually conflicting, and it is impossible to satisfy all the requirements simultaneously. To resolve this, the requirements are usually broken up into "hard" constraints for which any violation is prohibited, and "soft" constraints for which violations are allowed. Typically, hard constraints are included in the formulation as explicit constraints, whereas soft constraints are incorporated into the objective function via some penalty that is imposed for their violation.

For example, the requirement that region S in the primary treatment volume will receive a minimum dose l and a maximum dose u could be treated as a hard constraint by explicitly requiring that

$$l \le d_i \le u \quad \text{for all } i \in S.$$

Alternatively the requirement could be treated as a soft constraint, where a violation is allowed, but with penalty. One approach is to include in the objective function the nonlinear term

$$w_l \sum_{i \in S} \max(0, l - d_i)^2 + w_u \sum_{i \in S} \max(0, d_i - u)^2,$$

which sums up the squared deviation from the desired bound for those voxels where the bounds are violated. The parameters w_l and w_u are weights representing the relative importance of the bounds on the doses and may differ by region. For instance, underdosing the tumor can be more harmful than overdosing it, so the weights for this region satisfy $w_l \ge w_u$. For an alternative way to impose a penalty for violating the bounds on the doses, see the Exercises.

The "dose-volume constraints" that specify that a fraction β of some volume must receive a dose of u or less (or a dose of l or more) are more difficult to incorporate. As an example, suppose that the bladder volume in our example has 10,000 voxels. Then at least 7,000 of the voxels must receive 32 Gy or less. To count the number of voxels that exceed 32 Gy we must define an indicator for each voxel that determines whether its dose meets 32 Gy or exceeds it. This can be done by defining for each voxel a variable y_i that is either zero or one, depending on whether the dose meets the desired upper limit or not. Then adding the constraints

$$d_i \le 32(1 - y_i) + 78 y_i, \quad \sum_{i \in S} y_i \le 3{,}000, \quad y_i \in \{0, 1\}$$

enforces the dose-volume constraints. The first constraint implies that if d_i exceeds 32 Gy, then y_i must be one; the second implies that the number of voxels where the dose exceeds 32 Gy is at most 3,000.

This formulation expresses the dose-volume requirements as hard constraints. However models with integer variables can be difficult to solve and may require a specialized implementation. For this reason, some researchers prefer other formulations. One way to use a soft constraint for the dose-volume requirement is to add to the objective function a penalty term of the form

$$w \sum_{i \in S(d)} \max(0, d_i - 32)^2,$$

where $S(d)$ is the set of 7,000 voxels (out of the 10,000) with the lowest dose, and w is the weight of the penalty. Unfortunately, we have traded one difficulty for another. In this alternative formulation, the penalty term does not have continuous derivatives (see the Exercises), which can create challenges for many optimization algorithms.

One may wonder why there are so many different models and formulations. There are several reasons. First, because the requirements are conflicting, there is no consensus among physicians as to what should be a hard constraint and what should be a soft constraint. Second, physicians have other desired objectives in the treatment that are extremely important yet cannot be adequately modeled. For example, they are concerned about the tumor control probability—the probability that the dose delivered will indeed kill the tumor. However models that incorporate these probabilities directly are computationally impractical. As another example, physicians obtain important information from the shape of the *dose-volume histogram*, a graph displaying for each dose level the percentage of the volume that receives at least that dose amount. Ideally one would like to include constraints that enforce the dose-volume histogram to have a "good" shape, but this would amount to including numerous dose-volume constraints, which again is computationally impractical. A third factor is the trade-off between solution time and solution quality. Most commercial systems use the weighted sum of penalties since these can typically be solved efficiently. However, because all the constraints are "soft," the solutions are not always adequate. The solutions can sometimes include undesirable features, such as regions of low dosage ("cold spots") within the tumor, or regions of high dosage ("hot spots") in healthy tissue.

The problem of optimizing the fluence map can be immense. The number of voxels may range from tens of thousands to hundreds of thousands. Typically a treatment may use 5–10 beams, and the number of beamlets per beam can run into the thousands. Even if the direction of the beams is prescribed, the problem can be challenging. The problem becomes even harder if one attempts to optimize the number of beams and their directions, in addition to their fluence.

There is an additional challenge. Recall that the beamlets are formed by the movement of the leaf collimators; the longer a leaf is open, the more dose it allows to pass through. It is also necessary to determine the sequence of leaf positions and length of their open times that creates the desired fluence map—or an approximation to it—in a total sequencing time that does not unduly prolong the patient's total treatment time.

Exercises

7.1. One possible way to allow some violation of the constraint $l \leq d_i$ in a region S is to introduce for each voxel i in S two new nonnegative variables s_i' and s_i'' satisfying

$$d_i - s_i' + s_i'' = l_i$$
$$s_i', s_i'' \geq 0,$$

and to include a penalty term of the form $w_l \sum_{i \in S} s_i''$ in the objective. Explain why this approach would work, and derive an equivalent approach for the constraint $d_i \leq u$.

7.2. The purpose of this exercise is to show that when the dose-volume requirements are included as soft constraints in the objective, the resulting penalty term may have discontinuous derivatives. Consider a region with only two voxels, and suppose that it is required that not more than half the voxels exceed a dose of u. Show that the approach described in this section for incorporating this requirement as a soft constraint adds a penalty term of the form $w \max(0, (\min_{i=1,2}\{d_i\} - u))^2$ to the objective. Evaluate the gradient of this penalty term at points where it exists. Determine whether the first derivatives are continuous on $d \geq 0$.

1.7.5 Positron Emission Tomography Image Reconstruction[6]

Positron emission tomography (PET) is a medical imaging technique that helps diagnose disease and assess the effect of treatment. Unlike other imaging techniques such as X-rays or CT-scans that directly study the anatomical structure of an organ, PET studies the physiology (blood flow or level of metabolism) of the organ. Metabolic activity is an important tool in diagnosis: cancerous cells have high metabolism or high activity, while tumor cells damaged by irradiation have low metabolism or low activity. Alzheimer's disease is indicated by regions of reduced activity in the brain, and coronary tissue damage is indicated by regions of reduced activity in the heart.

In a PET scan the patient is injected with a radioactively labeled compound (most commonly glucose, but sometimes water or ammonia) that is selected for its tendency to be absorbed in the organ of interest. Once the compound settles, it starts emitting radioactive emissions that are counted by the PET scanner. The level of emissions is proportional to the amount of drug absorbed, or in turn, to the level of cell activity. The emissions are counted using a PET scanner that surrounds the body. Based on the emissions counts obtained in the scanner, the goal is to determine the level of emissions from within the organ, and hence the level of metabolic activity. The output of the reconstruction is typically presented in a color image that reflects the different activity levels in the organ.

We describe the physics of PET in further detail. As the radioisotope decays, it emits positrons. Each positron annihilates with an electron, and produces two photons which move in nearly opposite directions, each hitting a tiny photodetector within the scanner at almost the same time. Any near-simultaneous detection of an event by two such detectors defines a *coincidence event* along a *coincidence line*. The number of coincidence events y_j detected along each of the possible coincidence lines j is the input to the image reconstruction.

Consider the situation depicted in Figure 1.13, where a grid of boxes or *voxels* has been imposed over the emitting object (for simplicity, the figure is depicted in two dimensions; the concept is readily extended to three dimensions). Given a set of measurements y_j along the coincidence lines $j = 1, \ldots, N$, we seek to estimate $x_i, i = 1, \ldots n$, the expected number of counts emitted from voxel i, where n is the number of voxels in the grid.

Most reconstruction methods are based on a technique known as filtered back projection. Although this technique yields fast reconstructions, the quality of the image can be poor in situations where the amount of radioactive substance used must be small. Under such situations it is necessary to use a statistical model of the emission process to determine the most likely image that fits the data. The approach is via the maximum likelihood estimation technique. The radioactive emissions from voxels $i = 1, \ldots, n$ are assumed to be statistically independent random variables that follow a Poisson distribution with mean x_i. Denote by $C_{i,j}$ the probability that an emission emanating from voxel i will hit detector pair (coincidence line) j. The $n \times N$ matrix $C = C_{i,j}$ depends on the geometry of the scanner and on the tissue being scanned, and is assumed to be known.

---

[6]This section requires some basic concepts from probability theory.

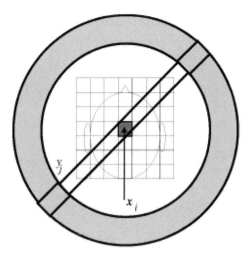

Figure 1.13. *PET.*

Using these assumptions one can show that the emissions emanating from voxel i and hitting detector pair j are also independent Poisson variables with mean rate $C_{i,j}x_i$, and the total emissions received by the detector pairs $j = 1, \ldots, N$ are independent Poisson distributed variables with mean rate $\sum_i C_{i,j}x_i$. Let $q = Ce_N$ where e_N is a vector of 1's. The vector q denotes the sum of the columns of C (which need not be 1). It is computationally easier if we write the optimization model using the logarithm of the likelihood function. If we ignore a constant term, the resulting logarithm is

$$f_{ML} = -q^T x + \sum_j y_j \log \left(C^T x \right)_j.$$

(See the Exercises.) Since the emission level is nonnegative, the final reconstruction problem becomes

$$\text{maximize} \quad f_{ML} = -q^T x + \sum_j y_j \log \left(C^T x \right)_j$$
$$\text{subject to} \quad x \geq 0.$$

The size of the problem can be enormous. If one wishes to reconstruct, say, a volume of, say, 5 cubic cm at a resolution of half a millimeter, then the size of the grid would be 100 by 100 by 100, corresponding to $n = 100{,}000$ variables. Problems of this size and even larger are not uncommon. The size of the data is also huge. The scanner may have thousands of photodetectors and since any pair of these can define a coincidence line, the number of coincidence lines N can be on the order of millions. Since every function evaluation requires the computation of a matrix product $C^T x$, and the matrix C is large, the function evaluations are time consuming.

The efficient solution of such large problems often requires understanding of their structure. By structure we mean special characteristics of the function, its gradient, and Hessian. Often structure is associated with the sparsity pattern of the Hessian, that is the number of zeros, and possibly their location. The special structure of f_{ML} and its derivatives can be used in designing effective methods for solving the problem. Here we will just give the formulas for the derivatives. Defining

$$\hat{y} = C^T x,$$

we can write the gradient and Hessian of the objective function, respectively, as

$$\nabla f_{ML}(x) = -q + C\hat{Y}^{-1}y,$$
$$\nabla^2 f_{ML}(x) = -CY\hat{Y}^{-2}C^T,$$

where $Y = \text{diag}(y)$ and $\hat{Y} = \text{diag}(\hat{y})$. The matrix C itself is sparse, and only a small fraction of its entries are nonzero. The diagonal matrices Y and $\hat{Y}$ are of course also sparse. Even so, the Hessian $\nabla^2 f_{ML}(x)$ is dense; almost all of its entries are likely to be nonzero. A key challenge in the design of effective algorithms is to exploit the sparsity of C.

Exercises

7.1. The goal of this exercise is to derive the maximum likelihood model for PET image reconstruction. Parts (a) and (b) require some basic background in stochastic methods.

    (i) Let Z_{ij} be the number of events emitted from voxel i and detected at coincidence line j, and let Y_j be the total emissions received by detector pair j, for $j = 1, \ldots, N$. Use the assumptions given in the section to prove that $\{ Z_{ij} \}$ are independent Poisson variables with mean $C_{i,j}x_i$, and that $\{ Y_j \}$ are independent Poisson distributed variables with mean rates $\hat{y}_j = \sum_i C_{i,j}x_i$.

    (ii) Prove that the likelihood may be written as

$$P\{y|x\} = \prod_j \frac{e^{-\hat{y}_j}\hat{y}_j^{y_j}}{y_j!} = \prod_j \frac{e^{-\sum_i C_{i,j}x_i}(\sum_i C_{i,j}x_i)^{y_j}}{y_j!}.$$

    (iii) Prove the final expression for the maximum likelihood estimation objective function f_{ML}. Hint: Take the logarithm of the likelihood and omit the constant term that does not depend on x.

7.2. Derive the formulas for the gradient and Hessian matrix of f_{ML}.

7.3. The purpose of this exercise is to show that the Hessian of f_{ML} may be dense, even when its matrix factors are sparse. Suppose that $C = (I \quad I \quad e_n)$ and $y = \hat{y} = e_{2n+1}$ where I is the identity matrix, and e_k is a vector of ones of size k. Show that every element of the Hessian is nonzero.

7.4. The purpose of this problem is to write a program in the modeling language of your choice to solve a PET image reconstruction problem. Your model should not only be correct, but also efficient and clear. Try and make your model as general as possible.

    (i) Develop the model and test it on a problem with $n = 9$ variables corresponding to a 3×3 grid, and with $N = 33$ detector pairs. The data are

$$C = (\; B \quad B \quad B \;),$$

where B is a sparse $n \times (n + 2)$ matrix with the following nonzero entries:

$$B_{i,i} = a, \quad B_{i,i+1} = b, \quad B_{i,i+2} = a, \quad i = 1, \ldots, n,$$

where

$$a = 0.18, \quad b = 0.017,$$

and

$$y^T = (\; 0 \quad 0 \quad 1 \quad 19 \quad 27 \quad 30 \quad 40 \quad 50 \quad 35 \quad 15 \quad 1 \quad \ldots$$
$$0 \quad 0 \quad 1 \quad 7 \quad 20 \quad 38 \quad 56 \quad 55 \quad 38 \quad 20 \quad 7 \quad \ldots$$
$$1 \quad 0 \quad 1 \quad 3 \quad 17 \quad 38 \quad 40 \quad 20 \quad 7 \quad 1 \quad 0 \;).$$

(ii) Test your software on a problem with $n = 1080$ variables corresponding to a 36×30 grid, and with $N = 1444$ detector pairs. The data are

$$C = (\; B \quad 2B \;),$$

where B is defined as in part (a) with the parameter values $a = 0.15$ and $b = 0.05$. The vector y can be downloaded in text format from the Web page for this book (http://www.siam.org/books/ot108). Display the values of the first row of the reconstructed image.

(iii) Identify the image you obtained in (ii). You will need software for displaying intensity images.

1.7.6 Shape Optimization

In this section we show how nonlinear optimization can address a problem of finding the shape of a hanging cable, which in equilibrium minimizes the potential energy of the cable. This problem often is called the catenary problem (from the Latin word "catena" meaning a chain).

The solution to the simplest case of the hanging cable problem, when the mass of the cable is uniformly distributed along the cable, was found at the end of the 18th century independently by John Bernoulli, Christian Huygens, and Gottfried Leibniz.

More recently, the catenary has played an important role in civil engineering. The solution to the catenary problem helps understand the effects on suspended cables of external applied forces arising from the live loads on a suspension bridge.

Here we demonstrate how a general hanging cable problem can be modeled as an optimization problem. We present several optimization models to illustrate that sometimes a physical problem can have multiple equivalent mathematical formulations, some of which are numerically tractable while others are not.

First, for simplicity we assume that the mass of the cable is distributed uniformly. The objective will be to minimize the potential energy of the cable

$$\underset{y(x)}{\text{minimize}} \int_{x_a}^{x_b} y(x)\sqrt{1 + y'(x)^2}\,dx.$$

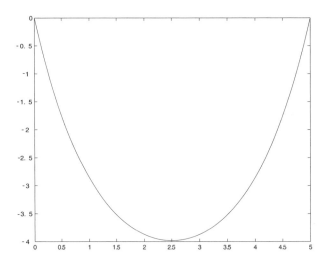

Figure 1.14. *Hanging cable with uniformly distributed mass.*

Here $y(x)$ is the height of the cable measured from some zero level, and $\sqrt{1 + y'(x)^2}dx$ is the arc length, which is proportional to mass since the mass is distributed uniformly. The model also has constraints: the cable has a specified length L

$$\int_{x_a}^{x_b} \sqrt{1 + y'(x)^2}dx = L,$$

and the ends of the cable are fixed

$$y(x_a) = y_a, \quad y(x_b) = y_b.$$

It can be shown that the solution to this problem is a hyperbolic cosine

$$y(x) = C_0 \cosh\left(\frac{x+C_1}{C_0}\right) + C_2,$$

where $\cosh(x) = (e^x + e^{-x})/2$ and the values of C_0, C_1, and C_2 are determined by the constraints. Figure 1.14 shows the graphical representation of $y(x)$.

In contrast to our previous optimization models where we had a finite number of variables, here we are seeking an optimal *function*, that is, an infinite continuum of values. In order to solve such a problem using nonlinear optimization algorithms, we *discretize* the function by approximating it at a finite number of points, as shown in Figure 1.15.

Here we describe the simplest method for discretizing such problems. If $x_a = x_0 < x_1 < \cdots < x_{N-1} < x_N = x_b$ is a uniform discretization of segment $[x_a, x_b]$ such that $\Delta x = x_1 - x_0 = x_2 - x_1 = \cdots = x_N - x_{N-1}$, a simple approximation to an integral of a function $f(x)$ is

$$\int_a^b f(x)dx \approx \sum_{i=0}^{N-1} f(x_i)\Delta x.$$

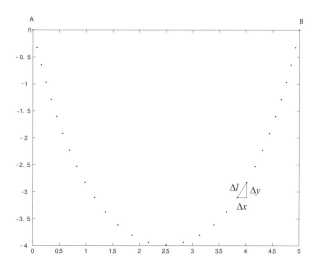

Figure 1.15. *Discretized hanging cable with uniformly distributed mass.*

The function values used to approximate the integral for the shape optimization problem are $f(x_i) = y(x_i)\sqrt{1 + y'(x_i)^2}$. We will approximate the values of the derivative $y'(x)$ at the discretization points x_i by

$$y_i' = \frac{y_{i+1} - y_i}{\Delta x}, \quad i = 0, 1, \ldots, N - 1,$$

where $y_i = y(x_i)$. The discretized problem consists of finding variables y_i, $i = 1, \ldots, N-1$, and y_i', $i = 0, \ldots, N - 1$, that solve the problem

$$\text{minimize} \quad E(y, y') = \sum_{i=0}^{N-1} y_i \sqrt{1 + (y_i')^2} \, \Delta x$$

$$\text{subject to} \quad y_{i+1} = y_i + y_i' \Delta x, \quad i = 0, \ldots, N - 1$$

$$\sum_{i=0}^{N-1} \sqrt{1 + (y_i')^2} \, \Delta x = L$$

$$y_0 = y_a, \quad y_N = y_b.$$

We refer to this as *optimization model 1*.

The greater the number of discretizations points N, the better the solution to optimization model 1 approximates the solution of the original problem. However for very large N, the optimization model 1 is difficult to solve. The constraint $\sum_{i=1}^{N} \sqrt{1 + (y_i')^2} \Delta x = L$ is nonlinear and can be a source of numerical difficulties for optimization algorithms. In the two-dimensional case this constraint defines the perimeter shown in Figure 1.16 (left). The point x_0 is on the perimeter and hence is feasible, but almost any perturbation of x_0 will move off the perimeter and hence out of the feasible region. Fortunately, there is another formulation of the catenary problem that leads to a more tractable model.

Rather than representing the cable as a function $y(x)$ of the variable x, we parameterize it as a function of its length with respect to its left end point. The points on the cable will

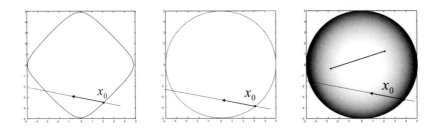

Figure 1.16. *Feasible regions.*

now have the form $(x(l), y(l)), l \in [0, L]$. This representation leads to a model that is simpler to analyze both mathematically and numerically.

Now we look for $(x(l), y(l)), l \in [0, L]$, which minimizes the potential energy

$$\min \int_0^L m(l) y(l) dl$$

subject to a constraint based on the Pythagorean theorem that defines the relations between dx, dy, and dl (see Figure 1.15),

$$dx^2 + dy^2 = dl^2,$$

and the ends of the cable are fixed

$$x(0) = x_a, \quad y(0) = y_a, \quad x(L) = x_b, \quad y(L) = y_b.$$

Here $m(l)$ is a mass distribution function such that $\int_0^L m(l) dl = M$ is the total mass of the cable.

The discretization of this problem with the uniform distribution of mass and the total mass of the cable M consists of finding variables x_l, $l = 1, \ldots, N - 1$, and y_i, $i = 1, \ldots, N - 1$, using the following *optimization model* 2:

$$\text{minimize} \quad E(y) = \frac{M}{N} \sum_{l=0}^{N} y_l$$

$$\text{subject to} \quad (x_l - x_{l-1})^2 + (y_l - y_{l-1})^2 = \left(\tfrac{L}{N}\right)^2, \quad l = 1, \ldots, N$$
$$x_0 = x_a, \quad x_N = x_b$$
$$y_0 = y_a, \quad y_N = y_b,$$

where the mass distribution function is $m = \text{const} = M/N$. This optimization model also has N nonlinear constraints:

$$(x_l - x_{l-1})^2 + (y_l - y_{l-1})^2 = \left(\tfrac{L}{N}\right)^2, \quad l = 1, \ldots, N,$$

which again can be a potential source of difficulties for optimization algorithms if N is large (the two-dimensional case is shown in Figure 1.16 (center)). However, the optimization model 2 can be simplified substantially by relaxing these constraints into inequalities:

$$(x_l - x_{l-1})^2 + (y_l - y_{l-1})^2 \leq \left(\tfrac{L}{N}\right)^2, \quad l = 1, \ldots, N.$$

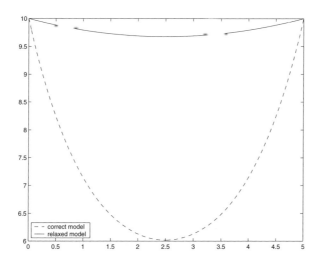

Figure 1.17. *Constraints cannot always be relaxed.*

Of course, we changed the formulation, which is legitimate only if we can prove that the new formulation has the same solution as the original one. In other words, we have to prove that both optimization model 2 and its relaxation have the same solution. We can prove this by contradiction. Suppose that the optimal solution of the relaxed model satisfies at least one constraint as a strict inequality. Then we can lower the discretized components of the solution corresponding to this constraint and still remain feasible. But lowering part of the cable decreases the potential energy, i.e., it decreases the objective function, so our solution could not have been optimal. This contradicts our original assumption.

Thus optimization model 2 and its relaxation have the same optimal solutions. But the two models are not equivalent computationally, since the feasible region for the relaxation has properties that make it easier for optimization algorithms to handle. In the two-dimensional case, the feasible region of the relaxed optimization model 2 is shown in Figure 1.16 (right). It is the entire circle, not just its perimeter. If x_0 is a feasible point in the interior of the feasible region, any small perturbation of x_0 is also in the interior. This feasible region has a convex shape; i.e., if we connect any two points from the feasible set, all the points between them are also feasible. This property of the interior of feasible set helps some optimization algorithms, later described in the book, efficiently find the solution.

It is apparently not possible to relax the constraints of optimization model 1 without changing the optimal solution, but it is easy to do so with optimization model 2. Relaxation of the nonlinear equality in optimization model 1 to an inequality

$$\sum_{i=1}^{N} \sqrt{1 + (y_i')^2} \Delta x \leq L$$

gives a model that is not equivalent and can result in an incorrect solution as shown in Figure 1.17. In this example, the length of an optimal cable for the relaxed model is less than L.

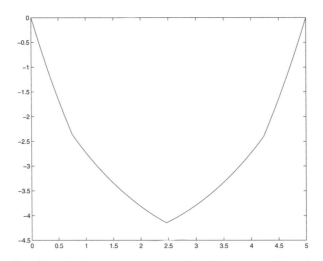

Figure 1.18. *Hanging cable with a nonuniform mass distribution.*

Optimization model 2 has another attractive property. Mathematicians in the 18th century assumed that the string is flexible and uniform, which implies that every segment of equal length has equal mass. This assumption is too restrictive for modern engineering. In many practical problems the total weight of the cable is not uniformly distributed along the cable.

If the mass distribution function is not uniform along the cable but instead is a general known function $m(l)$, then it is still easy to obtain a solution of a hanging cable problem using optimization model 2. We just have to replace the objective function $\frac{M}{N} \sum_{i=0}^{N+1} y_i$ with a more general linear objective function $\sum_{i=0}^{N+1} m_i y_i$ with appropriately selected coefficients m_i corresponding to a certain distribution of mass along the cable. For example, if the mass of most nodes is much smaller than that of three special nodes—the center node and the two nodes one quarter of the length away from both end points—then it is still easy to find the shape of such a cable (see Figure 1.18). We would not be able to easily model such a case using optimization model 1, for which the assumption of uniformly distributed mass is essential.

We conclude the section by emphasizing the importance of proper modeling of a problem. It is the responsibility of a modeler not to make the formulation more difficult than it need be. A problem that is computationally challenging in one formulation may become much easier to solve in a different formulation. It is up to the modeler to carefully consider the merits of a formulation prior to solving the problem.

1.8   Notes

Further information on integer programming can be found in the book by Wolsey (1998). References on global optimization are listed in the Notes for Chapter 2.

Overviews of the crew scheduling, fleet assignment problem, and other airline scheduling problems are given in the articles by Barnhart et al. (1999) and Gopalan and Talluri (1998);

methods for solving the related linear program are described in the paper by Bixby et al. (1992). The portfolio problem is described in the book by Markowitz and Todd (2000). An innovative approach to calculating the entire efficient frontier by solving just one linear programming problem using a specialized parametric method was developed by Ruszczynski and Vanderbei (2003).

The concept of support vector machines was initially developed by Vapnik (1998) in the late 1970s. A comprehensive overview on the subject is found in the tutorial by Burges (1988). More recent research is discussed in the books by Cristianini and Shawe-Taylor (2000), and by Schökopf et al. (1999).

Overviews of IMRT planning can be found in the articles by Shepard et al. (1999) and by Lee and Deasy (2006). The book by Herman (1980) and the papers by Shepp and Vardi (1982) and Lange and Carson (1984) are among the pioneering works pertaining to PET. Figure 1.13 is due to Calvin Johnson, and was taken from the paper by Johnson and Sofer (2001). Further applications of optimization can be found in the books by Vanderbei (2007), and by Fourer, Gay, and Kernighan (2003).

The hanging cable or catenary problem was first posed in the *Acta Eruditorium* in 1690 by Jacob Bernoulli. Simple catenary problems can be solved analytically. More complicated cases, those with nonuniformly distributed mass, may have to be solved numerically. More details about how to find shapes of a hanging cable analytically and numerically can be found in the paper of Griva and Vanderbei (2005) and many books on variational calculus; see, e.g., Gelfand and Fomin (1963, reprinted 2000).

Chapter 2

Fundamentals of Optimization

2.1 Introduction

This chapter discusses basic optimization topics that are relevant to both linear and nonlinear problems. Sections 2.2–2.4 discuss local and global optima, convexity, and the general form of an optimization algorithm. These topics have traditionally been considered as fundamental topics in all areas of optimization. The later sections of the chapter, discussing rates of convergence, series approximations to nonlinear functions, and Newton's method for nonlinear equations, are most relevant to nonlinear optimization. In fact, Part II on linear programming can be understood without these later sections.

The later topics are basic to discussions of nonlinear optimization, since they allow us to derive optimality conditions and develop and analyze algorithms for optimization problems involving nonlinear functions.

Although not essential, these topics give a fuller understanding of linear programming as well. For example, "interior-point" methods apply nonlinear optimization techniques to linear programming. They might use Newton's method to find a solution to the optimality conditions for a linear program, or use a nonlinear optimization algorithm on a linear programming problem. The tools from this chapter underlie the interior-point methods derived in Chapter 10.

2.2 Feasibility and Optimality

There are a variety of terms that are used to describe feasible and optimal points. We first discuss the terms associated with feasibility.

We consider a set of constraints of the form

$$g_i(x) = 0, \quad i \in \mathcal{E},$$
$$g_i(x) \geq 0, \quad i \in \mathcal{I}.$$

Here $\{g_i\}$ are given functions that define the constraints in the model, $\mathcal{E}$ is an index set for the equality constraints, and $\mathcal{I}$ is an index set for the inequality constraints. Any set of equations and inequalities can be rearranged in this form. For example, the equation $3x_1^2 + 2x_2 = 3x_3 - 9$ could be written as

$$g_1(x) = 3x_1^2 + 2x_2 - 3x_3 + 9 = 0,$$

and the inequality $\sin x_1 \le \cos x_2$ is equivalent to

$$g_2(x) = -\sin x_1 + \cos x_2 \ge 0.$$

Such transformations are merely cosmetic, but they simplify the notation for describing the constraints.

A point that satisfies all the constraints is said to be *feasible*. The set of all feasible points is termed the *feasible region* or *feasible set*. We shall denote it by S.

At a feasible point $\bar{x}$, an inequality constraint $g_i(x) \ge 0$ is said to be *binding* or *active* if $g_i(\bar{x}) = 0$, and *nonbinding* or *inactive* if $g_i(\bar{x}) > 0$. The point $\bar{x}$ is said to be on the *boundary* of the constraint in the former case, and in the *interior* of the constraint in the latter. All equality constraints are regarded as active at any feasible point. The *active set* at a feasible point is defined as the set of all constraints that are active at that point. The set of feasible points for which at least one inequality is binding is called the *boundary* of the feasible region. All other feasible points are *interior points*. (Interior points are only "interior" to the inequality constraints. If equality constraints are present, any feasible point will satisfy them. Since it is not possible to be interior to an equality constraint, some authors use the term *relative interior points*.)

Figure 2.1 illustrates the feasible region defined by the constraints

$$g_1(x) = x_1 + 2x_2 + 3x_3 - 6 = 0$$
$$g_2(x) = x_1 \ge 0$$
$$g_3(x) = x_2 \ge 0$$
$$g_4(x) = x_3 \ge 0.$$

At the feasible point $x_a = (0, 0, 2)^T$, the first two inequality constraints $x_1 \ge 0$ and $x_2 \ge 0$ are active, while the third is inactive. At the point $x_b = (3, 0, 1)^T$ only the second inequality is active, while at the interior point $x_c = (1, 1, 1)^T$ none of the inequalities are active. The boundary of the feasible region is indicated by bold lines.

Let us now look at terms associated with optimality. It may seem surprising that there is any question about what is meant by a "solution" to an optimization problem. The confusion arises because there are a variety of conditions associated with an optimal point and each of these conditions gives rise to a slightly different notion of a "solution."

Let us consider the n-dimensional problem

$$\underset{x \in S}{\text{minimize}}\ f(x).$$

There is no fundamental difference between minimization and maximization problems. We can maximize f by solving

$$\underset{x \in S}{\text{minimize}}\ (-f(x)),$$

and then multiplying the optimal objective value by -1. For this reason, it is sufficient to discuss minimization problems only.

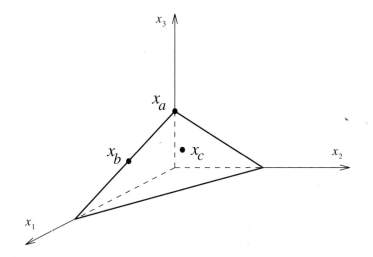

Figure 2.1. *Example of feasible region.*

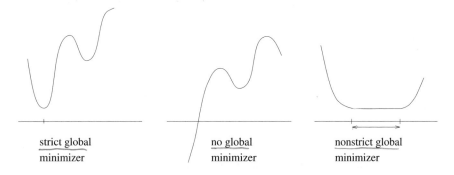

Figure 2.2. *Examples of global minimizers.*

The set S of feasible points is usually defined by a set of constraints, as above. For problems without constraints, the set S would be $\Re^n$, the set of vectors of length n whose components are real numbers.

The most basic definition of a solution is that x_* minimizes f if

$$f(x_*) \leq f(x) \quad \text{for all } x \in S.$$

The point x_* is referred to as a *global minimizer* of f in S. If in addition x_* satisfies

$$f(x_*) < f(x) \quad \text{for all } x \in S \text{ such that } x \neq x_*,$$

then x_* is a *strict global minimizer*. Not all functions have a finite global minimizer, and even if a function has a global minimizer there is no guarantee that it will have a strict global minimizer; see Figure 2.2.

It would be satisfying theoretically, and important practically, to be able to find global minimizers. However, many of the methods that we will study are based on the Taylor

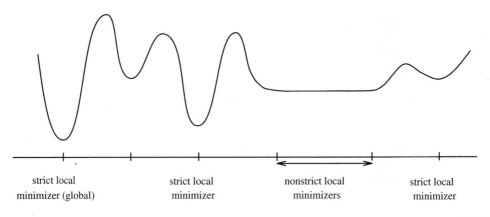

strict local   strict local   nonstrict local   strict local
minimizer (global)   minimizer   minimizers   minimizer

Figure 2.3. *Examples of local minimizers.*

series; that is, they are based on information about the function at a single point, and this information is normally only be valid within a small neighborhood of that point (see Section 2.6). Without additional information or assumptions about the problem it will not be possible to guarantee that a global solution has been found. An important exception is in the case where the function f and the set S are convex (see Section 2.3), which is true for linear programming problems.

   If we cannot find the global solution, then at the least we would like to find a point that is better than its surrounding points. More precisely, we would like to find a *local minimizer* of f in S, a point satisfying

$$f(x_*) \leq f(x) \quad \text{for all } x \in S \text{ such that } \|x - x_*\| < \epsilon.$$

Here ϵ is some small positive number that may depend on x_*. The point x_* is a *strict local minimizer* if

$$f(x_*) < f(x) \quad \text{for all } x \in S \text{ such that } x \neq x_* \text{ and } \|x - x_*\| < \epsilon.$$

Various one-dimensional examples are illustrated in Figure 2.3.

   In many important cases, strict local minimizers can be identified using first and second derivative values at $x = x_*$, and hence they can be identified by algorithms that compute first and second derivatives of the problem functions. (A local minimizer that is not a strict local minimizer is a degenerate case and is often considered to be a special situation.) Many algorithms, in particular those that only compute first derivative values, are only guaranteed to find a *stationary point* for the problem. (For unconstrained problems a stationary point is a point where the first derivatives of f are equal to zero. For constrained problems the definition is more complicated; see Chapter 14.) A local minimizer of f is also a stationary point of f but the reverse need not be true.

   Having all these various definitions of what is meant by a solution may seem perverse, but it merely reflects the fact that if we only have limited information, then we can draw only limited conclusions. The definitions are not without merit, though. In the case where all these various types of solutions are defined and where the function has several continuous derivatives, a global solution will also be both a local solution and a stationary point. In important special cases such as linear programming the reverse will also be true. In our experience, it is unusual for an algorithm to converge to a point that is a stationary point but not a local minimum. However, it is common for an algorithm to converge to a local minimum that is not a global minimum.

It may seem troubling that a local but not global solution is often found, but in many practical situations this can be acceptable if the local minimizer produces a satisfactory reduction in the value of the objective function. For example, if the objective function represented the costs of running a business, a 10% reduction in these costs would be a valuable saving, even if it did not correspond to the global solution to the optimization problem. Local optimization techniques are a valuable tool even if global solutions are desired, since techniques for global optimization typically solve a sequence of local optimization problems.

Exercises

2.1. Consider the feasible region defined by the constraints

$$1 - x_1^2 - x_2^2 \geq 0, \quad \sqrt{2} - x_1 - x_2 \geq 0, \quad \text{and} \quad x_2 \geq 0.$$

For each of the following points, determine whether the point is feasible or infeasible, and (if it is feasible) whether it is interior to or on the boundary of each of the constraints: $x_a = (\frac{1}{2}, \frac{1}{2})^T$, $x_b = (1, 0)^T$, $x_c = (-1, 0)^T$, $x_d = (-\frac{1}{2}, 0)^T$, and $x_e = (1/\sqrt{2}, 1/\sqrt{2})^T$.

2.2. Consider the one-variable function

$$f(x) = (x + 1)x(x - 2)(x - 5) = x^4 - 6x^3 + 3x^2 + 10x.$$

Graph this function and locate (approximately) the stationary points, local minima, and global minima.

2.3. Consider the problem

$$\begin{array}{ll} \text{minimize} & f(x) = x_1 \\ \text{subject to} & x_1^2 + x_2^2 \leq 4 \\ & x_1^2 \geq 1. \end{array}$$

Graph the feasible set. Use the graph to find all local minimizers for the problem, and determine which of those are also global minimizers.

2.4. Consider the problem

$$\begin{array}{ll} \text{minimize} & f(x) = x_1 \\ \text{subject to} & (x_1 - 1)^2 + x_2^2 = 1 \\ & (x_1 + 1)^2 + x_2^2 = 1. \end{array}$$

Graph the feasible set. Are there local minimizers? Are there global minimizers?

2.5. Give an example of a function that has no global minimizer and no global maximizer.

2.6. Provide definitions for a global maximizer, a strict global maximizer, a local maximizer, and a strict local maximizer.

2.7. Consider minimizing $f(x)$ for $x \in S$ where S is the set of integers. Prove that every point in S is a local minimizer of f.

2.8. Let $S = \{\, x : g_i(x) \geq 0, i = 1, \ldots, m \,\}$ and assume that the functions $\{\, g_i \,\}$ are continuous. Prove that if $g_i(\hat{x}) > 0$ for all i, then $\{\, x : \|x - \hat{x}\| < \epsilon \,\} \subset S$ for some $\epsilon > 0$.

2.9. Let S be the feasible region in Figure 2.1. Show that S can be represented by equality and inequality constraints in such a way that it has no interior points. Thus the interior of a set may depend on the way it is represented.

2.10. Let $S = \{\, x : g_i(x) \geq 0, i = 1, \ldots, m \,\}$ and assume that the functions $\{\, g_i \,\}$ are continuous. Assume that there exists a point $\hat{x}$ such that $g_i(\hat{x}) > 0$ for all i. Prove that S has a nonempty interior regardless of how S is represented.

2.3   Convexity

There is one important case where global solutions can be found, the case where the objective function is a convex function and the feasible region is a convex set. Let us first talk about the feasible region.

A set S is *convex* if, for any elements x and y of S,

$$\alpha x + (1 - \alpha)y \in S \quad \text{for all } 0 \leq \alpha \leq 1.$$

In other words, if x and y are in S, then the line segment connecting x and y is also in S. Examples of convex and nonconvex sets are given in Figure 2.4. More generally, every set defined by a system of linear constraints is a convex set; see the Exercises.

A function f is *convex* on a convex set S if it satisfies

$$f(\alpha x + (1 - \alpha)y) \leq \alpha f(x) + (1 - \alpha)f(y)$$

for all $0 \leq \alpha \leq 1$ and for all $x, y \in S$. This definition says that the line segment connecting the points $(x, f(x))$ and $(y, f(y))$ lies on or above the graph of the function; see Figure 2.5. Intuitively, the graph of the function is bowl shaped.

Analogously, a function is concave on S if it satisfies

$$f(\alpha x + (1 - \alpha)y) \geq \alpha f(x) + (1 - \alpha)f(y)$$

convex   nonconvex

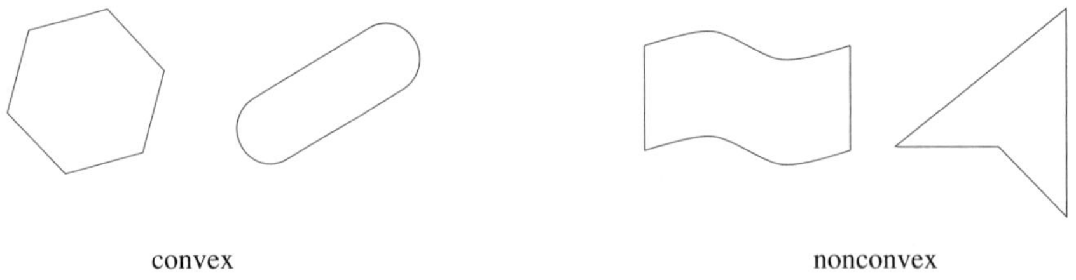

Figure 2.4. *Convex and nonconvex sets.*

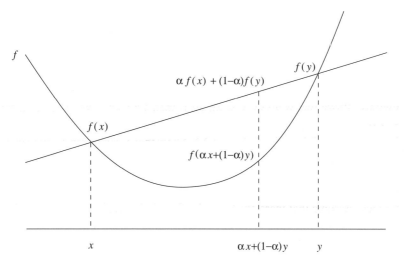

Figure 2.5. *Convex function.*

for all $0 \leq \alpha \leq 1$ and for all $x, y \in S$. Concave functions are explored in the Exercises below. Linear functions are both convex and concave.

We say that a function is *strictly convex* if

$$f(\alpha x + (1 - \alpha)y) < \alpha f(x) + (1 - \alpha)f(y)$$

for all $x \neq y$ and $0 < \alpha < 1$ where $x, y \in S$.

Let us now return to the discussion of local and global solutions. We define a *convex optimization problem* to be a problem of the form

$$\underset{x \in S}{\text{minimize}} \ f(x),$$

where S is a convex set and f is a convex function on S. A problem

$$\begin{aligned} \text{minimize} \quad & f(x) \\ \text{subject to} \quad & g_i(x) \geq 0, \ i = 1, \dots, m, \end{aligned}$$

is a convex optimization problem if f is convex and the functions $\{ g_i \}$ are concave; see the Exercises.

The following theorem shows that any local solution of such a problem is also a global solution. This result is important to linear programming, since every linear program is a convex optimization problem.

Theorem 2.1 (Global Solutions of Convex Optimization Problems). *Let x_* be a local minimizer of a convex optimization problem. Then x_* is also a global minimizer. If the objective function is strictly convex, then x_* is the unique global minimizer.*

Proof. The proof is by contradiction. Let x_* be a local minimizer and suppose, by contradiction, that it is not a global minimizer. Then there exists some point $y \in S$ satisfying

$f(y) < f(x_*)$. If $0 < \alpha < 1$, then

$$f(\alpha x_* + (1 - \alpha)y) \leq \alpha f(x_*) + (1 - \alpha) f(y)$$
$$< \alpha f(x_*) + (1 - \alpha) f(x_*) = f(x_*).$$

This shows that there are points arbitrarily close to x_* (i.e., when α is arbitrarily close to 1) whose function values are strictly less than $f(x_*)$. These points are in S because S is convex. This contradicts the definition of a local minimizer. Hence a point such as y cannot exist, and x_* must be a global minimizer.

If the objective function is strictly convex, then a similar argument can be used to show that x_* is the unique global minimizer; see the Exercises. □

For general problems it may be as difficult to determine if the function f and the region S are convex as it is to find a global solution, so this result is not always useful. However, there are important practical problems, such as linear programs, where convexity can be guaranteed.

We conclude this section by defining a *convex combination* (weighted average) of a finite set of points. A convex combination is a linear combination whose coefficients are nonnegative and sum to one. Algebraically, the point y is a convex combination of the points $\{ x_i \}_{i=1}^{k}$ if

$$y = \sum_{i=1}^{k} \alpha_i x_i,$$

where

$$\sum_{i=1}^{k} \alpha_i = 1 \quad \text{and} \quad \alpha_i \geq 0, \quad i = 1, \ldots, k.$$

There will normally be many ways in which y can be expressed as a convex combination of $\{ x_i \}$.

As an example, consider the points $x_1 = (0, 0)^T$, $x_2 = (1, 0)^T$, $x_3 = (0, 1)^T$, and $x_4 = (1, 1)^T$. If $y = (\frac{1}{2}, \frac{1}{2})^T$, then y can be expressed as a convex combination of $\{ x_i \}$ in the following ways:

$$y = 0x_1 + \tfrac{1}{2}x_2 + \tfrac{1}{2}x_3 + 0x_4$$
$$= \tfrac{1}{2}x_1 + 0x_2 + 0x_3 + \tfrac{1}{2}x_4$$
$$= \tfrac{1}{4}x_1 + \tfrac{1}{4}x_2 + \tfrac{1}{4}x_3 + \tfrac{1}{4}x_4,$$

and so forth.

2.3.1   Derivatives and Convexity

If a one-dimensional function f has two continuous derivatives, then an alternative definition of convexity can be given that is often easier to check. Such a function is convex if and only if

$$f''(x) \geq 0 \quad \text{for all } x \in S;$$

see the Exercises in Section 2.6. For example, the function $f(x) = x^4$ is convex on the entire real line because $f''(x) = 12x^2 > 0$ for all x. The function $f(x) = \sin x$ is neither convex nor concave on the real line because $f''(x) = -\sin x$ can be both positive and negative.

In the multidimensional case the Hessian matrix of second derivatives must be positive semidefinite; that is, at every point $x \in S$

$$y^T \nabla^2 f(x) y \geq 0 \quad \text{for all } y;$$

see the Exercises in Section 2.6. (The Hessian matrix is defined in Appendix B.4.) Notice that the vector y is not restricted to lie in the set S. The quadratic function

$$f(x_1, x_2) = 4x_1^2 + 12x_1 x_2 + 9x_2^2$$

is convex over any subset of $\Re^2$ since

$$y^T \nabla^2 f(x) y = (y_1 \quad y_2) \begin{pmatrix} 8 & 12 \\ 12 & 18 \end{pmatrix} \begin{pmatrix} y_1 \\ y_2 \end{pmatrix}$$
$$= 8y_1^2 + 24y_1 y_2 + 18y_2^2$$
$$= 2(2y_1 + 3y_2)^2 \geq 0.$$

Alternatively, it would have been possible to show that the eigenvalues of the Hessian matrix were all greater than or equal to zero.

In the one-dimensional case, if a function satisfies

$$f''(x) > 0 \quad \text{for all } x \in S,$$

then it is strictly convex on S. In the multidimensional case, if the Hessian matrix $\nabla^2 f(x)$ is positive definite for all $x \in S$, then the function is strictly convex on S. This is not an "if and only if" condition, since the Hessian of a strictly convex function need not be positive definite everywhere (see the Exercises).

Now we consider another characterization of convexity that can be applied to functions that have one continuous derivative. In this case a function f is convex over a convex set S if and only if it satisfies

$$f(y) \geq f(x) + \nabla f(x)^T (y - x)$$

for all $x, y \in S$. This property states that the function is on or above any of its tangents. (See Figure 2.6.)

To prove this property, note that if f is convex, then for any x and y in S and for any $0 < \alpha \leq 1$,

$$f(\alpha y + (1 - \alpha)x) \leq \alpha f(y) + (1 - \alpha) f(x),$$

so that

$$\frac{f(x + \alpha(y - x)) - f(x)}{\alpha} \leq f(y) - f(x).$$

If we let α approach 0 from above, we can conclude that $f(y) \geq f(x) + \nabla f(x)^T (y - x)$.

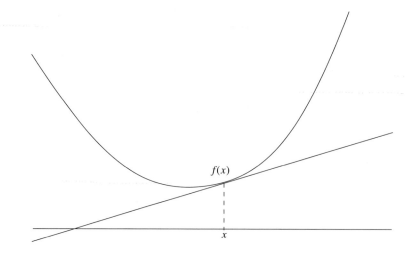

Figure 2.6. *Convex function with continuous first derivative.*

Conversely, suppose that the function f satisfies $f(y) \geq f(x) + \nabla f(x)^T(y - x)$ for all x and y in S. Let $t = \alpha x + (1 - \alpha)y$. Then t is also in the set S, so

$$f(x) \geq f(t) + \nabla f(t)^T(x - t)$$

and

$$f(y) \geq f(t) + \nabla f(t)^T(y - t).$$

Multiplying the two inequalities by α and $1 - \alpha$, respectively, and then adding yields the desired result. See the Exercises for details.

Exercises

3.1. Prove that the intersection of a finite number of convex sets is also a convex set.

3.2. Let $S_1 = \{x : x_1 + x_2 \leq 1, x_1 \geq 0\}$ and $S_2 = \{x : x_1 - x_2 \geq 0, x_1 \leq 1\}$, and let $S = S_1 \cup S_2$. Prove that S_1 and S_2 are both convex sets but S is not a convex set. This shows that the union of convex sets is not necessarily convex.

3.3. Consider a feasible region S defined by a set of linear constraints

$$S = \{x : Ax \leq b\}.$$

Prove that S is convex.

3.4. Prove that a function f is concave if and only if $-f$ is convex.

3.5. Let $f(x)$ be a function on $\Re^n$. Prove that f is both convex and concave if and only if $f(x) = c^T x$ for some constant vector c.

3.6. Prove that a convex combination of convex functions all defined on the same convex set S is also a convex function on S.

3.7. Let f be a convex function on a convex set $S \in \mathfrak{R}^n$. Let k be a nonzero scalar, and define $g(x) = kf(x)$. Prove that if $k > 0$, then g is a convex function on S, and if $k < 0$, then g is a concave function on S.

3.8. (Jensen's Inequality.) Let f be a function on a convex set $S \in \mathfrak{R}^n$. Prove that f is convex if and only if

$$f\left(\sum_{i=1}^{k} \alpha_i x_i\right) \le \sum_{i=1}^{k} \alpha_i f(x_i)$$

for all $x_1, \ldots, x_m \in S$ and $0 \le \alpha_i \le 1$ where $\sum_{i=1}^{k} \alpha_i = 1$.

3.9. Prove the well-known inequality between the arithmetic mean and the geometric mean of a set of positive numbers:

$$(x_1 + \cdots + x_k)/k \ge (x_1 \cdots x_k)^{1/k}.$$

Hint: Apply the previous problem to the function $f(x) = -\log(x)$.

3.10. Consider the function $f(x_1, x_2) = \alpha x_1^p x_2^q$, defined on $S = \{x : x > 0\}$. For what values of α, p, and q is the function convex? Strictly convex? For what values is it concave? Strictly concave?

3.11. Consider the problem

$$\underset{x \in S}{\text{maximize}} \ f(x),$$

where S is a convex set and f is a concave function. Prove that any local maximizer is also a global maximizer.

3.12. Let $g_1, \ldots, g_m$ be concave functions on $\mathfrak{R}^n$. Prove that the set

$$S = \{x : g_i(x) \ge 0, i = 1, \ldots, m\}$$

is convex.

3.13. Let f be a convex function on the convex set S. Prove that the level set

$$T = \{x \in S : f(x) \le k\}$$

is convex for all real number k.

3.14. A function f is said to be *quasi convex* on the convex set S if every level set of f in S is convex, that is, if $\{x \in S : f(x) \le k\}$ is convex for all k.

(i) Prove that $f(x) = \sqrt{x}$ is a quasi-convex function on $S = \{x \in \mathfrak{R}^1, x \ge 0\}$ but it is not convex on S.

(ii) Prove that f is quasi convex on a convex set S if and only if for every x and y in S and every $0 \le \alpha \le 1$,

$$f(\alpha x + (1 - \alpha)x) \le \max\{f(x), f(y)\}.$$

(iii) Prove that any local minimizer of a quasi-convex function on a convex set is also a global minimizer.

3.15. Let $g_1, \ldots, g_m$ be concave functions on $\Re^n$. Prove that the set

$$S = \{\, x : g_i(x) \geq 0, i = 1, \ldots, m \,\}$$

is convex.

3.16. Let $f : \Re^n \to \Re^1$ be a convex function, and let $g : \Re^1 \to \Re^1$ be a convex nondecreasing function. (The notation $f : \Re^n \to \Re^1$ means that f is a real-valued function of n variables; g is a real-valued function of one variable.) Prove that the composite function $h : \Re^n \to \Re^1$ defined by $h(x) = g(f(x))$ is convex.

3.17. Complete the proof of Theorem 2.1 for the case when the objective function is strictly convex.

3.18. Express $(2, 2)^T$ as a convex combination of $(0, 0)^T$, $(1, 4)^T$, and $(3, 1)^T$.

3.19. For each of the following functions, determine if it is convex, concave, both, or neither on the real line. If the function is convex or concave, indicate if it is strictly convex or strictly concave.

(i) $f(x) = 3x^2 + 4x - 5$
(ii) $f(x) = \exp(x^2)$
(iii) $f(x) = 7x - 15$
(iv) $f(x) = \sqrt{1 + x^2}$
(v) $f(x) = 4 - 5x + 3x^2$
(vi) $f(x) = 2x^4 + 3x^3 + 4x^2$
(vii) $f(x) = x/(1 + x^4)$.

3.20. Determine if
$$f(x_1, x_2) = 2x_1^2 - 3x_1x_2 + 5x_2^2 - 2x_1 + 6x_2$$

is convex, concave, both, or neither for $x \in \Re^2$.

3.21. Give an example of a one-dimensional function f that is strictly convex on the real line even though $f''(\hat{x}) = 0$ at some point $\hat{x}$.

3.22. Let $g_1, \ldots, g_m$ be concave functions on $\Re^n$, let f be a convex function on $\Re^n$, and let μ be a positive constant. Prove that the function

$$\beta(x) = f(x) - \mu \sum_{i=1}^{m} \log g_i(x)$$

is convex on the set $S = \{\, x : g_i(x) > 0, i = 1, \ldots, m \,\}$.

2.4   The General Optimization Algorithm

More algorithms for solving optimization problems have been proposed than could possibly be discussed in a single book. This has happened in part because optimization problems can come in so many forms, but even for particular problems such as one-variable unconstrained minimization problems, there are many different algorithms that one could use.

Despite this diversity of both algorithms and problems, all of the algorithms that we will discuss in any detail in this book will have the same general form.

ALGORITHM 2.1.
General Optimization Algorithm I

1. Specify some initial guess of the solution x_0.
2. For $k = 0, 1, \ldots$
   - (i) If x_k is optimal, stop.
   - (ii) Determine x_{k+1}, a new estimate of the solution.

This algorithm is so simple that it almost conveys no information at all. However, as we discuss ever more complex algorithms for ever more elaborate problems, it is often helpful to keep in mind that we are still working within this simple and general framework.

The algorithm suggests that testing for optimality and determining a new point x_{k+1} are separate ideas, but this is usually not true. Often the information obtained from the optimality test is the basis for the computation of the new point. For example, if we are trying to solve the one-dimensional problem without constraints

$$\text{minimize } f(x),$$

then the optimality test will often be based on the condition

$$f'(x) = 0.$$

If $f'(x_k) \neq 0$, then x_k is not optimal, and the sign and value of $f'(x_k)$ indicate whether f is increasing or decreasing at the point x_k, as well as how rapidly f is changing. Such information is valuable in selecting x_{k+1}.

Many of our algorithms will have a more specific form.

ALGORITHM 2.2.
General Optimization Algorithm II

1. Specify some initial guess of the solution x_0.
2. For $k = 0, 1, \ldots$
   - (i) If x_k is optimal, stop.
   - (ii) Determine a *search direction* p_k.
   - (iii) Determine a *step length* α_k that leads to an improved estimate of the solution:
     $$x_{k+1} = x_k + \alpha_k p_k.$$

In this algorithm, p_k is a *search direction* that we hope points in the general direction of the solution, or that "improves" our solution in some sense. The scalar α_k is a *step length* that determines the point x_{k+1}; once the search direction p_k has been computed, the step length α_k is found by solving some auxiliary one-dimensional problem; see Figure 2.7.

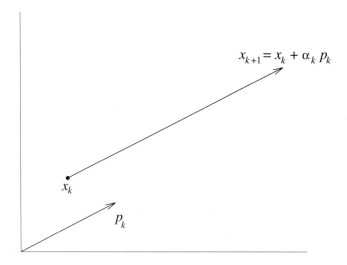

Figure 2.7. *General optimization algorithm.*

Why do we not just solve for the solution directly? Except for the simplest optimization problems, formulas for the solution do not exist. For example, consider the problem

$$\text{minimize } f(x) = e^x + x^2.$$

The optimality condition $f'(x) = 0$ has the form

$$e^x + 2x = 0,$$

but there is no simple formula for the solution to this equation. Hence for many problems some form of iterative method must be employed to determine a solution. (Any finite sequence of calculations is a formula of some sort, and so the solution of a general optimization problem can only be found as the limit of an infinite sequence. When we refer to computing a "solution" we most always mean an approximate solution, an element of this sequence that has sufficient accuracy. Determining the exact solution, or the limit of such a sequence, would be an "infinite" calculation.)

Why do we split the computation of x_{k+1} into two calculations? Ideally we would like to have $x_{k+1} = x_k + p_k$ where p_k solves

$$\underset{p}{\text{minimize }} f(x_k + p),$$

but this is equivalent to our original problem

$$\underset{x}{\text{minimize }} f(x).$$

Instead a compromise is employed. For an unconstrained problem of the form here, we will typically require that the search direction p_k be a *descent direction* for the function f at the point x_k. This means that for "small" steps taken along p_k the function value is guaranteed to decrease:

$$\rightarrow \quad f(x_k + \alpha p_k) < f(x_k) \quad \text{for } 0 < \alpha \leq \epsilon$$

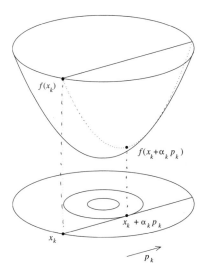

Figure 2.8. *Line search.*

for some ϵ. For a linear function $f(x) = c^T x$, p_k is a descent direction if

$$c^T(x_k + \epsilon p_k) = c^T x_k + \epsilon c^T p_k < c^T x_k,$$

or in other words if $c^T p_k < 0$. Techniques for computing descent directions for nonlinear functions are discussed in Chapter 11.

With p_k available, we would ideally like to determine the step length α_k so as to minimize the function in that direction:

$$\underset{\alpha \geq 0}{\text{minimize}} \ f(x_k + \alpha p_k).$$

This is a problem only involving one variable, the parameter α. The restriction $\alpha \geq 0$ is imposed because p_k is a descent direction.

Even for this one-dimensional problem there may not be a simple formula for the solution, so it too cannot normally be solved exactly. Instead, an α_k is computed that either "sufficiently decreases" the value of f or yields an "approximate minimizer" of the function f in the direction p_k. Both these terms have precise theoretical meanings that will be specified in later chapters, and computational techniques are available that allow α_k to be determined at reasonable cost. The calculation of α_k is called a *line search* because it corresponds to a search along the line $x_k + \alpha p_k$ defined by α. The line search is illustrated in Figure 2.8.

Algorithm II with its three major steps (the optimality test, computation of p_k, and computation of α_k) has been the basis for a great many of the most successful optimization algorithms ever developed. It has been used to develop many software packages for nonlinear optimization, and it is also present implicitly as part of the simplex method for linear programming. It is not the only approach possible (see Section 11.6), but it is the approach that we will emphasize in this book.

Using the concept of descent directions, we can establish an important condition for optimality for the constrained problem

$$\underset{x \in S}{\text{minimize}} \ f(x).$$

We define p to be a *feasible descent direction* at a point $x_k \in S$ if, for some $\epsilon > 0$,

$$x_k + \alpha p \in S \quad \text{and} \quad f(x_k + \alpha p) < f(x_k)$$

for all $0 < \alpha \leq \epsilon$. If a feasible descent direction exists at a point x_k, then it is possible to move a short distance along this direction to a feasible point with a better objective value. Then x_k cannot be a local minimizer for this problem. Hence, if x_* is a local minimizer, there cannot exist any feasible descent directions at x_*. This result will be used to derive optimality conditions for a variety of optimization problems.

Exercises

4.1. Let $x_k = (2, 1)^T$ and $p_k = (-1, 3)^T$. Plot the set $\{ x : x = x_k + \alpha p_k, \alpha \geq 0 \}$.

4.2. Find all descent directions for the linear function $f(x) = x_1 - 2x_2 + 3x_3$. Does your answer depend on the value of x?

4.3. Consider the problem

$$\begin{array}{ll} \text{minimize} & f(x) = -x_1 - x_2 \\ \text{subject to} & x_1 + x_2 \leq 2 \\ & x_1, x_2 \geq 0. \end{array}$$

(i) Determine the feasible directions at $x = (0, 0)^T$, $(0, 1)^T$, $(1, 1)^T$, and $(0, 2)^T$.

(ii) Determine whether there exist feasible descent directions at these points, and hence determine which (if any) of the points can be local minimizers.

2.5   Rates of Convergence

Many of the algorithms discussed in this book do not find a solution in a finite number of steps. Instead these algorithms compute a sequence of approximate solutions that we hope get closer and closer to a solution. When discussing such an algorithm, the following two questions are often asked:

- Does it converge?
- How fast does it converge?

It is the second question that is the topic of this section.

If an algorithm converges in a finite number of steps, the cost of that algorithm is often measured by counting the number of steps required, or by counting the number of arithmetic operations required. For example, if Gaussian elimination is applied to a system of n linear equations, then it will require about n^3 operations. This cost is referred to as the *computational complexity* of the algorithm. This concept is discussed in more detail in Chapter 9 in the context of linear programming.

For many optimization methods, the number of operations or steps required to find an exact solution will be infinite, so some other measure of efficiency must be used. The rate of convergence is one such measure. It describes how quickly the estimates of the solution approach the exact solution.

Let us assume that we have a sequence of points x_k converging to a solution x_*. We define the sequence of errors to be

$$e_k = x_k - x_*.$$

Note that

$$\lim_{k \to \infty} e_k = 0. \quad \text{since } \lim_{k \to \infty} x_k = x_*$$

We say that the sequence $\{x_k\}$ converges to x_* with rate r and rate constant C if

$$\lim_{k \to \infty} \frac{\|e_{k+1}\|}{\|e_k\|^r} = C \quad \text{if } |e_{k+1}| = \mathcal{O}(|e_k|^r)$$

and $C < \infty$. To understand this idea better, let us look at some examples.

Initially let us assume that we have ideal convergence behavior

$$\|e_{k+1}\| = C \|e_k\|^r \quad \text{for all } k,$$

so that we can avoid having to deal with limits. When $r = 1$ this is referred to as *linear* convergence:

$$\|e_{k+1}\| = C \|e_k\|.$$

If $0 < C < 1$, then the norm of the error is reduced by a constant factor at every iteration. If $C > 1$, then the sequence diverges. (What can happen when $C = 1$?) If we choose $C = 0.1 = 10^{-1}$ and $\|e_0\| = 1$, then the norms of the errors are

$$1, 10^{-1}, 10^{-2}, 10^{-3}, 10^{-4}, 10^{-5}, 10^{-6}, 10^{-7},$$

and seven-digit accuracy is obtained in seven iterations, a good result. On the other hand, if $C = 0.99$, then the norms of the errors take on the values

$$1, 0.99, 0.9801, 0.9703, 0.9606, 0.9510, 0.9415, 0.9321, \ldots,$$

and it would take about 1600 iterations to reduce the error to 10^{-7}, a less impressive result.

If $r = 1$ and $C = 0$, the convergence is called *superlinear*. Superlinear convergence includes all cases where $r > 1$ since if

$$\lim_{k \to \infty} \frac{\|e_{k+1}\|}{\|e_k\|^r} = C < \infty,$$

then

$$\lim_{k \to \infty} \frac{\|e_{k+1}\|}{\|e_k\|} = \lim_{k \to \infty} \frac{\|e_{k+1}\|}{\|e_k\|^r} \|e_k\|^{r-1} = C \times \lim_{k \to \infty} \|e_k\|^{r-1} = 0.$$

When $r = 2$, the convergence is called *quadratic*. As an example, let $r = 2$, $C = 1$, and $\|e_0\| = 10^{-1}$. Then the sequence of error norms is

$$10^{-1}, 10^{-2}, 10^{-4}, 10^{-8},$$

and so three iterations are sufficient to achieve seven-digit accuracy. In this form of quadratic convergence the error is squared at each iteration. Another way of saying this is that the number of correct digits in x_k doubles at every iteration. Of course, if the constant $C \neq 1$, then this is not an accurate statement, but it gives an intuitive sense of the attractions of a quadratic convergence rate.

For optimization algorithms there is one other important case, and that is when $1 < r < 2$. This is another special case of superlinear convergence. This case is important because (a) it is qualitatively similar to quadratic convergence for the precision of common computer calculations, and (b) it can be achieved by algorithms that only compute first derivatives, whereas to achieve quadratic convergence it is often necessary to compute second derivatives as well. To get a sense of what this form of superlinear convergence looks like, let $r = 1.5$, $C = 1$, and $\|e_0\| = 10^{-1}$. Then the sequence of error norms is

$$1 \times 10^{-1}, 3 \times 10^{-2}, 6 \times 10^{-3}, 4 \times 10^{-4}, 9 \times 10^{-6}, 3 \times 10^{-8},$$

and five iterations are required to achieve single-precision accuracy.

Example 2.2 (Rate of Convergence of a Sequence). Consider the sequence

$$2, 1.1, 1.01, 1.001, 1.0001, 1.00001, \ldots$$

with general term $x_k = 1 + 10^{-k}$. This sequence converges to $x_* = 1$ and $e_k = x_k - x_* = 10^{-k}$. Hence

$$\lim_{k \to \infty} \frac{\|e_{k+1}\|}{\|e_k\|} = \lim_{k \to \infty} \frac{10^{-(k+1)}}{10^{-k}} = \frac{1}{10},$$

so that the sequence converges linearly with rate constant $\frac{1}{10}$.

Now consider the sequence

$$4, 2.5, 2.05, 2.00060975, \ldots$$

defined by the formula

$$x_{k+1} = \frac{1}{2}\left(x_k + \frac{4}{x_k}\right) = \frac{x_k}{2} + \frac{2}{x_k}$$

with $x_0 = 4$. It can be shown that $x_k \to 2$. Also

$$
\begin{aligned}
e_{k+1} &= x_{k+1} - x_* \\
&= \frac{x_k}{2} + \frac{2}{x_k} - 2 \\
&= \frac{1}{2x_k}(x_k^2 + 4 - 4x_k) \\
&= \frac{1}{2x_k}(x_k - 2)^2 = \frac{1}{2x_k}e_k^2.
\end{aligned}
$$

From this it follows that

$$\lim_{k\to\infty} \frac{\|e_{k+1}\|}{\|e_k\|^2} = \frac{1}{2|x_*|} = \frac{1}{4}.$$

Hence this sequence converges quadratically with rate constant $\frac{1}{4}$.   ■

In practical situations ideal convergence behavior is not always observed. The rate of convergence is only observed in the limit, so at the initial iterations there is no guarantee that the norm of the error will be reduced at all, let alone at any predictable rate. In fact, it is not uncommon for an algorithm to expend almost all of its effort far from the solution, with this asymptotic convergence rate only becoming apparent at the last few iterations. In addition, the algorithm will be terminated after a finite number of iterations when the error in the solution is below some tolerance, and so the limiting behavior described here may be only imperfectly observed.

There is ambiguity in the definition of the rate of convergence. For instance, any sequence that converges quadratically also converges linearly, but with rate constant equal to zero. It is common when discussing algorithms to refer to the *fastest* rate at which the algorithm *typically* converges. For example, in Section 2.7 we show that a certain sequence $\{x_k\}$ satisfies

$$x_{k+1} - x_* \approx \left(\frac{f''(x_*)}{2f'(x_*)}\right)(x_k - x_*)^2,$$

where $x_* = \lim x_k$ and f is a function used to define the sequence. Based on this formula, the sequence $\{x_k\}$ is said to converge quadratically. However, if $f'(x_*) = 0$ the right-hand side is not defined. On the other hand, if $f'(x_*) \neq 0$ but $f''(x_*) = 0$, then the sequence can converge faster than quadratically. "Typically" these things do not happen.

In many situations people use a sort of shorthand and only refer to the convergence rate without mention of the rate constant. For quadratic rates of convergence this is not too misleading, since the ideal behavior and the observed behavior are similar unless the rate constant is exceptionally large or small. However, in the linear case the rate constant plays an important role. It is not uncommon to see rate constants that are close to one, and more unusual to see rate constants near zero. As a result, linear convergence rates are often considered to be inferior. However, if the rate constant is small, then there is little practical difference between linear and higher rates of convergence at the level of precision common on many computers. In summary, even though it is generally true that higher rates of convergence often represent improvements in performance, this is not guaranteed, and an algorithm with a linear rate of convergence can sometimes be effective in a practical setting.

Exercises

5.1. For each of the following sequences, prove that the sequence converges, find its limit, and determine the rate of convergence and the rate constant.

(i) The sequence

$$\frac{1}{2}, \frac{1}{4}, \frac{1}{8}, \frac{1}{16}, \frac{1}{32}, \ldots$$

with general term $x_k = 2^{-k}$, for $k = 1, 2, \ldots$.

(ii) The sequence

$$1.05, 1.0005, 1.000005, \ldots$$

with general term $x_k = 1 + 5 \times 10^{-2k}$, for $k = 1, 2, \ldots$.

(iii) The sequence with general term $x_k = 2^{-2^k}$.

(iv) The sequence with general term $x_k = 3^{-k^2}$.

(v) The sequence with general term $x_k = 1 - 2^{-2^k}$ for k odd, and $x_k = 1 + 2^{-k}$ for k even.

5.2. Consider the sequence defined by $x_0 = a > 0$ and

$$x_{k+1} = \frac{1}{2}\left(x_k + \frac{a}{x_k}\right).$$

Prove that this sequence converges to $x_* = \sqrt{a}$ and that the convergence rate is quadratic, and determine the rate constant.

5.3. Consider a convergent sequence $\{x_k\}$ and define a second sequence $\{y_k\}$ with $y_k = cx_k$ where c is some nonzero constant. What is the relationship between the convergence rates and rate constants of the two sequences?

5.4. Let $\{x_k\}$ and $\{c_k\}$ be convergent sequences, and assume that

$$\lim_{k \to \infty} c_k = c \neq 0.$$

Consider the sequence $\{y_k\}$ with $y_k = c_k x_k$. Is this sequence guaranteed to converge? If so, can its convergence rate and rate constant be determined from the rates and rate constants for the sequences $\{x_k\}$ and $\{c_k\}$?

2.6   Taylor Series

The Taylor series is a tool for approximating a function f near a specified point x_0. The approximation obtained is a polynomial, i.e., a function that is easy to manipulate. The Taylor series is a general tool—it can be applied whenever the function has derivatives—and it has many uses:

- It allows you to estimate the value of the function near the given point (when the function is difficult to evaluate directly).

- The derivatives and integral of the approximation can be used to estimate the derivatives and integral of the original function.

- It is used to derive many algorithms for finding zeroes of functions (see below), for minimizing functions, etc.

Since many problems are difficult to solve exactly, and an approximate solution is often adequate (the data for the problem may be inaccurate), the Taylor series is widely used, both theoretically and practically. Even if the data are exact, an approximate solution may be adequate, and in any case it is all we can hope for under most circumstances.

How does it work? We first consider the case of a one-dimensional function f with n continuous derivatives. Let x_0 be a specified point (say $x_0 = 17.5$ or $x_0 = 0$). Then the nth order Taylor series approximation is

$$f(x_0 + p) \approx f(x_0) + pf'(x_0) + \frac{1}{2}p^2 f''(x_0) + \cdots + \frac{p^n}{n!}f^{(n)}(x_0).$$

Here $f^{(n)}(x_0)$ is the nth derivative of f at the point x_0, and $n! = n(n-1)(n-2)\cdots 3 \cdot 2 \cdot 1$. Notice that $\frac{1}{2}p^2 f''(x_0) = (p^2/2!)f^{(2)}(x_0)$. In this formula, p is a variable; we will decide later what values p will take. The approximation will normally only be accurate for small values of p.

Example 2.3 (Taylor Series). Let $f(x) = \sqrt{x}$ and let $x_0 = 1$. Then

$$f(x_0) = \sqrt{x_0} = \sqrt{1} = 1$$

$$f'(x_0) = \tfrac{1}{2}x_0^{-\frac{1}{2}} = \tfrac{1}{2}1^{-\frac{1}{2}} = \tfrac{1}{2}$$

$$f''(x_0) = -\tfrac{1}{4}x_0^{-\frac{3}{2}} = -\tfrac{1}{4}1^{-\frac{3}{2}} = -\tfrac{1}{4}$$

$$f'''(x_0) = \tfrac{3}{8}x_0^{-\frac{5}{2}} = \tfrac{3}{8}1^{-\frac{5}{2}} = \tfrac{3}{8}$$

$$\vdots$$

Hence, substituting into the formula for the Taylor series,

$$\begin{aligned}
\sqrt{1+p} &= f(x_0 + p) \\
&\approx f(x_0) + pf'(x_0) + \tfrac{1}{2}p^2 f''(x_0) + \tfrac{1}{6}p^3 f'''(x_0) \\
&= 1 + p(\tfrac{1}{2}) + \tfrac{1}{2}p^2(-\tfrac{1}{4}) + \tfrac{1}{6}p^3(\tfrac{3}{8}) \\
&= 1 + \tfrac{1}{2}p - \tfrac{1}{8}p^2 + \tfrac{1}{16}p^3.
\end{aligned}$$

How do we use this? Suppose we want to approximate $f(1.6)$. Then $x_0 + p = 1 + p = 1.6$, and so $p = 0.6$:

$$\begin{aligned}
\sqrt{1.6} &= \sqrt{1+0.6} \\
&\approx 1 + \tfrac{1}{2}(0.6) - \tfrac{1}{8}(0.6)^2 + \tfrac{1}{16}(0.6)^3 \approx 1.2685.
\end{aligned}$$

The true value is $1.264911\ldots$; the approximation is accurate to three digits.   ∎

The first two terms of the Taylor series give us the formula for the tangent line for the function f at the point x_0. We commonly define the tangent line in terms of a general point x, and not in terms of p. Since $x_0 + p = x$, we can rearrange to get $p = x - x_0$. Substitute this into the first two terms of the series to get the tangent line:

$$y = f(x_0) + (x - x_0)f'(x_0).$$

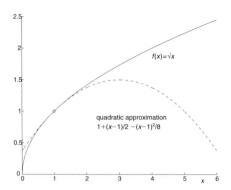

Figure 2.9. *Taylor series approximation.*

For the example above we get

$$y = 1 + (x - 1)\tfrac{1}{2} \quad \text{or} \quad y = \tfrac{1}{2}(x + 1).$$

The first three terms of the Taylor series give a quadratic approximation to the function f at the point x_0. This is illustrated in Figure 2.9.

So far we have only considered a Taylor series for a function of one variable. The Taylor series can also be derived for real-valued functions of many variables. If we use matrix and vector notation, then there is an obvious analogy between the two cases:

$$\text{1-variable:} \quad f(x_0 + p) = f(x_0) + pf'(x_0) + \tfrac{1}{2}p^2 f''(x_0) + \cdots$$

$$n\text{-variables:} \quad f(x_0 + p) = f(x_0) + p^T \nabla f(x_0) + \tfrac{1}{2}p^T \nabla^2 f(x_0)p + \cdots.$$

In the second line above x_0 and p are both vectors. The notation $\nabla f(x_0)$ refers to the gradient of the function f at the point $x = x_0$. The notation $\nabla^2 f(x_0)$ represents the Hessian of f at the point $x = x_0$. (See Appendix B.4.) The higher-order terms of the Taylor series can also be written down, but the notation is more complex and they will not be required in this book.

Example 2.4 (Multidimensional Taylor Series)**.** Consider the function

$$f(x_1, x_2) = x_1^3 + 5x_1^2 x_2 + 7x_1 x_2^2 + 2x_2^3$$

at the point

$$x_0 = (-2, 3)^T.$$

The gradient of this function is

$$\nabla f(x) = \begin{pmatrix} 3x_1^2 + 10x_1 x_2 + 7x_2^2 \\ 5x_1^2 + 14x_1 x_2 + 6x_2^2 \end{pmatrix}$$

and the Hessian matrix is

$$\nabla^2 f(x) = \begin{pmatrix} 6x_1 + 10x_2 & 10x_1 + 14x_2 \\ 10x_1 + 14x_2 & 14x_1 + 12x_2 \end{pmatrix}.$$

At the point $x_0 = (-2, 3)^T$ these become

$$\nabla f(x_0) = \begin{pmatrix} 15 \\ -10 \end{pmatrix} \quad \text{and} \quad \nabla^2 f(x_0) = \begin{pmatrix} 18 & 22 \\ 22 & 8 \end{pmatrix}.$$

If $p = (p_1, p_2)^T = (0.1, 0.2)^T$, then

$$
\begin{aligned}
f(-1.9, 3.2) &= f(-2 + 0.1, 3 + 0.2) \\
&= f(x_0 + p) \\
&\approx f(x_0) + p^T \nabla f(x_0) + \frac{1}{2} p^T \nabla^2 f(x_0) p \\
&= -20 + (0.1 \quad 0.2) \begin{pmatrix} 15 \\ -10 \end{pmatrix} + \frac{1}{2} (0.1 \quad 0.2) \begin{pmatrix} 18 & 22 \\ 22 & 8 \end{pmatrix} \begin{pmatrix} 0.1 \\ 0.2 \end{pmatrix} \\
&= -20 - 0.5 + 0.69 = -19.81.
\end{aligned}
$$

The true value is $f(-1.9, 3.2) = -19.755$, so the approximation is accurate to three digits. ■

The Taylor series for multidimensional problems can also be derived using summations rather than matrix-vector notation:

$$f(x_0 + p) = f(x_0) + \sum_{i=1}^{n} p_i \left. \frac{\partial f(x)}{\partial x_i} \right|_{x=x_0} + \frac{1}{2} \sum_{i=1}^{n} \sum_{j=1}^{n} p_i p_j \left. \frac{\partial^2 f(x)}{\partial x_i \partial x_j} \right|_{x=x_0} + \cdots.$$

The formula is the same as before; only the notation has changed.

There is an alternate form of the Taylor series that is often used, called the *remainder form*. If three terms are used it looks like

$$\text{1-variable:} \quad f(x_0 + p) = f(x_0) + pf'(x_0) + \tfrac{1}{2} p^2 f''(\xi)$$
$$\text{n-variables:} \quad f(x_0 + p) = f(x_0) + p^T \nabla f(x_0) + \tfrac{1}{2} p^T \nabla^2 f(\xi) p.$$

The point ξ is an *unknown* point lying between x_0 and $x_0 + p$. In this form the series is exact, but it involves an unknown point, so it cannot be evaluated. This form of the series is often used for theoretical purposes, or to derive bounds on the accuracy of the series. The accuracy of the series can be analyzed by establishing bounds on the final "remainder" term.

If the remainder form of the series is used, but with only two terms, then we obtain

$$\text{1-variable:} \quad f(x_0 + p) = f(x_0) + pf'(\xi)$$
$$\text{n-variables:} \quad f(x_0 + p) = f(x_0) + p^T \nabla f(\xi).$$

This result is known as the *mean-value theorem*.

Exercises

6.1. Find the first four terms of the Taylor series for

$$f(x) = \log(1 + x)$$

about the point $x_0 = 0$. Evaluate the series for $p = 0.1$ and $p = 0.01$ and compare with the value of $f(x_0 + p)$. Derive the remainder form of the Taylor series using five terms (the four terms you already derived plus a remainder term). Derive a bound on the accuracy of the four-term series. Compare the bound you derived with the actual errors for $p = 0.1$ and $p = 0.01$.

6.2. Find the first three terms of the Taylor series for the following functions.

　　(i) $f(x) = \sin x$ about the point $x_0 = \pi$.
　　(ii) $f(x) = 2/(3x + 5)$ about the point $x_0 = -1$.
　　(iii) $f(x) = e^x$ about the point $x_0 = 0$.

6.3. Determine the general term in the Taylor series for the function

$$f(x) = \begin{cases} e^{-1/x} & \text{if } x > 0, \\ 0 & \text{if } x \leq 0, \end{cases}$$

about the point $x_0 = 0$. Compare this with the Taylor series for the function $f(x) = 0$ about the same point. What can you conclude about the limitations of the Taylor series as a tool for approximating functions?

6.4. Find the first three terms of the Taylor series for

$$f(x_1, x_2) = 3x_1^4 - 2x_1^3 x_2 - 4x_1^2 x_2^2 + 5x_1 x_2^3 + 2x_2^4$$

at the point

$$x_0 = (1, -1)^T.$$

Evaluate the series for $p = (0.1, 0.01)^T$ and compare with the value of $f(x_0 + p)$.

6.5. Find the first three terms of the Taylor series for

$$f(x_1, x_2) = \sqrt{x_1^2 + x_2^2}$$

about the point $x_0 = (3, 4)^T$.

6.6. Prove that if $p^T \nabla f(x_k) < 0$, then $f(x_k + \epsilon p) < f(x_k)$ for $\epsilon > 0$ sufficiently small. Hint: Expand $f(x_k + \epsilon p)$ in a Taylor series about the point x_k and look at $f(x_k + \epsilon p) - f(x_k)$.

6.7. (The results of this and the next problem show that a function f is convex on a convex set S if the Hessian matrix $\nabla^2 f(x)$ is positive semidefinite for all $x \in S$.) Let f be a real-valued function of n variables x with continuous first derivatives. Prove that f is convex on the convex set S if and only if

$$f(y) \geq f(x) + \nabla f(x)^T (y - x)$$

for all $x, y \in S$.

6.8. Let f be a real-valued function of n variables x with continuous second derivatives. Use the result of the previous problem to prove that f is convex on the convex set S if $\nabla^2 f(x)$ is positive semidefinite for all $x \in S$.

2.7   Newton's Method for Nonlinear Equations

Let us now consider methods for solving

$$f(x) = 0.$$

We first consider the one-dimensional case where x is a scalar and f is a real-valued function. Later we will look at the n-dimensional case where $x = (x_1, \ldots, x_n)^T$ and $f(x) = (f_1(x), \ldots, f_n(x))^T$. Note that both x and $f(x)$ are vectors of the same length n. Throughout this section we assume that the function f has two continuous derivatives.

If $f(x)$ is a linear function, it is possible to find a solution if the system is nonsingular. The cost of finding the solution is predictable—it is the cost of applying Gaussian elimination. Except for a few isolated special cases, such as quadratic equations in one variable, in the nonlinear case it is not possible to guarantee that a solution can be found, nor is it possible to predict the cost of finding a solution. However, the situation is not totally bleak. There are effective algorithms that work much of the time, and that are efficient on a wide variety of problems. They are based on solving a *sequence* of linear equations. As a result, if the function f is linear, they can be as efficient as the techniques for linear systems. Also, we can apply our knowledge about linear systems in the nonlinear case.

The methods that we will discuss are based on Newton's method. Given an estimate of the solution x_k, the function f is approximated by the linear function consisting of the first two terms of the Taylor series for the function f at the point x_k. The resulting linear system is then solved to obtain a new estimate of the solution x_{k+1}.

To derive the formulas for Newton's method, we first write out the Taylor series for the function f at the point x_k:

$$f(x_k + p) \approx f(x_k) + p f'(x_k).$$

If $f'(x_k) \neq 0$, then we can solve the equation

$$f(x_*) \approx f(x_k) + p f'(x_k) = 0$$

for p to obtain

$$p = -f(x_k)/f'(x_k).$$

The new estimate of the solution is then $x_{k+1} = x_k + p$ or

$$x_{k+1} = x_k - f(x_k)/f'(x_k).$$

This is the formula for Newton's method.

Example 2.5 (Newton's Method). As an example, consider the one-dimensional problem

$$f(x) = 7x^4 + 3x^3 + 2x^2 + 9x + 4 = 0.$$

Then

$$f'(x) = 28x^3 + 9x^2 + 4x + 9$$

and the formula for Newton's method is

$$x_{k+1} = x_k - \frac{7x_k^4 + 3x_k^3 + 2x_k^2 + 9x_k + 4}{28x_k^3 + 9x_k^2 + 4x_k + 9}.$$

Table 2.1. *Newton's method for a one-dimensional problem.*

k	x_k	$f(x_k)$	$|x_k - x_*|$
0	0	4×10^0	5×10^{-1}
1	-0.4444444444444444	4×10^{-1}	7×10^{-2}
2	-0.5063255748934088	3×10^{-2}	5×10^{-3}
3	-0.5110092428604380	2×10^{-4}	3×10^{-5}
4	-0.5110417864454134	9×10^{-9}	2×10^{-9}
5	-0.5110417880368663	0	0

If we start with the initial guess $x_0 = 0$, then

$$
\begin{aligned}
x_1 &= x_0 - \frac{7x_0^4 + 3x_0^3 + 2x_0^2 + 9x_0 + 4}{28x_0^3 + 9x_0^2 + 4x_0 + 9} \\
&= 0 - \frac{7 \times 0^4 + 3 \times 0^3 + 2 \times 0^2 + 9 \times 0 + 4}{28 \times 0^3 + 9 \times 0^2 + 4 \times 0 + 9} \\
&= 0 - \frac{4}{9} = -4/9 = -0.4444\ldots.
\end{aligned}
$$

At the next iteration we substitute $x_1 = -4/9$ into the formula for Newton's method and obtain $x_2 \approx -0.5063$. The complete iteration is given in Table 2.1. ∎

Newton's method corresponds to approximating the function f by its tangent line at the point x_k. The point where the tangent line crosses the x-axis (i.e., a zero of the tangent line) is taken as the new estimate of the solution. This geometric interpretation is illustrated in Figure 2.10.

The performance of Newton's method in Example 2.5 is considered to be typical for this method. It converges rapidly and, once x_k is close to the solution x_*, the error is approximately squared at every iteration. It has a quadratic rate of convergence as we now show.

It is not difficult to analyze the convergence of Newton's method using the Taylor series. Define the error in x_k by $e_k = x_k - x_*$. Using the remainder form of the Taylor series:

$$0 = f(x_*) = f(x_k - e_k) = f(x_k) - e_k f'(x_k) + \tfrac{1}{2}e_k^2 f''(\xi).$$

Dividing by $f'(x_k)$ and rearranging gives

$$e_k - \frac{f(x_k)}{f'(x_k)} = \frac{1}{2}e_k^2 \frac{f''(\xi)}{f'(x_k)}.$$

Since $e_k = x_k - x_*$ we obtain

$$x_k - \frac{f(x_k)}{f'(x_k)} - x_* = \frac{1}{2}(x_k - x_*)^2 \frac{f''(\xi)}{f'(x_k)},$$

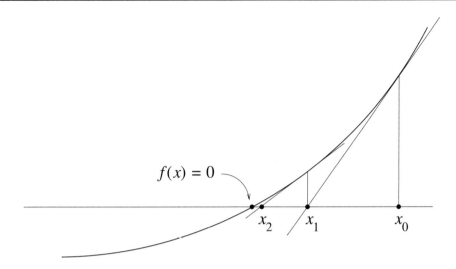

Figure 2.10. *Newton's method—geometric interpretation.*

which is the same as

$$x_{k+1} - x_* = \frac{1}{2}(x_k - x_*)^2 \frac{f''(\xi)}{f'(x_k)}.$$

If the sequence $\{x_k\}$ converges, then $\xi \to x_*$, and hence when x_k is sufficiently close to x_*,

$$x_{k+1} - x_* \approx \frac{1}{2}\left(\frac{f''(x_*)}{f'(x_*)}\right)(x_k - x_*)^2$$

indicating that the error in x_k is approximately squared at every iteration, assuming that the rate constant $\frac{1}{2}f''(x_*)/f'(x_*)$ is not ridiculously large or small. These results are summarized in the following theorem.

Theorem 2.6 (Convergence of Newton's Method). *Assume that the function $f(x)$ has two continuous derivatives. Let x_* be a zero of f with $f'(x_*) \neq 0$. If $|x_0 - x_*|$ is sufficiently small, then the sequence defined by*

$$x_{k+1} = x_k - f(x_k)/f'(x_k)$$

converges quadratically to x_* with rate constant

$$C = |f''(x_*)/2f'(x_*)|.$$

Proof. See the Exercises.   □

Example 2.5 also shows that the function values $f(x_k)$ converge quadratically to zero. This also follows from the Taylor series:

$$0 = f(x_*) = f(x_k + e_k) = f(x_k) + e_k f'(\xi).$$

This can be rearranged to obtain

$$f(x_k) = -e_k f'(\xi) = -f'(\xi)(x_* - x_k)$$

so that $f(x_k)$ is proportional to $(x_* - x_k)$. Hence they converge at the same rate if $f'(x_*) \neq 0$.

In the argument above we have assumed that $\{ f'(x_k) \}$ and $f'(x_*)$ are all nonzero. If $f'(x_k) = 0$ for some k, then Newton's method fails (there is a division by zero in the formula). Geometrically this means that the tangent line is horizontal, parallel to the x-axis, and so it does not have a zero. If on the other hand $f'(x_k) \neq 0$ for all k, $f''(x_*) \neq 0$, but $f'(x_*) = 0$, then the coefficient in the convergence analysis

$$\frac{f''(\xi)}{2f'(x_k)}$$

tends to infinity, and the algorithm does not have a quadratic rate of convergence. If f is a polynomial, this corresponds to f having a multiple zero at the point x_*; this case is illustrated in Example 2.7.

Example 2.7 (Newton's Method; $f'(x_*) = 0$). We now apply Newton's method to the example

$$f(x) = x^4 - 7x^3 + 17x^2 - 17x + 6$$
$$= (x - 1)^2 (x - 2)(x - 3) = 0.$$

This function has a multiple zero at $x_* = 1$ and at this point $f(x_*) = f'(x_*) = 0$. The derivative of f is

$$f'(x) = 4x^3 - 21x^2 + 34x - 17$$

and the formula for Newton's method is

$$x_{k+1} = x_k - \frac{x^4 - 7x^3 + 17x^2 - 17x + 6}{4x^3 - 21x^2 + 34x - 17}.$$

If we start with the initial guess $x_0 = 1.1$, then the method converges to $x_* = 1$ at a linear rate, whereas if we start with $x_0 = 2.1$, then the method converges to $x_* = 2$ at a quadratic rate. The results for these iterations are given in Tables 2.2 and 2.3. (In the final lines of both tables the function value $f(x_k)$ is zero; this is the value calculated by the computer and is a side effect of using finite-precision arithmetic.)   ∎

In Example 2.7 the slow convergence only occurs when the method converges to a solution where $f'(x_*) = 0$. Quadratic convergence is obtained at the other roots, where $f'(x_*) \neq 0$.

It should also be noticed that the accuracy of the solution was worse at a multiple root. This too can be explained by the Taylor series, although this time we expand about the point x_*:

$$f(x_k) = f(x_* + e_k) = f(x_*) + e_k f'(x_*) + \tfrac{1}{2} e_k^2 f''(\xi).$$

At the solution, $f(x_*) = 0$, and since this is assumed to be a multiple zero, $f'(x_*) = 0$ as well. Hence

$$f(x_k) = \tfrac{1}{2} e_k^2 f''(\xi) = (\tfrac{1}{2} f''(\xi))(x_k - x_*)^2.$$

Table 2.2. *Newton's method:* $f'(x_*) = 0$ $(x_0 = 1.1)$.

k	x_k	$f(x_k)$	$|x_k - x_*|$
0	1.100000000000000	2×10^{-2}	1×10^{-1}
1	1.045541401273894	4×10^{-3}	5×10^{-2}
2	1.021932395992710	9×10^{-4}	2×10^{-2}
3	1.010779316995807	2×10^{-4}	1×10^{-2}
4	1.005345328998912	6×10^{-5}	5×10^{-3}
5	1.002661858321646	1×10^{-5}	3×10^{-3}
6	1.001328260855184	4×10^{-6}	1×10^{-3}
7	1.000663467429195	9×10^{-7}	7×10^{-4}
8	1.000331568468827	2×10^{-7}	3×10^{-4}
9	1.000165742989413	6×10^{-8}	2×10^{-4}
10	1.000082861192927	1×10^{-8}	8×10^{-5}
$\vdots$			
24	1.000000075780004	1×10^{-14}	8×10^{-8}
25	1.000000040618541	0	4×10^{-8}

Table 2.3. *Newton's method:* $f'(x_*) \neq 0$ $(x_0 = 2.1)$.

k	x_k	$f(x_k)$	$|x_k - x_*|$
0	2.100000000000000	-1×10^{-1}	1×10^{-1}
1	2.006603773584894	-7×10^{-3}	7×10^{-3}
2	2.000042472785593	-4×10^{-5}	4×10^{-5}
3	2.000000001803635	-2×10^{-9}	2×10^{-9}
4	2.000000000000001	0	9×10^{-16}

The function value $f(x_k)$ is now proportional to the *square* of the error $(x_k - x_*)$. So, for example, if $f(x_k) = 10^{-16}$ (about the level of machine precision in typical double precision arithmetic), and $\frac{1}{2} f''(\xi) = 1$, then $x_k - x_* = 10^{-8}$. In this case the point x_k is only accurate to half precision.

The proof of convergence for Newton's method requires that the initial point x_0 be sufficiently close to a zero. If not, the method can fail to converge, even when there is no division by zero in the formula for the method. This is illustrated in the example below. In Chapter 11 we discuss safeguards that can be added to Newton's method that prevent this from happening.

Example 2.8 (Failure of Newton's Method). Consider the problem

$$f(x) = \frac{e^x - e^{-x}}{e^x + e^{-x}} = 0.$$

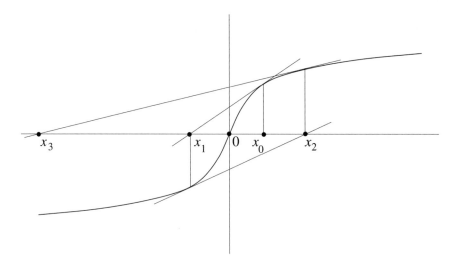

Figure 2.11. *Failure of Newton's method.*

If Newton's method is used with the initial guess $x_0 = 1$, then the sequence of approximate solutions is

$$x_0 = 1, \qquad x_1 = -0.8134, \qquad x_2 = 0.4094$$
$$x_3 = -0.0473, \quad x_4 = 7.06 \times 10^{-5}, \quad x_5 = -2.35 \times 10^{-13}$$

and at the final point $f(x_5) = -2.35 \times 10^{-13}$, so the method converges to a solution.

However if $x_0 = 1.1$, then

$$x_0 = 1.1, \qquad x_1 = -1.1286, \qquad x_2 = 1.2341$$
$$x_3 = -1.6952, \qquad x_4 = 5.7154, \qquad x_5 = -2.30 \times 10^4$$

and at the next iteration an overflow results. At the final point $f(x_5) = 1$, so the sequence is not converging to a solution.

A graph of the function is given in Figure 2.11. This function is also called the hyperbolic tangent function, $f(x) = \tanh x$.   ∎

2.7.1   Systems of Nonlinear Equations

Much of the discussion in the one-dimensional case can be transferred with only minor changes to the n-dimensional case. Suppose now that we are solving

$$f(x) = 0,$$

where this represents

$$f_1(x_1, \ldots, x_n) = 0,$$
$$f_2(x_1, \ldots, x_n) = 0,$$
$$\vdots$$
$$f_n(x_1, \ldots, x_n) = 0.$$

Define the matrix $\nabla f(x)$ with columns $\nabla f_1(x), \ldots, \nabla f_n(x)$. This is the transpose of the Jacobian of f at the point x. (The Jacobian is discussed in Appendix B.4.) As before, we write out the Taylor series approximation for the function f at the point x_k:

$$f(x_k + p) \approx f(x_k) + \nabla f(x_k)^T p,$$

where p is now a vector. Now we solve the equation

$$f(x_*) \approx f(x_k) + \nabla f(x_k)^T p = 0$$

for p to obtain

$$p = -\nabla f(x_k)^{-T} f(x_k).$$

The new estimate of the solution is then

$$x_{k+1} = x_k + p = x_k - \nabla f(x_k)^{-T} f(x_k).$$

This is the formula for Newton's method in the n-dimensional case.

Example 2.9 (Newton's Method in n Dimensions). As an example, consider the two-dimensional problem

$$f_1(x_1, x_2) = 3x_1 x_2 + 7x_1 + 2x_2 - 3 = 0,$$
$$f_2(x_1, x_2) = 5x_1 x_2 - 9x_1 - 4x_2 + 6 = 0.$$

Then

$$\nabla f(x_1, x_2) = \begin{pmatrix} 3x_2 + 7 & 5x_2 - 9 \\ 3x_1 + 2 & 5x_1 - 4 \end{pmatrix},$$

and the formula for Newton's method is

$$x_{k+1} = x_k - \begin{pmatrix} 3x_2 + 7 & 5x_2 - 9 \\ 3x_1 + 2 & 5x_1 - 4 \end{pmatrix}^{-T} \begin{pmatrix} 3x_1 x_2 + 7x_1 + 2x_2 - 3 \\ 5x_1 x_2 - 9x_1 - 4x_2 + 6 \end{pmatrix}.$$

If we start with the initial guess $x_0 = (1, 2)^T$, then

$$x_1 = x_0 - \begin{pmatrix} 3x_2 + 7 & 5x_2 - 9 \\ 3x_1 + 2 & 5x_1 - 4 \end{pmatrix}^{-T} \begin{pmatrix} 3x_1 x_2 + 7x_1 + 2x_2 - 3 \\ 5x_1 x_2 - 9x_1 - 4x_2 + 6 \end{pmatrix}$$

$$= \begin{pmatrix} 1 \\ 2 \end{pmatrix} - \begin{pmatrix} 13 & 1 \\ 5 & 1 \end{pmatrix}^{-T} \begin{pmatrix} 14 \\ -1 \end{pmatrix}$$

$$= \begin{pmatrix} 1 \\ 2 \end{pmatrix} - \begin{pmatrix} 2.375 \\ -3.375 \end{pmatrix} = \begin{pmatrix} -1.375 \\ 5.375 \end{pmatrix}.$$

The complete iteration is given in Table 2.4. ∎

In the n-dimensional case, Newton's method corresponds to approximating the function f by a linear function at the point x_k. The zero of this linear approximation is the new estimate x_{k+1}. As in the one-dimensional case, the method typically converges with a

Table 2.4. *Newton's method for an n-dimensional problem.*

k	x_1^k	x_2^k	$\|f\|_2$	$\|x - x_*\|_2$
0	1.0000000×10^0	2.0000000	1×10^1	1×10^0
1	-1.3749996×10^0	5.3749991	5×10^1	4×10^0
2	$-5.4903371 \times 10^{-1}$	3.0472771	1×10^1	2×10^0
3	$-1.6824928 \times 10^{-1}$	1.9741571	2×10^0	5×10^{-1}
4	$-2.7482068 \times 10^{-2}$	1.5774495	3×10^{-1}	8×10^{-2}
5	$-1.0090199 \times 10^{-3}$	1.5028436	1×10^{-2}	3×10^{-3}
6	$-1.4637396 \times 10^{-6}$	1.5000041	2×10^{-5}	4×10^{-6}
7	$-3.0852447 \times 10^{-12}$	1.5000000	4×10^{-11}	9×10^{-12}
8	$-2.0216738 \times 10^{-18}$	1.5000000	0	2×10^{-18}

quadratic rate of convergence, as the theorem below indicates. A proof of quadratic convergence for Newton's method can be found in the book by Ortega and Rheinboldt (1970, reprinted 2000).

Theorem 2.10 (Convergence of Newton's Method in *n* Dimensions). *Assume that the function $f(x)$ has two continuous derivatives. Assume that x_* satisfies $f(x_*) = 0$ with $\nabla f(x_*)$ nonsingular. If $\|x_0 - x_*\|$ is sufficiently small, then the sequence defined by*

$$x_{k+1} = x_k - (\nabla f(x_k))^{-1} f(x_k)$$

converges quadratically to x_*.

Our discussion has implicitly assumed that every Jacobian matrix $\nabla f(x_k)^T$ is nonsingular, that is, the system of linear equations that defines the new point x_{k+1} has a unique solution. If this assumption is not satisfied, Newton's method fails. If the Jacobian matrix at the solution $\nabla f(x_*)^T$ is singular, then there is no guarantee of quadratic convergence.

The proof of convergence assumes that x_k is "sufficiently close" to x_*, as in the one-dimensional case. If it is not, the method can diverge.

Exercises

7.1. Apply Newton's method to find all three solutions of

$$f(x) = x^3 - 5x^2 - 12x + 19 = 0.$$

You will have to use several different initial guesses.

7.2. Let a be some positive constant. It is possible to use Newton's method to calculate $1/a$ to any desired accuracy without doing division. Determine a function f such that $f(1/a) = 0$, and for which the formula for Newton's method only uses the arithmetic operations of addition, subtraction, and multiplication. For what initial values does Newton's method converge for this function?

7.3. Apply Newton's method to

$$f(x) = (x - 2)^4 + (x - 2)^5$$

with initial guess $x_0 = 3$. You should observe that the sequence converges linearly with rate constant $\frac{3}{4}$. Now apply the iterative method

$$x_{k+1} = x_k - 4f(x_k)/f'(x_k).$$

This method should converge more rapidly for this problem. Prove that the new method converges quadratically, and determine the rate constant.

7.4. A function f has a root of multiplicity $m > 1$ at the point x_* if

$$f(x_*) = f'(x_*) = \cdots = f^{(m-1)}(x_*) = 0.$$

Assume that Newton's method with initial guess x_0 converges to such a root. Prove that Newton's method converges linearly but not quadratically. Assume that the iteration

$$x_{k+1} = x_k - mf(x_k)/f'(x_k)$$

converges to x_*. If $f^{(m)}(x_*) \neq 0$, prove that this sequence converges quadratically.

7.5. Apply Newton's method to solve $f(x) = x^2 - a = 0$, where $a > 0$. This is a good way to compute $\pm\sqrt{a}$. How does the iteration behave if $a \leq 0$? What happens if you choose x_0 as a complex number?

7.6. Prove that your iteration from the previous problem converges to a root if $x_0 \neq 0$. When does the iteration converge to $+\sqrt{a}$ and when does it converge to $-\sqrt{a}$?

7.7. For the iteration in the previous problem, can you efficiently determine a good initial guess x_0 using the value of a and the elementary operations of addition, subtraction, multiplication, and division? Can you determine an upper bound on how many elementary operations are required to determine a root to within a specified accuracy?

7.8. Newton's method was derived by approximating the general function f by the first two terms of its Taylor series at the current point x_k. Derive another method for finding zeroes by approximating f with the first *three* terms of its Taylor series at the current point, and finding a zero of this approximation. Determine the rate of convergence for this new method (you may assume that the method converges). Apply the method to the functions in Examples 2.5 and 2.7.

7.9. Prove Theorem 2.10.

7.10. Apply Newton's method to the system of nonlinear equations

$$f_1(x_1, x_2) = x_1^2 + x_2^2 - 1 = 0$$
$$f_2(x_1, x_2) = 5x_1^2 - x_2 - 2 = 0.$$

There are four solutions to this system of equations. Can you find all four of them by using different initial guesses?

7.11. (Extended Project) Apply Newton's method to solve $f(x) = x^n - a = 0$ for various values of n. Experiment with the method in an attempt to understand its properties. Under what circumstances will the method converge to a root? Can you, by using complex-valued initial guesses, determine all n roots of this equation? What can you prove about the convergence of the iteration? What happens if n is not an integer?

7.12. Suppose that Newton's method is applied to a system of nonlinear equations, where some of the equations are linear. Prove that the linear equations are satisfied at every iteration, except possibly at the initial point.

2.8 Notes

Global Optimization—Techniques for global optimization are discussed in the books by Hansen (1992) and Floudas and Pardalos (1992, reprinted 2007); Hansen and Walster (2003); Horst et al. (2000); and Liberti and Maculan (2006). A survey of results can be found in article by Rinnooy Kan and Timmer (1989).

Newton's Method—If a function is known to have a multiple root, and if the multiplicity of the root is known (e.g., if it is known to be a double root), then it is possible to adjust the formula for Newton's method to restore the quadratic rate of convergence. (See the Exercises above.) However, on a general problem it is unlikely that this information will be available, so this is not normally a practical alternative.

Chapter 3

Representation of Linear Constraints

3.1 Basic Concepts

In this chapter we examine ways of representing linear constraints. The goal is to write the constraints in a form that makes it easy to move from one feasible point to another. The constraints specify interrelationships among the variables so that, for example, if we increase the first variable, retaining feasibility might require making a complicated sequence of changes to all the other variables. It is much easier if we express the constraints using a coordinate system that is "natural" for the constraints. Then the interrelationships among the variables are taken care of by the coordinate system, and moves between feasible points are almost as simple as for a problem without constraints.

In the general case these constraints may be either equalities or inequalities. Since any inequality of the "less than or equal" type may be transformed to an equivalent constraint of the "greater or equal" type, any problem with linear constraints may be written as follows:

$$\begin{array}{ll} \text{minimize} & f(x) \\ \text{subject to} & a_i^T x = b_i, \quad i \in \mathcal{E} \\ & a_i^T x \geq b_i, \quad i \in \mathcal{I}. \end{array}$$

Each a_i here is a vector of length n and each b_i is a scalar. $\mathcal{E}$ is an index set for the equality constraints and $\mathcal{I}$ is an index set for the inequality constraints. We denote by A the matrix whose rows are the vectors a_i^T and denote by b the vector of right-hand side coefficients b_i. Let S be the set of feasible points. A set of this form, defined by a finite number of linear constraints, is sometimes called a *polyhedron* or a *polyhedral set*. In this chapter we are not concerned with the properties of the objective function f.

Example 3.1 (Problem with Linear Constraints). Consider the problem

$$\begin{array}{ll} \text{minimize} & f(x) = x_1^2 + x_2^3 x_3^4 \\ \text{subject to} & x_1 + 2x_2 + 3x_3 = 6 \\ & x_1, x_2, x_3 \geq 0. \end{array}$$

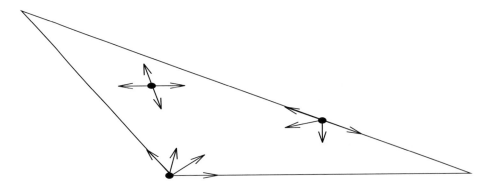

Figure 3.1. *Feasible directions.*

For this example $\mathcal{E} = \{1\}$ and $\mathcal{I} = \{2, 3, 4\}$. The vectors $\{a_i\}$ that determine the constraints are

$$a_1 = (1 \quad 2 \quad 3)^T, \quad a_2 = (1 \quad 0 \quad 0)^T$$
$$a_3 = (0 \quad 1 \quad 0)^T, \quad a_4 = (0 \quad 0 \quad 1)^T$$

and the right-hand sides are

$$b_1 = 6, \quad b_2 = 0, \quad b_3 = 0, \quad \text{and} \quad b_4 = 0. \quad \blacksquare$$

We start by taking a closer look at the relation between a feasible point and its neighboring feasible points. We shall be interested in determining how the function value changes as we move from a feasible point $\bar{x}$ to nearby feasible points.

First let us look at the direction of movement. We define p to be a *feasible direction* at the point $\bar{x}$ if a small step taken along p leads to a feasible point in the set. Mathematically, p is a feasible direction if there exists some $\epsilon > 0$ such that $\bar{x} + \alpha p \in S$ for all $0 \le \alpha \le \epsilon$. Thus, a small movement from $\bar{x}$ along a feasible direction maintains feasibility. In addition, since the feasible set is convex, any feasible point in the set can be reached from $\bar{x}$ by moving along some feasible direction. Examples of feasible directions are shown in Figure 3.1.

In many applications, it is useful to maintain feasibility at every iteration. For example, the objective function may only be defined at feasible points. Or if the algorithm is terminated before an optimal solution has been found, only a feasible point may have practical value. These considerations motivate a class of methods called *feasible-point methods*. These methods have the following form.

ALGORITHM 3.1.
Feasible-Point Method

1. Specify some initial feasible guess of the solution x_0.
2. For $k = 0, 1, \ldots$

    (i) Determine a feasible direction of descent p_k at the point x_k. If none exists, stop.
    (ii) Determine a new feasible estimate of the solution: $x_{k+1} = x_k + \alpha_k p_k$, where $f(x_{k+1}) < f(x_k)$.

In this chapter we are mainly concerned with representing feasible directions with respect to S in terms of the constraint vectors a_i. We begin by characterizing feasible directions with respect to a single constraint. Specifically, we determine conditions that ensure that small movements away from a feasible point $\bar{x}$ will keep the constraint satisfied.

Consider first an equality constraint $a_i^T x = b_i$. Let us examine the effect of taking a small positive step α in the direction p. Since $a_i^T \bar{x} = b_i$, then $a_i^T (\bar{x} + \alpha p) = b_i$ will hold if and only if $a_i^T p = 0$.

Example 3.2 (An Equality Constraint). Suppose that we wished to solve

$$\text{minimize} \quad f(x_1, x_2)$$
$$\text{subject to} \quad x_1 + x_2 = 1.$$

For this constraint $a_1 = (1, 1)^T$ and $b_1 = 1$. Let $\bar{x} = (0, 1)^T$ so that $\bar{x}$ satisfies the constraint. Then $\bar{x} + \alpha p$ will satisfy the constraint if and only if $a_1^T p = 0$, that is,

$$p_1 + p_2 = 0.$$

For this example

$$a_1^T(\bar{x} + \alpha p) = (\bar{x}_1 + \bar{x}_2) + \alpha(p_1 + p_2) = (1) + \alpha(0) = 1,$$

as expected.

The original problem is equivalent to

$$\underset{\alpha}{\text{minimize}} \ f(\bar{x} + \alpha p),$$

where $\bar{x} = (0, 1)^T$, as before, and where $p = (1, -1)^T$ is a vector satisfying $a_1^T p = 0$. Expressing feasible points in the form $\bar{x} + \alpha p$ will be a way for us to transform constrained problems to equivalent problems without constraints. ∎

Continuing to inequality constraints, consider first some constraint $a_i^T x \geq b_i$ which is inactive at $\bar{x}$. Since $a_i^T \bar{x} > b_i$, then $a_i^T (\bar{x} + \alpha p) > b_i$ for all α sufficiently small. Thus, we can move a small distance in any direction p without violating the constraint.

If the inequality constraint is active at $\bar{x}$, we have $a_i^T \bar{x} = b_i$. Then to guarantee that $a_i^T (\bar{x} + \alpha p) \geq b_i$ for small positive step lengths α, the direction p must satisfy $a_i^T p \geq 0$.

Example 3.3 (An Inequality Constraint). Suppose that we wished to solve

$$\text{minimize} \quad f(x_1, x_2)$$
$$\text{subject to} \quad x_1 + x_2 \geq 1.$$

For this constraint $a_1 = (1, 1)^T$ and $b_1 = 1$. If $\bar{x} = (0, 2)^T$, then the constraint is inactive and any nearby point is feasible.

If $\bar{x} = (0, 1)^T$, then the constraint is active and nearby points can be expressed in the form $\bar{x} + \alpha p$ with $a_1^T p \geq 0$. For this example this corresponds to the condition $p_1 + p_2 \geq 0$, or $p_1 \geq -p_2$. ∎

In summary, we conclude that the feasible directions at a point $\bar{x}$ are determined by the equality constraints and the active inequalities at that point. Let $\hat{\mathcal{I}}$ denote the set of active inequality constraints at $\bar{x}$. Then p is a feasible direction with respect to the feasible set at $\bar{x}$ if and only if

$$a_i^T p = 0, \quad i \in \mathcal{E}, \qquad a_i^T p \geq 0, \quad i \in \hat{\mathcal{I}}.$$

In the following, it will be convenient to consider separately problems that have only equality constraints, or only inequality constraints.

The general form of the *equality-constrained problem* is

$$\begin{aligned} \text{minimize} \quad & f(x) \\ \text{subject to} \quad & Ax = b. \end{aligned}$$

It is evident from our discussion above that a vector p is a feasible direction for the linear equality constraints if and only if

$$Ap = 0.$$

We call the set of all vectors p such that $Ap = 0$ the *null space* of A. A direction p is a feasible direction for the linear equality constraints if and only if it lies in the null space of A.

The general form of the *inequality-constrained problem* is

$$\begin{aligned} \text{minimize} \quad & f(x) \\ \text{subject to} \quad & Ax \geq b. \end{aligned}$$

Let $\bar{x}$ be a feasible point for this problem. We have observed already that the inactive constraints at $\bar{x}$ do not influence the feasible directions at this point. Let $\hat{A}$ be the submatrix of A corresponding to the rows of the active constraints at $\bar{x}$. Then a direction p is a feasible direction for S at $\bar{x}$ if and only if

$$\hat{A}p \geq 0.$$

Since the inactive constraints at a point have no impact on its feasible directions, such constraints can be ignored when testing whether the point is locally optimal. In particular, if we had prior knowledge of which constraints are active at the optimum, we could cast aside the inactive constraints and treat the active constraints as equalities. A solution of the inequality-constrained problem is a solution of the equality-constrained problem defined by the active constraints.

The theory for inequality-constrained problems draws on the theory for equality-constrained problems. For this reason, it is important to study problems with only equality constraints. In particular, it will be useful to study ways to represent all the vectors in the null space of a matrix. This is the topic of Sections 3.2 and 3.3.

Once a feasible direction p is determined, the new estimate of the solution is of the form $\bar{x} + \alpha p$ where $\alpha \geq 0$. Since the new point must be feasible, in general there is an upper limit on how large α can be.

For an equality constraint we have $a_i^T p = 0$, and so

$$a_i^T(\bar{x} + \alpha p) = a_i^T \bar{x} = b_i$$

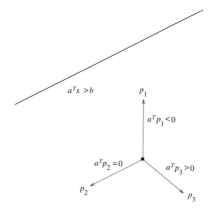

Figure 3.2. *Movement to and away from the boundary.*

for *all* values of α. For an active inequality constraint we have $a_i^T p \geq 0$, and so

$$a_i^T(\bar{x} + \alpha p) \geq a_i^T \bar{x} \geq b_i$$

for all values of $\alpha \geq 0$. Thus only the inactive constraints are relevant when determining an upper bound on α.

Because $\bar{x}$ is feasible, $a_i^T \bar{x} > b_i$ for all inactive constraints. Thus, if $a_i^T p \geq 0$, the constraint remains satisfied for all $\alpha \geq 0$. As α increases, the movement is *away* from the boundary of the constraint. On the other hand, if $a_i^T p < 0$, the inequality will remain valid only if $\alpha \leq (a_i^T \bar{x} - b_i)/(-a_i^T p)$. A positive step along p is a move towards the boundary, and any step larger than this bound will violate the constraint. (See Figure 3.2.) The maximum step length $\bar{\alpha}$ that maintains feasibility is obtained from a *ratio test*:

$$\bar{\alpha} = \min\left\{ (a_i^T \bar{x} - b_i)/(-a_i^T p) : a_i^T p < 0 \right\},$$

where the minimum is taken over all inactive constraints. If $a_i^T p \geq 0$ for all inactive constraints, then an arbitrarily large step can be taken without violating feasibility.

Example 3.4 (Ratio Test). Let $\bar{x} = (1, 1)^T$ and $p = (4, -2)^T$. Suppose that there are three inactive constraints with

$$a_1^T = (1 \quad 4) \quad \text{and} \quad b_1 = 3$$
$$a_2^T = (0 \quad 3) \quad \text{and} \quad b_2 = 2$$
$$a_3^T = (5 \quad 1) \quad \text{and} \quad b_3 = 4.$$

Then

$$a_1^T p = -4 < 0, \quad a_2^T p = -6 < 0, \quad \text{and} \quad a_3^T p = 18 > 0,$$

so only the first two constraints are used in the ratio test:

$$\bar{\alpha} = \min\left\{ (a_i^T \bar{x} - b_i)/(-a_i^T p) : a_i^T p < 0 \right\}$$
$$= \min\left\{ (5 - 3)/4, (3 - 2)/6 \right\} = 1/6.$$

Notice that the point $\bar{x} + \bar{\alpha} p = (\frac{5}{3}, \frac{2}{3})^T$ is on the boundary of the second constraint. ∎

Exercises

1.1. Find the sets of all feasible directions at points $x_a = (0, 0, 2)^T$, $x_b = (3, 0, 1)^T$, and $x_c = (1, 1, 1)^T$ for Example 3.1.

1.2. Consider the set defined by the constraints $x_1 + x_2 = 1$, $x_1 \geq 0$, and $x_2 \geq 0$. At each of the following points determine the set of feasible directions: (a) $(0, 1)^T$; (b) $(1, 0)^T$; (c) $(0.5, 0.5)^T$.

1.3. Consider the system of inequality constraints $Ax \geq b$ with

$$A = \begin{pmatrix} 9 & 4 & 1 & 9 & -7 \\ 6 & -7 & 8 & -4 & -6 \\ 1 & 6 & 3 & -7 & 6 \end{pmatrix} \quad \text{and} \quad b = \begin{pmatrix} -15 \\ -30 \\ -20 \end{pmatrix}.$$

For the given values of x and p, perform a ratio test to determine the maximum step length $\bar{\alpha}$ such that $x + \bar{\alpha}p$ remains feasible.

   (i) $x = (8, 4, -3, 4, 1)^T$ and $p = (1, 1, 1, 1, 1)^T$,
   (ii) $x = (7, -4, -3, -3, 3)^T$ and $p = (3, 2, 0, 1, -2)^T$,
   (iii) $x = (5, 0, -6, -8, -3)^T$ and $p = (5, 0, 5, 1, 3)^T$,
   (iv) $x = (9, 1, -1, 6, 3)^T$ and $p = (-4, -2, 4, -2, 2)^T$.

1.4. What are the potential consequences of miscalculating $\bar{\alpha}$ in the ratio test?

1.5. Let $S = \{x : Ax \leq b\}$. Derive the conditions that must be satisfied by a feasible direction at a point $\bar{x} \in S$.

1.6. On a computer, there is a danger that an overflow can occur during the ratio test if, in a particular ratio, the numerator is large and the denominator is small. How can the ratio test be implemented so that this danger is removed?

3.2 Null and Range Spaces

Let A be an $m \times n$ matrix with $m \leq n$. We denote the *null space* of A by

$$\mathcal{N}(A) = \{p \in \Re^n : Ap = 0\}.$$

↗constraints

The null space of a matrix is the set of vectors orthogonal to the rows of the matrix. Recall that the null space represents the set of feasible directions for the constraints $Ax = b$. It is easy to see that any linear combination of two vectors in $\mathcal{N}(A)$ is also in $\mathcal{N}(A)$, and thus the null space is a subspace of $\Re^n$. It can be shown that the dimension of this subspace is $n - \text{rank}(A)$. When A has full row rank (i.e., its rows are linearly independent), this is just $n - m$.

Another term that will be important to our discussions is the *range space* of a matrix. This is the set of vectors spanned by the columns of the matrix (that is, the set of all linear combinations of these columns). In particular, we are interested in the range space of A^T, defined by

$$\mathcal{R}(A^T) = \{q \in \Re^n : q = A^T\lambda \quad \text{for some } \lambda \in \Re^m\}.$$

↳ ie, all linear combinations of the rows

$$q^T = \lambda^T A$$

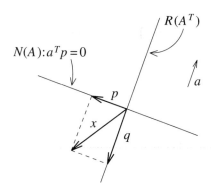

Figure 3.3. *Null space and range space of $A = (a^T)$.*

Throughout this text, if we mention a range space without specifying a matrix, it refers to the range space of A^T. The dimension of the range space is the same as the rank of A^T, or equivalently the rank of A.

There is an important relationship between $\mathcal{N}(A)$ and $\mathcal{R}(A^T)$: they are *orthogonal subspaces*. This means that any vector in one subspace is orthogonal to any vector in the other. To verify this statement, we note that any vector $q \in \mathcal{R}(A^T)$ can be expressed as $q = A^T\lambda$ for some $\lambda \in \Re^m$, and therefore, for any vector $p \in \mathcal{N}(A)$ we have

$$q^Tp = \lambda^T Ap = 0.$$

There is more. Because the null and range spaces are orthogonal subspaces whose dimensions sum to n, any n-dimensional vector x can be written uniquely as the sum of a null-space and a range-space component:

$$x = p + q,$$

where $p \in \mathcal{N}(A)$ and $q \in \mathcal{R}(A^T)$. Figure 3.3 illustrates the null and range spaces for $A = (a^T)$, where a is a two-dimensional nonzero vector. Notice that the vector a is orthogonal to the null space and that any range-space vector is a scalar multiple of a. The decomposition of a vector x into null-space and range-space components is also shown in Figure 3.3.

How can we represent vectors in the null space of A? For this purpose, we define a matrix Z to be a *null-space matrix* for A if any vector in $\mathcal{N}(A)$ can be expressed as a linear combination of the columns of Z. The representation of a null-space matrix is not unique. If A has full row rank m, any matrix Z of dimension $n \times r$ and rank $n - m$ that satisfies $AZ = 0$ is a null-space matrix. The column dimension r must be at least $(n - m)$. In the special case where r is equal to $n - m$, the columns of Z are linearly independent, and Z is then called a *basis* matrix for the null space of A. If Z is an $n \times r$ null-space matrix, the null space can be represented as

$$\mathcal{N}(A) = \{\, p : p = Zv \quad \text{for some } v \in \Re^r \,\},$$

thus $\mathcal{N}(A) = \mathcal{R}(Z)$. This representation of the null space gives us a practical way to generate feasible points. If $\bar{x}$ is any point satisfying $Ax = b$, then all other feasible points can be written as

$$x = \bar{x} + Zv$$

for some vector v.

As an example consider the rank-two matrix

$$A = \begin{pmatrix} 1 & -1 & 0 & 0 \\ 0 & 0 & 1 & 1 \end{pmatrix}.$$

The null space of A is the set of all vectors p such that

$$Ap = \begin{pmatrix} 1 & -1 & 0 & 0 \\ 0 & 0 & 1 & 1 \end{pmatrix} \begin{pmatrix} p_1 \\ p_2 \\ p_3 \\ p_4 \end{pmatrix} = \begin{pmatrix} p_1 - p_2 \\ p_3 + p_4 \end{pmatrix} = \begin{pmatrix} 0 \\ 0 \end{pmatrix};$$

that is, the vector must satisfy $p_1 = p_2$ and $p_3 = -p_4$. Thus any null-space vector must have the form

$$p = \begin{pmatrix} v_1 \\ v_1 \\ v_2 \\ -v_2 \end{pmatrix}$$

for some scalars v_1 and v_2. A possible basis matrix for the null space of A is

$$Z = \begin{pmatrix} 1 & 0 \\ 1 & 0 \\ 0 & 1 \\ 0 & -1 \end{pmatrix}$$

and the null space can be expressed as

$$\mathcal{N}(A) = \left\{ p : p = Zv \quad \text{for some } v \in \Re^2 \right\}.$$

The matrix

$$\bar{Z} = \begin{pmatrix} 1 & 0 & 2 \\ 1 & 0 & 2 \\ 0 & 1 & -1 \\ 0 & -1 & 1 \end{pmatrix}$$

is also a null-space matrix for A, but it is not a basis matrix since its third column is a linear combination of the first two columns. The null space of A can be expressed in terms of $\bar{Z}$ as

$$\mathcal{N}(A) = \left\{ p : p = \bar{Z}\bar{v} \text{ for some } \bar{v} \in \Re^3 \right\}.$$

Exercises

2.1. In each of the following cases, compute a basis matrix for the null space of the matrix A and express the points x_i as $x_i = p_i + q_i$ where p_i is in the null space of A and q_i is in the range space of A^T.

(i)

$$A = \begin{pmatrix} 1 & 1 & 1 & 1 \\ 1 & -1 & -1 & 1 \\ 0 & 1 & 0 & 1 \end{pmatrix}, \quad x_1 = \begin{pmatrix} 1 \\ 3 \\ 1 \\ 2 \end{pmatrix}, \quad x_2 = \begin{pmatrix} 0 \\ -2 \\ -3 \\ 4 \end{pmatrix}.$$

(ii)

$$A = \begin{pmatrix} 1 & 1 & 1 & 1 \end{pmatrix}, \quad x_1 = \begin{pmatrix} -2 \\ 4 \\ 5 \\ -2 \end{pmatrix}, \quad x_2 = \begin{pmatrix} 7 \\ 5 \\ -13 \\ 1 \end{pmatrix}.$$

(iii)

$$A = \begin{pmatrix} 1 & 1 & 1 & 1 \\ 1 & -1 & -1 & 1 \end{pmatrix}, \quad x_1 = \begin{pmatrix} 4 \\ 3 \\ 4 \\ 0 \end{pmatrix}, \quad x_2 = \begin{pmatrix} -1 \\ 1 \\ 5 \\ -5 \end{pmatrix}.$$

(iv)

$$A = \begin{pmatrix} 1 & 1 & 1 & 1 \\ 2 & 0 & 0 & 2 \\ 1 & -1 & -1 & 1 \end{pmatrix}, \quad x_1 = \begin{pmatrix} 3 \\ 1 \\ 1 \\ 2 \end{pmatrix}, \quad x_2 = \begin{pmatrix} 8 \\ 9 \\ -2 \\ -4 \end{pmatrix}.$$

2.2. Let Z be an $n \times r$ null-space matrix for the matrix A. If Y is any invertible $r \times r$ matrix, prove that $\hat{Z} = ZY$ is also a null-space matrix for A.

2.3. Let A be a given $m \times n$ matrix and let Z be a null-space matrix for A. Let X be an invertible $m \times m$ matrix and let Y be an invertible $n \times n$ matrix. If a change of variable is made to transform A into $\hat{A} = XAY$, how can Z be transformed into $\hat{Z}$, a null-space matrix for $\hat{A}$?

2.4. Let A be a full-rank $m \times n$ matrix and let Z be a basis matrix for the null space of A. Use the results of the previous problem to prove that, for appropriate choices of X and Y, $\hat{A}$ and $\hat{Z}$ have the form

$$\hat{A} = \begin{pmatrix} 0 & I_m \end{pmatrix} \quad \text{and} \quad \hat{Z} = \begin{pmatrix} I_{n-m} \\ 0 \end{pmatrix},$$

where I_m and I_{n-m} are identity matrices of the appropriate size. What is the corresponding result in the case where A is not of full rank and Z is any null-space matrix?

2.5. Let A be an $m \times n$ matrix with $m < n$. Prove that any n-dimensional vector x can be written uniquely as the sum of a null-space and a range-space component:

$$x = p + q,$$

where $p \in \mathcal{N}(A)$ and $q \in \mathcal{R}(A^T)$.

2.6. Suppose that you are given a matrix A and a vector p and are told that p is in the null space of A. On a computer, you cannot expect that Ap will be exactly equal to zero because of rounding errors. How large would the computed value of $\|Ap\|$ have to be before you could conclude that p was not in the null space of A? (Your answers should incorporate the values of the machine precision and the components of A and p.) If the computed value of $\|Ap\|$ is zero, can you conclude that p is in the null space of A?

3.3   Generating Null-Space Matrices

We present here four commonly used methods for deriving a null-space matrix for A. The discussion assumes that A is an $m \times n$ matrix of full row rank (and hence $m \leq n$). Two of the approaches, the variable reduction method and the QR factorization, yield an $n \times (n - m)$ basis matrix for $\mathcal{N}(A)$. The other two methods yield an $n \times n$ null-space matrix.

3.3.1   Variable Reduction Method

This method is the approach used by the simplex algorithm for linear programming. It is also used in nonlinear optimization (see Section 15.6). We start with an example.

Consider the linear system of equations:

$$\begin{aligned}
p_1 + \ p_2 - p_3 &= 0 \\
- 2p_2 + p_3 &= 0.
\end{aligned}$$

This system has the form $Ap = 0$. We wish to generate all solutions to this system.

We can solve for any two variables whose associated columns in A are linearly independent in terms of the third variable. For example, we can solve for p_1 and p_3 in terms of p_2 as follows:

$$\begin{aligned}
p_1 &= \ p_2 \\
p_3 &= 2p_2.
\end{aligned}$$

The set of all solutions to the system can be written as

$$p = \begin{pmatrix} 1 \\ 1 \\ 2 \end{pmatrix} p_2,$$

where p_2 is chosen arbitrarily. Thus $Z = (1, 1, 2)^T$ is a basis for the null space of A.

Since the values of p_1 and p_3 depend on p_2, they are called *dependent variables*. They are also sometimes called *basic* variables. The variable p_2 which can take on any value is called an *independent variable*, or a *nonbasic variable*.

To generalize this, consider the $m \times n$ system $Ap = 0$. Select any set of m variables whose corresponding columns are linearly independent—these will be the basic variables. Denote by B the $m \times m$ matrix defined by these columns. The remaining variables will be the nonbasic variables; we denote the $m \times (n - m)$ matrix of their respective columns by N. The general solution to the system $Ap = 0$ is obtained by expressing the basic variables in terms of the nonbasic variables, where the nonbasic variables can take on any arbitrary value.

For ease of notation we assume here that the first m variables are the basic variables. Thus

$$Ap = (\begin{array}{cc} B & N \end{array}) \begin{pmatrix} p_B \\ p_N \end{pmatrix} = Bp_B + Np_N = 0.$$

Premultiplying the last equation by B^{-1} we get

$$p_B = -B^{-1}Np_N.$$

Thus the set of solutions to the system $Ap = 0$ is

$$p = \begin{pmatrix} p_B \\ p_N \end{pmatrix} = \begin{pmatrix} -B^{-1}N \\ I \end{pmatrix} p_N,$$

and the $n \times (n - m)$ matrix

$$Z = \begin{pmatrix} -B^{-1}N \\ I \end{pmatrix}$$

is a basis for the null space of A.

Consider now the system $Ax = b$. One feasible solution is

$$\bar{x} = \begin{pmatrix} B^{-1}b \\ 0 \end{pmatrix}.$$

If x is *any* point that satisfies $Ax = b$, then x can be written in the form

$$x = \bar{x} + p = \bar{x} + Zp_N = \begin{pmatrix} B^{-1}b \\ 0 \end{pmatrix} + \begin{pmatrix} -B^{-1}N \\ I \end{pmatrix} p_N.$$

If the basis matrix B is chosen differently, then the representation of the feasible points changes, but the set of feasible points does not.

In this derivation we assumed that the first m variables were the basic variables. If this is not true, the rows in Z must be reordered to correspond to the ordering of the basic and nonbasic variables. This technique is illustrated in the following example.

Example 3.5 (Variable Reduction). Consider the system of constraints $Ax = b$ with

$$A = \begin{pmatrix} 1 & -2 & 1 & 3 \\ 0 & 1 & 1 & 4 \end{pmatrix} \quad \text{and} \quad b = \begin{pmatrix} 5 \\ 6 \end{pmatrix}.$$

Let B consist of the first two columns of A, and let N consist of the last two columns:

$$B = \begin{pmatrix} 1 & -2 \\ 0 & 1 \end{pmatrix} \quad \text{and} \quad N = \begin{pmatrix} 1 & 3 \\ 1 & 4 \end{pmatrix}.$$

Then

$$\bar{x} = \begin{pmatrix} B^{-1}b \\ 0 \end{pmatrix} = \begin{pmatrix} 17 \\ 6 \\ 0 \\ 0 \end{pmatrix}$$

and

$$Z = \begin{pmatrix} -B^{-1}N \\ I \end{pmatrix} = \begin{pmatrix} -3 & -11 \\ -1 & -4 \\ 1 & 0 \\ 0 & 1 \end{pmatrix}.$$

It is easy to verify that $A\bar{x} = b$ and $AZ = 0$. Every point satisfying $Ap = 0$ is of the form

$$Zp_N = \begin{pmatrix} -3 & -11 \\ -1 & -4 \\ 1 & 0 \\ 0 & 1 \end{pmatrix} \begin{pmatrix} p_3 \\ p_4 \end{pmatrix} = \begin{pmatrix} -3p_3 - 11p_4 \\ -p_3 - 4p_4 \\ p_3 \\ p_4 \end{pmatrix}.$$

If instead B is chosen as columns 4 and 3 of A (in that order), and N as columns 2 and 1, then

$$B = \begin{pmatrix} 3 & 1 \\ 4 & 1 \end{pmatrix} \quad \text{and} \quad N = \begin{pmatrix} -2 & 1 \\ 1 & 0 \end{pmatrix}.$$

Care must be taken in defining $\bar{x}$ and Z to ensure that their components are positioned correctly. In this case

$$B^{-1}b = \begin{pmatrix} 1 \\ 2 \end{pmatrix} \quad \text{and} \quad \bar{x} = \begin{pmatrix} 0 \\ 0 \\ 2 \\ 1 \end{pmatrix}.$$

Notice that the components of $B^{-1}b$ are at positions 4 and 3 in $\bar{x}$, corresponding to the columns of A that were used to define B. Similarly

$$-B^{-1}N = \begin{pmatrix} -3 & 1 \\ 11 & -4 \end{pmatrix} \quad \text{and} \quad Z = \begin{pmatrix} 0 & 1 \\ 1 & 0 \\ 11 & -4 \\ -3 & 1 \end{pmatrix}.$$

The rows of $-B^{-1}N$ are placed in rows 4 and 3 of Z, and the rows of I are placed in rows 2 and 1. As before, $A\bar{x} = b$ and $AZ = 0$. Every point satisfying $Ap = 0$ is of the form

$$Zp_N = \begin{pmatrix} 0 & 1 \\ 1 & 0 \\ 11 & -4 \\ -3 & 1 \end{pmatrix} \begin{pmatrix} p_2 \\ p_1 \end{pmatrix} = \begin{pmatrix} p_1 \\ p_2 \\ 11p_2 - 4p_1 \\ -3p_2 + p_1 \end{pmatrix}. \quad \blacksquare$$

In practice the matrix Z itself is rarely formed explicitly, since the inverse of B should not be computed. This is not a limitation; Z is only needed to provide matrix-vector products of the form $p = Zv$, or the form $Z^T g$. These computations do not require Z explicitly. For example, the vector $p = Zv$ may be computed as follows. First we compute $t = Nv$. Next we compute $u = -B^{-1}t$, by solving the system $Bu = -t$. (This should be done via a numerically stable method such as the LU factorization.) The vector $p = Zv$ is now given by $p = (u^T, v^T)^T$.

The variable reduction approach for representing the null space is the method used in the simplex algorithm for linear programming. This approach has been enhanced so that ever larger problems can be solved. These enhancements exploit the sparsity that is often present in large problems, in order to reduce computational effort and increase accuracy. A more detailed exposition of these techniques is given in Chapter 7.

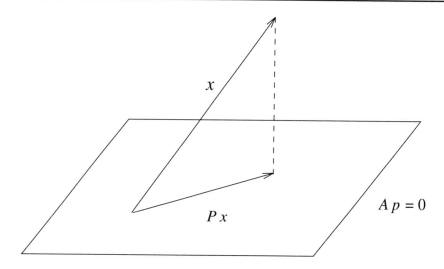

Figure 3.4. *Orthogonal projection.*

3.3.2 Orthogonal Projection Matrix

Let x be an n-dimensional vector, and let A be an $m \times n$ matrix of full row rank. Then x can be expressed as a sum of two components, one in $\mathcal{N}(A)$ and the other in $\mathcal{R}(A^T)$:

$$x = p + q,$$

where $Ap = 0$, and $q = A^T\lambda$ for some m-dimensional vector λ. Multiplying this equation on the left by A gives $Ax = AA^T\lambda$, from which we obtain $\lambda = (AA^T)^{-1}Ax$. Substituting for q gives the null-space component of x:

$$p = x - A^T(AA^T)^{-1}Ax = (I - A^T(AA^T)^{-1}A)x.$$

The $n \times n$ matrix

$$P = I - A^T(AA^T)^{-1}A$$

is called an *orthogonal projection matrix* into $\mathcal{N}(A)$. The null-space component of the vector x can be found by premultiplying x by P; the resulting vector Px is also termed the orthogonal projection of x onto $\mathcal{N}(A)$ (see Figure 3.4).

The orthogonal projection matrix is the unique matrix with the following properties:

- It is a null-space matrix for A;
- $P^2 = P$, which means repeated application of the orthogonal projection has no further effect;
- $P^T = P$ (P is symmetric).

The name "orthogonal projection" may be misleading—unless P is the identity matrix it is *not* orthogonal.

There are a number of ways to compute the projection matrix. Selection of the method depends in general on the application, the size of m and n, as well as the sparsity of A. We point out that by "computing the matrix" we mean representing the matrix so that a matrix-vector product of the form Px can be formed for any vector x. The projection matrix itself is rarely formed explicitly.

To demonstrate this point, suppose that A consists of a single row: $A = a^T$, where a is an n-vector. Then

$$P = I - \frac{1}{a^T a} a a^T.$$

Forming P explicitly would require approximately $n^2/2$ multiplications and $n^2/2$ storage locations. Forming the product Px for some vector x would require n^2 additional multiplications. These costs can be reduced dramatically if only the vector a and the scalar $a^T a$ are stored. "Forming" P this way only requires n multiplications in the calculation of $(a^T a)$. The matrix-vector product is computed as $Px = x - a(a^T x)/(a^T a)$. This requires only $2n$ multiplications.

In the example above the matrix $A A^T$ is the scalar $a^T a$, which is easy to invert. In the more general case where A has several rows, the task of "inverting" $A A^T$ becomes expensive, and care must be taken to perform this in a numerically stable manner. Often, this is done by the Cholesky factorization. However, if A is dense it is not advisable to form the matrix $A A^T$ explicitly, since it can be shown that its condition number is the square of that of A. A more stable approach is to use a QR factorization of A^T (see Appendix A.7.3 and Section 3.3.4 below).

For the case when A is large and sparse, the QR factorization may be too expensive, since it tends to produce dense factors. Special techniques that attempt to exploit the sparsity structure of A have been developed for this situation.

3.3.3 Other Projections

As before, let A be an $m \times n$ matrix of full row rank. Let D be a positive-definite $n \times n$ matrix, and consider the $n \times n$ matrix

$$P_D = I - D A^T (A D A^T)^{-1} A.$$

It is easy to show that P_D is a null-space matrix for A. Also, $P_D P_D = P_D$. An $n \times n$ matrix with these two properties is called a *projection matrix*. An orthogonal projection is therefore a symmetric projection matrix.

Many of the new interior point algorithms for optimization use projections of this form. In the case of linear programming, the matrix D is generally a diagonal matrix with positive diagonal terms. This matrix D changes from iteration to iteration, while A remains unchanged. Special techniques for computing and updating these projections have been developed.

3.3.4 The QR Factorization

Again let A be an $m \times n$ matrix with full row rank. We perform an orthogonal factorization of A^T :

$$A^T = QR.$$

Let $Q = (Q_1, Q_2)$, where Q_1 consists of the first m columns of Q, and Q_2 consists of the last $n - m$ columns. Also denote the top $m \times m$ triangular submatrix of R by R_1. The rest of R is an $(n - m) \times m$ zero matrix. Since Q is an orthogonal matrix, it follows that $AQ = R^T$, or

$$AQ_1 = R_1^T \quad \text{and} \quad AQ_2 = 0.$$

Thus

$$Z = Q_2$$

is a basis for the null space of A. This basis is also known as an *orthogonal basis*, since $Z^T Z = I$.

Example 3.6 (Generating a Basis Matrix Using the QR Factorization). Consider the matrix

$$A = \begin{pmatrix} 1 & -1 & 0 & 0 \\ 0 & 0 & 1 & 1 \end{pmatrix}.$$

An orthogonal factorization of A^T yields

$$Q = \begin{pmatrix} -\sqrt{2}/2 & 0 & -1/2 & -1/2 \\ \sqrt{2}/2 & 0 & -1/2 & -1/2 \\ 0 & -\sqrt{2}/2 & 1/2 & -1/2 \\ 0 & -\sqrt{2}/2 & -1/2 & 1/2 \end{pmatrix}, \qquad R = \begin{pmatrix} -\sqrt{2} & 0 \\ 0 & -\sqrt{2} \\ 0 & 0 \\ 0 & 0 \end{pmatrix},$$

hence

$$Z = \begin{pmatrix} -1/2 & -1/2 \\ -1/2 & -1/2 \\ 1/2 & -1/2 \\ -1/2 & 1/2 \end{pmatrix}$$

is a basis for the null space of A. ∎

The QR factorization method has the important advantage that the basis Z can be formed in a numerically stable manner. Moreover, computations performed with respect to the resulting basis Z are numerically stable. (For further information, see the references cited in the Notes.) However, this numerical stability comes at a price, since computing the QR factorization is relatively expensive. If m is small relative to n, some savings may be gained by not forming Q explicitly. An additional drawback of the QR method is that the basis Z can be dense even when A is sparse. As a result it may be unsuitable for large sparse problems.

Exercises

3.1. For each of the following matrices, compute a basis for the null space using variable reduction (with A written in the form (B, N)).

(i)

$$A = \begin{pmatrix} 1 & 1 & 1 & 1 \\ 1 & -1 & -1 & 1 \\ 0 & 1 & 0 & 1 \end{pmatrix}.$$

(ii)
$$A = (1 \quad 1 \quad 1 \quad 1).$$

(iii)
$$A = \begin{pmatrix} 1 & 1 & 1 & 1 \\ 1 & -1 & -1 & 1 \end{pmatrix}.$$

(iv)
$$A = \begin{pmatrix} 1 & 1 & 1 & 1 \\ 2 & 0 & 0 & 2 \\ 1 & -1 & -1 & 1 \end{pmatrix}.$$

3.2. Compute the orthogonal projection matrix for each of the matrices in the previous problem.

3.3. Consider the system $Ap = 0$, where
$$A = \begin{pmatrix} 1 & 2 & 0 & 2 \\ 2 & 1 & 2 & 4 \end{pmatrix}.$$

Compute a basis for the null-space matrix of A using p_2 and p_3 as the basic variables. Use this to write a general expression for all solutions to this system. Could you do the same if p_1 and p_4 were the basic variables?

3.4. Let A be an $m \times n$ matrix of full row rank. Prove that the matrix AA^T is positive definite, and hence its inverse exists.

3.5. Let A be an $m \times n$ matrix of full column rank. Prove that the matrix $A^T A$ is positive definite, and hence its inverse exists.

3.6. Let A be an $m \times n$ full row rank matrix and let Z be a basis for its null space. Prove that
$$I - A^T(AA^T)^{-1}A = Z(Z^TZ)^{-1}Z^T.$$

3.7. Let P be the orthogonal projection matrix associated with an $m \times n$ full row rank matrix A. Prove that P has $n - m$ linearly independent eigenvectors associated with the eigenvalue 1, and m linearly independent eigenvectors associated with the eigenvalue 0.

3.8. Prove that if P is the orthogonal projection matrix associated with $\mathcal{N}(A)$, then $I - P$ is the orthogonal projection matrix associated with $\mathcal{R}(A^T)$.

3.9. Let $A = (1, 3, 2, -1)^T$ and let $x = (6, 8, -2, 1)^T$. Compute the orthogonal projection of x into the null space of A without explicitly forming the projection matrix.

3.10. Prove that an orthogonal projection matrix is positive semidefinite.

3.11. Let A be an $m \times n$ matrix of full row rank, and let P be the orthogonal projection matrix corresponding to A. Let a be an n-dimensional vector and suppose that a is not a linear combination of the rows of A.

   (i) Prove that $a^T P a \neq 0$.

   (ii) Let
$$\hat{A} = \begin{pmatrix} A \\ a^T \end{pmatrix},$$

and let $\hat{P}$ be the orthogonal projection matrix corresponding to $\hat{A}$. Prove that $\hat{P} = P - Pa(a^T Pa)^{-1}a^T P$.

3.12. Let A be an $m \times n$ full row rank matrix and D an $n \times n$ positive-definite matrix.

    (i) Prove that the matrix ADA^T is positive definite, and hence its inverse exists.

    (ii) Let $P_D = I - DA^T(ADA^T)^{-1}A$. Prove that $P_D x = 0$ if and only if $x = DA^T\eta$ for some m-dimensional vector η.

    (iii) Prove that the matrix $P_D D$ is positive semidefinite, and $x^T P_D D x = 0$, if and only if $x = A^T\eta$ for some vector η.

3.13. Compute an orthogonal basis matrix for the matrices in Exercise 3.1.

3.14. Consider the QR factorization of a full row rank matrix A. Prove that $Q_1 Q_1^T + Q_2 Q_2^T = I$.

3.15. Consider the problem of forming the orthogonal projection matrix associated with a matrix A. One approach to avoid the potential ill-conditioning of the matrix AA^T is to use the QR factorization for the matrix AA^T. Assume that A has full row rank.

    (i) Prove that the $AA^T = R_1^T R_1$ and hence R_1^T is the lower triangular matrix of the Cholesky factorization for AA^T.

    (ii) Prove that the resulting orthogonal projection is $P = Q_2 Q_2^T$.

    (iii) Prove that $A^T(AA^T)^{-1} = Q_1 R_1^{-T}$.

3.16. Let A be a matrix with full row rank, and let Z be an orthogonal basis matrix for A. Prove that the orthogonal projection matrix associated with A satisfies $P = ZZ^T$.

3.17. Let P be an orthogonal projection. Prove that P is unique. Hint: Let $P = ZZ^T$ where Z is an orthogonal basis matrix for the null space. Suppose that P_1 is another orthogonal projection. Then $P_1 = ZV^T$ for some full-rank matrix V. Now prove that $V = Z$.

3.18. Compute an orthogonal projection matrix for

$$A = \begin{pmatrix} 1 & 1 & 1 & 1 \\ 2 & 2 & 2 & 2 \end{pmatrix}.$$

3.4 Notes

Further information on these topics can be found in the books by Gill, Murray, and Wright (1991); Golub and Van Loan (1996); and Trefethen and Bau (1997).

Part II

Linear Programming

Chapter 4

Geometry of Linear Programming

4.1 Introduction

Linear programs can be studied both algebraically and geometrically. The two approaches are equivalent, but one or the other may be more convenient for answering a particular question about a linear program.

The algebraic point of view is based on writing the linear program in a particular way, called *standard form*. Then the coefficient matrix of the constraints of the linear program can be analyzed using the tools of linear algebra. For example, we might ask about the rank of the matrix, or for a representation of its null space. It is this algebraic approach that is used in the simplex method, the topic of the next chapter.

The geometric point of view is based on the geometry of the feasible region and uses ideas such as convexity to analyze the linear program. It is less dependent on the particular way in which the constraints are written. Using geometry (particularly in two-dimensional problems where the feasible region can be graphed) makes many of the concepts in linear programming easy to understand, because they can be described in terms of intuitive notions such as moving along an edge of the feasible region.

There is a direct correspondence between these two points of view. This chapter will explore several aspects of this correspondence.

Before giving an outline of the chapter, we show how a two-dimensional linear program can be solved graphically. Consider the problem

$$\begin{aligned} \text{minimize} \quad & z = -x_1 - 2x_2 \\ \text{subject to} \quad & -2x_1 + x_2 \leq 2 \\ & -x_1 + x_2 \leq 3 \\ & x_1 \leq 3 \\ & x_1, x_2 \geq 0. \end{aligned}$$

The feasible region is graphed in Figure 4.1.

The figure also includes lines corresponding to various values of the objective function. For example, the line $z = -2 = -x_1 - 2x_2$ passes through the points $(2, 0)^T$ and $(0, 1)^T$,

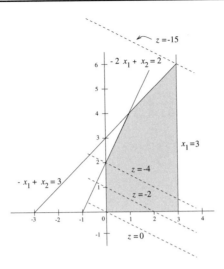

Figure 4.1. *Graphical solution of a linear program.*

and the parallel line $z = 0$ passes through the origin. The goal of the linear program is to minimize the value of z. As the figure illustrates, z decreases as these lines move upward and to the right. The objective z cannot be decreased indefinitely, however. Eventually the z line ceases to intersect the feasible region, indicating that there are no longer any feasible points corresponding to that particular value of z. The minimum occurs when $z = -15$ at the point $(3, 6)^T$, that is, at the last point where an objective line intersects the feasible region. This is a corner of the feasible region.

It is no coincidence that the solution occurred at a corner or *extreme point*. Proving this result will be the major goal of this chapter.

To achieve this goal, we will first describe *standard form*, a particular way of writing a system of linear constraints. Standard form will be used to define a *basic feasible solution*. We will then show that the algebraic notion of a basic feasible solution is equivalent to the geometric notion of an extreme point. This equivalence is of value because, in higher dimensions, basic feasible solutions are easier to generate and identify than extreme points. It will then be shown how to represent any feasible point in terms of extreme points and *directions of unboundedness* (directions used in the description of unbounded feasible regions). Finally, this representation of feasible points will be used to prove that any linear program with a finite optimal solution has an optimal extreme point. This last result will, in turn, motivate our discussion of the simplex method, a method that solves linear programs by examining a sequence of basic feasible solutions, that is, extreme points.

Exercises

1.1. Solve the following linear programs graphically.

(i)

$$\begin{aligned}
\text{minimize} \quad & z = 3x_1 + x_2 \\
\text{subject to} \quad & x_1 - x_2 \le 1 \\
& 3x_1 + 2x_2 \le 12 \\
& 2x_1 + 3x_2 \le 3 \\
& -2x_1 + 3x_2 \ge 9 \\
& x_1, x_2 \ge 0.
\end{aligned}$$

(ii)

$$\begin{aligned}
\text{maximize} \quad & z = x_1 + 2x_2 \\
\text{subject to} \quad & 2x_1 + x_2 \ge 12 \\
& x_1 + x_2 \ge 5 \\
& -x_1 + 3x_2 \le 3 \\
& 6x_1 - x_2 \ge 12 \\
& x_1, x_2 \ge 0.
\end{aligned}$$

(iii)

$$\begin{aligned}
\text{minimize} \quad & z = x_1 - 2x_2 \\
\text{subject to} \quad & x_1 - 2x_2 \ge 4 \\
& x_1 + x_2 \le 8 \\
& x_1, x_2 \ge 0.
\end{aligned}$$

(iv)

$$\begin{aligned}
\text{minimize} \quad & z = -x_1 - x_2 \\
\text{subject to} \quad & x_1 - x_2 \ge 1 \\
& x_1 - 2x_2 \ge 2 \\
& x_1, x_2 \ge 0.
\end{aligned}$$

(v)

$$\begin{aligned}
\text{minimize} \quad & z = x_1 - x_2 \\
\text{subject to} \quad & x_1 - x_2 \ge 2 \\
& 2x_1 + x_2 \ge 1 \\
& x_1, x_2 \ge 0.
\end{aligned}$$

(vi)

$$\begin{aligned}
\text{minimize} \quad & z = 4x_1 - x_2 \\
\text{subject to} \quad & x_1 + x_2 \le 6 \\
& x_1 - x_2 \ge 3 \\
& -x_1 + 2x_2 \ge 2 \\
& x_1, x_2 \ge 0.
\end{aligned}$$

(vii)

$$\begin{aligned}
\text{maximize} \quad & z = 6x_1 - 3x_2 \\
\text{subject to} \quad & 2x_1 + 5x_2 \ge 10 \\
& 3x_1 + 2x_2 \le 40 \\
& x_1, x_2 \le 15.
\end{aligned}$$

(viii)

$$\begin{aligned}
\text{minimize} \quad & z = x_1 + 9x_2 \\
\text{subject to} \quad & 2x_1 + x_2 \le 100 \\
& x_1 + x_2 \le 80 \\
& x_1 \le 40 \\
& x_1, x_2 \ge 0.
\end{aligned}$$

(ix)

$$\begin{aligned}
\text{minimize} \quad & z = 2x_1 + 13x_2 \\
\text{subject to} \quad & x_1 + x_2 \le 5 \\
& x_1 + 2x_2 \le 6 \\
& x_1, x_2 \ge 0.
\end{aligned}$$

(x)

$$\begin{aligned}
\text{minimize} \quad & z = -5x_1 - 7x_2 \\
\text{subject to} \quad & -3x_1 + 2x_2 \le 30 \\
& -2x_1 + x_2 \le 12 \\
& x_1, x_2 \ge 0.
\end{aligned}$$

1.2. Find graphically all the values of the parameter a such that $(-3, 4)^T$ is the optimal solution of the following problem:

$$\begin{aligned}
\text{maximize} \quad & z = ax_1 + (2 - a)x_2 \\
\text{subject to} \quad & 4x_1 + 3x_2 \le 0 \\
& 2x_1 + 3x_2 \le 7 \\
& x_1 + x_2 \le 1.
\end{aligned}$$

1.3. Find graphically all the values of the parameter a such that the following systems define nonempty feasible sets.

(i)

$$\begin{aligned}
5x_1 + \ x_2 + x_3 + 3x_4 &= a \\
8x_1 + 3x_2 + x_3 + 2x_4 &= 2 - a \\
x_1, x_2, x_3, x_4 &\ge 0.
\end{aligned}$$

(ii)

$$\begin{aligned}
ax_1 + x_2 + 3x_3 - x_4 &= 2 \\
x_1 - x_2 - x_3 - 2x_4 &= 2 \\
x_1, x_2, x_3, x_4 &\ge 0.
\end{aligned}$$

1.4. Suppose that the linear program

$$\begin{aligned}
\text{minimize} \quad & z = c^T x \\
\text{subject to} \quad & Ax = b \\
& x \ge 0
\end{aligned}$$

has an optimal objective z_*. Discuss how the optimal objective would change if (a) a constraint is added to the problem; and (b) a constraint is deleted from the problem.

4.2   Standard Form

There are many different ways to represent a linear program. It is sometimes more convenient to use one instead of another, at times to make a property of the linear program more apparent, at other times to simplify the description of an algorithm. One such representation, called *standard form*, will be used to describe the simplex method.

In matrix-vector notation, a linear program in standard form will be written as

$$\text{minimize} \quad z = c^T x$$
$$\text{subject to} \quad Ax = b$$
$$x \geq 0$$

with $b \geq 0$. Here x and c are vectors of length n, b is a vector of length m, and A is an $m \times n$ matrix called the *constraint matrix*. The important things to notice are (i) it is a minimization problem, (ii) all the variables are constrained to be nonnegative, (iii) all the other constraints are represented as equations, and (iv) the components of the right-hand side vector b are all nonnegative. This will be the form of a linear program used within the simplex method. In other settings, other forms of a linear program may be more convenient.

Example 4.1 (Standard Form). The linear program

$$\text{minimize} \quad z = 4x_1 - 5x_2 + 3x_3$$
$$\text{subject to} \quad 3x_1 - 2x_2 + 7x_3 = 7$$
$$8x_1 + 6x_2 + 6x_3 = 5$$
$$x_1, x_2, x_3 \geq 0$$

is in standard form. In terms of the matrix-vector notation,

$$x = \begin{pmatrix} x_1 \\ x_2 \\ x_3 \end{pmatrix}, \quad c = \begin{pmatrix} 4 \\ -5 \\ 3 \end{pmatrix}, \quad A = \begin{pmatrix} 3 & -2 & 7 \\ 8 & 6 & 6 \end{pmatrix}, \quad b = \begin{pmatrix} 7 \\ 5 \end{pmatrix}.$$

There are $n = 3$ variables and $m = 2$ constraints. ∎

All linear programs can be converted to standard form. The rules for doing this are simple and can be performed automatically by software. Most linear programming software packages allow the user to represent a linear program in any convenient way and then the software performs the conversion internally. We illustrate these techniques via examples. Justification for these rules is left to the Exercises.

If the original problem is a maximization problem:

$$\text{maximize } z = 4x_1 - 3x_2 + 6x_3 = c^T x,$$

then the objective can be multiplied by -1 to obtain

$$\text{minimize } \hat{z} = -4x_1 + 3x_2 - 6x_3 = -c^T x.$$

After the problem has been solved, the optimal objective value must be multiplied by -1, so that $z_* = -\hat{z}_*$. The optimal values of the variables are the same for both objective functions.

If any of the components of b are negative, then those constraints should be multiplied by -1. This will cause a constraint of the "$\leq$" form to be converted to a "$\geq$" constraint and vice versa.

If a variable has a lower bound other than zero, say

$$x_1 \geq 5,$$

then the variable can be replaced in the problem by

$$x_1' = x_1 - 5.$$

The constraint $x_1 \geq 5$ is equivalent to $x_1' \geq 0$. An upper bound on a variable (say, $x_1 \leq 7$) can be treated as a general constraint, that is, as one of the constraints included in the coefficient matrix A. This is inefficient but satisfactory for explaining the simplex method. More efficient techniques for handling upper bounds are described in Section 7.2.

A variable without specified lower or upper bounds, called a *free* or *unrestricted variable*, can be replaced by a pair of nonnegative variables. For example, if x_2 is a free variable, then throughout the problem it will be replaced by

$$x_2 = x_2' - x_2'' \quad \text{with} \quad x_2', x_2'' \geq 0.$$

Intuitively, x_2' will record positive values of x_2, and x_2'' will record negative values. So if $x_2 = 7$, then $x_2' = 7$ and $x_2'' = 0$, and if $x_2 = -4$, then $x_2' = 0$ and $x_2'' = 4$. The properties of the simplex method ensure that at most one of x_2' and x_2'' will be nonzero at a time (see the Exercises in Section 4.3). This is only one way of handling a free variable; an alternative is given in the Exercises; another is given in Section 7.6.6.

The remaining two transformations are used to convert general constraints into equations. A constraint of the form

$$2x_1 + 7x_2 - 3x_3 \leq 10$$

is converted to an equality constraint by including a *slack variable* s_1:

$$2x_1 + 7x_2 - 3x_3 + s_1 = 10$$

together with the constraint $s_1 \geq 0$. The slack variable just represents the difference between the left- and right-hand sides of the original constraint. Similarly a constraint of the form

$$6x_1 - 2x_2 + 4x_3 \geq 15$$

is converted to an equality by including an *excess variable* e_2:

$$6x_1 - 2x_2 + 4x_3 - e_2 = 15$$

together with the constraint $e_2 \geq 0$. (For emphasis, the slack and excess variables are labeled here as s_1 and e_2 to distinguish them from the variables used in the original formulation of the linear program. In other settings it may be more convenient to label them like the other variables, for example as x_4 and x_5. Of course, the choice of variable names does not affect the properties of the linear program.)

Example 4.2 (Transformation to Standard Form). To illustrate these transformation rules, we consider the example

$$
\begin{aligned}
\text{maximize} \quad & z = -5x_1 - 3x_2 + 7x_3 \\
\text{subject to} \quad & 2x_1 + 4x_2 + 6x_3 = 7 \\
& 3x_1 - 5x_2 + 3x_3 \leq 5 \\
& -4x_1 - 9x_2 + 4x_3 \leq -4 \\
& x_1 \geq -2, \ 0 \leq x_2 \leq 4, \ x_3 \text{ free.}
\end{aligned}
$$

To convert to a minimization problem, we multiply the objective by -1:

$$\text{minimize } \hat{z} = 5x_1 + 3x_2 - 7x_3.$$

The third constraint is multiplied by -1 so that all the right-hand sides of the constraints are nonnegative:

$$4x_1 + 9x_2 - 4x_3 \geq 4.$$

The variable x_1 will be transformed to

$$x_1' = x_1 + 2.$$

The upper bound $x_2 \leq 4$ will be treated here as one of the general constraints and the variable x_3 will be transformed to

$$x_3 = x_3' - x_3'',$$

because it is a free variable. When these substitutions have been made we obtain

$$
\begin{aligned}
\text{minimize } \quad & \hat{z} = 5x_1' + 3x_2 - 7x_3' + 7x_3'' - 10 \\
\text{subject to } \quad & 2x_1' + 4x_2 + 6x_3' - 6x_3'' = 11 \\
& 3x_1' - 5x_2 + 3x_3' - 3x_3'' \leq 11 \\
& 4x_1' + 9x_2 - 4x_3' + 4x_3'' \geq 12 \\
& \hspace{4.5cm} x_2 \leq 4 \\
& x_1', x_2, x_3', x_3'' \geq 0.
\end{aligned}
$$

The constant term in the objective, "-10," is usually removed via a transformation of the form $z' = \hat{z} + 10$ so that we obtain the revised objective

$$\text{minimize } z' = 5x_1' + 3x_2 - 7x_3' + 7x_3''.$$

The final step in the conversion is to add slack and excess variables to convert the general constraints to equalities:

$$
\begin{aligned}
\text{minimize } \quad & z' = 5x_1' + 3x_2 - 7x_3' + 7x_3'' \\
\text{subject to } \quad & 2x_1' + 4x_2 + 6x_3' - 6x_3'' = 11 \\
& 3x_1' - 5x_2 + 3x_3' - 3x_3'' + s_2 = 11 \\
& 4x_1' + 9x_2 - 4x_3' + 4x_3'' - e_3 = 12 \\
& \hspace{4cm} x_2 + s_4 = 4 \\
& x_1', x_2, x_3', x_3'', s_2, e_3, s_4 \geq 0.
\end{aligned}
$$

With this the original linear program has been converted to an equivalent one in standard form.

In matrix-vector form it would be represented as

$$
\begin{aligned}
\text{minimize } \quad & z = c^T x \\
\text{subject to } \quad & Ax = b \\
& x \geq 0
\end{aligned}
$$

with $c = (5, 3, -7, 7, 0, 0, 0)^T$, $b = (11, 11, 12, 4)^T$, and

$$A = \begin{pmatrix} 2 & 4 & 6 & -6 & 0 & 0 & 0 \\ 3 & -5 & 3 & -3 & 1 & 0 & 0 \\ 4 & 9 & -4 & 4 & 0 & -1 & 0 \\ 0 & 1 & 0 & 0 & 0 & 0 & 1 \end{pmatrix}.$$

The vector of variables is $x = (x_1', x_2, x_3', x_3'', s_2, e_3, s_4)^T$.

It can be shown that the solution to the problem in standard form is

$$z' = -0.12857, \ x_1' = 0, \ x_2 = 1.65714, \ x_3' = 0.728571,$$
$$x_3'' = 0, \ s_2 = 17.1, \ e_3 = 0, \ s_4 = 2.34286,$$

so that the solution to the original problem is

$$z = 10.12857, \ x_1 = -2, \ x_2 = 1.65714, \ x_3 = 0.728571. \quad \blacksquare$$

One of the reasons that the general constraints in the problem are converted to equalities is that it allows us to use the techniques of elimination to manipulate and simplify the constraints. For example, the system

$$x_1 = 1$$
$$x_1 + x_2 = 2$$

can be reduced to the equivalent system

$$x_1 = 1$$
$$x_2 = 1$$

by subtracting the first constraint from the second. However, if we erroneously apply the same operation to

$$x_1 \geq 1$$
$$x_1 + x_2 \geq 2,$$

then it results in

$$x_1 \geq 1$$
$$x_2 \geq 1,$$

a system of constraints that defines a *different* feasible region. The two regions are illustrated in Figure 4.2. Elimination is not a valid way to manipulate systems of inequalities because it can alter the set of solutions to such systems.

It might seem that the rules for transforming a linear program to standard form could greatly increase the size of a linear program, particularly if a large number of slack and excess variables must be added to obtain a problem in standard form. However, these new variables only appear in the problem in a simple way so that the additional variables do not make the problem significantly harder to solve.

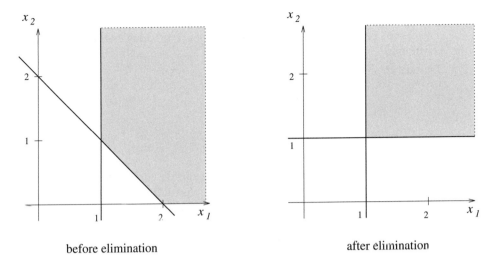

before elimination   after elimination

Figure 4.2. *Elimination and inequalities.*

Exercises

2.1. Convert the following linear program to standard form:

$$\begin{aligned}
\text{maximize} \quad & z = 3x_1 + 5x_2 - 4x_3 \\
\text{subject to} \quad & 7x_1 - 2x_2 - 3x_3 \geq 4 \\
& -2x_1 + 4x_2 + 8x_3 = -3 \\
& 5x_1 - 3x_2 - 2x_3 \leq 9 \\
& x_1 \geq 1, \; x_2 \leq 7, \; x_3 \geq 0.
\end{aligned}$$

2.2. Convert the following linear program to standard form:

$$\begin{aligned}
\text{minimize} \quad & z = x_1 - 5x_2 - 7x_3 \\
\text{subject to} \quad & 5x_1 - 2x_2 + 6x_3 \geq 5 \\
& 3x_1 + 4x_2 - 9x_3 = 3 \\
& 7x_1 + 3x_2 + 5x_3 \leq 9 \\
& x_1 \geq -2, \; x_2, x_3 \text{ free.}
\end{aligned}$$

2.3. Convert the following linear program to standard form:

$$\begin{aligned}
\text{maximize} \quad & z = 6x_1 - 3x_2 \\
\text{subject to} \quad & 2x_1 + 5x_2 \geq 10 \\
& 3x_1 + 2x_2 \leq 40 \\
& x_1, x_2 \leq 15.
\end{aligned}$$

2.4. Consider the linear program in Example 4.2. Convert it to standard form, except do not make the substitution $x_3 = x_3' - x_3''$. Show that the problem can be replaced by an equivalent problem with one less variable and one less constraint by eliminating

x_3 using the equality constraints. (This is a general technique for handling free variables.) Why cannot this technique be used to eliminate variables with nonnegativity constraints?

2.5. Consider the linear program

$$
\begin{aligned}
\text{minimize} \quad & z = c^T x \\
\text{subject to} \quad & Ax \le b \\
& e^T x = 1 \\
& x_1, \dots, x_{n-1} \ge 0, \ x_n \text{ free},
\end{aligned}
$$

where $e = (1, \dots, 1)^T$, b and c are arbitrary vectors of length n, and A is the matrix with entries $a_{i,i} = a_{i,n} = 1$ for $i = 1, \dots, n$ and all other entries zero. Use the constraint $e^T x = 1$ to eliminate the free variable x_n from the linear program (as in the previous problem). Is this a good approach when n is large?

2.6. Prove that each of the transformation rules used to convert a linear program to standard form produces an equivalent linear programming problem. Hint: For each of the rules, prove that a solution to the original problem can be used to obtain a solution to the transformed problem, and vice versa.

2.7. Consider the linear program

$$
\begin{aligned}
\text{minimize} \quad & z = c^T x \\
\text{subject to} \quad & Ax = b \\
& x \ge 0.
\end{aligned}
$$

Transform it into an equivalent standard-form problem for which the right-hand-side vector is zero. Hint: You can achieve this by introducing an additional variable and an additional constraint.

4.3 Basic Solutions and Extreme Points

In this section we examine the relationship between the geometric notion of an extreme point of the feasible region and the algebraic notion of a basic feasible solution. First, it is necessary to give a precise definition of both these terms. To do this, let us consider a linear programming problem in standard form

$$
\begin{aligned}
\text{minimize} \quad & z = c^T x \\
\text{subject to} \quad & Ax = b \\
& x \ge 0.
\end{aligned}
$$

In this problem x is a vector of length n and A is an $m \times n$ matrix with $m \le n$. We will assume that the matrix A has full rank, that is, the rows of A are linearly independent.

The full-rank assumption is not unreasonable. If A is not of full rank, then either the constraints are inconsistent or there are redundant constraints, depending on the right-hand-side vector b. If the constraints are inconsistent, then the problem has no solution and the feasible region is empty, so there are no extreme points. If there are redundant constraints,

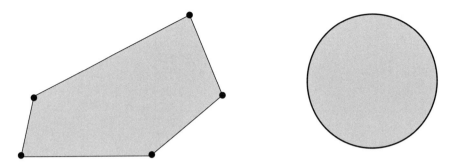

Figure 4.3. *Definition of an extreme point.*

then theoretically they could be removed from the problem without changing either the solution or the feasible region.

If $m = n$, then the constraints $Ax = b$ would completely determine x, and the feasible region would consist of either a single point (if $x \geq 0$) or would be empty (otherwise). If $m > n$, then in most cases the constraints $Ax = b$ would have no solution.

An extreme point is defined geometrically using convexity. A point $x \in S$ is an *extreme point* or *vertex* of a convex set S if it *cannot* be expressed in the form

$$x = \alpha y + (1 - \alpha)z$$

with $y, z \in S$, $0 < \alpha < 1$, and $y, z \neq x$. That is, x cannot be expressed as a convex combination of feasible points y and z different from x. See Figure 4.3. Notice that the values $\alpha = 0$ and $\alpha = 1$ are excluded in this definition. If $\alpha = 0$ then $x = z$, and if $\alpha = 1$ then $x = y$. Since y and z are supposed to be different from x, these two cases are ruled out.

The definition of an extreme point applies to any convex set. In particular, since a system of linear constraints defines a convex set (see Section 2.3), it applies to the feasible region of a linear programming problem.

A *basic solution* is defined algebraically using the standard form of the constraints. A point x is a basic solution if

- x satisfies the equality constraints of the linear program, and

- the columns of the constraint matrix corresponding to the nonzero components of x are linearly independent.

Since the matrix A has full row rank, it is possible to separate the components of x into two subvectors, one consisting of $n - m$ *nonbasic* variables x_N all of which are zero, and the other consisting of m *basic* variables x_B whose constraint coefficients correspond to an invertible $m \times m$ *basis matrix* B. In cases where more than $n - m$ components of x are zero there may be more than one way to choose x_B and x_N. The set of basic variables is called the *basis*.

A point x is a *basic feasible solution* if in addition it satisfies the nonnegativity constraint $x \geq 0$. It is an *optimal basic feasible solution* if it is also optimal for the linear

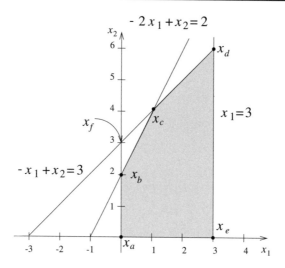

Figure 4.4. *Feasible region.*

program. The word "solution" in these definitions refers only to the equality constraints for the linear program in standard form, and has no connection with the value of the objective function. In Section 4.4 we show that if a linear program has an optimal solution, then it has an optimal basic feasible solution. For this reason it will be sufficient to examine just the basic feasible solutions when solving a linear programming problem.

Example 4.3 (Basic Feasible Solutions). Consider the linear program from Section 4.1:

$$\begin{aligned}
\text{minimize} \quad & z = -x_1 - 2x_2 \\
\text{subject to} \quad & -2x_1 + x_2 \leq 2 \\
& -x_1 + x_2 \leq 3 \\
& x_1 \leq 3 \\
& x_1, x_2 \geq 0.
\end{aligned}$$

The feasible region for this problem is illustrated in Figure 4.4, and the optimal value of this problem is $z_* = -15$ at the point $x_* = (3, 6)^T$. The graph will be used to examine the extreme points.

The boundaries of the feasible region are defined by the lines

$$\begin{aligned}
-2x_1 + x_2 &= 2 \\
-x_1 + x_2 &= 3 \\
x_1 &= 3 \\
x_1 &= 0 \\
x_2 &= 0
\end{aligned}$$

and each corner of the feasible region corresponds to the intersection of two of these lines. There are ten potential intersections of this type, but only five of them (x_a, x_b, x_c, x_d, x_e) are corners of the feasible region. Four others lie outside the feasible region, and one pairing is impossible since the lines $x_1 = 0$ and $x_1 = 3$ do not intersect.

In standard form this linear program is written as

$$
\begin{aligned}
\text{minimize} \quad & z = -x_1 - 2x_2 \\
\text{subject to} \quad & -2x_1 + x_2 + s_1 = 2 \\
& -x_1 + x_2 + s_2 = 3 \\
& x_1 + s_3 = 3 \\
& x_1, x_2, s_1, s_2, s_3 \geq 0.
\end{aligned}
$$

Standard form will be used to describe the basic feasible solutions. In this form the problem has five variables.

In our example, the basis $\{x_2, s_1, s_3\}$ produces the basic solution

$$
(x_1 \quad x_2 \quad s_1 \quad s_2 \quad s_3)^T = (0 \quad 3 \quad -1 \quad 0 \quad 3)^T;
$$

it corresponds to the infeasible *corner* x_f. The basis $\{s_1, s_2, s_3\}$ produces the basic feasible solution

$$
(x_1 \quad x_2 \quad s_1 \quad s_2 \quad s_3)^T = (0 \quad 0 \quad 2 \quad 3 \quad 3)^T;
$$

it corresponds to the corner x_a. If the basis $\{x_1, x_2, s_1\}$ is chosen, we obtain the optimal basic feasible solution

$$
(x_1 \quad x_2 \quad s_1 \quad s_2 \quad s_3)^T = (3 \quad 6 \quad 2 \quad 0 \quad 0)^T;
$$

it corresponds to the corner x_d. We will show how to determine basic feasible and optimal basic feasible solutions when we discuss the simplex method in Chapter 5.

Two different bases can correspond to the same point. To see this, consider the constraints defined by

$$
Ax = \begin{pmatrix} 2 & 1 & 0 & 0 \\ 3 & 0 & 1 & 0 \\ 4 & 0 & 0 & 1 \end{pmatrix} \begin{pmatrix} x_1 \\ x_2 \\ x_3 \\ x_4 \end{pmatrix} = \begin{pmatrix} 6 \\ 13 \\ 12 \end{pmatrix} = b.
$$

If $x = (3, 0, 4, 0)^T \geq 0$, then there is ambiguity about the choice of x_B and x_N. If $x_B = (x_1, x_2, x_3)^T$ and $x_N = (x_4)$, then the coefficient matrix for the nonzero components of x_B

$$
\begin{pmatrix} 2 & 0 \\ 3 & 1 \\ 4 & 0 \end{pmatrix}
$$

has linearly independent columns, so x is a basic feasible solution. In this example the coefficient matrix for x_B

$$
B = \begin{pmatrix} 2 & 1 & 0 \\ 3 & 0 & 1 \\ 4 & 0 & 0 \end{pmatrix}
$$

is invertible. The same basic feasible solution is obtained using

$$
x_B = (x_1 \quad x_3 \quad x_4)^T \quad \text{and} \quad x_N = (x_2)
$$

with invertible basis matrix

$$B = \begin{pmatrix} 2 & 0 & 0 \\ 3 & 1 & 0 \\ 4 & 0 & 1 \end{pmatrix}.$$

Because of this ambiguity, the point $(3, 0, 4, 0)^T$ is called a *degenerate basic feasible solution.* ∎

Let x be any basic feasible solution. Once a set of basic variables has been selected it is possible to reorder the variables so that the basic variables are listed first:

$$x = \begin{pmatrix} x_B \\ x_N \end{pmatrix}.$$

The constraint matrix can then be written as

$$A = (B \quad N),$$

where B is the coefficient matrix for x_B and N is the coefficient matrix for x_N. For a basic solution we have $x_N = 0$, so that the set of constraints $Ax = b$ simplifies to $Bx_B = b$:

$$Ax = (B \quad N) \begin{pmatrix} x_B \\ x_N \end{pmatrix} = Bx_B + Nx_N = Bx_B = b.$$

Thus x_B, and hence x, is determined by B and b.

The number of basic feasible solutions is finite and is bounded by the number of ways that the m variables x_B can be selected from among the n variables x. This number is a binomial coefficient

$$\binom{n}{m} = \frac{n!}{m!(n-m)!},$$

where

$$n! = n(n-1)(n-2)\cdots 3 \cdot 2 \cdot 1.$$

Not all choices of x_B will necessarily correspond to feasible points, so this number can be an overestimate.

The concept of an extreme point is equivalent to the concept of a basic feasible solution, as is proved in the following theorem.

Theorem 4.4. *A point x is an extreme point of the set $\{ x : Ax = b, x \geq 0 \}$ if and only if it is a basic feasible solution.*

Proof. We first show that if x is a basic feasible solution, then it is also an extreme point. If x is a basic feasible solution, then it is a feasible point. For convenience we may assume that the last $n - m$ variables of x are nonbasic so that

$$x = \begin{pmatrix} x_B \\ x_N \end{pmatrix} = \begin{pmatrix} x_B \\ 0 \end{pmatrix}.$$

Let B be the invertible basis matrix corresponding to x_B. The proof will be by contradiction: If x is not an extreme point, then there exist two distinct feasible points y and z satisfying $x = \alpha y + (1 - \alpha)z$ with $0 < \alpha < 1$. We will write y and z in terms of the same basis

$$y = \begin{pmatrix} y_B \\ y_N \end{pmatrix} \quad \text{and} \quad z = \begin{pmatrix} z_B \\ z_N \end{pmatrix}.$$

Both y and z are feasible, so that $y_N \geq 0$ and $z_N \geq 0$. Since $0 = x_N = \alpha y_N + (1 - \alpha)z_N$ and $0 < \alpha < 1$, all the terms on the right-hand side are nonnegative, and we can conclude that $y_N = z_N = 0$. Also, because x, y, and z are feasible they satisfy the equality constraints of the problem, so that

$$B x_B = B y_B = B z_B = b.$$

Since B is invertible, $x_B = y_B = z_B$, contradicting our assumption that y and z were distinct from x. Hence x is an extreme point.

The more difficult part of the proof is to show that if x is an extreme point then it is a basic feasible solution. This will also be proved by contradiction. An extreme point x must be feasible so that $Ax = b$ and $x \geq 0$. By reordering the variables if necessary so that the zero variables are last, x can be written as

$$x = \begin{pmatrix} x_B \\ x_N \end{pmatrix},$$

where $x_N = 0$ and $x_B > 0$. We write $A = (B, N)$ where B and N are the coefficients corresponding to x_B and x_N, respectively. (B may not be a square matrix.) If the columns of B are linearly independent, then x is a basic feasible solution, and nothing needs to be proved. So we will suppose that the columns of B are linearly dependent and construct distinct feasible points y and z that satisfy $x = \frac{1}{2}y + \frac{1}{2}z$, hence showing that x cannot be an extreme point.

Let B_i be the ith column of B. If the columns of B are linearly dependent, then there exist real numbers $p_1, \ldots, p_k$, not all of which are zero, such that

$$B_1 p_1 + B_2 p_2 + \cdots + B_k p_k = 0.$$

If we define $p = (p_1, \ldots, p_k)^T$, then the above equation can be written as $Bp = 0$. Note that

$$B(x_B \pm \alpha p) = B x_B \pm \alpha B p = B x_B \pm 0 = B x_B = b$$

for all values of α. Since $x_B > 0$, for small positive values of ϵ we will have

$$x_B + \epsilon p > 0$$
$$x_B - \epsilon p > 0.$$

Let

$$y = \begin{pmatrix} x_B + \epsilon p \\ x_N \end{pmatrix} \quad \text{and} \quad z = \begin{pmatrix} x_B - \epsilon p \\ x_N \end{pmatrix}.$$

Then y and z are feasible and distinct from x. Since $x = \frac{1}{2}y + \frac{1}{2}z$, this contradicts our assumption that x was an extreme point. This completes the proof. $\quad\square$

It is possible that one or more of the basic variables in a basic feasible solution will be zero. If this occurs, then the point is called a *degenerate vertex*, and the linear program is said to be *degenerate*. At a degenerate vertex several different bases may correspond to the same basic feasible solution. This was illustrated in the latter part of Example 4.3, where the basic feasible solution $(x_1, x_2, x_3, x_4)^T = (3, 0, 4, 0)^T$ could be represented using either $x_B = (x_1, x_2, x_3)^T$ or $x_B = (x_1, x_3, x_4)^T$.

Degeneracy can arise when a linear program contains a redundant constraint. For example, the constraints in Example 4.3 arose when slack variables were added to the constraints

$$
\begin{aligned}
2x_1 &\leq 6 \\
3x_1 &\leq 13 \\
4x_1 &\leq 12.
\end{aligned}
$$

In this form, the first and third constraints are equivalent, and so either of them could be removed from the problem without changing its solution.

There are several more definitions that will be useful when discussing the simplex method. Geometrically, two extreme points are *adjacent* if they are connected by an edge of the feasible region. For example, in Figure 4.4 the extreme points x_a and x_b are adjacent, but x_a and x_c are not. For a linear program in standard form with m equality constraints, two bases will be *adjacent* if they have $m - 1$ variables in common. Adjacent bases define *adjacent basic feasible solutions*. (Note that adjacent bases may not define distinct basic feasible solutions; see Example 4.3.)

One further concept is needed to describe the feasible region geometrically, the concept of a *direction of unboundedness*. (Some authors use the term *direction of a set*.) If S is a convex set, then $d \neq 0$ is a direction of unboundedness if

$$ x + \gamma d \in S \quad \text{for all } x \in S \text{ and } \gamma \geq 0. $$

As we will show in the next section, every feasible point can be represented as a convex combination of extreme points plus, if applicable, a direction of unboundedness.

Example 4.5 (Direction of Unboundedness). We obtain an unbounded feasible region by deleting one constraint from our example:

$$
\begin{aligned}
\text{minimize} \quad & z = -x_1 - 2x_2 \\
\text{subject to} \quad & -2x_1 + x_2 \leq 2 \\
& -x_1 + x_2 \leq 3 \\
& x_1, x_2 \geq 0.
\end{aligned}
$$

The feasible region for this new problem is illustrated in Figure 4.5.

Now there are only three extreme points, $x_a = (0, 0)^T$, $x_b = (0, 2)^T$, and $x_c = (1, 4)^T$. The point $y = (2, 1)^T$ cannot be represented as a convex combination of these extreme points. This follows from the conditions

$$
\begin{aligned}
\alpha_1 x_a + \alpha_2 x_b + \alpha_3 x_c &= y \\
\alpha_1 + \alpha_2 + \alpha_3 &= 1 \\
\alpha_1, \alpha_2, \alpha_3 &\geq 0.
\end{aligned}
$$

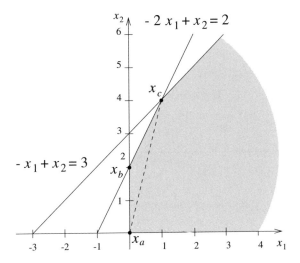

Figure 4.5. *Unbounded feasible region.*

The first condition represents two linear equations, one for each component of y. Combined with the second condition, it gives the linear system

$$\begin{pmatrix} 0 & 0 & 1 \\ 0 & 2 & 4 \\ 1 & 1 & 1 \end{pmatrix} \begin{pmatrix} \alpha_1 \\ \alpha_2 \\ \alpha_3 \end{pmatrix} = \begin{pmatrix} 2 \\ 1 \\ 1 \end{pmatrix}$$

whose unique solution is $\alpha_1 = 5/2, \alpha_2 = -7/2, \alpha_3 = 2$. Since $\alpha_2 < 0$, this is not a convex combination. The triangular area in Figure 4.5 shows which points are convex combinations of extreme points.

For this example, if x is any feasible point and $\gamma \geq 0$, then any point

$$x + \gamma \begin{pmatrix} 1 \\ 0 \end{pmatrix}$$

is also feasible. The direction $(1, 0)^T$ is a direction of unboundedness because it is possible to move arbitrarily far in that direction and remain feasible. In this example it is possible to select two linearly independent directions of unboundedness, such as $d_1 = (1, 0)^T$ and $d_2 = (1, 1)^T$. It is not difficult to show that any feasible point can be written as a convex combination of the extreme points x_a, x_b, and x_c, plus some multiple of either of these directions of unboundedness. ∎

Let x be a feasible point for the linear program in standard form ($Ax = b, x \geq 0$) and let d be a direction of unboundedness. Then both x and $x + \gamma d$ must be feasible for all $\gamma \geq 0$, so that

$$Ax = b, \quad x \geq 0,$$
$$A(x + \gamma d) = b, \quad x + \gamma d \geq 0.$$

Together these conditions show that a direction of unboundedness must satisfy

$$Ad = 0$$
$$d \geq 0.$$

In addition, any nonzero vector d satisfying these two conditions will be a direction of unboundedness; see the Exercises.

Exercises

3.1. Consider the system of linear constraints

$$
\begin{aligned}
2x_1 + x_2 &\leq 100 \\
x_1 + x_2 &\leq 80 \\
x_1 &\leq 40 \\
x_1, x_2 &\geq 0.
\end{aligned}
$$

(i) Write this system of constraints in standard form, and determine all the basic solutions (feasible and infeasible).

(ii) Determine the extreme points of the feasible region (corresponding to both the standard form of the constraints, as well as the original version).

3.2. Consider the following system of inequalities:

$$
\begin{aligned}
x_1 + x_2 &\leq 5 \\
x_1 + 2x_2 &\leq 6 \\
x_1, x_2 &\geq 0.
\end{aligned}
$$

(i) Find the extreme points of the region defined by these inequalities.

(ii) Does this set have any directions of unboundedness? Either prove that none exist, or give an example of a direction of unboundedness.

3.3. Consider the feasible region in Figure 4.5.

(i) Show that $d_1 = (1, 0)^T$ and $d_2 = (1, 1)^T$ are directions of unboundedness. Determine the corresponding directions of unboundedness for the problem written in standard form, and verify that the conditions $Ad = 0$ and $d \geq 0$ are satisfied for both directions.

(ii) Prove that d is a direction of unboundedness if and only if d is a nonnegative combination of d_1 and d_2.

3.4. Consider the linear program

$$
\begin{aligned}
\text{minimize} \quad & z = -5x_1 - 7x_2 \\
\text{subject to} \quad & -3x_1 + 2x_2 \leq 30 \\
& -2x_1 + x_2 \leq 12 \\
& x_1, x_2 \geq 0.
\end{aligned}
$$

     (i) Draw a graph of the feasible region.

    (ii) Determine the extreme points of the feasible region.

   (iii) Determine two linearly independent directions of unboundedness.

   (iv) Convert the linear program to standard form and determine the basic feasible solutions and two linearly independent directions of unboundedness for this version of the problem. Verify that the directions of unboundedness satisfy $Ad = 0$ and $d \geq 0$.

3.5. Consider a linear program with the constraints in standard form

$$Ax = b \quad \text{and} \quad x \geq 0.$$

Prove that if $d \neq 0$ satisfies

$$Ad = 0 \quad \text{and} \quad d \geq 0,$$

then d is a direction of unboundedness.

3.6. Consider the system of constraints

$$2x_1 + x_2 \leq 3$$
$$3x_1 + x_2 \leq 4$$
$$4x_1 + x_2 \leq 5$$
$$5x_1 + x_2 \leq 6$$
$$x_1, x_2 \geq 0.$$

     (i) Determine the extreme points for the feasible region.

    (ii) Convert the problem to standard form, and determine the basic feasible solutions.

   (iii) Which basic feasible solution corresponds to the extreme point $(1, 1)^T$? How many different bases can be used to generate this basic feasible solution? Which of these bases are adjacent?

3.7. Find all the vertices of the region defined by the following system:

$$3x_1 + x_2 + x_3 + x_4 = 1$$
$$x_1 + 6x_2 - 2x_3 + x_4 = 1$$
$$x_1, x_2, x_3, x_4 \geq 0.$$

Does the system have degenerate vertices?

3.8. Find all the values of the parameter a such that the regions defined by the following systems have degenerate vertices.

     (i)

$$x_1 + x_2 \leq 8$$
$$6x_1 + x_2 \leq 12$$
$$2x_1 + x_2 \leq a$$
$$x_1, x_2 \geq 0.$$

(ii)
$$ax_1 + x_2 \geq 1$$
$$2x_1 + x_2 \leq 6$$
$$-x_1 + x_2 \leq 6$$
$$x_1 + 2x_2 \geq 6$$
$$x_1, x_2 \geq 0.$$

3.9. Consider a linear program with the following constraints:

$$4x_1 + 7x_2 + 2x_3 - 3x_4 + x_5 + 4x_6 = \quad 4$$
$$-x_1 - 2x_2 + x_3 + x_4 - x_6 = -1$$
$$x_2 - 3x_3 - x_4 - x_5 + 2x_6 = \quad 0$$
$$x_i \geq 0, \quad i = 1, \ldots, 6.$$

Determine every basis that corresponds to the basic feasible solution $(0, 1, 0, 1, 0, 0)^T$.

3.10. Consider the feasible region in Figure 4.4. Determine formulas for the points on the edges of the feasible region. What are the corresponding formulas for the problem in standard form? The formulas you determine should be of the form

$$\text{(extreme point)} + \alpha(\text{direction}) \quad \text{for } 0 \leq \alpha \leq \alpha_{max}.$$

3.11. Repeat the previous problem for the feasible region in Figure 4.5. Note that in some cases there will be no upper bound on α.

3.12. Consider the system of constraints $Ax = b$, $x \geq 0$ with

$$A = \begin{pmatrix} 1 & 4 & 7 & 1 & 0 & 0 \\ 2 & 5 & 8 & 0 & 1 & 0 \\ 3 & 6 & 9 & 0 & 0 & 1 \end{pmatrix} \quad \text{and} \quad b = \begin{pmatrix} 12 \\ 15 \\ 18 \end{pmatrix}.$$

Is $x = (1, 1, 1, 0, 0, 0)^T$ a basic feasible solution? Explain your answer.

3.13. Suppose that a linear program includes a free variable x_i. In converting this problem to standard form, x_i is replaced by a pair of nonnegative variables:

$$x_i = x_i' - x_i'', \quad x_i', x_i'' \geq 0.$$

Prove that no basic feasible solution can include both x_i' and x_i'' as basic variables.

3.14. Let the $m \times n$ matrix A be the coefficient matrix for a linear program in standard form. The upper bound

$$\binom{n}{m} = \frac{n!}{m!(n-m)!}$$

on the number of basic feasible solutions can sometimes be precise, but it can also be a considerable overestimate.

(i) Construct an example with $n = 4$ and $m = 2$ where the number of basic feasible solutions is equal to $\binom{n}{m}$.

(ii) Construct examples of arbitrary size where the number of basic feasible solutions is equal to zero.

3.15. Prove that the set $S = \{ x : Ax < b \}$ does not contain any extreme points.

3.16. Let $S = \{ x : x^T x \le 1 \}$. Prove that the extreme points of S are the points on its boundary.

3.17. Consider the set $S = \{ x : x_1 \ge x_2 \ge \cdots \ge x_n \ge 0 \}$.

    (i) Prove that if $x \in S$ then so is $\alpha x \in S$ for all $\alpha \ge 0$. A set with this property is called a *cone*.

    (ii) Prove that the origin is the only extreme point of S.

    (iii) Find n linearly independent directions of unboundedness for this set.

3.18. Give an example of a degenerate linear program that does not contain a redundant constraint.

3.19. Give an example of a linear program where a degenerate basic feasible solution only corresponds to a single basis.

4.4   Representation of Solutions; Optimality

The first goal of this section is to prove that any feasible point can be represented as a convex combination of extreme points plus, possibly, a direction of unboundedness. Then this result will be used to prove that any linear program with a finite optimal solution has an optimal basic feasible solution.

    The idea behind the representation theorem is straightforward and will first be illustrated using two examples of feasible sets, one bounded and one unbounded. The examples will be in two dimensions so they can be graphed, but the techniques used in the examples are the same as those used in the proof.

    We will use the examples from Section 4.3. First we consider a bounded problem with the constraints

$$\begin{aligned} -2x_1 + x_2 &\le 2 \\ -x_1 + x_2 &\le 3 \\ x_1 &\le 3 \\ x_1, x_2 &\ge 0. \end{aligned}$$

We would like to show that if x is any feasible point, then it can be expressed as a convex combination of extreme points of the feasible region. Our discussion will be based on Figure 4.6.

    Let us choose the feasible point $x = (2, 1)^T$. We would like to express x as a convex combination of the extreme points $x_a, \dots, x_e$. Consider the direction $p = (1, 1)^T$. Since x is in the interior of the feasible region, $x + \gamma p$ will be feasible for small values of γ. (By "small" we mean small in absolute value.) However, since the region is bounded, as we move along p or $-p$ eventually we will hit the boundary of the region. In this example this occurs at the points $y_1 = x + p = (3, 2)^T$ and $y_2 = x - p = (1, 0)^T$, that is, for $\gamma = 1$ and $\gamma = -1$. Notice that $x = \frac{1}{2} y_1 + \frac{1}{2} y_2$ so that x is a convex combination of y_1 and y_2.

    Neither y_1 nor y_2 is an extreme point; both are along an edge of the feasible region. For small values of γ the points $y_1 + \gamma p_1$ and $y_2 + \gamma p_2$ will be feasible, where $p_1 = (0, 1)^T$ and $p_2 = (1, 0)^T$. However, as γ is increased in magnitude, we will eventually hit

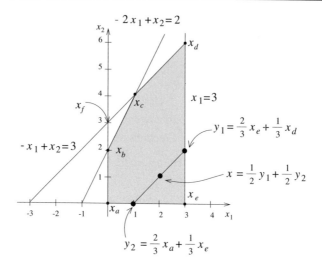

Figure 4.6. *Representation via extreme points: Bounded case.*

another boundary of the region. For y_1 this occurs at $y_{11} = y_1 + 4p_1 = (3, 6)^T = x_d$ and $y_{12} = y_1 - 2p_1 = (3, 0)^T = x_e$, and for y_2 this occurs at $y_{21} = y_2 + 2p_2 = (3, 0)^T$ and $y_{22} = y_2 - 1p_2 = (0, 0)^T$. The points y_1 and y_2 can be written as

$$y_1 = \tfrac{2}{3}y_{12} + \tfrac{1}{3}y_{11}$$
$$y_2 = \tfrac{2}{3}y_{22} + \tfrac{1}{3}y_{21}.$$

The points on the right-hand side are extreme points.

Since $x = \tfrac{1}{2}y_2 + \tfrac{1}{2}y_1$ we can combine these results to obtain

$$x = \tfrac{1}{3}y_{22} + \tfrac{1}{6}y_{21} + \tfrac{1}{3}y_{12} + \tfrac{1}{6}y_{11}$$
$$= \tfrac{1}{3}x_a + \tfrac{1}{6}x_e + \tfrac{1}{3}x_e + \tfrac{1}{6}x_d$$
$$= \tfrac{1}{3}x_a + \tfrac{1}{2}x_e + \tfrac{1}{6}x_d.$$

Thus we have expressed x as a convex combination of extreme points.

Now we will consider the unbounded region obtained by deleting one of the constraints:

$$-2x_1 + x_2 \le 2$$
$$-x_1 + x_2 \le 3$$
$$x_1, x_2 \ge 0.$$

We would like to show that if x is any feasible point, then it can be expressed as a convex combination of extreme points plus, if required, a direction of unboundedness. Our discussion will be based on Figure 4.7.

Let us again choose the feasible point $x = (2, 1)^T$ and the direction $p = (1, 1)^T$. As before, $x + \gamma p$ will be feasible for small values of γ. For $\gamma < 0$, the boundary is

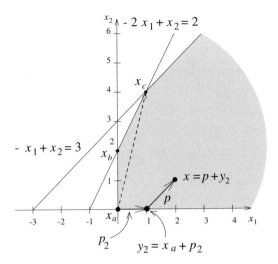

Figure 4.7. *Representation via extreme points: Unbounded case.*

encountered at the point $y_2 = x - p = (1, 0)^T$. However, the direction p is a direction of unboundedness so that $x + \gamma p$ is feasible for all positive values of γ. In this case we will represent x as the sum of a direction of unboundedness and a point on the boundary, that is, $x = p + y_2$.

The point y_2 is not an extreme point so we will represent it in terms of other points along the same edge. For $p_2 = (1, 0)^T$ we examine points of the form $y_2 + \gamma p_2$. Another boundary is encountered at the point $y_{22} = y_2 - p_2 = (0, 0)^T$. In the direction p_2 the region is unbounded and $y_2 = p_2 + y_{22}$, that is, y_2 is the sum of a direction of unboundedness and an extreme point.

Combining these two results we obtain

$$x = p + y_2 = p + (p_2 + y_{22})$$
$$= (p + p_2) + y_{22} = \hat{p} + x_a,$$

where $\hat{p} = p + p_2 = (2, 1)^T$, another direction of unboundedness. In this way we have expressed x as the sum of a direction of unboundedness and a (trivial) convex combination of extreme points.

The representation theorem is given below. For the examples above, the constraints were not in standard form; this was so the examples could be graphed easily. The theorem works with a problem expressed in standard form. This is not an essential detail—it merely eliminates ambiguity about how the constraints are represented. The argument is the same.

To point out the connection between the two approaches, we write the constraints for the unbounded example in standard form, that is, $S = \{ x : Ax = b, x \geq 0 \}$ with

$$A = \begin{pmatrix} -2 & 1 & 1 & 0 \\ -1 & 1 & 0 & 1 \end{pmatrix} \quad \text{and} \quad b = \begin{pmatrix} 2 \\ 3 \end{pmatrix}.$$

The point $x = (2, 1)^T$ is transformed into $\bar{x} = (2, 1, 5, 4)^T$, where $\bar{x}_3 = 5$ and $\bar{x}_4 = 4$ are the slack variables for the two constraints. The direction $p = (1, 1)^T$ is transformed into the direction $\bar{p} = (1, 1, 1, 0)^T$, with the last two components chosen so that $A(\bar{x} + \bar{p}) = b$ or, equivalently, $A\bar{p} = 0$.

Theorem 4.6 (Representation Theorem). *Consider the set*

$$S = \{x : Ax = b, x \ge 0\},$$

representing the feasible region for a linear program in standard form. Let

$$V = \{v_1, v_2, \ldots, v_k\}$$

be the set of extreme points (vertices) of S. If S is nonempty, then V is nonempty, and every feasible point $x \in S$ can be written in the form

$$x = d + \sum_{i=1}^{k} \alpha_i v_i,$$

where

$$\sum_{i=1}^{k} \alpha_i = 1 \quad and \quad \alpha_i \ge 0, \quad i = 1, \ldots, k,$$

and d satisfies $Ad = 0$ and $d \ge 0$, i.e., either $d = 0$ or d is a direction of unboundedness of S.

Proof. The proof will make repeated use of the equivalence between extreme points and basic feasible solutions. We will assume that A is of full row rank, since if A is not of full row rank it can be replaced by a smaller full-rank matrix.

We will first consider the case where the set S is bounded, so that there are no directions of unboundedness and $d = 0$. Let $x \in S$ be any feasible point. If x is an extreme point, then $x = v_i$ for some i and the theorem is true with $\alpha_i = 1$ and $\alpha_j = 0$ for $j \ne i$.

If x is not an extreme point, then, by the results in the last section, x is not a basic feasible solution. Hence the columns of A corresponding to the nonzero entries are linearly dependent and we can find a feasible direction p, that is, a vector $p \ne 0$ satisfying

$$\begin{aligned} Ap &= 0 \\ p_i &= 0 \quad \text{if } x_i = 0. \end{aligned}$$

If ϵ is small in magnitude,

$$\begin{aligned} A(x + \epsilon p) &= b \\ x + \epsilon p &\ge 0 \\ (x + \epsilon p)_i &= 0 \quad \text{if } x_i = 0. \end{aligned}$$

Hence $x + \epsilon p \in S$. Since S is bounded, as ϵ increases in magnitude (either positive or negative) eventually points are encountered where some additional component of $x + \epsilon p$ becomes zero. Let y_1 be the point obtained with $\epsilon > 0$ and y_2 be the point obtained with $\epsilon < 0$. Then x is a convex combination of y_1 and y_2 and both y_1 and y_2 have at least one more zero component than x does; see the Exercises.

The argument is now completed by induction. If y_1 and y_2 are both extreme points, then we are finished. Otherwise, the same reasoning is applied as necessary to one or both of y_1 and y_2 to express them as convex combinations of points with one more zero component. This is repeated until eventually a representation is obtained in terms of extreme points. (There is one detail that must be checked: it must be shown that if y_1 and y_2 are convex combinations of extreme points, then so is x; see the Exercises.) This argument also shows that the set of extreme points is nonempty. Because the number of nonzero components is decreasing by one at each step, and is bounded below by 0, eventually the points generated by this scheme must be basic feasible solutions, that is, extreme points.

The unbounded case is proved similarly. Choose $x \in S$. If x is not an extreme point we can form $x + \epsilon p$ for a vector p chosen as before. However, it is possible that either p or $-p$ is a direction of unboundedness if either $p \geq 0$ or $p \leq 0$, respectively. (They cannot both be directions of unboundedness because of the nonnegativity constraints $x \geq 0$; see the Exercises.) Suppose that p is a direction of unboundedness, so a move in the direction $-p$ will hit the boundary at some point y_2, that is, $x - \gamma p = y_2$ with $\gamma > 0$. (Analogous remarks apply if $-p$ is a direction of unboundedness.) Then

$$x = d + 1 \cdot y_2,$$

where $d = \gamma p$ for some γ, so that x is the sum of a direction of unboundedness and a (trivial) convex combination of y_2. As before, y_2 has at least one more zero entry than x does.

Now the same argument can be applied inductively to y_2 to show that it can be expressed as a convex combination of extreme points plus a nonnegative linear combination of directions of unboundedness with nonnegative coefficients. Since such a combination of directions of unboundedness is again a direction of unboundedness (see the Exercises), this completes the proof. $\square$

So far our main concern has been the constraints in a linear program. We now examine the objective function and show that a solution to a linear program, if one exists, can always be chosen from among the extreme points of the feasible region.

Theorem 4.7. *If a linear program in standard form has a finite optimal solution, then it has an optimal basic feasible solution.*

Proof. Let x be a finite optimal solution for the linear program represented in standard form. Using the representation theorem we can write x as

$$x = d + \sum_{i=1}^{k} \alpha_i v_i,$$

where

$$\sum_{i=1}^{k} \alpha_i = 1 \quad \text{and} \quad \alpha_i \geq 0, \quad i = 1, \ldots, k.$$

As before $\{v_i\}$ is the set of extreme points of the feasible region, and d is either zero or a direction of unboundedness. The objective function has the value

$$c^T x = c^T d + \sum_{i=1}^{k} \alpha_i c^T v_i.$$

We first show that $c^T d = 0$. If $c^T d < 0$, then the objective function is unbounded below, since it is straightforward to verify that

$$x_\gamma = \gamma d + \sum_{i=1}^{k} \alpha_i v_i$$

will be feasible for any $\gamma > 0$ and $c^T(\gamma d) = \gamma c^T d$ will be unbounded below as γ increases. This in turn implies that $c^T x_\gamma$ is unbounded below. Since x was assumed to be a finite optimal solution, this is a contradiction and so $c^T d \geq 0$. Now if $c^T d > 0$, then $c^T x > c^T y$ where

$$y = \sum_{i=1}^{k} \alpha_i v_i$$

is a feasible point. This shows that x would not be optimal in this case. Hence $c^T d = 0$ and $c^T x = c^T y$, showing that y is also an optimal solution.

Now pick an index j for which $c^T v_j = \min_i \{ c^T v_i \}$. Then for any convex combination of the v_i's,

$$c^T y = \sum_{i=1}^{k} \alpha_i c^T v_i \geq \sum_{i=1}^{k} \alpha_i c^T v_j$$
$$= c^T v_j \sum_{i=1}^{k} \alpha_i = c^T v_j.$$

Since y is optimal it must be true that $c^T y = c^T v_j$, showing that there is an optimal extreme point, namely v_j, or equivalently an optimal basic feasible solution.   □

One of the conditions for an optimal solution is that the objective function cannot decrease if we move in any feasible direction. Consider the linear program from Section 4.1:

$$
\begin{array}{ll}
\text{minimize} & z = -x_1 - 2x_2 \\
\text{subject to} & -2x_1 + x_2 \leq 2 \\
& -x_1 + x_2 \leq 3 \\
& x_1 \leq 3 \\
& x_1, x_2 \geq 0.
\end{array}
$$

If we select the optimal point $x = (3, 6)^T$ and move some small distance $\epsilon > 0$ in the feasible direction $d = (-2, -3)^T$, then

$$c^T(x + \epsilon d) = c^T x + \epsilon c^T d$$
$$= -15 + 8\epsilon > -15,$$

so that the objective function increases in this direction.

We can also represent this idea algebraically. Suppose that we have a linear program in standard form

$$\text{minimize} \quad z = c^T x$$
$$\text{subject to} \quad Ax = b$$
$$x \geq 0,$$

and that x is an optimal basic feasible solution. If p is a feasible direction, then $x + \epsilon p$ must be feasible for small $\epsilon > 0$. In addition, because x is optimal, $c^T(x + \epsilon p) \geq c^T x$. Hence p must satisfy

$$c^T p \geq 0$$
$$Ap = 0$$
$$p_i \geq 0 \quad \text{if } x_i = 0.$$

We will use these conditions when deriving the simplex method in the next chapter.

Exercises

4.1. Let x be a feasible point for the constraints

$$Ax = b, \quad x \geq 0$$

that is not an extreme point. Prove that there exists a vector $p \neq 0$ satisfying

$$Ap = 0$$
$$p_i = 0 \quad \text{if } x_i = 0.$$

4.2. Let x be an element of a convex set S. Assume that $x_1 = x + \epsilon_1 p \in S$ and $x_2 = x - \epsilon_2 p \in S$, where $p \neq 0$ and $\epsilon_1, \epsilon_2 > 0$. Prove that x is a convex combination of x_1 and x_2. That is, prove that

$$x = \alpha x_1 + (1 - \alpha) x_2,$$

where $0 < \alpha < 1$, and determine the value of α.

4.3. Let x be a convex combination of $\{ y_1, \ldots, y_k \}$. Assume in turn that each y_i is a convex combination of $\{ y_{i,1}, \ldots, y_{i,k_i} \}$. Prove that x is a convex combination of the vectors $\{ y_{i,j} \}$.

4.4. Let p be a direction of unboundedness for the constraints

$$Ax = b, \quad x \geq 0.$$

Prove that $-p$ cannot be a direction of unboundedness for these constraints.

4.5. Let $\{ d_1, \ldots, d_k \}$ be directions of unboundedness for the constraints

$$Ax = b, \quad x \geq 0.$$

Prove that

$$d = \sum_{i=1}^{k} \alpha_i d_i \quad \text{with } \alpha_i \geq 0$$

is also a direction of unboundedness for these constraints.

4.6. Consider the linear program

$$
\begin{aligned}
\text{minimize} \quad & z = 2x_1 - 3x_2 \\
\text{subject to} \quad & 4x_1 + 3x_2 \le 12 \\
& x_1 - 2x_2 \le 2 \\
& x_1, x_2 \ge 0.
\end{aligned}
$$

Represent the point $x = (1, 1)^T$ as a convex combination of extreme points plus, if applicable, a direction of unboundedness. Find three different representations.

4.7. Consider the linear program

$$
\begin{aligned}
\text{minimize} \quad & z = 3x_1 + x_2 \\
\text{subject to} \quad & x_1 - x_2 \ge 2 \\
& -2x_1 + x_2 \le 4 \\
& x_1, x_2 \ge 0.
\end{aligned}
$$

Represent the point $x = (5, 2)^T$ as a convex combination of extreme points plus, if applicable, a direction of unboundedness. Find three different representations.

4.8. Suppose that a linear program with bounded feasible region has ℓ optimal extreme points $v_1, \ldots, v_\ell$. Prove that a point is optimal for the linear program if and only if it can be expressed as a convex combination of $\{ v_i \}$.

4.9. Complete the proof of Theorem 4.6 in the case where S is bounded by showing that x is a convex combination of y_1 and y_2.

4.5   Notes

The material in this chapter is well known and is discussed in a number of books on linear programming such as the books of Dantzig (1963, reprinted 1998), Chvátal (1983), Murty (1983), and Schrijver (1986, reprinted 1998).

Chapter 5

The Simplex Method

5.1 Introduction

The simplex method is the most widely used method for linear programming and one of the most widely used of all numerical algorithms. It was developed in the 1940's at the same time as linear programming models came to be used for economic and military planning. It had competitors at that time, but these competitors could not match the efficiency of the simplex method and they were soon discarded. Even as problems have become larger and computers more powerful, the simplex method has been able to adapt and remain the method of choice for many people. It is only in recent years with the development of interior-point methods (see Chapter 10) that the simplex method has had a serious challenge for primacy in the realm of linear programming.

Even though the simplex method only solves linear programming problems, its techniques are of more general interest. The same techniques can be used to handle linear constraints in nonlinear optimization problems and can be generalized to handle nonlinear constraints. This is discussed in Chapter 15. The ways that constraints are represented are used in other settings, as are the methods for computing Lagrange multipliers (dual variables; see Chapter 6). Our study of the simplex method will also provide a good setting for discussing degeneracy and a number of other topics.

The simplex method has important historic ties to economics, and this has influenced the terminology associated with the method. For example, it is common to speak of reduced "costs" and shadow "prices." For many applications these terms are useful and suggestive of the interpretations that will be given to the linear programming model.

In this chapter we describe the basic form of the simplex method. We apply the method to a linear program in standard form, show how to find an initial feasible point, and adapt the simplex method to solve degenerate problems. Our emphasis will be on the general properties of the method. The details that make up a modern implementation of the method are delayed until Chapter 7.

The results of Chapter 4 provide the major motivation for the simplex method. We proved that if a linear program has a finite optimal solution, then it has an optimal basic feasible solution. This implies that we need only examine basic feasible solutions to solve a linear program. The simplex method is a systematic and effective way to do just this.

5.2 The Simplex Method

The simplex method is an iterative method for solving a linear programming problem written in standard form. When applied to nondegenerate problems, it moves from one basic feasible solution (extreme point) to another. The simplex method is an example of a feasible-point method (see Section 3.1). What distinguishes the simplex method from a general feasible-point method is that every estimate of the solution is a basic feasible solution. At each iteration the method tests to see if the current basis is optimal. If it is not, the method selects a feasible direction along which the objective function improves and moves to an adjacent basic feasible solution along that direction. Then everything repeats.

Here we present the simplex method using explicit matrix inverses. Modern computer implementations of the simplex method do not do this, but rather use matrix factorizations and related techniques (see Section 7.5). The main reason is that explicit matrix inverses are not suitable for sparse problems. However, many important ideas about linear programming and about the simplex method can be explained without reference to the specific representation of the inverse matrix.

The simplex method will be illustrated using the linear program

$$
\begin{aligned}
\text{minimize} \quad & z = -x_1 - 2x_2 \\
\text{subject to} \quad & -2x_1 + x_2 \le 2 \\
& -x_1 + 2x_2 \le 7 \\
& x_1 \le 3 \\
& x_1, x_2 \ge 0.
\end{aligned}
$$

Slack variables are added to put it in standard form:

$$
\begin{aligned}
\text{minimize} \quad & z = -x_1 - 2x_2 \\
\text{subject to} \quad & -2x_1 + x_2 + x_3 = 2 \\
& -x_1 + 2x_2 + x_4 = 7 \\
& x_1 + x_5 = 3 \\
& x_1, x_2, x_3, x_4, x_5 \ge 0.
\end{aligned}
$$

As usual, we denote the objective function by $z = c^T x$ and the constraints by $Ax = b$, with $x \ge 0$. The feasible region for the original form of the problem is illustrated in Figure 5.1.

In this problem each of the constraints has a slack variable. This makes it easy to find a basic feasible solution, that is, $x_B = (x_3, x_4, x_5)^T$ and $x_N = (x_1, x_2)^T$. The coefficient matrix associated with a complete set of slack variables will always be the identity matrix I, whose columns are linearly independent. Since the nonbasic variables will be zero, the basic variables will satisfy

$$
I x_B = x_B = b.
$$

In standard form the right-hand side b will be nonnegative, so $x \ge 0$ and is feasible. For this example, the initial basic feasible solution is

$$
(x_1 \quad x_2 \quad x_3 \quad x_4 \quad x_5)^T = (0 \quad 0 \quad 2 \quad 7 \quad 3)^T.
$$

This corresponds to the extreme point x_a in Figure 5.1.

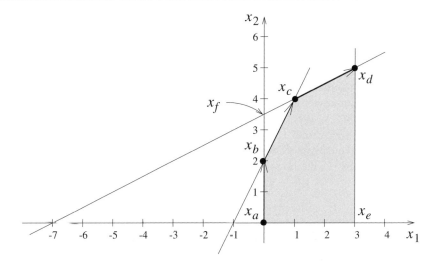

Figure 5.1. *The simplex method.*

We now test if this point is optimal. To do this, we determine if there exist any feasible descent directions. The constraints can be written with the basic variables expressed in terms of the nonbasic variables:

$$x_3 = 2 + 2x_1 - x_2$$
$$x_4 = 7 + x_1 - 2x_2$$
$$x_5 = 3 - x_1.$$

All other feasible points can be found by varying the values of the nonbasic variables x_1 and x_2 and using the constraints to determine the values of the basic variables x_3, x_4, and x_5. Because the nonbasic variables are currently zero and all variables must be nonnegative, it is only valid to *increase* a nonbasic variable so that it becomes positive.

Our goal is to minimize the objective function

$$z = -x_1 - 2x_2,$$

and its current value is $z = 0$. If either x_1 or x_2 is increased from zero, then z will decrease. This shows that nearby feasible points obtained by increasing either x_1 or x_2 give lower values of the objective function, so that the current basis is not optimal.

The simplex method moves from one basis to an adjacent basis, deleting and adding just one variable from the basis. This corresponds to moving between adjacent basic feasible solutions. In geometric terms, the simplex method moves along edges of the feasible region. It is not difficult to calculate how the objective function changes along an edge of the feasible region, and this contributes to the simplicity of the method.

For this example, moving to an adjacent basic feasible solution corresponds to increasing either x_1 or x_2, but not both. The coefficient of x_2 is greater in absolute value than the coefficient of x_1, so z decreases more rapidly when x_2 is increased. In the hope of making more rapid progress toward the solution, we choose to increase x_2 rather than x_1.

Every unit increase of x_2 decreases the objective value by two, so the more x_2 is increased, the better the value of the objective. However, the value of x_2 cannot be increased indefinitely because the region is bounded. The constraint equations show that as x_2 increases and x_1 is kept fixed at zero, $x_3 = 2 - x_2$ and $x_4 = 7 - 2x_2$ decrease but $x_5 = 3$ is unaffected. To maintain nonnegativity of the variables, x_2 can be increased only until one of x_3 or x_4 becomes zero. The first constraint shows that $x_3 = 0$ when $x_2 = 2$ (and $x_4 = 3 > 0$); this corresponds to the point x_b in Figure 5.1. The second constraint shows that $x_4 = 0$ when $x_2 = \frac{7}{2}$ (and $x_3 = -\frac{3}{2} < 0$); this corresponds to the infeasible point x_f in the figure. Consequently x_2 can only be increased to the value $x_2 = 2$. At this point x_3 becomes zero and leaves the basis, and x_2 has entered the basis. The new basic feasible solution is x_b:

$$(x_1 \quad x_2 \quad x_3 \quad x_4 \quad x_5)^T = (0 \quad 2 \quad 0 \quad 3 \quad 3)^T,$$

where $x_B = (x_2, x_4, x_5)^T$ and $x_N = (x_1, x_3)^T$.

The final step in the iteration is to make the transition to the new basic feasible solution. One way to do this is to rewrite the problem so that the new basic variables are expressed in terms of the new nonbasic variables. We want only the nonbasic variables to appear in the objective, and we want the coefficient matrix for the basic variables in the equality constraints to be the identity matrix. Writing the constraints in this way will allow us to repeat the same analysis at the new basic feasible solution. It will make it easy to determine if the current basis is optimal, and if not, how the basis can be changed to improve the value of the objective.

Since x_2 is replacing x_3 in the basis, we use the first constraint (the one that defines x_3 in terms of the other variables) to express x_2 in terms of the nonbasic variables x_1 and x_3:

$$x_2 = 2 + 2x_1 - x_3.$$

We then use this equation to make substitutions for x_2 in the remaining constraints and the objective function. After simplification the linear program has the form

$$\text{minimize } z = -4 - 5x_1 + 2x_3$$

subject to the constraints

$$x_2 = 2 + 2x_1 - x_3$$
$$x_4 = 3 - 3x_1 + 2x_3$$
$$x_5 = 3 - x_1$$

and with all variables nonnegative. Since $x_1 = x_3 = 0$ the current objective value is $z = -4$ and the basic variables have the values $x_2 = 2$, $x_4 = 3$, and $x_5 = 3$. This completes one iteration of the simplex method.

We can now begin again by testing for optimality, examining how the objective changes when we increase the nonbasic variables from zero. This basic feasible solution is not optimal and we can improve the objective by increasing x_1. And so forth. At each iteration, we identify a nonbasic variable that can improve the objective (if one exists). This variable is increased until some basic variable decreases to zero. This gives a new basic feasible solution, and the process repeats.

For this example, the simplex method moves from x_a to x_b to x_c to x_d, the optimal point. We will go through the remaining iterations in Example 5.2.

5.2.1 General Formulas

Let us now consider a general linear program and derive general formulas for the steps in the simplex method. Assume that the problem has n variables and m linearly independent equality constraints.

We derive the formulas in matrix-vector form for the linear program

$$\text{minimize} \quad z = c^T x$$
$$\text{subject to} \quad Ax = b$$
$$x \geq 0.$$

Let x be a basic feasible solution with the variables ordered so that

$$x = \begin{pmatrix} x_B \\ x_N \end{pmatrix},$$

where x_B is the vector of basic variables and x_N is the (currently zero) vector of nonbasic variables. The objective function can be written as

$$z = c_B^T x_B + c_N^T x_N,$$

where the coefficients for the basic variables are in c_B and the coefficients for the nonbasic variables are in c_N. Similarly, we write the constraints as

$$B x_B + N x_N = b.$$

The constraints can be rewritten as

$$x_B = B^{-1} b - B^{-1} N x_N.$$

By varying the values of the nonbasic variables we can obtain all possible solutions to $Ax = b$.

If this formula is substituted into the formula for z, we obtain

$$z = c_B^T B^{-1} b + (c_N^T - c_B^T B^{-1} N) x_N.$$

If we define $y = (c_B^T B^{-1})^T = B^{-T} c_B$, then z can be written as

$$z = y^T b + (c_N^T - y^T N) x_N.$$

This formula is efficient computationally. The vector y is the vector of *simplex multipliers*. The current values of the basic variables and the objective are obtained by setting $x_N = 0$. We denote these by

$$x_B = \hat{b} = B^{-1} b \quad \text{and} \quad \hat{z} = c_B^T B^{-1} b.$$

Example 5.1 (General Formulas). For our sample linear program,

$$A = \begin{pmatrix} -2 & 1 & 1 & 0 & 0 \\ -1 & 2 & 0 & 1 & 0 \\ 1 & 0 & 0 & 0 & 1 \end{pmatrix}, \quad b = \begin{pmatrix} 2 \\ 7 \\ 3 \end{pmatrix}, \quad \text{and} \quad c = \begin{pmatrix} -1 \\ -2 \\ 0 \\ 0 \\ 0 \end{pmatrix}.$$

If $x_B = (x_1, x_2, x_3)^T$ and $x_N = (x_4, x_5)^T$, then

$$B = \begin{pmatrix} -2 & 1 & 1 \\ -1 & 2 & 0 \\ 1 & 0 & 0 \end{pmatrix}, \quad B^{-1} = \begin{pmatrix} 0 & 0 & 1 \\ 0 & \frac{1}{2} & \frac{1}{2} \\ 1 & -\frac{1}{2} & \frac{3}{2} \end{pmatrix}, \quad N = \begin{pmatrix} 0 & 0 \\ 1 & 0 \\ 0 & 1 \end{pmatrix},$$

$c_B^T = (-1, -2, 0)$, and $c_N^T = (0, 0)$. The current values of the variables are

$$x_B = \hat{b} = B^{-1}b = \begin{pmatrix} 3 \\ 5 \\ 3 \end{pmatrix}$$

and $x_N = (0, 0)^T$. The objective value is

$$\hat{z} = c_B^T B^{-1} b = -13.$$

If we define

$$y^T = c_B^T B^{-1} = (0 \quad -1 \quad -2),$$

then the objective value could also be computed as $\hat{z} = y^T b = -13$. If $x_N \neq 0$, then the general formula for the basic variables is

$$x_B = B^{-1}b - B^{-1}N x_N = \begin{pmatrix} 3 \\ 5 \\ 3 \end{pmatrix} - \begin{pmatrix} 0 & 1 \\ \frac{1}{2} & \frac{1}{2} \\ -\frac{1}{2} & \frac{3}{2} \end{pmatrix} \begin{pmatrix} x_4 \\ x_5 \end{pmatrix},$$

and the general formula for the objective value is

$$z = y^T b + (c_N^T - y^T N) x_N = -13 + (1 \quad 2) \begin{pmatrix} x_4 \\ x_5 \end{pmatrix}. \quad \blacksquare$$

Let $\hat{c}_j$ be the entry in the vector $\hat{c}_N^T \equiv (c_N^T - c_B^T B^{-1} N)$ corresponding to x_j. The coefficient $\hat{c}_j$ is called the *reduced cost* of x_j. Then

$$z = \hat{z} + \hat{c}_N^T x_N.$$

(In Example 5.1, $\hat{c}_4 = 1$ and $\hat{c}_5 = 2$.) If the nonbasic variable x_j is assigned some nonzero value ϵ, then the objective function will change by $\hat{c}_j \epsilon$.

To test for optimality we examine what would happen to the objective function if each of the nonbasic variables were increased from zero. If $\hat{c}_j > 0$ the objective function will increase, if $\hat{c}_j = 0$ the objective will not change, and if $\hat{c}_j < 0$ the objective will decrease. Hence if $\hat{c}_j < 0$ for some j, then the objective function can be improved if x_j is increased from zero. If the current basis is not optimal, then a variable x_t with $\hat{c}_t < 0$ can be selected to enter the basis.

Once the entering variable x_t has been selected, we must then determine how much it can be increased before a nonnegativity constraint is violated. This determines which variable (if any) will leave the basis. The basic variables are defined by

$$x_B = B^{-1}b - B^{-1}N x_N,$$

and, with the exception of x_t, all components of x_N are zero. Thus

$$x_B = \hat{b} - \hat{A}_t x_t,$$

where $\hat{A}_t$ is the vector $B^{-1} A_t$ and A_t is the tth column of A.

We examine this equation componentwise:

$$(x_B)_i = \hat{b}_i - \hat{a}_{i,t} x_t.$$

If $\hat{a}_{i,t} > 0$, then $(x_B)_i$ will decrease as the entering variable x_t increases, and $(x_B)_i$ will equal zero when $x_t = \hat{b}_i / \hat{a}_{i,t}$. If $\hat{a}_{i,t} < 0$, then $(x_B)_i$ will increase, and if $\hat{a}_{i,t} = 0$, then $(x_B)_i$ will remain unchanged.

The variable x_t can be increased as long as all the variables remain nonnegative, that is, until it reaches the value

$$\bar{x}_t = \min_{1 \le i \le m} \left\{ \frac{\hat{b}_i}{\hat{a}_{i,t}} : \hat{a}_{i,t} > 0 \right\}.$$

This is a ratio test (see Section 3.1), but of an especially simple form. The minimum ratio from the ratio test identifies the new nonbasic variable, and hence determines the new basic feasible solution, with x_t as the new basic variable. The formulas

$$x_B \leftarrow x_B - \hat{A}_t \bar{x}_t \quad \text{and} \quad \hat{z} \leftarrow \hat{z} + \hat{c}_t \bar{x}_t$$

can be used to determine the new values of the objective function and the basic variables in the current basis. The variable x_t is assigned the value $\bar{x}_t$; the remaining nonbasic variables are still zero.

If $\hat{a}_{i,t} \le 0$ for all values of i, then none of the basic variables will decrease in value as x_t is increased from zero, and so x_t can be made arbitrarily large. In this case, the objective function will decrease without bound as $x_t \to \infty$, indicating that the linear program does not have a finite minimum. Such a problem is said to be "unbounded."

We can now outline the simplex algorithm. The method starts with a basis matrix B corresponding to a basic feasible solution $x_B = \hat{b} = B^{-1}b \ge 0$. The steps of the algorithm are given below.

1. *The Optimality Test*—Compute the vector $y^T = c_B^T B^{-1}$. Compute the coefficients $\hat{c}_N^T = c_N^T - y^T N$. If $\hat{c}_N^T \ge 0$, then the current basis is optimal. Otherwise, select a variable x_t that satisfies $\hat{c}_t < 0$ as the entering variable.

2. *The Step*—Compute $\hat{A}_t = B^{-1} A_t$, the constraint coefficients corresponding to the entering variable. Find an index s that satisfies

$$\frac{\hat{b}_s}{\hat{a}_{s,t}} = \min_{1 \le i \le m} \left\{ \frac{\hat{b}_i}{\hat{a}_{i,t}} : \hat{a}_{i,t} > 0 \right\}.$$

This ratio test determines the leaving variable and the "pivot entry" $\hat{a}_{s,t}$. If $\hat{a}_{i,t} \le 0$ for all i, then the problem is unbounded.

3. *The Update*—Update the basis matrix B and the vector of basic variables x_B.

The optimality test is a *local* test since it only involves the reduced costs in the current basis. Since linear programming problems are convex optimization problems, however, any local solution is also a global solution. (See Section 2.3.) Thus this test identifies a global solution to a linear program.

Example 5.2 (Simplex Algorithm). We will illustrate the simplex method on our example linear program:

$$A = \begin{pmatrix} -2 & 1 & 1 & 0 & 0 \\ -1 & 2 & 0 & 1 & 0 \\ 1 & 0 & 0 & 0 & 1 \end{pmatrix}, \quad b = \begin{pmatrix} 2 \\ 7 \\ 3 \end{pmatrix}, \quad \text{and} \quad c = \begin{pmatrix} -1 \\ -2 \\ 0 \\ 0 \\ 0 \end{pmatrix}.$$

If we use the slack variables as the initial basis, then $x_B = (x_3, x_4, x_5)^T$, $x_N = (x_1, x_2)^T$, $B = I = B^{-1}$, $c_B^T = (0, 0, 0)$, $c_N^T = (-1, -2)$, and

$$N = \begin{pmatrix} -2 & 1 \\ -1 & 2 \\ 1 & 0 \end{pmatrix}.$$

Thus $x_B = \hat{b} = B^{-1}b = (2, 7, 3)^T$. With this basis

$$y^T = c_B^T B^{-1} = (0 \quad 0 \quad 0) \quad \text{and} \quad \hat{c}_N^T = c_N^T - y^T N = (-1 \quad -2).$$

Both components of $\hat{c}_N$ are negative, so this basis is not optimal. Since $(\hat{c}_N)_2$ is the more negative component, we select x_2 (the second nonbasic variable) as the entering variable.

For the ratio test we compute the entering column

$$\hat{A}_2 = B^{-1} A_2 = \begin{pmatrix} 1 \\ 2 \\ 0 \end{pmatrix},$$

so that the ratios (corresponding to the first two components of $\hat{A}_2$) are

$$\frac{\hat{b}_1}{\hat{a}_{1,2}} = 2 \quad \text{and} \quad \frac{\hat{b}_2}{\hat{a}_{2,2}} = \frac{7}{2}.$$

The first ratio is smaller, so x_3 (the first basic variable) is the variable that leaves the basis.

At the next iteration x_2 replaces x_3 in the basis, so that $x_B = (x_2, x_4, x_5)^T$, $x_N = (x_1, x_3)^T$,

$$B = \begin{pmatrix} 1 & 0 & 0 \\ 2 & 1 & 0 \\ 0 & 0 & 1 \end{pmatrix}, \quad B^{-1} = \begin{pmatrix} 1 & 0 & 0 \\ -2 & 1 & 0 \\ 0 & 0 & 1 \end{pmatrix}, \quad N = \begin{pmatrix} -2 & 1 \\ -1 & 0 \\ 1 & 0 \end{pmatrix},$$

$c_B^T = (-2, 0, 0)$, and $c_N^T = (-1, 0)$. Thus

$$x_B = \hat{b} = B^{-1}b = (2 \quad 3 \quad 3)^T$$
$$y^T = c_B^T B^{-1} = (-2 \quad 0 \quad 0)$$
$$\hat{c}_N^T = c_N^T - y^T N = (-5 \quad 2).$$

The first reduced cost is negative, so this basis is not optimal, and x_1 (the first nonbasic variable) is the entering variable. The entering column is

$$\hat{A}_1 = B^{-1}A_1 = \begin{pmatrix} -2 \\ 3 \\ 1 \end{pmatrix}$$

and the candidate ratios are $\hat{b}_2/\hat{a}_{2,1} = 1$ and $\hat{b}_3/\hat{a}_{3,1} = 3$, so that x_4 (the second basic variable) is the leaving variable.

At the third iteration $x_B = (x_2, x_1, x_5)^T$, $x_N = (x_3, x_4)^T$,

$$B = \begin{pmatrix} 1 & -2 & 0 \\ 2 & -1 & 0 \\ 0 & 1 & 1 \end{pmatrix}, \quad B^{-1} = \begin{pmatrix} -\frac{1}{3} & \frac{2}{3} & 0 \\ -\frac{2}{3} & \frac{1}{3} & 0 \\ \frac{2}{3} & -\frac{1}{3} & 1 \end{pmatrix}, \quad N = \begin{pmatrix} 1 & 0 \\ 0 & 1 \\ 0 & 0 \end{pmatrix},$$

$c_B^T = (-2, -1, 0)$, and $c_N^T = (0, 0)$. Then

$$\begin{aligned}
x_B &= \hat{b} = B^{-1}b = (4 \quad 1 \quad 2)^T \\
y^T &= c_B^T B^{-1} = (\tfrac{4}{3} \quad -\tfrac{5}{3} \quad 0) \\
\hat{c}_N^T &= c_N^T - y^T N = (-\tfrac{4}{3} \quad \tfrac{5}{3}).
\end{aligned}$$

This basis is not optimal and x_3 is the entering variable. The entering column is

$$\hat{A}_3 = B^{-1}A_3 = \begin{pmatrix} -\frac{1}{3} \\ -\frac{2}{3} \\ \frac{2}{3} \end{pmatrix}$$

and the only candidate ratio is $\hat{b}_3/\hat{a}_{3,1} = 3$, so x_5 is the leaving variable.

At the fourth iteration, $x_B = (x_2, x_1, x_3)^T$, $x_N = (x_4, x_5)^T$,

$$B = \begin{pmatrix} 1 & -2 & 1 \\ 2 & -1 & 0 \\ 0 & 1 & 0 \end{pmatrix}, \quad B^{-1} = \begin{pmatrix} 0 & \frac{1}{2} & \frac{1}{2} \\ 0 & 0 & 1 \\ 1 & -\frac{1}{2} & \frac{3}{2} \end{pmatrix}, \quad N = \begin{pmatrix} 0 & 0 \\ 1 & 0 \\ 0 & 1 \end{pmatrix},$$

$c_B^T = (-2, -1, 0)$, and $c_N^T = (0, 0)$. Then

$$\begin{aligned}
x_B &= \hat{b} = B^{-1}b = (5 \quad 3 \quad 3)^T \\
y^T &= c_B^T B^{-1} = (0 \quad -1 \quad -2) \\
\hat{c}_N^T &= c_N^T - y^T N = (1 \quad 2).
\end{aligned}$$

This basis is optimal. ∎

In the optimality test of the simplex method there is an ambiguity about the choice of the entering variable. In the example, we selected the entering variable corresponding to the most negative $\hat{c}_j < 0$. If x_j is increased by ϵ, then z will change by $\hat{c}_j\epsilon$, so this choice achieves the best rate of decrease in z. This choice does not take into account the results of the ratio test, so it is possible that only a tiny step will be taken and that z will only decrease by a small amount. It also does not take into account the scaling of the variables in the problem. Other more sophisticated ways of choosing the entering variable are possible, but they may require additional computations and can be more expensive to use. They are discussed in Section 7.6.1.

Had we chosen x_1 to enter the basis at the first iteration, the method would have moved from x_a through x_e to the optimal point x_d, requiring only two iterations; see Figure 5.1. For general problems, however, there is no practical way to predict which choice of entering variable would lead to the least number of iterations.

5.2.2 Unbounded Problems

In step 2 of the simplex method there is the possibility that the problem will be unbounded. If $\hat{a}_{i,t} > 0$, then basic variable $(x_B)_i$ will decrease as the entering variable x_t increases, and $(x_B)_i$ will equal zero when $x_t = \hat{b}_i / \hat{a}_{i,t}$. If $\hat{a}_{i,t} \leq 0$ for all i, then none of the basic variables will decrease as x_t increases, implying that x_t can be increased without bound, and hence the feasible region is unbounded. The objective function will change by an amount equal to $\hat{c}_t x_t$ as x_t increases. Since the entering variable was chosen because $\hat{c}_t < 0$, the objective function can be decreased indefinitely. Thus the linear program will not have a finite minimum value. Unboundedness is illustrated in the following example.

Example 5.3 (Unbounded Linear Program). Consider the linear program

$$\begin{array}{ll}
\text{minimize} & z = -x_1 - 2x_2 \\
\text{subject to} & -x_1 + x_2 \leq 2 \\
& -2x_1 + x_2 \leq 1 \\
& x_1, x_2 \geq 0.
\end{array}$$

After two iterations, the basis is $x_B = (x_1, x_2)^T$ with $x_N = (x_3, x_4)^T$,

$$B = \begin{pmatrix} -1 & 1 \\ -2 & 1 \end{pmatrix} \quad \text{and} \quad B^{-1} = \begin{pmatrix} 1 & -1 \\ 2 & -1 \end{pmatrix}.$$

At this iteration, $x_B = (1, 3)^T$ and the reduced costs for the nonbasic variables are $\hat{c}_N^T = (5, -3)$, so the current basis is not optimal, and x_4 (the second nonbasic variable) is the entering variable. The entering column is

$$\hat{A}_4 = \begin{pmatrix} -1 \\ -1 \end{pmatrix},$$

so there are no candidates for the ratio test. The entering variable x_4 can be increased without limit, so the objective function can be decreased without limit, and there is no finite solution to this linear program. This can also be seen by looking at a graph of the feasible region; see Figure 5.2. (The figure represents the two-variable version of the problem, not the problem in standard form.)

The current basic feasible solution is $(x_1, x_2, x_3, x_4)^T = (1, 3, 0, 0)^T$. From the equation $x_B = \hat{b} - \hat{A}_4 x_4$ we conclude that all points of the form

$$\begin{pmatrix} x_1 \\ x_2 \\ x_3 \\ x_4 \end{pmatrix} = \begin{pmatrix} 1 \\ 3 \\ 0 \\ 0 \end{pmatrix} + \begin{pmatrix} 1 \\ 1 \\ 0 \\ 1 \end{pmatrix} x_4$$

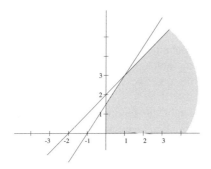

Figure 5.2. *Unbounded linear program.*

are feasible. Let $d = (1, 1, 0, 1)^T$. It is easy to check that $d \geq 0$ and $Ad = 0$, where

$$A = \begin{pmatrix} -1 & 1 & 1 & 0 \\ -2 & 1 & 0 & 1 \end{pmatrix}$$

is the coefficient matrix for the constraints of the problem in standard form. Hence d is a direction of unboundedness. Since $c^T d < 0$, the objective decreases as x_4 is increased, showing that the problem is unbounded.

In this example it would have been possible to stop at the first iteration. Our rule for choosing the entering variable picked x_2 to enter the basis, but any variable x_j with $\hat{c}_j < 0$ can be the entering variable since such a variable will lead to an improvement in the objective function. If x_1 were chosen as the entering variable, then the entering column would be

$$\hat{A}_1 = \begin{pmatrix} -1 \\ -2 \end{pmatrix},$$

and there would be no candidates for the ratio test, again indicating that the problem is unbounded. At the first iteration, all points of the form

$$\begin{pmatrix} x_1 \\ x_2 \\ x_3 \\ x_4 \end{pmatrix} = \begin{pmatrix} 0 \\ 0 \\ 2 \\ 1 \end{pmatrix} + \begin{pmatrix} 1 \\ 0 \\ 1 \\ 2 \end{pmatrix} x_1$$

are feasible, where $d = (1, 0, 1, 2)^T$ is a direction of unboundedness along which the objective decreases. ∎

5.2.3 Notation for the Simplex Method (Tableaus)

Although we have presented the formulas for the simplex method already, these formulas are not always convenient for classroom and explanatory use because they require the calculation of matrix inverses or solving systems of equations. (If software is available for performing the necessary matrix calculations, however, they are satisfactory.) In this section we present a notational device called a "tableau" for representing the calculations in the simplex method. The tableau uses the inverse of the basis matrix, but updates it at every iteration of the simplex method, rather than calculating it anew. This makes it possible to solve small linear programs "by hand."

The tableau is also a convenient and compact format to present examples. For this reason we will sometimes use tableaus to discuss examples. To understand our examples it is only necessary to be able to extract information from the tableaus, not to manipulate them.

We emphasize that the tableaus (and the use of explicit matrix inverses) are merely notational devices that assist our explanations of the simplex method. Computer implementations of the simplex method use other techniques more suitable for large sparse problems (see Chapter 7).

For our example

$$\begin{array}{ll}
\text{minimize} & z = -x_1 - 2x_2 \\
\text{subject to} & -2x_1 + x_2 + x_3 = 2 \\
& -x_1 + 2x_2 + x_4 = 7 \\
& x_1 + x_5 = 3 \\
& x_1, x_2, x_3, x_4, x_5 \geq 0
\end{array}$$

the initial tableau looks like

basic	x_1	x_2	x_3	x_4	x_5	rhs
$-z$	-1	-2	0	0	0	0
x_3	-2	1	1	0	0	2
x_4	-1	2	0	1	0	7
x_5	1	0	0	0	1	3

The lower part of the tableau contains the coefficients of the constraints of the linear program in standard form. For example, the last row corresponds to the third constraint $x_1 + x_5 = 3$. The top row of the tableau contains the coefficients in the objective function. It corresponds to writing the objective function in the form of an equality constraint

$$-z - x_1 - 2x_2 + 0x_3 + 0x_4 + 0x_5 = 0,$$

where the right-hand side is the negative of the current value of the objective function. To emphasize that the objective value is multiplied by -1, the top row of the tableau is labeled $-z$. The first column of the tableau lists the basic variables and the column labeled "rhs" for "right-hand side" records the values of $-z$ and the basic variables. (The nonbasic variables are zero.)

We will again solve the example problem, this time using the tableau. Because the initial basis matrix is $B = I$, the entries in the lower part of the "rhs" column are $x_B = \hat{b} = B^{-1}b$ and the entries in the top row are the current reduced costs $\hat{c}$. At every iteration, the entries in the tableau will be represented in terms of the current basis, so that the "rhs" column will include $\hat{b}$ and the top row will include $\hat{c}$.

Before proceeding with the example, we give the general formulas for the tableau. Consider a linear program in standard form with n variables and m equality constraints. Let us assume that at the current iteration the vectors of basic and nonbasic variables are $x_B = (x_1, \ldots, x_m)^T$ and $x_N = (x_{m+1}, \ldots, x_n)^T$, respectively.

The original linear program corresponds to the tableau

basic	x_B	x_N	rhs
$-z$	c_B^T	c_N^T	0
x_B	B	N	b

and the tableau for the problem in the current basis is

basic	x_B	x_N	rhs
$-z$	0	$c_N^T - c_B^T B^{-1} N$	$-c_B^T B^{-1} b$
x_B	I	$B^{-1} N$	$B^{-1} b$

These are the matrix-vector formulas for the tableau.

The simplex iteration begins with the optimality test. For the basic variables the reduced costs are zero. In our example, at the first iteration the reduced costs for the nonbasic variables are negative, so the current basis is not optimal. The reduced cost for x_2 is larger in magnitude, so we select x_2 as the entering variable.

We determine the leaving variable using the ratio test. The ratios are computed using the "rhs" values and the values in the entering column, where the ratio is computed only if the coefficient of the entering variable is positive. The smallest nonnegative ratio will correspond to the leaving variable. In the tableau, only the first two constraint coefficients for x_2 are greater than zero, giving the ratios $2/1 = 2$ and $7/2 = \frac{7}{2}$. Hence, x_3 is the leaving variable. In the tableau we mark the entering variable as well as the *pivot* entry in the x_2 column and the x_3 row:

basic	x_1	x_2 $\Downarrow$	x_3	x_4	x_5	rhs
$-z$	-1	-2	0	0	0	0
x_3	-2	$\boxed{1}$	1	0	0	2
x_4	-1	2	0	1	0	7
x_5	1	0	0	0	1	3

The final step is to transform the tableau to express the coefficients in terms of the new basis. This step is sometimes called *pivoting*. This can be done using the matrix-vector formulas for the tableau using the new basis. It can also be done directly from the tableau by applying elementary row operations to transform the x_2 column into

$$\begin{pmatrix} 0 \\ 1 \\ 0 \\ 0 \end{pmatrix},$$

that is, into a column of the identity matrix with a one as the pivot entry and zeroes elsewhere. The result of this transformation is that the new basic variables are represented in terms of the new nonbasic variables.

In this case we add 2 times the x_3 row to the $-z$ row, and subtract 2 times the x_3 row from the x_4 row to obtain the new tableau:

basic	x_1	x_2	x_3	x_4	x_5	rhs
$-z$	-5	0	2	0	0	4
x_2	-2	1	1	0	0	2
x_4	3	0	-2	1	0	3
x_5	1	0	0	0	1	3

Notice that the "basic" column has been modified to reflect the change in the basis. This is the tableau corresponding to the transformed linear program at the basic feasible solution x_b that we derived earlier.

We now perform the second iteration of the simplex method. In the top row of the tableau the reduced cost of x_1 is $-5 < 0$, so this basis is not optimal and x_1 will be the entering variable. The ratio test indicates that x_4 will leave the basis:

basic	$\Downarrow$ x_1	x_2	x_3	x_4	x_5	rhs
$-z$	-5	0	2	0	0	4
x_2	-2	1	1	0	0	2
x_4	$\boxed{3}$	0	-2	1	0	3
x_5	1	0	0	0	1	3

We then apply elimination operations to get the next tableau. The tableaus for the remaining iterations are

basic	x_1	x_2	$\Downarrow$ x_3	x_4	x_5	rhs
$-z$	0	0	$-\frac{4}{3}$	$\frac{5}{3}$	0	9
x_2	0	1	$-\frac{1}{3}$	$\frac{2}{3}$	0	4
x_1	1	0	$-\frac{2}{3}$	$\frac{1}{3}$	0	1
x_5	0	0	$\boxed{\frac{2}{3}}$	$-\frac{1}{3}$	1	2

and

basic	x_1	x_2	x_3	x_4	x_5	rhs
$-z$	0	0	0	1	2	13
x_2	0	1	0	$\frac{1}{2}$	$\frac{1}{2}$	5
x_1	1	0	0	0	1	3
x_3	0	0	1	$-\frac{1}{2}$	$\frac{3}{2}$	3

At the fourth iteration the reduced costs of the nonbasic variables are all positive, so the current basis is optimal. The solution can be read from the "rhs" column of the tableau: $z = -13$, $x_2 = 5$, $x_1 = 3$, and $x_3 = 3$. The nonbasic variables, x_4 and x_5, are zero. This is the same as in Section 5.2.

5.2.4 Deficiencies of the Tableau

In the tableau form of the simplex method, many of the computations performed in a given iteration are not used in that iteration. For example, the tableau columns of *all* nonbasic variables are computed, even though only the column of the entering variable is needed in order to determine the new solution.

Implementations of the simplex method generate at each iteration only the information that is specifically required for that iteration. The result is a version of the method which requires less storage and less computation. It also makes it possible to utilize the sparsity of the matrix A to reduce the number of computations. Historically this approach was named the *revised* simplex method to distinguish it from the tableau form.

The version of the simplex method presented in Section 5.2.1 is of this type. Here we discuss some of the advantages of this approach.

As before, we work with a problem in standard form

$$\text{minimize} \quad z = c^T x$$
$$\text{subject to} \quad Ax = b$$
$$x \geq 0,$$

where A is an $m \times n$ matrix of full row rank. Let B be the basis matrix at some iteration. In matrix-vector notation, the current tableau is

basic	x_B	x_N	rhs
$-z$	0	$c_N^T - c_B^T B^{-1} N$	$-c_B^T B^{-1} b$
x_B	I	$B^{-1} N$	$B^{-1} b$

The information required for the simplex method can be generated directly from B^{-1} and the original data. This allows us to compute information only as needed. More specifically, suppose that some representation of the $m \times m$ inverse of a basis matrix for some iteration is available. Then the only other information that would be needed at that iteration is the current solution vector, the reduced costs, and the column of the entering variable. This information may be computed from the formulas for the method.

If the basis matrix inverse B^{-1} is available, then

$$x_B = \hat{b} = B^{-1} b$$

and the associated objective value is

$$\hat{z} = c_B^T B^{-1} b = c_B^T x_B.$$

The columns of the current tableau, $\hat{A}_j$, are obtained from

$$\hat{A}_j = B^{-1} A_j,$$

where A_j is the jth column of A. If we define

$$y^T = c_B^T B^{-1},$$

we can compute the reduced costs from

$$\hat{c}_j = c_j - y^T A_j.$$

The process of computing the coefficients $\hat{c}_j$ is called *pricing*.

Some mechanism is needed to update the representation of the inverse matrix for the next iteration. The inverse could be computed anew at every iteration, but more efficient techniques are discussed in Chapter 7.

We can be more precise about the computational differences between the simplex method via formulas and via the tableau.

The simplex method, whether implemented using the formulas from Section 5.2.1 or using the tableau, will go through the same sequence of bases, provided that the same criteria are used for selecting the entering variable and for breaking ties in selecting the leaving variable. Thus, on a given problem the two versions perform the same number of iterations. The difference between the versions of the method is in the organization of the computation. In the following we compare the formulas with the tableau for a problem in standard form with m constraints and n variables.

Consider first the storage requirements, beyond those required for storing the problem itself. The tableau requires an $(m + 1) \times (n + 1)$ array. The formulas require

- an array of length m to store the value of x_B,

- an array of length m to store the entering column,

- an array of length $n - m$ to store the reduced costs, and

- a representation of B^{-1}.

If B^{-1} is represented explicitly, then an $m \times m$ array is required. If B is a sparse matrix, then its inverse can typically be represented using storage proportional to the number of nonzero entries in B. Thus if n is much larger than m (as is frequently the case), the formulas achieve significant savings in storage requirements as compared to the tableau.

Consider now the computational effort required by the two approaches. One measure of this effort is the number of operations required per iteration. For simplicity we shall only count the number of multiplications and divisions. The number of additions and subtractions is roughly the same. We start by examining the work required to solve a dense problem.

The main computational effort in the tableau method is in pivoting. Each pivot updates $n - m + 1$ tableau columns, corresponding to the $(n - m)$ nonbasic variables plus the right-hand-side vector. First, the pivot row is divided by the pivot term; this requires $n - m + 1$ operations. Next, a multiple of the updated pivot row is added to each of the remaining m rows (including the top row); this requires $m(n - m + 1)$ multiplications. In total, each pivot requires

$$(n - m + 1) + m(n - m + 1) = mn + n + 1 - m^2$$

multiplications. The only other calculations occur in the ratio test, where at most m divisions are performed. The effort in the ratio test is negligible compared to the effort in pivoting, and for simplicity we shall ignore it.

The computational effort in an iteration with the formulas includes: computation of $\hat{c}_j$ (pricing); computation of the entering column; the ratio test; and the update. Each computation $\hat{c}_j = c_j - y^T A_j$ requires m multiplications in the inner product. Since there are $n - m$ nonbasic variables, pricing will require $m(n - m)$ multiplications. Computing the entering column $\hat{A}_t$ involves a matrix-vector product $B^{-1}A_t$, or m^2 multiplications. In the update step, the representation of B^{-1} must be updated, along with the reduced costs and the value of x_B. If Gaussian elimination is used to do this (see Section 7.5), then the cost is $m + 1$ multiplications per row, or a total cost of $(m + 1)^2$. Summing up (and again ignoring the cost of the ratio test), we conclude that the formula-based simplex method requires

$$m(n - m) + m^2 + (m + 1)^2 = mn + (m + 1)^2$$

multiplications per iteration.

It appears that unless n is substantially larger than m, the tableau method will require less computation. However, our operation count has not taken into account the effects of sparsity. To examine this, consider for example a sparse problem, where each column of A has exactly 5 nonzero elements. Then, if we are using the formulas, each inner product $y^T A_j$ will now only require 5 multiplications, hence full pricing will require $5(n - m)$ multiplications. The matrix-vector product $B^{-1}A_t$ will require $5m$ multiplications. The update step will still require $(m + 1)^2$ operations. In total, the number of operations will be

$$5(n - m) + 5m + (m + 1)^2 = 5n + (m + 1)^2.$$

In contrast, the tableau will still require the number of computations given above. When m and n are large, the savings offered by the revised simplex method are dramatic. For example, if $m = 1000$ and $n = 100{,}000$, then each iteration of the tableau simplex method will require about 99 million operations, while each iteration of the formula-based simplex method will only require about 1.5 million operations. Such savings might reduce the total solution time from days to just hours, or from hours to just minutes.

In the operation count, we assumed that B^{-1} is dense. This is often the case, even when the matrix B is sparse. Thus, if m is large the $(m + 1)^2$ operations required to update a dense matrix B^{-1} may become expensive. In Chapter 7 we describe a variant of the simplex method that represents B^{-1} as a product of factors which tend to be sparse. This can further reduce the work and storage required by the simplex method.

Exercises

2.1. Verify the computational results in Example 5.2.

2.2. Solve the following linear programs using the simplex method. If the problem is two dimensional, graph the feasible region, and outline the progress of the algorithm.

(i)

$$
\begin{aligned}
\text{minimize} \quad & z = -5x_1 - 7x_2 - 12x_3 + x_4 \\
\text{subject to} \quad & 2x_1 + 3x_2 + 2x_3 + x_4 \leq 38 \\
& 3x_1 + 2x_2 + 4x_3 - x_4 \leq 55 \\
& x_1, x_2, x_3, x_4 \geq 0.
\end{aligned}
$$

(ii)

$$\text{maximize} \quad z = 5x_1 + 3x_2 + 2x_3$$
$$\text{subject to} \quad 4x_1 + 5x_2 + 2x_3 + x_4 \le 20$$
$$3x_1 + 4x_2 - x_3 + x_4 \le 30$$
$$x_1, x_2, x_3, x_4 \ge 0.$$

(iii)

$$\text{minimize} \quad z = 3x_1 + 9x_2$$
$$\text{subject to} \quad -5x_1 + 2x_2 \le 30$$
$$-3x_1 + x_2 \le 12$$
$$x_1, x_2 \ge 0.$$

(iv)

$$\text{minimize} \quad z = 3x_1 - 2x_2 - 4x_3$$
$$\text{subject to} \quad 4x_1 + 5x_2 - 2x_3 \le 22$$
$$x_1 - 2x_2 + x_3 \le 30$$
$$x_1, x_2, x_3 \ge 0$$

(v)

$$\text{maximize} \quad z = 7x_1 + 8x_2$$
$$\text{subject to} \quad 4x_1 + x_2 \le 100$$
$$x_1 + x_2 \le 80$$
$$x_1 \le 40$$
$$x_1, x_2 \ge 0$$

(vi)

$$\text{minimize} \quad z = -6x_1 - 14x_2 - 13x_3$$
$$\text{subject to} \quad x_1 + 4x_2 + 2x_3 \le 48$$
$$x_1 + 2x_2 + 4x_3 \le 60$$
$$x_1, x_2, x_3 \ge 0.$$

2.3. Consider the linear program

$$\text{minimize} \quad z = x_1 - x_2$$
$$\text{subject to} \quad -x_1 + x_2 \le 1$$
$$x_1 - 2x_2 \le 2$$
$$x_1, x_2 \ge 0.$$

Derive an expression for the set of optimal solutions to this problem, and show that this set is unbounded.

2.4. Find all the values of the parameter a such that the following linear program has a finite optimal solution:

$$\text{minimize} \quad z = -ax_1 + 4x_2 + 5x_3 - 3x_4$$
$$\text{subject to} \quad 2x_1 + x_2 - 7x_3 - x_4 = 2$$
$$x_1, x_2, x_3, x_4 \ge 0.$$

2.5. Using the optimality test to find all the values of the parameter a such that $x_* = (0, 1, 1, 3, 0, 0)^T$ is the optimal solution of the following linear program:

$$\begin{aligned}
\text{minimize} \quad & z = -x_1 - a^2 x_2 + 2x_3 - 2ax_4 - 5x_5 + 10x_6 \\
\text{subject to} \quad & -2x_1 - x_2 + x_4 + 2x_6 = 2 \\
& 2x_1 + x_2 + x_3 = 2 \\
& -2x_1 - x_3 + x_4 + 2x_5 = 2 \\
& x_1, x_2, x_3, x_4, x_5, x_6 \geq 0.
\end{aligned}$$

2.6. The reduced costs are given by the formula $\hat{c}_N^T = c_N^T - c_B^T B^{-1} N$, and a basic feasible solution is optimal if $\hat{c}_N^T \geq 0$. Construct an example involving a degenerate basic feasible solution that corresponds to two different bases, where in one basis the basic feasible solution is optimal, but in the other basis it is not.

2.7. Prove that the set of optimal solutions to a linear programming problem is a convex set.

2.8. Prove that in the simplex method a variable which has just left the basis cannot re-enter the basis in the following iteration.

2.9. Consider the linear program

$$\begin{aligned}
\text{minimize} \quad & z = c^T x \\
\text{subject to} \quad & Ax \leq b \\
& x \geq 0,
\end{aligned}$$

where $x = (x_1, \ldots, x_n)^T$, $c = (0, \ldots, 0, -\alpha)^T$, $b = (1, \ldots, 1)^T$, and

$$A = \begin{pmatrix}
1 & & & & \\
-2 & 1 & & & \\
& -2 & 1 & & \\
& & \ddots & \ddots & \\
& & & -2 & 1
\end{pmatrix}.$$

Here α is small positive number, say $\alpha = 2^{-50}$.

    (i) Consider the basic feasible solution where the slacks are the basic variables. Compute the reduced costs for this basis. By how much does this basis violate the optimality conditions? What is the current value of the objective?

    (ii) Consider now the solution where $\{ x_1, \ldots, x_n \}$ is the set of basic variables. Prove that this is a basic feasible solution.

    (iii) Prove that the solution defined in (ii) is optimal.

    (iv) What is the optimal objective value? (Find a closed-form solution, if possible.)

2.10. Consider a linear program with a single constraint

$$\begin{aligned}
\text{minimize} \quad & z = c_1 x_1 + c_2 x_2 + \cdots + c_n x_n \\
\text{subject to} \quad & a_1 x_1 + a_2 x_2 + \cdots + a_n x_n \leq b \\
& x_1, x_2, \ldots, x_n \geq 0.
\end{aligned}$$

    (i) Under what conditions is the problem feasible?

    (ii) Develop a simple rule to determine an optimal solution, if one exists.

2.11. Solve the linear programs in Exercise 2.2 using the tableau.

2.12. This problem concerns the number of additions/subtractions in the various versions of the simplex method for a problem with n variables and $m < n$ equality constraints.

(i) Compute the number of additions/subtractions required in each iteration of the tableau version of the simplex method.

(ii) Compute the number of additions/subtractions required in each iteration of the Simplex method implemented using the formulas in Section 5.2.4.

(iii) Assume now that each column of the constraint matrix has $l < m$ nonzero entries. Repeat part (ii).

2.13. The following tableau corresponds to an iteration of the simplex method:

basic	x_1	x_2	x_3	x_4	x_5	x_6	rhs
$-z$	0	a	0	b	c	3	d
	0	-2	1	e	0	2	f
	1	g	0	-2	0	1	1
	0	0	0	h	1	4	3

Find conditions on the parameters $a, b, \ldots, h$ so that the following statements are true.

(i) The current basis is optimal.

(ii) The current basis is the unique optimal basis.

(iii) The current basis is optimal but alternative optimal bases exist.

(iv) The problem is unbounded.

(v) The current solution will improve if x_4 is increased. When x_4 is entered into the basis, the change in the objective is zero.

5.3 The Simplex Method (Details)

In Section 5.2, a general discussion of the simplex method was given, and a small linear program was solved, but this does not give a complete description of the method. For example, it was not shown how to initialize the method, and there were no guarantees that it would terminate. The rest of this chapter will fill some of these gaps. In this section, we show how to detect if the linear program has multiple solutions. In Section 5.4, techniques for initializing the simplex method are described. And in Section 5.5, we give conditions under which the simplex method will be guaranteed to terminate when applied to any linear program.

5.3.1 Multiple Solutions

A linear program can have more than one optimal solution. This can occur when the reduced cost of a nonbasic variable is equal to zero in the optimal basis.

Example 5.4 (Multiple Solutions). Consider the linear program

$$
\begin{aligned}
\text{minimize} \quad & z = -x_1 \\
\text{subject to} \quad & -2x_1 + x_2 \leq 2 \\
& -x_1 + x_2 \leq 3 \\
& x_1 \leq 3 \\
& x_1, x_2 \geq 0.
\end{aligned}
$$

After one iteration of the simplex method,

basic	x_1	x_2	x_3	x_4	x_5	rhs
$-z$	0	0	0	0	1	3
x_3	0	1	1	0	2	8
x_4	0	1	0	1	1	6
x_1	1	0	0	0	1	3

This basis is optimal, but the reduced cost for the nonbasic variable x_2 is zero. This indicates that if this variable entered the basis, then the objective would change by $\hat{c}_2 \times$ (new value of x_2) $= 0$, so the objective value would not be altered and would remain optimal. If we perform this update we obtain

basic	x_1	x_2	x_3	x_4	x_5	rhs
$-z$	0	0	0	0	1	3
x_3	0	0	1	-1	1	2
x_2	0	1	0	1	1	6
x_1	1	0	0	0	1	3

This basis is also optimal, and again the reduced cost of a nonbasic variable (x_4) is zero.

This problem has two optimal basic feasible solutions. Any convex combination of these two solutions is also optimal. These points correspond to an edge of the feasible region. This is illustrated in Figure 5.3. ■

5.3.2 Feasible Directions and Edge Directions

The simplex method is an example of a feasible point method. It moves from one extreme point to another along a sequence of feasible descent directions. For nondegenerate linear programs, these directions correspond to edges of the feasible region.

We can determine formulas for the feasible directions in the simplex method. Let $A = (B, N)$ be the constraint matrix, and let

$$
\hat{x} = \begin{pmatrix} B^{-1}b \\ 0 \end{pmatrix}
$$

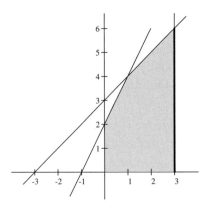

Figure 5.3. *Multiple solutions.*

be the corresponding basic feasible solution. (It may be necessary to reorder the variables, with the basic variables listed first.) Any feasible point can be represented as

$$x = \begin{pmatrix} x_B \\ x_N \end{pmatrix} = \begin{pmatrix} B^{-1}b - B^{-1}Nx_N \\ x_N \end{pmatrix}$$

$$= \begin{pmatrix} B^{-1}b \\ 0 \end{pmatrix} + \begin{pmatrix} -B^{-1}N \\ I \end{pmatrix} x_N$$

$$= \hat{x} + Zx_N$$

for some nonnegative value of x_N. The matrix

$$Z = \begin{pmatrix} -B^{-1}N \\ I \end{pmatrix}$$

is the null-space matrix for A obtained via variable reduction with this basis.

Alternatively, following the discussion in Section 3.1, x may be written as $\hat{x} + p$ where p is a feasible direction. Since

$$A(\hat{x} + p) = b \quad \text{and} \quad \hat{x} + p \geq 0,$$

it follows that

$$Ap = 0 \quad \text{and} \quad p_N \geq 0.$$

Hence p is in the null space of A and can be written as $p = Zv$ for some v. Comparing this with the result above shows that we can set $v = p_N = x_N$. Any nonnegative choice of x_N will correspond to a feasible direction.

In the simplex method, only one nonbasic variable is allowed to enter the basis at a time. This implies that only one component of x_N will be nonzero during an update. In turn, this implies that the feasible directions considered at an iteration of the simplex method are the columns of Z. If $(x_N)_k$ is the entering variable and Z_k is the kth column of Z, then an update step in the simplex method corresponds to a step of the form

$$x = \hat{x} + (x_N)_k Z_k.$$

The edges of the feasible region at the point $\hat{x}$ are the nonnegative points x that can be written in this form, and the vectors Z_k are called *edge directions*. The step in the simplex method is of the form

$$\hat{x} + \alpha p$$

for a search direction $p = Z_k$ and a step length $\alpha = (x_N)_k$, and so the simplex method fits into the framework of our general optimization algorithm.

If some column Z_i of Z satisfies $Z_i \geq 0$, then $d = Z_i$ is a direction of unboundedness for the problem. It is easy to verify that $Ad = 0$ and $d \geq 0$, i.e., that the conditions for a direction of unboundedness are satisfied.

Example 5.5 (Feasible Directions and Edge Directions). We look again at the linear program

$$
\begin{array}{ll}
\text{minimize} & z = -x_1 - 2x_2 \\
\text{subject to} & -2x_1 + x_2 \leq 2 \\
& -x_1 + 2x_2 \leq 7 \\
& x_1 \leq 3 \\
& x_1, x_2 \geq 0.
\end{array}
$$

Its feasible region is depicted in Figure 5.1.

At the first iteration $\hat{x} = x_a = (0, 0, 2, 7, 3)^T$. The basic and nonbasic variables are $x_B = (x_3, x_4, x_5)^T$ and $x_N = (x_1, x_2)^T$, respectively. The corresponding null-space matrix is

$$
Z = \begin{pmatrix} I \\ -B^{-1}N \end{pmatrix} = \begin{pmatrix} \overbrace{\qquad\qquad}^{I} \\ -\begin{pmatrix} 1 & 0 & 0 \\ 0 & 1 & 0 \\ 0 & 0 & 1 \end{pmatrix}^{-1} \begin{pmatrix} -2 & 1 \\ -1 & 2 \\ 1 & 0 \end{pmatrix} \end{pmatrix} = \begin{pmatrix} 1 & 0 \\ 0 & 1 \\ 2 & -1 \\ 1 & -2 \\ -1 & 0 \end{pmatrix}.
$$

At this iteration, x_2 entered the basis. This is the second nonbasic variable and so the feasible direction is

$$p = Z_2 = (0 \quad 1 \quad -1 \quad -2 \quad 0)^T.$$

The step procedure determines that the new value of x_2 is 2, and hence the step length is $\alpha = 2$. The new basic feasible solution can be written as

$$
\begin{aligned}
x = \hat{x} + \alpha p &= (0 \quad 0 \quad 2 \quad 7 \quad 3)^T + 2(0 \quad 1 \quad -1 \quad -2 \quad 0)^T \\
&= (0 \quad 2 \quad 0 \quad 3 \quad 3)^T.
\end{aligned}
$$

It would also have been possible to take a step in the feasible direction $p = Z_1$. Both Z_1 and Z_2 are edge directions. They correspond to the edges connecting x_a to x_e and x_a to x_b in Figure 5.1. ∎

The null-space matrix Z can also be used to derive a formula for the reduced costs:

$$c^T Z = (c_B^T \quad c_N^T) \begin{pmatrix} -B^{-1}N \\ I \end{pmatrix}$$
$$= -c_B^T B^{-1} N + c_N^T = c_N^T - c_B^T B^{-1} N.$$

Thus the reduced cost of the kth nonbasic variable is just $c^T Z_k$.

Exercises

3.1. Consider the linear program

$$\begin{array}{ll} \text{minimize} & z = -x_1 + 2x_2 - x_3 \\ \text{subject to} & x_1 + 2x_2 + x_3 \le 12 \\ & 2x_1 + \ x_2 - x_3 \le 6 \\ & -x_1 + 3x_2 \le 9 \\ & x_1, x_2, x_3 \ge 0. \end{array}$$

Add slack variables x_4, x_5, and x_6 to put the problem in standard form.

(i) Consider the basis $\{x_1, x_4, x_6\}$. Use the formulas for the simplex method to represent the linear program in terms of this basis.

(ii) Perform an iteration of the simplex method, constructing the null-space matrix Z, and computing the search direction d and the step length α so that $x_{k+1} = x_k + \alpha d$, where x_k is the current vector of variables and x_{k+1} is the new vector of variables.

3.2. Derive an expression for the family of optimal solutions to the linear program in Example 5.5.

3.3. At each iteration of the simplex method $x_{k+1} = x_k + \alpha p$. Determine α and p for each iteration in the solution of the linear program from Section 5.2:

$$\begin{array}{ll} \text{minimize} & z = -x_1 - 2x_2 \\ \text{subject to} & -2x_1 + x_2 \le 2 \\ & -x_1 + 2x_2 \le 7 \\ & x_1 \le 3 \\ & x_1, x_2 \ge 0 \end{array}$$

and verify that $Ap = 0$ where A is the coefficient matrix for the equality constraints in the problem.

3.4. Suppose that the optimal solution to a linear program has been found, and a reduced cost associated with a nonbasic variable is zero. Must the linear program have multiple solutions? Explain your answer.

3.5. Let $\hat{x}$ be an optimal basic feasible solution to a linear program in standard form with $m \times n$ constraint matrix A of full row rank. Let Z be the null-space matrix for A obtained via variable reduction using the basis corresponding to $\hat{x}$. Suppose that

$c^T Z_k = 0$ for $k = 1, \ldots, \ell \leq n - m$, where Z_k is the kth column of Z and c is the vector of objective coefficients. Show that every nonnegative vector

$$x = \hat{x} + \sum_{j=1}^{\ell} \alpha_j Z_j$$

is also optimal for the linear program.

5.4 Getting Started—Artificial Variables

The simplex method moves from one basic feasible solution to another until either a solution is found or until it is determined that the problem is unbounded. In the example in Section 5.2, an initial basic feasible solution was obtained by choosing the slack variables as a basis. In problems where every constraint has a slack variable this will always be a valid choice.

General problems will not have this property, raising the question of how to find a basic feasible solution. Sometimes the person posing the problem will be able to provide one. In cases where a sequence of similar linear programs is solved, such as a weekly budget prediction where the data vary slightly from week to week, the optimal basis from the previous linear program may be feasible for the new linear program. Or, say, if the linear program is designed to optimize the operations of a factory, the current setup at the factory may represent a basic feasible solution. The use of a specified initial basis was illustrated in Example 5.1.

This still leaves problems for which no obvious initial feasible point is available. One could guess at a basis but there is no guarantee that it would correspond to a point that satisfied the nonnegativity constraints. Randomly trying one basis after another until a basic feasible solution is found can be time consuming; if the problem were infeasible, every basis would have to be examined before this could be concluded.

When no initial point is provided, some general technique for getting started is required. We describe two standard approaches. The first (called the *two-phase method*) solves an auxiliary linear program to find an initial basic feasible solution. The second (called the *big-M method*) adds terms to the objective function that penalize for infeasibility. Although these are usually considered to be two separate approaches for finding a feasible point, they are closely related. They both use artificial variables as an algorithmic device, and in fact the two-phase method is the limit of the big-M method as the magnitude of the penalty goes to infinity. There are differences, however, in the way in which these methods are implemented in software, and for this reason it is worthwhile to consider them separately.

We will study these approaches using the following example:

$$
\begin{aligned}
\text{minimize} \quad & z = 2x_1 + 3x_2 \\
\text{subject to} \quad & 3x_1 + 2x_2 = 14 \\
& 2x_1 - 4x_2 \geq 2 \\
& 4x_1 + 3x_2 \leq 19 \\
& x_1, x_2 \geq 0.
\end{aligned}
$$

In standard form this becomes

$$\begin{array}{ll}
\text{minimize} & z = 2x_1 + 3x_2 \\
\text{subject to} & 3x_1 + 2x_2 = 14 \\
& 2x_1 - 4x_2 - x_3 = 2 \\
& 4x_1 + 3x_2 + x_4 = 19 \\
& x_1, x_2, x_3, x_4 \geq 0.
\end{array}$$

The first constraint contains no obvious candidate for a basic variable. The second constraint contains an excess variable, but if it were a basic variable it would take on the infeasible value $-2 < 0$. Only the third constraint has a slack variable suitable as a member of the initial basis.

Both initialization techniques use the device of *artificial* variables, that is, extra variables that are temporarily added to the problem. An artificial variable is added to every constraint that does not contain a slack variable:

$$\begin{array}{ll}
\text{minimize} & z = 2x_1 + 3x_2 \\
\text{subject to} & 3x_1 + 2x_2 + a_1 = 14 \\
& 2x_1 - 4x_2 - x_3 + a_2 = 2 \\
& 4x_1 + 3x_2 + x_4 = 19 \\
& x_1, x_2, x_3, x_4, a_1, a_2 \geq 0.
\end{array}$$

Now it is possible to initialize the simplex method using $x_B = (a_1, a_2, x_4)^T$ with values $a_1 = 14$, $a_2 = 2$, and $x_4 = 19$. This choice of x_B has coefficient matrix B that is a permutation of the identity matrix I.

Since the artificial variables are not part of the original problem, this choice of basis *does not correspond* to a basic feasible solution to the original problem; it is not even feasible for the original problem. The methods discussed below try to move to a basic feasible solution which does not include artificial variables. If this is possible, then the new basis will only include variables from the original problem and will represent a feasible point for the linear program. If the artificial variables cannot be driven to zero, then the constraints for the original problem are infeasible and the problem has no solution.

5.4.1   The Two-Phase Method

In the two-phase method the artificial variables are used to create an auxiliary linear program, called the *phase-1 problem*, whose only purpose is to determine a basic feasible solution for the original set of constraints. The objective function for the phase-1 problem is

$$\text{minimize } z' = \sum_i a_i,$$

where $\{a_i\}$ are the artificial variables. For our example $z' = a_1 + a_2$. The constraints for the phase-1 problem are the constraints of the original problem put in standard form, with artificial variables added as necessary. If the constraints for the original linear program are feasible, then the phase-1 problem will have optimal value $z'_* = 0$. If the original constraints are infeasible, then $z'_* > 0$.

We will illustrate this approach using the tableau. The tableau for the problem with artificial variables is

basic	x_1	x_2	x_3	x_4	a_1	a_2	rhs
$-z'$	0	0	0	0	1	1	0
a_1	3	2	0	0	1	0	14
a_2	2	-4	-1	0	0	1	2
x_4	4	3	0	1	0	0	19

The top-row entries for a_1 and a_2 are not zero, so z' is not expressed only in terms of the nonbasic variables. If we write the linear program in terms of the current basis, we obtain

basic	x_1	x_2	x_3	x_4	a_1	a_2	rhs
$-z'$	-5	2	1	0	0	0	-16
a_1	3	2	0	0	1	0	14
a_2	2	-4	-1	0	0	1	2
x_4	4	3	0	1	0	0	19

This transformation is necessary whenever the entries for the initial basic variables are not zero; it can be performed using the general formulas for the simplex method or by using elimination within the tableau. If we use the general formulas, the reduced costs for the nonbasic variables are

$$c_N^T - c_B^T B^{-1} N = (0 \quad 0 \quad 0) - (0 \quad 1 \quad 1) \begin{pmatrix} 0 & 0 & 1 \\ 1 & 0 & 0 \\ 0 & 1 & 0 \end{pmatrix} \begin{pmatrix} 3 & 2 & 0 \\ 2 & -4 & -1 \\ 4 & 3 & 0 \end{pmatrix}$$
$$= (-5 \quad 2 \quad 1).$$

The reduced costs for the basic variables are zero. Also, the objective value for the initial basis is obtained either via elimination or from the formula $-z = -c_B^T B^{-1} b$.

At the first iteration, the reduced cost for x_1 is negative so this basis is not optimal. The ratio test indicates that a_2 is the leaving variable. We would like to remove a_2 from the problem completely. The artificial variables were added to constraints where there was no obvious choice for a basic variable. In the current basis x_1 serves that function for the second constraint, and a_2 is no longer required. For this reason a_2 (or any other artificial variable that has left the basis) can be removed from the problem. The new basic solution is

		⇓				
basic	x_1	x_2	x_3	x_4	a_1	rhs
$-z'$	0	-8	$-\frac{3}{2}$	0	0	-11
a_1	0	8	$\frac{3}{2}$	0	1	11
x_1	1	-2	$-\frac{1}{2}$	0	0	1
x_4	0	$\boxed{11}$	2	1	0	15

At iteration 2 the reduced costs for x_2 and x_3 are negative, so this basis is not optimal. Since the coefficient of x_2 is larger in magnitude, x_2 will be selected as the entering variable. Then x_4 is the leaving variable. After pivoting we obtain

basic	x_1	x_2	$\Downarrow$ x_3	x_4	a_1	rhs
$-z'$	0	0	$-\frac{1}{22}$	$\frac{8}{11}$	0	$-\frac{1}{11}$
a_1	0	0	$\boxed{\frac{1}{22}}$	$-\frac{8}{11}$	1	$\frac{1}{11}$
x_1	1	0	$-\frac{3}{22}$	$\frac{2}{11}$	0	$\frac{41}{11}$
x_2	0	1	$\frac{2}{11}$	$\frac{1}{11}$	0	$\frac{15}{11}$

At iteration 3 the reduced cost for x_3 is negative so this basis is not optimal and x_3 is the entering variable. The ratio test shows that a_1 is the leaving variable. Pivoting (and removing the a_1 column because it is irrelevant) gives the new tableau:

basic	x_1	x_2	x_3	x_4	rhs
$-z'$	0	0	0	0	0
x_3	0	0	1	-16	2
x_1	1	0	0	-2	4
x_2	0	1	0	3	1

The current basis does not involve any artificial variables and the objective value is zero, so this is a feasible point for the constraints of the original problem.

The solution of the phase-1 problem only gives a basic feasible solution for the original problem; it is not optimal. It can be used as an initial basic feasible solution for the original problem with objective $z = 2x_1 + 3x_2$. This is called the *phase-2 problem*, with the following data:

basic	x_1	x_2	x_3	x_4	rhs
$-z$	2	3	0	0	0
x_3	0	0	1	-16	2
x_1	1	0	0	-2	4
x_2	0	1	0	3	1

If the simplex method is implemented without a tableau, then all that is necessary is to retain the final basis from the phase-1 problem as the initial basis of the phase-2 problem.

The reduced costs for the basic variables x_1 and x_2 are not zero, so the problem must be expressed in standard form before the simplex method can be used. If we represent the linear program in terms of the current basis, we obtain

basic	x_1	x_2	x_3	x_4 ⇓	rhs
$-z$	0	0	0	-5	-11
x_3	0	0	1	-16	2
x_1	1	0	0	-2	4
x_2	0	1	0	$\boxed{3}$	1

The reduced cost for x_4 is negative so this basis is not optimal. Only x_2 is a candidate for the ratio test, so it is the leaving variable. Pivoting gives

basic	x_1	x_2	x_3	x_4	rhs
$-z$	0	$\frac{5}{3}$	0	0	$-\frac{28}{3}$
x_3	0	$\frac{16}{3}$	1	0	$\frac{22}{3}$
x_1	1	$\frac{2}{3}$	0	0	$\frac{14}{3}$
x_4	0	$\frac{1}{3}$	0	1	$\frac{1}{3}$

This basis is optimal.

Much of the time, the two-phase method will work as indicated. The phase-1 problem will be set up and solved via the simplex method. If the constraints for the original problem have a basic feasible solution, at the end of phase 1 the artificial variables will all be nonbasic, and the final basis from phase 1 can be used as an initial basis for the original linear program. However, there are several exceptional cases that can arise at the end of phase 1, all associated with artificial variables remaining in the basis. We will discuss these using examples. These exceptional cases occur in an analogous way when a big-M approach is used; see the Exercises.

In the following examples, intermediate results of the simplex method are omitted so we can focus on the exceptional cases.

Example 5.6 (Infeasible Problem). Consider the linear program

$$\begin{array}{ll}
\text{minimize} & z = -x_1 \\
\text{subject to} & x_1 + x_2 \geq 6 \\
& 2x_1 + 3x_2 \leq 4 \\
& x_1, x_2 \geq 0.
\end{array}$$

An artificial variable will be used in the first constraint; the second constraint will have a slack variable and so will not need an artificial variable. The optimal phase-1 basic solution is

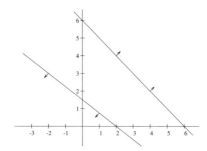

Figure 5.4. *Infeasible problem.*

basic	x_1	x_2	x_3	x_4	a_1	rhs
$-z'$	0	$\frac{1}{2}$	1	$\frac{1}{2}$	0	-4
a_1	0	$-\frac{1}{2}$	-1	$-\frac{1}{2}$	1	4
x_1	1	$\frac{3}{2}$	0	$\frac{1}{2}$	0	2

The objective function is nonzero, and the artificial variable is still in the basis with value $a_1 = 4 > 0$. There is no solution to the phase-1 problem that has $a_1 = 0$, indicating that there is no feasible solution to the constraints. This can be seen from Figure 5.4.   ∎

In the next example an artificial variable remains in the basis at the end of phase 1, but with value 0, and the phase-1 objective function also has value 0. In this case a basic feasible solution has been found, but additional update steps are required to remove the artificial variables from the basis before phase 2 can begin.

Example 5.7 (Removing Artificial Variables). Consider the linear program

$$\begin{aligned}
\text{minimize} \quad & z = x_1 + x_2 \\
\text{subject to} \quad & 2x_1 + x_2 + x_3 = 4 \\
& x_1 + x_2 + 2x_3 = 2 \\
& x_1, x_2, x_3 \geq 0.
\end{aligned}$$

If artificial variables are added to both equations, and x_1 replaces a_2 in the basis at the first iteration, then the optimal phase-1 basic solution is

basic	x_1	x_2	x_3	a_1	rhs
$-z'$	0	1	3	0	0
a_1	0	-1	-3	1	0
x_1	1	1	2	0	2

The value of the objective function is zero, and the point $(x_1, x_2, x_3)^T = (2, 0, 0)^T$ is a feasible point for the original set of constraints. However, the artificial variable a_1 is still in the basis so it is not possible to proceed with phase 2, since an appropriate basis for the original problem has not yet been found.

At the current point the values of the variables are

$$(x_1 \quad x_2 \quad x_3 \quad a_1)^T = (2 \quad 0 \quad 0 \quad 0)^T$$

and so there are three possible choices of basis that would lead to the same solution (assuming that the corresponding coefficient matrices are of full rank): $\{ x_1, x_2 \}$, $\{ x_1, x_3 \}$, and $\{ x_1, a_1 \}$. The last involves an artificial variable and is not of interest.

If $\{ x_1, x_2 \}$ is selected as the basis, then

basic	x_1	x_2	x_3	a_1	rhs
$-z'$	0	0	0	1	0
x_2	0	1	3	-1	0
x_1	1	0	-1	1	2

This basis is optimal for the phase-1 problem. If $\{ x_1, x_3 \}$ is selected, then

basic	x_1	x_2	x_3	a_1	rhs
$-z'$	0	0	0	1	0
x_3	0	$\frac{1}{3}$	1	$-\frac{1}{3}$	0
x_1	1	$\frac{1}{3}$	0	$\frac{2}{3}$	2

This basis is also optimal for the phase-1 problem. In both these cases it is now possible to proceed with phase 2.

For both of these choices we ignored the usual rules for the simplex method. The reduced costs for the entering variables were positive, and the rules for the ratio test were violated. It was only possible to do this because the artificial variable was zero. The purpose here was to find a feasible basis that did not include artificial variables, having found a solution to the phase-1 problem with objective value zero. We were only interested in changing the way the solution was represented, that is, in changing the basis to an equivalent one. ∎

There is a general rule for cases where the phase-1 problem has optimal value zero but the final basis includes artificial variables equal to zero. If the ith basic variable is a zero-valued artificial variable, then it can be replaced in the basis by any nonbasic variable x_j from the original linear program for which $\hat{a}_{i,j} \neq 0$.

The final example is a linear program with linearly dependent constraints. We have assumed up to now that such constraints would be removed. This example shows what can happen if they are not.

Example 5.8 (Linearly Dependent Constraints). Consider the linear program

$$\begin{aligned}
\text{minimize} \quad & z = x_1 + 2x_2 \\
\text{subject to} \quad & x_1 + x_2 = 2 \\
& 2x_1 + 2x_2 = 4 \\
& x_1, x_2 \geq 0.
\end{aligned}$$

The second constraint is twice the first constraint. If artificial variables are added, and x_1 replaces a_1 in the basis at the first iteration, we obtain the optimal phase-1 basis

basic	x_1	x_2	a_2	rhs
$-z'$	0	0	0	0
x_1	1	1	0	2
a_2	0	0	1	0

This basis is optimal with objective value zero, but the artificial variable a_2 is still in the basis. It is not possible to choose the basis $\{\, x_1, x_2 \,\}$ because the entry in column x_2 and row a_2 is zero, so it cannot be a pivot entry.

To resolve the difficulty we look more carefully at the last row of the tableau. It corresponds to the equation

$$a_2 = 0.$$

Since this equation has no influence on the original problem it can be removed (together with the a_2 column), leaving the reduced problem

basic	x_1	x_2	rhs
$-z'$	0	0	0
x_1	1	1	2

This basis can now be used to start phase 2. If the second constraint had been removed from the problem to begin with, we would have obtained the same result.   ∎

This is another case where the phase-1 problem has optimal value zero but the final basis includes artificial variables equal to zero. In general, if the ith basic variable is a zero-valued artificial variable and if $\hat{a}_{i,j} = 0$ for every variable x_j from the original linear program, then the constraints in the original program must have been linearly dependent.

On a computer it can be difficult to identify linear dependence. Small rounding errors will be introduced making it unlikely that any computed value will be exactly zero. It is then necessary to decide if a small number should be considered to be zero. A wrong decision can lead to a dramatic change in the solution to the linear program. This topic is discussed further in Chapter 7.

5.4.2   The Big-M Method

In the Big-M method penalty terms are added to the objective function that are designed to push artificial variables out of the basis. We will again use the example

$$
\begin{aligned}
\text{minimize} \quad & z = 2x_1 + 3x_2 \\
\text{subject to} \quad & 3x_1 + 2x_2 = 14 \\
& 2x_1 - 4x_2 \geq 2 \\
& 4x_1 + 3x_2 \leq 19 \\
& x_1, x_2 \geq 0
\end{aligned}
$$

to illustrate the method. As before, the problem is put in standard form and artificial variables are added. But in this case, instead of setting up an auxiliary phase-1 problem, the objective function will be changed to

$$\text{minimize } z' = 2x_1 + 3x_2 + Ma_1 + Ma_2,$$

where M is a symbol representing some large positive number. In general, there will be one penalty term for each artificial variable. For pencil-and-paper calculations M is left as a symbol and no specific value is given for it. In a computer calculation M would be set large enough to dominate all other numbers arising during the solution of the linear program.

If M is large, then any basis that includes a positive artificial variable will lead to a large positive value of the objective function z'. If there is any basic feasible solution to the constraints of the *original* linear program, then the corresponding basis will not include any artificial variables and its objective value will be much smaller. Because the artificial variables have a high cost associated with them, the simplex method will eventually remove them from the basis if this is at all possible. Any basic feasible solution to the penalized problem in which all the artificial variables are nonbasic (and hence zero) is also a basic feasible solution to the original problem. The corresponding basis can be used as an initial basis for the original problem.

The objective function in the phase-1 problem can be obtained as a limit of the objective function in the big-M method. The big-M method has objective function

$$z' = c^T x + M \sum_i a_i.$$

This is equivalent to using the objective

$$\hat{z} = M^{-1} c^T x + \sum_i a_i.$$

Taking the limit as $M \to \infty$ gives the phase-1 objective. As a consequence, the tableaus for the phase-1 problem will only differ from the big-M tableaus in the top row. For this reason we will go through the simplex method for the example more quickly than we did when examining the two-phase method.

In our example, the initial basis for the penalized problem with artificial variables gives

basic	x_1	x_2	x_3	x_4	a_1	a_2	rhs
$-z'$	2	3	0	0	M	M	0
a_1	3	2	0	0	1	0	14
a_2	2	-4	-1	0	0	1	2
x_4	4	3	0	1	0	0	19

As before, the reduced costs for the artificial variables are not zero and the problem must be written in terms of the current basis:

basic	x_1 ⇓	x_2	x_3	x_4	a_1	a_2	rhs
$-z'$	$-5M+2$	$2M+3$	M	0	0	0	$-16M$
a_1	3	2	0	0	1	0	14
a_2	$\boxed{2}$	-4	-1	0	0	1	2
x_4	4	3	0	1	0	0	19

At the first iteration, x_1 is the entering variable and a_2 is the leaving variable. As in the two-phase method, once an artificial variable leaves the basis it becomes irrelevant and can be removed from the problem. After pivoting (and removing a_2), we obtain the basic solution

basic	x_1	x_2 ⇓	x_3	x_4	a_1	rhs
$-z'$	0	$-8M+7$	$-\frac{3}{2}M+1$	0	0	$-11M-2$
a_1	0	8	$\frac{3}{2}$	0	1	11
x_1	1	-2	$-\frac{1}{2}$	0	0	1
x_4	0	$\boxed{11}$	2	1	0	15

At iteration 2, x_2 is the entering variable, and x_4 is the leaving variable. After pivoting we obtain

basic	x_1	x_2	x_3 ⇓	x_4	a_1	rhs
$-z'$	0	0	$-\frac{M+6}{22}$	$\frac{8M-7}{11}$	0	$-\frac{M+127}{11}$
a_1	0	0	$\boxed{\frac{1}{22}}$	$-\frac{8}{11}$	1	$\frac{1}{11}$
x_1	1	0	$-\frac{3}{22}$	$\frac{2}{11}$	0	$\frac{41}{11}$
x_2	0	1	$\frac{2}{11}$	$\frac{1}{11}$	0	$\frac{15}{11}$

At iteration 3, x_3 is the entering variable and a_1 is the leaving variable. After pivoting (and removing the a_1 column because it is irrelevant) we obtain the new basic solution:

basic	x_1	x_2	x_3	x_4 ⇓	rhs
$-z'$	0	0	0	-5	-11
x_3	0	0	1	-16	2
x_1	1	0	0	-2	4
x_2	0	1	0	$\boxed{3}$	1

The current basis does not involve any artificial variables, so this is a feasible point for the original problem. With the artificial variables gone, the objective function is now that of the original linear program.

At iteration 4 the reduced cost for x_4 is negative so this basis is not optimal. In the column for x_4, x_2 is the only possible exiting variable. Pivoting gives

basic	x_1	x_2	x_3	x_4	rhs
$-z$	0	$\frac{5}{3}$	0	0	$-\frac{28}{3}$
x_3	0	$\frac{16}{3}$	1	0	$\frac{22}{3}$
x_1	1	$\frac{2}{3}$	0	0	$\frac{14}{3}$
x_4	0	$\frac{1}{3}$	0	1	$\frac{1}{3}$

This basis is optimal. As expected, it is the same as the optimal basis obtained using the two-phase method.

In a software implementation it can be challenging to select an appropriate value for the penalty M. M must be large enough to dominate the other values in the problem, but if it is too large it can introduce serious computational errors through rounding. This topic is discussed further in Section 16.3.

Exercises

4.1. Use the simplex method (via a phase-1 problem) to find a basic feasible solution to the following system of linear inequalities:

$$2x_1 - 3x_2 + 2x_3 \geq 3$$
$$-x_1 + x_2 + x_3 \geq 5$$
$$x_1, x_2, x_3 \geq 0.$$

4.2. Solve the problem

$$\begin{aligned}
\text{minimize} \quad & z = -4x_1 - 2x_2 - 8x_3 \\
\text{subject to} \quad & 2x_1 - x_2 + 3x_3 \leq 30 \\
& x_1 + 2x_2 + 4x_3 = 40 \\
& x_1, x_2, x_3 \geq 0,
\end{aligned}$$

using (a) the two-phase method; (b) the big-M method.

4.3. Solve the problem

$$\begin{aligned}
\text{minimize} \quad & z = -4x_1 - 2x_2 \\
\text{subject to} \quad & 3x_1 - 2x_2 \geq 4 \\
& -2x_1 + x_2 = 2 \\
& x_1, x_2, \geq 0,
\end{aligned}$$

using (a) the two-phase method; (b) the big-M method.

4.4. Solve the following problem using the two-phase or big-M method:

$$\text{minimize} \quad z = 2x_1 - 2x_2 - x_3 - 2x_4 + 3x_5$$
$$\text{subject to} \quad - 2x_1 + x_2 - x_3 - x_4 = 1$$
$$x_1 - x_2 + 2x_3 + x_4 + x_5 = 4$$
$$-x_1 + x_2 - x_5 = 4$$
$$x_1, x_2, x_3, x_4, x_5 \geq 0.$$

4.5. Consider the phase-1 problem for a linear program with the constraints

$$x_1 \geq 5$$
$$x_2 \geq 1$$
$$x_1 + 2x_2 \geq 4$$
$$x_1, x_2 \geq 0.$$

Consider the following sequence of points $(x_1, x_2)^T$:

$$\begin{pmatrix} 0 \\ 0 \end{pmatrix}, \begin{pmatrix} 0 \\ 1 \end{pmatrix}, \begin{pmatrix} 2 \\ 1 \end{pmatrix}, \begin{pmatrix} 4 \\ 0 \end{pmatrix}, \begin{pmatrix} 5 \\ 0 \end{pmatrix}, \text{ and } \begin{pmatrix} 5 \\ 1 \end{pmatrix}.$$

Show that these points could correspond to successive basic feasible solutions if the simplex method were applied to the phase-1 problem. Hence show that it is possible for artificial variables to leave and then re-enter the basis if they are retained throughout the solution of the phase-1 problem.

4.6. Apply the big-M method to the linear programs in Examples 5.7, 5.8, and 5.9.

4.7. The following are the final phase-1 basic solutions for four different linear programming problems. In each problem a_1 and a_2 are the artificial variables for the two constraints, and the objective of each of the problems is

$$\text{minimize } z = x_1 + x_2 + x_3.$$

For each of the problems, determine whether the problem is feasible; and if it is feasible, find the initial basis for phase 2 and write the linear program in terms of that basis.

(i)

basic	x_1	x_2	x_3	a_1	a_2	rhs
$-z'$	0	0	0	1	1	0
	3	0	1	-1	2	0
	2	1	0	0	1	5

(ii)

basic	x_1	x_2	x_3	a_1	a_2	rhs
$-z'$	1	0	1	0	0	0
	3	1	0	0	1	2
	-1	0	-1	1	1	0

(iii)

basic	x_1	x_2	x_3	a_1	a_2	rhs
$-z'$	0	1	2	0	0	-1
	0	1	-2	-3	1	1
	1	3	4	1	0	3

(iv)

basic	x_1	x_2	x_3	a_1	a_2	rhs
$-z'$	0	0	0	3	0	0
	1	2	12	1	0	3
	0	0	0	-2	1	0

4.8. The following is the final basic solution for phase 1 in a linear programming problem, where a_1 and a_2 are the artificial variables for the two constraints:

basic	x_1	x_2	x_3	x_4	x_5	a_1	a_2	rhs
$-z'$	0	a	0	0	b	c	1	d
	-2	0	4	1	0	0	-2	1
	e	f	g	0	h	i	1	j

Find conditions on the parameters $a, b, c, d, e, f, g, h, i$, and j such that the following statements are true. You need not mention those parameters that can take on arbitrary positive or negative values. You should attempt to find the most general conditions possible.

(i) A basic feasible solution to the original problem has been found.
(ii) The problem is infeasible.
(iii) The problem is feasible but some artificial variables are still in the basis. However, by performing update operations a basic feasible solution to the original problem can be found.
(iv) The problem is feasible but it has a redundant constraint.
(v) For the case $a = 4$, $b = 1$, $c = 0$, $d = 0$, $e = 0$, $f = -4$, $g = 0$, $h = -1$, $i = 1$, and $j = 0$, determine whether the system is feasible. If so, find an initial basic solution for phase 2. Assume that the objective is to minimize $z = x_1 + x_4$.

4.9. Consider the linear programming problem

$$\begin{aligned}
\text{minimize} \quad & z = c^T x \\
\text{subject to} \quad & Ax = b \\
& x \geq 0.
\end{aligned}$$

Let $a_1, \ldots, a_m$ be the artificial variables, and suppose that at the end of phase 1 a basic feasible solution to the problem has been found (no artificial variables are in the basis). Prove that, in the final phase-1 basis, the reduced costs are zero for the original variables $x_1, \ldots, x_n$ and are one for the artificial variables.

4.10. Describe how you would use the big-M method to solve a maximization problem.

4.11. Consider the linear programming problem

$$\begin{aligned}
\text{minimize} \quad & z = c^T x \\
\text{subject to} \quad & Ax \geq b \\
& x \geq 0,
\end{aligned}$$

where $b \geq 0$. It is possible to use a single artificial variable to obtain an initial basic feasible solution to this problem. Let s be the vector of excess variables, $e = (1, \ldots, 1)^T$ and a be an artificial variable. Consider the phase-1 problem

$$\begin{aligned}
\text{minimize} \quad & z' = a \\
\text{subject to} \quad & Ax - s + ae = b \\
& x, s, a \geq 0.
\end{aligned}$$

(i) Assume for simplicity that $b_1 = \max \{ b_i \}$. Prove that $\{ a, s_2, s_3, \ldots, s_m \}$ is a feasible basis for the new problem.

(ii) Prove that if the original problem is feasible, then the phase-1 problem will have optimal objective value $z'_* = 0$, and if the original problem is infeasible it will have optimal objective value $z'_* > 0$.

5.5   Degeneracy and Termination

The version of the simplex method that we have described can fail, cycling endlessly without any improvement in the objective and without finding a solution. This can only happen on *degenerate* problems, problems where a basic variable is equal to zero in some basis.

On a degenerate problem, an iteration of the simplex method need not improve the value of the objective function. Suppose that at some iteration, x_t is the entering variable and x_s is the leaving variable. Then the formulas for the simplex method in Section 5.2 indicate that

$$\bar{x}_t = \frac{\hat{b}_s}{\hat{a}_{s,t}} \quad \text{and} \quad \bar{z} = \hat{z} + \hat{c}_t \bar{x}_t,$$

where $\bar{z}$ is the new objective value. On a degenerate problem it is possible that $\hat{b}_s = 0$ and $\bar{x}_t = 0$, the entering variable will have value 0 in the new basis (the same value it had as a nonbasic variable), and the objective value will not change ($\bar{z} = \hat{z}$).

Example 5.9 (Degeneracy). Consider the problem

$$\begin{aligned}
\text{minimize} \quad & z = -x_1 - x_2 \\
\text{subject to} \quad & x_1 \leq 2 \\
& x_1 + x_2 \leq 2 \\
& x_1, x_2 \geq 0.
\end{aligned}$$

The successive bases for this problem are

basic	x_1 ⇓	x_2	x_3	x_4	rhs
$-z$	-1	-1	0	0	0
x_3	$\boxed{1}$	0	1	0	2
x_4	1	1	0	1	2

basic	x_1	x_2 ⇓	x_3	x_4	rhs
$-z$	0	-1	1	0	2
x_1	1	0	1	0	2
x_4	0	$\boxed{1}$	-1	1	0

basic	x_1	x_2	x_3	x_4	rhs
$-z$	0	0	0	1	2
x_1	1	0	1	0	2
x_2	0	1	-1	1	0

The degeneracy arises because of the tie in the ratio test at the first iteration. At the second iteration, x_2 enters the basis but its new value is zero. As a result, the values of the variables and the objective function are unchanged. ∎

If the problem is not degenerate, then $\hat{b}_s > 0$ and so $\bar{x}_t > 0$ and $\bar{z} < \hat{z}$. This fact will be used to prove that, if the problem is not degenerate, then our version of the simplex method is guaranteed to terminate.

The "linear program" mentioned in the theorem might be a phase-1 problem or might include big-M terms for problems where an initial basic feasible solution is not available. In the case of a phase-1 problem (say), the "optimal basic feasible solution" would only be a solution to the phase-1 problem and, if the optimal objective value were positive, would indicate that the original problem were infeasible.

Theorem 5.10 (Finite Termination; Nondegenerate Case). *Suppose that the simplex method is applied to a linear program, and that at every iteration every basic variable is strictly positive. Then in a finite number of iterations the method either terminates at an optimal basic feasible solution or determines that the problem is unbounded.*

Proof. Consider an iteration of the simplex method. If all the reduced costs satisfy $\hat{c}_j \geq 0$, then the current basis is optimal and the method terminates. Otherwise, it is possible to choose an entering variable x_t with $\hat{c}_t < 0$. The ratio test for this variable computes

$$\min_{1 \leq i \leq m} \left\{ \frac{\hat{b}_i}{\hat{a}_{i,t}} : \hat{a}_{i,t} > 0 \right\}.$$

We have assumed that at every iteration every basic variable is strictly positive, so that $\hat{b}_i > 0$ for all i. If $\hat{a}_{i,t} \leq 0$ for all i, then there is no valid ratio to consider in the ratio test, and the problem is unbounded. Otherwise, the minimum ratio from the ratio test will be strictly positive; let its value be α. The new value of the entering variable will be $x_t = \alpha$ and the objective will change by $\alpha \hat{c}_t < 0$, so that the new value of the objective will be strictly less than the current value.

The value of the objective is completely determined by the choice of basis (the values of the basic variables are determined from the equality constraints, and the nonbasic variables are equal to zero). Since the objective is strictly decreased at every iteration, no basis can reoccur. Since there are only finitely many bases, the simplex method must terminate in a finite number of iterations. $\square$

Termination is not guaranteed for degenerate problems. Consider the linear program

$$\begin{aligned}
\text{minimize} \quad & z = -\tfrac{3}{4}x_1 + 150x_2 - \tfrac{1}{50}x_3 + 6x_4 \\
\text{subject to} \quad & \tfrac{1}{4}x_1 - 60x_2 - \tfrac{1}{25}x_3 + 9x_4 \leq 0 \\
& \tfrac{1}{2}x_1 - 90x_2 - \tfrac{1}{50}x_3 + 3x_4 \leq 0 \\
& x_3 \leq 1 \\
& x_1, x_2, x_3, x_4 \geq 0.
\end{aligned}$$

We will apply the simplex method to this problem, using the most negative reduced cost to select the entering variable, and breaking ties in the ratio test by selecting the first candidate row. If this is done, then the simplex method cycles—endlessly repeating the same sequence of bases with no improvement in the objective and without finding the optimal solution. It leads to the following sequence of basic solutions:

basic	x_1 ⇓	x_2	x_3	x_4	x_5	x_6	x_7	rhs
$-z$	$-\tfrac{3}{4}$	150	$-\tfrac{1}{50}$	6	0	0	0	0
x_5	$\boxed{\tfrac{1}{4}}$	-60	$-\tfrac{1}{25}$	9	1	0	0	0
x_6	$\tfrac{1}{2}$	-90	$-\tfrac{1}{50}$	3	0	1	0	0
x_7	0	0	1	0	0	0	1	1

basic	x_1	x_2 ⇓	x_3	x_4	x_5	x_6	x_7	rhs
$-z$	0	-30	$-\tfrac{7}{50}$	33	3	0	0	0
x_1	1	-240	$-\tfrac{4}{25}$	36	4	0	0	0
x_6	0	$\boxed{30}$	$\tfrac{3}{50}$	-15	-2	1	0	0
x_7	0	0	1	0	0	0	1	1

basic	x_1	x_2	$\Downarrow$ x_3	x_4	x_5	x_6	x_7	rhs
$-z$	0	0	$-\frac{2}{25}$	18	1	1	0	0
x_1	1	0	$\boxed{\frac{8}{25}}$	-84	-12	8	0	0
x_2	0	1	$\frac{1}{500}$	$-\frac{1}{2}$	$-\frac{1}{15}$	$\frac{1}{30}$	0	0
x_7	0	0	1	0	0	0	1	1

basic	x_1	x_2	x_3	$\Downarrow$ x_4	x_5	x_6	x_7	rhs
$-z$	$\frac{1}{4}$	0	0	-3	-2	3	0	0
x_3	$\frac{25}{8}$	0	1	$-\frac{525}{2}$	$-\frac{75}{2}$	25	0	0
x_2	$-\frac{1}{160}$	1	0	$\boxed{\frac{1}{40}}$	$\frac{1}{120}$	$-\frac{1}{60}$	0	0
x_7	$-\frac{25}{8}$	0	0	$\frac{525}{2}$	$\frac{75}{2}$	-25	1	1

basic	x_1	x_2	x_3	x_4	$\Downarrow$ x_5	x_6	x_7	rhs
$-z$	$-\frac{1}{2}$	120	0	0	-1	1	0	0
x_3	$-\frac{125}{2}$	10500	1	0	$\boxed{50}$	-150	0	0
x_4	$-\frac{1}{4}$	40	0	1	$\frac{1}{3}$	$-\frac{2}{3}$	0	0
x_7	$\frac{125}{2}$	-10500	0	0	-50	150	1	1

basic	x_1	x_2	x_3	x_4	x_5	$\Downarrow$ x_6	x_7	rhs
$-z$	$-\frac{7}{4}$	330	$\frac{1}{50}$	0	0	-2	0	0
x_5	$-\frac{5}{4}$	210	$\frac{1}{50}$	0	1	-3	0	0
x_4	$\frac{1}{6}$	-30	$-\frac{1}{150}$	1	0	$\boxed{\frac{1}{3}}$	0	0
x_7	0	0	1	0	0	0	1	1

basic	$\Downarrow$ x_1	x_2	x_3	x_4	x_5	x_6	x_7	rhs
$-z$	$-\frac{3}{4}$	150	$-\frac{1}{50}$	6	0	0	0	0
x_5	$\boxed{\frac{1}{4}}$	-60	$-\frac{1}{25}$	9	1	0	0	0
x_6	$\frac{1}{2}$	-90	$-\frac{1}{50}$	3	0	1	0	0
x_7	0	0	1	0	0	0	1	1

The final basis is the same as the initial basis, so the simplex method has made no progress and will continue to cycle through these six bases indefinitely.

A variety of techniques have been developed that guarantee termination of the simplex method even on degenerate problems. One of these, discovered by Bland (1977) and often referred to as *Bland's rule*, is described here. It is a rule for determining the entering and leaving variables, and it depends on an ordering of all the variables in the problem. Suppose that we have chosen the natural ordering: $x_1, x_2, \ldots$. Then at each iteration of the simplex method choose the entering variable as the first variable from this list for which the reduced cost is strictly negative. Then, among all the potential leaving variables that give the minimum ratio in the ratio test, choose the one that appears first in this list. Bland's rule determines how to break ties in the ratio test.

If Bland's rule is applied to this example, then the first few bases are the same. The first change occurs with the fifth basic solution

basic	x_1 ⇓	x_2	x_3	x_4	x_5	x_6	x_7	rhs
$-z$	$-\frac{1}{2}$	120	0	0	-1	1	0	0
x_3	$-\frac{125}{2}$	10500	1	0	50	-150	0	0
x_4	$-\frac{1}{4}$	40	0	1	$\frac{1}{3}$	$-\frac{2}{3}$	0	0
x_7	$\boxed{\frac{125}{2}}$	-10500	0	0	-50	150	1	1

The rest of the basic solutions are

basic	x_1	x_2	x_3	x_4	x_5 ⇓	x_6	x_7	rhs
$-z$	0	36	0	0	$-\frac{7}{5}$	$\frac{11}{5}$	$\frac{1}{125}$	$\frac{1}{125}$
x_3	0	0	1	0	0	0	1	1
x_4	0	-2	0	1	$\boxed{\frac{2}{15}}$	$-\frac{1}{15}$	$\frac{1}{250}$	$\frac{1}{250}$
x_1	1	-168	0	0	$-\frac{4}{5}$	$\frac{12}{5}$	$\frac{2}{125}$	$\frac{2}{125}$

basic	x_1	x_2	x_3	x_4	x_5	x_6	x_7	rhs
$-z$	0	15	0	$\frac{21}{2}$	0	$\frac{3}{2}$	$\frac{1}{20}$	$\frac{1}{20}$
x_3	0	0	1	0	0	0	1	1
x_4	0	-15	0	$\frac{15}{2}$	1	$-\frac{1}{2}$	$\frac{3}{100}$	$\frac{3}{100}$
x_1	1	-180	0	6	0	2	$\frac{2}{50}$	$\frac{2}{50}$

As hoped, with Bland's rule the simplex method terminates.

Bland's rule can be inefficient if applied at every simplex iteration since it may select entering variables that do not greatly improve the value of the objective function. To rectify this, Bland's rule need only be used at degenerate vertices where there is a danger of cycling.

At other iterations a more effective pivot rule should be used. An alternative is to use the perturbation method described in Section 5.5.1.

It can be shown that if the simplex method uses Bland's rule it will always terminate.

Theorem 5.11 (Termination with Bland's Rule). *If the simplex method is implemented using Bland's rule to select the entering and leaving variables, then the simplex method is guaranteed to terminate.*

Proof. See the paper by Bland (1977). $\square$

5.5.1 Resolving Degeneracy Using Perturbation

Another way to resolve degeneracy in the simplex method is to introduce small perturbations into the right-hand sides of the constraints. These perturbations remove the degeneracy, so the method makes progress at every iteration and hence is guaranteed to terminate. In some software packages explicit perturbations are introduced. However, in the technique that we describe here, the perturbations are merely symbolic. They are used to derive a pivot rule for the simplex method that prevents cycling. This approach is also referred to as the *lexicographic* method of resolving degeneracy. A related technique can be applied to network problems in a particularly efficient manner (see Section 8.5).

If the simplex method is applied to a degenerate problem, then it is possible that at some iteration the minimum ratio from the ratio test will be zero, and thus there is a risk of cycling. (Even if cycling does not occur, the simplex method may perform a long sequence of degenerate updates, a phenomenon known as *stalling*, and only make slow progress toward a solution.) Suppose that each basic variable were perturbed:

$$(x_B)_i \leftarrow (x_B)_i + \epsilon_i,$$

where $\{\,\epsilon_i\,\}$ is a set of small positive numbers. Then none of the perturbed basic variables would be zero, and the risk of cycling would be removed (at least at the current iteration). The method we will describe is a more elaborate version of this simple idea.

Consider a linear program where the constraints have been perturbed to

$$Ax = b + \epsilon,$$

where

$$\epsilon = (\,\epsilon_0 \quad \epsilon_0^2 \quad \cdots \quad \epsilon_0^m\,)^T$$

and $\epsilon_0 > 0$ is some "sufficiently small" positive number. (There will not be any need to specify ϵ_0; it will only need to be "small enough" for certain inequalities to hold.) The simplex method will be applied to this perturbed problem and, once the solution has been found, ϵ_0 will be set equal to zero to obtain the solution to the original problem.

Let x_B be some basic feasible solution to the perturbed problem corresponding to a basis matrix B, and denote the entries in B^{-1} by $(\beta_{i,j})$. Then $x_B = B^{-1}(b + \epsilon) = B^{-1}b + B^{-1}\epsilon$ and so

$$(x_B)_i = \hat{b}_i + \beta_{i,1}\epsilon_0 + \beta_{i,2}\epsilon_0^2 + \cdots + \beta_{i,m}\epsilon_0^m,$$

where $\hat{b}_i = (B^{-1}b)_i$.

We will say that $(x_B)_i$ is *lexicographically positive* if the first nonzero term in the above formula is positive. This is equivalent to saying that $(x_B)_i$ is positive for all sufficiently small ϵ_0. To see this, first consider the case where $\hat{b}_i > 0$. Then

$$(x_B)_i = \hat{b}_i + \epsilon_0(\beta_{i,1} + \beta_{i,2}\epsilon_0 + \cdots)$$

and so if ϵ_0 is small enough, $(x_B)_i > 0$. Now suppose that $\hat{b}_i = 0$, $\beta_{i,j} = 0$ for $j = 1, \ldots, k-1$, and $\beta_{i,k} > 0$. Then

$$(x_B)_i = \beta_{i,k}\epsilon_0^k + \beta_{i,k+1}\epsilon_0^{k+1} + \cdots + \beta_{i,m}\epsilon_0^m,$$

or

$$\frac{(x_B)_i}{\epsilon_0^k} = \beta_{i,k} + \epsilon_0(\beta_{i,k+1} + \beta_{i,k+2}\epsilon_0 + \cdots).$$

Once again, $(x_B)_i > 0$ for small enough ϵ_0.

Correspondingly, we will say that $(x_B)_j$ is lexicographically smaller than $(x_B)_i$ if $(x_B)_i - (x_B)_j$ is lexicographically positive. For sufficiently small ϵ_0, this is the same as $(x_B)_i > (x_B)_j$. This will be true if and only if the first nonzero term in the formula for $(x_B)_i - (x_B)_j$ is positive. It is possible to test these lexicographic conditions without specifying a value for ϵ_0.

Example 5.12 (Lexicographic Ordering). Let

$$B^{-1} = \begin{pmatrix} 1 & -2 & 2 \\ 1 & -2 & 3 \\ 0 & 3 & 4 \end{pmatrix} \quad \text{and} \quad \hat{b} = \begin{pmatrix} 0 \\ 0 \\ 1 \end{pmatrix}.$$

Then

$$(x_B)_1 = 0 + 1\epsilon_0 - 2\epsilon_0^2 + 2\epsilon_0^3$$
$$(x_B)_2 = 0 + 1\epsilon_0 - 2\epsilon_0^2 + 3\epsilon_0^3$$
$$(x_B)_3 = 1 + 0\epsilon_0 + 3\epsilon_0^2 + 4\epsilon_0^3.$$

All three components of x_B are lexicographically positive since the first nonzero term in each expression is nonzero. Also, $(x_B)_1$ is smaller than $(x_B)_2$ since

$$(x_B)_2 - (x_B)_1 = 0 + 0\epsilon_0 + 0\epsilon_0^2 + 1\epsilon_0^3,$$

and the first nonzero coefficient in this expression is positive. In addition, $(x_B)_2$ is lexicographically smaller than $(x_B)_3$.   ∎

For general ϵ_0 it is not possible for two components $(x_B)_i$ and $(x_B)_j$ to be lexicographically equal (i.e., all the terms in their formulas have identical coefficients). This would imply that

$$\beta_{i,k} = \beta_{j,k} \quad \text{for } k = 1, \ldots, m$$

and hence that B^{-1} had two identical rows. This is impossible since the rows of an invertible matrix must be linearly independent. It is this property that guarantees that the perturbed linear program will never have a degenerate basic feasible solution.

We will now prove that the simplex method applied to the perturbed problem is guaranteed to terminate. For simplicity, we will assume that the linear program has a complete set of slack variables. The application of the technique to linear programs in standard form will be considered in the Exercises.

Theorem 5.13. *Consider a linear program of the form*

$$\begin{aligned}\text{minimize} \quad & z = c^T x \\ \text{subject to} \quad & Ax \le b \\ & x \ge 0\end{aligned}$$

with $b \ge 0$. Assume that the constraints are perturbed to

$$Ax \le b + \epsilon,$$

where

$$\epsilon = (\epsilon_0 \quad \cdots \quad \epsilon_0^m)^T$$

and ϵ_0 is sufficiently small. Then the simplex method applied to the perturbed problem is guaranteed to terminate.

Proof. We will show by induction that the components of x_B are lexicographically positive at every iteration, and hence (by Theorem 5.10) the simplex method is guaranteed to terminate.

For the linear program with slack variables, at the first iteration we can select $B = I$ and

$$(x_B)_i = b_i + \epsilon_0^i.$$

Since $b_i \ge 0$, each component of the initial x_B is lexicographically positive.

At a general iteration with basis matrix B, assume that the components of x_B are lexicographically positive. If the current basis is not optimal, let x_t be the entering variable. The only way that the next basic feasible solution can be degenerate is if there is a tie in the minimum ratio test:

$$\frac{(x_B)_i}{\hat{a}_{i,t}} = \frac{(x_B)_j}{\hat{a}_{j,t}}.$$

The left-hand and right-hand sides of this equation would then be lexicographically equal, implying that rows i and j of B^{-1} were multiples of each other, and hence B^{-1} would not be invertible (which is impossible). Hence the ratio test must identify a unique leaving variable, say $(x_B)_s$.

We will now show that the new basic feasible solution is lexicographically positive. In the pivot row

$$(x_B)_s \leftarrow (x_B)_s / \hat{a}_{s,t},$$

where $\hat{a}_{s,t} > 0$, so $(x_B)_s$ remains lexicographically positive. In the other rows of the tableau

$$(x_B)_j \leftarrow (x_B)_j - \frac{\hat{a}_{j,t}}{\hat{a}_{s,t}}(x_B)_s.$$

If $\hat{a}_{j,t} \leq 0$, then this is the sum of a lexicographically positive term and a term that is either lexicographically positive or zero, so the result is lexicographically positive. If $\hat{a}_{j,t} > 0$, then the update can be rewritten as

$$(x_B)_j \leftarrow \hat{a}_{j,t} \left[\frac{(x_B)_j}{\hat{a}_{j,t}} - \frac{(x_B)_s}{\hat{a}_{s,t}} \right].$$

Since the right-hand side is the difference of two ratios from the ratio test, and $(x_B)_s$ produced the minimum ratio, the new value of $(x_B)_j$ is lexicographically positive. $\square$

The number ϵ_0 can be considered merely as a symbol. It need not be assigned a specific value. To determine whether a component $(x_B)_i$ is lexicographically positive, it is only necessary to know the coefficients $\{ \beta_{i,k} \}$, i.e., to know the coefficients in the corresponding row of B^{-1}. In fact, we only need to know the first nonzero coefficient in this set. Similarly, in the ratio test we only need to compare the leading terms in the formulas for $(x_B)_i$ and $(x_B)_j$ to determine the minimum ratio. For nondegenerate problems, the coefficients $\{ \beta_{i,k} \}$ would never have to be examined.

Exercises

5.1. Suppose that at the current iteration of the simplex method the basic feasible solution is degenerate. Is the objective value guaranteed to remain unchanged?

5.2. Consider the system of equations $Bx = b + \epsilon$ where

$$B^{-1} = \begin{pmatrix} 1 & 2 & 1 \\ 1 & 1 & 2 \\ 1 & 1 & 3 \end{pmatrix}, \quad b = \begin{pmatrix} 5 \\ 5 \\ 5 \end{pmatrix}, \quad \text{and} \quad \epsilon = \begin{pmatrix} \epsilon_0 \\ \epsilon_0^2 \\ \epsilon_0^3 \end{pmatrix}.$$

Sort the components of $x = B^{-1}(b + \epsilon)$ lexicographically.

5.3. Apply the perturbation method to the linear program from this section:

$$\begin{aligned} \text{minimize} \quad & z = -\tfrac{3}{4}x_1 + 150x_2 - \tfrac{1}{50}x_3 + 6x_4 \\ \text{subject to} \quad & \tfrac{1}{4}x_1 - 60x_2 - \tfrac{1}{25}x_3 + 9x_4 \leq 0 \\ & \tfrac{1}{2}x_1 - 90x_2 - \tfrac{1}{50}x_3 + 3x_4 \leq 0 \\ & x_3 \leq 1 \\ & x_1, x_2, x_3, x_4 \geq 0. \end{aligned}$$

5.4. When the simplex method was applied to the sample linear program (see the previous problem) and cycling occurred, ties in the ratio test were broken by choosing the first candidate variable. Does cycling occur in this example when the *last* candidate variable is chosen? (Be sure that you choose the first candidate entering variable in the optimality test, just as before.)

5.5. Show how to apply the perturbation technique to a linear program in standard form. (In the proof of Theorem 5.13, the linear program had a complete set of slack variables. This is not true in general.)

5.6. Consider a linear program in standard form with exactly two variables. Prove that cycling cannot occur.

5.7. Consider a linear program in standard form with exactly one equality constraint. Prove that cycling cannot occur.

5.6 Notes

The Simplex Method—From the 1940s to the present, George Dantzig's work on linear programming has been immensely influential. Dantzig's book (1963, reprinted 1998) contains a vast amount of relevant material. More recent reference works include the books by Chvátal (1983), Murty (1983), Schrijver (1986, reprinted 1998), and Bazaraa, Jarvis, and Sherali (1990). Early discussions of what would later be called linear programming can be found in the works of Kantorovich (1939) and von Neumann (1937).

The revised simplex method was first described by Dantzig (1953) and Orchard-Hays (1954). An extensive discussion can be found in the book by Dantzig (1963).

Degeneracy—The first example of cycling was constructed by Hoffman (1953). Our smaller example is due to Beale (1955). The perturbation method was described by Charnes (1952), and the lexicographic method by Dantzig, Orden, and Wolfe (1955). Bland's rule is, not surprisingly, found in a paper by Bland (1977).

Chapter 6

Duality and Sensitivity

6.1 The Dual Problem

For every linear programming problem there is a companion problem, called the "dual" linear program, in which the roles of variables and constraints are reversed. That is, for every variable in the original or "primal" linear program there is a constraint in the dual problem, and for every constraint in the primal there is a variable in the dual.

In an application, the variables in the primal problem might represent products, and the objective coefficients might represent the profits associated with manufacturing those products. Hence the objective in the primal indicates directly how an increase in production affects profit. The constraints in the primal problem might represent the availability of raw materials. An increase in the availability of raw materials might allow an increase in production, and hence an increase in profit, but this relationship is not as easy to deduce from the primal problem. One of the effects of duality theory is to make explicit the effect of changes in the constraints on the value of the objective. It is because of this interpretation that the variables in the dual problem are sometimes called "shadow prices," since they measure the implicit "costs" associated with the constraints.

Duality can also be used to develop efficient linear programming methods. For example, at the current time, the most successful interior-point software relies on a combination of primal and dual information.

While it is possible to define a dual to any linear program, the symmetry of the two problems is most obvious when the linear program is in *canonical form*. A minimization problem is in canonical form if all problem constraints are of the "≥" type, and all variables are nonnegative:

$$\text{minimize} \quad z = c^T x$$
$$\text{subject to} \quad Ax \geq b$$
$$x \geq 0.$$

We shall refer to this original problem as the *primal* linear program. The corresponding *dual* linear program will have the form

$$\text{maximize} \quad w = b^T y$$
$$\text{subject to} \quad A^T y \leq c$$
$$y \geq 0.$$

173

If the primal problem has n variables and m constraints, then the dual problem will have m variables (one dual variable for each primal constraint) and n constraints (one dual constraint for each primal variable). The coefficients in the objective of the primal are the coefficients on the right-hand side of the dual, and vice versa. The constraint matrix in the dual is the transpose of the matrix in the primal.

The dual problem is a maximization problem, where all constraints are of the "$\leq$" type, and all variables are nonnegative. This form is referred to as the *canonical form* for a maximization problem.

Example 6.1 (Canonical Dual Linear Program). Consider the primal problem, a linear program in canonical form

$$
\begin{aligned}
\text{minimize} \quad & z = 6x_1 + 2x_2 - x_3 + 2x_4 \\
\text{subject to} \quad & 4x_1 + 3x_2 - 2x_3 + 2x_4 \geq 10 \\
& 8x_1 + x_2 + 2x_3 + 4x_4 \geq 18 \\
& x_1, x_2, x_3, x_4 \geq 0.
\end{aligned}
$$

Then its dual is

$$
\begin{aligned}
\text{maximize} \quad & w = 10y_1 + 18y_2 \\
\text{subject to} \quad & 4y_1 + 8y_2 \leq 6 \\
& 3y_1 + y_2 \leq 2 \\
& -2y_1 + 2y_2 \leq -1 \\
& 2y_1 + 4y_2 \leq 2 \\
& y_1, y_2 \geq 0.
\end{aligned}
$$

Here y_1 is the dual variable corresponding to the first primal constraint and y_2 is the dual variable corresponding to the second primal constraint. The first dual constraint ($4y_1 - 8y_2 \leq 8$) corresponds to the primal variable x_1; similarly the second, third, and fourth constraints in the dual correspond to the primal variables x_2, x_3, and x_4, respectively. ∎

Any linear program can be transformed to an equivalent problem in canonical form. A "$\leq$" constraint can simply be multiplied by -1. An equality constraint can be written as two inequalities, since the equation $a = b$ is equivalent to the simultaneous inequalities $a \geq b$ and $-a \geq -b$. The requirement that all variables be nonnegative can be handled in the same way that conversion to standard form was handled (see Section 4.2). And a maximization problem can be converted to a minimization problem by multiplying the objective by -1.

The next lemma shows that the role of the primal and dual could be interchanged. It also indicates that the dual of a maximization problem in canonical form is a minimization problem in canonical form.

Lemma 6.2. *The dual of the dual linear program is the primal linear program.*

Proof. We need only consider a canonical minimization problem

$$
\begin{aligned}
\text{minimize} \quad & z = c^T x \\
\text{subject to} \quad & Ax \geq b \\
& x \geq 0,
\end{aligned}
$$

since any linear program can be transformed to this form. The dual program is

$$\text{maximize} \quad w = b^T y$$
$$\text{subject to} \quad A^T y \leq c$$
$$y \geq 0.$$

This is equivalent to the following minimization problem in canonical form:

$$\text{minimize} \quad w' = -b^T y$$
$$\text{subject to} \quad -A^T y \geq -c$$
$$y \geq 0.$$

The dual of this problem is

$$\text{maximize} \quad z' = -c^T x$$
$$\text{subject to} \quad -Ax \leq -b$$
$$x \geq 0.$$

This linear program is equivalent to the program

$$\text{minimize} \quad z = c^T x$$
$$\text{subject to} \quad Ax \geq b$$
$$x \geq 0,$$

which is the primal linear program. $\square$

Although it is possible to determine the dual of *any* linear program simply by converting it to canonical form, there are easy rules for obtaining the dual problem from the primal problem directly. These rules can be deduced by considering some general linear programs.

First consider a primal problem which has a mix of "$\geq$" constraints, "$\leq$" constraints, and "$=$" constraints:

$$\text{minimize} \quad z = c^T x$$
$$\text{subject to} \quad A_1 x \geq b_1$$
$$A_2 x \leq b_2$$
$$A_3 x = b_3$$
$$x \geq 0.$$

We can convert it to an equivalent problem in canonical form:

$$\text{minimize} \quad z = c^T x$$
$$\text{subject to} \quad A_1 x \geq b_1$$
$$-A_2 x \geq -b_2$$
$$A_3 x \geq b_3$$
$$-A_3 x \geq -b_3$$
$$x \geq 0.$$

If we define y_1, y_2', y_3', and y_3'' to be the vectors of dual variables corresponding to the four groups of constraints, then the dual problem is

$$\text{maximize} \quad w = b_1^T y_1 - b_2^T y_2' + b_3^T y_3' - b_3^T y_3''$$
$$\text{subject to} \quad A_1^T y_1 - A_2^T y_2' + A_3^T y_3' - A_3^T y_3'' \leq c$$
$$y_1 \geq 0, \ y_2' \geq 0, \ y_3' \geq 0, \ y_3'' \geq 0.$$

Defining $y_2 = -y_2'$ and $y_3 = y_3' - y_3''$, the dual problem can be rewritten in the form

$$\text{maximize} \quad w = b_1^T y_1 + b_2^T y_2 + b_3^T y_3$$

$$\text{subject to} \quad A_1^T y_1 + A_2^T y_2 + A_3^T y_3 \leq c$$

$$y_1 \geq 0,\ y_2 \leq 0,\ y_3 \text{ unrestricted}.$$

Notice that the directions of the constraints in the original primal are not in canonical form. Likewise, the signs of the variables in the final dual are not in canonical form. Let us examine these anomalies. The dual variables associated with primal "$\geq$" constraints are nonnegative, but the dual variables associated with "$\leq$" constraints are nonpositive, and the dual variables associated with the "$=$" constraints are unrestricted. This could be restated as follows: If the direction of a primal constraint is consistent with canonical form, the corresponding dual variable is nonnegative; if the direction of the constraint is reversed with respect to canonical form, the corresponding dual variable is nonpositive; and if the constraint is an equality, the corresponding dual variable is unrestricted. This is a general rule. It also applies to maximization problems which have a mix of "$\geq$" constraints, "$\leq$" constraints, and "$=$" constraints (see the Exercises). The direction of a constraint in a problem will be "consistent with respect to canonical form" if it is of the "$\geq$" type in a minimization problem, or if it is of the "$\leq$" type in a maximization problem.

Now consider a primal linear program, which has a mix of nonnegative, nonpositive, and unrestricted variables:

$$\text{minimize} \quad z = c_1^T x_1 + c_2^T x_2 + c_3^T x_3$$

$$\text{subject to} \quad A_1 x_1 + A_2 x_2 + A_3 x_3 \geq b$$

$$x_1 \geq 0,\ x_2 \leq 0,\ x_3 \text{ unrestricted}.$$

If we put this problem in canonical form, and then simplify the dual problem, we obtain

$$\text{maximize} \quad w = b^T y$$

$$\text{subject to} \quad A_1^T y \leq c_1$$

$$A_2^T y \geq c_2$$

$$A_3^T y = c_3$$

$$y \geq 0.$$

Here the signs of the variables in the primal are not in canonical form, and neither are the directions of the constraints in the dual. If a primal variable is nonnegative, the direction of the corresponding dual constraint will be consistent with (the dual's) canonical form; if it is nonpositive, the direction of the dual constraint will be reversed with respect to canonical form; and if the variable is unrestricted, the corresponding constraint will be an equality. This is a general rule, both for minimization and maximization problems. Notice that it is symmetric (or "dual") to the rule that we obtained earlier.

We can summarize the relationship between the constraints and variables in the primal and dual problems as follows:

primal/dual constraint   **dual/primal variable**

consistent with canonical form $\Longleftrightarrow$ variable ≥ 0

reversed from canonical form $\Longleftrightarrow$ variable ≤ 0

equality constraint $\Longleftrightarrow$ variable unrestricted

Example 6.3 (General Dual Linear Problem). Consider the primal problem

$$\begin{aligned}
\text{maximize} \quad & z = 6x_1 + x_2 + x_3 \\
\text{subject to} \quad & 4x_1 + 3x_2 - 2x_3 = 1 \\
& 6x_1 - 2x_2 + 9x_3 \geq 9 \\
& 2x_1 + 3x_2 + 8x_3 \leq 5 \\
& x_1 \geq 0, x_2 \leq 0, \ x_3 \text{ unrestricted.}
\end{aligned}$$

Then its dual is

$$\begin{aligned}
\text{minimize} \quad & w = y_1 + 9y_2 + 5y_3 \\
\text{subject to} \quad & 4y_1 + 6y_2 + 2y_3 \geq 6 \\
& 3y_1 - 2y_2 + 3y_3 \leq 1 \\
& -2y_1 + 9y_2 + 8y_3 = 1 \\
& y_1 \text{ unrestricted}, y_2 \leq 0, y_3 \geq 0.
\end{aligned}$$

The primal problem is a maximization problem. Its first constraint is an equality, and its second constraint and third constraint are, respectively, reversed and consistent with respect to the canonical form of a maximization problem. For this reason y_1 is unrestricted, $y_2 \leq 0$, and $y_3 \geq 0$. Now the dual problem is a minimization problem. Because $x_1 \geq 0$ and $x_2 \leq 0$, the first and second dual constraints are, respectively, consistent and reversed with respect to the canonical form of a minimization problem. Because x_3 is unrestricted, the third dual constraint is an equality.

It is easy to verify that the dual of the dual is the primal.   ■

In the following sections it will be useful to consider the dual of a problem in standard form. If the primal problem is

$$\begin{aligned}
\text{minimize} \quad & z = c^T x \\
\text{subject to} \quad & Ax = b \\
& x \geq 0,
\end{aligned}$$

then its dual is

$$\begin{aligned}
\text{maximize} \quad & w = b^T y \\
\text{subject to} \quad & A^T y \leq c.
\end{aligned}$$

The dual variables y are unrestricted.

The concept of a dual problem applies not only to linear programs, but also to a wide range of problems from a wide variety of fields such as engineering, physics, and mathematics. For example it is also possible to define a dual problem for nonlinear optimization problems (see Chapter 14). There, the dual variables are often called Lagrange multipliers.

Exercises

1.1. Find the dual of

$$\begin{aligned}
\text{minimize} \quad & z = 3x_1 - 9x_2 + 5x_3 - 6x_4 \\
\text{subject to} \quad & 4x_1 + 3x_2 + 5x_3 + 8x_4 \geq 24 \\
& 2x_1 - 7x_2 - 4x_3 - 6x_4 \geq 17 \\
& x_1, x_2, x_3, x_4 \geq 0.
\end{aligned}$$

1.2. Find the dual of

$$\text{minimize} \quad z = -2x_1 + 4x_2 - 3x_3$$
$$\text{subject to} \quad 9x_1 - 2x_2 - 8x_3 = 5$$
$$3x_1 + 3x_2 + 3x_3 = 7$$
$$7x_1 - 5x_2 + 2x_3 = 9$$
$$x_1, x_2, x_3 \geq 0.$$

1.3. Find the dual of

$$\text{maximize} \quad z = 6x_1 - 3x_2 - 2x_3 + 5x_4$$
$$\text{subject to} \quad 4x_1 + 3x_2 - 8x_3 + 7x_4 = 11$$
$$3x_1 + 2x_2 + 7x_3 + 6x_4 \geq 23$$
$$7x_1 + 4x_2 + 3x_3 + 2x_4 \leq 12$$
$$x_1, x_2 \geq 0, x_3 \leq 0$$
$$(x_4 \text{ unrestricted}).$$

Verify that the dual of the dual is the primal.

1.4. Obtain the dual to the problem

$$\text{minimize} \quad z = c_1^T x_1 + c_2^T x_2 + c_3^T x_3$$
$$\text{subject to} \quad A_1 x_1 + A_2 x_2 + A_3 x_3 \geq b$$
$$x_1 \geq 0, x_2 \leq 0, x_3 \text{ unrestricted}$$

by converting the problem to canonical form, finding its dual, and then simplifying the result.

1.5. Find the dual to the problem

$$\text{minimize} \quad z = c^T x$$
$$\text{subject to} \quad Ax = b$$
$$l \leq x \leq u,$$

where l and u are vectors of lower and upper bounds on x.

1.6. Find the dual to the problem

$$\text{minimize} \quad z = c^T x$$
$$\text{subject to} \quad b_1 \leq Ax \leq b_2$$
$$x \geq 0.$$

1.7. Can you find a linear program which is its own dual? (We will say that the two problems are the same if one can be obtained from the other merely by multiplying the objective, any of the constraints, or any of the variables by -1.)

1.8. Write a computer program that, when given a linear program (not necessarily in standard form), will generate the dual linear program automatically.

1.9. If you have linear programming software available, experiment with the properties of a pair of primal and dual linear programs. What is the relationship between their optimal values? Change the coefficients in the objective or on the right-hand side of the constraints and observe what happens to the optimal values of the linear programs. Examine the relationship between the ith variable in one problem and the ith constraint in the other.

6.2 Duality Theory

There are two major results relating the primal and dual problems. The first, called "weak" duality, is easier to prove. It states that primal objective values provide bounds for dual objective values, and vice versa. This weak duality property can be extended to nonlinear optimization problems and other more general settings. The second, called "strong" duality, states that the optimal values of the primal and dual problems are equal, provided that they exist. For nonlinear problems there may not be a strong duality result.

In the theoretical results below we work with primal linear programs in *standard form*. In Section 4.2 it was shown that every linear program can be converted to standard form. Hence working with problems in standard form is primarily a matter of convenience. It makes it unnecessary to examine a great many different cases corresponding to linear programs in a variety of forms. We begin with a simple theorem.

Theorem 6.4 (Weak Duality). *Let x be a feasible point for the primal problem in standard form, and let y be a feasible point for the dual problem. Then*

$$z = c^T x \geq b^T y = w.$$

lower bound for $c^T x$

upper bound for $b^T y$

Proof. The constraints for the dual show that $c^T \geq y^T A$. Since $x \geq 0$,

$$z = c^T x \geq y^T A x = y^T b = b^T y = w. \quad \square$$

We have stated and proved the weak duality theorem in the case where the primal problem is a minimization problem. For a primal problem in general form, the weak duality result would say that the objective value corresponding to a feasible point for the maximization problem would always be less than or equal to the objective value corresponding to a feasible point for the minimization problem.

Example 6.5 (Weak Duality). Consider the primal and dual linear programs in Example 6.1. It is easy to check that the point $x = (4, 0, 0, 0)^T$ is feasible for the primal and that the point $y = (\frac{1}{2}, 0)^T$ is feasible for the dual. At these points

$$z = c^T x = 24 > 5 = b^T y = w,$$

so that the weak duality theorem is satisfied. ∎

There are several simple consequences of the weak duality theorem. For proofs, see the Exercises.

Corollary 6.6. *If the primal is unbounded, then the dual is infeasible. If the dual is unbounded, then the primal is infeasible.*

Corollary 6.7. *If x is a feasible solution to the primal, y is a feasible solution to the dual, and $c^T x = b^T y$, then x and y are optimal for their respective problems.*

Corollary 6.7 is used in the proof of strong duality. It shows that it is possible to check if the points x and y are optimal without solving the corresponding linear programs. By the way, it is possible for both the primal and dual problems to be infeasible.

Example 6.8 (Primal/Dual Relationships). First consider the primal problem

$$\begin{aligned} \text{maximize} \quad & z = x_1 + x_2 \\ \text{subject to} \quad & x_1 - x_2 \leq 1 \\ & x_1, x_2 \geq 0, \end{aligned}$$

and its dual problem

$$\begin{aligned} \text{minimize} \quad & w = y_1 \\ \text{subject to} \quad & y_1 \geq 1 \\ & -y_1 \geq 1 \\ & y_1 \geq 0. \end{aligned}$$

Here, the primal problem is unbounded, and the dual is infeasible.

Next consider the infeasible problem

$$\begin{aligned} \text{maximize} \quad & z = 2x_1 - x_2 \\ \text{subject to} \quad & x_1 + x_2 \geq 1 \\ & -x_1 - x_2 \geq 1. \end{aligned}$$

In general, the dual of an infeasible problem could be either infeasible or unbounded. Here the dual problem is

$$\begin{aligned} \text{minimize} \quad & z = y_1 + y_2 \\ \text{subject to} \quad & y_1 - y_2 = 2 \\ & y_1 - y_2 = -1, \end{aligned}$$

which is infeasible.   ∎

Theorem 6.9 (Strong Duality). *Consider a pair of primal and dual linear programs. If one of the problems has an optimal solution then so does the other, and the optimal objective values are equal.*

Proof. For convenience, we can assume that (a) the primal problem has an optimal solution (since the roles of primal and dual could be interchanged), (b) the primal problem is in standard form, and (c) x_*, the solution to the primal, is an optimal basic feasible solution. By reordering the variables we can write x_* in terms of basic and nonbasic variables:

$$x_* = \begin{pmatrix} x_B \\ x_N \end{pmatrix}$$

and correspondingly we write

$$A = (B \quad N) \quad \text{and} \quad c = \begin{pmatrix} c_B \\ c_N \end{pmatrix}.$$

Then $x_B = B^{-1}b$. If x_* is optimal, the reduced costs satisfy $c_N^T - c_B^T B^{-1} N \geq 0$ or

$$c_B^T B^{-1} N \leq c_N^T.$$

Let y_* be the vector of simplex multipliers corresponding to this basic feasible solution: $y_* = B^{-T}c_B$ or

$$y_*^T = c_B^T B^{-1}.$$

We will show that y_* is feasible for the dual and that $b^T y_* = c^T x_*$. Then Corollary 6.7 will show that y_* is optimal for the dual. We first check feasibility:

$$y_*^T A = c_B^T B^{-1} (B \quad N)$$
$$= (c_B^T \quad c_B^T B^{-1}N) \le (c_B^T \quad c_N^T) = c^T;$$

hence $A^T y_* \le c$ and y satisfies the dual constraints. We now compute the objective values for the primal and the dual:

$$z = c^T x = c_B^T x_B = c_B^T B^{-1} b$$
$$w = b^T y = y^T b = c_B^T B^{-1} b = z.$$

So y_* is feasible for the dual and has dual value equal to the optimal primal value. Hence by Corollary 6.7, y_* is optimal for the dual. □

The proof of the strong duality theorem provides the optimal dual solution. If we write x_* in terms of basic and nonbasic variables

$$x_* = \begin{pmatrix} x_B \\ x_N \end{pmatrix}$$

and write

$$A = (B \quad N) \quad \text{and} \quad c = \begin{pmatrix} c_B \\ c_N \end{pmatrix},$$

then the optimal values of the dual variables are given by the corresponding vector of simplex multipliers

$$y_* = B^{-T}c_B.$$

It also follows from the proof that at any iteration, if y is the vector of simplex multipliers, then the vector of reduced costs is

$$\hat{c} = c - A^T y.$$

Thus the reduced costs are the dual slack variables. If they are all nonnegative, then y is dual feasible and the solution is optimal. (In such cases, the basis is said to be *dual feasible*.) At any intermediate step the reduced costs are not all nonnegative and the vector of simplex multipliers is dual infeasible. Thus the simplex method generates a sequence of primal feasible solutions x and dual infeasible solutions y with $c^T x = b^T y$, terminating when y is dual feasible.

If the original linear program has a complete set of slack variables, then the reduced costs for the slack variables are given by

$$c_N^T - c_B^T B^{-1} N = 0^T - c_B^T B^{-1} I = -(B^{-T}c_B)^T = -y_*^T,$$

because the objective coefficients (c_N^T) for the slack variables are zero, and their constraint coefficients (N) are given by I. In this case the values of the optimal dual variables are the same as the reduced costs of the slack variables (except for the sign). This is also true when there are excess or artificial variables; see the Exercises.

Example 6.10 (Linear Program with Slack Variables). Consider the example from Section 5.2:

$$
\begin{aligned}
\text{minimize} \quad & z = -x_1 - 2x_2 \\
\text{subject to} \quad & -2x_1 + x_2 \leq 2 \\
& -x_1 + 2x_2 \leq 7 \\
& x_1 \leq 3 \\
& x_1, x_2 \geq 0.
\end{aligned}
$$

The optimal basic solution is

basic	x_1	x_2	x_3	x_4	x_5	rhs
$-z$	0	0	0	1	2	13
x_2	0	1	0	$\frac{1}{2}$	$\frac{1}{2}$	5
x_1	1	0	0	0	1	3
x_3	0	0	1	$-\frac{1}{2}$	$\frac{3}{2}$	3

The optimal dual variables are

$$
y_*^T = c_B^T B^{-1} = (-2 \quad -1 \quad 0) \begin{pmatrix} 0 & \frac{1}{2} & \frac{1}{2} \\ 0 & 0 & 1 \\ 1 & -\frac{1}{2} & \frac{3}{2} \end{pmatrix} = (0 \quad -1 \quad -2).
$$

These are the negatives of the reduced costs corresponding to the slack variables. The dual objective function is

$$
\text{maximize} \ w = 2y_1 + 7y_2 + 3y_3
$$

and so $w_* = 2(0) + 7(-1) + 3(-2) = -13 = z_*$, as expected. It is straightforward to verify that the constraints of the dual problem are satisfied. ∎

6.2.1 Complementary Slackness

We discuss here a further relationship between a pair of primal and dual problems that have optimal solutions. There is an interdependence between the nonnegativity constraints in the primal ($x \geq 0$) and the constraints in the dual ($A^T y \leq c$). At optimal solutions to both problems it is not possible to have both $x_j > 0$ and $(A^T y)_j < c_j$. At least one of these constraints must be binding: either x_j is zero or the jth dual slack variable is zero. This property, called *complementary slackness*, can be summarized in the equation

$$
x^T(c - A^T y) = 0.
$$

This equation is the same as $\sum_j x_j(c - A^T y)_j = 0$. Since the primal and dual constraints ensure that each of the terms in the summation must be nonnegative, if the entire sum is zero, then every term is zero. The complementary slackness property is established in the following theorem. The theorem states that complementary slackness will hold between *any* pair of optimal primal and optimal dual solutions; these solutions need not correspond to a basis.

Theorem 6.11 (Complementary Slackness). *Consider a pair of primal and dual linear programs, with the primal problem in standard form. If x is optimal for the primal and y is optimal for the dual, then $x^T(c - A^T y) = 0$. If x is feasible for the primal, y is feasible for the dual, and $x^T(c - A^T y) = 0$, then x and y are optimal for their respective problems.*

Proof. As in the proof of weak duality (Theorem 6.4), if x and y are feasible, then

$$z = c^T x \geq y^T A x = y^T b = w.$$

If x and y are optimal, then $w = z$ so that $c^T x = y^T A x = x^T A^T y$. Rearranging this final formula gives the first result.

If $x^T(c - A^T y) = 0$, then $z = w$ and Corollary 6.7 shows that x and y are then optimal. $\square$

Example 6.12 (Complementary Slackness). We look again at the linear program

$$\begin{array}{ll}
\text{minimize} & z = -x_1 - 2x_2 \\
\text{subject to} & -2x_1 + x_2 + x_3 = 2 \\
& -x_1 + 2x_2 + x_4 = 7 \\
& x_1 + x_5 = 3 \\
& x_1, x_2, x_3, x_4, x_5 \geq 0.
\end{array}$$

The optimal solutions are $x = (x_1, x_2, x_3, x_4, x_5)^T = (3, 5, 3, 0, 0)^T$ and $y = (y_1, y_2, y_3)^T = (0, -1, -2)^T$. The dual constraints are

$$\begin{array}{r}
-2y_1 - y_2 + y_3 \leq -1 \\
y_1 + 2y_2 \leq -2 \\
y_1 \leq 0 \\
y_2 \leq 0 \\
y_3 \leq 0.
\end{array}$$

In the primal the last two nonnegativity constraints are binding ($x_4 \geq 0$ and $x_5 \geq 0$). In the dual the first three constraints are binding. So the complementary slackness condition is satisfied. ■

It is possible to have both $x_j = 0$ and $c_j - (A^T y)_j = 0$, for example, when the problem is degenerate and one of the basic variables is zero. If this does *not* happen, that is, if exactly one of these two quantities is zero for all j, then the problem is said to satisfy a *strict* complementary slackness condition. If a linear programming problem has an optimal solution, then there always exists a strictly complementary optimal pair of solutions to the primal and the dual problems. This pair of solutions need not be basic solutions, however. (See the Exercises.)

In the simplex method the complementary slackness conditions hold between *any* basic feasible solution and its associated vector of simplex multipliers: If $x_j > 0$, then x_j is a basic variable and its reduced cost (or dual slack variable) is zero. Conversely, if a dual slack variable (reduced cost) is nonzero, then the associated primal variable is nonbasic and hence zero. Thus the simplex method maintains primal feasibility and complementary slackness and strives to achieve dual feasibility.

If a linear program is not in standard form, then a complementary slackness condition holds between any restricted (nonnegative or nonpositive) variable and its corresponding dual constraint, as well as between any inequality constraint and its associated dual variable. Thus for the pair of primal and dual canonical linear programs

$$\begin{array}{llll} \text{minimize} & z = c^T x & \text{maximize} & w = b^T y \\ \text{subject to} & Ax \geq b & \text{subject to} & A^T y \leq c \\ & x \geq 0 & & y \geq 0 \end{array}$$

the complementary slackness conditions are

$$x^T(c - A^T y) = 0 \quad \text{and} \quad y^T(Ax - b) = 0.$$

(See the Exercises.)

6.2.2 Interpretation of the Dual

The dual linear program can be used to gain practical insight into the properties of a model. We will examine this idea via an example. Although the exact interpretation of the dual will vary from application to application, the approach we use (looking at the optimal values of the dual variables, as well as the dual problem as a whole) is general.

Let us consider a baker who makes and sells two types of cakes, one simple and one elaborate. Both cakes require basic ingredients (flour, sugar, eggs, and so forth), as well as fancier ingredients such as nuts and fruit for decoration and flavor, with the elaborate cake using more of the fancier ingredients. There are also greater labor costs associated with the elaborate cake. The baker would like to maximize profit.

A linear programming model for this situation might be

$$\begin{array}{ll} \text{maximize} & z = 24x_1 + 14x_2 \\ \text{subject to} & 3x_1 + 2x_2 \leq 120 \\ & 4x_1 + x_2 \leq 100 \\ & 2x_1 + x_2 \leq 70 \\ & x_1, x_2 \geq 0. \end{array}$$

Here x_1 and x_2 represent the number of batches of the elaborate and simple cakes produced per day. The objective records the profit. The first constraint represents the daily limits on the availability of basic ingredients (in pounds), where a batch of the elaborate cake requires 3 pounds, and a batch of the simple cake requires 2 pounds. The second constraint similarly records the limits on fancier ingredients. The third constraint records the limits on labor (measured in hours) where a batch of the elaborate cakes uses 2 hours of labor, and a batch of the simple cakes uses 1 hour of labor.

The dual linear program is

$$\begin{array}{ll} \text{minimize} & w = 120y_1 + 100y_2 + 70y_3 \\ \text{subject to} & 3y_1 + 4y_2 + 2y_3 \geq 24 \\ & 2y_1 + y_2 + y_3 \geq 14 \\ & y_1, y_2, y_3 \geq 0. \end{array}$$

The optimal solution to the primal problem is $z = 888$, $x_1 = 16$, and $x_2 = 36$. The optimal solution to the dual problem is $w = 888$, $y_1 = 6.4$, $y_2 = 1.2$, and $y_3 = 0$. Note that the two objective values are equal, and that the complementary slackness conditions are satisfied.

In this problem the limiting factors are the availability of basic and fancy ingredients. (There are 2 hours of excess labor available; the bakery might employ one of the bakers part time or give additional tasks to this baker.) The baker might be able to purchase additional quantities of these ingredients. How much should the baker be willing to pay? Since the optimal primal and dual objective values are equal, and the dual objective is

$$w = 120y_1 + 100y_2 + 70y_3,$$

each extra pound of basic ingredients will be worth $y_1 = 6.4$ dollars in profit, and each extra pound of fancy ingredients will be worth $y_2 = 1.2$ dollars. Hence the dual variables determine the marginal values of these raw materials. Additional labor is of no value to the baker ($y_3 = 0$) because excess labor is already available. (There are limits to this argument, however; if too many cakes are made the 2 excess hours of labor will be used up, and additional analysis of the model will be required.)

There is an additional interpretation of the dual problem. Suppose that some other company would like to take over the baker's business. What price should be offered? A price could be determined by setting values on the baker's assets (plain ingredients, fancy ingredients, and labor); call these values y_1, y_2, and y_3. The other company would like to minimize the amount paid to the baker:

$$\text{minimize} \quad w = 120y_1 + 100y_2 + 70y_3.$$

These values would be fair to the baker if they represented a profit at least as good as could be obtained by producing cakes, that is, if

$$3y_1 + 4y_2 + 2y_3 \geq 24$$
$$2y_1 + y_2 + y_3 \geq 14$$

These are the objective and constraints for the dual problem. Thus the dual problem allows us to determine the daily value of the baker's business.

Another interpretation of the dual problem arises in game theory. This is discussed in Section 14.8.

Exercises

2.1. Consider the linear program

$$\begin{aligned}
\text{maximize} \quad & z = -x_1 - x_2 \\
\text{subject to} \quad & -x_1 + x_2 \geq 1 \\
& 2x_1 - x_2 \leq 2 \\
& x_1, x_2 \geq 0.
\end{aligned}$$

Find the dual to the problem. Solve the primal and the dual graphically, and verify that the results of the strong duality theorem hold. Verify that the optimal dual solution satisfies $y^T = c_B^T B^{-1}$ where B is the optimal basis matrix.

2.2. Prove that if both the primal and the dual problems have feasible solutions, then both have optimal solutions, and the optimal objective values of the two problems are equal.

2.3. Prove Corollary 6.6.

2.4. Prove Corollary 6.7.

2.5. Prove that if an excess variable has been included in the ith constraint, then the optimal reduced cost for this variable is the ith optimal dual variable y_i.

2.6. Prove that if an artificial variable has been added to the ith constraint within a big-M approach, then the optimal reduced cost for this variable is $y_i - M$, where y_i is the ith optimal dual variable.

2.7. Consider a linear program with a single constraint

$$\begin{aligned} \text{minimize} \quad & z = c_1 x_1 + c_2 x_2 + \cdots + c_n x_n \\ \text{subject to} \quad & a_1 x_1 + a_2 x_2 + \cdots + a_n x_n \leq b \\ & x_1, x_2, \ldots, x_n \geq 0. \end{aligned}$$

Using duality develop a simple rule to determine an optimal solution, if the latter exists.

2.8. Using duality theory find the solution to the following linear program:

$$\begin{aligned} \text{minimize} \quad & z = x_1 + 2x_2 + \cdots + n x_n \\ \text{subject to} \quad & x_1 \geq 1 \\ & x_1 + x_2 \geq 2 \\ & \qquad \vdots \\ & x_1 + x_2 + x_3 + \cdots + x_n \geq n \\ & x_1, x_2, x_3, \ldots, x_n \geq 0. \end{aligned}$$

2.9. Consider the primal linear programming problem

$$\begin{aligned} \text{minimize} \quad & z = c^T x \\ \text{subject to} \quad & Ax = b \\ & x \geq 0. \end{aligned}$$

Assume that this problem and its dual are both feasible. Let x_* be an optimal solution to the primal and let y_* be an optimal solution to the dual. For each of the following changes, describe what effect they have on x_* and y_*, if any. These changes should be considered individually—they are not cumulative.

(i) The vector c is multiplied by λ, where $\lambda > 0$.

(ii) The kth equality constraint is multiplied by λ.

(iii) The ith equality constraint is modified by adding to it λ times the kth equality constraint.

(iv) The right-hand side b is multiplied by λ.

2.10. Consider the following linear programming problems:

$$\begin{aligned} \text{maximize} \quad & z = c^T x \\ \text{subject to} \quad & Ax \leq b \end{aligned}$$

and
$$\text{minimize} \quad z = c^T x$$
$$\text{subject to} \quad Ax \geq b.$$

(i) Write the duals to these problems.

(ii) If both of these problems are feasible, prove that if one of these problems has a finite optimal solution then so does the other.

(iii) If both of these problems are feasible, prove that the first objective is unbounded above if and only if the second objective is unbounded below.

(iv) Assume that both of these problems have finite optimal solutions. Let x be feasible for the first problem and let $\hat{x}$ be feasible for the second. Prove that
$$c^T x \leq c^T \hat{x}.$$

2.11. Prove that if the problem
$$\text{minimize} \quad z = c^T x$$
$$\text{subject to} \quad Ax = b$$
$$x \geq 0$$
has a finite optimal solution, then the new problem
$$\text{minimize} \quad z = c^T x$$
$$\text{subject to} \quad Ax = \hat{b}$$
$$x \geq 0$$
cannot be unbounded for any choice of the vector $\hat{b}$.

2.12. Consider the linear programming problem
$$\text{minimize} \quad z = c^T x$$
$$\text{subject to} \quad Ax = b$$
$$x \geq 0.$$
Let B be the optimal basis, and suppose that $B^{-1}b > 0$. Consider the problem
$$\text{minimize} \quad z = c^T x$$
$$\text{subject to} \quad Ax = b + \epsilon$$
$$x \geq 0,$$
where ϵ is a vector of perturbations. Prove that if the elements of ϵ are sufficiently small in absolute value, then B is also the optimal basis for the perturbed problem, and that the optimal dual solution is unchanged. What is the optimal objective value in this case?

2.13. Prove that if the system
$$Ax \leq b$$
has a solution, then the system
$$A^T y = 0$$
$$b^T y < 0$$
$$y \geq 0$$
has no solution.

2.14. (Farkas' Lemma) Use the duality theorems to prove that the system

$$A^T y \leq 0$$
$$b^T y > 0$$

has a solution if and only if the system

$$Ax = b$$
$$x \geq 0$$

has no solution.

2.15. Consider the linear program

$$\begin{aligned}
\text{maximize} \quad & z = 2x_1 + 9x_2 + 3x_3 \\
\text{subject to} \quad & -2x_1 + 2x_2 + x_3 \geq 1 \\
& x_1 + 4x_2 - x_3 \geq 1 \\
& x_1, x_2, x_3 \geq 0.
\end{aligned}$$

(i) Find the dual to this problem and solve it graphically.

(ii) Use complementarity slackness to obtain the solution to the primal.

2.16. Suppose that in the previous problem the first constraint is replaced by the constraint $-3x_1 + 2x_2 + x_3 \geq 1$. Find the dual to the problem and solve it graphically. Can you use complementary slackness to obtain the dual solution?

2.17. Use a combination of duality theory, elimination of variables, and graphical solution to solve the following linear programs. Do not use the simplex method.

(i)

$$\begin{aligned}
\text{minimize} \quad & z = -3x_1 + 2x_2 + x_3 \\
\text{subject to} \quad & -3x_2 - x_3 \leq 2 \\
& -x_1 - x_2 \geq -3 \\
& -x_1 - 2x_2 - x_3 \geq 1 \\
& x_1, x_2 \geq 0.
\end{aligned}$$

(ii)

$$\begin{aligned}
\text{minimize} \quad & z = -2x_1 - 4x_2 + x_3 + x_4 \\
\text{subject to} \quad & 2x_1 - 2x_2 + x_3 + x_4 \geq 2 \\
& -x_1 + x_2 - x_3 \geq -1 \\
& 3x_1 + x_2 + x_4 = 5 \\
& x_1, x_2, x_4 \geq 0.
\end{aligned}$$

2.18. Derive the complementary slackness conditions for a pair of primal and dual linear programs in canonical form.

2.19. Consider the primal linear programming problem

$$\begin{aligned}
\text{minimize} \quad & z = c^T x \\
\text{subject to} \quad & Ax \leq b \\
& x \geq 0.
\end{aligned}$$

Assume that this problem and its dual are both feasible. Let x_* be an optimal solution vector to the primal, let z_* be its associated objective value, and let y_* be an optimal solution vector to the dual problem. Show that

$$z_* = y_*^T A x_*.$$

2.20. Let x_* be an optimal solution to a linear program in standard form. Let y_* be an optimal solution to the dual problem and let s_* be the associated vector of dual slack variables. Prove that the solutions satisfy strict complementarity if and only if $x_*^T s_* = 0$ and $x_* + s_* > 0$.

2.21. In the next two exercises we will prove that if both primal and dual linear programs have feasible solutions, then there exist feasible solutions to these problems that satisfy strict complementarity. We will assume that the primal problem is given in standard form, and will denote the primal by (P) and its dual by (D). To start, we will prove in this exercise that there exists a feasible solution $\bar{x}$ to the primal and a feasible solution $\bar{y}$ to the dual with slack variables $\bar{s}$, such that $\bar{x} + \bar{s} > 0$.

(i) Suppose that every feasible solution to the primal satisfies $x_j = 0$ for some index j. Consider the linear programming problem (P')

$$\begin{aligned} \text{maximize} \quad & z' = e_j^T x \\ \text{subject to} \quad & Ax = b \\ & x \geq 0, \end{aligned}$$

where e_j is a vector with an entry of 1 in location j and zeros elsewhere. Prove that (P') is feasible and has an optimal objective value of zero.

(ii) Formulate the dual (D') to (P') and prove that it has an optimal solution with optimal objective value of zero.

(iii) Let y' be an optimal solution to (D') and let s' be the associated vector of slack variables. Prove that for any feasible solution y to (D) and corresponding slack variables $s = c - A^T y$, the vector $y + y'$ is feasible to (D), with corresponding slack variables $s + s' + e_j$, so that the jth dual slack variable is at least 1.

(iv) Prove that by taking appropriate strictly convex combinations of solutions to the primal (P) and to the dual (D) we can obtain a feasible solution $\bar{x}$ to the primal and a feasible solution $\bar{y}$ to the dual with slack variables $\bar{s}$, such that $\bar{x} + \bar{s} > 0$.

2.22. Prove that any linear program with a finite optimal value has a strictly complementary primal-dual optimal pair. Hint: Let z_* be the optimal objective value for a problem in standard form and consider the problem

$$\begin{aligned} \text{minimize} \quad & z = c^T x \\ \text{subject to} \quad & Ax = b \\ & c^T x \leq z_* \\ & x \geq 0. \end{aligned}$$

Apply the results of the previous exercise to this problem.

6.3 The Dual Simplex Method

The version of the simplex method that we have been using, which we now refer to as the *primal* simplex method, begins with a basic feasible solution to the primal linear program and iterates until the primal optimality conditions are satisfied. It is also possible to apply the simplex method to the dual problem, starting with a feasible solution to the dual program and iterating until the dual optimality conditions are satisfied.

The optimality conditions for the primal correspond to the feasibility conditions for the dual. This result was derived as part of the proof of Theorem 6.9, where it was shown that the primal optimality condition

$$c_N^T - c_B^T B^{-1} N \geq 0$$

is equivalent to the dual feasibility condition

$$A^T y \leq c,$$

where $y = B^{-T} c_B$ is the vector of simplex multipliers corresponding to the basis B. Thus the primal simplex method moves through a sequence of primal feasible but dual infeasible bases, at each iteration trying to reduce dual infeasibility until the dual feasibility conditions are satisfied.

The dual simplex method works in a "dual" manner. It goes through a sequence of dual feasible but primal infeasible bases, trying to reduce primal infeasibility until the primal feasibility conditions are satisfied. Although the dual simplex method can be viewed as the simplex method applied to the dual problem, it can be implemented directly in terms of the primal problem, if an initial dual feasible solution is available. The practical importance of the dual simplex method is discussed in Section 6.4.

We assume that an initial dual-feasible basis has been specified; i.e., the reduced costs are nonnegative. As described here, the algorithm uses B^{-1}, $x_B = \hat{b} = B^{-1} b$, and the current values of the reduced costs $\{\hat{c}_j\}$. (If the full tableau is used, this information can be read from the tableau.)

The dual simplex method terminates when the current basis is primal feasible, so an iteration of the method begins by checking if

$$x_B \geq 0.$$

If not, some entry $(x_B)_s < 0$ is used to select the pivot row.

We now describe an iteration of the dual simplex method, using an argument similar to that used to derive the primal simplex method. Suppose that a variable $(x_B)_s$ is infeasible so that its right-hand-side entry $\hat{b}_s < 0$. In terms of the current basis the sth constraint has the form

$$(x_B)_s + \sum_{j \in \mathcal{N}} \hat{a}_{s,j} x_j = \hat{b}_s < 0,$$

where $\mathcal{N}$ is the set of indices of the nonbasic variables, and $\{\hat{a}_{s,j}\}$ are the entries in row s of $B^{-1} A$. If some entry $\hat{a}_{s,j} < 0$ and nonbasic variable x_j were to replace $(x_B)_s$ in the basis, then the new value of x_j would be

$$\frac{\hat{b}_s}{\hat{a}_{s,j}} > 0,$$

that is, the new basic variable would be feasible. Not all such nonbasic variables can enter the basis because the dual feasibility (primal optimality) conditions must remain satisfied.

If x_j is the entering variable, the new reduced costs will satisfy

$$\bar{c}_l = \hat{c}_l - \hat{c}_j \frac{\hat{a}_{s,l}}{\hat{a}_{s,j}} \quad \text{for } l = 1, \ldots, n.$$

(If $l = j$ then $\bar{c}_l = 0$.) Since each $\bar{c}_l$ must be nonnegative, the smallest ratio $\{\, |\hat{c}_j / \hat{a}_{s,j}| \,\}$ with $\hat{a}_{s,j} < 0$ determines which reduced cost goes to zero first. (See the Exercises.) This ratio determines the pivot entry $\hat{a}_{s,t}$.

In our example below, the leaving variable is the one that is most negative. Any negative variable may be chosen as the leaving variable, and so other selection rules are possible. See Section 7.6.1.

The ratio test requires the computation of

$$\frac{\hat{c}_j}{\hat{a}_{s,j}}$$

for any nonbasic variable j for which $\hat{a}_{s,j} < 0$. Thus it is necessary to know the entries in the pivot row. If the full tableau is used, this information is available. Otherwise, the pivot row must be computed. The nonbasic entries in the entering row are given by

$$e_s^T B^{-1} A_j,$$

where e_s is column s of the $m \times m$ identity matrix. These entries can be computed by first letting $\sigma^T = e_s^T B^{-1}$, that is, computing row s of B^{-1}, and then forming

$$\sigma^T A_j$$

for all nonbasic variables j. The costs of this last calculation are almost the same as the pricing step that computes $\{\, \hat{c}_j \,\}$ in the primal method. (The vector e_s is a sparse vector, and this can be exploited to make the computations more efficient.)

The update step is performed just as in the (primal) simplex method. If the full tableau is used, elimination operations are applied to transform the pivot column to a column of the identity matrix. Otherwise, the pivot column is computed using

$$\hat{A}_t = B^{-1} A_t,$$

and the reduced costs are updated using

$$\hat{c}_j \leftarrow \hat{c}_j - \frac{\hat{c}_t}{\hat{a}_{s,t}} \hat{a}_{s,j}.$$

Finally x_B and B^{-1} are updated.

We now summarize the dual simplex method. At the initial basis, the reduced costs must satisfy $\hat{c}_j \geq 0$. There are three major steps: the feasibility test, the step, and the update.

1. *The Feasibility Test*—If $x_B = \hat{b} = B^{-1}b \geq 0$, then the current basis is a solution. Otherwise, choose $(x_B)_s$ as the leaving variable, where $\hat{b}_s < 0$.

2. *The Step*—In the pivot row (the row with entries $\hat{a}_{s,j} = e_s^T B^{-1} A_j$, where e_s is column s of the identity matrix) find an index t that satisfies

$$\left| \frac{\hat{c}_t}{\hat{a}_{s,t}} \right| = \min_{1 \leq j \leq n} \left\{ \left| \frac{\hat{c}_j}{\hat{a}_{s,j}} \right| : \hat{a}_{s,j} < 0, x_j \text{ nonbasic} \right\}.$$

This determines the entering variable x_t and the pivot entry $\hat{a}_{s,t}$. If no such index t exists, then the primal problem is infeasible and the dual problem is unbounded.

3. *The Update*—Represent the linear program in terms of the new basis. (Compute the pivot column $\hat{A}_t = B^{-1} A_t$, and update B^{-1}, x_B, and the reduced costs $\hat{c}$.)

The next example illustrates the dual simplex method.

Example 6.13 (Dual Simplex Method). Consider the linear program

$$\begin{array}{rl} \text{minimize} & z = 2x_1 + 3x_2 \\ \text{subject to} & 3x_1 - 2x_2 \geq 4 \\ & x_1 + 2x_2 \geq 3 \\ & x_1, x_2 \geq 0. \end{array}$$

We will describe the dual simplex method using the full tableau.

If excess variables but not artificial variables are added, then the tableau for this problem is

basic	x_1	x_2	x_3	x_4	rhs
$-z$	2	3	0	0	0
	3	-2	-1	0	4
	1	2	0	-1	3

Consider the initial basis $x_B = (x_3, x_4)^T$. If we multiply the constraints by -1, we obtain

basic	x_1	x_2	x_3	x_4	rhs
$-z$	2	3	0	0	0
x_3	-3	2	1	0	-4
x_4	-1	-2	0	1	-3

This basis is primal infeasible since both x_3 and x_4 are negative, but the primal optimality conditions are satisfied (the reduced costs are positive).

The dual problem is

$$\begin{array}{rl} \text{maximize} & w = 4y_1 + 3y_2 \\ \text{subject to} & 3y_1 + y_2 \leq 2 \\ & -2y_1 + 2y_2 \leq 3 \\ & y_1, y_2 \geq 0. \end{array}$$

Although not necessary for the algorithm, the corresponding dual solution can be found from the formula $y^T = c_B^T B^{-1}$. (We will compute the sequence of dual solutions in this example to emphasize that the dual simplex method is moving through a sequence of dual feasible solutions.) For this basis, the dual variables are $y_1 = y_2 = 0$ with dual objective $w = 0$. This point is dual feasible but not dual optimal. Throughout the execution of the dual simplex method, complementary slackness will be maintained and the primal and dual objectives will be equal.

The current basis is not (primal) feasible. The most negative variable is x_3, so it will be the leaving variable. In the ratio test there is only one valid ratio, in the x_1 column:

basic	x_1	x_2	x_3	x_4	rhs
$-z$	2	3	0	0	0
x_3	$\boxed{-3}$	2	1	0	-4 $\Leftarrow$
x_4	-1	-2	0	1	-3

We now apply elimination operations to obtain the new basic solution:

basic	x_1	x_2	x_3	x_4	rhs
$-z$	0	$\frac{13}{3}$	$\frac{2}{3}$	0	$-\frac{8}{3}$
x_1	1	$-\frac{2}{3}$	$-\frac{1}{3}$	0	$\frac{4}{3}$
x_4	0	$-\frac{8}{3}$	$-\frac{1}{3}$	1	$-\frac{5}{3}$

The corresponding dual feasible solution is $y_1 = \frac{2}{3}$, $y_2 = 0$.

At the next iteration, x_4 is the only negative variable, so it will be the leaving variable. There are two ratios to consider in the ratio test:

$$\left| \frac{\frac{13}{3}}{-\frac{8}{3}} \right| = \frac{13}{8} \quad \text{and} \quad \left| \frac{\frac{2}{3}}{-\frac{1}{3}} \right| = 2.$$

The first of these is smaller so x_2 is the entering variable:

basic	x_1	x_2	x_3	x_4	rhs
$-z$	0	$\frac{13}{3}$	$\frac{2}{3}$	0	$-\frac{8}{3}$
x_1	1	$-\frac{2}{3}$	$-\frac{1}{3}$	0	$\frac{4}{3}$
x_4	0	$\boxed{-\frac{8}{3}}$	$-\frac{1}{3}$	1	$-\frac{5}{3}$ $\Leftarrow$

After applying elimination operations we obtain

basic	x_1	x_2	x_3	x_4	rhs
$-z$	0	0	$\frac{1}{8}$	$\frac{13}{8}$	$-\frac{43}{8}$
x_1	1	0	$-\frac{1}{4}$	$-\frac{1}{4}$	$\frac{7}{4}$
x_2	0	1	$\frac{1}{8}$	$-\frac{3}{8}$	$\frac{5}{8}$

This basis is optimal and feasible, so we stop. The dual solution is $y_1 = \frac{1}{8}$, $y_2 = \frac{13}{8}$.

    If the full tableau were not used, then the same sequence of bases would be obtained. The only difference would be that the pivot rows and columns would be computed at every iteration using the current basis matrix B. ■

Exercises

3.1. Use the dual simplex method to solve

$$\begin{array}{rl} \text{minimize} & z = 5x_1 + 4x_2 \\ \text{subject to} & 4x_1 + 3x_2 \geq 10 \\ & 3x_1 - 5x_2 \geq 12 \\ & x_1, x_2 \geq 0. \end{array}$$

3.2. Use the dual simplex method to solve

$$\begin{array}{rl} \text{minimize} & z = 5x_1 + 2x_2 + 8x_3 \\ \text{subject to} & 2x_1 - 3x_2 + 2x_3 \geq 3 \\ & -x_1 + x_2 + x_3 \geq 5 \\ & x_1, x_2, x_3 \geq 0. \end{array}$$

3.3. Use the dual simplex method to solve

$$\begin{array}{rl} \text{maximize} & z = -2x_1 - 7x_2 - 6x_3 - 5x_4 \\ \text{subject to} & 2x_1 - 3x_2 - 5x_3 - 4x_4 \geq 20 \\ & 7x_1 + 2x_2 + 6x_3 - 2x_4 \leq 35 \\ & 4x_1 + 5x_2 - 3x_3 - 2x_4 \geq 15 \\ & x_1, x_2, x_3, x_4 \geq 0. \end{array}$$

3.4. In step 2 of the dual simplex method, explain why the dual problem is unbounded if there is no admissible entering variable in the ratio test. Also, find a direction of unboundedness for the dual problem in such a case.

3.5. Prove that at each iteration of the dual simplex method,

$$\Delta w = \Delta z = \frac{\hat{c}_t \hat{b}_s}{\hat{a}_{s,t}} \geq 0,$$

where Δw is the change in the dual objective function, and Δz is the change in the primal objective function.

3.6. Prove that in the step procedure of the dual simplex method the smallest ratio

$$\left\{ \left| \frac{\hat{c}_j}{\hat{a}_{s,j}} \right| : \hat{a}_{s,j} < 0, x_j \text{ nonbasic} \right\}$$

determines which reduced cost goes to zero first.

3.7. Is it possible for a basic variable that is nonnegative to become negative in the course of the dual simplex?

3.8. The following is a tableau obtained when solving a minimization linear programming problem via the dual simplex algorithm.

basic	x_1	x_2	x_3	x_4	x_5	x_6	x_7	rhs
$-z$	0	a	0	0	3	b	c	2
	0	-1	1	3	-1	0	1	1
	1	1	0	d	e	0	f	g
	0	h	0	-2	-2	1	-1	2

Find conditions on the parameters a, b, c, d, e, f, g, h such that the following are true. State the most general conditions that apply. (You do not have to mention those parameters that can take on any value from $-\infty$ to $+\infty$.)

(i) The above tableau is a valid tableau for the dual simplex algorithm.

(ii) A basic feasible solution to the problem has been found.

(iii) The problem is infeasible.

(iv) The problem is unbounded.

(v) The current solution is not feasible. According to the dual simplex method, the variable to enter the basis is x_4. (Assume that there are no ties.)

(vi) x_7 enters the basis, and the resulting solution is still infeasible.

3.9. In Exercises 3.1 and 3.2, apply the primal simplex method to the dual linear program, using bases that correspond to the iterations of the dual simplex method. Show that the two approaches are equivalent in these cases.

3.10. Suppose that the primal and dual simplex methods are implemented using the revised simplex tableau. Compare the operation counts for an iteration of both methods, if they are applied to a problem with n variables and m constraints. How are these operation counts affected when sparsity is taken into account?

3.11. Define dual degeneracy. Show (via an example) that degeneracy can cause the objective value to remain unchanged during an iteration of the dual simplex method.

3.12. Devise a "phase-1" procedure for the dual simplex method that would allow the method to be applied to any linear program.

3.13. Devise a "big-M" procedure for the dual simplex method that would allow the method to be applied to any linear program. Hint: Add a new variable and constraint.

6.4 Sensitivity

The purpose of sensitivity analysis is to determine how the solution of a linear program changes when changes are made to the data in the problem. This is an important practical technique, since it is rare that the data in a model are known exactly. For example, the linear program might represent a model of the economy, and one of the entries in the model might be the predicted inflation rate six months in the future. This rate can only be guessed at, and so it would be worrisome if the solution to the linear program was especially sensitive to its estimated value. The developer of the model might wish to know the effect on the objective value of a change in the right-hand side of one of the constraints. Or what happens when the cost coefficients change, or when a new constraint is added to the problem. Another possibility is that a particular entry in the model lies in some interval, and hence it would be desirable to know the solution of the linear program for all permissible values of this parameter. This situation might arise, for example, if the model of the economy included the number of unemployed workers, a number that might be estimated in the form

$$7,000,000 \pm 450,000.$$

Sensitivity analysis is designed to answer such questions.

Sensitivity analysis attempts to answer these questions without having to re-solve the problem. The idea is to start from the information provided by the optimal basis to answer these "what if" questions.

When performing sensitivity analysis it is also possible to determine the *range* of values that a perturbation can take without changing the optimal basis. Within this range, the values of the variables may change, but the basis will remain constant. In particular, the nonbasic variables will remain equal to zero. This could be of value, for example, if each variable represented the number of hours that an employee were assigned to a task, so that the optimal basis would determine the staffing required, even if the actual number of hours worked by each employee might vary. Most software packages for linear programming provide sensitivity information as well as range information for each objective coefficient and for the right-hand side of each constraint.

Our techniques depend on the feasibility and optimality conditions for a linear program. The current basis is feasible if

$$B^{-1}b \geq 0.$$

It is optimal if

$$c_N^T - c_B^T B^{-1} N \geq 0.$$

From a mathematical point of view, all of sensitivity analysis can be considered as a consequence of these formulas.

We will consider only the simpler cases, but the approach is general:

- Do the changes in the data affect the optimality conditions? How much can the data change before the optimality conditions are violated? If the current basis is no longer optimal, apply the primal simplex method to restore optimality.

- Do the changes in the data affect the feasibility conditions? How much can the data change before the feasibility conditions are violated? If the current basis is no longer feasible, apply the dual simplex method to restore feasibility.

We will examine these ideas via an example.

Example 6.14 (Sensitivity Analysis). Consider the linear program

$$
\begin{aligned}
\text{minimize} \quad & z = -x_1 - 2x_2 \\
\text{subject to} \quad & -2x_1 + x_2 \leq 2 \\
& -x_1 + 2x_2 \leq 7 \\
& x_1 \leq 3 \\
& x_1, x_2 \geq 0.
\end{aligned}
$$

The optimal solution is given by

basic	x_1	x_2	x_3	x_4	x_5	rhs
$-z$	0	0	0	1	2	13
x_2	0	1	0	$\frac{1}{2}$	$\frac{1}{2}$	5
x_1	1	0	0	0	1	3
x_3	0	0	1	$-\frac{1}{2}$	$\frac{3}{2}$	3

The current basis is $x_B = (x_2, x_1, x_3)^T$, and

$$B = \begin{pmatrix} 1 & -2 & 1 \\ 2 & -1 & 0 \\ 0 & 1 & 0 \end{pmatrix}, \quad B^{-1} = \begin{pmatrix} 0 & \frac{1}{2} & \frac{1}{2} \\ 0 & 0 & 1 \\ 1 & -\frac{1}{2} & \frac{3}{2} \end{pmatrix},$$

$$N = \begin{pmatrix} 0 & 0 \\ 1 & 0 \\ 0 & 1 \end{pmatrix}, \quad B^{-1}N = \begin{pmatrix} \frac{1}{2} & \frac{1}{2} \\ 0 & 1 \\ -\frac{1}{2} & \frac{3}{2} \end{pmatrix},$$

$$c_B = \begin{pmatrix} -2 \\ -1 \\ 0 \end{pmatrix}, \quad c_N = \begin{pmatrix} 0 \\ 0 \end{pmatrix}, \quad y^T = c_B^T B^{-1} = (0 \quad -1 \quad -2),$$

$$B^{-1}b = \begin{pmatrix} 5 \\ 3 \\ 3 \end{pmatrix}, \quad \hat{c}_N^T = c_N^T - y^T N = (1 \quad 2).$$

Here y is the vector of optimal dual variables.

We now perturb the linear program in various ways. Each of these changes will be independent and will be applied to the original linear program.

Suppose that the right-hand side of the second constraint is perturbed. We will denote this by $\bar{b}_2 = b_2 + \delta$, where $\bar{b}_2$ is the new right-hand-side value, and δ is the perturbation. This is a change of the form $\bar{b} = b + \Delta b$ for some vector Δb. In this case $\Delta b = (0, \delta, 0)^T$. This change has no effect on the optimality conditions since they do not involve the right-hand side. However, it does affect the feasibility conditions: $B^{-1}\bar{b} \geq 0$. The feasibility condition will remain satisfied as long as $B^{-1}(b + \Delta b) \geq 0$, or equivalently, as long as $B^{-1}b \geq -B^{-1}\Delta b$. For this example, this condition is

$$\begin{pmatrix} 5 \\ 3 \\ 3 \end{pmatrix} \geq \begin{pmatrix} -\frac{1}{2}\delta \\ 0 \\ \frac{1}{2}\delta \end{pmatrix}.$$

That is, the basis does not change if $-10 \leq \delta \leq 6$.

The new value of the objective function will be $\bar{z} = c_B^T B^{-1} \bar{b} = y^T (b + \Delta b) = z + y^T \Delta b$. This shows that $\Delta z = y^T \Delta b$. For this example, $\bar{z} = z + y_2 \delta = -13 - \delta$.

If $\delta = -4$, then the basis will not change, and

$$\bar{x}_B = x_B + B^{-1}\Delta b = \begin{pmatrix} 5 \\ 3 \\ 3 \end{pmatrix} + \begin{pmatrix} -2 \\ 0 \\ 2 \end{pmatrix} = \begin{pmatrix} 3 \\ 3 \\ 5 \end{pmatrix}$$

$$\bar{z} = z + y^T \Delta b = -13 - 1(-4) = -9.$$

No other values are affected.

If $\delta = 8$, then the basis changes, since

$$\bar{x}_B = x_B + B^{-1}\Delta b = \begin{pmatrix} 5 \\ 3 \\ 3 \end{pmatrix} + \begin{pmatrix} 4 \\ 0 \\ -4 \end{pmatrix} = \begin{pmatrix} 9 \\ 3 \\ -1 \end{pmatrix} \not\geq 0$$

$$\bar{z} = z + y^T\Delta b = -13 - 1(8) = -21$$

is infeasible. In terms of the current basis the perturbed problem is

basic	x_1	x_2	x_3	x_4	x_5	rhs
$-z$	0	0	0	1	2	21
x_2	0	1	0	$\frac{1}{2}$	$\frac{1}{2}$	9
x_1	1	0	0	0	1	3
x_3	0	0	1	$\boxed{-\frac{1}{2}}$	$\frac{3}{2}$	-1 $\Leftarrow$

The reduced costs are unchanged, and hence the optimality conditions remain satisfied. The dual simplex method can be applied to obtain the new solution

basic	x_1	x_2	x_3	x_4	x_5	rhs
$-z$	0	0	2	0	5	19
x_2	0	1	1	0	2	8
x_1	1	0	0	0	1	3
x_4	0	0	-2	1	-3	2

Suppose now that the coefficient of x_2 in the objective is changed: $\bar{c}_2 = c_2 + \delta$. This is a change in the cost coefficient of a basic variable, that is, a change of the form $\bar{c}_B = c_B + \Delta c_B$ with $\Delta c_B = (\delta, 0, 0)^T$. This affects the optimality condition $c_N^T - \bar{c}_B^T B^{-1} N \geq 0$, but not the feasibility condition. The optimality condition will remain satisfied if the new reduced costs are nonnegative:

$$c_N^T - (c_B + \Delta c_B)^T B^{-1} N = \hat{c}_N^T - \Delta c_B^T B^{-1} N \geq 0,$$

that is, if

$$\hat{c}_N^T \geq (\Delta c_B)^T B^{-1} N.$$

Substituting the data listed at the beginning of this example, we obtain

$$(1 \quad 2) \geq (\delta \quad 0 \quad 0)\begin{pmatrix} \frac{1}{2} & \frac{1}{2} \\ 0 & 1 \\ -\frac{1}{2} & \frac{3}{2} \end{pmatrix}$$

or

$$(1 \quad 2) \geq (\tfrac{1}{2}\delta \quad \tfrac{1}{2}\delta).$$

This will be satisfied if $\delta \leq 2$.

If $\delta = 1$, the current basis remains optimal and the reduced costs become

$$\hat{c}_N^T - \Delta c_B^T B^{-1} N = (1 - \tfrac{1}{2}\delta \quad 2 - \tfrac{1}{2}\delta) = (\tfrac{1}{2} \quad \tfrac{3}{2}) \geq 0.$$

The new value of the objective is

$$\bar{z} = \bar{c}_B^T B^{-1} b = (c_B^T + \Delta c_B^T)x_B = z + (\Delta c_B)^T x_B.$$

In this case $\bar{z} = z + \delta x_2 = z + 5\delta = -13 + 5(1) = -8$.

If $\delta = 4$, the current basis is no longer optimal. The reduced costs for x_4 and x_5 become

$$(1 - \tfrac{1}{2}\delta \quad 2 - \tfrac{1}{2}\delta) = (-1 \quad 0) \ngeq 0.$$

The new value of the objective is

$$\bar{z} = z + 5\delta = -13 + 5(4) = 7.$$

We apply the primal simplex method to the perturbed problem:

basic	x_1	x_2	x_3	$\Downarrow$ x_4	x_5	rhs
$-z$	0	0	0	-1	0	-7
x_2	0	1	0	$\boxed{\tfrac{1}{2}}$	$\tfrac{1}{2}$	5
x_1	1	0	0	0	1	3
x_3	0	0	1	$-\tfrac{1}{2}$	$\tfrac{3}{2}$	3

The new optimal basic solution is

basic	x_1	x_2	x_3	x_4	x_5	rhs
$-z$	0	2	0	0	1	3
x_4	0	2	0	1	1	10
x_1	1	0	0	0	1	3
x_3	0	1	1	0	2	8

As a final illustration we consider the addition of a new variable x_3 to the problem. Suppose that its coefficient in the objective is c_3 and its coefficients in the constraints are

$$A_3 = \begin{pmatrix} a_{1,3} \\ a_{2,3} \\ a_{3,3} \end{pmatrix}.$$

We will also assume that the new variable x_3 is constrained to be nonnegative.

The current basic feasible solution is also a basic feasible solution to the augmented problem (the problem that includes x_3) if x_3 is included as a nonbasic variable. Is the current basis optimal? We must check if the reduced cost for x_3 satisfies $\hat{c}_3 \geq 0$. This condition has the form

$$\hat{c}_3 = c_3 - y^T A_3 \geq 0.$$

If $\hat{c}_3 \geq 0$, then no further work is necessary: the current basis will remain optimal with $x_3 = 0$. If $\hat{c}_3 < 0$, then the current basis is not optimal. A new column will be added to the problem (corresponding to x_3) and the primal simplex method will be applied to determine the new optimal basis.

If $c_3 = 4$ and $A_3 = (5, -3, 4)^T$, then the optimality condition for x_3 is

$$4 - (0 \quad -1 \quad -2) \begin{pmatrix} 5 \\ -3 \\ 4 \end{pmatrix} = 9 \geq 0$$

so the new variable does not affect the solution.

If $c_3 = 2$ and $A_3 = (4, -5, 1)^T$, then the optimality condition for x_3 is

$$2 - (0 \quad -1 \quad -2) \begin{pmatrix} 4 \\ -5 \\ 1 \end{pmatrix} = -1 \not\geq 0$$

so the current solution is no longer optimal. The entries in the new column are obtained by computing $B^{-1} A_3$:

$$B^{-1} A_3 = \begin{pmatrix} 0 & \frac{1}{2} & \frac{1}{2} \\ 0 & 0 & 1 \\ 1 & -\frac{1}{2} & \frac{3}{2} \end{pmatrix} \begin{pmatrix} 4 \\ -5 \\ 1 \end{pmatrix} = \begin{pmatrix} -2 \\ 1 \\ 8 \end{pmatrix}$$

so that the augmented problem becomes

basic	x_1	x_2	x_3	x_3	x_4	x_5	rhs
$-z$	0	0	-1	0	1	2	13
x_2	0	1	-2	0	$\frac{1}{2}$	$\frac{1}{2}$	5
x_1	1	0	1	0	0	1	3
x_3	0	0	$\boxed{8}$	1	$-\frac{1}{2}$	$\frac{3}{2}$	3

The new optimal basic solution is

basic	x_1	x_2	x_3	x_3	x_4	x_5	rhs
$-z$	0	0	0	$\frac{1}{8}$	$\frac{15}{16}$	$\frac{35}{16}$	$\frac{107}{8}$
x_2	0	1	0	$\frac{1}{4}$	$\frac{3}{8}$	$\frac{7}{8}$	$\frac{23}{4}$
x_1	1	0	0	$-\frac{1}{8}$	$\frac{1}{16}$	$\frac{13}{16}$	$\frac{21}{8}$
x_3	0	0	1	$\frac{1}{8}$	$-\frac{1}{16}$	$\frac{3}{16}$	$\frac{3}{8}$

Some general rules can be stated for doing sensitivity analysis. In the following we denote the reduced costs by $\hat{c}_N^T = c_N^T - c_B^T B^{-1} N$.

- *Change in the right-hand side*—If $\bar{b} = b + \Delta b$, then determine if the current basis is still feasible by checking if $\hat{b} \geq -B^{-1}\Delta b$. If so, then $\bar{x}_B = x_B + B^{-1}\Delta b$ and $\bar{z} = z + y^T\Delta b$. (The vector y is the vector of dual variables.) If not, apply the dual simplex method to the perturbed problem to restore feasibility.

- *Change to an objective coefficient (nonbasic variable)*—If $\bar{c}_N = c_N + \Delta c_N$, then determine if the basis is still optimal by checking if $\hat{c}_N^T \geq -(\Delta c_N)^T$. If the basis does not change, then there are no changes to the variables or to the objective. If the basis does change, then apply the primal simplex method to the perturbed problem to restore optimality.

- *Change to an objective coefficient (basic variable)*—If $\bar{c}_B = c_B + \Delta c_B$, determine if the basis is still optimal by checking if $\hat{c}_N^T \geq (\Delta c_B)^T B^{-1} N$. If the basis does not change, then $\bar{y} = y + B^{-T}\Delta c_B$ and $\bar{z} = z + (\Delta c_B)^T x_B$. If the basis does change, apply the primal simplex method to the perturbed problem to restore optimality.

- *New constraint coefficients (nonbasic variable)*—If $\bar{N} = N + \Delta N$, then determine if the current basis is still optimal by checking if $\hat{c}_N^T \geq c_B^T B^{-1}\Delta N$. If the basis does not change, then there are no changes to the variables or to the objective. If the basis does change, then apply the primal simplex method to restore optimality.

- *New variable*—If x_t is a new variable with objective coefficient c_t and constraint coefficients $A_t = (a_{1,t}, \ldots, a_{m,t})^T$, then determine if the current basis is still optimal by testing if $c_t - y^T A_t \geq 0$. If it is, then there are no changes to the variables or to the objective. If it is not, then apply the primal simplex method to restore optimality.

- *New constraint*—See the Exercises.

It is also possible to change the coefficients of the constraints corresponding to a basic variable, or to have a combination of changes of the above forms. In these cases it might happen that the current basis would be neither feasible nor optimal for the new problem, so that neither the primal nor the dual simplex method could be applied directly to find the new solution. It would be necessary to use some sort of phase-1 procedure to find a basic feasible solution to the new problem before applying the primal simplex method. This might not be any faster than solving the new problem from scratch.

Exercises

4.1. The following questions apply to the linear program in Example 6.14. Each of the questions is independent.

    (i) By how much can the right-hand side of the first constraint change before the current basis ceases to be optimal?

    (ii) What would the new solution be if the right-hand side of the third constraint were increased by 5?

    (iii) What would the new solution be if the coefficient of x_1 in the objective were decreased by 2? Increased by 2?

(iv) Would the current basis remain optimal if a new variable x_3 were added to the model with objective coefficient $c_3 = 5$ and constraint coefficients $A_3 = (-2, 4, 5)^T$?

4.2. Show how to update the solution to a linear program when a new constraint is added.

(i) Consider first a constraint of the form

$$a_1 x_1 + a_2 x_2 + \cdots + a_n x_n \leq \beta.$$

There are two cases: (a) when the current optimal solution satisfies the new constraint, and (b) when the current optimal solution violates the new constraint.

(ii) Next, consider a constraint of the form

$$a_1 x_1 + a_2 x_2 + \cdots + a_n x_n = \beta.$$

In this case, an artificial variable may have to be added to the constraint, and a big-M term may have to be added to the objective.

(iii) How can a constraint of the form

$$a_1 x_1 + a_2 x_2 + \cdots + a_n x_n \geq \beta$$

be handled?

4.3. The following questions below apply to the linear program

$$\begin{aligned}
\text{maximize} \quad & z = 3x_1 + 13x_2 + 13x_3 \\
\text{subject to} \quad & x_1 + x_2 \leq 7 \\
& x_1 + 3x_2 + 2x_3 \leq 15 \\
& 2x_2 + 3x_3 \leq 9 \\
& x_1, x_2, x_3 \geq 0
\end{aligned}$$

with optimal basis $\{x_1, x_2, x_3\}$ and

$$B^{-1} = \begin{pmatrix} 5/2 & -3/2 & 1 \\ -3/2 & 3/2 & -1 \\ 1 & -1 & 1 \end{pmatrix}.$$

All of the questions are independent.

(i) What is the solution to the problem? What are the optimal dual variables?

(ii) What is the solution of the linear program obtained by decreasing the right-hand side of the second constraint by 5?

(iii) By how much can the right-hand side of the first constraint increase and decrease without changing the optimal basis?

(iv) What is the solution of the linear program obtained by increasing the coefficient of x_2 in the objective by 15?

(v) By how much can the objective coefficient of x_1 increase and decrease without changing the optimal basis?

(vi) Would the current basis remain optimal if a new variable x_4 were added to the model with objective coefficient $c_4 = 5$ and constraint coefficients $A_4 = (2, -1, 5)^T$?

(vii) Determine the solution of the linear program obtained by adding the constraint

$$x_1 - x_2 + 2x_3 \leq 10.$$

(viii) Determine the solution of the linear program obtained by adding the constraint

$$x_1 - x_2 + x_3 \geq 6.$$

(ix) Determine the solution of the linear program obtained by adding the constraint

$$x_1 + x_2 + x_3 = 10.$$

4.4. The following questions below apply to the linear program

$$\begin{aligned}
\text{minimize} \quad & z = -101x_1 + 87x_2 + 23x_3 \\
\text{subject to} \quad & 6x_1 - 13x_2 - 3x_3 \leq 11 \\
& 6x_1 + 11x_2 + 2x_3 \leq 45 \\
& x_1 + 5x_2 + x_3 \leq 12 \\
& x_1, x_2, x_3 \geq 0
\end{aligned}$$

with optimal basic solution

basic	x_1	x_2	x_3	x_3	x_4	x_5	rhs
$-z$	0	0	0	12	4	5	372
x_1	1	0	0	1	-2	7	5
x_2	0	1	0	-4	9	-30	1
x_3	0	0	1	19	-43	144	2

All of the questions are independent.

(i) What is the solution of the linear program obtained by decreasing the right-hand side of the second constraint by 15?

(ii) By how much can the right-hand side of the second constraint increase and decrease without changing the optimal basis?

(iii) What is the solution of the linear program obtained by increasing the coefficient of x_1 in the objective by 25?

(iv) By how much can the objective coefficient of x_3 increase and decrease without changing the optimal basis?

(v) Would the current basis remain optimal if a new variable x_4 were added to the model with objective coefficient $c_4 = 46$ and constraint coefficients $A_4 = (12, -14, 15)^T$?

(vi) Determine the solution of the linear program obtained by adding the constraint

$$5x_1 + 7x_2 + 9x_3 \leq 50.$$

   (vii) Determine the solution of the linear program obtained by adding the constraint

$$12x_1 - 15x_2 + 7x_3 \geq 10.$$

   (viii) Determine the solution of the linear program obtained by adding the constraint

$$x_1 + x_2 + x_3 = 30.$$

6.5   Parametric Linear Programming

Parametric linear programming is a form of sensitivity analysis, but one in which a *range* of values of the objective or the right-hand side is analyzed. For the case of the objective, we will examine problems of the form

$$\begin{aligned} \text{minimize} \quad & z = (c + \alpha \Delta c)^T x \\ \text{subject to} \quad & Ax = b \\ & x \geq 0, \end{aligned}$$

where the parameter α is allowed to range over all positive and negative values. If the right-hand side were allowed to vary, then the problems would be of the form

$$\begin{aligned} \text{minimize} \quad & z = c^T x \\ \text{subject to} \quad & Ax = b + \alpha \Delta b \\ & x \geq 0. \end{aligned}$$

We will concentrate on the case where the objective coefficients are varied.

   Parametric programming can be of value in applications where the coefficients in the model are uncertain and are only known to lie within particular intervals. It can also be valuable when there are two conflicting objective functions, for example, one representing minimum cost $(c^T x)$ and the other representing minimum time $(\bar{c}^T x)$. To understand the trade-offs between the two, a compromise objective function might be used:

$$z = (1 - \alpha)c^T x + \alpha \bar{c}^T x = c^T x + \alpha(\bar{c} - c)^T x.$$

This function is of the desired form with $\Delta c = c - \bar{c}$. In this application, only values of α in the interval [0, 1] would be relevant.

   We will assume that the linear program has been solved with $\alpha = 0$, that is, with objective function $z = (c + \alpha \Delta c)^T x = c^T x$. Techniques from sensitivity analysis will be used to examine how the solution changes as α is varied from zero. If the current basis remains optimal, then the current basic feasible solution $x_B = B^{-1}b$ will not change. Hence only the optimality conditions need be examined. For the perturbed problem they are

$$(c_N + \alpha \Delta c_N)^T - (c_B + \alpha \Delta c_B)^T B^{-1} N \geq 0$$

or

$$\alpha(\Delta c_N^T - \Delta c_B^T B^{-1} N) \geq -(c_N^T - c_B^T B^{-1} N),$$

where Δc_N and Δc_B represent the perturbations to c_N and c_B, respectively. This inequality must be satisfied for every component in the optimality test.

The coefficients on the right-hand side (that is, the reduced costs from the simplex method) satisfy

$$\hat{c}_N^T = c_N^T - c_B^T B^{-1} N \geq 0$$

since the current basis is assumed to be optimal. For $\alpha > 0$, the inequality is of interest only when

$$(\Delta c_N^T - \Delta c_B^T B^{-1} N)_i < 0.$$

As a result, α can be increased up to the value

$$\bar{\alpha} = \min_i \left\{ -\frac{(c_N^T - c_B^T B^{-1} N)_i}{(\Delta c_N^T - \Delta c_B^T B^{-1} N)_i} : (\Delta c_N^T - \Delta c_B^T B^{-1} N)_i < 0 \right\}$$

before the current basis ceases to be optimal. For $\alpha > \bar{\alpha}$ the basis changes, and the index i that determines $\bar{\alpha}$ specifies the entering variable for the simplex method. Similarly, for $\alpha < 0$, it is possible to decrease α up to the value

$$\underline{\alpha} = \max_i \left\{ -\frac{(c_N^T - c_B^T B^{-1} N)_i}{(\Delta c_N^T - \Delta c_B^T B^{-1} N)_i} : (\Delta c_N^T - \Delta c_B^T B^{-1} N)_i > 0 \right\}$$

before the current basis ceases to be optimal. Again, the index i that determines $\underline{\alpha}$ determines the entering variable.

For $\alpha \in [\underline{\alpha}, \bar{\alpha}]$ the reduced costs for the nonbasic variables are given by the formula

$$(c_N^T - c_B^T B^{-1} N) + \alpha(\Delta c_N^T - \Delta c_B^T B^{-1} N).$$

The parametric objective value is given by

$$z(\alpha) = z(0) + \alpha \Delta c_B^T x_B,$$

where $z(0)$ is the objective value for the problem with $\alpha = 0$.

If, when attempting to calculate $\bar{\alpha}$, there is no index that satisfies

$$(\Delta c_N^T - \Delta c_B^T B^{-1} N)_i < 0,$$

then α can be increased without bound with the current basis remaining optimal. If, when applying the simplex method to determine the new basis at $\bar{\alpha}$, there is no leaving variable, then the linear program is unbounded for $\alpha > \bar{\alpha}$. Similarly, if there is no index that satisfies

$$(\Delta c_N^T - \Delta c_B^T B^{-1} N)_i > 0,$$

then α can be decreased without bound with the current basis remaining optimal, and if there is no leaving variable at $\underline{\alpha}$, then the linear program is unbounded for $\alpha < \underline{\alpha}$.

Parametric linear programming is illustrated in the following example.

Example 6.15 (Parametric Linear Programming). We will examine the linear program from Example 6.12:

$$\begin{array}{ll}
\text{minimize} & z = -x_1 - 2x_2 \\
\text{subject to} & -2x_1 + x_2 \leq 2 \\
& -x_1 + 2x_2 \leq 7 \\
& x_1 \leq 3 \\
& x_1, x_2 \geq 0
\end{array}$$

with optimal basic solution

basic	x_1	x_2	x_3	x_4	x_5	rhs
$-z$	0	0	0	1	2	13
x_2	0	1	0	$\frac{1}{2}$	$\frac{1}{2}$	5
x_1	1	0	0	0	1	3
x_3	0	0	1	$-\frac{1}{2}$	$\frac{3}{2}$	3

Consider the parametric objective function $z(\alpha) = (c + \alpha \Delta c)^T x$ with

$$c = (-1 \quad -2 \quad 0 \quad 0 \quad 0)^T \quad \text{and} \quad \Delta c = (2 \quad 3 \quad 0 \quad 0 \quad 0)^T,$$

so that $\Delta c_B = (3, 2, 0)^T$ and $\Delta c_N = (0, 0)^T$. For the original problem with $\alpha = 0$ the optimal basis is $x_B = (x_2, x_1, x_3)^T$. The values of B, c_B, etc. are listed in Example 6.12.

To determine how much α can be varied without changing the basis, we calculate

$$\Delta c_N^T - \Delta c_B^T B^{-1} N = \begin{pmatrix} -\frac{3}{2} \\ -\frac{7}{2} \end{pmatrix} \quad \text{and} \quad \hat{c}_N^T = c_N^T - c_B^T B^{-1} N = \begin{pmatrix} 1 \\ 2 \end{pmatrix}.$$

Since there is no entry satisfying

$$(\Delta c_N^T - \Delta c_B^T B^{-1} N)_i > 0,$$

α can be decreased without bound, with the current basis remaining optimal. However, we can compute an upper bound on the range of α:

$$\bar{\alpha} = \min \left\{ \tfrac{2}{3}, \tfrac{4}{7} \right\} = \tfrac{4}{7}$$

and x_5 is the entering variable for this value of α. The nonbasic reduced costs are

$$\begin{pmatrix} 1 \\ 2 \end{pmatrix} + \alpha \begin{pmatrix} -\frac{3}{2} \\ -\frac{7}{2} \end{pmatrix} = \begin{pmatrix} 1 - \frac{3}{2}\alpha \\ 2 - \frac{7}{2}\alpha \end{pmatrix}$$

and the objective value is

$$z(\alpha) = -13 + \alpha c_B^T x_B = -13 + 21\alpha.$$

Hence for $\alpha = \bar{\alpha} = \tfrac{4}{7}$, the coefficients in terms of the current basis are

basic	x_1	x_2	x_3	x_4	$\Downarrow$ x_5	rhs
$-z$	0	0	0	$\frac{1}{7}$	0	1
x_2	0	1	0	$\frac{1}{2}$	$\frac{1}{2}$	5
x_1	1	0	0	0	1	3
x_3	0	0	1	$-\frac{1}{2}$	$\boxed{\frac{3}{2}}$	3

After pivoting, the new optimal basic solution is

basic	x_1	x_2	x_3	x_4	x_5	rhs
$-z$	0	0	0	$\frac{1}{7}$	0	1
x_2	0	1	$-\frac{1}{3}$	$\frac{2}{3}$	0	4
x_1	1	0	$-\frac{2}{3}$	$\frac{1}{3}$	0	1
x_5	0	0	$\frac{2}{3}$	$-\frac{1}{3}$	1	2

The whole process can now be repeated using the new basis.

To determine how much α can be increased we use the new basis to calculate

$$\Delta c_N^T - \Delta c_B^T B^{-1} N = \begin{pmatrix} \frac{7}{3} \\ -\frac{8}{3} \end{pmatrix} \quad \text{and} \quad \hat{c}_N^T = c_N^T - c_B^T B^{-1} N = \begin{pmatrix} 0 \\ \frac{1}{7} \end{pmatrix}.$$

Then

$$\bar{\alpha} = \frac{3}{56}$$

and x_4 is the entering variable for $\alpha = \frac{4}{7} + \bar{\alpha}$. (Note that we are calculating how much further α can be increased from its current value of $\frac{4}{7}$.)

The nonbasic reduced costs are

$$\begin{pmatrix} 0 \\ \frac{1}{7} \end{pmatrix} + (\alpha - \frac{4}{7}) \begin{pmatrix} \frac{7}{3} \\ -\frac{8}{3} \end{pmatrix}$$

and the objective value is

$$z(\alpha) = -1 + 14(\alpha - \frac{4}{7}).$$

For $\alpha - \frac{4}{7} = \frac{3}{56}$ we obtain

basic	x_1	x_2	x_3	$\Downarrow$ x_4	x_5	rhs
$-z$	0	0	$\frac{1}{8}$	0	0	$\frac{1}{4}$
x_2	0	1	$-\frac{1}{3}$	$\frac{2}{3}$	0	4
x_1	1	0	$-\frac{2}{3}$	$\boxed{\frac{1}{3}}$	0	1
x_5	0	0	$\frac{2}{3}$	$-\frac{1}{3}$	1	2

After pivoting, the new optimal basic solution is

basic	x_1	x_2	x_3	x_4	x_5	rhs
$-z$	0	0	$\frac{1}{8}$	0	0	$\frac{1}{4}$
x_2	-2	1	1	0	0	2
x_4	3	0	-2	1	0	3
x_5	1	0	0	0	1	3

To determine how much further α can be increased we calculate

$$\Delta c_N^T - \Delta c_B^T B^{-1} N = \begin{pmatrix} 8 \\ -3 \end{pmatrix} \quad \text{and} \quad \hat{c}_N^T = c_N^T - c_B^T B^{-1} N = \begin{pmatrix} 0 \\ \frac{1}{8} \end{pmatrix}.$$

Then

$$\bar{\alpha} = \tfrac{1}{24}$$

and x_3 is the entering variable for $\alpha = \frac{4}{7} + \frac{3}{56} + \bar{\alpha} = \frac{5}{8} + \bar{\alpha}$.

The nonbasic reduced costs are

$$\begin{pmatrix} 0 \\ \frac{1}{8} \end{pmatrix} + (\alpha - \tfrac{5}{8}) \begin{pmatrix} 8 \\ -3 \end{pmatrix}$$

and the objective value is

$$z(\alpha) = -\tfrac{1}{4} + 6(\alpha - \tfrac{5}{8}).$$

For $\alpha - \frac{5}{8} = \frac{1}{24}$ we obtain

basic	x_1	x_2	$\Downarrow$ x_3	x_4	x_5	rhs
$-z$	$\frac{1}{3}$	0	0	0	0	0
x_2	-2	1	$\boxed{1}$	0	0	2
x_4	3	0	-2	1	0	3
x_5	1	0	0	0	1	3

After pivoting, the new optimal basic solution is

basic	x_1	x_2	x_3	x_4	x_5	rhs
$-z$	$\frac{1}{3}$	0	0	0	0	0
x_3	-2	1	1	0	0	2
x_4	-1	2	0	1	0	7
x_5	1	0	0	0	1	3

For this basis

$$\Delta c_N^T - \Delta c_B^T B^{-1} N = \begin{pmatrix} 2 \\ 3 \end{pmatrix} > 0$$

so there is no entering variable and the current basis remains optimal for all larger values of α. Also, since $\Delta c_B = (0, 0, 0)^T$ for this basis, the objective value remains constant as α increases.

To summarize: If $\alpha \in (-\infty, \frac{4}{7}]$, then

$$x_B = (x_1, x_2, x_3)^T$$
$$z(\alpha) = -13 + 21\alpha.$$

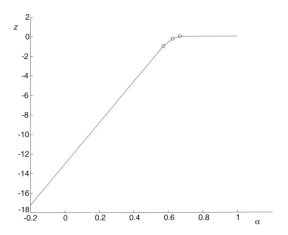

Figure 6.1. *Parametric objective function.*

If $\alpha \in [\frac{4}{7}, \frac{5}{8}]$, then

$$x_B = (x_1, x_2, x_5)^T$$
$$z(\alpha) = -1 + 14(\alpha - \tfrac{4}{7}) = -9 + 14\alpha.$$

If $\alpha \in [\frac{5}{8}, \frac{2}{3}]$, then

$$x_B = (x_1, x_4, x_5)^T$$
$$z(\alpha) = -\tfrac{1}{4} + 6(\alpha - \tfrac{5}{8}) = -4 + 6\alpha.$$

If $\alpha \in [\frac{2}{3}, +\infty)$, then

$$x_B = (x_3, x_4, x_5)^T$$
$$z(\alpha) = 0.$$

The graph of the objective value as a function of α is plotted in Figure 6.1. ∎

For the example, the value of the parametric objective function is piecewise linear and concave. This result is true in general for parametric linear programs (see the Exercises).

A similar technique can be developed for solving problems of the form

$$\begin{aligned} \text{minimize} \quad & z = c^T x \\ \text{subject to} \quad & Ax = b + \alpha \Delta b \\ & x \geq 0. \end{aligned}$$

The technique can be derived in one of two ways: either directly using sensitivity analysis, or by applying parametric analysis to the dual linear program. Regardless of how it is derived, the dual simplex method is used to find the new basis for each critical value of α. See the Exercises.

Parametric linear programming can be used as a general technique for solving linear programming problems. To develop this idea we assume that an initial basic feasible solution is available. (If not, a two-phase or big-M approach could be used to initialize the method.) This basis is used to define an artificial objective function $\bar{c}$ with respect to which the initial basis is optimal. One possible choice would be

$$\bar{c}_N = (1 \quad \cdots \quad 1)^T$$
$$\bar{c}_B = (0 \quad \cdots \quad 0)^T,$$

so that

$$\bar{c}_N^T - \bar{c}_B^T B^{-1} N = (1 \quad \cdots \quad 1)^T \geq 0.$$

Then the parametric programming method is applied to the linear program with objective function

$$z = (1 - \alpha)\bar{c}^T x + \alpha c^T x = \bar{c}^T x + \alpha(c - \bar{c})^T x,$$

where c is the original vector of objective coefficients. The solution for $\alpha = 1$ is the solution to the original problem. This technique is sometimes called the *shadow vertex method*.

Exercises

5.1. Apply parametric linear programming to the linear program in Example 6.13. The original objective uses $c = (2, 3, 0, 0)^T$. Use $\Delta c = (4, 1, 0, 0)^T$.

5.2. Consider a linear program with parametric objective function

$$\text{minimize} \quad z = (c + \alpha \Delta c)^T x.$$

Prove that the optimal value $z(\alpha)$ is a concave, piecewise linear function of α.

5.3. Derive a parametric linear programming algorithm to solve

$$\begin{aligned} \text{minimize} \quad & z = c^T x \\ \text{subject to} \quad & Ax = b + \alpha \Delta b \\ & x \geq 0 \end{aligned}$$

for $\alpha \geq 0$. Assume that an optimal basis is known for the problem with $\alpha = 0$.

5.4. Apply the algorithm obtained in the previous problem to the linear program in Example 6.12. Use $\Delta b = (4, 1, 1)^T$.

5.5. Use the shadow vertex method to solve the linear program in Example 6.14. Use the initial basis $x_B = (x_3, x_4, x_5)^T$ and $x_N = (x_1, x_2)^T$, and let the artificial objective function have coefficients

$$\bar{c}_N = (1, 1)^T \quad \text{and} \quad \bar{c}_B = (0, 0, 0)^T.$$

6.6 Notes

Duality—Duality theory for linear programming was first developed by von Neumann (1947), but the first published result is in the paper by Gale, Kuhn, and Tucker (1951). Von Neumann's result built upon his earlier work in game theory. Farkas' lemma (Exercise 2.14) was proved (in a slightly different form) by Julius Farkas in 1901 and was used in the work of Gale, Kuhn, and Tucker. The existence of a strictly complementary primal-dual optimal pair for any linear program with a finite optimum is due to Goldman and Tucker (1956). Further historical discussion of duality theory can be found in Section 14.9.

The Dual Simplex Method—The dual simplex method was first described in the papers of Lemke (1954) and Beale (1954).

Parametric Programming—Parametric programming was first developed in the paper by Gass and Saaty (1955). A more recent survey of works on this topic can be found in the book by Gal (1979). On degenerate problems there is a possibility that the method described here can cycle but, as with the simplex method, it is possible to modify the method so that it is guaranteed to terminate. The papers by Dantzig (1989) and Magnanti and Orlin (1988) describe techniques for doing this. The paper by Klee and Kleinschmidt (1990) explains when cycling can occur.

The shadow vertex method is not widely used for practical computations, but it is used theoretically to study the average-case behavior of the simplex method, that is, the expected performance of the simplex method on a random problem. (See Section 9.5.)

Chapter 7

Enhancements of the Simplex Method

7.1 Introduction

In previous chapters, we made use of the formulas for the simplex method but paid less attention to the computational details of the method. In particular, we did not explain how the basis matrix was represented. One possibility would be to use the explicit inverse of the basis matrix. This can be a sensible choice when solving small problems using hand calculations; however, it is inefficient for solving large- or even moderate-size problems. It is also inflexible—it is less able to take advantage of linear programs that have special structure, and thereby less able to achieve computational savings within the simplex method.

An important goal of this chapter is to focus on the essentials of the simplex method and to move away from the obvious interpretations of its formulas. The simplex method consists of three major steps: the optimality test that identifies, if the current basis is not optimal, the entering variable; the step procedure that determines the leaving variable and the new basis; and the update that changes the basis. As long as these calculations can be performed, the simplex method can be used, regardless of how the calculations are organized.

Many of the techniques we describe are designed to permit the solution of large problems, for which matrix inverses are particularly ill suited. Consider, for example, a problem with $m = 10,000$ equalities. Practical problems of this size or larger are not uncommon. The overwhelming majority of such large problems are sparse, with typically only a handful of nonzero elements in each column of the constraint matrix and possibly the right-hand side vector as well. Suppose our problem has, say, a total of 50,000 nonzero elements in the basis matrix. Then this matrix can be stored in a compact form by recording only the values of the nonzero elements and their row indices. If the basis matrix were inverted, its inverse might be dense, and updating this inverse would require updating all $m^2 = 100,000,000$ nonzero entries. The computational effort required to perform these updates over thousands of iterations would be prohibitive. This is the major disadvantage of using matrix inverses in the simplex method: they do not exploit sparsity.

Section 7.2 discusses a type of problem with special structure. Many linear programming models include upper bounds on the variables. In earlier chapters we treated these

213

upper bounds as general constraints. However, it is possible to handle these upper bounds in much the same way as the nonnegativity constraints on the variables, with considerable computational savings. The resulting simplex method is only slightly more complicated than for a problem in standard form.

In some linear programming models it can be inconvenient or expensive to generate the coefficients associated with a variable. It is possible to implement the simplex method in such a way that these coefficients are only generated as needed. The hope is that the linear program can be optimized by examining only a subset of the coefficients, and hence avoid unnecessary calculations. This technique is called *column generation* and is the subject of Section 7.3.

Column generation is applied in Section 7.4 to problems whose constraints are divided into two groups: one group of "easy" constraints, and another (usually small) group of "hard" constraints. This special problem structure can be exploited using the *decomposition principle*.

No matter how the simplex method is described, its ultimate effectiveness depends on how it is implemented in software. Details of the algorithm that are mathematically routine may require considerable transformation to turn them into efficient software. These ideas are discussed in Sections 7.5 and 7.6. Section 7.5 discusses the representation of the basis matrix. If sparsity is handled effectively, the storage requirements and computational effort required to solve large linear programs can be dramatically reduced.

The many topics discussed in this chapter have a joint goal. They aim to generalize our view of the simplex method to obtain a more powerful method capable of solving problems with millions of variables. There are only a few basic steps that are necessary to define the simplex method, and a focus on these basics serves as a unifying theme running through these seemingly disparate topics.

For the most part, the sections in this chapter can be read independently of each other. The only exception is Section 7.4 on the decomposition principle, which is easier to understand if Section 7.3 on column generation has already been read.

7.2 Problems with Upper Bounds

It is common in linear programming models to include upper bound constraints on the variables. These might represent upper limits on demand for a product, or perhaps just limits on allowable values (for example, a probability cannot be greater than one). In integer programming, where some or all of the variables are constrained to be integers, upper bound constraints are included to reduce the size of the feasible region, and hence reduce the amount of time required to compute a solution.

An upper bound constraint can be treated as a general linear constraint, and in fact we have used this approach in earlier sections. This is computationally wasteful, however. It increases the size of the problem by one general constraint and by one slack variable, and it does not take full advantage of the special form of these constraints. Upper bound constraints can be handled within the simplex method almost as easily as nonnegativity constraints. We will discuss how to do this below. In fact, general bound constraints

$$\ell \leq x \leq u$$

can be incorporated. We restrict our attention to constraints of the form

$$0 \leq x \leq u$$

so as to simplify the presentation. We also assume that $u > 0$, and that all the components of u are finite. Techniques for more general problems are left to the Exercises.

To develop the method, we require a more general definition of a basic feasible solution. Up to now we have assumed that the nonbasic variables are set equal to zero, their lower bound. With upper bounds present, we will allow the nonbasic variables to be equal to their lower bound (zero) or equal to their upper bound. For example, for the constraints

$$x_1 + 2x_2 = 4$$
$$0 \leq x_1 \leq 5, \; 0 \leq x_2 \leq 1$$

one basic feasible solution would be

$$x_B = (x_1) = (4)$$
$$x_N = (x_2) = (0)$$

and another would be

$$x_B = (x_1) = (2)$$
$$x_N = (x_2) = (1).$$

The same basis can lead to two different basic feasible solutions. Since $x_N \neq 0$ when upper bounds are present, some of the formulas for the simplex method will become more complicated.

Let us define this new form of basic feasible solution more precisely. Consider a linear program of the form

$$\text{minimize} \quad z = c^T x$$
$$\text{subject to} \quad Ax = b$$
$$0 \leq x \leq u,$$

where $u > 0$ and A is an $m \times n$ matrix of full row rank. A point x will be an (extended) basic feasible solution to this problem if (i) x satisfies the constraints of the linear program, and (ii) the columns of the constraint matrix corresponding to the components of x that are strictly between their bounds are linearly independent. The new definition of basic feasible solution is consistent with the old definition applied to the standard form of the problem, as the next lemma shows.

Lemma 7.1. *An extended basic feasible solution for the bounded-variable problem is equivalent to a basic feasible solution to the problem in standard form*

$$\text{minimize} \quad z = c^T x$$
$$\text{subject to} \quad Ax = b$$
$$x + s = u$$
$$x, s \geq 0.$$

Proof. Consider a feasible solution (x, s) for the standard form. We may assume that x can be split into the following three pieces: the first k components of x strictly between their bounds, the next ℓ components at their upper bounds, and the remaining $n-k-\ell$ components equal to zero. Then the first k components of s are positive, the next ℓ components are zero, and the remaining $n - k - \ell$ components are positive. Let A_1 and A_2 be the submatrices of A corresponding to the first two pieces of x, respectively.

Let $\hat{B}$ be the matrix consisting of the constraint coefficients corresponding to the positive components of x and s:

$$\hat{B} = \begin{pmatrix} A_1 & A_2 & & \\ I_k & & I_k & \\ & I_\ell & & \\ & & & I_{n-k-\ell} \end{pmatrix}.$$

(Here I_k denotes a $k \times k$ identity matrix, etc.) The columns of $\hat{B}$ are linearly independent (and hence x is a basic feasible solution in the old sense) if and only if there are no nontrivial solutions to

$$A_1\alpha_1 + A_2\alpha_2 = 0$$
$$\alpha_1 + \alpha_3 = 0$$
$$\alpha_2 = 0$$
$$\alpha_4 = 0$$

and hence there are no nontrivial solutions to

$$A_1\alpha_1 = 0$$
$$\alpha_2 = \alpha_4 = 0$$
$$\alpha_3 = -\alpha_1.$$

This shows that the columns of $\hat{B}$ are linearly independent if and only if the columns of A_1 are linearly independent. Since A_1 is also the coefficient matrix corresponding to the components of x that are strictly between their bounds, then x is a basic feasible solution in the new sense if and only if (x, s) is a basic feasible solution in the old sense.   □

Let us now return to the problem with upper bounds

$$\text{minimize} \quad z = c^T x$$
$$\text{subject to} \quad Ax = b$$
$$0 \leq x \leq u$$

and assume that an initial feasible basis is provided. (A two-phase or big-M approach can be used to find an initial basis.) Corresponding to the basis, we can identify m basic variables x_B and $n - m$ nonbasic variables x_N so that the constraints take the form $Bx_B + Nx_N = b$, where B is an $m \times m$ invertible matrix. Hence,

$$x_B = B^{-1}b - B^{-1}Nx_N.$$

The nonbasic variables will be either zero or at their upper bound.

If this formula is substituted into the objective function we obtain

$$z = c_B^T x_B + c_N^T x_N$$
$$= y^T b + (c_N^T - y^T N)x_N$$
$$= y^T b + \sum_{j \in \mathcal{N}} \hat{c}_j x_j,$$

where $y^T = c_B^T B^{-1}$, and $\mathcal{N}$ is the index set for the nonbasic variables. As before, the optimality test is based on the reduced costs

$$\hat{c}_j = c_j - y^T A_j,$$

although the test is more complicated in this case. If the nonbasic variable x_j is zero and if $\hat{c}_j \geq 0$, then the solution will not improve if x_j enters the basis. (If x_j is increased from zero, then the objective function will not decrease.) In addition, if x_j is at its upper bound and if $\hat{c}_j \leq 0$, then again the solution will not improve if x_j enters the basis. (If x_j is decreased from its upper bound, then the objective function will not decrease.)

If the optimality test is not satisfied, then any violation can be used to determine the entering variable x_t. As before, the entering column is

$$\hat{A}_t = B^{-1} A_t.$$

The ratio test for determining the leaving variable is also more complicated than before. One of three things can happen as the entering variables is changed: (a) the "entering" variable can move from one bound to another with the basis unchanged, (b) a basic variable can increase and leave the basis by going to its upper bound, or (c) a basic variable can decrease and leave the basis by going to zero. Cases (b) and (c) can be further refined depending on whether the entering variable is equal to zero or to its upper bound. The ratio test must determine which of these things happens first. Since every variable has finite upper and lower bounds, the problem cannot be unbounded. In more general problems with infinite upper bounds, unboundedness would be a possibility.

We now derive the ratio test, and hence determine the leaving variable. Suppose that the entering variable x_t is changed by α. Then the vector of basic variables will change from its current value $\hat{b}$ to $x_B = \hat{b} - \alpha \hat{A}_t$. To maintain feasibility with respect to the bounds, the following condition must remain satisfied:

$$0 \leq (x_B)_i = \hat{b}_i - \alpha \hat{a}_{i,t} \leq \hat{u}_i,$$

where $\hat{u}_i$ is the upper bound for the ith basic variable and $\hat{a}_{i,t}$ is the ith component of $\hat{A}_t$. If the entering variable is equal to zero (its lower bound), then $\alpha > 0$. If $\hat{a}_{i,t} > 0$, then $(x_B)_i$ will decrease towards zero. Thus the ratio test

$$\min_{1 \leq i \leq m} \left\{ \frac{\hat{b}_i}{\hat{a}_{i,t}} : \hat{a}_{i,t} > 0 \right\}$$

determines which (if any) is the first basic variable to go to zero. If on the other hand $\hat{a}_{i,t} < 0$, then $(x_B)_i$ will increase towards its upper bound. The corresponding ratio test

$$\min_{1 \leq i \leq m} \left\{ \frac{\hat{u}_i - \hat{b}_i}{-\hat{a}_{i,t}} : \hat{a}_{i,t} < 0 \right\}$$

determines which (if any) is the first basic variable to reach its upper bound. Similar ratio tests can be derived in the case where $x_t = u_t$, its upper bound, and where $\alpha < 0$. (See the Exercises.)

The overall algorithm is summarized below. The method starts with a basis matrix B corresponding to a basic feasible solution

$$x_B = \hat{b} = B^{-1}b - \sum_{j \in \mathcal{N}} B^{-1}A_jx_j.$$

The steps of the algorithm are given below.

ALGORITHM 7.1.
Bounded-Variable Simplex Method

1. *The Optimality Test*—Compute the vector of simplex multipliers $y^T = c_B^T B^{-1}$. Compute the reduced costs $\hat{c}_j = c_j - y^T A_j$ for the nonbasic variables x_j. If for all nonbasic variables either (a) $x_j = 0$ and $\hat{c}_j \geq 0$, or (b) $x_j = u_j$ and $\hat{c}_j \leq 0$, then the current basis is optimal. Otherwise, select a variable x_t that violates the optimality test as the entering variable.

2. *The Step*—Compute the entering column $\hat{A}_t = B^{-1}A_t$. Find an index s that corresponds to the minimum value θ of the following quantities (if any of the quantities is undefined, its value should be taken to be $+\infty$):

    (i) the distance between the bounds for the entering variable x_t:

    $$u_t;$$

    (ii) if $x_t = 0$,

    $$\min_{1 \leq i \leq m} \left\{ \frac{\hat{b}_i}{\hat{a}_{i,t}} : \hat{a}_{i,t} > 0 \right\},$$

    $$\min_{1 \leq i \leq m} \left\{ \frac{\hat{u}_i - \hat{b}_i}{-\hat{a}_{i,t}} : \hat{a}_{i,t} < 0 \right\},$$

    where $\hat{u}_i$ is the upper bound for the ith basic variable;

    (iii) if $x_t = u_t$,

    $$\min_{1 \leq i \leq m} \left\{ \frac{\hat{b}_i}{-\hat{a}_{i,t}} : \hat{a}_{i,t} < 0 \right\},$$

    $$\min_{1 \leq i \leq m} \left\{ \frac{\hat{u}_i - \hat{b}_i}{\hat{a}_{i,t}} : \hat{a}_{i,t} > 0 \right\}.$$

    Here $\hat{b}$ is the vector of current values of the basic variables, $\hat{u}_i$ is the upper bound for the ith basic variable, and $\hat{a}_{i,t}$ is the ith component of $\hat{A}_t$. The ratio test determines the leaving variable.

3. *The Update*—Update the basis matrix B and the vector of basic variables x_B. If x_t is both the entering and leaving variable, then B does not change.

We now give formulas for step 3 of the above algorithm; the formulas use the result θ of the ratio test. Let α be the amount by which x_t changes; $\alpha = \theta$ if x_t was zero, and $\alpha = -\theta$ if x_t was at its upper bound. Then in terms of the *current* basis, the variables can be updated using the formula

$$\begin{pmatrix} x_B \\ x_N \end{pmatrix} \leftarrow \begin{pmatrix} x_B \\ x_N \end{pmatrix} + \alpha \begin{pmatrix} -\hat{A}_t \\ e_t \end{pmatrix}.$$

Hence the basic variables can be updated via $x_B = \hat{b} - \alpha\hat{A}_t$. The new value of the entering variable is $x_t + \alpha$. The objective value z will decrease by $\hat{c}_t\alpha$. An example illustrating the method is given below.

Example 7.2 (Upper Bounded Variables). Consider the linear program

$$\begin{aligned} \text{minimize} \quad & z = -4x_1 + 5x_2 \\ \text{subject to} \quad & 3x_1 - 2x_2 + x_3 = 6 \\ & -2x_1 - 4x_2 + x_4 = 4 \\ & 0 \le x_1 \le 4 \\ & 0 \le x_2 \le 3 \\ & 0 \le x_3 \le 20 \\ & 0 \le x_4 \le 20. \end{aligned}$$

We use the initial basis $x_B = (x_3, x_4)^T$ and $x_N = (x_1, x_2)^T$ and set the nonbasic variables at the values $x_1 = 0$ and $x_2 = 3$. (The choice of basis does not uniquely determine the values of the basic variables.) Hence

$$c_B = \begin{pmatrix} 0 \\ 0 \end{pmatrix}, \quad c_N = \begin{pmatrix} -4 \\ 5 \end{pmatrix},$$

$$B = \begin{pmatrix} 1 & 0 \\ 0 & 1 \end{pmatrix}, \quad N = \begin{pmatrix} 3 & -2 \\ -2 & -4 \end{pmatrix}, \quad \text{and} \quad x_N = \begin{pmatrix} 0 \\ 3 \end{pmatrix}.$$

The basic variables are computed from

$$x_B = B^{-1}b - B^{-1}Nx_N = \begin{pmatrix} 12 \\ 16 \end{pmatrix}.$$

At the first iteration of the simplex method, the simplex multipliers are

$$y^T = c_B^T B^{-1} = (0 \quad 0)$$

and the coefficients for the optimality test are

$$\hat{c}_1 = -4 \quad \text{and} \quad \hat{c}_2 = 5.$$

Both components fail the optimality test. We will use the larger violation to select x_2 as the entering variable.

The entering column is

$$\hat{A}_2 = B^{-1}A_2 = \begin{pmatrix} -2 \\ -4 \end{pmatrix}.$$

The entering variable will be decreased from its upper bound. In the ratio test the distance between the bounds for the entering variable is 3. Both components of the entering column are negative, so the rest of the ratio test is based on

$$\min_{1 \le i \le m} \left\{ \frac{\hat{b}_i}{-\hat{a}_{i,2}} : \hat{a}_{i,2} < 0 \right\} = \min_{1 \le i \le m} \left\{ \frac{12}{2}, \frac{16}{4} \right\} = 4.$$

The result of the ratio test is that the entering variable moves to its lower bound. (The other possible ratio tests are irrelevant.) The basis does not change, so B and B^{-1} do not change. The change in the entering variable is $\alpha = -3$, and the new basic feasible solution is

$$x_B \leftarrow x_B + \alpha(-\hat{A}_2) = x_B + 3\hat{A}_2 = \begin{pmatrix} 6 \\ 4 \end{pmatrix} = \begin{pmatrix} x_3 \\ x_4 \end{pmatrix} \quad \text{and} \quad x_N = \begin{pmatrix} 0 \\ 0 \end{pmatrix} = \begin{pmatrix} x_1 \\ x_2 \end{pmatrix}.$$

This completes the first iteration.

At the second iteration, the dual variables and the reduced costs are unchanged:

$$y = (0, 0)^T, \quad \hat{c}_1 = -4, \quad \text{and} \quad \hat{c}_2 = 5$$

because the basis has not changed. However, now the variable x_2 is at its lower bound and $\hat{c}_2$ satisfies the optimality test. The optimality test fails for $\hat{c}_1$, so x_1 is the entering variable.

The entering column is

$$\hat{A}_1 = \begin{pmatrix} 3 \\ -2 \end{pmatrix}.$$

Variable x_1 will be increased from zero. In the ratio test, the distance between bounds for x_1 is 4. The ratio test for the first component is based on

$$\frac{\hat{b}_1}{\hat{a}_{1,1}} = \frac{6}{3} = 2.$$

The ratio test for the second component is

$$\frac{\hat{u}_2 - \hat{b}_2}{-\hat{a}_{2,1}} = \frac{20 - 4}{2} = 8.$$

(Note that $\hat{u}_2 = u_4 = 20$.) The smallest of these values is 2, so x_3 is the leaving variable. The change in the entering variable is $\alpha = 2$.

The new basis corresponds to $x_B = (x_1, x_4)^T$ and $x_N = (x_2, x_3)^T$. In terms of this basis

$$c_B = \begin{pmatrix} -4 \\ 0 \end{pmatrix}, \quad c_N = \begin{pmatrix} -5 \\ 0 \end{pmatrix},$$

$$B = \begin{pmatrix} 3 & 0 \\ -2 & 1 \end{pmatrix}, \quad N = \begin{pmatrix} 2 & 1 \\ 4 & 0 \end{pmatrix}, \quad \text{and} \quad x_N = \begin{pmatrix} 0 \\ 0 \end{pmatrix}.$$

The basic variables are

$$x_B = \begin{pmatrix} x_1 \\ x_4 \end{pmatrix} = \begin{pmatrix} 2 \\ 4 - 2(-2) \end{pmatrix} = \begin{pmatrix} 2 \\ 8 \end{pmatrix}.$$

At the third iteration, the dual variables are

$$y = \begin{pmatrix} -\frac{4}{3} \\ 0 \end{pmatrix}$$

and the coefficients in the optimality test are

$$\hat{c}_2 = \tfrac{7}{3} \quad \text{and} \quad \hat{c}_3 = \tfrac{4}{3}.$$

Since x_2 and x_3 are at their lower bounds, this basis is optimal.
        At the solution,

$$x = (2 \quad 0 \quad 0 \quad 8)^T$$

and $z = c^T x = -8$ is the optimal objective value.   ∎

Exercises

2.1. Solve the following linear programs with the upper bound form of the simplex method. Use the explicit representation of the inverse of the basis matrix. Use the slack variables to form an initial basic feasible solution (note that there will be no upper bounds on the slack variables).

(i)

$$\begin{aligned}
\text{minimize} \quad & z = -5x_1 - 10x_2 - 15x_3 \\
\text{subject to} \quad & 2x_1 + 4x_2 + 2x_3 \le 50 \\
& 3x_1 + 5x_2 + 4x_3 \le 80 \\
& 0 \le x_1, x_2, x_3 \le 20.
\end{aligned}$$

(ii)

$$\begin{aligned}
\text{minimize} \quad & z = x_1 - x_2 \\
\text{subject to} \quad & -x_1 + x_2 \le 5 \\
& x_1 - 2x_2 \le 9 \\
& 0 \le x_1 \le 6, 0 \le x_2 \le 8.
\end{aligned}$$

(iii)

$$\begin{aligned}
\text{maximize} \quad & z = 6x_1 - 3x_2 \\
\text{subject to} \quad & 2x_1 + 5x_2 \le 20 \\
& 3x_1 + 2x_2 \le 40 \\
& 0 \le x_1, x_2 \le 15.
\end{aligned}$$

(iv)

$$\begin{aligned}
\text{minimize} \quad & z = -5x_1 - 7x_2 \\
\text{subject to} \quad & -3x_1 + 2x_2 \le 30 \\
& -2x_1 + x_2 \le 12 \\
& 0 \le x_1, x_2 \le 20.
\end{aligned}$$

2.2. Repeat Exercise 2.1(i) using the initial values $x_1 = 20$, $x_2 = 0$, and $x_3 = 0$.

2.3. Repeat Exercise 2.1(ii) using the initial values $x_1 = 0$ and $x_2 = 5$.

2.4. Repeat Exercise 2.1(iv) using the initial values $x_1 = 20$ and $x_2 = 0$.

2.5. Consider the bounded-variable linear program

$$\begin{aligned}
\text{minimize} \quad & z = x_1 - 4x_3 + x_4 - 3x_5 + 2x_6 \\
\text{subject to} \quad & x_1 + x_2 - x_3 - 2x_5 = 0 \\
& x_1 - 2x_3 - x_4 + x_6 = 3 \\
& 0 \le x_1, x_2, x_3 \le 5 \\
& 0 \le x_4, x_5, x_6 \le 2.
\end{aligned}$$

At a certain iteration of the bounded simplex algorithm, the basic variables are x_1 and x_2, with basis inverse matrix

$$B^{-1} = \begin{pmatrix} 0 & 1 \\ 1 & -1 \end{pmatrix}.$$

The nonbasic variables x_3 and x_4 are at their lower bound (zero), while x_5 and x_6 are at their upper bound. Now do the following:

(i) Determine the corresponding basic solution.

(ii) Determine which of the nonbasic variables will yield an improvement in the objective value if chosen to enter the basis.

(iii) For each of the candidate variables that you found in part (ii), determine the corresponding leaving variable and the resulting basic solution.

2.6. Derive the remaining portions of the ratio test for the case where the entering variable x_t is at its upper bound.

2.7. Derive a simplex method for linear programming problems with general bounds on the variables $\ell \le x \le u$.

2.8. Determine how the simplex method for problems with upper bounds would be modified if some of the variables had upper and lower bounds both equal to 0.

2.9. A basic feasible solution to the bounded-variable problem is said to be *degenerate* if one of the basic variables is equal to its upper or lower bound. Prove that, in the absence of degeneracy, the bounded-variable simplex method will terminate in a finite number of iterations.

2.10. Consider the bounded-variable problem

$$\begin{aligned}
\text{minimize} \quad & z = c^T x \\
\text{subject to} \quad & Ax = b \\
& 0 \le x \le u.
\end{aligned}$$

What is the dual to this problem? What are the complementary slackness conditions at the optimum?

7.3 Column Generation

One of the important properties of the simplex method is that it does not require that the constraint matrix A be explicitly available. Indeed, if we review the steps of the simplex method, we see that the columns of A are used only to determine whether the reduced costs

$\hat{c}_j = c_j - y^T A_j$ are nonnegative for every j, and if not, to generate a column A_t that violates this condition. (A_t is then used to obtain the entering column $\hat{A}_t = B^{-1} A_t$.) All that is needed is some technique that determines whether a basic solution is optimal, and if not, produces a column that violates the optimality conditions. When the constraint matrix A has some specific known structure, it is sometimes possible to find such an A_t without explicit knowledge of the columns of A. This technique, which generates columns of A only as needed, is called *column generation*.

One application where column generation is possible is the *cutting stock problem*, discussed below. Practical cutting stock problems may involve many millions of variables—so many that it is impossible even to form their columns in reasonable time. Yet by using column generation, the simplex algorithm can be used to solve such problems. At each iteration of the algorithm, an auxiliary problem determines the column of A that yields the largest coefficient $\hat{c}_j$. Remarkably, this auxiliary problem can be solved without generating the columns of A.

To present the cutting stock problem, we start with an example. A manufacturer produces sheets of material (such as steel, paper, or foil) of standard width of $50''$. To satisfy customer demand, the sheets must be cut into sections of the same length but of smaller widths. Suppose the manufacturer has orders for 25 rolls of width $20''$, 120 rolls of width $14''$, and 20 rolls of width $8''$. To fill these orders, the manufacturer can cut the standard sheets in a variety of ways. For instance, a standard sheet could be cut into two sections of width $20''$ and one section of width $8''$, with a waste of $2''$; or it could be cut into two sections of width $14''$ and two sections of width $8''$, with a waste of $6''$ (see Figure 7.1). Each such alternative is called a *pattern*, and clearly there are many possible patterns. The problem is to determine how many sheets need to be cut and into which patterns. Assuming that waste material is thrown away, our objective is to minimize the total number of sheets that need to be cut.

In the general case, a manufacturer produces sheets of standard width W. These sheets must then be cut to smaller widths to meet customer demand. Specifically, the manufacturer has to supply b_i sections of width $w_i < W$, for $i = 1, \ldots, m$. To formulate this problem, we represent each cutting pattern by a vector of length m, whose ith component indicates the number of sections of width w_i that are used in that pattern. For example, the two patterns described above are represented by

$$\begin{pmatrix} 2 \\ 0 \\ 1 \end{pmatrix} \quad \text{and} \quad \begin{pmatrix} 0 \\ 2 \\ 2 \end{pmatrix}.$$

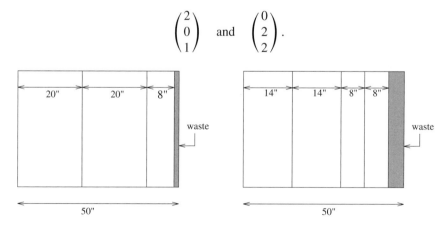

Figure 7.1. *Cutting patterns.*

The first component in each vector indicates the number of sections of width $20''$ used in a pattern, the second component indicates the number of sections of width $14''$, and the third component indicates the number of sections of width $8''$. A vector $a = (a_1, \ldots, a_m)^T$ represents a cutting pattern if and only if

$$w_1 a_1 + w_2 a_2 + \cdots + w_m a_m \leq W,$$

where $\{a_i\}$ are nonnegative integers.

Let x_i denote the number of sheets to be cut into pattern i, and let n denote the number of all possible cutting patterns. Even for small values of m, this number may be enormous. In practical problems, m may be of the order of a few hundred. In such cases, n may be of the order of hundreds of millions. Due to its sheer size, the matrix defined by the various patterns will not be available explicitly. We denote this conceptual matrix by A. The problem of minimizing the number of sheets used to satisfy the demands becomes

$$
\begin{aligned}
\text{minimize} \quad & z = \sum_{i=1}^{n} x_i \\
\text{subject to} \quad & Ax \geq b \\
& x \geq 0.
\end{aligned}
$$

The variables x_i must also be integer. Here we shall solve the linear program, ignoring the integrality restrictions, and then round the solution variables appropriately. Although rounding the solution of a linear program does not necessarily give an optimal solution of the associated program with integrality constraints, in the case of cutting stock problems, rounding is often appropriate because it is applied in large-scale production settings.

Finding an initial basic feasible solution is not difficult. For example, in the problem above we can use the initial basis matrix

$$B = \begin{pmatrix} 2 & 0 & 0 \\ 0 & 3 & 0 \\ 0 & 0 & 6 \end{pmatrix}.$$

The columns of B corresponds to the patterns that cut as many $20''$, $14''$, and $8''$ sections as possible. For convenience we denote these columns by A_1, A_2, and A_3 and denote the number of sheets cut according to each of these patterns by x_1, x_2, and x_3. The solution corresponding to this basis matrix is

$$x_B = \begin{pmatrix} x_1 \\ x_2 \\ x_3 \end{pmatrix} = B^{-1} b = \begin{pmatrix} \frac{1}{2} & 0 & 0 \\ 0 & \frac{1}{3} & 0 \\ 0 & 0 & \frac{1}{6} \end{pmatrix} \begin{pmatrix} 25 \\ 120 \\ 20 \end{pmatrix} = \begin{pmatrix} \frac{25}{2} \\ 40 \\ \frac{10}{3} \end{pmatrix}.$$

Suppose now that at some iteration we have a basic feasible solution with corresponding basis matrix B. To determine whether the solution is optimal we first compute the vector of simplex multipliers

$$y^T = c_B^T B^{-1}.$$

Next we must determine whether the solution is optimal. Suppose first that some component of y, say y_i, is negative. Then the reduced cost of the ith excess variable (x_{n+i}) is $\hat{c}_{n+i} = 0$

$-y^T(-e_i) = y_i < 0$ (here e_i denotes a vector of length m with a 1 in its ith position, and zeroes elsewhere). We can therefore choose this excess variable to be the variable that enters the basis. No further computation of the coefficients $\hat{c}_j$ will be needed.

Suppose now that $y \geq 0$. Then the excess variables will all have nonnegative reduced costs. We must now determine whether the variables in the original problem satisfy the optimality conditions, that is, whether

$$\hat{c}_j = 1 - y^T A_j \geq 0, \quad j = 1, \ldots, n.$$

Because A has so many columns, and because these columns are not explicitly available, it is virtually impossible to perform this computation variable by variable. Fortunately, we know the structure of these columns. We will now show how we can use this knowledge either to verify optimality or to select an entering variable.

The idea is simple but clever. To determine whether the optimality conditions are satisfied, we will find a column $A_t = (a_1, a_2, \ldots, a_m)^T$ that corresponds to the most negative reduced cost $\hat{c}_t$. This in turn is the column t with the largest value of $y^T A_j$. Since each column corresponds to some cutting pattern, A_t will be the solution to the problem

$$\begin{aligned} \text{maximize} \quad & \bar{z} = y_1 a_1 + y_2 a_2 + \cdots + y_m a_m \\ \text{subject to} \quad & w_1 a_1 + w_2 a_2 + \cdots + w_m a_m \leq W \\ & a_i \geq 0 \text{ and integer}, i = 1, \ldots, m. \end{aligned}$$

This problem has a linear objective and a single linear constraint; all variables are required to be nonnegative integers. A problem of this form is called a *knapsack problem*. It can be solved efficiently by special-purpose algorithms. Since finding the variable x_t with the most negative reduced cost $\hat{c}_t$ can be formulated as a knapsack problem, it can be accomplished without explicit knowledge of each column of A.

The solution of the knapsack problem is a pattern A_t corresponding to the largest value of $y^T A_j$. If $y^T A_t > 1$, this column violates the optimality condition and is selected to enter the basis. If $y^T A_t \leq 1$, the current basic feasible solution is optimal, and the algorithm is terminated.

Example 7.3 (Cutting Stock Problem). Consider the example discussed in this section. Suppose that the initial basis matrix is given as above. The vector y is given by

$$y^T = c_B^T B^{-1} = (1 \quad 1 \quad 1) \begin{pmatrix} \frac{1}{2} & 0 & 0 \\ 0 & \frac{1}{3} & 0 \\ 0 & 0 & \frac{1}{6} \end{pmatrix} = (\tfrac{1}{2} \quad \tfrac{1}{3} \quad \tfrac{1}{6}).$$

To find the column with the most negative reduced cost $\hat{c}_j$, we solve the problem

$$\begin{aligned} \text{maximize} \quad & \tfrac{1}{2}a_1 + \tfrac{1}{3}a_2 + \tfrac{1}{6}a_3 \\ \text{subject to} \quad & 20a_1 + 14a_2 + 8a_3 \leq 50 \\ & a_1, a_2, a_3 \geq 0 \text{ and integer}, \end{aligned}$$

using a special-purpose algorithm for solving knapsack problems. The solution is $a = (0, 3, 1)^T$, with a knapsack objective value of $y^T a = 7/6$. For convenience we label a by

A_4, and label the corresponding variable (representing the number of sheets to be cut in pattern A_4) by x_4. Then $\hat{c}_4$ is the most negative of the coefficients $\hat{c}_j$. Since $\hat{c}_4 = 1 - 7/6 = -1/6 < 0$, the current solution is not optimal, and x_4 enters the basis. To determine which variable leaves the basis we compute

$$\hat{A}_4 = B^{-1}A_4 = \begin{pmatrix} \frac{1}{2} & 0 & 0 \\ 0 & \frac{1}{3} & 0 \\ 0 & 0 & \frac{1}{6} \end{pmatrix} \begin{pmatrix} 0 \\ 3 \\ 1 \end{pmatrix} = \begin{pmatrix} 0 \\ 1 \\ \frac{1}{6} \end{pmatrix}.$$

Recall that $x_B = (\frac{25}{4}, 40, \frac{10}{3})^T$. The admissible ratios in the ratio test are the second ratio $40/1 = 40$, and the third ratio $(10/3)/(1/6) = 20$. The latter is smaller, so the third basic variable leaves the basis. The new basis matrix and its inverse are

$$B = \begin{pmatrix} 2 & 0 & 0 \\ 0 & 3 & 3 \\ 0 & 0 & 1 \end{pmatrix} \quad \text{and} \quad B^{-1} = \begin{pmatrix} \frac{1}{2} & 0 & 0 \\ 0 & \frac{1}{3} & -1 \\ 0 & 0 & 1 \end{pmatrix},$$

and the vector of basic variables is

$$x_B = \begin{pmatrix} x_1 \\ x_2 \\ x_4 \end{pmatrix} = B^{-1}b = \begin{pmatrix} \frac{1}{2} & 0 & 0 \\ 0 & \frac{1}{3} & -1 \\ 0 & 0 & 1 \end{pmatrix} \begin{pmatrix} 25 \\ 120 \\ 20 \end{pmatrix} = \begin{pmatrix} \frac{25}{2} \\ 20 \\ 20 \end{pmatrix}.$$

The vector of simplex multipliers is

$$y^T = c_B^T B^{-1} = (1 \quad 1 \quad 1) B^{-1} = (\frac{1}{2} \quad \frac{1}{3} \quad 0),$$

and the new knapsack problem is

$$\begin{aligned} \text{maximize} \quad & \tfrac{1}{2}a_1 + \tfrac{1}{3}a_2 \\ \text{subject to} \quad & 20a_1 + 14a_2 + 8a_3 \le 50 \\ & a_1, a_2, a_3 \ge 0 \text{ and integer.} \end{aligned}$$

The solution to the knapsack problem is the pattern $a = (1, 2, 0)^T$ with knapsack objective value $y^T a = \frac{7}{6}$. We label a by A_5 and the corresponding variable by x_5. Then $\hat{c}_5 = 1 - y^T A_5 = 1 - 7/6 = -1/6 < 0$, and hence the current solution is still not optimal, and x_5 enters the basis. We compute

$$\hat{A}_5 = B^{-1}A_5 = \begin{pmatrix} \frac{1}{2} & 0 & 0 \\ 0 & \frac{1}{3} & -1 \\ 0 & 0 & 1 \end{pmatrix} \begin{pmatrix} 1 \\ 2 \\ 0 \end{pmatrix} = \begin{pmatrix} \frac{1}{2} \\ \frac{2}{3} \\ 0 \end{pmatrix}.$$

The ratio test compares the ratios $(\frac{25}{2})/(\frac{1}{2})$ and $(20)/(\frac{2}{3})$. The first of these ratios is smaller, and the leaving variable is x_1. The new basis matrix and its inverse are

$$B = \begin{pmatrix} 1 & 0 & 0 \\ 2 & 3 & 3 \\ 0 & 0 & 1 \end{pmatrix} \quad \text{and} \quad B^{-1} = \begin{pmatrix} 1 & 0 & 0 \\ -\frac{2}{3} & \frac{1}{3} & -1 \\ 0 & 0 & 1 \end{pmatrix},$$

and the vector of basic variables is

$$x_B = \begin{pmatrix} x_5 \\ x_2 \\ x_4 \end{pmatrix} = B^{-1}b = \begin{pmatrix} 25 \\ \frac{10}{3} \\ 20 \end{pmatrix}.$$

The vector of simplex multipliers is

$$y^T = c_B^T B^{-1} = (1 \quad 1 \quad 1) B^{-1} = (\tfrac{1}{3} \quad \tfrac{1}{3} \quad 0).$$

To find the column with the largest coefficient $\hat{c}_j$ we solve the knapsack problem

$$\begin{aligned} \text{maximize} \quad & \tfrac{1}{3}a_1 + \tfrac{1}{3}a_2 \\ \text{subject to} \quad & 20a_1 + 14a_2 + 8a_3 \le 50 \\ & a_1, a_2, a_3 \ge 0 \text{ and integer.} \end{aligned}$$

The solution to this problem is $a = (0, 3, 0)^T$, which is the column of the basic variable x_2. The knapsack objective value is $y^T a = 1$, indicating (as expected) that the reduced cost of x_2 is zero. Since $\hat{c}_2 = \min \{ \hat{c}_j \} = 0$, the current solution is optimal. Our basic variables are $x_5 = 25$, $x_2 = 3.33$, and $x_4 = 20$. In practice we are interested in an integer solution. For this problem, rounding upwards in fact gives the optimal integer solution. The solution is to cut 25 sections according to the pattern $(1, 2, 0)^T$, 4 sections according to the pattern $(0, 3, 0)^T$, and 20 sections according to the pattern $(0, 3, 1)^T$, using a total of 49 standard sheets. ∎

Exercises

3.1. Find all feasible patterns in the example discussed in this section.

3.2. Consider the cutting stock problem that arises when a company manufactures sheets of standard width $100''$ and has commitments to supply 40 sections of width $40''$, 60 sections of width $24''$, and 80 sections of width $18''$. Find an initial basic feasible solution to this problem, and formulate the knapsack problem that will determine whether this solution is optimal.

7.4   The Decomposition Principle

In some applications of linear programming the constraints of the problem are divided into two groups, one group of "easy" constraints and another of "hard" constraints. This can happen in network problems where the constraints that describe the network (the easy constraints) are augmented by additional constraints of a more general form (the hard constraints). This can also happen in "block angular" problems (see below) where there are a small number of constraints that involve all the variables (the hard constraints), but if these are removed the problem decomposes into several independent smaller problems, each of which is easier to solve. (On a parallel computer these smaller problems could be solved simultaneously on separate processors.)

Referring to the constraints as "easy" and "hard" may be a bit deceptive. The "hard" constraints need not be in themselves intrinsically difficult, but rather they can complicate the linear program, making the overall problem more difficult to solve. If these "complicating" constraints could be removed from the problem, then more efficient techniques could be applied to solve the resulting linear program.

The decomposition principle is a tool for solving linear programs having this structure. It is another example of column generation. The decomposition principle uses a change of variables to transform the original linear program into a new linear program that involves only the hard constraints. If the number of hard constraints is small, then it is likely that this new linear program can be solved by the simplex method in few iterations. However, the optimality test for the new linear program will require solving an auxiliary linear program involving the easy constraints, and so will be expensive. The cost of performing the optimality test will determine whether using the decomposition principle is more effective than applying the simplex method directly to the original problem.

Consider a linear program of the form

$$\begin{aligned} \text{minimize} \quad & z = c^T x \\ \text{subject to} \quad & A_H x = b_H \\ & A_E x = b_E \\ & x \geq 0, \end{aligned}$$

where A_H is the constraint matrix for the hard constraints and A_E is the matrix for the easy constraints. We will assume that the set

$$\{\, x : A_E x = b_E, x \geq 0 \,\}$$

is bounded. (The unbounded case will be discussed later in this section.) Then every feasible point x for the easy constraints can be represented as a convex combination of the extreme points $\{\, v_i \,\}$ of this set:

$$x = \sum_{i=1}^{k} \alpha_i v_i,$$

where

$$\sum_{i=1}^{k} \alpha_i = 1 \quad \text{and} \quad \alpha_i \geq 0, \quad i = 1, \ldots, k$$

(see Theorem 4.6 of Chapter 4). If the matrix A_E really does represent easy constraints, then in principle it should not be difficult to generate the associated extreme points (basic feasible solutions). The decomposition principle only generates individual extreme points as needed and does not normally generate the entire set of extreme points.

This representation in terms of extreme points can be used to rewrite the linear program:

$$\text{minimize} \quad z = c^T \left(\sum_i \alpha_i v_i \right)$$

$$\text{subject to} \quad A_H \left(\sum_i \alpha_i v_i \right) = b_H$$

$$\sum_i \alpha_i = 1$$

$$\alpha \geq 0.$$

If we define

$$c_M = (c^T v_1 \quad \cdots \quad c^T v_k)^T$$

$$A_M = \begin{pmatrix} A_H v_1 & \cdots & A_H v_k \\ 1 & \cdots & 1 \end{pmatrix}$$

$$b_M = \begin{pmatrix} b_H \\ 1 \end{pmatrix},$$

then the linear program becomes

$$\underset{\alpha}{\text{minimize}} \quad z = c_M^T \alpha$$

$$\text{subject to} \quad A_M \alpha = b_M$$

$$\alpha \geq 0.$$

This is the linear program that we will solve using the simplex method, with the coefficients α as the variables. It is sometimes referred to as the *master problem* to distinguish it from the auxiliary linear program that will be solved as part of the optimality test in the simplex method. Its coefficients are denoted with a subscript M because of this name.

Example 7.4 (Transformation of a Linear Program). Consider the linear program

$$\begin{aligned}
\text{minimize} \quad & z = 3x_1 - 5x_2 - 7x_3 + 2x_4 \\
\text{subject to} \quad & x_1 + 2x_2 - x_3 + 2x_4 = 2.75 \\
& 3x_1 - x_2 + 4x_3 - 5x_4 = 14.375 \\
& 2x_1 + 3x_2 \leq 9 \\
& x_1 + 2x_2 \leq 5 \\
& 3x_3 + 5x_4 \leq 15 \\
& x_3 - x_4 \leq 3 \\
& x \geq 0.
\end{aligned}$$

We will consider the first two constraints to be the hard constraints and the last four to be the easy constraints. If the hard constraints are removed, then the problem can be divided into two independent linear programs—one involving the variables x_1 and x_2, the other involving x_3 and x_4:

$$\begin{aligned}
\text{minimize} \quad & 3x_1 - 5x_2 \\
\text{subject to} \quad & 2x_1 + 3x_2 \leq 9 \\
& x_1 + 2x_2 \leq 5 \\
& x_1, x_2 \geq 0
\end{aligned}$$

and

$$\text{minimize} \quad -7x_3 + 2x_4$$
$$\text{subject to} \quad 3x_3 + 5x_4 \leq 15$$
$$x_3 - x_4 \leq 3$$
$$x_3, x_4 \geq 0.$$

These linear programs involve only two variables and can be solved graphically.

If slack variables x_5, x_6, x_7, and x_8 are added to the easy constraints to convert them into equations, then

$$c = (3 \quad -5 \quad -7 \quad 2 \quad 0 \quad 0 \quad 0 \quad 0)^T,$$

$$A_H = \begin{pmatrix} 1 & 2 & -1 & 2 & 0 & 0 & 0 & 0 \\ 3 & -1 & 4 & -5 & 0 & 0 & 0 & 0 \end{pmatrix}, \quad b_H = \begin{pmatrix} 2.750 \\ 14.375 \end{pmatrix},$$

$$A_E = \begin{pmatrix} 2 & 3 & 0 & 0 & 1 & 0 & 0 & 0 \\ 1 & 2 & 0 & 0 & 0 & 1 & 0 & 0 \\ 0 & 0 & 3 & 5 & 0 & 0 & 1 & 0 \\ 0 & 0 & 1 & -1 & 0 & 0 & 0 & 1 \end{pmatrix}, \quad b_E = \begin{pmatrix} 9 \\ 5 \\ 15 \\ 3 \end{pmatrix}.$$

The extreme points for the first linear program (with the slack variables included) are

$$\begin{pmatrix} x_1 \\ x_2 \\ x_5 \\ x_6 \end{pmatrix} = \begin{pmatrix} 3 \\ 1 \\ 0 \\ 0 \end{pmatrix}, \quad \begin{pmatrix} 4.5 \\ 0 \\ 0 \\ 0.5 \end{pmatrix}, \quad \begin{pmatrix} 0 \\ 2.5 \\ 1.5 \\ 0 \end{pmatrix}, \quad \text{and} \quad \begin{pmatrix} 0 \\ 0 \\ 9 \\ 5 \end{pmatrix}.$$

The extreme points for the second linear program are

$$\begin{pmatrix} x_3 \\ x_4 \\ x_7 \\ x_8 \end{pmatrix} = \begin{pmatrix} 3.75 \\ 0.75 \\ 0 \\ 0 \end{pmatrix}, \quad \begin{pmatrix} 3 \\ 0 \\ 6 \\ 0 \end{pmatrix}, \quad \begin{pmatrix} 0 \\ 3 \\ 0 \\ 6 \end{pmatrix}, \quad \text{and} \quad \begin{pmatrix} 0 \\ 0 \\ 15 \\ 3 \end{pmatrix}.$$

The extreme points for the set of easy constraints $\{ x : A_E x = b_E, x \geq 0 \}$ are obtained by combining extreme points from the two smaller linear programs. The resulting matrix of extreme points $V = (v_1 \quad \cdots \quad v_k)$ is

$$V = \begin{pmatrix} 3 & 3 & 3 & 3 & 4.5 & 4.5 & 4.5 & 4.5 & 0 & 0 & 0 & 0 & 0 & 0 & 0 & 0 \\ 1 & 1 & 1 & 1 & 0 & 0 & 0 & 0 & 2.5 & 2.5 & 2.5 & 2.5 & 0 & 0 & 0 & 0 \\ 3.75 & 3 & 0 & 0 & 3.75 & 3 & 0 & 0 & 3.75 & 3 & 0 & 0 & 3.75 & 3 & 0 & 0 \\ 0.75 & 0 & 3 & 0 & 0.75 & 0 & 3 & 0 & 0.75 & 0 & 3 & 0 & 0.75 & 0 & 3 & 0 \\ 0 & 0 & 0 & 0 & 0 & 0 & 0 & 0 & 1.5 & 1.5 & 1.5 & 1.5 & 9 & 9 & 9 & 9 \\ 0 & 0 & 0 & 0 & 0.5 & 0.5 & 0.5 & 0.5 & 0 & 0 & 0 & 0 & 5 & 5 & 5 & 5 \\ 0 & 6 & 0 & 15 & 0 & 6 & 0 & 15 & 0 & 6 & 0 & 15 & 0 & 6 & 0 & 15 \\ 0 & 0 & 6 & 3 & 0 & 0 & 6 & 3 & 0 & 0 & 6 & 3 & 0 & 0 & 6 & 3 \end{pmatrix}.$$

The master linear program has constraint matrix A_M, whose first two rows are equal to $A_H V$, and whose last row is $(1, \ldots, 1)$:

$$A_M = \begin{pmatrix} 2.75 & 2 & 11 & 5 & 2.25 & 1.5 & 10.5 & 4.5 \\ 19.25 & 20 & -7 & 8 & 24.75 & 25.5 & -1.5 & 13.5 \\ 1 & 1 & 1 & 1 & 1 & 1 & 1 & 1 \end{pmatrix}$$

$$\begin{pmatrix} 2.75 & 2 & 11 & 5 & -2.25 & -3 & 6 & 0 \\ 8.75 & 9.5 & -17.5 & -2.5 & 11.25 & 12 & -15 & 0 \\ 1 & 1 & 1 & 1 & 1 & 1 & 1 & 1 \end{pmatrix}.$$

The objective coefficients in the master problem are

$$c_M = V^T c = \begin{pmatrix} -20.75 & -17 & 10 & 4 & -11.25 & -7.5 & 19.5 & 13.5 \end{pmatrix}$$
$$\begin{pmatrix} -37.25 & -33.5 & -6.5 & -12.5 & -24.75 & -21 & 6 & 0 \end{pmatrix}^T.$$

The right-hand side for the master problem is

$$b_M = \begin{pmatrix} 2.75 \\ 14.375 \\ 1 \end{pmatrix}. \quad \blacksquare$$

It is neither necessary nor desirable to write down the master linear program explicitly. The number of extreme points can be immense—much larger than the number of variables in the original problem. Fortunately, it is possible to apply the simplex method to the master problem without explicitly generating all the columns of A_M.

To describe the method we assume that an initial basic feasible solution has been specified. (Initialization procedures are discussed below.) If there are m hard constraints in A_H, the basis will be of size $m + 1$ because of the additional constraint $\sum \alpha_i = 1$. Let B be the basis matrix (B is a submatrix of A_M). As usual, the dual variables are computed via

$$y^T = (c_M)_B^T B^{-1},$$

where $(c_M)_B$ is the subvector of c_M corresponding to the current basis. The optimality test is carried out by computing the components of

$$c_M^T - y^T A_M$$

corresponding to the nonbasic variables. If any of these entries is negative, then the current basis is not optimal, and the most negative of these entries can be used to select the entering variable. Hence the optimality test can be carried out by determining

$$\min_i (c_M^T)_i - (y^T A_M)_i$$

and checking to see if the optimal value is negative. All values of i can be considered, since this expression will be zero if α_i is a basic variable.

This minimization problem is equivalent to

$$\min_i c^T v_i - \bar{y}^T A_H v_i - y_{m+1} \cdot 1,$$

where $\bar{y} = (y_1, \ldots, y_m)^T$. Since $\{v_i\}$ is the set of extreme points, and since any bounded linear program always has an optimal extreme point (see Section 4.4), the optimality test can be written as

$$\begin{array}{ll} \underset{x}{\text{minimize}} & z = (c - A_H^T \bar{y})^T x - y_{m+1} \\ \text{subject to} & A_E x = b_E \\ & x \geq 0. \end{array}$$

This is a linear program involving only the easy constraints, so presumably it is easy to solve. (The term y_{m+1} in the objective is a constant. It can be ignored when solving this linear program for x but must be included when determining the optimal value of z.)

If the solution to this linear program has optimal objective value zero, then the current basis for the master problem is optimal. If the optimal objective value is negative, then the current basis is not optimal. Note that the optimal basic feasible solution (the optimal basic feasible solution for the easy problem in the optimality test) is one of the extreme points in $\{v_i\}$. Denote it by v.

The rest of the simplex method for solving the master problem is much as before. The entering column is computed using the formula

$$(\hat{A}_M)_t = B^{-1} \begin{pmatrix} A_H v \\ 1 \end{pmatrix}.$$

A ratio test is performed to determine the leaving variable, and then B^{-1} and α_B are updated. (Recall that α is the vector of variables in the master problem.)

We now summarize the steps in an iteration of the method. A representation of B^{-1} must be provided, where the basis matrix B consists of the basic columns of the matrix A_M. The corresponding basic feasible solution to the master problem is $\alpha_B = B^{-1} b_M$. The vector $(c_M)_B$ contains the basic components of the vector c_M. Note that A_M and c_M are not normally available explicitly; only B and $(c_M)_B$ may be available.

1. *The Optimality Test*—Compute the dual variables $y^T = (c_M)_B^T B^{-1}$ and set $\bar{y} = (y_1, \ldots, y_m)^T$. Solve the linear program

$$\begin{array}{ll} \underset{x}{\text{minimize}} & z = (c - A_H^T \bar{y})^T x - y_{m+1} \\ \text{subject to} & A_E x = b_E \\ & x \geq 0 \end{array}$$

for an optimal extreme point v. If the optimal value is zero, then the current basis is optimal. Otherwise, use v to define the entering column.

2. *The Step*—Compute the entering column

$$(\hat{A}_M)_t = B^{-1} \begin{pmatrix} A_H v \\ 1 \end{pmatrix}.$$

Find an index s that satisfies

$$\frac{(\hat{b}_M)_s}{(\hat{A}_M)_{s,t}} = \min_{1 \leq i \leq m+1} \left\{ \frac{(\hat{b}_M)_i}{(\hat{A}_M)_{i,t}} : (\hat{A}_M)_{i,t} > 0 \right\}.$$

(Here $(\hat{A}_M)_{i,t}$ denotes the ith component of $(\hat{A}_M)_t$.) The ratio test determines the leaving variable and the pivot entry $(\hat{A}_M)_{s,t}$. If $(\hat{A}_M)_{i,t} \le 0$ for all i, then the problem is unbounded.

3. *The Update*—Update the inverse matrix B^{-1} and the vector of basic variables α_B (for example, by performing elimination operations that transform $(\hat{A}_M)_t$ into the sth column of the identity matrix).

Once an optimal solution to the master problem has been found, the solution to the original problem is obtained from

$$x = V_B \alpha_B,$$

where V_B is the matrix whose columns are the vertices corresponding to the optimal basis for the master problem.

We now illustrate the decomposition principle with an example. The set of extreme points for this example was given in Example 7.4, but it is not used here, and would not normally be available. Only information associated with the current basis is used in the calculations. We assume that an initial basic feasible solution for the master problem is available; initialization procedures are discussed later in this section.

Example 7.5 (Decomposition Principle). Consider the linear program from Example 7.4. As before, we will consider the first two constraints to be the hard constraints and the last four to be the easy constraints. An initial feasible point for the master problem can be obtained using the extreme points

$$v_1 = \begin{pmatrix} 3 \\ 1 \\ 3 \\ 0 \\ 0 \\ 0 \\ 6 \\ 0 \end{pmatrix}, \quad v_2 = \begin{pmatrix} 0 \\ 2.5 \\ 3 \\ 0 \\ 1.5 \\ 0 \\ 6 \\ 0 \end{pmatrix}, \quad \text{and} \quad v_3 = \begin{pmatrix} 0 \\ 2.5 \\ 0 \\ 3 \\ 1.5 \\ 0 \\ 0 \\ 6 \end{pmatrix}.$$

For convenience, we have labeled the extreme points as v_1, v_2, and v_3; this merely reflects the fact that they are the initial extreme points and does not correspond to their location in the matrix V in Example 7.4. For this initial basis,

$$(c_M)_B = (c^T v_1 \quad c^T v_2 \quad c^T v_3)^T = (-17 \quad -33.5 \quad -6.5)^T$$

$$B = \begin{pmatrix} A_H v_1 & A_H v_2 & A_H v_3 \\ 1 & 1 & 1 \end{pmatrix} = \begin{pmatrix} 2 & 2 & 11 \\ 20 & 9.5 & -17.5 \\ 1 & 1 & 1 \end{pmatrix}$$

$$\alpha_B = \hat{b}_M = B^{-1} b_M = (0.6786 \quad 0.2381 \quad 0.0833)^T$$

$$z = -20.0536.$$

The dual variables are

$$y^T = (c_M)_B^T B^{-1} = (7.7143 \quad 1.5714 \quad -63.8571),$$

so $\bar{y} = (7.7143 \quad 1.5714)^T$. The coefficients of the objective in the linear program for the optimality test are

$$c - A_{H.}^T \bar{y} = (-9.4286 \quad -18.8571 \quad -5.5714 \quad -5.5714 \quad 0 \quad 0 \quad 0 \quad 0)^T.$$

Recall from Example 7.4 that, because of the special structure of the easy constraint matrix A_E, this linear program splits into two smaller linear programs:

$$
\begin{aligned}
\text{minimize} \quad & -9.4286x_1 - 18.8571x_2 \\
\text{subject to} \quad & 2x_1 + 3x_2 \le 9 \\
& x_1 + 2x_2 \le 5 \\
& x_1, x_2 \ge 0
\end{aligned}
$$

and

$$
\begin{aligned}
\text{minimize} \quad & -5.5714x_3 - 5.5714x_4 \\
\text{subject to} \quad & 3x_3 + 5x_4 \le 15 \\
& x_3 - x_4 \le 3 \\
& x_3, x_4 \ge 0.
\end{aligned}
$$

These small linear programs can be solved graphically.

The first linear program has two optimal basic feasible solutions (here the optimal slack variables are also listed):

$$(x_1 \quad x_2 \quad x_5 \quad x_6)^T = (3 \quad 1 \quad 0 \quad 0)^T \quad \text{and} \quad (0 \quad 2.5 \quad 1.5 \quad 0)^T$$

with optimal objective value -47.1429. The second linear program has the solution

$$(x_3 \quad x_4 \quad x_7 \quad x_8)^T = (3.75 \quad 0.75 \quad 0 \quad 0)^T$$

with objective value -25.0714. The objective value for the linear program in the optimality test is obtained by subtracting y_3 from these two objective values:

$$-47.1429 - 25.0714 - y_3 = -47.1429 - 25.0714 - (-63.8571) = -8.3571 < 0,$$

so the current basis is not optimal. If the first solution to the first linear program is used, then

$$v_4 = (3 \quad 1 \quad 3.75 \quad 0.75 \quad 0 \quad 0 \quad 0 \quad 0)^T$$

is the entering vertex.

The entering column is given by

$$(\hat{A}_M)_t = B^{-1} \begin{pmatrix} A_H v_4 \\ 1 \end{pmatrix} = \begin{pmatrix} 1.1429 \\ -0.2262 \\ 0.0833 \end{pmatrix}.$$

With the right-hand side $\hat{b}_M = (0.6786 \quad 0.2381 \quad 0.0833)^T$, the ratios in the ratio test are

$$\begin{pmatrix} 0.5938 \\ - \\ 1 \end{pmatrix},$$

so that α_1 is the leaving variable. The new basis consists of $\{\alpha_4, \alpha_2, \alpha_3\}$.

At the second iteration,

$$(c_M)_B = (-20.75 \quad -33.5 \quad -6.5)^T$$

$$B = \begin{pmatrix} 2.75 & 2 & 11 \\ 19.25 & 9.5 & -17.5 \\ 1 & 1 & 1 \end{pmatrix}$$

$$\alpha_B = (0.5938 \quad 0.3724 \quad 0.0339)^T$$

$$z = -25.0156.$$

The dual variables are

$$y^T = (c_M)_B^T B^{-1} = (5.625 \quad 0.875 \quad -53.0625).$$

The objective coefficients for the linear program in the optimality test are

$$c - A_H^T \bar{y} = (-5.25 \quad -15.375 \quad -4.875 \quad -4.875 \quad 0 \quad 0 \quad 0 \quad 0)^T.$$

The resulting two small linear programs are solved graphically.

The first has the solution

$$(0 \quad 2.5 \quad 1.5 \quad 0)^T$$

with optimal objective value -38.4275. The second has the solution

$$(3.75 \quad 0.75 \quad 0 \quad 0)^T$$

with objective value -21.9375. For the optimality test the objective value is

$$-38.4275 - 21.9375 - (-53.0625) = -7.3125 < 0,$$

so the current basis is not optimal, and

$$v_5 = (0 \quad 2.5 \quad 3.75 \quad 0.75 \quad 1.5 \quad 0 \quad 0 \quad 0)^T$$

is the entering vertex.

The entering column is

$$(\hat{A}_M)_t = B^{-1} \begin{pmatrix} A_H v_5 \\ 1 \end{pmatrix} = \begin{pmatrix} 0.1250 \\ 0.8021 \\ 0.0729 \end{pmatrix},$$

and the ratios in the ratio test are

$$\begin{pmatrix} 4.7500 \\ 0.4643 \\ 0.4643 \end{pmatrix}.$$

There is a tie; we will (arbitrarily) choose α_2 as the leaving variable. The new basis consists of $\{\alpha_4, \alpha_5, \alpha_3\}$.

At the third iteration,

$$(c_M)_B = (-20.75 \quad -37.25 \quad -6.5)^T$$

$$B = \begin{pmatrix} 2.75 & 2.75 & 11 \\ 19.25 & 8.75 & -17.5 \\ 1 & 1 & 1 \end{pmatrix}$$

$$\alpha_B = (0.5357 \quad 0.4643 \quad 0)^T$$

$$z = -28.4107.$$

The dual variables are

$$y^T = (c_M)_B^T B^{-1} = (8.7273 \quad 1.5714 \quad -75).$$

The objective coefficients for the linear program in the optimality test are

$$c - A_H^T \bar{y} = (-10.4416 \quad -20.8831 \quad -4.5584 \quad -7.5974 \quad 0 \quad 0 \quad 0 \quad 0)^T.$$

The resulting two small linear programs are solved graphically.
   The first has two solutions:

$$(3 \quad 1 \quad 0 \quad 0)^T \quad \text{and} \quad (0 \quad 2.5 \quad 1.5 \quad 0)^T$$

with optimal objective value -52.2078. The second also has two solutions:

$$(3.75 \quad 0.75 \quad 0 \quad 0)^T \quad \text{and} \quad (0 \quad 3 \quad 0 \quad 6)^T$$

with objective value -22.7922. For the optimality test the objective value is

$$-52.2078 - 22.7922 - (-75) = 0,$$

so the current basis for the master problem is optimal.
   The optimal objective value for the original linear program is $z = -28.4107$, the same as for the master linear program. The optimal values of the original variables are

$$x = \alpha_4 v_4 + \alpha_5 v_5 + \alpha_3 v_3 = (1.6071 \quad 1.6964 \quad 3.75 \quad 0.75 \quad 0.6964 \quad 0 \quad 0 \quad 0)^T.$$

It is straightforward to check that this point is feasible for the original problem, and that it gives the correct objective value.   ∎

   Some procedure is needed to find an initial basic feasible solution for the simplex method. In some cases, an initial basis is readily available. For example, suppose that the original linear program is of the form

$$\begin{aligned}
\text{minimize} \quad & z = c^T x \\
\text{subject to} \quad & A_H x \leq b_H \\
& A_E x \leq b_E \\
& x \geq 0
\end{aligned}$$

with $b_H \geq 0$ and $b_E \geq 0$. Then $v_0 = (0, \dots, 0)^T$ is an extreme point for the set $\{x : A_E x \leq b_E, x \geq 0\}$. Let V be the matrix whose columns are the extreme points of this set, with v_0 as its first column. Then the master problem will have the form

$$\begin{aligned}
\underset{\alpha}{\text{minimize}} \quad & z = c_M^T \alpha \\
\text{subject to} \quad & (A_H V)\alpha + s = b_H \\
& e^T \alpha = 1 \\
& \alpha \geq 0,
\end{aligned}$$

where s is a vector of slack variables and $e = (1, \ldots, 1)^T$. The constraint matrix for the master problem will be

$$A_M = \begin{pmatrix} A_H V & I \\ e^T & 0 \end{pmatrix}.$$

An initial basis can be obtained using the slack variables s (the final columns of A_M) together with the variable corresponding to the extreme point $v_0 = (0, \ldots, 0)^T$ (the initial column of A_M). Since $A_H v_0 = (0, \ldots, 0)^T$, the basis matrix will be $B = I$. This idea can also be used for problems of the form

$$\begin{aligned}
\text{minimize} \quad & z = c^T x \\
\text{subject to} \quad & A_H x \leq b_H \\
& A_E x = b_E \\
& x \geq 0
\end{aligned}$$

if we know an extreme point v_0 for the easy constraints (see the Exercises).

In cases where there is no obvious initial basis for the master problem, artificial variables must be added. Then either a two-phase or big-M approach can be used to initialize the method. For example, if a two-phase approach is used and $b_M \geq 0$, then the phase-1 problem would be

$$\begin{aligned}
\underset{\alpha, a}{\text{minimize}} \quad & z = e^T a \\
\text{subject to} \quad & A_M \alpha + a = b_M \\
& \alpha, a \geq 0,
\end{aligned}$$

where a is a vector of artificial variables and $e = (1, \ldots, 1)^T$. The initial basis consists entirely of the artificial variables and $B = I$. Then the algorithm for the decomposition principle is used to solve this linear program to obtain an initial basic feasible solution for the original problem.

Up to this point we have assumed that the set $\{ x : A_E x = b_E, x \geq 0 \}$ is bounded. This is not necessary. In general every feasible point x can be represented in terms of extreme points $\{ v_i \}$ and directions of unboundedness $\{ d_j \}$:

$$x = \sum_{i=1}^{k} \alpha_i v_i + \sum_{j=1}^{\ell} \beta_j d_j,$$

where

$$\sum_{i=1}^{k} \alpha_i = 1$$
$$\alpha_i \geq 0, \quad i = 1, \ldots, k$$
$$\beta_j \geq 0, \quad j = 1, \ldots, \ell.$$

(See Theorem 4.6 of Chapter 4.) This representation can then be used to derive a similar simplex method for the decomposed problem. For further details, see the book by Chvátal (1983).

One of the most important applications of the decomposition principle is to *block angular* problems. These are problems where A_E, the matrix for the easy constraints, is block diagonal. That is,

$$A_E = \begin{pmatrix} (A_E)_{(1)} & & & \mathbf{0} \\ & (A_E)_{(2)} & & \\ & & \ddots & \\ \mathbf{0} & & & (A_E)_{(r)} \end{pmatrix},$$

where each $(A_E)_{(j)}$ is itself a matrix. In this case the linear program for the optimality test

$$\begin{aligned} \underset{x}{\text{maximize}} \quad & z = (A_H^T \bar{y} - c)^T x + y_{m+1} \\ \text{subject to} \quad & A_E x = b_E \\ & x \geq 0 \end{aligned}$$

splits into r disjoint linear programs of smaller size:

$$\begin{aligned} \underset{x_{(j)}}{\text{maximize}} \quad & z_{(j)} = (A_H^T \bar{y} - c)_{(j)}^T x_{(j)} \\ \text{subject to} \quad & (A_E)_{(j)} x_{(j)} = (b_E)_{(j)} \\ & x_{(j)} \geq 0. \end{aligned}$$

(The subscript (j) indicates the components of a vector corresponding to the submatrix $(A_E)_{(j)}$.) On a parallel computer, it would be possible to solve these smaller problems simultaneously on a set of processors.

The linear program in Example 7.4 is of this type, with

$$(A_E)_{(1)} = \begin{pmatrix} 2 & 3 \\ 1 & 2 \end{pmatrix}, \quad (b_E)_{(1)} = \begin{pmatrix} 9 \\ 5 \end{pmatrix},$$

$$(A_E)_{(2)} = \begin{pmatrix} 3 & 5 \\ 1 & -1 \end{pmatrix}, \quad \text{and} \quad (b_E)_{(2)} = \begin{pmatrix} 15 \\ 3 \end{pmatrix}.$$

Problems with this structure arise when modeling an organization consisting of many separate divisions. Each of the blocks in the easy constraints corresponds to the portion of the model for a particular division. The hard constraints correspond to the linkages connecting one division with another, and with the allocation of activities and resources among the divisions. The overall objective is to optimize some "benefit" for the entire organization.

Exercises

4.1. Set up and solve the phase-1 problem for the master problem in Example 7.4.

4.2. Consider a linear program of the form

$$\text{minimize} \quad z = c^T x$$
$$\text{subject to} \quad A_H x \leq b_H$$
$$A_E x = b_E$$
$$x \geq 0,$$

where A_H represents the hard constraints and A_E represents the easy constraints. Assume that an extreme point v_0 for the easy constraint set

$$\{ x : A_E x = b_E, x \geq 0 \}$$

is known. Show how to find an initial basic feasible solution for the decomposition principle. What is the basis matrix B? Show how to compute B^{-1} efficiently.

4.3. Solve the following linear program using the decomposition principle:

$$\text{minimize} \quad z = -2x_1 - 5x_2$$
$$\text{subject to} \quad x_1 + 2x_2 = 13.5$$
$$x_1 + 3x_2 = 18.0$$
$$x_1 \leq 9$$
$$x_2 \leq 5$$
$$x_1, x_2 \geq 0.$$

Let the first two constraints be the "hard" constraints and the remaining constraints be the "easy" constraints. As an initial basis use the extreme points $(0, 0)^T$, $(0, 5)^T$, and $(9, 5)^T$. (These are extreme points for the easy constraints in their original form, without slack variables.)

4.4. Use the decomposition principle to solve the problem

$$\text{minimize} \quad z = -x_1 - 2x_2 - 4y_1 - 3y_2$$
$$\text{subject to} \quad x_1 + x_2 + 2y_1 \leq 4$$
$$x_2 + y_1 + y_2 \leq 3$$
$$2x_1 + x_2 \leq 4$$
$$x_1 + x_2 \leq 2$$
$$y_1 + y_2 \leq 2$$
$$3y_1 + 2y_2 \leq 5$$
$$x \geq 0, \quad y \geq 0.$$

4.5. At each iteration of the decomposition principle the current objective value provides an upper bound on the optimal objective value. It is also possible to determine a lower bound on the objective. Let x be a feasible point for the original linear program, and let y be the vector of dual variables at the current iteration of the decomposition principle. Prove that

$$(c - A_H^T \bar{y})^T x - y_{m+1} \geq \bar{z}_*,$$

where $\bar{z}_*$ is the optimal value of the linear program in the optimality test. Use this formula to prove that

$$z_* \geq \bar{y}^T b_H + y_{m+1} + \bar{z}_*,$$

where z_* is the optimal objective value for the original linear program.

4.6. Use the formula from the previous problem to compute lower bounds on the objective at each iteration for the linear program in Example 7.4.

7.5   Representation of the Basis

In previous chapters we described the formulas that govern the simplex method. We have seen that all of the information needed for an iteration can be obtained from the set of variables that are basic and from the corresponding basic matrix B. Thus the vectors x_b, y, and $\hat{A}_s$ can be computed directly from the formulas

$$x_b = B^{-1}b, \quad y^T = c_B^T B^{-1}, \quad \hat{A}_s = B^{-1}A_s.$$

In this section we discuss approaches for efficient implementation of these formulas. In Section 7.5.1 we describe the approach known as the "product form of the inverse" in which the basis inverse B^{-1} is represented as a product of *elementary matrices*. These elementary matrices are quite sparse, so matrix-vector multiplications with respect to the basis matrix inverse can be performed at relatively low cost. This method, however, has been superseded by a more efficient approach that represents the LU factorization of the basis B as a product of elementary matrices; this approach is described in Section 7.5.2. Using this approach the vectors x_b, y, and $\hat{A}_s$ are obtained by solving a system of equations with respect to B:

$$Bx_b = b, \quad y^T B = c_B^T, \quad B\hat{A}_s = A_s.$$

The LU factorization is superior to the product form of the inverse both in its utilization of sparsity and its control of roundoff errors. However, it is more complicated both in notation and in the operations involved. For this reason we have chosen to set the background for key ideas in the use of products of elementary matrices by first describing the product form of the inverse, and only then discussing the LU factorization. If the background ideas are familiar to the reader, it is possible to skip Section 7.5.1 and turn directly to Section 7.5.2.

7.5.1   The Product Form of the Inverse

To develop the product form of the inverse, consider a simplex iteration with basis matrix B. Suppose that at the end of this iteration, the sth basic variable is replaced by x_t. The new basis matrix $\bar{B}$ is obtained from B by replacing its sth column by A_t. Since $\hat{A}_t = B^{-1}A_t$ (or $B\hat{A}_t = A_t$), it follows that

$$\bar{B} = BF,$$

where

$$F = \begin{pmatrix} 1 & & & & & \\ & 1 & & & & \\ & & \ddots & \begin{bmatrix} \\ \hat{A}_t \\ \\ \end{bmatrix} & & \\ & & & & \ddots & \\ & & & & & 1 \end{pmatrix}$$

is obtained from the identity matrix by replacing its sth column by $\hat{A}_t$.

Denoting $E = F^{-1}$, the new inverse matrix is obtained by multiplying B^{-1} from the left by E:

$$\bar{B}^{-1} = E B^{-1}.$$

It is easy to verify that E also differs from the identity matrix only in its sth column

$$E = \begin{pmatrix} 1 & & & & & & \\ & 1 & & & & & \\ & & \ddots & & & & \\ & & & \eta & & & \\ & & & & \ddots & & \\ & & & & & 1 & \\ & & & & & & 1 \end{pmatrix},$$

where

$$\eta = \begin{pmatrix} -\hat{a}_{1,t}/\hat{a}_{s,t} \\ \vdots \\ -\hat{a}_{s-1,t}/\hat{a}_{s,t} \\ 1/\hat{a}_{s,t} \\ -\hat{a}_{s+1,t}/\hat{a}_{s,t} \\ \vdots \\ -\hat{a}_{m,t}/\hat{a}_{s,t} \end{pmatrix}$$

and $\{\hat{a}_{j,t}\}$ are the entries in $\hat{A}_t$ (see the Exercises). The matrix E is called an *elementary matrix* and the vector η is called an *eta vector*; one can show that the pivot operations performed on $\hat{A}$ are achieved by multiplying $\hat{A}$ on the left by E (see the Exercises).

Suppose that we start the simplex algorithm with an initial basis matrix $B_1 = I$. Let E_1 denote the elementary matrix corresponding to the pivot operations in the first iteration. Then at the second iteration the inverse basis matrix is $B_2^{-1} = E_1 B_1^{-1} = E_1$. Similarly, in the third iteration, $B_3^{-1} = E_2 B_2^{-1} = E_2 E_1$, and in general at the kth iteration the inverse basis matrix is

$$B_k^{-1} = E_{k-1} E_{k-2} \cdots E_2 E_1,$$

where E_i is the elementary matrix corresponding to iteration i. This representation of the basis matrix inverse is known as the *product form of the inverse*. The basis matrix inverse is not formed explicitly, but kept as a product of its factors. Since each elementary matrix is uniquely defined by its eta vector and its column position, it may be stored compactly. For historic reasons, the collection of eta vectors corresponding to $E_1, E_2, \ldots, E_k$ is known as the *eta file*.

How can we perform the simplex computations without explicitly forming B^{-1}? The matrix is not needed explicitly, but only to provide matrix-vector products. In the simplex method these operations occur in two forms: (a) premultiplication—multiplication of a column vector from the left ($\hat{b} = B^{-1}b$ and $\hat{A}_t = B^{-1}A_t$); and (b) postmultiplication—multiplication of a row vector from the right ($y^T = c_B^T B^{-1}$). We now show how these operations can be performed.

A computation of the form $B_k^{-1}a$ is performed sequentially via

$$B_k^{-1}a = E_{k-1}(E_{k-2}(E_{k-3}(\cdots (E_2(E_1 a)) \cdots))),$$

by first premultiplying a by E_1, next premultiplying the resulting vector by E_2, and so forth. This operation is called a forward transformation (FTRAN), because it corresponds to a forward scan of the eta file.

A computation of the form $c^T B_k^{-1}$ is performed sequentially via

$$c^T B_k^{-1} = ((\cdots(((c^T E_{k-1})E_{k-2})E_{k-3})\cdots)E_2)E_1$$

by first postmultiplying c^T by E_{k-1}, then postmultiplying the result by E_{k-2}, and so forth. This operation is called a backward transformation (BTRAN), because it corresponds to a backward scan of the eta file.

Each matrix operation with respect to an elementary matrix is fast and simple. To see this, let E be an elementary matrix with eta vector η in its sth column. Then a matrix-vector product of the form Ea for some vector a is computed as

$$Ea = \begin{pmatrix} 1 & & \eta_1 & & \\ & \ddots & \vdots & & \\ & & \eta_s & & \\ & & \vdots & \ddots & \\ & & \eta_m & & 1 \end{pmatrix} \begin{pmatrix} a_1 \\ \vdots \\ a_s \\ \vdots \\ a_m \end{pmatrix}$$

$$= \begin{pmatrix} a_1 + \eta_1 a_s \\ \vdots \\ \eta_s a_s \\ \vdots \\ a_m + \eta_m a_s \end{pmatrix} = \begin{pmatrix} a_1 \\ \vdots \\ 0 \\ \vdots \\ a_m \end{pmatrix} + a_s \begin{pmatrix} \eta_1 \\ \vdots \\ \eta_s \\ \vdots \\ \eta_m \end{pmatrix}.$$

The rule is, replace the sth term of a by zero, and add to that a_s times η. The matrix E need not be formed explicitly.

Example 7.6 (Premultiplication of a Column Vector). Consider the 4×4 elementary matrix E defined by

$$s = 3, \quad \eta = \begin{pmatrix} -\frac{3}{2} \\ 1 \\ \frac{1}{2} \\ -3 \end{pmatrix}.$$

This matrix is obtained from the identity matrix by replacing its third column by η. Let $a = (7, -3, 4, 2)^T$. Then the matrix-vector product Ea is computed as

$$Ea = \begin{pmatrix} a_1 \\ a_2 \\ 0 \\ a_4 \end{pmatrix} + a_3 \eta = \begin{pmatrix} 7 \\ -3 \\ 0 \\ 2 \end{pmatrix} + 4 \begin{pmatrix} -\frac{3}{2} \\ 1 \\ \frac{1}{2} \\ -3 \end{pmatrix} = \begin{pmatrix} 1 \\ 1 \\ 2 \\ -10 \end{pmatrix}.$$

The computation was carried out without explicitly forming the matrix E. ∎

A vector product of the form $c^T E$ is computed as

$$c^T E = (c_1 \quad c_2 \quad \cdots \quad c_s \quad \cdots \quad c_m) \begin{pmatrix} 1 & & & & \eta_1 & & \\ & 1 & & & \vdots & & \\ & & \ddots & & \vdots & & \\ & & & & \eta_s & & \\ & & & & \vdots & \ddots & \\ & & & & \eta_m & & 1 \end{pmatrix}$$

$$= (c_1 \quad c_2 \quad \cdots \quad c_{s-1} \quad (c_1\eta_1 + \cdots + c_m\eta_m) \quad c_{s+1} \quad \cdots \quad c_m)$$

$$= (c_1 \quad c_2 \quad \cdots \quad c_{s-1} \quad c^T\eta \quad c_{s+1} \quad \cdots \quad c_m).$$

Thus, the computation leaves c unchanged except for its sth component which is replaced by $c^T\eta$.

Example 7.7 (Postmultiplication of a Row Vector). Consider the matrix E in the previous example, and let $c^T = (-1, 2, -3, 4)$. The matrix-vector product $c^T E$ is computed as follows:

$$c^T E = (c_1 \quad c_2 \quad c^T\eta \quad c_4) = (-1 \quad 2 \quad -10 \quad 4),$$

since

$$c^T\eta = (-1 \quad 2 \quad -3 \quad 4) \begin{pmatrix} -\frac{3}{2} \\ 1 \\ \frac{1}{2} \\ -3 \end{pmatrix} = -10. \qquad \blacksquare$$

We can now outline the steps of the kth iteration of the product-form version of the simplex method. Available at this iteration is a basis matrix inverse B_k^{-1}, represented as a product of elementary matrices $E_{k-1} \cdots E_1$. Each elementary matrix E_i is represented by its eta vector η_i and its row index s_i. We also have available $x_B = \hat{b} = B_k^{-1}b$. The steps of the algorithm are given below.

1. *The Optimality Test*—Compute

$$y^T = c_B^T E_{k-1} E_{k-2} \cdots E_1.$$

Compute the coefficients $\hat{c}_j = c_j - y^T A_j$ for the nonbasic variables x_j. If $\hat{c}_j \geq 0$ for all nonbasic variables, then the current basis is optimal. Otherwise, select a variable x_t that satisfies $\hat{c}_t < 0$ as the entering variable.

2. *The Step*—Compute the entering column $\hat{A}_t = E_{k-1} E_{k-2} \cdots E_1 A_t$. Find an index $s = s_k$ that satisfies

$$\frac{\hat{b}_s}{\hat{a}_{s,t}} = \min_{1 \leq i \leq m} \left\{ \frac{\hat{b}_i}{\hat{a}_{i,t}} : \hat{a}_{i,t} > 0 \right\}.$$

This ratio test determines the leaving variable and the pivot entry $\hat{a}_{s,t}$. If $\hat{a}_{i,t} \leq 0$ for all i, then the problem is unbounded.

3. *The Update*—Update the inverse matrix B_{k+1}^{-1}: Form the eta vector η_k that transforms $\hat{A}_t$ to the sth column of the identity matrix. Update the solution vector $x_B = E_k \hat{b}$.

In the computation of the new basic variables some savings can be achieved, since $x_B = E_k B_k^{-1} b = E_k \hat{b}$.

In the following example we solve a problem with the simplex method using the product form of the inverse. The same problem was solved in Sections 5.2 and 5.3.

Example 7.8 (Simplex Method Using the Product Form of the Inverse). Consider the problem

$$
\begin{array}{rl}
\text{minimize} & z = -x_1 - 2x_2 \\
\text{subject to} & -2x_1 + x_2 \leq 2 \\
& -x_1 + 2x_2 \leq 7 \\
& x_1 \leq 3 \\
& x_1, x_2 \geq 0.
\end{array}
$$

Slack variables are added to put it in standard form:

$$
\begin{array}{rl}
\text{minimize} & z = -x_1 - 2x_2 \\
\text{subject to} & -2x_1 + x_2 + x_3 = 2 \\
& -x_1 + 2x_2 + x_4 = 7 \\
& x_1 + x_5 = 3 \\
& x_1, x_2, x_3, x_4, x_5 \geq 0.
\end{array}
$$

We start with $x_B = (x_3, x_4, x_5)^T$. The basis matrix B_1 is the identity matrix.

Iteration 1. Since $c_B^T = (0, 0, 0)$, our initial vector of simplex multipliers is $y^T = (0, 0, 0)$. Pricing the nonbasic variables, we obtain

$$
\hat{c}_1 = c_1 - y^T A_1 = -1 - 0 = -1, \qquad \hat{c}_2 = c_2 - y^T A_2 = -2 - 0 = -2,
$$

and we choose x_2 as the entering variable (corresponding to the most negative reduced cost). The entering column is

$$
\hat{A}_2 = B_1^{-1} A_2 = I A_2 = \begin{pmatrix} 1 \\ 2 \\ 0 \end{pmatrix}.
$$

The ratio test with the right-hand-side vector $\hat{b}^T = (2, 7, 3)^T$ gives the first basic variable x_3 as the leaving variable. We can now update B_2^{-1} by updating the eta vector for E_1:

$$
\eta_1 = \begin{pmatrix} 1/\hat{a}_{1,2} \\ -\hat{a}_{2,2}/\hat{a}_{1,2} \\ -\hat{a}_{3,2}/\hat{a}_{1,2} \end{pmatrix} = \begin{pmatrix} 1/1 \\ -2/1 \\ 0/1 \end{pmatrix} = \begin{pmatrix} 1 \\ -2 \\ 0 \end{pmatrix}.
$$

We also record the position of η_1 in E_1: $s_1 = 1$. Updating the resulting solution, we obtain

$$
x_B = \begin{pmatrix} x_2 \\ x_4 \\ x_5 \end{pmatrix} = B_2^{-1} b = E_1 b
$$

$$
= E_1 \begin{pmatrix} 2 \\ 7 \\ 3 \end{pmatrix} = \begin{pmatrix} 0 \\ 7 \\ 3 \end{pmatrix} + 2 \begin{pmatrix} 1 \\ -2 \\ 0 \end{pmatrix} = \begin{pmatrix} 2 \\ 3 \\ 3 \end{pmatrix}.
$$

Iteration 2. Updating the vector of multipliers gives

$$y^T = c_B^T B_2^{-1} = c_B^T E_1 = (-2 \quad 0 \quad 0)\, E_1 = (-2 \quad 0 \quad 0).$$

Pricing the nonbasic variables, we obtain

$$\hat{c}_1 = c_1 - y^T A_1 = -1 - (-2 \quad 0 \quad 0) \begin{pmatrix} -2 \\ -1 \\ 1 \end{pmatrix} = -5$$

$$\hat{c}_3 = c_3 - y^T A_3 = 0 - (-2 \quad 0 \quad 0) \begin{pmatrix} 1 \\ 0 \\ 0 \end{pmatrix} = 2,$$

and hence the solution is not optimal and x_1 will enter the basis. Computing the entering column, we obtain

$$\hat{A}_1 = B_2^{-1} A_1 = E_1 A_1$$

$$= E_1 \begin{pmatrix} -2 \\ -1 \\ 1 \end{pmatrix} = \begin{pmatrix} 0 \\ -1 \\ 1 \end{pmatrix} + (-2) \begin{pmatrix} 1 \\ -2 \\ 0 \end{pmatrix} = \begin{pmatrix} -2 \\ 3 \\ 1 \end{pmatrix}.$$

Performing the ratio test with respect to the right-hand-side vector, we conclude that the second basic variable x_4 will leave and will be replaced by x_1. The new eta vector is

$$\eta_2 = \begin{pmatrix} -\hat{a}_{1,1}/\hat{a}_{2,1} \\ 1/\hat{a}_{2,1} \\ -a_{3,1}/\hat{a}_{2,1} \end{pmatrix} = \begin{pmatrix} -(-2)/3 \\ 1/3 \\ -(1)/3 \end{pmatrix} = \begin{pmatrix} 2/3 \\ 1/3 \\ -1/3 \end{pmatrix}.$$

The eta file now includes the following information:

$$s_1 = 1 \quad \text{and} \quad \eta_1 = \begin{pmatrix} 1 \\ -2 \\ 0 \end{pmatrix}$$

$$s_2 = 2 \quad \text{and} \quad \eta_2 = \begin{pmatrix} \frac{2}{3} \\ \frac{1}{3} \\ -\frac{1}{3} \end{pmatrix}.$$

Updating the solution, we obtain

$$x_B = \begin{pmatrix} x_2 \\ x_1 \\ x_5 \end{pmatrix} = B_3^{-1} b = E_2 E_1 b$$

$$= E_2 \hat{b} = \begin{pmatrix} 2 \\ 0 \\ 3 \end{pmatrix} + 3 \begin{pmatrix} \frac{2}{3} \\ \frac{1}{3} \\ -\frac{1}{3} \end{pmatrix} = \begin{pmatrix} 4 \\ 1 \\ 2 \end{pmatrix}.$$

Iteration 3. To test for optimality we first compute

$$y^T = c_B^T B_3^{-1} = ((-2 \quad -1 \quad 0)\, E_2)\, E_1 = (-2 \quad -\tfrac{5}{3} \quad 0)\, E_1 = (\tfrac{4}{3} \quad -\tfrac{5}{3} \quad 0).$$

Pricing yields

$$\hat{c}_3 = c_3 - y^T A_3 = 0 - (\tfrac{4}{3} \quad -\tfrac{5}{3} \quad 0) \begin{pmatrix} 1 \\ 0 \\ 0 \end{pmatrix} = -\tfrac{4}{3}$$

$$\hat{c}_4 = c_4 - y^T A_4 = 0 - (\tfrac{4}{3} \quad -\tfrac{5}{3} \quad 0) \begin{pmatrix} 0 \\ 1 \\ 0 \end{pmatrix} = \tfrac{5}{3},$$

and hence the solution may be improved by letting x_3 enter the basis. To determine the leaving variable, we compute the entering column

$$\hat{A}_3 = B_3^{-1} A_3 = E_2 E_1 A_3$$

$$= E_2 E_1 \begin{pmatrix} 1 \\ 0 \\ 0 \end{pmatrix} = E_2 \left[\begin{pmatrix} 0 \\ 0 \\ 0 \end{pmatrix} + 1 \begin{pmatrix} 1 \\ -2 \\ 0 \end{pmatrix} \right]$$

$$= E_2 \begin{pmatrix} 1 \\ -2 \\ 0 \end{pmatrix} = \begin{pmatrix} 1 \\ 0 \\ 0 \end{pmatrix} - 2 \begin{pmatrix} \tfrac{2}{3} \\ \tfrac{1}{3} \\ -\tfrac{1}{3} \end{pmatrix} = \begin{pmatrix} -\tfrac{1}{3} \\ -\tfrac{2}{3} \\ \tfrac{2}{3} \end{pmatrix}.$$

The ratio test results in the third basic variable x_5 leaving the basis. At the end of this iteration we update the eta file with

$$s_3 = 3 \quad \text{and} \quad \eta_3 = \begin{pmatrix} \tfrac{1}{2} \\ 1 \\ \tfrac{3}{2} \end{pmatrix}.$$

Updating the solution gives

$$x_B = \begin{pmatrix} x_2 \\ x_1 \\ x_3 \end{pmatrix} = B_4^{-1} b = E_3 E_2 E_1 b$$

$$= E_3 \hat{b} = \begin{pmatrix} 4 \\ 1 \\ 0 \end{pmatrix} + 2 \begin{pmatrix} \tfrac{1}{2} \\ 1 \\ \tfrac{3}{2} \end{pmatrix} = \begin{pmatrix} 5 \\ 3 \\ 3 \end{pmatrix}.$$

Iteration 4. Updating the multiplier vector yields

$$y^T = c_B^T B_4^{-1} = (((-2 \quad -1 \quad 0) E_3) E_2) E_1$$
$$= ((-2 \quad -1 \quad -2) E_2) E_1 = (-2 \quad -1 \quad -2) E_1 = (0 \quad -1 \quad -2).$$

Pricing yields

$$\hat{c}_4 = c_4 - y^T A_4 = 0 - (0 \quad -1 \quad -2) \begin{pmatrix} 0 \\ 1 \\ 0 \end{pmatrix} = 1$$

$$\hat{c}_5 = c_5 - y^T A_5 = 0 - (0 \quad -1 \quad -2) \begin{pmatrix} 0 \\ 0 \\ 1 \end{pmatrix} = 2,$$

and the optimality conditions are satisfied. Our solution is $x_1 = 3$ and $x_2 = 5$, with slack variables $x_3 = 3$, $x_4 = 0$, $x_5 = 0$, and objective $z = -x_1 - 2x_2 = -13$.   ∎

From the example above, the product form of the simplex method may appear to be cumbersome. Indeed, this problem is too small and dense to afford any computational savings. The major savings of the product form occur when the problem is large and sparse. In such problems, the eta vectors corresponding to the elementary matrix tend to be sparse. The result is reduced storage and fewer computations.

Example 7.9 (Sparsity of Eta Vectors). Consider the basis matrix

$$B = \begin{pmatrix} 1 & 0 & 0 & 0 & 1 \\ 1 & 1 & 0 & 0 & 0 \\ 0 & 1 & 1 & 0 & 0 \\ 0 & 0 & 1 & 1 & 0 \\ 0 & 0 & 0 & 1 & 1 \end{pmatrix}.$$

To obtain B^{-1}, we start with a 5×5 identity matrix and sequentially replace its ith column by the ith column of B. At each step we update the inverse of the resulting matrix, using the diagonal elements as the pivots ($s_i = i$).

Let $B_0 = I$, and let b_i be the ith column of B. We transform B_0 in five stages to $B_5 = B$. The first eta vector is obtained from b_1. The second eta vector is obtained from $E_1 b_2$. The last eta vector is obtained from $E_4 E_3 E_2 E_1 b_5$. This procedure yields the following eta vectors:

$$\eta_1 = \begin{pmatrix} 1 \\ -1 \\ 0 \\ 0 \\ 0 \end{pmatrix}, \quad \eta_2 = \begin{pmatrix} 0 \\ 1 \\ -1 \\ 0 \\ 0 \end{pmatrix}, \quad \eta_3 = \begin{pmatrix} 0 \\ 0 \\ 1 \\ -1 \\ 0 \end{pmatrix}, \quad \eta_4 = \begin{pmatrix} 0 \\ 0 \\ 0 \\ 1 \\ -1 \end{pmatrix}, \quad \eta_5 = \begin{pmatrix} -\frac{1}{2} \\ \frac{1}{2} \\ -\frac{1}{2} \\ \frac{1}{2} \\ \frac{1}{2} \end{pmatrix}.$$

Thus most of the eta vectors are sparse. In contrast, the explicit inverse

$$B^{-1} = \begin{pmatrix} \frac{1}{2} & \frac{1}{2} & -\frac{1}{2} & \frac{1}{2} & -\frac{1}{2} \\ -\frac{1}{2} & \frac{1}{2} & \frac{1}{2} & -\frac{1}{2} & \frac{1}{2} \\ \frac{1}{2} & -\frac{1}{2} & \frac{1}{2} & \frac{1}{2} & -\frac{1}{2} \\ -\frac{1}{2} & \frac{1}{2} & -\frac{1}{2} & \frac{1}{2} & \frac{1}{2} \\ \frac{1}{2} & -\frac{1}{2} & \frac{1}{2} & -\frac{1}{2} & \frac{1}{2} \end{pmatrix}$$

is completely dense.

Consider now a dense vector a. The matrix-vector product $B^{-1} a$ will require 13 multiplications if the product form of the inverse is used; however, it will require 25 multiplications if the explicit inverse is used (see the Exercises).   ∎

7.5.2   Representation of the Basis—The *LU* Factorization

In the previous subsection we represented B^{-1} as a product of elementary matrices and showed how to compute the vector of basic variables $x_b = B^{-1}b$, the simplex multipliers $y^T = c_B^T B^{-1}$, and the entering column $\hat{A}_t = B^{-1}A_t$ using this representation. This method is no longer in wide use, since an approach based on Gaussian elimination is superior in terms of its numerical accuracy, the overall number of operations required, and the greater flexibility in utilizing sparsity and controlling the fill-in of nonzeroes.

The application of Gaussian elimination modifies the computations within the simplex algorithm. Rather than using the formulas based on B^{-1}, we solve a system of equations with respect to the matrix B. Thus x_b, y^T, and $\hat{A}_t$ are computed as solutions of the systems

$$Bx_b = b, \quad y^T B = c_B^T, \quad B\hat{A}_t = A_t.$$

The key idea in the method is to reduce the system via elementary row operations to an equivalent system where the matrix is upper triangular.

One of the main techniques for utilizing sparsity in Gaussian elimination is the switching of rows or switching of columns. Intuitively, we would like to get a matrix that is "almost" upper triangular to begin with. Judicious switching of rows can also help control the roundoff error. Our discussion below assumes that we are using row permutations; for example we might perform partial pivoting where rows are switched so that the pivot element is the element of largest magnitude in the noneliminated part of its column. We will ignore the effect of column permutations, but note that a change in the order of the columns is simply a change in the order of the variables.

To utilize sparsity in Gaussian elimination, it is advantageous to represent the sequence of operations required for the triangularization in product form:

$$L_r P_r \cdots L_1 P_1 B = U,$$

where the matrices L_i are lower triangular pivot matrices and the matrices P_i are permutation matrices (see Appendix A.6). Each of these matrices can be stored in a compact form. The number r here represents the number of Gaussian pivots used to transform B into an upper triangular matrix, and U is the transformed upper triangular system matrix. If we write

$$\bar{L} = L_r P_r \cdots L_1 P_1,$$

then $\bar{L}B = U$. When no row permutations are required (that is, $P_i = I$), then $\bar{L}$ is a lower triangular matrix, and so is its inverse. Letting $L = \bar{L}^{-1}$, we can write B as a product of lower and upper triangular matrices

$$B = LU.$$

Because of this representation, the method is also called the *LU* factorization or *LU* decomposition. In the more general case, where row permutations are used, the matrix $\bar{L}$ and hence L may no longer be lower triangular, but the method is still known as the *LU* factorization.

Our overview of the implementation of the *LU* factorization will begin with a discussion of the method for computing the factors of the *LU* decomposition, storing them in compact form, and performing computations using this compact form. Next we will discuss how to solve the systems of equations that arise in the course of a simplex iteration. Finally, we will discuss how to update the factorization following a simplex iteration, when one variable leaves the basis and another variable enters.

We start with the triangulization of the matrix B:

$$L_r P_r \cdots L_1 P_1 B = U.$$

For simplicity we denote the intermediate ("partially" upper triangular) matrices generated in the course of the factorization (regardless of the step) by $\hat{U}$. The P_i matrices are *elementary permutation matrices* formed by switching two rows of the identity. Specifically, if P is an elementary permutation matrix obtained by switching rows j and k of the identity, then the operation $P\hat{U}$ switches rows j and k of $\hat{U}$ (see the Exercises). In general the permutation matrix need not be formed explicitly. Only the indices of the rows being interchanged need be stored. If P, say, interchanges rows j and k, then multiplying P from the right by a column vector, or from the left by a row vector, simply switches elements j and k of the vector.

The matrices L_i are the matrices that perform the elementary row operations involved in Gaussian elimination. The matrix L_i that pivots on column s of $\hat{U}$ is the identity matrix with its sth column replaced by

$$\eta = \begin{pmatrix} 0 \\ \vdots \\ 0 \\ 1 \\ -\hat{u}_{s+1,s}/\hat{u}_{s,s} \\ \vdots \\ -\hat{u}_{m,s}/\hat{u}_{s,s} \end{pmatrix} \leftarrow s,$$

where $\{\hat{u}_{j,s}\}$ are the entries in $\hat{U}$ (see the Exercises). Denoting column s of the identity matrix by e_s, we can write L_i as

$$L_i = I + (\eta - e_s)e_s^T.$$

The elementary matrices need not be formed explicitly. Only the index s of the pivot term and the lower part of the eta vector containing entries $s + 1, \ldots, m$ need be stored. Using just this information, it is easy to premultiply or postmultiply the matrix by a vector. For example,

$$L_i a = (I + (\eta - e_s)e_s^T)a = a + (\eta - e_s)a_s = a + a_s \begin{pmatrix} 0 \\ \vdots \\ 0 \\ \eta_{s+1} \\ \vdots \\ \eta_m \end{pmatrix}.$$

Thus, to compute $L_i a$ we take the vector a and add to its subdiagonal portion (components $s + 1, \ldots, m$) the corresponding portion of a_s times η.

Forming $c^T L_i$ is also easy, since

$$c^T L_i = c^T (I + (\eta - e_s)e_s^T) = c^T + (c^T \eta - c_s)e_s^T$$
$$= (\, c_1 \quad c_2 \quad \ldots \quad c_{s-1} \quad c^T \eta \quad c_{s+1} \quad \ldots \quad c_m \,).$$

Thus, the computation $c^T L_i$ leaves c unchanged, except for its sth component which is replaced by $c^T \eta$.

Factorization of the matrix B using the product form is done a column at a time, starting with its first column. At the beginning of step k we typically have an "eta file" consisting of the eta vectors from the previous iterations, their associated pivot index, and the permutations applied. Available also are columns $1, \ldots, k - 1$ of U. In the course of triangularizing column k of B we first compute the effect of all previous row operations on the column. After a possible row interchange, the top part of the resulting vector will be column k of U, while the subdiagonal portion will define the elimination eta vector.

Example 7.10 (*LU* Factorization). We will illustrate the factorization using the 3×3 example

$$B = \begin{pmatrix} 1.6 & -4.2 & -0.8 \\ 4.0 & 1.5 & 3.0 \\ 8.0 & -1.0 & 1.0 \end{pmatrix}$$

from Appendix A.6. We will use partial pivoting which chooses at each iteration k the pivot term with largest magnitude from among those available.

At the first step, rows 1 and 3 are switched:

$$P_1 : \quad 1 \leftrightarrow 3;$$

therefore P_1 is defined by the pair $(1, 3)$ indicating that P_1 is obtained by switching the first and third rows of the identity matrix. The effect of this on the first column of B is

$$P_1 B_1 = \begin{pmatrix} 8.0 \\ 4.0 \\ 1.6 \end{pmatrix}.$$

We now record the first column of U, the eta vector, and the row index of the pivot:

$$U_1 = \begin{pmatrix} 8 \\ 0 \\ 0 \end{pmatrix}, \quad \eta_1 = \begin{pmatrix} 1 \\ -0.5 \\ -0.2 \end{pmatrix}, \quad s_1 = 1.$$

Only the second and third components of η_1 need be stored since the first component will always be 1. At the second step we first reconstruct the effect of $L_1 P_1$ on the second column of B:

$$L_1 P_1 B_2 = L_1 P_1 \begin{pmatrix} -4.2 \\ 1.5 \\ -1 \end{pmatrix} = L_1 \begin{pmatrix} -1 \\ 1.5 \\ -4.2 \end{pmatrix} = \begin{pmatrix} -1 \\ 1.5 \\ -4.2 \end{pmatrix} - \begin{pmatrix} 0 \\ -0.5 \\ 0.2 \end{pmatrix} = \begin{pmatrix} -1 \\ 2 \\ -4 \end{pmatrix}.$$

Since the third element is greater than the second we define P_2 by the pair $(2, 3)$:

$$P_2 : \quad 2 \leftrightarrow 3.$$

Since

$$P_2 \hat{B}_2 = \begin{pmatrix} -1 \\ -4 \\ 2 \end{pmatrix},$$

we obtain

$$U_2 = \begin{pmatrix} -1 \\ -4 \\ 0 \end{pmatrix}, \quad \eta_2 = \begin{pmatrix} 0 \\ 1 \\ 0.5 \end{pmatrix}, \quad s_2 = 2.$$

Here only the third element of η_2 need be stored. At the third and last step we reconstruct the effect of $L_2 P_2 L_1 P_1$ on the third column of B:

$$L_2 P_2 L_1 P_1 B_3 = L_2 P_2 L_1 P_1 \begin{pmatrix} -0.8 \\ 3 \\ 1 \end{pmatrix} = L_2 P_2 L_1 \begin{pmatrix} 1 \\ 3 \\ -0.8 \end{pmatrix}$$

$$= L_2 P_2 \left[\begin{pmatrix} 1 \\ 3 \\ -0.8 \end{pmatrix} + \begin{pmatrix} 0 \\ -0.5 \\ -0.2 \end{pmatrix} \right] = L_2 P_2 \begin{pmatrix} 1 \\ 2.5 \\ -1 \end{pmatrix}$$

$$= L_2 \begin{pmatrix} 1 \\ -1 \\ 2.5 \end{pmatrix} = \begin{pmatrix} 1 \\ -1 \\ 2.5 \end{pmatrix} - \begin{pmatrix} 0 \\ 0 \\ 0.5 \end{pmatrix} = \begin{pmatrix} 1 \\ -1 \\ 2 \end{pmatrix}.$$

We obtain

$$U_3 = \begin{pmatrix} 1 \\ -1 \\ 2 \end{pmatrix}. \quad \blacksquare$$

To solve a system of the form $Bx = a$ we use the fact that $\bar{L} B = U$ to obtain $\bar{L} a = Ux$. We compute the vector

$$w = \bar{L} b = L_r P_r \cdots L_1 P_1 a$$

and then solve

$$Ux = a.$$

Likewise, to solve a system $y^T B = c^T$, we use the fact that $\bar{L} B = U$ to obtain $y^T \bar{L}^{-1} U = c^T$. Defining $u^T = y^T \bar{L}^{-1}$ (so that $u^T \bar{L} = y^T$) we first solve for u in

$$u^T U = c^T,$$

and then compute y:

$$y^T = u^T \bar{L} = u^T L_r P_r \cdots L_1 P_1.$$

Example 7.11 (Solution of System of Equations). Consider the system $Bx = a$, where B is the matrix in the previous example and $a = (0, 10, 10)^T$. We first compute

$$w = L_2 P_2 L_1 P_1 b = L_2 P_2 L_1 P_1 \begin{pmatrix} 0 \\ 10 \\ 10 \end{pmatrix} = L_2 P_2 L_1 \begin{pmatrix} 10 \\ 10 \\ 0 \end{pmatrix}$$

$$= L_2 P_2 \begin{pmatrix} 10 \\ 5 \\ -2 \end{pmatrix} = L_2 \begin{pmatrix} 10 \\ -2 \\ 5 \end{pmatrix} = \begin{pmatrix} 10 \\ -2 \\ 4 \end{pmatrix}.$$

Solving the system $Ux = w$, we obtain through backsubstitution that

$$
\begin{aligned}
2x_3 &= 4 \rightarrow x_3 = 2 \\
-4x_2 - 2 &= -2 \rightarrow x_2 = 0 \\
8x_1 + 2 &= 10 \rightarrow x_1 = 1
\end{aligned}
$$

so the solution is $x = (1, 0, 2)^T$.

Consider now the system $y^T B = c^T$ where $c = (32, -20, 4)^T$. We first solve for u in $u^T U = c^T$ to obtain

$$
\begin{aligned}
8u_1 &= 32 \rightarrow u_1 = 4 \\
-4 - 4u_2 &= -20 \rightarrow u_2 = 4 \\
4 - 4 + 2u_3 &= 4 \rightarrow u_3 = 2
\end{aligned}
$$

so that $u = (4, 4, 2)^T$. Next we compute $y^T = u^T \bar{L}$:

$$
\begin{aligned}
y^T = u^T L_2 P_2 L_1 P_1 &= (4 \quad 4 \quad 2) L_2 P_2 L_1 P_1 = (4 \quad 5 \quad 2) P_2 L_1 P_1 \\
&= (4 \quad 2 \quad 5) L_1 P_1 = (2 \quad 2 \quad 5) P_1 = (5 \quad 2 \quad 2). \quad \blacksquare
\end{aligned}
$$

So far we have described how to factor the initial basis matrix B. In most simplex iterations—when one variable leaves the basis and another variable enters the basis—we do not factor B from scratch. Instead the existing factorization is updated by performing additional elementary row operations and permutations. As a result, the number of factors in the LU decompositions gradually increases from iteration to iteration. After some number of iterations this ceases to be efficient, since the effort to update and use the factorization grows with each iteration. It may also be true that the accuracy of the factorization has deteriorated. At this point a new LU factorization of the current basis matrix is computed using the techniques we have just described. This step is called a *refactorization*. A refactorization is also typically performed at the final iteration (see Section 7.6.3).

We now describe the technique for updating the factorization when the basis changes. We will describe here the technique proposed by Bartels and Golub (1969). Suppose that $\bar{L}B = U$, where $\{ B_i \}$ and $\{ U_i \}$ are the columns of B and U, respectively. Then

$$
\bar{L}B = \bar{L} (B_1 \quad \cdots \quad B_m) = (U_1 \quad \cdots \quad U_m) = U.
$$

Let $a = A_t$ and B_i be the columns associated with the entering and leaving variables, respectively. Instead of replacing B_i by a, we will delete B_i from B, shift columns $B_{i+1}, \ldots, B_m$ one position to the left, and insert the new column a at the end. This will give the updated basis matrix $\bar{B}$ with

$$
\begin{aligned}
\bar{L}\bar{B} = \bar{L} (B_1 \quad \cdots \quad B_{i-1} \quad B_{i+1} \quad \cdots \quad B_m \quad a) \\
= (U_1 \quad \cdots \quad U_{i-1} \quad U_{i+1} \quad \cdots \quad U_m \quad w) \equiv \hat{U},
\end{aligned}
$$

where

$$
w = \bar{L}a.
$$

The reordering of the columns of $\bar{B}$ corresponds to a reordering of the basic variables x_B. The vector w can be obtained as a by-product of the computation of the entering column $\hat{A}_t$, since it is computed in the first step of the solution of the system $B\hat{A}_t = a$.

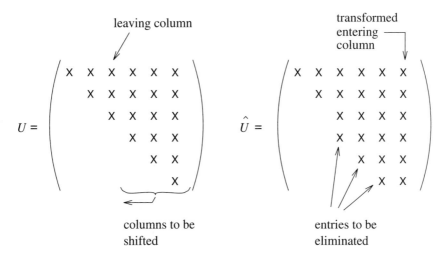

$$U = \begin{pmatrix} \times & \times & \times & \times & \times & \times \\ & \times & \times & \times & \times & \times \\ & & \times & \times & \times & \times \\ & & & \times & \times & \times \\ & & & & \times & \times \\ & & & & & \times \end{pmatrix} \qquad \hat{U} = \begin{pmatrix} \times & \times & \times & \times & \times & \times \\ & \times & \times & \times & \times & \times \\ & & \times & \times & \times & \times \\ & & & \times & \times & \times \\ & & & & \times & \times \\ & & & & & \times \end{pmatrix}$$

leaving column

transformed entering column

columns to be shifted

entries to be eliminated

Figure 7.2. *Updating a factorization.*

The columns of $\bar{B}$ are reordered so as to simplify the updating of the LU factorization. As is seen in Figure 7.2, the matrix $\hat{U}$ is almost upper triangular—the only entries that need to be eliminated are just below the main diagonal in columns $i, \ldots, m - 1$.

The subdiagonal entries in $\hat{U}$ are eliminated using Gaussian elimination with partial pivoting. For column j ($j = i, \ldots, m - 1$) the values $|\hat{u}_{j,j}|$ and $|\hat{u}_{j+1,j}|$ are compared. If $|\hat{u}_{j,j}| < |\hat{u}_{j+1,j}|$, then rows j and $j + 1$ of $\hat{U}$ are interchanged. Then the entry $\hat{U}_{j+1,j}$ in the resulting matrix is eliminated. The resulting eta vector will only have one element other than the diagonal, so storage is minimal. The update procedure corresponds to an LU factorization of $\hat{U}$, resulting in an updated upper triangular matrix $\bar{U}$.

Example 7.12 (Updating the LU Factorization). Consider the basis matrix

$$B = \begin{pmatrix} 2 & -1 & 0 & 1 & -2 \\ -1 & \frac{3}{2} & 2 & \frac{1}{2} & 1 \\ -\frac{1}{2} & \frac{1}{2} & \frac{5}{2} & -1 & \frac{1}{2} \\ 1 & -1 & -1 & 2 & -2 \\ 0 & 1 & 0 & 1 & \frac{5}{2} \end{pmatrix}.$$

In the course of the factorization we find that no permutations are required, that the eta vectors corresponding to pivot indices $s_1 = 1$, $s_2 = 2$, $s_3 = 3$, and $s_4 = 4$ are

$$\eta_1 = \begin{pmatrix} 1 \\ \frac{1}{2} \\ \frac{1}{4} \\ -\frac{1}{2} \\ 0 \end{pmatrix}, \quad \eta_2 = \begin{pmatrix} 0 \\ 1 \\ -\frac{1}{4} \\ \frac{1}{2} \\ 1 \end{pmatrix}, \quad \eta_3 = \begin{pmatrix} 0 \\ 0 \\ 1 \\ 0 \\ 1 \end{pmatrix}, \quad \eta_4 = \begin{pmatrix} 0 \\ 0 \\ 0 \\ 1 \\ \frac{1}{2} \end{pmatrix},$$

and that the resulting upper triangular matrix is

$$
U = \begin{pmatrix}
2 & -1 & 0 & 1 & -2 \\
0 & 1 & 2 & 1 & 0 \\
0 & 0 & 2 & -1 & 0 \\
0 & 0 & 0 & 2 & -1 \\
0 & 0 & 0 & 0 & 2
\end{pmatrix}.
$$

Although the matrix $\bar{L}L_4L_3L_2L_1$ is not formed explicitly, we will construct it here to facilitate the exposition:

$$
\bar{L} = \begin{pmatrix}
1 & 0 & 0 & 0 & 0 \\
\frac{1}{2} & 1 & 0 & 0 & 0 \\
\frac{1}{4} & -\frac{1}{4} & 1 & 0 & 0 \\
-\frac{1}{2} & \frac{1}{2} & 0 & 1 & 0 \\
0 & -1 & 1 & \frac{1}{2} & 1
\end{pmatrix}.
$$

Suppose that the entering column for the new basis is

$$
a = \begin{pmatrix}
2 \\
0 \\
-1 \\
4 \\
3
\end{pmatrix}
$$

and that

$$
w = \bar{L}a = \begin{pmatrix}
2 \\
1 \\
-\frac{3}{4} \\
\frac{7}{2} \\
3
\end{pmatrix}.
$$

The new basis matrix (with column 2 deleted and a inserted at the end) is

$$
\bar{B} = \begin{pmatrix}
2 & 0 & 1 & -2 & 2 \\
-1 & 2 & \frac{1}{2} & 1 & 0 \\
-\frac{1}{2} & \frac{5}{2} & -1 & \frac{1}{2} & -1 \\
1 & -1 & 2 & -2 & 4 \\
\frac{1}{2} & 0 & 1 & \frac{5}{2} & 3
\end{pmatrix}
$$

and thus U is transformed into (with column 2 deleted and w at the end)

$$
\hat{U} = \begin{pmatrix}
2 & 0 & 1 & -2 & 2 \\
0 & 2 & 1 & 0 & 1 \\
0 & 2 & -1 & 0 & -\frac{3}{4} \\
0 & 0 & 2 & -1 & \frac{7}{2} \\
0 & 0 & 0 & 2 & 3
\end{pmatrix}.
$$

Gaussian elimination is then applied to this matrix.

We start with column 2. No row interchange is required, so $P_5 = I$. The elementary matrix L_5 corresponding to pivot term $s_5 = 2$ and eta vector

$$\eta_5 = \begin{pmatrix} 0 \\ 1 \\ -1 \\ 0 \\ 0 \end{pmatrix}$$

is used to eliminate the $(3, 2)$ entry. The effect of this on the third column of $\hat{U}$ is

$$L_5 \begin{pmatrix} 1 \\ 1 \\ -1 \\ 2 \\ 0 \end{pmatrix} = \begin{pmatrix} 1 \\ 1 \\ -2 \\ 0 \\ 0 \end{pmatrix}.$$

Again no interchange is required (so $P_6 = I$), and the elementary matrix L_6 corresponding to pivot term $s_6 = 3$ and eta vector

$$\eta_6 = \begin{pmatrix} 0 \\ 0 \\ 1 \\ 1 \\ 0 \end{pmatrix}$$

eliminates the $(4, 3)$ entry. The effect of these operations on the fourth column of $\hat{U}$ is

$$L_6 L_5 \begin{pmatrix} -2 \\ 0 \\ 0 \\ -1 \\ 2 \end{pmatrix} = L_6 \begin{pmatrix} -2 \\ 0 \\ 0 \\ -1 \\ 2 \end{pmatrix} = \begin{pmatrix} -2 \\ 0 \\ 0 \\ -1 \\ 2 \end{pmatrix}.$$

Finally, a permutation matrix

$$P_7: \quad 4 \leftrightarrow 5$$

is used to interchange rows 4 and 5, and the elementary matrix L_7 corresponding to pivot term $s_7 = 4$ and eta vector

$$\eta_7 = \begin{pmatrix} 0 \\ 0 \\ 0 \\ \frac{1}{2} \\ 0 \end{pmatrix}$$

eliminates the $(5, 4)$ entry. The effect of these operations on the fifth column of U are computed via

$$L_7 P_7 L_6 L_5 \begin{pmatrix} 2 \\ 1 \\ -\frac{3}{4} \\ \frac{7}{2} \\ 3 \end{pmatrix} = L_7 P_7 L_6 \begin{pmatrix} 2 \\ 1 \\ -\frac{7}{4} \\ \frac{7}{2} \\ 3 \end{pmatrix} = L_7 P_7 \begin{pmatrix} 2 \\ 1 \\ -\frac{7}{4} \\ \frac{7}{4} \\ 3 \end{pmatrix} = L_7 \begin{pmatrix} 2 \\ 1 \\ -\frac{7}{4} \\ 3 \\ \frac{7}{4} \end{pmatrix} = \begin{pmatrix} 2 \\ 1 \\ -\frac{7}{4} \\ 3 \\ \frac{13}{4} \end{pmatrix}.$$

The resulting upper triangular matrix is

$$
\bar{U} = \begin{pmatrix}
2 & 0 & 1 & -2 & 2 \\
0 & 2 & 1 & 0 & 1 \\
0 & 0 & -2 & 0 & -\frac{7}{4} \\
0 & 0 & 0 & 2 & 3 \\
0 & 0 & 0 & 0 & \frac{13}{4}
\end{pmatrix}.
$$

This corresponds to the transformation

$$
L_7 P_7 L_6 L_5 \hat{U} = \bar{U}
$$

or in turn

$$
L_7 P_7 L_6 L_5 L_4 L_3 L_2 L_1 \bar{B} = \bar{U}. \quad \blacksquare
$$

Although this example may seem daunting, when the problem is large and sparse, the sparsity of the eta vectors can be used to great advantage. Various other schemes have been developed to further accelerate the simplex iterations, either by attempting to reduce fill-in, or by devising approaches that have fast access to data in computer memory, based on the scheme for storing sparse data (see Appendix A.6.1). As an example, we note that the LU factorization we described uses row interchanges to maintain numerical stability. Unfortunately these interchanges can interfere with the sparse storage schemes used to represent the basis matrix. A related updating scheme has been proposed by Forrest and Tomlin (1972) that alleviates some of these difficulties. In this alternative, a row interchange is performed at every step of the elimination for $\hat{U}$, regardless of the values of $|\hat{u}_{j,j}|$ and $|\hat{u}_{j+1,j}|$. Because there is no choice about the interchange, this approach is less numerically stable. Nevertheless, the Forrest–Tomlin update can be superior computationally.

Exercises

5.1. Use the simplex method to solve the linear programs in Exercise 2.2 of Chapter 5. Use the product form of the inverse.

5.2. Consider the problem

$$
\begin{aligned}
\text{minimize} \quad & z = 34x_1 + 5x_2 + 10x_3 + 9x_4 \\
\text{subject to} \quad & 2x_1 + x_2 + x_3 + x_4 = 9 \\
& 4x_1 - 2x_2 + 5x_3 + x_4 \le 8 \\
& 4x_1 - x_2 + 3x_3 + x_4 \ge 5 \\
& x_1, x_2, x_3, x_4 \ge 0.
\end{aligned}
$$

Let x_5 be the slack variable corresponding to the second constraint, x_6 the surplus variable corresponding to the third constraint, and let x_7 and x_8 be the artificial variables corresponding to the first and third constraints, respectively. Assume that the problem was solved via the simplex method, using a two-phase approach. The following information is available at the end of phase 1:

$$x_B = (x_2, x_5, x_4)^T,$$

$$s_1 = 3, \quad \eta_1 = \begin{pmatrix} -1 \\ -1 \\ 1 \end{pmatrix}$$

$$s_2 = 1, \quad \eta_2 = \begin{pmatrix} \frac{1}{2} \\ \frac{1}{2} \\ \frac{1}{2} \end{pmatrix}.$$

Find the current basic solution. Determine if it is optimal for phase 2, and if not, find the optimal solution. Use the product form of the simplex method.

5.3. Let $\hat{A}$ be an $m \times n$ matrix. Suppose that $\hat{a}_{s,t} \neq 0$. Let η be a vector of length m such that $\eta_s = 1/\hat{a}_{s,t}$, and $\eta_i = -\hat{a}_{i,t}/\hat{a}_{s,t}$ for $i \neq s$. Let E be the elementary matrix obtained by replacing the sth column of the $m \times m$ identity matrix by the vector η.

   (i) Prove that multiplying $\hat{A}$ on the left by E is equivalent to pivoting on $\hat{A}$ with $\hat{a}_{s,t}$ as the pivot.
   (ii) Let F be a matrix obtained by replacing the sth column of an $m \times m$ identity matrix by the tth column of $\hat{A}$. Prove that $F^{-1} = E$.

5.4. Let E be an $m \times m$ elementary matrix with eta vector η. Suppose that η has $l < m$ nonzero elements. Let a and c be dense vectors of length m.

   (i) Show that the matrix-vector product Ea requires l multiplications and $l - 1$ additions.
   (ii) Show that the matrix-vector product $c^T E$ requires l multiplications and $l - 1$ additions.

5.5. Compute the number of multiplications/divisions required in the kth iteration of the product form of the simplex method. (You may assume that the problem is dense.) Also compute the total number of multiplications/divisions required in k iterations of the product form of the simplex method.

5.6. Verify the calculations in Example 7.8.

5.7. Compute the inverse of the matrix

$$B = \begin{pmatrix} 1 & 0 & 0 & 0 & 1 \\ 0 & 1 & 0 & 0 & 1 \\ 0 & 0 & 1 & 0 & 1 \\ 0 & 0 & 0 & 1 & 1 \\ 1 & 1 & 1 & 1 & 6 \end{pmatrix}$$

by sequentially replacing the ith column of the identity matrix by the ith column of B and performing the required pivot operations. Show that most of the eta vectors obtained are sparse. Compute B^{-1} explicitly, and verify that it is a dense matrix.

5.8. Consider the matrix

$$B = \begin{pmatrix} 4 & 0 & 0 & 2 \\ 0 & 1 & 3 & 0 \\ 0 & 0 & 2 & 1 \\ 1 & 1 & 1 & 1 \end{pmatrix}.$$

(i) Represent B^{-1} using the product form of the inverse. Incorporate the columns in order (first column 1, then column 2, and so on).

(ii) Solve the linear system $Bx = b$ with $b = (1, -1, 1, -1)^T$ using the product-form representation from (i).

5.9. Assume that B is the current basis matrix, $\bar{B}$ is the new basis matrix, A_s is the column corresponding to the leaving variable (and is in column s in B), and that A_t is the column corresponding to the entering variable. Show that $\bar{B}^{-1} = EB^{-1}$ where

$$E = I + (\eta - e_s)e_s^T$$

and e_s is column s of the identity matrix. Also show that $\bar{B} = BF$ where $F = I + (\hat{A}_t - \hat{A}_s)e_s^T$.

5.10. Prove that the product of two lower triangular matrices is lower triangular.

5.11. Prove that if the inverse of a lower triangular matrix exists, it is lower triangular.

5.12. Let L_s be an $m \times m$ lower triangular elementary matrix formed by replacing the sth column of the identity matrix by η. Prove that L_s^{-1} is obtained from the identity matrix by replacing entries $j = s + 1, \ldots, m$ of column s by $-\eta_j$.

5.13. Suppose that L_i and L_j are two lower triangular elementary matrices used in Gaussian elimination, formed by replacing the ith and jth columns of the identity matrix, respectively, by η_i and η_j. Prove that their product $L_i L_j$ is the identity matrix with columns i and j replaced by η_i and η_j, respectively.

5.14. Consider the matrix factorization

$$b = \begin{pmatrix} 2 & 4 & -2 \\ 1 & 6 & 5 \\ 0 & 2 & 11 \end{pmatrix} = \begin{pmatrix} 1 & 0 & 0 \\ \frac{1}{2} & 1 & 0 \\ 0 & \frac{1}{2} & 1 \end{pmatrix} \begin{pmatrix} 2 & 4 & -2 \\ 0 & 4 & 6 \\ 0 & 0 & 8 \end{pmatrix} = LU.$$

Replace the first column of B by $a = (1, 3, 4)^T$ and compute the updated factorization.

5.15. Find an LU factorization of the matrix

$$B = \begin{pmatrix} 0 & 2 & 4 \\ 1 & 0 & 5 \\ -2 & 2 & 0 \end{pmatrix}.$$

Use partial pivoting. Note: Find the eta vectors for the factorization, but do not form the eta matrices. Use the result to solve the system $y^T B = c_B^T$, where $c_B^T = (2, 11, -8)$.

5.16. In a certain iteration of the simplex algorithm, the basic variables are x_1, x_2, and x_3, and the basis matrix is

$$B = \begin{pmatrix} 0 & -2 & 2 \\ 1 & -2 & 0 \\ -3 & 0 & 3 \end{pmatrix}.$$

(i) Find the LU decomposition of the basis matrix using partial pivoting. *Note:* Find the eta vectors for the factorization, but do not form the eta matrices.

(ii) Use the results of (i) to solve the system $y^T B = c_B^T$, where $c_B^T = (8, -10, 4)$.

(iii) Suppose now that the second basic variable is replaced by x_4, whose constraint coefficients are

$$A_4 = (0 \quad -2 \quad 3)^T.$$

Let $\hat{B}$ be the new basis matrix. Compute the updates required to obtain the LU decomposition of $\hat{B}$.

(iv) Solve the system of equations

$$\hat{B}x = \begin{pmatrix} 10 \\ 0 \\ 12 \end{pmatrix}.$$

5.17. Compute the LU factorization of the matrix

$$B = \begin{pmatrix} 1 & 0 & 0 & 0 & 1 \\ 0 & 1 & 0 & 0 & 1 \\ 0 & 0 & 1 & 0 & 1 \\ 0 & 0 & 0 & 1 & 1 \\ 1 & 1 & 1 & 1 & 6 \end{pmatrix}.$$

5.18. Consider the matrix

$$B = \begin{pmatrix} 4 & 0 & 0 & 2 \\ 0 & 1 & 3 & 0 \\ 0 & 0 & 2 & 1 \\ 1 & 1 & 1 & 1 \end{pmatrix}.$$

(i) Find the LU decomposition of B using partial pivoting. *Note:* Find the eta vectors for the factorization, but do not form the eta matrices.

(ii) Solve the linear system $Bx = b$ with $b = (1, -1, 1, -1)^T$ using the product-form representation from (i).

7.6 Numerical Stability and Computational Efficiency

A great deal of effort must be expended to translate the simplex method into a high-quality piece of software. Part of this effort is concerned with efficiency, making every step of the method run as efficiently as possible. But there are other issues, such as reliability and flexibility. Linear programming software should work effectively on a computer, where the arithmetic is subject to rounding errors, and where the problems may not satisfy the assumptions we have been routinely making (for example, that the constraint matrix has full row rank). In addition, the software should be able to solve problems that are not specified in standard form, but rather in a form that is more convenient to the user of the software.

This section will describe some of the ideas and techniques that arise in the development of software for the simplex method. Ideally, we would like to say, "This is the best way to implement the simplex method," but this is not possible. On different sets of problems, on different computers, and on different variants of the simplex method, the choices can be (and often are) different. Even subtle changes in the computer hardware can influence the way the software is written. If we tried to indicate the "best" techniques, there is a good chance that our description would almost immediately be out of date, or would be invalid in many contexts.

There is another reason that limits our discussion. Many details of linear programming software have never been published. They are implemented in the computer software, but this software is often proprietary. Even if the corresponding algorithms have been published, the algorithmic descriptions may be less precise than the software, with many small but important details omitted. Such details of software craftsmanship are rarely mentioned in research publications.

In this section we discuss a number of implementation issues: (a) the choice of the entering variable (pricing), (b) the choice of an initial basis, (c) tolerances for rounding errors, (d) scaling, (e) preprocessing, and (f) alternate model formats. The first issue will occupy most of our attention.

Although many of our comments are motivated by specific software packages, our discussion here avoids any such identifications and frequently uses words like "often" and "usually." This is an attempt to be accurate, as well as to avoid having our comments go quickly out of date. Software packages may include alternative choices for specific steps in the simplex method, with default choices selected to achieve good performance on a large class of problems. The alternatives may be invoked at the request of a particular user, or perhaps when the software itself identifies that the alternative might be preferable. Thus, from problem to problem, the behavior of the software may change.

7.6.1   Pricing

One of the most expensive operations in the simplex method is pricing, i.e., the optimality test. Because it is so expensive, simplex algorithms try to either reduce the costs of this step, or to make better use of all the calculations to select a more promising entering variable.

Partial pricing is one technique for reducing the costs of the optimality test. Instead of computing all the coefficients $\{\,\hat{c}_j\,\}$ (*full pricing*), only a subset of these coefficients is computed (100, say). If one of the optimality conditions is violated, then this violation identifies an entering variable. If not, then another subset of the coefficients is computed. This continues until either an entering variable is identified, or until it is determined that the current basis is optimal.

A much different approach to pricing is in fact to do *extra* calculations to identify a "better" entering variable. We will examine one such approach, called *steepest-edge* pricing. Suppose that at the current iteration, $\hat{c}_t$ has been used to select the entering column $\hat{A}_t$. Then the variables x will be updated using the formula

$$\begin{pmatrix} x_B \\ x_N \end{pmatrix} \leftarrow \begin{pmatrix} x_B \\ x_N \end{pmatrix} + \alpha p_t,$$

where α is the minimum ratio from the ratio test, and p_t is an edge direction, that is, p_t is a column of the matrix

$$Z = \begin{pmatrix} -B^{-1}N \\ I \end{pmatrix}.$$

(See Section 5.4.2.) The initial portion of p_t is the vector $-\hat{A}_t = -B^{-1}A_t$, and the corresponding reduced cost can be computed using

$$\hat{c}_t = c_t - c_B^T B^{-1} A_t = (c_B^T \quad c_N^T) p_t = c^T p_t.$$

In our descriptions of the simplex method we have been selecting the entering variable as the variable with the most negative reduced cost:

$$\hat{c}_t = \min_j c^T p_j,$$

where p_j is the column of Z corresponding to A_j. We will refer to this as the *steepest-descent pricing rule*. If the entering variable is increased from zero to ϵ, then the objective decreases by $\hat{c}_t \epsilon$, suggesting that this choice may give the greatest reduction ("steepest descent") in the objective value.

The steepest-descent rule has a drawback. It measures improvement in the objective per unit change in the variable x_j. If the feasible region is rotated, then this measure would change, even though rotating the feasible region is merely a cosmetic change to the problem. It would be preferable to have a rule that was insensitive to transformations of this type. We would like to measure improvement in the objective per unit movement along an edge of the feasible region.

The steepest-edge rule does this. It selects the entering variable using

$$\min_j \frac{c^T p_j}{\|p_j\|}.$$

The rule determines how the objective function is changing in the *direction* determined by the vector p_j, without regard to the particular coordinate system used to represent it.

A disadvantage of the steepest-edge rule is that it requires the computation of

$$\|p_j\| = \sqrt{1 + \left\|\hat{A}_j\right\|^2}$$

for all nonbasic columns j. Computing these norms in the obvious way would require computing $\{\hat{A}_j\}$, which would be prohibitively expensive. However, it is possible to update the values of the norms inexpensively as the basis changes.

We will first show how to update $\hat{A}_j$. (This is just an intermediate step in the derivation of the steepest-edge rule; only the norm values are actually calculated by the algorithm.) To derive this, we need a formula for updating the inverse of the basis matrix. Let us assume that B is the current basis matrix, $\bar{B}$ is the new basis matrix, A_s is the column corresponding to the leaving variable (and, for simplicity, is in column s in B), and A_t is the column corresponding to the entering variable. Then (see the Exercises)

$$\bar{B}^{-1} = EB^{-1},$$

where E is a matrix of the form

$$E = I + (\eta - e_s)e_s^T.$$

and e_s is column s of an $m \times m$ identity matrix. More specifically,

$$
\eta - e_s \equiv
\begin{pmatrix}
-\hat{a}_{1,t}/\hat{a}_{s,t} \\
\vdots \\
-\hat{a}_{s-1,t}/\hat{a}_{s,t} \\
1/\hat{a}_{s,t} \\
-\hat{a}_{s+1,t}/\hat{a}_{s,t} \\
\vdots \\
-\hat{a}_{m,t}/\hat{a}_{s,t}
\end{pmatrix}
-
\begin{pmatrix}
0 \\
\vdots \\
0 \\
1 \\
0 \\
\vdots \\
0
\end{pmatrix}
= \frac{e_s - \hat{A}_t}{\hat{a}_{s,t}},
$$

where $\hat{A}_t$ is the entering column and $\hat{a}_{s,t}$ is the pivot entry at the current iteration. We also define

$$
\sigma \equiv B^{-T} e_s,
$$

that is, σ is equal to row s of B^{-1}. Expanding the formula for $\bar{B}^{-1}$ gives

$$
\begin{aligned}
\bar{B}^{-1} &= E B^{-1} \\
&= B^{-1} + (\eta - e_s) e_s^T B^{-1} \\
&= B^{-1} + \frac{1}{\hat{a}_{s,t}} (e_s - \hat{A}_t) \sigma^T.
\end{aligned}
$$

Let A_j be a nonbasic column in both the current and the new basis. Then

$$
\begin{aligned}
\bar{B}^{-1} A_j &= B^{-1} A_j + \frac{1}{\hat{a}_{s,t}} (e_s - \hat{A}_t) \sigma^T A_j \\
&= \hat{A}_j + \frac{\sigma^T A_j}{\hat{a}_{s,t}} (e_s - \hat{A}_t).
\end{aligned}
$$

This formula can be used to update the norms of the vectors $\{ p_j \}$. First notice that

$$
\| p_j \|^2 = 1 + \left\| \hat{A}_j \right\|^2 .
$$

If we define $\gamma_j \equiv \left\| \hat{A}_j \right\|^2 = \hat{A}_j^T \hat{A}_j = (B^{-1} A_j)^T (B^{-1} A_j)$ and $\bar{\gamma}_j \equiv \left\| \bar{B}^{-1} A_j \right\|^2$, then

$$
\begin{aligned}
\bar{\gamma}_j &= (\bar{B}^{-1} A_j)^T (\bar{B}^{-1} A_j) \\
&= \gamma_j + \left(\frac{\sigma^T A_j}{\hat{a}_{s,t}} \right)^2 (1 - 2\hat{a}_{s,t} + \gamma_t) + 2 \left(\frac{\sigma^T A_j}{\hat{a}_{s,t}} \right) (\hat{a}_{s,j} - \hat{A}_j^T \hat{A}_t).
\end{aligned}
$$

This formula can be reorganized to make it more suitable for computation. The final term can be adjusted using the identity

$$
\hat{A}_j^T \hat{A}_t = (B^{-1} A_j)^T \hat{A}_t = A_j^T (B^{-T} \hat{A}_t).
$$

Also, since

$$2\left(\frac{\sigma^T A_j}{\hat{a}_{s,t}}\right)^2 \hat{a}_{s,t} = 2\left(\frac{e_s^T B^{-1} A_j}{\hat{a}_{s,t}}\right)^2 \hat{a}_{s,t}$$

$$= 2\left(\frac{\hat{a}_{s,j}}{\hat{a}_{s,t}}\right)^2 \hat{a}_{s,t}$$

$$= 2\left(\frac{\hat{a}_{s,j}}{\hat{a}_{s,t}}\right) \hat{a}_{s,j}$$

$$= 2\left(\frac{\sigma^T A_j}{\hat{a}_{s,t}}\right) \hat{a}_{s,j},$$

two of the terms involving $\hat{a}_{s,t}$ cancel. As a result we obtain

$$\bar{\gamma}_j = \gamma_j + \left(\frac{\sigma^T A_j}{\hat{a}_{s,t}}\right)^2 (1 + \gamma_t) - 2\left(\frac{\sigma^T A_j}{\hat{a}_{s,t}}\right) A_j^T (B^{-T} \hat{A}_t).$$

A slightly different formula can be derived for the coefficient of the leaving variable. (See the Exercises.)

This final formula is the basis for the steepest-edge rule. It requires the calculation of $B^{-T} \hat{A}_t$, some additional inner products, as well as storage for $\{\gamma_j\}$ and the initialization of these quantities. (The other calculations are by-products of the simplex method.) For sparse problems, many of the $\sigma^T A_j$ terms can be zero, so the number of extra calculations required to implement this technique may not be excessive.

Example 7.13 (Steepest Edge Update Formula). Consider the constraint matrix

$$A = \begin{pmatrix} 1 & 2 & 0 & 4 & 1 & 5 \\ 0 & 1 & 2 & 2 & 5 & 4 \\ 0 & 0 & 1 & 1 & 3 & 5 \end{pmatrix}.$$

We will begin with the basis $x_B = (x_1, x_2, x_3)^T$ so that

$$B = \begin{pmatrix} 1 & 2 & 0 \\ 0 & 1 & 2 \\ 0 & 0 & 1 \end{pmatrix} \quad \text{and} \quad B^{-1} = \begin{pmatrix} 1 & -2 & 4 \\ 0 & 1 & -2 \\ 0 & 0 & 1 \end{pmatrix}.$$

For the nonbasic columns,

$$\hat{A}_4 = B^{-1} A_4 = \begin{pmatrix} 4 \\ 0 \\ 1 \end{pmatrix}, \quad \hat{A}_5 = \begin{pmatrix} 3 \\ -1 \\ 3 \end{pmatrix}, \quad \text{and} \quad \hat{A}_6 = \begin{pmatrix} 17 \\ -6 \\ 5 \end{pmatrix}.$$

The new basis will be $\bar{x}_B = (x_1, x_2, x_4)^T$ so that $s = 3$, $t = 4$, $e_s = (0, 0, 1)^T$, and

$$\bar{B} = \begin{pmatrix} 1 & 2 & 4 \\ 0 & 1 & 2 \\ 0 & 0 & 1 \end{pmatrix} \quad \text{and} \quad \bar{B}^{-1} = \begin{pmatrix} 1 & -2 & 0 \\ 0 & 1 & -2 \\ 0 & 0 & 1 \end{pmatrix}.$$

We will concentrate on the final two columns of A, the columns that remain nonbasic. Then $\hat{a}_{s,t} = 1$,

$$\sigma = B^{-T}e_s = (0 \quad 0 \quad 1)^T,$$

and

$$\kappa_5 \equiv \sigma^T A_5 / \hat{a}_{s,t} = 3$$
$$\kappa_6 \equiv \sigma^T A_6 / \hat{a}_{s,t} = 5.$$

Now it would be possible to update $\hat{A}_5$ and $\hat{A}_6$ using

$$\hat{A}_5 \leftarrow \hat{A}_5 + \kappa_5(e_s - \hat{A}_t)$$
$$\hat{A}_6 \leftarrow \hat{A}_6 + \kappa_6(e_s - \hat{A}_t),$$

although this is not required by the steepest-edge rule.

Now let us examine the update formula for the squares of the norms of these vectors. Initially

$$\gamma_4 = 17, \quad \gamma_5 = 19, \quad \text{and} \quad \gamma_6 = 350.$$

We compute

$$B^{-T}\hat{A}_t = B^{-T}\hat{A}_4 = \begin{pmatrix} 4 \\ -8 \\ 17 \end{pmatrix}.$$

Then

$$\bar{\gamma}_5 = \gamma_5 + \kappa_5^2(1 + \gamma_4) - 2\kappa_5 A_5^T(B^{-T}\hat{A}_4) = 91$$
$$\bar{\gamma}_6 = \gamma_6 + \kappa_6^2(1 + \gamma_4) - 2\kappa_6 A_6^T(B^{-T}\hat{A}_4) = 70,$$

and these are the squares of the norms of the vectors in $\bar{p}$. ∎

The steepest-edge rule can dramatically decrease the overall number of simplex iterations required to solve a linear program. It is especially valuable within the dual simplex method since in that setting a separate calculation is not required to obtain the vector σ. For further details, see the paper by Forrest and Goldfarb (1992).

There are other pricing rules that attempt to choose a better entering variable than given by the steepest-descent rule. One of these, called Devex, only *approximates* the norms that are computed exactly by the steepest-edge rule. The costs per simplex iteration are lower, but more iterations are typically required. For further information on the Devex and other approximate forms of steepest-edge pricing, see the papers by Harris (1973), Goldfarb and Reid (1977), and Świetanowski (1998).

Currently, steepest-edge pricing is considered to be a good choice for the dual simplex method, whereas for the primal simplex method it is preferable to start with partial pricing and later switch to Devex or some other form of approximate steepest-edge pricing.

7.6.2 The Initial Basis

Many software packages can take advantage of a specified initial basis, that is, if the user is able to provide one. More commonly, the package will have to determine an initial basis automatically. The simplex method can be initialized with a basis consisting of slack and/or artificial variables. However, if the model is feasible, the optimal basis need not contain any artificial variables. Also, if a slack variable is in the optimal basis, then the corresponding constraint is redundant. Hence an initial basis consisting of artificial and slack variables might have little in common with the optimal basis, and the simplex algorithm would then have to perform a great many pivots before reaching the optimal solution.

For these reasons, some packages attempt to find an initial basis that avoids using artificial variables and (to a lesser extent) slack variables. This operation is sometimes referred to as a *crash* procedure. One such strategy is described in the paper by Bixby (1993).

Crash procedures attempt to choose an initial basis B according to criteria such as (a) the columns of B do not correspond to artificial variables, (b) the columns of B are sparse, (c) the columns of B form an (approximately) upper or lower triangular matrix, (d) the diagonal entries of B are suitable pivot entries for Gaussian elimination, (e) the matrix B is not "too ill conditioned," and (f) the columns of B are "likely" to be in the optimal basis.

A crash procedure can reduce the number of simplex iterations required to find an optimal solution. However, the initial basis that results will be less sparse than for a slack/artificial basis (where $B = I$) so the early simplex iterations will be more expensive, and the resulting savings in computer time may not be as dramatic.

7.6.3 Tolerances; Degeneracy

Ideally, linear programming software would return a solution that exactly satisfied the constraints, all of whose variables were nonnegative, and where all the optimality conditions were satisfied. Unfortunately, due to the realities of finite-precision arithmetic, this is not always possible. Instead, the computed solution will only satisfy these conditions to within certain tolerances related to machine epsilon ϵ_{mach} (the accuracy of the computer arithmetic; see Appendix B.2). The tolerances indicated below are based on a value of $\epsilon_{mach} \approx 10^{-16}$. Many software packages allow the user to modify the tolerances used by the algorithm.

Not all of these conditions are equally difficult to satisfy. If x_B is computed using an LU factorization of B with partial pivoting, then

$$\frac{\| Bx_B - b \|}{\| B \| \cdot \| x_B \|} = O(\epsilon_{mach}).$$

This indicates that (under these assumptions) the constraints $Ax = b$ will be satisfied to near the limits of machine accuracy. These assumptions are not fully satisfied in linear programming software, however. The pivoting strategy may be modified to enhance sparsity of the factorization, and updates to the factorization can lead to additional deterioration. In fact, if $\| Bx_B - b \|$ (scaled as above) becomes "too large" (larger than 10^{-6}, say), then this is an indication that it is time to refactor the basis matrix. Also, it is common to refactor the basis matrix at the optimal point to enhance the accuracy of the computed solution.

The computed solution may violate the nonnegativity constraints $x \geq 0$ (primal feasibility) or the optimality conditions (dual feasibility), but only by a small amount. For example, violations in these conditions of up to 10^{-6} might be tolerated.

In addition, small coefficients in the model might be ignored. For example, any entry in A satisfying (say) $|A_{i,j}| \leq 10^{-12}$ might be replaced by zero.

These tolerances for zero can be exploited as a technique for resolving degeneracy. Suppose that at some iteration of the simplex method the step procedure resulted in a step of zero. Then the feasibility tolerance for the corresponding basic variable could be randomly perturbed. The simplex algorithm would allow this variable to become slightly negative, and a nonzero step would be taken. The perturbed problem would not be degenerate. Similar perturbations could be incorporated whenever a degeneracy was detected. Later, when the solution to the perturbed problem had been found, the perturbations would be removed. The current basis might then be infeasible, and additional calculations would be required to restore feasibility with respect to the original problem. These further calculations would correspond to a phase-1 problem.

It is common to encounter degeneracy when solving large practical linear programming problems. Strategies such as these are important enhancements to software for the simplex method.

The tolerances can be used in a similar manner within the ratio test to expand the list of potential leaving variables. This can be of value in controlling ill-conditioning in the basis matrix, since leaving variables associated with small pivot entries can perhaps be avoided. For further details, see the paper by Bixby (1993).

7.6.4   Scaling

It is possible to make a problem ill conditioned merely by changing the units in which the model is specified. For example, consider the constraints

$$\begin{pmatrix} 3 & 1 \\ 1 & 3 \end{pmatrix} \begin{pmatrix} x_1 \\ x_2 \end{pmatrix} = \begin{pmatrix} 5 \\ 9 \end{pmatrix}.$$

The matrix has condition number equal to 2. Suppose that the first constraint measures kilograms of flour, say. If this first constraint is changed so that it measures grams of flour, then the system of constraints becomes

$$\begin{pmatrix} 3000 & 1000 \\ 1 & 3 \end{pmatrix} \begin{pmatrix} x_1 \\ x_2 \end{pmatrix} = \begin{pmatrix} 5000 \\ 9 \end{pmatrix},$$

and the condition number of the transformed matrix is approximately 1250, which is about 1000 times worse than before. A similar situation would occur if the variable x_1 had its units changed via a change of variables of the form

$$\hat{x}_1 = 1000x_1,$$

causing a column of the matrix (as well as the cost coefficient c_1) to be modified.

Transformations such as these are cosmetic changes to the model and lead to new models that are mathematically equivalent to the originals. However, on a computer where finite-precision arithmetic is used, they can alter the behavior of the simplex method and lead to a deterioration in performance.

Scaling problems can also arise if the model includes a large upper bound for a variable that need not have an upper bound. For example, the model might replace the constraint $0 \le x_5$ with

$$0 \le x_5 \le 10^{12},$$

where the value 10^{12} is out of scale with the remaining data in the model. This can cause difficulties if at some iteration the variable is set equal to its upper bound.

It is not uncommon to encounter such "poorly scaled" problems. Large models are developed over a long period of time, often by a changing team of people, making it difficult to ensure that all the constraints are measured in consistent units. Or perhaps data might be collected from a variety of agencies whose reporting schemes did not conform to any common standard.

Linear programming software attempts to cope with such difficulties by scaling of the variables and constraints. (In some codes this is done by default; in others it is optional.) A simple scaling rule divides the ith constraint (including the right-hand side) by

$$\max_{j} |a_{i,j}|$$

to obtain $\hat{A}$. Then the jth column of $\hat{A}$ and the cost coefficient c_j are divided by

$$\max_{i} |\hat{a}_{i,j}|$$

to obtain $\bar{A}$. Then the simplex method is applied to the transformed problem. If this scaling is used, then the largest entry (in absolute value) in any nonzero row or column of $\bar{A}$ is equal to 1 (see the Exercises).

This scaling strategy is heuristic in the sense that it is not guaranteed to improve the performance or accuracy of the simplex method. Ideally the scaling would be chosen so as to minimize, or at least reduce, the condition numbers of the basis matrices B. Such a strategy is not practical, even for finding an ideal strategy for a single basis, let alone a set of bases. For more information on scaling, see the paper by Skeel (1979).

7.6.5 Preprocessing

Since large models are often created by teams of people or automatically by software, it is common for these models to contain redundancies. These redundancies are generally harmless, so there is little motivation for the creator of the model to examine a model in detail in an attempt to eliminate them. Even though they are harmless, these redundancies do increase the size of a model, and this can lead to computational inefficiencies.

Some software packages attempt to eliminate redundancies by preprocessing the model before applying the simplex method. We will list some of these techniques here. Further ideas can be found in the papers by Lustig, Marsten, and Shanno (1994); Brearly, Mitra, and Williams (1975); and Andersen and Andersen (1995).

If all the entries in a row of A are equal to zero, then either the constraint is redundant (if the right-hand side is zero) and the constraint can be deleted, or it is inconsistent and the problem is infeasible. A similar technique can be applied if a column of A is zero, in which case the dual problem might be infeasible.

A row of the matrix might represent a simple bound on a variable that had been written as a general constraint. It is better to handle such a constraint explicitly as a bound (see Section 7.2).

The upper and lower bounds on a variable might be equal (see Section 7.6.6), for example,

$$3 \le x_5 \le 3.$$

Then 3 could be substituted for x_5 throughout the model, and x_5 eliminated.

A more sophisticated technique uses the bounds on a variable to identify redundant constraints. For example, suppose the model contained the constraints

$$x_1 + x_2 \le 20$$
$$0 \le x_1 \le 10$$
$$0 \le x_2 \le 5.$$

Then the first constraint could be removed from the model since the upper bounds on the variables indicate that $x_1 + x_2 \le 15$.

These rules could be applied repeatedly to a model since one set of reductions might reveal new possibilities. Once this process had stabilized, the simplex method would be applied to the reduced problem. Then, after the solution had been found, the transformations would be reversed to find the solution to the original problem.

7.6.6   Model Formats

We have assumed that the linear program being solved is in standard form:

$$\text{minimize} \quad z = c^T x$$
$$\text{subject to} \quad Ax = b$$
$$x \ge 0.$$

We have discussed how to convert any linear program into standard form, so this is not a restrictive assumption, but it can increase the size of a model unnecessarily. Linear programming software usually allows more general models, such as

$$\text{minimize} \quad z = c^T x$$
$$\text{subject to} \quad Ax = b$$
$$\ell \le x \le u,$$

or even

$$\text{minimize} \quad z = c^T x$$
$$\text{subject to} \quad b_1 \le Ax \le b_2$$
$$\ell \le x \le u,$$

where b_1, b_2, ℓ, and u are (possibly infinite) upper and lower bounds on the constraints and the variables. Models of these types can be solved using straightforward variants of the simplex method (see Section 7.2).

This flexibility permits models that might seem eccentric or even perverse. For example, it would allow a *free constraint*:

$$-\infty \le 7x_1 + 9x_2 \le +\infty.$$

This might arise if a user was interested in solving a linear program, and then knowing the values of alternate objective functions at the optimal point x_*. Each of these objective functions could be included in the original model as a free constraint.

It would allow a *fixed variable*:

$$1 \le x_3 \le 1.$$

This might arise if a general, open-ended model were being developed for a company, but currently there was no flexibility for certain terms in the model. It would also allow a *free variable*:

$$-\infty \le x_4 \le \infty,$$

a variable that can take on any value.

All of these cases can be handled by the simplex method. A free constraint is always satisfied, so it can be ignored until the problem is solved and the solution is presented to the user. Fixed variables are just constants in the model.

Free variables could be handled by the software in several ways. One way would be to eliminate them from the problem. For example, if x_4 appeared in the constraint

$$5x_4 + x_5 - 3x_6 = 2,$$

then the equivalent formula

$$x_4 = \tfrac{1}{5}(2 - x_5 + 3x_6)$$

could be substituted for x_4 everywhere in the model. However, this approach can destroy some of the sparsity in the model. Instead, it may be preferable to retain x_4 in the model and add it to the basis as soon as possible. Once a free variable enters the basis it will never leave the basis, since a change in the value of the entering variable will not cause the free variable to violate a bound. The only way that a free variable will not be part of the optimal basis is if its reduced cost is zero at every iteration.

Exercises

6.1. Let

$$A = \begin{pmatrix} 1 & 3 & 2 & 5 & 8 & 7 \\ 0 & 2 & 5 & 1 & 4 & 6 \\ 0 & 0 & 3 & 5 & 2 & 1 \end{pmatrix}.$$

Suppose that the current basis uses the basic variables $x_B = (x_1, x_2, x_3)^T$ and the new basis $\bar{x}_B = (x_1, x_2, x_6)^T$. Use the formulas for the steepest-edge pricing scheme to compute the updated vector $\bar{\gamma}$ corresponding to columns 4 and 5 of A.

6.2. In a linear program, suppose that the jth variable is transformed via $x'_j = \theta x_j$, where $\theta > 0$. Is the steepest-edge pricing rule affected by this change? Explain your answer.

6.3. The discussion of the steepest-edge pricing rule did not derive the formulas corresponding to the leaving variable. Using the notation of Section 7.6.1, show that

$$\bar{B}^{-1}A_s = e_s + (1/\hat{a}_{s,t})(e_s - \hat{A}_t) = (1 + 1/\hat{a}_{s,t})e_s - (1/\hat{a}_{s,t})\hat{A}_t$$

and

$$\bar{\gamma}_s = (1/\hat{a}_{s,t}^2)(1 + \gamma_t) - 1.$$

Verify your result using the data in Example 7.13.

6.4. Another possible pricing scheme would be to determine, for each potential entering variable, what the new value of the objective function would be if that variable were to enter the basis. Why would this scheme be expensive within the simplex method?

6.5. Suppose that the scaling strategy in Section 7.6.4 is applied to an $m \times n$ matrix A to obtain a scaled matrix $\bar{A}$. Assume that no row or column of A has all its entries equal to zero. Prove that

$$\max_i |\bar{a}_{i,j}| = 1 \quad \text{for } 1 \le j \le n,$$
$$\max_j |\bar{a}_{i,j}| = 1 \quad \text{for } 1 \le i \le m.$$

7.7   Notes

Product Form—The product form of the inverse was developed by Dantzig and Orchard-Hays (1954).

Column Generation—The technique of column generation is described in the papers of Eisemann (1957), Ford and Fulkerson (1958), and Manne (1958). The cutting stock problem is discussed in the papers of Gilmore and Gomory (1961, 1963, 1965). The knapsack problem is discussed in the book by Nemhauser and Wolsey (1988, reprinted 1999).

Decomposition—The decomposition principle is due to Dantzig and Wolfe (1960). The implications for parallel computing are discussed in the paper by Ho, Lee, and Sundarraj (1988).

In our description of the decomposition principle, a point x is represented as a convex combination of all the extreme points for the set described by the easy constraints. In cases such as our example, where these constraints can be decomposed into independent subproblems, the convex combinations can also be decomposed, and this can lead to computational efficiencies; see the paper by Jones et al. (1993).

Numerical Stability and Computational Efficiency—In addition to the references cited in this section, general discussions of computational issues for the simplex method can be found in the books by Nazareth (1987) and Vanderbei (2007).

Chapter 8

Network Problems

8.1 Introduction

Linear programming problems defined on networks have many special properties. These properties allow the simplex method to be implemented more efficiently, making it possible to solve large problems efficiently. The structure of a basis, as well as the steps in the simplex method, can be interpreted directly in terms of the network, providing further insight into the workings of the simplex method. These relationships between the simplex method and the network form one of the major themes of this chapter. We use them to derive the *network simplex method*, a refinement of the simplex method specific to network problems.

Network problems arise in many settings. The network might be a physical network, such as a road system or a network of telephone lines. Or the network might only be a modeling tool, perhaps reflecting the time restrictions in scheduling a complicated construction project. A number of these applications are discussed in Section 8.2.

8.2 Basic Concepts and Examples

The most general network optimization problem that we treat in this chapter is called the *minimum cost network flow problem*. It is a linear program of the form

$$\begin{aligned}
\text{minimize} \quad & z = c^T x \\
\text{subject to} \quad & Ax = b \\
& \ell \leq x \leq u,
\end{aligned}$$

where ℓ and u are vectors of lower and upper bounds on x. We allow components of ℓ and u to take on the values $-\infty$ and $+\infty$, respectively, to indicate that a variable can be arbitrarily small or large.

The notation for describing network problems is slightly different than for the linear programs we have discussed so far. Consider the network in Figure 8.1. You might think of this as a set of roads through a park. This network has seven *nodes* (the small black circles) and eleven *arcs* connecting the nodes. The nodes are numbered 1–7. An arc between nodes i and j is denoted as (i, j). So, for example, this network includes the arcs $(1, 2)$ and $(4, 5)$.

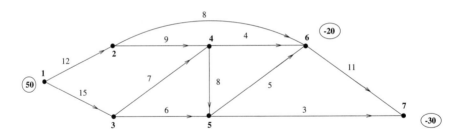

Figure 8.1. *Sample network.*

In this example, the existence of an arc (i, j) means that it is possible to drive from node i to node j, but not from node j to node i. There is a difference between arc (i, j) and arc (j, i). For each arc (i, j) the linear program will have a corresponding variable $x_{i,j}$ and cost coefficient $c_{i,j}$. The variable $x_{i,j}$ records the flow in arc (i, j) of the network, and for this application it might represent the number of cars on a road. In this problem there are eleven variables, one for each road. (The remaining information on the network is explained in Example 8.1.)

In the general network problem there will be a variable for each arc in the network, and an equality constraint for each node. We assume that there are m nodes and n arcs in the network, so that A is an $m \times n$ matrix.[7] The bounds on the variables represent the upper and lower limits on flow on an arc. Often the lower bound will be zero.

The ith row of the constraint matrix A corresponds to a constraint at the ith node:

$$(\text{flow out of node } i) - (\text{flow into node } i) = b_i,$$

or in algebraic terms

$$\sum_j x_{i,j} - \sum_k x_{k,i} = b_i,$$

where the respective summations are taken over all arcs leading out of and into node i. If $b_i > 0$, then node i is called a *source* since it adds flow to the network. If $b_i < 0$, then the node is called a *sink* since it removes flow from the network. If $b_i = 0$, then the node is called a *transshipment* node, a node where flow is conserved. A component $c_{i,j}$ of the cost vector c records the cost of shipping one unit of flow over arc (i, j).

Example 8.1 (Network Linear Program). Consider the network in Figure 8.1. We now use it to represent the flow of oil through pipes. Suppose that 50 barrels of oil are being produced at node 1, and that they must be shipped through a system of pipes to nodes 6 and 7 (20 barrels to node 6, and 30 barrels to node 7). The costs of pumping a barrel of oil along each arc are marked on the figure. The flow on each arc has a lower bound of zero and an upper bound of 30.

Node 1 is a source and nodes 6 and 7 are sinks. The other nodes are transshipment nodes. The cost of each arc is marked in the figure. The corresponding minimum cost linear

---

[7]This reverses the more common usage of n for the number of nodes and m for the number of arcs that is used in many references on network problems. This choice, however, provides consistency with the other chapters in this book, where n refers to the number of variables and m refers to the number of constraints.

program is

$$\begin{aligned}
\text{minimize} \quad & z = 12x_{1,2} + 15x_{1,3} + 9x_{2,4} + 8x_{2,6} + 7x_{3,4} + 6x_{3,5} \\
& \quad + 8x_{4,5} + 4x_{4,6} + 5x_{5,6} + 3x_{5,7} + 11x_{6,7} \\
\text{subject to} \quad & x_{1,2} + x_{1,3} = 50 \\
& x_{2,4} + x_{2,6} - x_{1,2} = 0 \\
& x_{3,4} + x_{3,5} - x_{1,3} = 0 \\
& x_{4,5} + x_{4,6} - x_{2,4} - x_{3,4} = 0 \\
& x_{5,6} + x_{5,7} - x_{3,5} - x_{4,5} = 0 \\
& x_{6,7} - x_{2,6} - x_{4,6} - x_{5,6} = -20 \\
& -x_{5,7} - x_{6,7} = -30 \\
& 0 \le x \le 30.
\end{aligned}$$

The constraints are listed in order of node number, where the left-hand side of the constraint corresponds to (flow out) − (flow in). If we order the variables as in the objective function, then in matrix-vector form the linear program could be written with cost vector

$$c = (\,12 \quad 15 \quad 9 \quad 8 \quad 7 \quad 6 \quad 8 \quad 4 \quad 5 \quad 3 \quad 11\,)^T,$$

right-hand side vector

$$b = (\,50 \quad 0 \quad 0 \quad 0 \quad 0 \quad -20 \quad -30\,)^T,$$

and coefficient matrix

$$A = \begin{pmatrix}
1 & 1 & & & & & & & & & \\
-1 & & 1 & 1 & & & & & & & \\
& -1 & & & 1 & 1 & & & & & \\
& & -1 & & -1 & & 1 & 1 & & & \\
& & & & & -1 & -1 & & 1 & 1 & \\
& & & -1 & & & & -1 & -1 & & 1 \\
& & & & & & & & & -1 & -1
\end{pmatrix}.$$

Each column of A has two nonzero entries, $+1$ and -1. Each column corresponds to an arc (i, j): $+1$ appears in row i and -1 in row j to indicate that the arc carries flow out of node i and into node j. ∎

The total *supply* in the network is given by the formula

$$S \equiv \sum_{\{i:b_i>0\}} b_i$$

and the total *demand* by

$$D \equiv - \sum_{\{i:b_i<0\}} b_i.$$

We will assume that $S = D$ (that total supply equals total demand). A network can always be modified to guarantee this. If $S > D$, that is, there is excess supply, then an artificial node is added to the network with demand $S - D$, and artificial arcs are added, connecting

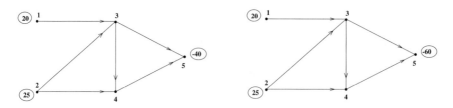

Figure 8.2. *Unbalanced networks.*

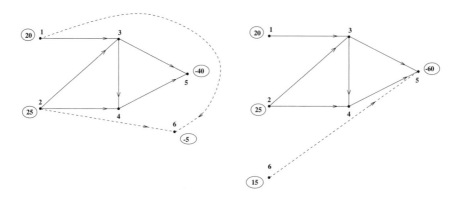

Figure 8.3. *Transformed networks.*

every source to this artificial node; each such arc has its associated cost coefficient equal to zero. (This assumes that there is no cost associated with excess production.) If there is excess demand, then an artificial node is added with supply $D - S$, together with artificial arcs connecting this artificial node with every sink. These new arcs have cost coefficients that correspond to the cost (if any) of unmet demand.

Example 8.2 (Ensuring that Total Supply Equals Total Demand). Consider the networks in Figure 8.2. The first has excess supply and the second has excess demand.

If artificial sources and sinks are added appropriately, together with the associated artificial arcs, then the networks are brought into balance. The results of these transformations are illustrated in Figure 8.3. No cost has been associated with either excess supply or excess demand.   ∎

We have written above that there are lower and upper bounds on every flow: $\ell \leq x \leq u$. For the sake of simplicity, when deriving the network simplex method in this chapter we will assume that the variables are only constrained to be nonnegative:

$$x \geq 0.$$

This simplifying assumption can be made without any loss of generality. (The reasons for this are outlined here, although their justification is left to the Exercises.) A simple change of variables can be used to transform $\ell \leq x$ into $0 \leq \hat{x}$. Upper bounds can also be eliminated.

The technique used to eliminate upper bounds does increase the size of the linear program, which can lead to an increase in solution time. An alternative is to develop a variant of the network simplex method that handles upper bounds, analogous to the bounded-variable simplex method developed in Section 7.2.

The constraint matrix A in a network problem is sparse. As observed in Example 8.1, each column of A has precisely two nonzero entries, $+1$ and -1. If (i, j) is an arc in the network, then the corresponding column in A has $+1$ in row i and -1 in row j. This implies that if all the rows of A are added together, then the result is a vector of all zeroes. Thus, the rank of A is at most $m - 1$, where m is the number of nodes in the network. We prove in the next section that the rank of A is exactly $m - 1$.

If we add together all the rows in the equality constraints $Ax = b$, then, by the above remarks, the left-hand side will be zero. The right-hand side will sum to $S - D$ (supply minus demand), and so it also will be zero. Therefore, any one of the constraints is redundant. The rank deficiency in A does not lead to an inconsistent system of linear equations.

There are a number of special forms of the minimum cost network flow problem that are of independent interest. Special-purpose algorithms have been developed for these problems, some of which are dramatically more efficient than the general-purpose simplex method. For this reason, giving them individualized treatment has resulted in many practical benefits. We will not discuss these special-purpose algorithms in this book, but we mention some of these special problems to give some idea of the range of applications of network models.

In a *transportation problem*, every node in the network is either a source or a sink, and every arc goes from a source node to a sink node. Hence the flow conservation constraints have one of two forms:

$$\sum_j x_{i,j} = b_i$$

for a source with $b_i > 0$, or

$$-\sum_k x_{k,i} = b_i$$

for a sink with $b_i < 0$. A transportation problem models the direct movement of goods from suppliers to customers, where some cost is associated with the shipments.

Example 8.3 (Transportation Problem). Suppose that a toy company imports dolls manufactured in Asia. Ships carrying the dolls arrive in either San Francisco or Los Angeles, and then the dolls are transported by truck to distribution centers in Chicago, New York, and Miami. We assume that the costs of the truck shipments are roughly proportional to the distances traveled. The corresponding transportation problem is given in Figure 8.4, with supplies and demands marked. Note that total supply equals total demand. ∎

An *assignment problem* is an optimization model for assigning people to jobs. Everyone must be assigned a job, and only one person can fill each job. It is a special case of a transportation problem, where $b_i = 1$ for a source and $b_i = -1$ for a sink. There are the same number of sources as sinks, because there are the same number of people as jobs. An assignment problem is frequently written as a maximization problem with the objective coefficients $c_{i,j}$ indicating the value of a person if assigned to a particular job

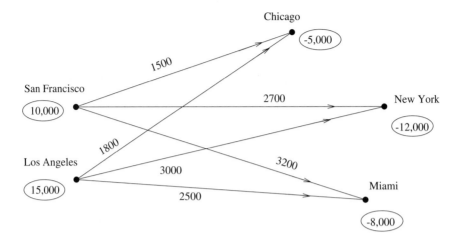

Figure 8.4. *Transportation problem.*

(perhaps based on their experience or education, as well as the skill requirements of the job). If required, the assignment problem can be expressed as an equivalent minimization problem by multiplying the objective function by -1. Assignment problems suffer from severe degeneracy, but special algorithms have been designed to solve them that are more efficient than general algorithms for the transportation problem.

It is not normally possible to assign a fraction of a person to a fraction of a job, so an assignment problem also includes the requirement that the variables take on integer values. This integrality constraint is common to many network problems. We show in the next section that, if the data for a network problem are integers, then any basic solution will be integer valued, and hence an optimal basic feasible solution will be integer valued. For this reason, we omit the integrality constraint from the model for an assignment problem.

Example 8.4 (Assignment Problem). Suppose that a company is planning to assign three people to three jobs. The jobs are accountant, budget director, and personnel manager. The first two people have degrees in business, but the second has ten years of corporate experience, while the first is just out of school. The second and third persons both have some management experience, but in different departments. The third person's degree was in anthropology. Based on this information, the personnel department has determined numerical values $\{\,c_{i,j}\,\}$ corresponding to each person's appropriateness for a particular job. The corresponding assignment problem is illustrated in Figure 8.5.

The corresponding linear program can be written as

$$\text{maximize} \quad z = 11x_{1,1} + 5x_{1,2} + 2x_{1,3} + 15x_{2,1} + 12x_{2,2}$$
$$+ 8x_{2,3} + 3x_{3,1} + 1x_{3,2} + 10x_{3,3}$$

subject to the constraints

$$x_{1,1} + x_{1,2} + x_{1,3} = 1$$
$$x_{2,1} + x_{2,2} + x_{2,3} = 1$$

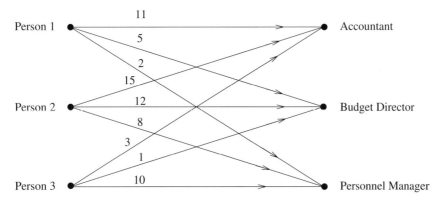

Figure 8.5. *Assignment problem.*

$$x_{3,1} + x_{3,2} + x_{3,3} = 1$$
$$-x_{1,1} - x_{2,1} - x_{3,1} = -1$$
$$-x_{1,2} - x_{2,2} - x_{3,2} = -1$$
$$-x_{1,3} - x_{2,3} - x_{3,3} = -1$$
$$0 \leq x \leq 1. \qquad \blacksquare$$

A *shortest path* problem determines the shortest or the fastest route between an origin and a destination. It can be represented as a minimum cost network flow problem with one source (the origin) with supply equal to 1, and one sink (the destination) with demand equal to 1. There are usually many transshipment nodes where flow is conserved. The cost coefficients $\{ c_{i,j} \}$ represent the length of an arc, the time required to traverse a particular arc, or the financial cost of using an arc.

In many applications the cost coefficients satisfy $c_{i,j} \geq 0$. This is a natural requirement if $c_{i,j}$ represents the length of arc (i, j) in the network. If this requirement is satisfied, then especially efficient algorithms are available to solve the shortest path problem. There exist applications, however, where it is sensible to allow $c_{i,j} < 0$. If negative costs are present, the special-purpose algorithms can break down, and the shortest path problem can be more difficult to solve.

Example 8.5 (Shortest Path Problem). The network in Figure 8.6 represents a road system. Some of the streets are one way, while others can be driven in both directions. The goal is to drive from the source (node 1) to the sink (node 11) via the shortest possible route. The travel times for each road segment are marked on the arcs. $\qquad \blacksquare$

A *maximum flow* problem determines the maximum amount of flow that can be moved through a network from the source to the sink. As in the shortest path problem, there is a single source and a single sink. This problem includes an additional variable f that records the flow in the network. For convenience, assume that node 1 is the source and node m is the sink.

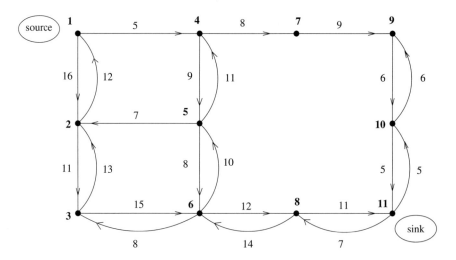

Figure 8.6. *Shortest path problem.*

This problem is normally written in the form

$$\underset{x,f}{\text{maximize}} \quad z = f$$

$$\text{subject to} \quad \sum_j x_{1,j} - \sum_k x_{k,1} = f$$

$$\sum_j x_{i,j} - \sum_k x_{k,i} = 0, \quad i = 2, \ldots, m-1$$

$$\sum_j x_{m,j} - \sum_k x_{k,m} = -f$$

$$0 \le x_{i,j} \le u_{i,j}.$$

(Note that f is a variable, even though it is written on the right-hand side of the constraints.) If the artificial arc $(m, 1)$ is added to the network, with unlimited capacity ($u_{m,1} = +\infty$), then the maximum flow problem can be converted to an equivalent minimum cost network flow problem:

$$\underset{x}{\text{minimize}} \quad z = -x_{m,1}$$

$$\text{subject to} \quad \sum_j x_{i,j} - \sum_k x_{k,i} = 0, \quad i = 1, \ldots, m$$

$$0 \le x_{i,j} \le u_{i,j}.$$

The maximum flow problem is illustrated in the following example.

Example 8.6 (Maximum Flow Problem). Suppose that you wish to transport a large number of military personnel between Seattle and New York by airplane. The network in Figure 8.7 indicates the available flights and the capacities (in hundreds) of the planes. The solution to the maximum flow problem determines the number of people that can be transported, as well as the routings that achieve this result. ∎

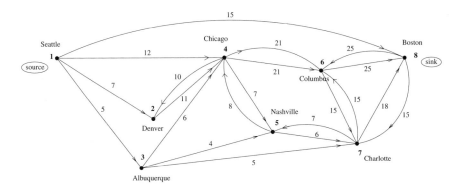

Figure 8.7. *Maximum flow problem.*

The dual of a maximum flow problem can be interpreted in terms of *cuts*. A cut is defined to be a division of the nodes into two disjoint sets, the first $\mathcal{N}_1$ containing the source, and the second $\mathcal{N}_2$ containing the sink. The capacity of the cut is the sum of the capacities of the arcs that lead from $\mathcal{N}_1$ to $\mathcal{N}_2$.

Example 8.7 (Cuts in a Network). Consider the maximum flow problem in Example 8.6. If we pick the cut defined by the node sets $\mathcal{N}_1 = \{\,1, 2, 3\,\}$ and $\mathcal{N}_2 = \{\,4, 5, 6, 7, 8\,\}$, then the capacity of the cut is 5300 (the sum of the capacities of arcs $(1, 4)$, $(1, 8)$, $(2, 4)$, $(3, 4)$, $(3, 5)$, and $(3, 7)$). If we pick the cut defined by the sets $\mathcal{N}_1 = \{\,1, 4, 5, 7\,\}$ and $\mathcal{N}_2 = \{\,2, 3, 6, 8\,\}$, then the capacity of the cut is 9100 (for arcs $(1, 2)$, $(1, 3)$, $(1, 8)$, $(4, 2)$, $(4, 6)$, $(7, 6)$, and $(7, 8)$).   ∎

A famous theorem states that the value of the maximum flow in a network is equal to the minimum of the capacities of all cuts in the network, a result that we will sketch here. This is a special case of the strong duality theorem for linear programming. For complete details, see the book by Ford and Fulkerson (1962).

The dual of the original form of the maximum flow problem is

$$
\begin{aligned}
\operatorname*{minimize}_{y,v} \quad & w = \sum u_{i,j} v_{i,j} \\
\text{subject to} \quad & y_m - y_1 = 1 \\
& y_i \; - y_j + v_{i,j} \geq 0 \quad \text{for all arcs } (i, j) \\
& v_{i,j} \geq 0.
\end{aligned}
$$

The dual variable y_i corresponds to the flow-conservation constraint for the ith node in the primal. The dual variable $v_{i,j}$ corresponds to the upper bound $x_{i,j} \leq u_{i,j}$ in the primal.

To show the relationship of the dual problem with cuts, let $y_i = 0$ if node i is in the set $\mathcal{N}_1$, and let $y_i = 1$ if node i is in the set $\mathcal{N}_2$. This ensures that $y_m - y_1 = 1$. Let $v_{i,j} = 1$ if arc (i, j) connects $\mathcal{N}_1$ with $\mathcal{N}_2$, and let $v_{i,j} = 0$ otherwise. It is straightforward to check that this produces a feasible solution to the dual, and that the dual objective value is equal to the capacity of the cut. The fact that the optimal solution to the dual corresponds to a cut is left to the Exercises.

Exercises

2.1. Write down the linear program for the transportation problem in Example 8.3.

2.2. Write down the linear program for the shortest path problem in Example 8.5.

2.3. Consider a network linear program where the variables have general lower bounds $\ell \leq x$. Show how to use a change of variables to convert to a problem with nonnegativity constraints $0 \leq \hat{x}$.

2.4. Consider a network linear program that includes upper bounds on the variables $0 \leq x \leq u$. Show, by adding an artificial node for every upper bound, how to convert to an equivalent network problem without upper bounds on the variables.

2.5. Use the technique of the previous problem to remove the upper bounds in the linear program in Example 8.1.

2.6. Consider an arbitrary minimum cost network flow problem with lower bounds $\ell = 0$. Show that a feasible point for this problem can be found by solving a related maximum flow problem. *Hint*: Add a new "super source" to the network that can supply all the given sources, and a new "super sink" that can absorb the demand of all the given sinks.

2.7. Verify that the rank of the matrix A in Example 8.1 is equal to 6, one less than the number of equality constraints.

2.8. Solve the linear program in Example 8.4 (either by using simplex software or by examining the network in Figure 8.5) and verify that there is an integer-valued optimal basic feasible solution.

2.9. Show (by constructing an example) that the objective value in a shortest path problem with negative cost coefficients can be unbounded below.

2.10. Consider the maximum flow problem in Example 8.6.

   (i) Write down the linear program for this problem.
   (ii) What is the dual of this linear program?
   (iii) What is the capacity of the cut corresponding to the sets $\mathcal{N}_1 = \{1, 3, 5, 6\}$ and $\mathcal{N}_2 = \{2, 4, 7, 8\}$?
   (iv) What is the dual feasible solution corresponding to this cut?

2.11. Consider a maximum flow problem and its dual, with $\mathcal{N}_1$ and $\mathcal{N}_2$ being the sets associated with a cut. Let $y_i = 0$ if node i is in $\mathcal{N}_1$, and let $y_i = 1$ if node i is in $\mathcal{N}_2$. Let $v_{i,j} = 1$ if arc (i, j) connects $\mathcal{N}_1$ with $\mathcal{N}_2$, and let $v_{i,j} = 0$ otherwise. Verify that y and v are feasible for the dual.

2.12. Use duality results from linear programming to prove that an optimal basic feasible solution to a maximum flow problem corresponds to a minimum capacity cut.

8.3   Representation of the Basis

Many of the efficiencies in the network simplex method come about because of the special form of the basis in a network problem. As we shall prove below, a basis is equivalent to a spanning tree, a special subset of a network that will be defined below. Before we can

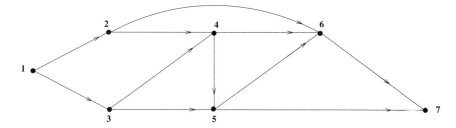

Figure 8.8. *Sample network.*

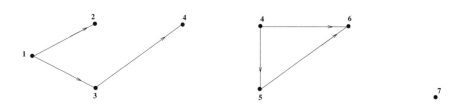

Figure 8.9. *Subnetworks.*

prove this important result, we shall need to define a number of terms relating to networks. The terms will be illustrated using the sample network in Figure 8.8. This same network was used in Example 8.1; the corresponding linear program will be referred to here also.

A *subnetwork* of a network is a subset of the nodes and arcs of the original network. The arcs in the subnetwork must connect nodes in the subnetwork and must not involve nodes that are not in the subnetwork. For example, if the subnetwork includes only nodes 1, 3, and 6, then an arc $(1, 2)$ could not be part of the subnetwork because node 2 is not part of the subnetwork; arc $(6, 3)$ could be included if it was present in the original network. Subnetworks are illustrated in Figure 8.9. A subnetwork is itself a network.

A *path* from node i_1 to node i_k is a subnetwork consisting of a sequence of nodes i_1, $i_2, \ldots, i_k$, together with a set of distinct arcs connecting each node in the sequence to the next. The arcs need not all point in the same direction. For example, the path could contain either arc (i_1, i_2) or arc (i_2, i_1). See Figure 8.10.

A network is said to be *connected* if there is a path between every pair of nodes in the subnetwork. See Figure 8.11.

A *cycle* is a path from a node i_1 to itself. That is, it consists of a sequence of nodes i_1, $i_2, \ldots, i_k = i_1$, together with arcs connecting them. See Figure 8.12.

A *tree* is a connected subnetwork containing no cycles. A *spanning tree* is a tree that includes every node in the network. A tree and spanning tree for the network in Figure 8.8 are shown in Figure 8.13.

We will examine further the properties of trees and spanning trees. These are established in a sequence of lemmas. For the remainder of this chapter we make the following two assumptions about any network that we will consider: (a) the network is connected (if not, the problem can be decomposed into two or more smaller problems), and (b) there are no arcs of the form (i, i) from a node to itself. We are now ready to prove our results.

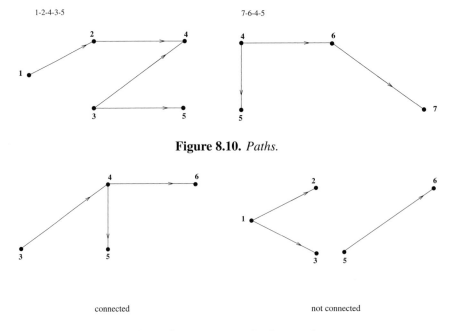

Figure 8.10. *Paths.*

Figure 8.11. *Connected subnetworks.*

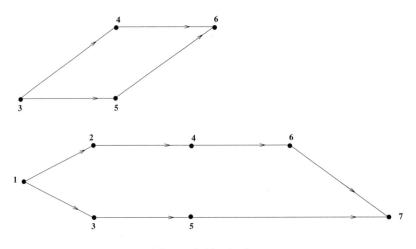

Figure 8.12. *Cycles.*

Lemma 8.8. *Every tree consisting of at least two nodes has at least one end (a node that is incident to exactly one arc).*

Proof. Pick some node i in the tree. Follow any path away from node i (one must exist since the tree is connected). Since there are no cycles in the tree, eventually the path must terminate at an end of the tree.   □

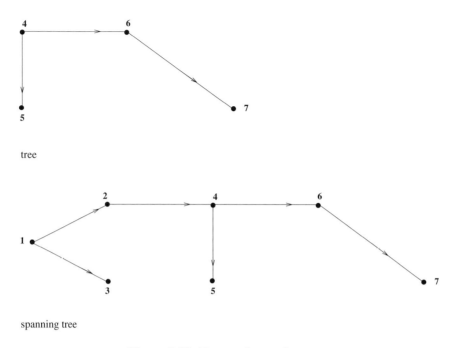

tree

spanning tree

Figure 8.13. *Trees and spanning trees.*

Lemma 8.9. *A spanning tree for a network with m nodes contains exactly $m - 1$ arcs.*

Proof. This is proved by induction on the number of nodes in the spanning tree. If the spanning tree consists of one node, then there are no arcs. If the spanning tree consists of $m \geq 2$ nodes, construct a subtree and subnetwork by removing an end node from the tree and the network, as well as the arc incident to it. Lemma 8.8 shows that such a node exists. The resulting tree has $m - 1$ nodes and (by induction) $m - 2$ arcs. Adding back the end node and the corresponding arc gives a spanning tree with m nodes and $m - 1$ arcs.   □

Lemma 8.10. *If a spanning tree is augmented by adding to it an additional arc of the network, then exactly one cycle is formed.*

Proof. Suppose that arc (i, j) is added to the spanning tree. Since the spanning tree already contains a path between nodes i and j, that path together with the arc (i, j) forms a cycle. So the augmented tree contains at least one cycle. Suppose that two distinct cycles were formed. They must both contain the new arc (i, j) because the spanning tree had no cycles. Then the union of the two cycles, minus the new arc (i, j), also contains a cycle, but consists only of arcs in the original tree. This is a contradiction, showing that exactly one cycle is formed.   □

Lemma 8.11. *Every connected network contains a spanning tree.*

Proof. If the network does not contain a cycle, then it is also a spanning tree since it is connected and contains all of the nodes. Otherwise, there exists a cycle. Deleting any arc from this cycle results in a subnetwork that is still connected. It is possible to continue deleting arcs in this way as long as the resulting subnetwork continues to contain a cycle. Ultimately a subnetwork is obtained that contains no cycle and is connected and contains all the nodes, that is, a spanning tree.   □

The submatrix of A corresponding to a spanning tree has special structure: it can be rearranged to form a full-rank lower triangular matrix. Let B be the submatrix of A corresponding to a spanning tree. For the network in Figure 8.8, the matrix A was derived in Example 8.1:

$$A = \begin{pmatrix} 1 & 1 & & & & & & & & \\ -1 & & 1 & 1 & & & & & & \\ & -1 & & & 1 & 1 & & & & \\ & & -1 & & -1 & & 1 & 1 & & \\ & & & & & -1 & -1 & & 1 & 1 \\ & & & -1 & & & & -1 & -1 & & 1 \\ & & & & & & & & & -1 & -1 \end{pmatrix}.$$

For the spanning tree in Figure 8.13, the matrix B is obtained by selecting the columns associated with the variables $x_{1,2}$, $x_{1,3}$, $x_{2,4}$, $x_{4,5}$, $x_{4,6}$, and $x_{6,7}$:

$$B = \begin{pmatrix} 1 & 1 & & & & \\ -1 & & 1 & & & \\ & -1 & & & & \\ & & -1 & 1 & 1 & \\ & & & -1 & & \\ & & & & -1 & 1 \\ & & & & & -1 \end{pmatrix}.$$

If the matrix B is rearranged so that the rows are listed in the order $(3, 1, 2, 7, 5, 6, 4)$, and the columns in the order $(2, 1, 3, 6, 4, 5)$, then B is transformed into

$$\hat{B} = \begin{pmatrix} -1 & & & & & \\ 1 & 1 & & & & \\ & -1 & 1 & & & \\ & & & -1 & & \\ & & & & -1 & \\ & & 1 & & & -1 \\ & & -1 & & 1 & 1 \end{pmatrix},$$

a lower triangular matrix with entries ± 1 along the diagonal. It is clear that the columns of B are linearly independent, and hence B is of full rank. The following lemma shows that this is always possible.

Lemma 8.12. *Let B be the submatrix of the constraint matrix A corresponding to a spanning tree with m nodes. Then B can be rearranged to form a full-rank lower triangular matrix of dimension $m \times (m-1)$ with diagonal entries ± 1.*

Proof. By Lemma 8.9, a spanning tree consists of m nodes and $m - 1$ arcs, so B is of dimension $m \times (m - 1)$. We will use induction to show that B can be rearranged into the required form. If $m = 1$, then B is empty. If $m = 2$, then the spanning tree consists of one arc so that either

$$B = \begin{pmatrix} 1 \\ -1 \end{pmatrix} \quad \text{or} \quad B = \begin{pmatrix} -1 \\ 1 \end{pmatrix}.$$

Both of these matrices are of the required form.

Suppose that the result is true for spanning trees of $m - 1$ nodes. Now consider a spanning tree with m nodes, and let node i be an end of the spanning tree. By Lemma 8.8 such a node exists. Since node i is only connected to one arc in the tree, row i of B has exactly one nonzero entry, with value ± 1. Suppose that this entry occurs in column j of B. Now interchange rows 1 and i of B, as well as columns 1 and j. Then B is transformed into

$$\hat{B} = \begin{pmatrix} \pm 1 & \mathbf{0} \\ v & B_1 \end{pmatrix},$$

where B_1 is the submatrix corresponding to the spanning tree with node i and the corresponding arc removed, and v consists of the remaining portion of column j of B with row i removed.

The matrix B_1 represents a spanning tree for the network with node i removed, and hence (by induction) it can be rearranged into a lower triangular matrix with diagonal entries ± 1. Hence $\hat{B}$ can also be rearranged into this form. Since all the diagonal entries of the rearranged matrix are nonzero, the matrix is full rank. $\qquad \square$

We are now in a position to show the relationship between a spanning tree and a basis. To do this we state two definitions. Given a spanning tree for a network, a *spanning tree solution* x is a set of flow values that satisfy the flow-balance constraints $Ax = b$ for the network, and for which $x_{i,j} = 0$ for any arc (i, j) that is not part of the spanning tree. A *feasible spanning tree solution* x is a spanning tree solution that satisfies the nonnegativity constraints $x \geq 0$. These definitions are analogous to the definitions of basic solution and basic feasible solution.

Recall from Section 4.3 that a point x is an extreme point for a linear program in standard form if and only if it is a basic feasible solution. Keep in mind, however, that in Chapter 4 we assumed that the constraint matrix A had full rank, whereas here one of the constraints is redundant. Hence the size of the basis will be $m - 1$, one less than the number of rows in A.

Theorem 8.13 (Equivalence of Spanning Tree and Basis). *A flow x is a basic feasible solution for the network flow constraints*

$$\{ x : Ax = b, x \geq 0 \}$$

if and only if it is a feasible spanning tree solution.

Proof. For the first half of the proof, let us assume that x corresponds to a feasible spanning tree solution. (That is, x is feasible, and the nonzero components of x together with, if necessary, a subset of the zero components of x are the variables for arcs that form a spanning tree.) Let B be the submatrix of A corresponding to the spanning tree. By the previous lemma the columns of B are linearly independent, and hence x is a basic feasible solution.

For the other half of the proof, consider the set of arcs corresponding to the strictly positive components of x. If these arcs do not contain a cycle, then they can be augmented with zero-flow arcs to form a spanning tree, showing that x is a feasible spanning tree solution. Otherwise, this set of arcs must contain a cycle. We may assume that, within the cycle, all the flows are strictly greater than zero (if not, any arc with zero flow could be removed from the subnetwork associated with x). If the flow on an arc (i, j) in the cycle is increased by some small $\epsilon > 0$, then the other flows in the cycle must be adjusted to maintain the flow-balance constraints. Any arc pointing in the same direction as arc (i, j) has its flow increased by ϵ, and any arc pointing in the opposite direction has its flow decreased by ϵ. If ϵ is sufficiently small, this can be done without violating the nonnegativity constraints. Call this new flow x_ϵ. Similarly, if the flow on $x_{i,j}$ is decreased by ϵ, we can obtain a new feasible flow $x_{-\epsilon}$. Since

$$x = \tfrac{1}{2}x_\epsilon + \tfrac{1}{2}x_{-\epsilon}$$

the flow x is not an extreme point, and hence not a basic feasible solution. Together these remarks show that if x is a basic feasible solution, then x corresponds to a feasible spanning tree solution.   □

If the right-hand-side entries $\{b_i\}$ for a network problem are all integers, then any basic feasible solution will also consist of integers. This is a consequence of the special form of the basis matrix B. Let $\bar{B}$ be the matrix obtained by deleting the (dependent) last row of the lower triangular rearrangement of B. A basic feasible solution can be obtained by solving

$$\bar{B}x_B = \bar{b},$$

where $\bar{b}$ is the correspondingly rearranged right-hand-side b with its last component removed. This linear system can be solved using forward substitution: $(x_B)_1 = \bar{B}_{1,1}^{-1}\bar{b}_1$, and for $i = 2, \ldots, m - 1$,

$$(x_B)_i = \bar{B}_{i,i}^{-1}\left(\bar{b}_i - \sum_{j=1}^{i-1}\bar{B}_{i,j}(x_B)_j\right).$$

Since $\bar{B}_{i,i} = \pm 1$ for all i, and $\bar{B}_{i,j} = 0$ or ± 1, then x_B must consist of integers if b consists of integers. By similar reasoning, if the cost coefficients $\{c_{i,j}\}$ are all integers, then the dual variables must also be integers (see the Exercises).

This property has important practical consequences. Linear programs in which the solutions are further constrained to take on integer values arise frequently. For example, it is difficult to build two-thirds of a warehouse or to send half of a soldier on a mission. Such problems are called *integer programming problems*. In general cases, they can be difficult to solve, requiring auxiliary search techniques beyond the simplex method, such as branch and bound or cutting plane methods (see the book by Nemhauser and Wolsey (1988, reprinted 1999)). For network problems, however, the basic feasible solutions will always take on integer values, and hence an integer solution can be obtained just by applying the simplex method to the linear program arising from the network.

Exercises

3.1. For each of the example networks in Section 8.2, identify a spanning tree and compute the corresponding spanning tree solution. For the network in Figure 8.5, identify the basis matrix B and show that it can be rearranged as a lower triangular matrix.

3.2. Show how to compute the spanning tree solution corresponding to any spanning tree by first determining the flow at an end node of the tree, and then traversing the tree along paths beginning at the end node. Use this technique to prove that, if the supplies and demands for a network are integers, then any spanning tree solution will consist of integers.

3.3. Prove that, if the cost coefficients $\{ c_{i,j} \}$ in a minimum cost network flow problem are all integers, then the simplex multipliers corresponding to any basic feasible solution must be integers. Hence prove that the values of the dual variables at an optimal basic feasible solution must also be integers.

3.4. Let A be the constraint matrix for a network linear program. Prove that the determinant of every square submatrix of A is equal to 0, 1, or -1. Such a matrix is called *totally unimodular*. (*Hint*: Use induction on the size of the square submatrix.)

3.5. Consider a linear program in standard form with constraint matrix A, right-hand-side vector b, and cost vector c. Assume that all the entries in A, b, and c are integers. Prove that if A is totally unimodular, then every basic feasible solution has integer entries. *Hint*: Use Cramer's rule.

3.6. For a connected network, prove that any tree can be augmented with additional arcs to form a spanning tree.

3.7. Prove that a set of arcs in a network does not contain a cycle if and only if the corresponding submatrix of A has full column rank.

3.8. Use the result of the previous problem to prove that a basic feasible solution is equivalent to a feasible spanning tree solution.

8.4   The Network Simplex Method

The network simplex method uses the same operations as the simplex method (see Section 5.2). The method takes advantage of the special form of the minimum cost network flow problem to reduce the operation count for the method, often performing the calculations directly on the network rather than using matrix operations. The exceptional efficiency of these operations has made the network simplex method an important tool for this special class of linear programming problems.

We describe the network simplex method, showing how each of the major operations (the optimality test, the step, and the update) can be performed using network techniques.

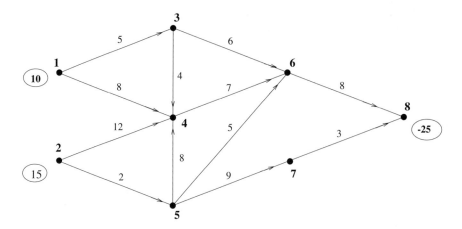

Figure 8.14. *A sample network.*

These operations will be related back to the formulas and algebraic techniques used in Chapter 5 to describe the simplex method.

Before we present the network simplex method, we will review the steps in the simplex method. Let B be the basis matrix at a given iteration, and assume that it corresponds to a basic feasible solution satisfying $Bx_B = b$. Let $A_{i,j}$ be the column of A associated with arc (i, j). Then the steps of the simplex method (adapted to the notation for the network problem) are as follows:

1. *The Optimality Test*—Compute the vector of simplex multipliers by solving $B^T y = c_B$. Compute the coefficients $\hat{c}_{i,j} = c_{i,j} - y^T A_{i,j}$ for the nonbasic variables $x_{i,j}$. If $\hat{c}_{i,j} \geq 0$ for all nonbasic variables, then the current basis is optimal. Otherwise, select a variable $x_{s,t}$ that satisfies $\hat{c}_{s,t} < 0$ as the entering variable.

2. *The Step*—Determine by how much the entering variable $x_{s,t}$ can be increased before one of the current basic variables is reduced to zero. If $x_{s,t}$ can be increased without bound, then the problem is unbounded.

3. *The Update*—Update the representation of the basis matrix B and the vector of basic variables x_B.

The simplifications in the method come about because of the special form of the basis matrix B (a lower triangular matrix with all entries equal to 0, 1, or -1) and its equivalent representation as a spanning tree.

To describe the method, we will use as an example the network problem in Figure 8.14. Nodes 1 and 2 are sources (with supplies equal to 10 and 15, respectively) and node 8 is a sink (with demand equal to 25). The costs of the arcs are indicated on the network.

We initialize the method with the basic feasible solution

$$x_{1,3} = 10, \quad x_{3,4} = 10, \quad x_{4,6} = 10, \quad x_{6,8} = 25$$
$$x_{2,5} = 15, \quad x_{5,6} = 15, \quad x_{7,8} = 0.$$

All other arcs are nonbasic, and the corresponding variables are zero. The value of the objective function is

$$z = 465.$$

It is easy to check that this flow satisfies all the constraints for the network, and that it is a feasible spanning tree solution.

To determine the simplex multipliers y we solve $B^T y = c_B$. If $x_{i,j}$ is a basic variable, and (i, j) is the corresponding arc in the network, then the corresponding equation for the simplex multipliers is

$$y_i - y_j = c_{i,j}.$$

There is a simplex multiplier associated with every node in the network. As has been mentioned, the rows of the matrix B are linearly dependent. This implies that one of the simplex multipliers is arbitrary. To determine the simplex multipliers, we will traverse the spanning tree (basis) starting at an end and set the first of the simplex multipliers equal to zero. Hence for this basis,

$$
\begin{aligned}
y_1 &= 0 \\
y_3 &= y_1 - 5 = -5 \\
y_4 &= y_3 - 4 = -9 \\
y_6 &= y_4 - 7 = -16 \\
y_8 &= y_6 - 8 = -24 \\
y_7 &= y_8 + 3 = -21 \\
y_5 &= y_6 + 5 = -11 \\
y_2 &= y_5 + 2 = -9.
\end{aligned}
$$

To determine the simplex multipliers we could have begun this process at any node. Also, we could have specified any value for the first simplex multiplier, not just zero. This would have resulted in different values for the simplex multipliers, but would not affect the optimality test, since this test only depends on the *differences* between pairs of simplex multipliers.

To perform the optimality test we compute

$$\hat{c}_{i,j} = c_{i,j} - y^T A_{i,j}$$

for the nonbasic variables $x_{i,j}$. Because each column of A contains only two nonzero entries, $+1$ and -1, we obtain the formula

$$\hat{c}_{i,j} = c_{i,j} - y_i + y_j.$$

If we carry out this calculation for all the nonbasic arcs, we get

$$
\begin{aligned}
\hat{c}_{1,4} &= 8 - y_1 + y_4 = -1 < 0 \\
\hat{c}_{3,6} &= 6 - y_3 + y_6 = -5 < 0 \\
\hat{c}_{5,4} &= 10 - y_5 + y_4 = 10 \\
\hat{c}_{2,4} &= 12 - y_2 + y_4 = 12 \\
\hat{c}_{5,7} &= 9 - y_5 + y_7 = -1 < 0.
\end{aligned}
$$

Since some of these entries are negative, this basis is not optimal.

The entry $\hat{c}_{3,6}$ is the most negative, and we choose the variable $x_{3,6}$ to enter the basis. By Lemma 8.10, adding this arc to the spanning tree creates a unique cycle, illustrated in

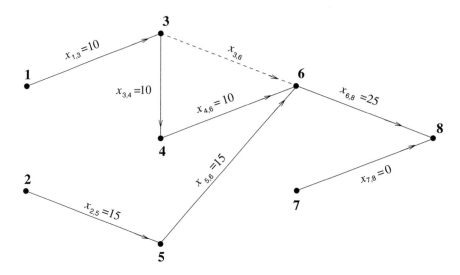

Figure 8.15. *Entering variable and cycle.*

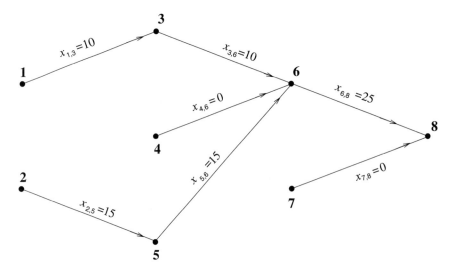

Figure 8.16. *Result of iteration* 1.

Figure 8.15. If $x_{3,6}$ is increased from zero, then the flows in the other arcs in the cycle must be adjusted to maintain the flow-balance constraints in the linear program.

In this case the flows in the other two arcs must decrease by one unit for every increase of one unit in $x_{3,6}$. In the current basis $x_{3,4} = x_{4,6} = 15$, so $x_{3,6}$ can be increased until it is equal to 15, at which point the other two flows are both equal to zero. One of the two arcs must be chosen to leave the basis. We pick $x_{3,4}$ to leave the basis. We obtain the new basic feasible solution shown in Figure 8.16.

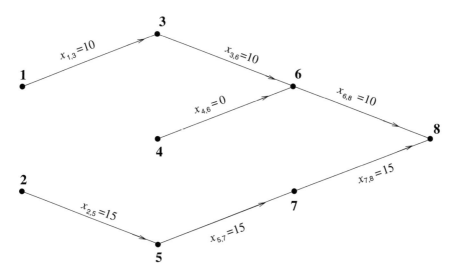

Figure 8.17. *Result of iteration 2.*

The new values of the basic variables are

$$x_{1,3} = 10, \quad x_{3,6} = 10, \quad x_{4,6} = 0, \quad x_{6,8} = 25$$
$$x_{2,5} = 15, \quad x_{5,6} = 15, \quad x_{7,8} = 0,$$

and the new value of the objective function is $z = 415$. Setting $y_1 = 0$, the simplex multipliers are

$$y = (\,0 \quad -4 \quad -5 \quad -4 \quad -6 \quad -11 \quad -16 \quad -19\,)^T.$$

In the optimality test,

$$\hat{c}_{5,7} = 9 - y_5 + y_7 = -1 < 0.$$

Hence this basis is not optimal, and $x_{5,7}$ is the entering variable. Adding arc $(5, 7)$ to the spanning tree produces a unique cycle.

To maintain the flow-balance constraints when $x_{5,7}$ is increased by one unit, $x_{7,8}$ will also increase by one unit, while $x_{5,6}$ and $x_{6,8}$ will both decrease by one unit. Since $x_{5,6} = 15$ and $x_{6,8} = 25$, $x_{5,6}$ will go to zero first and hence will leave the basis. The new basic feasible solution is illustrated in Figure 8.17.

The new values of the basic variables are

$$x_{1,3} = 10, \quad x_{3,6} = 10, \quad x_{4,6} = 0, \quad x_{6,8} = 10$$
$$x_{2,5} = 15, \quad x_{5,7} = 15, \quad x_{7,8} = 15,$$

and the new value of the objective function is $z = 400$. Computing the simplex multipliers gives

$$y = (\,0 \quad -5 \quad -5 \quad -4 \quad -7 \quad -11 \quad -16 \quad -19\,)^T.$$

In the optimality test,

$$\hat{c}_{1,4} = 4, \quad \hat{c}_{3,4} = 5, \quad \hat{c}_{2,4} = 13$$
$$\hat{c}_{5,4} = 11, \quad \hat{c}_{5,6} = 1.$$

Since these entries are all nonnegative, the current basis is optimal and the algorithm terminates.

We now summarize the steps in the network simplex method.

1. *The Optimality Test*

    (i) Compute the simplex multipliers y: Start at an end of the spanning tree and set the associated simplex multiplier to zero. Following the arcs (i, j) of the spanning tree, use the formula $y_i - y_j = c_{i,j}$ to compute the remaining simplex multipliers.

    (ii) Compute the reduced costs $\hat{c}$: For each nonbasic arc (i, j) compute $\hat{c}_{i,j} = c_{i,j} - y_i + y_j$. If $\hat{c}_{i,j} \geq 0$ for all nonbasic arcs, then the current basis is optimal. Otherwise, select an arc (s, t) that satisfies $\hat{c}_{s,t} < 0$ as the entering arc.

2. *The Step*—Identify the cycle formed by adding (s, t) to the spanning tree. Determine how much the flow on arc (s, t) can be increased before one of the other flows in the cycle is reduced to zero. If the flow in (s, t) can be increased without bound, then the problem is unbounded.

3. *The Update*—Update the spanning tree by adding arc (s, t) and removing an arc of the cycle whose flow has been reduced to zero.

It remains to show how to obtain an initial basic feasible solution.

In the example an initial basic feasible solution was provided, but in general cases a procedure for finding an initial point is required. The techniques for network problems are analogous to those used for general linear programs (see Section 5.5). In a network problem, artificial arcs (or, equivalently, artificial variables) can be added to the network in such a way that an "obvious" initial basic feasible solution is apparent. Then a phase-1 or big-M procedure is used to remove the artificial variables from the basis.

One way of doing this is to pick one node in the network to be labeled the *root* node. Then artificial arcs are added: one from each source node to the root node, and one from the root node to each sink and transshipment node. (No arc need be added from the root node to itself.) The costs associated with these arcs would be equal to 1 in a phase-1 problem, and equal to M in a big-M problem. The initial basic feasible solution would transmit all flow from the sources to the sinks via the root node, with zero flow from the root node to a transshipment node. This technique is illustrated in Figure 8.18, where the original arcs in the network are marked with solid lines and the artificial arcs with dotted lines.

The network simplex method performs the following arithmetic operations:

$$
\begin{array}{rl}
\text{computing } y: & m \text{ subtractions} \\
\text{computing } \hat{c}: & 2(n - m) \text{ subtractions} \\
\text{updating } x: & m \text{ additions/subtractions}
\end{array}
$$

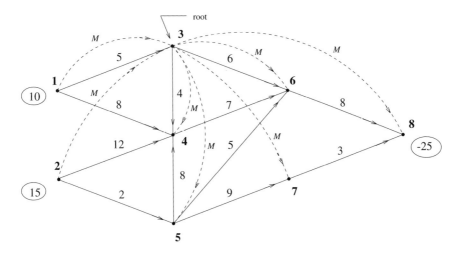

Figure 8.18. *Initial basic feasible solution.*

so that the total number of arithmetic operations is

$$m + 2(n - m) + m = 2n \text{ additions/subtractions.}$$

All of these calculations involve integers if the vectors b and c consist of integers.

We now compare this with the simplex method. The operation counts for the simplex method (see Sections 5.3 and 7.5) are harder to determine, since they depend on the sparsity of the constraint matrix A and the representation of the basis matrix B. We will assume that an LU factorization is used to represent B, and that the matrices A and B are sparse. An iteration of the simplex method involves the following steps:

$$\begin{aligned} \text{computing } y : \quad & \text{solving } B^T y = c_B \\ \text{computing } \hat{c} : \quad & \text{computing } c_j - y^T A_j \text{ for } (n - m) \text{ values of } j \\ \text{updating } B : \quad & \text{updating a sparse } LU \text{ factorization.} \end{aligned}$$

Each of these steps is more expensive than the corresponding step of the network simplex method. More operations are required, the operations involve multiplication and addition, and the operations involve real (decimal) numbers.

The network simplex method involves fewer operations, and each of those operations is faster (addition is usually faster than multiplication, and integer operations are usually faster than real operations). As a result, network simplex software is much faster than general-purpose simplex software.

In these operation counts we are ignoring the operations involved in maintaining the data structures for the algorithms. Much research has focused on ways of representing the network, as well as the spanning tree, within the network simplex algorithm. The data structure must allow all the simplex operations to be performed efficiently. It must also be designed so that it can be updated easily to reflect changes in the basis. For further details see the book by Ahuja, Magnanti, and Orlin (1993).

Many of the topics that have been discussed in other chapters for the simplex method have analogs for the network simplex method. For example, it is possible to do sensitivity analysis, but in the network case all the calculations can be done more efficiently. Also, for degenerate problems, it is possible that the network simplex method could cycle, and so some sort of anticycling procedure may be necessary.

Degeneracy in network problems is common. Even in cases where cycling does not occur, it is possible to have a large number of consecutive degenerate iterations. This is referred to as *stalling*. If the network simplex method is implemented with the basis represented and updated in a special way (using what is known as a "strongly feasible basis"), and if the entering variable is chosen appropriately, then at most nm consecutive degenerate iterations can occur. This approach guarantees that the network simplex method always terminates in a finite number of iterations, even on degenerate problems. It also improves the practical performance of the network simplex method. Details of this approach are discussed in Section 8.5.

Exercises

4.1. Apply the network simplex method to the linear programming problems in

    (i) Example 8.1.

    (ii) Example 8.3.

    (iii) Example 8.4.

    (iv) Example 8.6.

4.2. In the network simplex method, let r be the node whose simplex multiplier is set to zero. An arc (i, j) will be called a *forward* arc if the path from node r to node j along the spanning tree includes node i. Otherwise, arc (i, j) will be a *reverse* arc. Define the cost of a path from node r to node i as the sum of the costs of the reverse arcs in the path minus the sum of the costs of the forward arcs. Prove that the simplex multiplier y_i is equal to the cost of the path from node r to node i.

4.3. Consider the cycle created by adding a nonbasic arc (i, j) to a spanning tree. Define the cost of the cycle as the sum of the costs of the arcs in the cycle whose direction is the same as arc (i, j) minus the sum of the costs of the arcs whose direction is opposite to arc (i, j). Prove that the cost of this cycle is equal to $\hat{c}_{i,j}$, the reduced cost for the nonbasic arc.

4.4. The network simplex method can be made more efficient by updating the simplex multipliers y at each iteration, rather than recomputing them. Prove that the new simplex multipliers $\bar{y}$ satisfy either $\bar{y}_i = y_i$ or $\bar{y}_i = y_i - \hat{c}_{j,k}$, where (j, k) is the entering arc. What is the rule for determining which formula to use to compute $\bar{y}_i$?

4.5. Derive a variant of the network simplex method for upper bounded variables, analogous to the bounded-variable simplex method developed in Section 7.2.

4.6. Apply the initialization procedure in Figure 8.18 to the network problem in Figure 8.1. Use a phase-1 procedure to find an initial basic feasible solution to the original network.

4.7. Suppose that a minimum cost network flow problem has been modified by adding additional linear constraints. Show how to use the decomposition principle to solve the resulting problem, with the flow-balance constraints considered as the "easy" constraints.

4.8. Apply the method of the previous exercise to the linear program obtained by adding the constraint

$$x_{1,2} + x_{1,3} \geq 6$$

to the network problem in Figure 8.14.

8.5   Resolving Degeneracy

As with the regular simplex method, it is possible to solve degenerate linear programming problems and avoid cycling by using an appropriate pivot rule. The approach we will use here is a variant of the perturbation method (see Section 5.5.1). For network problems, this method can be realized in a particularly efficient manner. Much of our discussion will be specific to networks, and only at the end will the connections with the perturbation method be made clear.

The technique we describe uses a special form of basis, called a *strongly feasible basis* or *strongly feasible tree*. To define such a tree, we identify a particular node r as the *root* node. Then the tree is *strongly feasible* if any arc whose flow is zero points away from the root node r. (An arc (i, j) points away from the root if the path from node j to node r along the tree includes node i.) Any tree whose flows are all positive is strongly feasible. Strongly feasible trees are illustrated in Figure 8.19. The following theorem shows that they can be used to guarantee termination of the simplex method.

Theorem 8.14 (Guaranteed Termination). *If the basis at every iteration of the network simplex method is a strongly feasible basis, then the simplex method will terminate in a finite number of iterations.*

Proof. We will prove that the simplex method cannot cycle. Since there are only finitely many possible bases, and no basis can repeat, this will guarantee finite termination. We

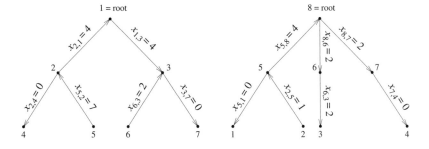

Figure 8.19. *Strongly feasible tree.*

denote by x and y the values of the variables at the current iteration, and by $\bar{x}$ and $\bar{y}$ the values at the next iteration.

As a preliminary step in the proof, we derive an update formula for the simplex multipliers y. Let (s, t) be the entering arc at the current iteration of the simplex method. Consider the subnetwork obtained by deleting the leaving arc from the current tree. This subnetwork consists of two trees, T_s containing node s and T_t containing node t. The new simplex multipliers $\bar{y}$ can be chosen as

$$\bar{y}_i = \begin{cases} y_i & \text{if node } i \text{ is in } T_s; \\ y_i - \hat{c}_{s,t} & \text{if node } i \text{ is in } T_t. \end{cases}$$

We must verify that $\bar{y}_i - \bar{y}_j = c_{i,j}$ for the new basis. If arc (i, j) is in T_s or in T_t, then

$$\bar{y}_i - \bar{y}_j = y_i - y_j = c_{i,j}.$$

To verify this for the entering arc (s, t) first recall that

$$\hat{c}_{s,t} = c_{s,t} - y_s + y_t.$$

Then

$$\begin{aligned} \bar{y}_s - \bar{y}_t &= y_s - (y_t - \hat{c}_{s,t}) \\ &= (c_{s,t} + y_t - \hat{c}_{s,t}) - y_t + \hat{c}_{s,t} = c_{s,t} \end{aligned}$$

as desired.

To prove that the simplex method does not cycle, we will show that "progress" is made at every iteration. Progress will be defined in terms of two functions:

$$\begin{aligned} f_1(x) &= c^T x \\ f_2(y) &= \sum_{i=1}^{m} (y_r - y_i), \end{aligned}$$

where r is the root node of the strongly feasible tree. (Even though the simplex multipliers are not uniquely determined, their *differences* are unique, and so the function f_2 is well defined.)

At a nondegenerate iteration there is strict improvement in the objective function, so

$$f_1(\bar{x}) < f_1(x)$$

and progress is made with respect to f_1.

At a degenerate iteration $f_1(\bar{x}) = f_1(x)$. The entering arc (s, t) will enter the basis with flow equal to zero, and hence it must point away from the root (by the definition of a strongly feasible tree). This implies that node r is in the subnetwork T_s, and that

$$f_2(\bar{y}) = f_2(y) + \hat{c}_{s,t} |T_t|,$$

where $|T_t|$ is the number of nodes in T_t. Since $\hat{c}_{s,t} < 0$ and $|T_t| > 0$,

$$f_2(\bar{y}) < f_2(y),$$

and so in this case progress is made with respect to f_2. Because progress is made with respect to one or the other of these functions at every iteration, a basis can never repeat and cycling cannot occur.   ☐

   If the network simplex method can be implemented so that at every iteration a strongly feasible basis is maintained, then it is guaranteed to terminate. The initialization scheme described in Section 8.4 produces an initial strongly feasible basis. (See the Exercises.) We now show that, if the leaving arc is chosen appropriately at every iteration, then every basis will be a strongly feasible basis.
   The rule for choosing the leaving arc will be based on the cycle created by the entering arc. If there is only one candidate for the leaving arc, then no choice is available. Otherwise, within this cycle define the *join* to be the node closest to the root of the tree, that is, the node whose path to the root consists of the fewest number of arcs. We will traverse the cycle, starting at the join, in the direction corresponding to the entering arc. (If (s, t) is the entering arc, this traversal will encounter node s just before node t.) The leaving arc will be chosen as the first candidate arc encountered during this traversal of the cycle. The following theorem shows that this rule has the desired property.

Theorem 8.15. *Assume that the network simplex method is initialized with a strongly feasible basis, and that at every iteration the leaving arc is chosen using the above rule. Then at every iteration the basis will be a strongly feasible basis.*

Proof. We need only prove that if the current basis is strongly feasible, then so is the new basis. There are two cases: nondegenerate and degenerate iterations. In both cases we need only examine the arcs in the cycle.
   For a nondegenerate iteration, all the candidate arcs must point in the opposite direction to the entering arc, since their flow will decrease towards zero. If the first arc encountered in traversing the cycle is selected as the leaving arc, then all the other candidate arcs will point away from the root in the new basis. Thus the new basis will be strongly feasible.
   For a degenerate iteration, all the arcs with zero flow will point away from the root, by the definition of a strongly feasible basis. As we traverse the cycle starting at the join, there will be no candidate arcs encountered until after we have traversed the entering arc. (The candidate arcs will all point in the opposite direction to the entering arc.) Hence the leaving arc will come after the entering arc, and once it is removed, all the arcs with zero flow in the new basis will point away from the root. This completes the proof.   ☐

   To conclude this section, we will show the relationship between strongly feasible bases and the perturbation method. This is the subject of the next theorem.

Theorem 8.16. *Consider a network linear program with equality constraints $Ax = b$ and a perturbed problem with constraints $Ax = b + \epsilon$, where*

$$\epsilon_i = \begin{cases} -(m-1)/m & \text{if } i = r; \\ 1/m & \text{otherwise.} \end{cases}$$

Here m is the number of nodes in the network, r is the index of the root node, and i is the index of a general node. Assume that the original problem has integer data, and that

*a feasible spanning tree solution for this problem has been specified. Then the tree for
the network is strongly feasible if and only if the corresponding flow is feasible for the
perturbed problem.*

Proof. We will define subsets of the nodes of the network relative to the root of the tree.
For a node i, $d(i)$ will be the set of nodes whose paths to the root include node i. Let $|d(i)|$
be the number of elements in $d(i)$.

Let x be the current basic feasible solution of the original problem, and let $\bar{x}$ be the
corresponding solution of the perturbed problem. If (i, j) is a basic arc, we will show that

$$\bar{x}_{i,j} = \begin{cases} x_{i,j} + |d(j)|/m & \text{if arc } (i, j) \text{ points away from node } r, \\ x_{i,j} - |d(i)|/m & \text{if arc } (i, j) \text{ points towards node } r \end{cases}$$

satisfies the flow constraints for the perturbed problem.

Suppose first that the arc (i, j) points away from node r. At node $j \neq r$ the perturbed
flow-balance equation must be satisfied:

$$\bar{x}_{i,j} + \sum_{k \neq i} \bar{x}_{k,j} - \sum_{\ell} \bar{x}_{j,\ell} = b_j + 1/m,$$

where these summations only include arcs that are in the tree. (Note that the nodes k and ℓ
satisfy $k, \ell \in d(j)$.) Substituting the proposed solution into the left-hand side gives

$$\bar{x}_{i,j} + \sum_{k \neq i} \bar{x}_{k,j} - \sum_{\ell} \bar{x}_{j,\ell}$$

$$= x_{i,j} + |d(j)|/m + \sum_{k \neq i}(x_{k,j} - |d(k)|/m) - \sum_{\ell}(x_{\ell,j} + |d(\ell)|/m)$$

$$= \left(x_{i,j} + \sum_{k \neq i} x_{k,j} - \sum_{\ell} x_{\ell,j} \right) + \left(|d(j)| - \sum_{k \neq i}|d(k)| - \sum_{\ell}|d(\ell)| \right)$$

$$= b_j + 1/m,$$

so the general constraints in the perturbed problem are satisfied. A similar argument can be
used when the arc (i, j) points towards node r, as well as at the root node (see the Exercises).

We now show that the perturbed solution is feasible if and only if the basis is strongly
feasible. On the basic arcs, the perturbed solution differs from the original solution by
$\pm|d(j)|/m$, a value that is less than one in magnitude. For a problem with integer data, the
only arcs that could become infeasible are those with zero flow in the original problem that
point towards the root node. If the perturbed solution is feasible, then no such nodes can
exist and so the basis is strongly feasible. Likewise, if the basis is strongly feasible, then
there are no such nodes and the perturbed solution is feasible.   □

In Chapter 5, perturbation was applied to general linear programs as a technique for
resolving degeneracy. In that setting it was shown that perturbation could be implemented
within the simplex method using a lexicographic technique. Here, in the context of network
problems, we have shown how perturbation can be implemented using strongly feasible

trees. This establishes an additional relationship between the properties of networks and the algebraic properties of the simplex method.

Exercises

5.1. Prove that the initialization scheme in Figure 8.18 produces a strongly feasible basis.

5.2. Prove that exactly two smaller trees are formed when an arc is deleted from a tree.

5.3. Complete the proof of Theorem 8.16 by showing that the proposed perturbed solution satisfies the flow constraints (a) at the root node, and (b) at node i when the arc (i, j) points towards the root node.

5.4. Solve the assignment problem in Example 8.4 with the perturbation method initialized using the technique in Figure 8.18. You may use either a big-M or a two-phase approach.

8.6   Notes

Network Models—Network models can sometimes be solved much faster than general linear programs. This can happen for one of two reasons: the problem might have a special structure that guarantees that the network simplex method will terminate in few iterations, or there might exist special algorithms for the problem. For example, the shortest path problem can be solved using an algorithm of Dijkstra (1959) that requires $O(m^2)$ operations. Fredman and Tarjan (1987) showed how to reduce this to $O(n + m \log m)$ by using appropriate data structures. There are other efficient algorithms with operation counts that depend on the magnitudes of the coefficients in the cost vector c. (In many applications, $m \ll n \ll m(m - 1)$.)

As with the shortest path problem, there are especially efficient algorithms for the maximum flow problem. Early results in this area can be found in the book of Ford and Fulkerson (1962). The first polynomial-time algorithm was described by Edmonds and Karp (1972), a method requiring $O(mn^2)$ operations. Cheriyan and Maheshwari (1989) have a method with an operation count of $O(m^2\sqrt{n})$, and there are a number of efficient methods with operation counts that depend on the magnitudes of the upper bounds u. An important class of algorithms for this problem is described in the papers of Goldberg (1985) and Goldberg and Tarjan (1988).

For further references, and for further information on algorithms for individual network problems, see the books by Murty (1992, reprinted 1998) and Ahuja, Magnanti, and Orlin (1993).

Network Simplex Method—The network simplex method continues to be a competitive method for solving network optimization problems, and it has the advantage that it is available in high-quality software packages. However, many alternative algorithms have been proposed for these problems that (a) have better theoretical properties, (b) have promising practical properties, or (c) take advantage of the special form of a particular problem (such as a shortest path problem).

The network simplex method requires few operations per iteration, but the number of iterations may be large. Technically, the number of iterations may be "exponential" in the size of the network. As a result, the total effort of solving the network problem can be large. It is desirable to have an algorithm that requires only a "polynomial" number of iterations.[8]

Some of these other algorithms are based on the simplex method. Before mentioning them, we should point out that the (primal) simplex method maintains primal feasibility (the constraints in the primal linear program are satisfied at every iteration) and complementary slackness, and it iterates until the dual feasibility (that is, primal optimality) conditions are satisfied.

The "primal-dual" method is one of these simplex-based methods. It starts with a dual feasible solution and uses the complementary slackness conditions to construct a "restricted" version of the primal problem. This restricted primal problem is then solved. If the restricted primal problem has optimal objective value zero, then the original network problem has been solved. Otherwise, the solution of the restricted primal can be used to improve the values of the dual variables, or to determine that no solution exists. The primal-dual method maintains dual feasibility and complementary slackness and strives for primal feasibility. The motivation for this method is that the restricted primal problem is a shortest path problem, for which special algorithms exist. For more information on the primal-dual method, see the book by Chvátal (1983).

Early tests of the primal-dual simplex method showed that it could be more efficient than the primal simplex method, but these conclusions were reversed when better implementations of the primal simplex method became available. Also, the primal-dual simplex method may require an exponential number of iterations, and so its theoretical behavior is not superior either.

It is also possible to apply the dual simplex method. In the version due to Orlin (1984), the dual simplex method requires only a polynomial number of iterations. (See also the paper by Orlin, Plotkin, and Tardos (1993).) Note, however, that not all versions of the dual simplex method may be polynomial-time methods. In fact, Zadeh (1979) has shown the equivalence of versions of the primal simplex method, the dual simplex method, the primal-dual simplex method, and the out-of-kilter method (see below), and in an earlier paper (1973) described an example where all of these methods require an exponential number of iterations.

There are a great many other network algorithms that are not based on the simplex method. One of the earliest, called the "out-of-kilter" algorithm, begins with an initial guess that satisfies the flow-balance constraints, but may violate the primal bound constraints as well as the dual feasibility constraints. It iterates, trying to find a point that satisfies the feasibility and optimality conditions, measuring progress in terms of a "kilter number" based on the optimality conditions for the problem. For further details, see the book by Ford and Fulkerson (1962).

Much recent research is concerned with the development of efficient polynomial-time methods for network problems. The earliest of these methods (derived from the primal-dual and out-of-kilter methods) had operation counts that depended on the magnitudes of the cost coefficients and upper bounds in the model. More recent work, starting with the paper by Tardos (1985), has developed "strongly" polynomial methods whose operation counts are independent of these magnitudes. For a survey of this work, see the book by Ahuja, Magnanti, and Orlin (1993).

---

[8]The terms "exponential" and "polynomial" are defined in Section 9.2.

Chapter 9

Computational Complexity of Linear Programming

9.1   Introduction

Almost as soon as it was developed, the simplex method was tested to determine how well it worked. Those early tests demonstrated that it was an effective method (at least on the examples it was applied to), and this conclusion was confirmed by numerous practical applications of the method to ever larger and more elaborate models.

This encouraging empirical experience was not supported by comparable theoretical results about the behavior of the simplex method, despite considerable effort to find such results. This raised a number of questions. Is the simplex method guaranteed to work well on all nondegenerate problems? Are there classes of problems on which the simplex method performs poorly? Is the simplex method the most efficient method possible for linear programming? These questions were answered (at least partially) in the 1970s and 1980s. In this chapter we present a brief survey of these results.

First, we discuss measures of performance of algorithms. Next we discuss the computational efficiency of the simplex method. We show that for some specially structured problems the number of iterations required by the simplex method grows exponentially with the size of the problem. Thus, measured by its worst-case performance, the simplex method is inefficient. This result spurred researchers to seek "polynomial algorithms" for which the computational effort required—even in the worst case—grows just polynomially with the size of the problem. We describe the ellipsoid method, the first method for linear programming shown to be polynomial. This discovery, in 1979, was received with much fanfare and optimism, soon to be followed by equally great disappointment. The method, while efficient in theory, is inefficient in practice, with its performance often matching the worst-case bound. These discouraging results led some researchers to focus on average-case rather than worst-case performance. In the last section of this chapter, we present results developed in the early 1980s that suggest that, measured by its average-case performance (albeit on a special set of problems), the simplex method is efficient.

The simplex algorithm remained the leading method for linear programming until 1984, when Karmarkar proposed a new polynommial method for linear programming that showed promising, if not stellar, computational results. Karmarkar's algorithm triggered research into a new class of methods called interior-point methods that, in some cases, have good theoretical properties and are also competitive computationally with the simplex method. These methods are the subject of the next chapter.

9.2   Computational Complexity

The purpose of computational complexity is to determine the number of arithmetic or other computational operations required to solve a particular problem using a specific algorithm. We will refer to this as the *cost* of solving a problem. For example, what is the cost of solving a nonsingular system of n linear equations, using Gaussian elimination with partial pivoting? In this chapter, we will mainly be concerned with the cost of solving a linear programming problem.

The cost of an algorithm can be measured in several ways. One measure is the "worst-case" cost: if some diabolical person were choosing an example so as to make the algorithm perform as poorly as possible, how many operations would the algorithm require to solve the problem? For algorithms such as Gaussian elimination applied to dense matrices, the worst-case behavior is also the typical behavior, so this is a useful measure of cost.

Another measure is the "average-case" behavior of an algorithm, that is, the number of arithmetic operations required when the algorithm is applied to an "average" problem. The simplex method has poor worst-case behavior (see Section 9.3) but good average-case behavior (see Section 9.5). In practice a worst-case analysis does not reflect the observed performance of the simplex method, so it is more plausible to consider average-case performance. However, it is often more difficult to analyze the average-case behavior of an algorithm than the worst-case behavior. One preliminary difficulty is the definition of an "average" problem. For example, most large linear programs that arise in applications have sparse constraint matrices, but randomly generated matrices (with common choices of the underlying probability distribution) are almost certain to be dense. Also, applied problems are often degenerate, whereas random problems are, with probability one, nondegenerate. Disagreements about the definition of an "average" problem can raise doubts about the applicability of the corresponding estimates of average-case performance.

There are other decisions that must be made in defining the cost of an algorithm, for example, defining what an "operation" is. In this book, we typically define an operation to be an arithmetic operation applied to two real numbers, such as an addition or a multiplication.

Arithmetic operations are not the only operations that could be counted. There is work associated with retrieving a number from memory, storing a result in memory, and printing the solution to a problem. These could also be included as part of the cost of an algorithm. The amount of storage required by an algorithm could also be counted, although this would be significant only if the algorithm required intermediate storage much greater than that used to store the problem data.

Computer implementations of the simplex method are almost always programmed using "real" arithmetic, meaning that the calculations are performed on floating-point numbers with a fixed number of digits. For the algorithms we discuss, the number of operations required to move numbers to and from memory is either proportional to, or overwhelmed by, the number of arithmetic operations, so these auxiliary operations are ignored. Finally, the algorithms discussed here have storage costs proportional to the size of the problem data. As a result, in assessing the cost of an algorithm we will only count arithmetic operations on floating-point numbers. For example, the cost of applying Gaussian elimination to a system of n linear equations is about $\frac{2}{3}n^3$ arithmetic operations.

To make all these cost measures precise, computer scientists often describe algorithms in terms of an associated "Turing machine." In his famous 1936 paper, Alan M. Turing described an imaginary computing device (since named after him) consisting of a processing unit together with an infinite tape divided into cells, each of which could record one of a finite set of symbols. The processing unit could be set in a finite number of "states," and at every time step of the algorithm, the processing unit could either (i) read the symbol on the current cell of the tape, (ii) write a symbol on the current cell of the tape, (iii) move the tape one cell to the left, (iv) move the tape one cell to the right, (v) change state, or (vi) stop. Despite the primitive nature of this device, Turing argued convincingly in his paper that every sequence of steps that might be considered as a calculation could be performed by a machine of this type. Turing also described a "universal" machine of this type that could mimic all other such machines, a forerunner of our modern general-purpose computer.

When Turing invented his machines, he was not interested in assessing the costs of algorithms, but instead used them as an intellectual tool to settle a famous problem about the axioms of arithmetic. They have since been used to define what is meant by an algorithm, or a step in an algorithm. We will not describe algorithms in terms of Turing machines, but will use more intuitive notions of computing. Even so, Turing machines will have a subtle influence on our discussions, particularly the notion of the "length of the input," a measure of the size of the problem with connections to the "tape" in a Turing machine.

When comparing algorithms, it is common to compare only the "order of magnitude" costs of the algorithms. For example, Gaussian elimination would cost $O(n^3)$ arithmetic operations, ignoring the constant $\frac{2}{3}$. If we say that the cost of an algorithm is $O(f(L))$, we mean that for sufficiently large L,

$$\text{number of arithmetic operations} \leq C f(L),$$

where C is some positive constant, L is a measure of the length of the input data for the problem, and f is some function. Because the constant is ignored, these order-of-magnitude estimates are mainly of value when L is large; for small L they can be deceptive, particularly if C is large.

What do we mean by "the length of the input data" for a problem? In the case of a linear program, we will consider it to be the number of bits required to store all the data for the problem. This would include the number of variables n, the number of general constraints m, and the coefficients in the matrix A and the vectors b and c. We will assume that these numbers are all integers,[9] so

$$L = \sum_{i,j} \lceil \log_2(|a_{i,j}| + 1) \rceil + \sum_{i} \lceil \log_2(|b_i| + 1) \rceil + \sum_{j} \lceil \log_2(|c_j| + 1) \rceil$$

$$+ \lceil \log_2(n + 1) \rceil + \lceil \log_2(m + 1) \rceil + (nm + n + m + 1).$$

---

[9]A problem involving fractions can be converted into one with integers; a problem involving general real numbers would require infinite space to store the binary representations of these numbers. Finite-precision numbers stored by a computer can be represented as fractions.

Table 9.1. *Polynomial and exponential growth rates.*

L	L^2	L^3	L^{100}	2^L	$L!$
2	4	8	1×10^{30}	4	2
5	25	1×10^2	8×10^{69}	32	1×10^2
10	1×10^2	1×10^3	1×10^{100}	1×10^3	4×10^6
50	3×10^3	1×10^5	8×10^{169}	1×10^{15}	3×10^{64}
100	1×10^4	1×10^6	1×10^{200}	1×10^{30}	9×10^{157}

(The notation $\lceil x \rceil$ denotes the smallest integer that is $\geq x$.) The final term $(nm + n + m + 1)$ represents the space needed to store the signs of all the numbers, plus an additional bit to indicate whether the linear program is a minimization or maximization problem.

The number L is a coarse measure of the size of a problem. In many cases it may be more convenient to use a different measure, such as the number of variables. For example, some of the algorithms for linear programming that we discuss have costs that are $O(n^3 L)$ in the worst case. Since $n < L$, we could have written that the costs were $O(L^4)$, but it is common to use a more precise cost estimate when one is available.

A distinction is made between "polynomial" and "exponential" algorithms. A polynomial algorithm has costs that are $O(f(L))$ in the worst case, where $f(L)$ is a polynomial in L. An exponential algorithm has costs that grow exponentially with L in the worst case. For example, an exponential algorithm might have costs proportional to 2^L. Exponential costs grow much more rapidly than polynomial costs as L increases, so exponential algorithms are often considered unacceptable for large problems. This is illustrated in Table 9.1. There are further categories of algorithms, between polynomial and exponential (with costs proportional to, say, $L^{\ln L}$) and beyond exponential (with costs proportional to, say, $L!$).

It is usually feasible to solve problems of size $L = 100$ if there is a polynomial-time algorithm, and the polynomial is of low degree. This may not be the case for exponential algorithms, where the costs increase rapidly with L. If $f(L)$ is a polynomial of high degree, say $f(L) = L^{100}$, then even a polynomial algorithm will be unworkable for large problems in the worst case. In a great many cases, however, polynomial algorithms have costs that are $O(L)$–$O(L^4)$, whereas exponential algorithms have costs that are $O(2^L)$ or worse, so the distinction between polynomial and exponential algorithms is a useful one.

Exercises

2.1. Determine L, the length of the input, for the linear program

$$
\begin{aligned}
\text{maximize} \quad & z = 5x_1 + 7x_2 + 9x_3 + 12x_4 \\
\text{subject to} \quad & 3x_1 - 9x_2 + 6x_3 - 4x_4 = 12 \\
& 2x_1 + 3x_2 - 2x_3 + 7x_4 = 21 \\
& x_1, x_2, x_3, x_4 \geq 0.
\end{aligned}
$$

2.2. Show that an iteration of the simplex algorithm in Section 5.2 has polynomial costs.

2.3. If the cost of one algorithm is 2^L and another is L^{100}, how large does L have to be before the polynomial algorithm becomes cheaper than the exponential algorithm?

2.4. Use Stirling's formula to show that if an algorithm requires $L!$ operations, then it is not a polynomial algorithm.

9.3   Worst-Case Behavior of the Simplex Method

Ever since it was invented, the simplex method has been considered a successful method for linear programming. In 1953 a paper by Hoffman et al. compared the simplex method with several other algorithms and concluded that it was much faster, even though the others were better suited to the computers available at that time.

It has been observed that the number of iterations required by the simplex method to find the optimal solution is often a small multiple of the number of general constraints. This is remarkable, since a problem with n variables and m constraints could have as many as

$$\binom{n}{m}$$

basic solutions (of course, many of these are likely to be infeasible). There was always the possibility that the simplex method might examine all of these bases before finding the optimal basis, but up until the 1970s no one had been able to exhibit a set of linear programs (with an arbitrary number of variables) where the simplex method took that many iterations to find a solution.

There is a considerable difference between m iterations and $\binom{n}{m}$ iterations. If we set $n = 2m$, then the values of these two quantities are

m	$\binom{2m}{m}$
1	2
5	252
10	184756
20	1×10^{11}
50	1×10^{29}
100	9×10^{58}
200	1×10^{119}
300	1×10^{179}
400	2×10^{239}
500	3×10^{299}

Even for small values of m the number of possible bases is huge. If we had a computer capable of performing one billion simplex iterations per second, then examining $\binom{100}{50}$ bases (that is, $m = 50$) would take 3,199,243,548,502.2 years.

If the number of simplex iterations were proportional to m, then the simplex method would be a polynomial algorithm. All the operations in a simplex iteration are simple matrix and vector calculations, with total costs of $O(mn)$ arithmetic operations if full pricing is done. Periodic refactorization of the basis matrix costs $O(m^3)$ arithmetic operations. So the costs of a simplex iteration can be as high as $O(m^3 + nm)$ operations, and if $O(m)$ iterations are performed, the overall costs of the simplex method are $O(m^4 + nm^2)$ arithmetic operations. This number is a polynomial in n and m. (These costs would be lower if the problem were sparse; the estimates here are based on dense-matrix computations.) On the other hand, if $\binom{n}{m}$ iterations were required, then the simplex method would be an exponential algorithm.

If a trial basis to a linear program has been proposed, it is possible to check if it is optimal in polynomial time. The optimality test in the simplex method is one way to do it, involving the computation of the reduced costs. The reduced costs can be computed using $O(m^3 + nm)$ arithmetic operations. (If the linear program is degenerate, it is possible that the basis may not be optimal, even though the corresponding point x is optimal.)

Together these comments show that the simplex method might be a polynomial algorithm (if the number of iterations were always $O(m)$) or an exponential algorithm (if the number of iterations were sometimes $\binom{n}{m}$), but in either case a solution can be verified in polynomial time. All these facts were known in the early 1970s.

In 1972 a paper by Klee and Minty showed that there exist problems of arbitrary size that cause the simplex method to examine every possible basis when the steepest-descent pricing rule is used, and hence showed that the simplex method is an exponential algorithm in the worst case.

We give here a variant of the original Klee–Minty problems:

$$\text{maximize} \quad z = \sum_{j=1}^{m} 10^{m-j} x_j$$

$$\text{subject to} \quad 2\sum_{j=1}^{i-1} 10^{i-j} x_j + x_i \leq 100^{i-1} \quad \text{for } i = 1, \ldots, m$$

$$x \geq 0.$$

When slack variables are added, there are $2m$ variables and m general constraints. These problems have 2^m feasible bases. If the simplex method chooses the entering variable as the one with the largest violation in the optimality test (as usual), then every basis will be examined.

Example 9.1 (Klee–Minty Problem). If $m = 3$, then the Klee–Minty problem has the form

$$\begin{aligned}
\text{maximize} \quad & z = 100x_1 + 10x_2 + x_3 \\
\text{subject to} \quad & x_1 \leq 1 \\
& 20x_1 + x_2 \leq 100 \\
& 200x_1 + 20x_2 + x_3 \leq 10000 \\
& x \geq 0.
\end{aligned}$$

The feasible region is illustrated schematically in Figure 9.1.

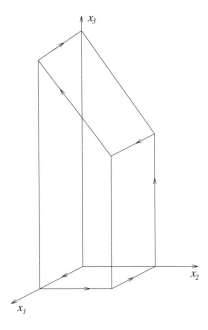

Figure 9.1. *Klee–Minty problem.*

When the simplex method is applied with the steepest-descent pricing rule, the sequence of basic feasible solutions is as follows:

Basis			z
$s_1 = 1$	$s_2 = 100$	$s_3 = 10000$	0
$x_1 = 1$	$s_2 = 80$	$s_3 = 9800$	100
$x_1 = 1$	$x_2 = 80$	$s_3 = 8200$	900
$s_1 = 1$	$x_2 = 100$	$s_3 = 8000$	1000
$s_1 = 1$	$x_2 = 100$	$x_3 = 8000$	9000
$x_1 = 1$	$x_2 = 80$	$x_3 = 8200$	9100
$x_1 = 1$	$s_2 = 80$	$x_3 = 9800$	9900
$s_1 = 1$	$s_2 = 100$	$x_3 = 10000$	10000

This problem has $2^3 = 8$ possible basic feasible solutions, and all are examined by the simplex method. ∎

In this example, the steepest-descent pricing rule causes the simplex method to examine all possible basic feasible solutions for the constraints. It would have been possible, however, to solve this problem at the first iteration if x_3 had been chosen as the entering variable and s_3 as the leaving variable. This raises the possibility that the simplex method, with a different pricing rule, might have better worst-case behavior. A number of researchers have examined this question and have shown that, for a variety of pricing rules, there are corresponding linear programming problems that cause the simplex method to examine an exponential number of basic feasible solutions. Although these results do not fully settle the question, they raise doubts that such an efficient pricing rule exists.

This is not the only question raised by the Klee–Minty example. For example, how common are linear programs that require exponentially many simplex iterations? How does the simplex method perform on an "average" problem? Are there algorithms for linear programming that require only a polynomial number of operations even in the worst case? These questions are discussed in the remaining sections of this chapter.

Exercise

3.1. Use linear programming software to solve Klee–Minty problems of various sizes. How many pivots are required? Are all basic feasible solutions examined?

9.4   The Ellipsoid Method

In 1979 the Soviet mathematician Leonid G. Khachiyan settled one of these questions by exhibiting a polynomial-time algorithm for linear programming. (The algorithm was not new, but Khachiyan's observations were.) His discovery received a great deal of attention, based on the hope that this algorithm might be a dramatically more efficient method for linear programming. Articles soon appeared in the *New York Times* and other general-interest publications, an indication of the immense practical importance of linear programming.

Khachiyan's method, based on a more general algorithm for convex programming and now called the ellipsoid method, is designed to find a point that *strictly* satisfies a system of linear inequalities. That is, it tries to find a point x such that

$$Ax < b.$$

Any linear programming problem can be transformed into a problem of this type, as we will now show.

Suppose we are given a linear program in canonical form

$$\begin{aligned} \text{minimize} \quad & z = c^T x \\ \text{subject to} \quad & Ax \geq b \\ & x \geq 0 \end{aligned}$$

together with its dual program

$$\begin{aligned} \text{maximize} \quad & w = b^T y \\ \text{subject to} \quad & A^T y \leq c \\ & y \geq 0. \end{aligned}$$

By duality theory (see Section 6.2), at the optimal solutions of both problems the two objective values will be equal, and the constraints to both problems will be satisfied:

$$c^T x - b^T y = 0$$
$$Ax \geq b$$
$$A^T y \leq c$$
$$x \geq 0$$
$$y \geq 0.$$

By weak duality, a pair of feasible points satisfies $b^T y \leq c^T x$, so these conditions are equivalent to a system of linear inequalities of the form

$$\hat{A}\hat{x} \leq \hat{b},$$

where

$$\hat{A} = \begin{pmatrix} c^T & -b^T \\ -A & 0 \\ 0 & A^T \\ -I & 0 \\ 0 & -I \end{pmatrix}, \quad \hat{x} = \begin{pmatrix} x \\ y \end{pmatrix}, \quad \text{and} \quad \hat{b} = \begin{pmatrix} 0 \\ -b \\ c \\ 0 \\ 0 \end{pmatrix}.$$

Because of this equivalence, we will assume in the rest of this section that we are solving a system of linear inequalities $Ax \leq b$, where A is an $m \times n$ matrix. Assume that the entries in A and b are all integers, and that L is the length of the input data for the system $Ax \leq b$:

$$L = \sum_{i,j} \lceil \log_2(|a_{i,j}| + 1) \rceil + \sum_i \lceil \log_2(|b_i| + 1) \rceil$$
$$+ \lceil \log_2(n) \rceil + \lceil \log_2(m) \rceil + (nm + m),$$

where n is the number of variables and m is the number of inequalities. Let $e = (1, \ldots, 1)^T$. It can be shown that if the system

$$Ax < b + 2^{-L} e$$

has a solution, then the system $Ax \leq b$ has a solution (see the paper by Gács and Lovász (1981)). These transformations allow us to solve a linear programming problem by solving a system of strict linear inequalities.

Before giving a precise description of the ellipsoid method, we will describe it intuitively. To begin, an *ellipsoid* is the higher-dimensional generalization of an ellipse. In n-dimensional space, it can be defined as the set of points

$$\left\{ x : (x - \bar{x})^T M^{-1}(x - \bar{x}) \leq 1 \right\},$$

where the vector $\bar{x}$ of length n is the "center" of the ellipsoid and the $n \times n$ positive definite matrix M defines the orientation and shape of the ellipsoid.

Example 9.2 (Ellipsoid). Let $\bar{x} = (5, 4)^T$ and

$$M = \begin{pmatrix} 3 & 1 \\ 1 & 3 \end{pmatrix}.$$

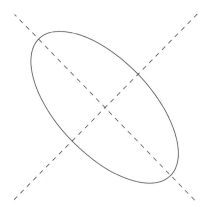

Figure 9.2. *Ellipsoid.*

Then

$$M^{-1} = \frac{1}{8} \begin{pmatrix} 3 & -1 \\ -1 & 3 \end{pmatrix}.$$

If we define $y = x - \bar{x}$, then the condition

$$(x - \bar{x})^T M^{-1} (x - \bar{x}) \leq 1$$

simplifies to

$$\tfrac{1}{4}(y_1 - y_2)^2 + \tfrac{1}{8}(y_1 + y_2)^2 \leq 1.$$

The ellipsoid is graphed in Figure 9.2. ■

The ellipsoid method begins by selecting an ellipsoid centered at the origin ($\bar{x} = x_0 = 0$) that contains part of the feasible region

$$S = \{ x : Ax < b \}.$$

The first ellipsoid is defined by a positive-definite matrix M_0 that is a multiple of the identity matrix. It is desirable to choose M_0 so that the initial ellipsoid is as small as possible, since this will reduce the bound on the number of iterations required by the method. In the absence of other information, it is possible to choose $M_0 = 2^L I$, a choice which defines an ellipsoid sufficiently large that it is guaranteed to contain part of the feasible region, if this region is nonempty. This is just a simple way to initialize the method. It would be possible to begin with any ellipsoid that contains some part $\bar{S}$ of the feasible region.

At the kth iteration, the method first checks if the center $\bar{x} = x_k$ of the current ellipsoid is feasible:

$$Ax_k < b.$$

If so, the method terminates with x_k as a solution. If not, then at least one of the constraints is violated. One of the violated constraints is used to determine a smaller ellipsoid with center x_{k+1} and matrix M_{k+1} that also contains the part $\bar{S}$ of the feasible region. Then the method repeats.

At each iteration the size of the ellipsoid shrinks by a constant factor. Because the data for the problem are all integers, it is possible to show that, eventually, either a solution has been found or the ellipsoid is so small that the feasible region must be empty. (Each ellipsoid in the sequence contains the same part $\bar{S}$ of the feasible region contained by the initial ellipsoid. It is possible to show that if the feasible region is nonempty, then there is a lower bound on the volume of $\bar{S}$. Eventually the volume of the ellipsoid will be smaller than this lower bound, implying that the feasible region must have been empty.)

Here then is the algorithm for finding a solution to $Ax < b$, where A is an $m \times n$ matrix, and L is the length of the input data.

ALGORITHM 9.1.
Ellipsoid Method

1. Set $x_0 = 0$, $M_0 = 2^L I$.
2. For $k = 0, 1, \ldots$

   (i) If $Ax_k < b$ stop. (A feasible point x_k has been found.)

   (ii) If $k > 6(n+1)^2 L$ stop. (The feasible region is empty.)

   (iii) Otherwise, find any inequality such that $a_i^T x_k \geq b_i$ (that is, an inequality that is violated by x_k). Then set

   $$x_{k+1} = x_k - \frac{1}{n+1} \frac{M_k a_i}{\sqrt{a_i^T M_k a_i}}$$

   $$M_{k+1} = \frac{n^2}{n^2 - 1} \left(M_k - \frac{2}{n+1} \frac{(M_k a_i)(M_k a_i)^T}{a_i^T M_k a_i} \right).$$

The form of the algorithm given here uses square roots, meaning that after the first iteration the numbers may not be representable as fractions. With more care, the limitations of finite-precision calculations can be taken into account.

Let E_k be the ellipsoid at the kth iteration. If $a_i^T x_k \geq b_i$, then any feasible point satisfies $a_i^T x \leq a_i^T x_k$. The formulas for x_{k+1} and M_{k+1} define an ellipsoid E_{k+1} that is the ellipsoid of minimum volume satisfying

$$E_{k+1} \supset E_k \cap \left\{ x : a_i^T x \leq a_i^T x_k \right\}$$

and

$$E_{k+1} \cap \left\{ x : a_i^T x = a_i^T x_k \right\} = E_k \cap \left\{ x : a_i^T x = a_i^T x_k \right\}.$$

It is clear that E_{k+1} contains the same portion of the feasible region that E_k does. This is illustrated in Figure 9.3.

It is possible to show that

$$\text{volume}\,(E_{k+1}) = c(n)\,\text{volume}\,(E_k),$$

where

$$c(n) = \left(\frac{n^2}{n^2 - 1} \right)^{(n-1)/2} \frac{n}{n+1} < e^{-1/2(n+1)} < 1,$$

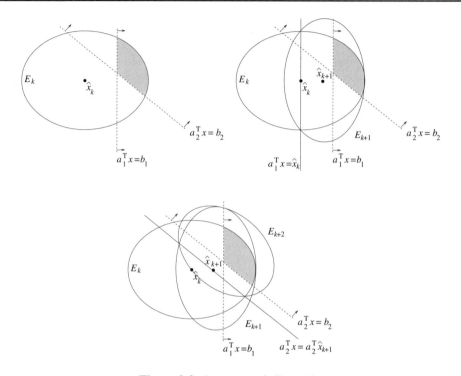

Figure 9.3. *Sequence of ellipsoids.*

so that the volume of the ellipsoid is reduced by a constant factor at every iteration. (Here the term $e^{-1/2(n+1)}$ involves the number $e \approx 2.71$.) Using the fact that the volume of the initial ellipsoid is bounded above by $2^{L(n+1)}$ and that the volume of the portion $\bar{S}$ of the feasible region (if the feasible region is nonempty) is bounded below by $2^{-(n+1)2L}$, it is straightforward to show that after at most $6(n+1)^2L$ iterations, either a solution is found or the feasible region is empty.

Example 9.3 (Ellipsoid Method). Consider the system of strict linear inequalities $Ax < b$:

$$\begin{pmatrix} -1 & 0 \\ 0 & -1 \end{pmatrix} \begin{pmatrix} x_1 \\ x_2 \end{pmatrix} < \begin{pmatrix} -1 \\ -1 \end{pmatrix}.$$

Here $m = n = 2$ and $L = 12$. At the first iteration, we will choose

$$x_0 = \begin{pmatrix} 0 \\ 0 \end{pmatrix} \quad \text{and} \quad M_0 = 4096 \begin{pmatrix} 1 & 0 \\ 0 & 1 \end{pmatrix}.$$

The first constraint is violated, so we can choose $a_i^T = a_1^T = (-1, 0)$. Then $M_0 a_1 = 4096(-1, 0)^T$, $a_1^T M_0 a_1 = 4096$,

$$x_1 = \begin{pmatrix} 0 \\ 0 \end{pmatrix} - \frac{1}{3} \frac{4096\binom{-1}{0}}{64} = \begin{pmatrix} 0 \\ 0 \end{pmatrix} - \frac{64}{3} \begin{pmatrix} -1 \\ 0 \end{pmatrix} = \begin{pmatrix} \frac{64}{3} \\ 0 \end{pmatrix},$$

and

$$M_1 = \frac{4}{3}\left[4096\begin{pmatrix}1 & 0\\0 & 1\end{pmatrix} - \frac{2}{3}\frac{(4096)^2\begin{pmatrix}1 & 0\\0 & 0\end{pmatrix}}{4096}\right]$$

$$= \frac{16384}{3}\left[\begin{pmatrix}1 & 0\\0 & 1\end{pmatrix} - \frac{2}{3}\begin{pmatrix}1 & 0\\0 & 0\end{pmatrix}\right] = \frac{16384}{9}\begin{pmatrix}1 & 0\\0 & 3\end{pmatrix}.$$

This completes the first iteration.

At the second iteration, $a_2^T x_1 = 0 > -1$, so the second constraint is violated. When the formulas for the ellipsoid method are applied with $a_i^T = a_2^T$, we obtain

$$x_2 = \begin{pmatrix}\frac{64}{3}\\\frac{128}{3\sqrt{3}}\end{pmatrix} \quad \text{and} \quad M_2 = \frac{65536}{27}\begin{pmatrix}1 & 0\\0 & 1\end{pmatrix}.$$

This completes the second iteration. The point x_2 satisfies the linear inequalities $Ax < b$, so the method terminates. ∎

The ellipsoid method requires at most $6(n + 1)^2 L$ iterations, where n is the number of variables. If A is an $m \times n$ matrix, then computing x_{k+1} from x_k requires $O(mn + n^2)$ arithmetic operations (the cost is dominated by the cost of finding a violated constraint and calculating $M_k a_i$). Computing M_{k+1} from M_k requires an additional $O(n^2)$ operations. Overall, the ellipsoid method requires at most $O((mn^3 + n^4)L)$ arithmetic operations, making it a polynomial-time algorithm.

Although the discovery of the ellipsoid method generated a great deal of excitement, the excitement quickly dissipated. It is true that in some cases (such as on the Klee–Minty problems) the simplex method would be much worse than the ellipsoid method. On many other problems, however, the simplex method is much better than the ellipsoid method. Computational experiments showed that the theoretical bounds on the performance of the ellipsoid method are qualitatively the same as its behavior on "typical" problems, whereas the performance of the simplex method is much better than its worst-case bounds. On practical problems the ellipsoid method is often slow to converge. It is not a practical alternative to the simplex method.

The ellipsoid method is not without its uses. Variants of it have been developed that provide polynomial-time algorithms for problems whose computational complexity had been previously unknown. Also, each iteration of the ellipsoid method requires only that a violated constraint be found. This does not require that the complete set of constraints be explicitly represented, which is a useful property in some settings (see the paper by Bland, Goldfarb, and Todd (1981)). Perhaps the most important contribution of the ellipsoid method was that it settled the question of whether linear programs can be solved in polynomial time. On the other hand, it left unanswered the question of whether a *practical* polynomial-time algorithm for linear programming could be found.

Exercises

4.1. Fill in the details of the calculations for the second iteration of the ellipsoid method in Example 9.3.

4.2. Determine a more precise count on the number of arithmetic operations required for an iteration of the ellipsoid method.

4.3. For the Klee–Minty linear programs in Section 9.3, how large does m have to be before the ellipsoid method becomes more efficient than the simplex method?

4.4. Consider the ellipsoid defined by

$$\left\{ x : (x - \bar{x})^T M^{-1} (x - \bar{x}) \leq 1 \right\},$$

where $\bar{x}$ is the center of the ellipsoid and M is a positive-definite matrix M. Define $y = x - \bar{x}$. Assume that M has been factored as

$$M = P \Lambda P^T,$$

where $\Lambda = \operatorname{diag}(\lambda_1, \ldots, \lambda_n)$ is the matrix of eigenvalues and

$$P = (p_1 \quad \cdots \quad p_n)$$

is the matrix of eigenvectors. Prove that the condition $(x - \bar{x})^T M^{-1}(x - \bar{x}) \leq 1$ is equivalent to

$$\sum_{i=1}^{n} \lambda_i^{-1} (p_i^T y)^2 \leq 1.$$

(The eigenvectors define the axes of the ellipsoid, and the eigenvalues determine the lengths of these axes.) Determine P and Λ for the ellipsoid in Example 9.2, and show that in this case the formula you have derived is equivalent to

$$\tfrac{1}{4}(y_1 - y_2)^2 + \tfrac{1}{8}(y_1 + y_2)^2 \leq 1.$$

9.5 The Average-Case Behavior of the Simplex Method[10]

By the early 1960s the "conventional wisdom" was that the simplex method requires between m and $3m$ iterations to find the solution of a linear program in standard form with m general constraints. This conclusion was based on a great deal of practical experience with the simplex method, and it continued to be accepted even as solutions of larger and larger problems were attempted. These iteration counts were low enough to make the simplex method an efficient and effective tool for everyday use. Even the pessimistic examples of Klee and Minty (see Section 9.3) were not enough to dim the enthusiasm for the simplex method (in part because no other competitive method was available).

Still, these pessimistic examples did raise doubts. Were such examples common? Would more bad examples show up as problems grew larger? Does there exist a large class of realistic problems that caused the simplex method to perform poorly? What is a "typical" linear programming problem? How did the simplex method behave on an "average" linear programming problem? We will concentrate on this last question.

This question needs to be phrased more precisely before it can be answered. In particular, three things must be specified: (1) the variant of the simplex method that will be

---

[10]This section uses ideas from statistics.

used, (2) the class of linear programming problems that will be solved, and (3) the stochastic model that will be used to define a "random" or "average" problem. We might hope to use the simplex method from Chapter 5, applied to linear programs in standard form, with some "simple" stochastic model. However, the results we present are not for this case.

Analyzing average-case behavior for the simplex method is difficult, and the results obtained are often influenced by the availability and tractability of appropriate mathematical tools. This has led researchers to study less familiar variants of the simplex method. In addition, care must be taken in the choice of a stochastic model, or else there is a possibility that an "average" problem may not be defined. These issues can quickly become complicated, while the discussion given here is brief; for further information see the book by Borgwardt (1987).

In this section we assume that the linear programs are in the form

$$\begin{aligned} \text{maximize} \quad & z = c^T x \\ \text{subject to} \quad & Ax \leq b, \end{aligned}$$

where A is an $m \times n$ matrix. The ith row of A is denoted by a_i^T. We also assume that a feasible initial point x_0 is provided. (For certain results, the right-hand-side vector will be chosen as $b = e \equiv (1, \ldots, 1)^T$ so that $x_0 = 0$ will automatically be feasible for the problems.)

A variant of the simplex method called the *shadow vertex* algorithm will be used to solve the linear programs. It is a form of parametric programming and is described in Section 6.5.

The first proof to show that, *on average*, a variant of the simplex method converged in a number of iterations that was a polynomial in m and n, was discovered by Borgwardt in 1982. His theorem is given below. It assumes that the coefficients in the linear program are chosen randomly in $\Re^n \setminus \{0\}$, meaning that none of the vectors in the problem can be equal to zero (although individual coefficients might be zero). The right-hand side in the "average" linear program considered in the theorem is not random—it is the vector $e = (1, \ldots, 1)^T$. Note that there are no explicit nonnegativity constraints on the variables.

Theorem 9.4. *Consider a linear program of the form*

$$\begin{aligned} \text{maximize} \quad & z = c^T x \\ \text{subject to} \quad & Ax \leq e, \end{aligned}$$

where $c, a_1, \ldots, a_m$ are independently and identically distributed in $\Re^n \setminus \{0\}$, and the distribution is symmetric under rotations. If the shadow vertex method is used with a feasible initial point, the expected number of iterations is bounded by

$$17n^3 m^{1/(n-1)}.$$

This result established the polynomial-time average behavior of the simplex method, but the bound obtained did not correspond to the observed practical behavior of the method (that is, between m and $3m$ iterations). A result with a more satisfying conclusion was obtained independently by Haimovich and Adler in 1983. It also uses the shadow vertex method, with an assumption about the existence of a "cooptimal path" (this term is explained in Borgwardt (1987)). The form of the linear program is slightly different (it allows a general right-hand-side vector b), and a different stochastic model is used.

Theorem 9.5. *Consider a linear program of the form*

$$maximize \quad z = c^T x$$
$$subject\ to \quad Ax \le b.$$

The coefficients c, A, and b are assumed to be chosen randomly in such a way that the problem is nondegenerate, and so that the constraints

$$a_i^T x \le b_i \quad and \quad -a_i^T x \le -b_i$$

are equally likely. If the shadow vertex method is used with a random initial basic feasible point x_0, and if a cooptimal path exists, then the expected number of iterations is bounded by

$$n \frac{m - n + 2}{m + 1}.$$

For the linear programs in Theorem 9.5 there need not be an obvious initial feasible point. Related results have shown that, if a two-phase approach is used in such cases, then the average number of iterations is bounded by

$$O\left(\min \left\{ (m - n)^2, n^2 \right\} \right)$$

for a related stochastic model as in the theorem, but without the assumption about the cooptimal path.

These two results are not straightforward to compare. Theorem 9.5 has a more optimistic conclusion, but its stochastic model can produce problems with a great many redundant constraints (so that m is an overestimate of the "effective size" of the linear program) and can produce many infeasible or unbounded, and hence irrelevant, problems. The stochastic model in Theorem 9.4 allows varying amounts of redundancy in the constraints, and all the problems generated are automatically feasible; moreover, the algorithm uses the known feasible solution at the origin to get started. However, this stochastic model rules out many practical problems since rotational symmetry implies that sparse problems are rare under these assumptions, whereas large, practical problems are almost always sparse.

9.6 Notes

Complexity—Background material on computational complexity can be found in the book by Papadimitriou and Steiglitz (1982, reprinted 1998).

Klee–Minty Problem—The variant of the Klee–Minty example that we use here is due to Chvátal (1983).

Ellipsoid Method—The survey papers of Bland, Goldfarb, and Todd (1981) and Schrader (1982, 1983) provide extensive background information on the ellipsoid method. The book of Schrijver (1986, reprinted 1998) is another useful source of information on this and other topics in this chapter. The ellipsoid method is derived from a method for solving nonsmooth convex programming problems discussed in the paper by Shor (1964). This method was improved by Yudin and Nemirovskii (1976). Khachiyan (1979) showed that this improved method, when specialized to linear programming problems, yielded a polynomial-time algorithm. Khachiyan's original publication was only an extended abstract, but was followed by a more detailed paper. The extended abstract was expanded upon in the paper of Gács and Lovász (1981), and this was the first detailed discussion of Khachiyan's work in English.

Average-Case Behavior—For an in-depth probabilistic analysis of the simplex method refer to the book by Borgwardt (1987). Recently, Spielman and Teng (2004) developed an approach called *smoothed analysis* that is a hybrid of worst-case and average-case analyses and can inherit the advantages of both.

Chapter 10

Interior-Point Methods for Linear Programming

10.1   Introduction

Interior-point methods are arguably the most significant development in linear optimization since the development of the simplex method. The methods can have good theoretical efficiency and good practical performance that are competitive with the simplex method. An important feature common to these algorithms is that the iterates are strictly feasible. (A strictly feasible point for the set $\{x : Ax = b, x \geq 0\}$ is defined as a point x such that $Ax = b$ and $x > 0$.) Thus, in contrast to the simplex algorithm, where the movement is along the boundary of the feasible region, the points generated by these new approaches lie in the *interior* of the inequality constraints. For this reason the methods are known as interior-point methods. This chapter discusses the underlying ideas in this important class of methods.

The seminal thrust in the development of interior-point methods for linear programming was the 1984 publication by Narendra Karmarkar of a new polynomial-time algorithm for linear programming. Five years earlier, the ellipsoid method—the first polynomial-time algorithm for linear programming—had been publicized and received with great excitement, soon to be followed with great disappointment due to its poor computational performance. Karmarkar's method, in contrast, was claimed from the outset to perform extraordinarily well on large linear programs. Although some initial assertions on the performance ("50 times faster than the simplex method") have not been established, it is now accepted that interior-point methods are an important tool in linear programming that can outperform the simplex method on many problems.

The publication of Karmarkar's new algorithm led to a flurry of research activity in the method and related methods. This activity was further increased with the surprising discovery in 1985 that Karmarkar's method is a specialized form of a class of algorithms for nonlinear optimization known as *barrier methods* (see Section 16.2).

Barrier methods solve a constrained problem by minimizing a sequence of unconstrained *barrier functions*. The methods were used and studied intensively in the 1960s. At that time, they were one of the few options for solving nonlinear optimization problems, but they were not seriously considered for solving linear programs because the simplex method was so effective. Eventually barrier methods were deemed by many researchers to be inefficient even for nonlinear optimization, mainly because they were thought to suffer from serious numerical difficulties. In the early 1970s barrier methods started to fall from grace, and by the early 1980s they were all but discarded. All this abruptly changed in 1984. Barrier methods, and in particular, those employing the *logarithmic barrier function*, became the focus of renewed research and spawned many new algorithms.

Most interior-point algorithms for linear programming fall into the following three main classes, each of which can be motivated by the logarithmic barrier function: path-following methods, potential-reduction methods, and affine-scaling methods. Path-following methods attempt to stay close to a "central trajectory" defined by the logarithmic barrier function; potential-reduction methods attempt to obtain a reduction in some merit or potential function that is related to the logarithmic barrier function; and affine-scaling methods sequentially transform the problem via an "affine scaling." This is only a rough classification, and algorithms often combine ideas from the various categories.

The algorithms may be characterized as either primal methods (maintaining primal feasibility), dual methods (maintaining dual feasibility), or primal-dual methods (maintaining feasibility to both problems). Throughout our discussion we will assume that the primal problem has the standard form

$$\begin{aligned} \text{minimize} \quad & z = c^T x \\ \text{subject to} \quad & Ax = b \\ & x \geq 0. \end{aligned}$$

In general, the iterates of primal interior-point methods satisfy the equality constraints and strictly satisfy (and hence are *interior* to) the nonnegativity constraints. Thus an iterate x_k satisfies $Ax_k = b$, with $x_k > 0$. Primal methods usually compute some estimate of the dual variables. Convergence is attained when the estimate is dual feasible and the duality gap is zero (to within specified tolerances).

Dual interior-point methods operate on the dual problem

$$\begin{aligned} \text{maximize} \quad & w = b^T y \\ \text{subject to} \quad & A^T y + s = c \\ & s \geq 0. \end{aligned}$$

Again, the iterates satisfy the equality constraints and strictly satisfy the nonnegativity constraints. Thus an iterate (y_k, s_k) satisfies $A^T y_k + s_k = c$, with $s_k > 0$. Dual methods usually compute some estimate of the optimal primal variables; as in primal methods, these are used to test for convergence.

Primal-dual methods attempt to solve the primal and dual problems simultaneously. In these methods, the primal and dual equality constraints are both satisfied exactly, while the nonnegativity constraints on x and s are strictly satisfied. Convergence is attained when the duality gap reaches zero (to within some tolerance). If x is feasible to the primal, and (y, s) is feasible to the dual, then

$$c^T x - b^T y = x^T s.$$

Thus a condition on the duality gap is equivalent to a condition on complementary slackness.

In terms of theoretical performance—complexity— there are many papers that give bounds on the number of iterations and arithmetic operations required by these methods to solve a linear programming problem. These bounds vary, depending on the particular algorithm used and the choice of the parameter settings. Karmarkar's original method requires at most $O(nL)$ iterations to solve a problem, with each iteration requiring $O(n^3)$ arithmetic operations, for a total of $O(n^4L)$ arithmetic operations. (As usual, n is the number of variables and L is the length of the input data.) Subsequent results have reduced this to $O(\sqrt{n}L)$ iterations and an average of $O(n^{2.5})$ operations per iteration, for a total of $O(n^3L)$ operations. The bound for the ellipsoid method is $O((mn^3 + n^4)L)$. The bounds for the two methods are surprisingly close, but while the practical performance of the ellipsoid method matches its theoretical bound, practical interior-point methods perform far better than the theory predicts.

It would be of great value if there were some theoretical criterion that predicted accurately whether a method would perform well practically, but unfortunately no such criterion is currently known. Some successful interior-point algorithms are polynomial algorithms, although the most successful implementations of these methods may not be. Also, algorithms such as the primal and dual affine methods are competitive, even though they are believed to be nonpolynomial. A great many algorithms with fine theoretical properties have never been tested on an extensive set of large problems.

Our focus in this chapter is on methods that have been successfully implemented in the solution of large-scale linear programs. We start with the primal-dual path-following method, the algorithm that has proved the most successful in practice. We discuss the method, as well as a technique for accelerating performance called the *predictor-corrector method*. We also discuss computational issues for the primal-dual method and interior-point methods in general. Next we present a "self-dual" formulation of the linear problem that allows for an efficient solution if a strictly feasible starting point is not known—and even if the problem is infeasible. We also present affine methods, and finally we give a more detailed discussion of path-following methods.

Interestingly, Karmarkar's method—the method that transformed the entire field of linear optimization—does not match the leading methods in its computational performance. For this reason we do not include a detailed discussion of Karmarkar's method, although one is available on the book's Web page, http://www.siam.org/books/ot108.

Interior-point methods are intrinsically based on nonlinear optimization methodology. However, it is possible to motivate the primal-dual path-following method using just concepts of linear optimization, and this is the approach taken in our presentation. The more theoretical aspects of path-following methods as well as affine-scaling methods are based on ideas from nonlinear optimization; a brief overview of the concepts required is given in Section 10.4.

10.2   The Primal-Dual Interior-Point Method

We describe here the primal-dual method, an interior-point method which has been particularly successful in practice. The method was originally developed as a theoretical, polynomial-time algorithm for linear programming, but it was quickly discovered that it could be adapted to give extraordinary practical performance.

To present the primal-dual method, we consider a linear program in standard form (the primal problem)

$$
\begin{aligned}
\text{minimize} \quad & z = c^T x \\
\text{subject to} \quad & Ax = b \\
& x \geq 0.
\end{aligned}
\tag{P}
$$

We assume that A is an $m \times n$ matrix of full row rank. This assumption is not always necessary, but it simplifies the discussion. If we denote by s the vector of dual slack variables, the corresponding dual problem can be written as

$$
\begin{aligned}
\text{maximize} \quad & w = b^T y \\
\text{subject to} \quad & A^T y + s = c \\
& s \geq 0.
\end{aligned}
\tag{D}
$$

Let $\bar{x}$ be a feasible solution to (P), and let $(\bar{y}, \bar{s})$ be a feasible solution to (D). These points will be optimal to (P) and (D) if and only if they satisfy the complementary slackness conditions

$$
x_j s_j = 0, \quad j = 1, \ldots, n.
$$

The main idea of the primal-dual method is to move through a sequence of strictly feasible primal and dual solutions that come increasingly closer to satisfying the complementary slackness conditions. Specifically, at each iteration we attempt to find vectors $x(\mu)$, $y(\mu)$, and $s(\mu)$ satisfying, for some $\mu > 0$,

$$
\begin{aligned}
Ax &= b \\
A^T y + s &= c \\
x_j s_j &= \mu, \quad j = 1, \ldots, n \\
x, s &\geq 0.
\end{aligned}
\tag{PD}
$$

The value of the parameter μ is then reduced and the process repeated until convergence is achieved (to within some tolerance). If $\mu > 0$, the condition $x_j s_j = \mu$ guarantees that $x > 0$ and $s > 0$; that is, the iterates are strictly feasible.

These conditions also constrain the duality gap since

$$
c^T x - b^T y = x^T s = n\mu.
$$

(See the Exercises.) Thus, the algorithm attempts to find a sequence of primal and dual feasible solutions with decreasing duality gaps. If the duality gap were zero, then the points would be optimal. Closeness to the solution will be measured by the size of the duality gap.

It is convenient to represent the conditions on complementary slackness in matrix-vector notation. Let $X = \operatorname{diag}(x)$ be a diagonal matrix whose jth diagonal term is x_j. Similarly, let $S = \operatorname{diag}(s)$, and let

$$
e = (1 \quad \cdots \quad 1)^T
$$

be the vector of length n whose entries are all equal to one. Thus we can write $x = Xe$, $s = Se$, and the complementary slackness condition may be written as

$$
X S e = \mu e.
$$

In the primal-dual algorithm the main computational effort is solving the primal-dual equations (PD). To save computation, we shall not solve them exactly but only approximately.

Thus at every iteration we shall only have an estimate of the solution to (PD). We now discuss how to obtain such an estimate.

Suppose that we have estimates $x > 0$, y, and $s > 0$ such that $Ax = b$ and $A^T y + s = c$, but $x_j s_j$ are not necessarily equal to μ. We shall find new estimates $x + \Delta x$, $y + \Delta y$, $s + \Delta s$ that are closer to satisfying these conditions. The requirements for primal feasibility are easy to state: we require that $A(x + \Delta x) = b$, and since $Ax = b$, it follows that

$$A \Delta x = 0,$$

which is a system of linear equations in Δx. Similarly, in order to maintain dual feasibility we must satisfy the linear system

$$A^T \Delta y + \Delta s = 0,$$

which is a system of linear equations in Δy and Δs. The equations for complementary slackness,

$$(x_j + \Delta x_j)(s_j + \Delta s_j) = \mu,$$

can be written as

$$s_j \Delta x_j + x_j \Delta s_j + \Delta x_j \Delta s_j = \mu - x_j s_j$$

and must be satisfied for all j. These are nonlinear equations in Δx_j and Δs_j. To obtain an approximate solution we ignore the term $\Delta x_j \Delta s_j$. If Δx_j and Δs_j are both small, then their product will be much smaller, justifying this action. The resulting system is now linear:

$$s_j \Delta x_j + x_j \Delta s_j = \mu - x_j s_j.$$

We have approximated the nonlinear system by a linear system. This technique for solving nonlinear equations by linearizing them is known as Newton's method (see Section 2.7). At each iteration of the primal-dual method we perform one iteration of Newton's method for solving (PD). We will refer to the search directions as the *Newton directions*.

In summary, the vectors Δx, Δy, Δs are obtained by solving the linear system

$$S \Delta x + X \Delta s = \mu e - X S e$$
$$A \Delta x = 0$$
$$A^T \Delta y + \Delta s = 0.$$

To solve this system, we use the third equation to obtain $\Delta s = -A^T \Delta y$. Substituting in the first equation we obtain

$$S \Delta x - X A^T \Delta y = \mu e - X S e.$$

Multiplying this equation by $A S^{-1}$ and using $A \Delta x = 0$, we get

$$-A S^{-1} X A^T \Delta y = A S^{-1}(\mu e - X S e).$$

Define the diagonal matrix $D \equiv S^{-1} X$, and let $v(\mu) = \mu e - X S e$. Then the solution vectors can be written as

$$\Delta y = -(A D A^T)^{-1} A S^{-1} v(\mu)$$
$$\Delta s = -A^T \Delta y$$
$$\Delta x = S^{-1} v(\mu) - D \Delta s.$$

If some x_j or s_j is zero, then these formulas are not defined.

The algorithm is as follows. Assume that we are given strictly feasible initial estimates $x > 0$, y, and $s > 0$ of the primal, dual, and dual slack variables. We compute the directions Δx, Δy, and Δs. We define the new estimates of the solutions as $x + \Delta x$, $y + \Delta y$, and $s + \Delta s$. Then we reduce μ and repeat.

If the parameter μ is updated as

$$\mu_{k+1} = \theta \mu_k,$$

with $0 < \theta < 1$, then under appropriate conditions on θ the new estimates are guaranteed to be strictly feasible ($x > 0$, $s > 0$), and the resulting algorithm is polynomial.

Unfortunately the values of θ for which this is true are close to 1. For example, the paper of Monteiro and Adler (1989) indicates that using $\theta = 1 - 3.5/\sqrt{n}$ will suffice. The resulting method requires at most $O(\sqrt{n}L)$ iterations, a good theoretical result. In practice, however, this value of θ is inefficient since μ decreases slowly, and the algorithm requires many iterations to reach a sufficiently low value of μ, a value where the difference between the primal and dual objectives is sufficiently close to zero.

To transform the method into a practical algorithm, we should decrease μ more rapidly. However, with these larger changes in μ, the new estimates of the solution may no longer be strictly feasible; that is, the variables may fail to satisfy the conditions $x > 0$ and $s > 0$. Therefore, the update rule for the new point is modified to

$$x(\alpha, \mu) = x + \alpha \Delta x$$
$$y(\alpha, \mu) = y + \alpha \Delta y$$
$$s(\alpha, \mu) = s + \alpha \Delta s,$$

where α is a step length chosen to ensure that x and s are positive. Even when a step length of 1 is strictly feasible, a larger step may be better. A strategy that has yielded good computational results is to take a large step which still maintains strict feasibility. For example, one may take α as 99.999% of the distance to boundary, that is,

$$\alpha = 0.99999 \alpha_{\max},$$

where $\alpha_{\max}$ is the largest step satisfying

$$x_j + \alpha_{\max} \Delta x_j \geq 0 \quad \text{and} \quad s_j + \alpha_{\max} \Delta s_j \geq 0$$

for all j. This value is explicitly given by

$$\alpha_{\max} = \min(\alpha_P, \alpha_D),$$

where

$$\alpha_P = \min_{\Delta x_j < 0} (-x_j/\Delta x_j)$$
$$\alpha_D = \min_{\Delta s_j < 0} (-s_j/\Delta s_j).$$

This is just a ratio test, as in the simplex method. (Practical algorithms often use different step lengths for the primal and dual variables, based on α_P and α_D, respectively. For simplicity, we do not do this in the example.)

Example 10.1 (Primal-Dual Method). We will apply the primal-dual method to the linear program

$$\begin{array}{rl}
\text{minimize} & z = -x_1 - 2x_2 \\
\text{subject to} & -2x_1 + x_2 \leq 2 \\
& -x_1 + 2x_2 \leq 7 \\
& x_1 + 2x_2 \leq 3 \\
& x_1, x_2 \geq 0.
\end{array}$$

Slack variables (labeled here x_3, x_4, and x_5) are added to put it into standard form

$$\begin{array}{rl}
\text{minimize} & z = -x_1 - 2x_2 \\
\text{subject to} & -2x_1 + x_2 + x_3 = 2 \\
& -x_1 + 2x_2 + x_4 = 7 \\
& x_1 + 2x_2 + x_5 = 3 \\
& x_1, x_2, x_3, x_4, x_5 \geq 0.
\end{array}$$

The dual of this problem, with slack variables s_j added, is

$$\begin{array}{rl}
\text{maximize} & w = 2y_1 + 7y_2 + 3y_3 \\
\text{subject to} & -2y_1 - y_2 + y_3 + s_1 = -1 \\
& y_1 + 2y_2 + 2y_3 + s_2 = -2 \\
& y_1 + s_3 = 0 \\
& y_2 + s_4 = 0 \\
& y_3 + s_5 = 0 \\
& s_1, s_2, s_3, s_4, s_5 \geq 0.
\end{array}$$

Hence the coefficient matrix for the constraints is

$$A = \begin{pmatrix} -2 & 1 & 1 & 0 & 0 \\ -1 & 2 & 0 & 1 & 0 \\ 1 & 2 & 0 & 0 & 1 \end{pmatrix}.$$

An initial set of strictly feasible points is given by

$$x = \begin{pmatrix} \frac{1}{2} \\ \frac{1}{2} \\ \frac{5}{2} \\ \frac{13}{2} \\ \frac{3}{2} \end{pmatrix}, \quad y = \begin{pmatrix} -1 \\ -1 \\ -5 \end{pmatrix}, \quad \text{and} \quad s = \begin{pmatrix} 1 \\ 11 \\ 1 \\ 1 \\ 5 \end{pmatrix}.$$

If we select, for instance, $\mu = 10$, then at the first iteration of the primal-dual method we set up the matrices and vectors

$$X = \begin{pmatrix} \frac{1}{2} & 0 & 0 & 0 & 0 \\ 0 & \frac{1}{2} & 0 & 0 & 0 \\ 0 & 0 & \frac{5}{2} & 0 & 0 \\ 0 & 0 & 0 & \frac{13}{2} & 0 \\ 0 & 0 & 0 & 0 & \frac{3}{2} \end{pmatrix}$$

$$S = \begin{pmatrix} 1 & 0 & 0 & 0 & 0 \\ 0 & 11 & 0 & 0 & 0 \\ 0 & 0 & 1 & 0 & 0 \\ 0 & 0 & 0 & 1 & 0 \\ 0 & 0 & 0 & 0 & 5 \end{pmatrix}$$

$$D = \begin{pmatrix} \frac{1}{2} & 0 & 0 & 0 & 0 \\ 0 & \frac{1}{22} & 0 & 0 & 0 \\ 0 & 0 & \frac{5}{2} & 0 & 0 \\ 0 & 0 & 0 & \frac{13}{2} & 0 \\ 0 & 0 & 0 & 0 & \frac{3}{10} \end{pmatrix}$$

$$e = (1 \quad 1 \quad 1 \quad 1 \quad 1)^T.$$

From these we compute

$$v = v(\mu) = \mu e - X S e = (9.5 \quad 4.5 \quad 7.5 \quad 3.5 \quad 2.5)^T$$

as well as

$$ADA^T = \begin{pmatrix} 4.5455 & 1.0909 & -0.9091 \\ 1.0909 & 7.1818 & -0.3182 \\ -0.9091 & -0.3182 & 0.9818 \end{pmatrix}$$

$$(ADA^T)^{-1} = \begin{pmatrix} 0.2767 & -0.0311 & 0.2461 \\ -0.0311 & 0.1448 & 0.0181 \\ 0.2461 & 0.0181 & 1.2523 \end{pmatrix}.$$

It is then possible to calculate

$$\Delta y = (0.2450 \quad 0.2092 \quad -10.7239)^T$$
$$\Delta s = (11.4231 \quad 20.7843 \quad -0.2450 \quad -0.2092 \quad 10.7239)^T$$
$$\Delta x = (3.7885 \quad -0.5356 \quad 8.1126 \quad 4.8598 \quad -2.7172)^T.$$

In the ratio test, α_P is determined by x_5

$$\alpha_P = -x_5/\Delta x_5 = 0.5520,$$

and α_D is determined by s_3

$$\alpha_D = -s_3/\Delta s_3 = 4.0812$$

so that $\alpha = 0.5520$. (Multiplying α by 0.99999 does not affect the first five digits of α.) For this value of α the new estimates of the solution are

$$x = (2.5914 \quad 0.2043 \quad 6.9785 \quad 9.1828 \quad 0.0000)^T$$
$$y = (-0.8647 \quad -0.8845 \quad -10.9200)^T$$
$$s = (7.3060 \quad 22.4738 \quad 0.8647 \quad 0.8845 \quad 10.9200)^T.$$

For these values of x and s the duality gap is $x^T s = 37.6812$ and the complementary slackness condition has the residual $x^T s - n\mu = -12.3188$.

At each of the subsequent iterations, we chose to reduce μ by 10. The values of the residual and the duality gap are

μ	$x^T s - n\mu$	$x^T s$
10^1	-1×10^1	4×10^1
10^0	-4×10^0	1×10^0
10^{-1}	-2×10^{-1}	3×10^{-1}
10^{-2}	-2×10^{-2}	3×10^{-2}
10^{-3}	-3×10^{-3}	2×10^{-3}
10^{-4}	-2×10^{-4}	3×10^{-4}
10^{-5}	-3×10^{-5}	2×10^{-5}
10^{-6}	-2×10^{-6}	3×10^{-6}
10^{-7}	-3×10^{-7}	2×10^{-7}

At this point the estimates of the solution are given by

$$x = (2.1794 \quad 0.4103 \quad 5.9486 \quad 8.3588 \quad 2 \times 10^{-8})^T$$
$$s = (5 \times 10^{-8} \quad 1 \times 10^{-7} \quad 2 \times 10^{-12} \quad 4 \times 10^{-9} \quad 1.0000)^T$$
$$y = (-2 \times 10^{-12} \quad -4 \times 10^{-9} \quad -1.0000)^T.$$

The primal and dual objective values are

$$z = c^T x = -3.0000$$
$$w = b^T y = -3.0000$$

and are equal (up to the number of digits displayed).

For this example the primal linear program (P) has multiple solutions, although the dual has a unique solution. The true solution of the dual is

$$s_* = (0 \quad 0 \quad 0 \quad 0 \quad 1)^T$$
$$y_* = (0 \quad 0 \quad -1)^T$$

and the solutions to the primal are all the points on the line segment connecting

$$x_1 = (0 \quad \tfrac{3}{2} \quad \tfrac{1}{2} \quad 4 \quad 0)^T$$
$$x_2 = (3 \quad 0 \quad 8 \quad 10 \quad 0)^T.$$

One of these points is

$$x_* = (2.1794 \quad 0.4103 \quad 5.9486 \quad 8.3588 \quad 0)^T$$

which is close to the solution given by the primal-dual method.

The final point x obtained by the algorithm is not a vertex of the feasible region. This is typical for interior-point methods whenever there are multiple solutions. Even so, the method converges to a point x_* that satisfies strict complementarity, as happens in this example for x_* and s_*. ∎

In our presentation of the primal-dual method we assumed that the vectors x, y, and s satisfied

$$Ax = b \quad \text{and} \quad A^Ty + s = c,$$

with $x > 0$ and $s > 0$. It might be difficult to find such points, and in such cases it is useful to modify the primal-dual method so that infeasible starting guesses can be used.

Assume that x, y, and s have been specified, with $x > 0$ and $s > 0$. Then we attempt to choose Δx, Δy, and Δs to satisfy

$$
\begin{aligned}
A(x + \Delta x) &= b \\
A^T(y + \Delta y) + (s + \Delta s) &= c \\
(x_j + \Delta x_j)(s_j + \Delta s_j) &= \mu.
\end{aligned}
$$

If, as before, we ignore the term $\Delta x_j \Delta s_j$, then these equations can be transformed into

$$
\begin{aligned}
S\Delta x + X\Delta s &= \mu e - XSe \equiv v(\mu) \\
A\Delta x &= b - Ax \equiv r_P \\
A^T\Delta y + \Delta s &= c - A^Ty - s \equiv r_D.
\end{aligned}
$$

Here r_P is the residual for the primal constraints $Ax = b$, and r_D is the residual for the dual constraints $A^Ty + s = c$. An analysis similar to that used earlier can be used to show that

$$
\begin{aligned}
\Delta y &= -(ADA^T)^{-1}[AS^{-1}v(\mu) - ADr_D - r_P] \\
\Delta s &= -A^T\Delta y + r_D \\
\Delta x &= S^{-1}v(\mu) - D\Delta s.
\end{aligned}
$$

These formulas are only a slight modification of those derived earlier, indicating that infeasible starting points can be handled in a straightforward manner within the primal-dual method. This modification is useful if the primal and dual have feasible solutions. If the primal or the dual is infeasible, this modification may fail to converge. An alternative approach that can detect whether the primal and dual are feasible is given in Section 10.3.

10.2.1   Computational Aspects of Interior-Point Methods

An important question is how interior-point methods and simplex methods compare on practical problems. Extensive computational tests indicate that interior-point methods in general—not just primal-dual path-following methods—require few iterations to solve a linear program, typically 20–60, even for large problems. Interior-point methods are affected little, if at all, by degeneracy. Each iteration of an interior-point method is expensive, requiring the solution of a system of linear equations involving the matrix ADA^T, where D is a diagonal matrix. (For the primal-dual method, $D = S^{-1}X$; the formulas for D differ among the various interior-point methods.) The matrix ADA^T changes in every entry at every iteration, so it must be factored at every iteration. This matrix can be much less sparse than A. (Since D is diagonal, the sparsity pattern of ADA^T does not change when D changes. Hence the sparsity pattern need only be analyzed once, and only a single *symbolic* factorization is required. The *numerical* factorization, however, is performed at every iteration. See Appendix A.6.1.)

Simplex methods typically require between m and $3m$ iterations, where m is the number of constraints. The number of iterations can increase dramatically on degenerate problems, or can decrease dramatically if a good initial basis is provided. Each iteration of the simplex method is cheap. The linear systems in the simplex method involve the basis matrix B. This matrix changes only in one column at each iteration, so that a factorization of B can be updated at every iteration, which is much less expensive than a refactorization. Since B consists of columns of A, it is as sparse as A.

A few conclusions can be drawn from these observations. If no good initial guess of the solution is available, or if ADA^T is sparse and easy to factor, or if the problem is degenerate, then interior-point methods are likely to perform well. However, if ADA^T is not sparse or is not easy to factor, or if a good initial guess is available (as in sensitivity analysis, or in applications where the linear program is solved repeatedly with slightly changing data), then the simplex method is likely to perform well.

Finally, interior-point methods do not produce an optimal basic feasible solution. Some auxiliary "basis recovery" procedure must be used to determine an optimal basis. In some applications, a basic feasible solution is of value, and this requirement can favor the use of the simplex method.

10.2.2   The Predictor-Corrector Algorithm

The predictor-corrector method is a modification of the primal-dual method that can reduce the number of iterations required by the method with only a modest increase in the cost per iteration. It is currently used in most interior-point software packages.

In our derivation we ignored the second-order terms of the form $\Delta x_j \Delta s_j$ in the equation

$$(x_j + \Delta x_j)(s_j + \Delta s_j) = \mu.$$

They were ignored under the assumption that they were small. This is likely to be true if x and s are near their optimal values, but it might be false at points far from the optimum.

The predictor-corrector method is designed to take these second-order terms into account by attempting to find a solution to the system

$$
\begin{aligned}
S\Delta x + X\Delta s &= \mu e - XSe - \Delta X\Delta Se \\
A\Delta x &= b - Ax \\
\Delta y + \Delta s &= c - A^T y - s,
\end{aligned}
\tag{PC}
$$

where $\Delta X = \mathrm{diag}\,(\Delta x_j)$ and $\Delta S = \mathrm{diag}\,(\Delta s_j)$. This is done in two steps. In the "predictor" step, a prediction of Δx and Δs is obtained, together with a prediction of a "good" value of μ (that is, a value of μ that is related to the values of x and s). In the "corrector" step, these predicted values are used to obtain an approximate solution to the (PC) system.

The predictor step solves the system (PC) with the term $\mu e - \Delta X\Delta Se$ ignored:

$$
\begin{aligned}
S\Delta x + X\Delta s &= -XSe \\
A\Delta x &= b - Ax \\
A^T\Delta y + \Delta s &= c - A^T y - s.
\end{aligned}
$$

This determines intermediate values $\Delta\hat{X}$ and $\Delta\hat{S}$. These are used to determine intermediate solutions $\hat{x}$ and $\hat{s}$, and in turn an updated value of μ based on the updated duality gap. (See the paper by Mehrotra (1992) for details.) The values of μ, $\Delta \hat{X}$, and $\Delta \hat{S}$ are then substituted into the (PC) equations,

$$S\Delta x + X\Delta s = \mu e - XSe - \Delta \hat{X}\Delta \hat{S}e$$
$$A\Delta x = b - Ax$$
$$A^T\Delta y + \Delta s = c - A^Ty - s,$$

which are then solved for Δx, Δy, and Δs.

This approach might seem to double the cost of an iteration of the primal-dual method. In fact, the corrector step uses the same factorization of the matrix ADA^T as the predictor step, so the predictor-corrector approach only slightly increases the cost of each iteration, and it offers the potential of decreasing the number of iterations required by the primal-dual method. Computational testing has shown that this combined approach forms one of the most effective of interior-point methods.

Exercises

2.1. Apply the primal-dual method to the linear program in Example 10.1, but using the infeasible initial guess $x = (1, 1, 1, 1, 1)^T$, $y = (0, 0, 0)^T$, and $s = (1, 1, 1, 1, 1)^T$.

2.2. Apply the predictor-corrector method to the linear program in Example 10.1. Choose μ so that
$$(x + \Delta \hat{x})^T(s + \Delta \hat{s}) = n\mu.$$

2.3. Apply the primal-dual method to the linear program in Example 10.1. Reduce μ by using the formula
$$\mu_{k+1} = \theta \mu_k$$
with $\theta = 1 - 1/\sqrt{n}$. Use $\alpha = 1$ at every iteration.

2.4. For the primal-dual pair of linear programs (P) and (D) prove that the duality gap satisfies
$$c^Tx - b^Ty = x^Ts,$$
and prove further that the solution to (PD) satisfies
$$c^Tx - b^Ty = x^Ts = n\mu.$$

2.5. Prove that, if D is a nonsingular diagonal matrix, then the sparsity pattern of ADA^T is in general unaffected by the values of D.

2.6. Prove that, if one of the columns of A is dense (has no zero entries), and D is a nonsingular diagonal matrix, then ADA^T will in general be a dense matrix.

2.7. Suppose that exactly one of the columns of A is dense (see the previous problem) and let $\hat{A}$ be the matrix obtained by replacing this column by a column of zeroes. Assume that $\hat{A}D\hat{A}^T$ is sparse and nonsingular, where D is a nonsingular diagonal matrix. Show how to use the Sherman–Morrison formula (see Appendix A.9) to efficiently solve linear systems involving the matrix ADA^T.

2.8. Suppose

$$\mu_{k+1} = \left(1 - \frac{\gamma}{\sqrt{n}}\right)\mu_k,$$

where $0 < \gamma < 1$. Given μ_0, find the number of iterations required to obtain $\mu_k \le \epsilon$, for any given $\epsilon > 0$.

10.3   Feasibility and Self-Dual Formulations

The path-following methods we have discussed have a drawback: they assume that the primal and dual problems are feasible. In practice, if either the primal or dual problem is infeasible, these methods can diverge, even when an adaptation for infeasible starting points is applied.

We describe here a method that overcomes this difficulty by reformulating the problem in a special way. The resulting formulation has several attractive properties. First, it does not require that an initial feasible solution to the original problem be known, nor does it assume that such a solution exists. Second, when an interior-point method such as the primal-dual algorithm is used to solve the reformulated problem, it will converge to an optimal solution if an optimum exists; otherwise, it will detect that either the primal or the dual is infeasible. Third, the size of the problem is only slightly larger than the original primal-dual pair. Finally, the problem can still be solved in $O(\sqrt{n}L)$ iterations.

The approach we describe embeds the problem in a *self-dual* linear optimization problem. A self-dual problem is a problem that is equivalent to its dual. In canonical form it can be written as

$$\begin{aligned}
&\underset{u}{\text{minimize}} && q^T u \\
&\text{subject to} && Mu \ge -q \\
& && u \ge 0,
\end{aligned}$$

where $q \ge 0$ and the matrix M is *skew symmetric*, that is, $M^T = -M$.

The problem has a feasible solution $u = 0$, and since $q \ge 0$, the optimal objective value is zero. Denote by v the slack vector for the constraints:

$$v = Mu + q,$$

where $v \ge 0$. The self-duality implies that any vectors u and v that are feasible to the primal are also feasible to the dual, with v corresponding to the dual slack variables. At the optimum, u and v will satisfy the complementary slackness condition, so $u^T v = 0$.

Suppose that we are given a linear program in canonical form

$$\begin{aligned}
&\text{minimize} && z = c^T x \\
&\text{subject to} && Ax \ge b \\
& && x \ge 0,
\end{aligned}$$

together with its dual program

$$\begin{aligned}
&\text{maximize} && w = b^T y \\
&\text{subject to} && A^T y \le c \\
& && y \ge 0.
\end{aligned}$$

By duality theory, if both problems have optimal solutions, then their solutions will satisfy
the system of inequalities

$$
\begin{aligned}
Ax \geq b, & \quad x \geq 0 \\
- A^T y \geq -c, & \quad y \geq 0 \\
b^T y - c^T x \geq 0. &
\end{aligned}
$$

Of course, if either the primal or the dual is infeasible, this system will not have a solution.
However, if we introduce a nonnegative scalar variable τ and define a new *homogeneous*
system of inequalities

$$
\begin{aligned}
Ax - b\tau &\geq 0 \\
- A^T y + c\tau &\geq 0 \\
b^T y - c^T x &\geq 0,
\end{aligned}
$$

where $x, y, \tau \geq 0$, then this new system always has a feasible solution since the vector of all
zeroes is feasible. Moreover, it is easy to see that if τ is positive at any feasible solution to
the system, then the vectors x/τ and y/τ solve the primal and dual problems, respectively.
Therefore, if we define

$$
\bar{M} = \begin{pmatrix} 0 & A & -b \\ -A^T & 0 & c \\ b^T & -c^T & 0 \end{pmatrix}, \quad \bar{u} = \begin{pmatrix} y \\ x \\ \tau \end{pmatrix},
$$

then we are interested in finding a solution of the homogeneous system

$$
\bar{M}\bar{u} \geq 0, \quad \bar{u} \geq 0
$$

for which $\tau > 0$. If such a solution exists, we can immediately obtain the solution of the
original linear program.

We still face two challenges. The first is to devise an approach that identifies whether
a solution with $\tau > 0$ exists. The second is that we would like to solve the system via an
interior-point method; however, the system has no feasible point that strictly satisfies all
inequalities. Indeed, any strictly feasible point $(x, y, \tau) > 0$ also satisfies $b^T y - c^T x = 0$
(see the Exercises), so the last inequality cannot be strictly satisfied.

To address this we will expand the dimension of the system by adding one nonnegative
variable and one constraint. The new variable, θ, is akin to an artificial variable that will
start with a value of 1 and converge to zero. The coefficients of θ in each of the $m + n + 1$
original rows are designed so that the slack in each of these rows will have an initial value
of 1. Assume for simplicity that the starting points x_0, y_0, and θ_0 are vectors of all 1's. Then
the coefficients of θ in the new column will be

$$
r = e - \bar{M}e,
$$

where e is a vector of 1's of length $m + n + 1$.

The new constraint is designed to keep the expanded matrix skew symmetric, so its
coefficients are $(-r^T \quad 0)$. The corresponding right-hand-side coefficient is set to $-(m + n + 2)$, so that the slack of the constraint at the initial iteration is equal to 1. We now obtain

our equivalent self-dual problem:

$$\begin{aligned}\text{minimize} \quad & q^T u \\ \text{subject to} \quad & Mu \geq -q \\ & u \geq 0,\end{aligned} \tag{SD}$$

where

$$M = \begin{pmatrix} \bar{M} & r \\ -r^T & 0 \end{pmatrix}, \quad u = \begin{pmatrix} \bar{u} \\ \theta \end{pmatrix}, \quad q = \begin{pmatrix} 0 \\ m+n+2 \end{pmatrix}.$$

The objective is to minimize $q^T u = (m+n+2)\theta$. Since $u = 0$ is a feasible solution and $\theta \geq 0$, any optimal solution u_* to (SD) will have $\theta_* = 0$.

The primal-dual algorithm for problem (SD) is straightforward. Since the primal is equivalent to the dual, the Newton step satisfies

$$\begin{aligned} U\Delta v + V\Delta u &= \mu e - UVe \\ M\Delta u + \Delta v &= 0, \end{aligned}$$

where $v = Mu + q$ is the primal (and dual) slack, $U = \text{diag}(u)$, and $V = \text{diag}(v)$. With appropriate reduction in the barrier parameter, the algorithm converges within $O(\sqrt{n}L)$ iterations.

Denote

$$u_* = \begin{pmatrix} y_* \\ x_* \\ \tau_* \\ \theta_* \end{pmatrix}, \quad v_* = \begin{pmatrix} v_{D*} \\ v_{P*} \\ \rho_* \\ \eta_* \end{pmatrix}.$$

The limiting solutions u_* and v_* satisfy strict complementarity, meaning that if $(u_*)_i = 0$, then $(v_*)_i > 0$. The proof of the latter point is quite technical and we will omit it; however, it is an important point. It allows us to prove the following lemma.

Lemma 10.2 (Strictly Complementary Solution to the Self-Dual Problem). *Let u_* and v_* be optimal solutions to the self-dual problem* (SD) *that satisfy strict complementarity. Then* (i) *if $\tau_* > 0$, then x_*/τ_* is an optimal solution to the primal and y_*/τ_* is an optimal solution to the dual;* (ii) *if $\tau_* = 0$, then either the primal problem or the dual problem is infeasible, or possibly both.*

Proof. Part (i) is straightforward and has already been discussed above. To prove part (ii) we note that, because $\theta_* = 0$, then if $\tau_* = 0$, we have

$$Ax_* \geq 0 \quad \text{and} \quad A^T y \leq 0.$$

Also, since $v_* = Mu_* + q$, we have that

$$\rho_* = b^T y_* - c^T x_*.$$

Consider now the strict complementarity between τ_* and its corresponding dual slack variable ρ_*. Since $\tau_* = 0$, then $\rho_* > 0$, and hence

$$b^T y_* - c^T x_* > 0.$$

This implies that either $b^T y_* > 0$, or $c^T x_* < 0$, or possibly both.

Suppose $b^T y_* > 0$. Then for any feasible solution x to the primal we have

$$0 \geq (y_*^T A)x = y_*^T (Ax) \geq y_*^T b > 0,$$

which is a contradiction, so the primal must be infeasible. Similarly if $c^T x_* < 0$, then for any feasible solution y to the dual we have

$$0 \leq (Ax_*)^T y = x_*^T (A^T y) \leq x_*^T c < 0,$$

which is a contradiction, and the dual must be infeasible. Finally, if both $b^T y_* > 0$ and $c^T x_* < 0$, then both the primal and dual problems are infeasible. $\Box$

The lemma shows that applying a primal-dual algorithm to a problem in self-dual form will yield either an optimal solution to the primal and dual problems, or a confirmation that one of these problems is infeasible.

Exercises

3.1. Let M be an $n \times n$ skew-symmetric matrix, and let q be a nonnegative n-dimensional vector. Prove that the problem

$$\begin{array}{ll} \underset{u}{\text{minimize}} & q^T u \\ \text{subject to} & Mu \geq -q \\ & u \geq 0 \end{array}$$

is self-dual.

3.2. Give an example for a two-dimensional self-dual linear program.

3.3. Prove that, if the $n \times n$ matrix M is skew symmetric, then $u^T M u = 0$ for any n-dimensional vector u.

3.4. Prove that, for any positive vector $\bar{u} = (y^T, x^T, \tau)^T$ satisfying $\bar{M}\bar{u} \geq 0$, the constraint $\hat{b}^T y - c^T x \geq 0$ is binding. (*Hint*: Use the fact that x/τ solves the primal and y/τ solves the dual.)

3.5. Set up the self-dual formulation corresponding to the problem

$$\begin{array}{ll} \text{minimize} & z = x_1 - x_2 \\ \text{subject to} & x_1 + x_2 \leq 2 \\ & x_1, x_2 \geq 0. \end{array}$$

10.4   Some Concepts from Nonlinear Optimization

Interior-point methods for linear programming are inherently based on nonlinear optimization techniques. Luckily, only a few results are needed to gain an understanding of these methods. Here we shall give a brief summary of the fundamental concepts needed. For a more detailed discussion, see the referenced sections of the book.

We start with the problem of minimizing an unconstrained function of n variables $f(x)$. At any local minimizer x_* the gradient (the vector of all partial derivatives) of f must satisfy $\nabla f(x_*) = 0$. If in addition the matrix of second derivatives $\nabla^2 f(x_*)$ is positive definite, then x_* is a strict local minimizer of f. (See Chapter 11.)

We will also be interested in minimizing a function subject to linear equality constraints. Consider the problem of minimizing a function f subject to m constraints of the form $a_i^T x = b_i$, that is,

$$\text{minimize} \quad f(x)$$
$$\text{subject to} \quad Ax = b.$$

To find the optimum we introduce an auxiliary function,

$$\mathcal{L}(x, \lambda) = f(x) - \sum_{i=1}^{m} \lambda_i (a_i^T x - b_i) = f(x) - \lambda^T (Ax - b),$$

called the Lagrangian. The variables λ_i are called the Lagrange multipliers, and, in the case of linear programming, correspond to the dual variables. For any optimal point x_* there exist values of $\lambda_1, \ldots, \lambda_m$ for which the gradient of the Lagrangian is zero. Thus the first-order optimality conditions can be written as

$$\nabla f(x_*) - A^T \lambda = 0$$
$$Ax_* = b.$$

More detail is given in Chapter 14.

The next concept is Newton's method for the solution of a system of nonlinear equations. Given a guess x_k of the solution, Newton's method replaces the nonlinear functions at $x_k + \Delta x$ by their linear approximation, obtained from the Taylor series, and solves the resulting linear system. The solution Δx is called the Newton direction and the resulting point $x_k + \Delta x$ will be the new guess of the solution of the nonlinear system. See Section 2.7.1 for further details.

An analogous idea is used in the minimization of an unconstrained function $f(x)$. Any local minimizer will satisfy $\nabla f(x) = 0$. The Newton direction is obtained by replacing the components of the gradient at x_k by their linear approximation obtained from the Taylor series. This leads to the linear system of equations

$$\nabla^2 f(x_k) \Delta x = -\nabla f(x_k),$$

and the solution is the Newton direction. See Section 11.3.

We can also give an alternative interpretation of Newton's method. The method approximates $f(x + \Delta x)$ by a quadratic function around $f(x_k)$ and then finds the step Δx_k that minimizes the quadratic approximation. Ignoring a constant term, we find that the resulting minimization problem is

$$\text{minimize} \quad \tfrac{1}{2} \Delta x \nabla^2 f(x_k) \Delta x + \nabla f(x_k)^T \Delta x.$$

In the case where the function to be minimized is subject to linear constraints $Ax = b$ and the current guess x_k satisfies $Ax_k = b$, a projected Newton search direction can be computed. The direction minimizes the quadratic approximation above and lies in the null space of A, so that $A\delta_x = 0$.

The final concept to be addressed is that of the logarithmic barrier method. We defer that to Section 10.6.

10.5   Affine-Scaling Methods

Affine methods were proposed soon after the publication of Karmarkar's method as a simplification of his method. It was later discovered, however, that the Russian scientist I. I. Dikin had proposed a primal affine method in 1967. Affine methods have been successful in solving large linear programs, although not quite as successful as the primal-dual path-following method. They are of interest because the directions they employ provide important insight into the search directions of path-following methods.

Affine-scaling methods are the simplest interior-point methods, yet they are effective at solving large problems. These methods transform the linear program into an equivalent one in which the current point is favorably positioned for a constrained form of the steepest-descent method.

Before we come to this idea, we discuss the search direction. Suppose that we are given some interior point for the primal problem. Since the nonnegativity constraints are nonbinding, we can move a small distance in a direction without violating these constraints. It would make sense to move in the *steepest-descent direction* $-c$ along which the function value decreases most rapidly. To maintain feasibility of the constraints $Ax = b$, we project this direction onto the null space of A. When A has full rank, this is achieved by premultiplying the steepest-descent direction by the orthogonal projection matrix

$$P = I - A^T(AA^T)^{-1}A.$$

The resulting direction

$$\Delta x = -Pc$$

is the *projected steepest-descent* direction. It is the feasible direction that produces the fastest rate of decrease in the objective (see the Exercises).

If the current point is close to the *center* of the feasible region (as is the point x_a in Figure 10.1), then considerable improvement could be made by moving in the projected steepest-descent direction. If the point is close to the boundary defined by the nonnegativity constraints, then possibly only a small step could be taken without losing feasibility, and thus little improvement achieved (see point x_b in the figure.) The steepest-descent direction will be effective if the current point is close to the center of the feasible region. Motivated by this, affine methods transform the linear problem into an equivalent problem that has the current point in a more "central" position; a step is then made along the projected steepest-descent direction for the transformed problem.

What is a "central" position? A plausible choice is the point $e = (1, 1, \ldots, 1)^T$, since all its variables are equally distant from their bounds. To transform the current point x_k into the point e we simply scale the variables, dividing them by the components of x_k. Let $X = \text{diag}(x_k)$ be the $n \times n$ diagonal matrix whose diagonal entries are the components of x_k. Then the scaling

$$\bar{x} = X^{-1}x \quad \text{or equivalently} \quad x = X\bar{x}$$

transforms the variables x into new variables $\bar{x}$, with x_k transformed into e. This transformation is an *affine-scaling* transformation, and X is the *scaling matrix*. Note that

$$e = X^{-1}x_k \quad \text{and} \quad x_k = Xe.$$

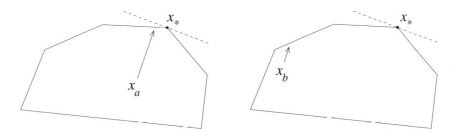

Figure 10.1. *Steepest-descent from "central" and "noncentral" points.*

Suppose that the original problem in "x-space" is written as

$$\text{minimize} \quad z = c^T x$$
$$\text{subject to} \quad Ax = b$$
$$x \geq 0.$$

Then the transformed problem in "$\bar{x}$-space" can be written as

$$\text{minimize} \quad z = \bar{c}^T \bar{x}$$
$$\text{subject to} \quad \bar{A}\bar{x} = b$$
$$\bar{x} \geq 0,$$

where

$$\bar{c} = Xc \quad \text{and} \quad \bar{A} = AX.$$

Our current position in "$\bar{x}$-space" is at the point e. We now make a move in the projected steepest-descent direction

$$\Delta\bar{x} = -\bar{P}\bar{c} = -(I - \bar{A}^T(\bar{A}\bar{A}^T)^{-1}\bar{A})\bar{c}$$
$$= -\left(I - XA^T(AX^2A^T)^{-1}AX\right)Xc.$$

The step in the transformed space along $\Delta\bar{x}$ yields

$$\bar{x}_{k+1} = e + \alpha\Delta\bar{x},$$

where α is a suitable step length. The final task in the iteration is to map $\bar{x}_{k+1}$ back to the original "x-space" to obtain

$$x_{k+1} = X\bar{x}_{k+1}.$$

We now summarize the *affine-scaling algorithm*. In each iteration, (i) the problem in "x-space" is transformed via affine scaling into an equivalent problem in "$\bar{x}$-space" so that the current point x_k is transformed into the point e, (ii) an appropriate step is taken from e along the projected steepest-descent direction in "$\bar{x}$-space," and (iii) the resulting point in "$\bar{x}$-space" is transformed back into the corresponding point in "x-space."

It is possible to express the algorithm entirely in terms of the original variables. To see this, note that

$$x_{k+1} = X\bar{x}_{k+1} = X(e + \alpha\Delta\bar{x}) = x_k + \alpha X\Delta\bar{x}.$$

Defining $\Delta x = X \Delta \bar{x}$, we obtain

$$x_{k+1} = x_k + \alpha \Delta x,$$

where

$$\Delta x = - \left(X^2 - X^2 A^T (AX^2 A^T)^{-1} AX^2 \right) c = -X \bar{P} X c$$

and $\bar{P} = I - XA^T (AX^2 A^T)^{-1} AX$ is the orthogonal projection matrix for AX. The direction Δx is the *primal affine-scaling direction*.

There are several ways to select the step length α. Because the function decreases at a constant rate along the search direction Δx, most implementations take as large a step as possible, stopping just short of the boundary. (On the boundary, some variable is zero and the method is not defined.) Commonly α is chosen as

$$\alpha = \gamma \alpha_{\max},$$

where $\alpha_{\max}$ is the step to the boundary, and $0 < \gamma < 1$; typically γ is set very close to 1, e.g., $\gamma = 0.99$. Since $\alpha_{\max}$ is the largest step satisfying

$$(x_k)_i + \alpha_{\max} \Delta x_i \geq 0, \quad i = 1, \ldots, n,$$

its value is given by a ratio test

$$\alpha_{\max} = \min_{\Delta x_i < 0} \left(-(x_k)_i / \Delta x_i \right).$$

If $\Delta x \geq 0$ and $\Delta x \neq 0$, then the problem is unbounded (see the Exercises).

We would like to terminate the algorithm when the optimality conditions are satisfied to within some tolerance, which occurs if a dual feasible solution with a near-zero duality gap is available. Under nondegeneracy assumptions, it is possible to estimate the solution to the dual. If we define

$$y_k = (AX^2 A^T)^{-1} AX^2 c \quad \text{and} \quad s_k = c - A^T y_k,$$

then (y_k, s_k) will converge to the optimal solution of the dual problem as x_k converges. Here, y_k is the solution to the least-squares problem

$$\min_y \left\| X(c - A^T y) \right\|,$$

and hence s_k is the solution to $\min_s \|Xs\|$; this is the vector s that yields the smallest "complementarity" (in norm) with respect to x_k. Since

$$\Delta x = -X^2 s_k,$$

the computation of the dual estimates is a by-product of the computation of the search direction and does not require extra work.

If the problem is degenerate, these dual estimates may not converge. A commonly used alternative is to stop when the relative improvement in the objective is small, that is, when

$$\frac{c^T x_k - c^T x_{k+1}}{\max \left\{ 1, |c^T x_k| \right\}} \leq \epsilon,$$

where ϵ is a small tolerance.

Example 10.3 (Primal Affine Method). We apply the primal affine method to the linear program

$$
\begin{aligned}
\text{minimize} \quad & z = -x_1 - 2x_2 \\
\text{subject to} \quad & -2x_1 + x_2 \leq 2 \\
& -x_1 + 2x_2 \leq 7 \\
& x_1 \leq 3 \\
& x_1, x_2 \geq 0.
\end{aligned}
$$

After slack variables are added to put the problem in standard form, the constraint matrix and cost vector are

$$
A = \begin{pmatrix} -2 & 1 & 1 & 0 & 0 \\ -1 & 2 & 0 & 1 & 0 \\ 1 & 0 & 0 & 0 & 1 \end{pmatrix} \quad \text{and} \quad c = (-1 \quad -2 \quad 0 \quad 0 \quad 0)^T.
$$

Suppose that we start from the initial feasible point

$$
x = (0.5 \quad 0.5 \quad 2.5 \quad 6.5 \quad 2.5)^T.
$$

Then the initial scaling matrix is

$$
X = \begin{pmatrix} 0.5 & 0 & 0 & 0 & 0 \\ 0 & 0.5 & 0 & 0 & 0 \\ 0 & 0 & 2.5 & 0 & 0 \\ 0 & 0 & 0 & 6.5 & 0 \\ 0 & 0 & 0 & 0 & 2.5 \end{pmatrix},
$$

and thus

$$
\bar{A} = AX = \begin{pmatrix} -1 & 0.5 & 2.5 & 0 & 0 \\ -0.5 & 1 & 0 & 6.5 & 0 \\ 0.5 & 0 & 0 & 0 & 2.5 \end{pmatrix}, \quad \bar{c} = (-0.5 \quad -1 \quad 0 \quad 0 \quad 0)^T.
$$

From these we compute

$$
AX^2 A^T = \begin{pmatrix} 7.50 & 1.00 & -0.50 \\ 1.00 & 43.50 & -0.25 \\ -0.50 & -0.25 & 6.50 \end{pmatrix}.
$$

Using this we compute

$$
y = (AX^2 A^T)^{-1} AX^2 c = \begin{pmatrix} -0.0003 \\ -0.0175 \\ -0.0392 \end{pmatrix}, \quad s = c - A^T y = \begin{pmatrix} -0.9789 \\ -1.9648 \\ 0.0003 \\ 0.0175 \\ 0.0392 \end{pmatrix},
$$

and

$$
\Delta x = X \Delta \bar{x} = \begin{pmatrix} 0.2447 \\ 0.4912 \\ -0.0018 \\ -0.7377 \\ -0.2447 \end{pmatrix}.
$$

The ratio test gives

$$\alpha_{max} = \min \left\{ \frac{2.5}{0.0018}, \frac{6.5}{0.7377}, \frac{2.5}{0.2477} \right\} = 8.8114.$$

Taking a step that is 0.9999 of the maximum step to the boundary yields $\alpha = 8.8105$. The new point is

$$x_1 = x_0 + \alpha \Delta x = (2.6561 \quad 4.8277 \quad 2.4845 \quad 0.0001 \quad 0.3439)^T.$$

After three additional iterations, the estimate of the solution is

$$x = (3.0000 \quad 5.0000 \quad 3.0000 \quad 7 \times 10^{-8} \quad 3 \times 10^{-9})^T.$$

The solution of this problem is $x_* = (3, 5, 3, 0, 0)^T$. The estimates of the dual variables at this iteration are

$$y = (3 \times 10^{-10} \quad -1.0000 \quad -2.0000)^T,$$
$$s = (-2 \times 10^{-10} \quad -4 \times 10^{-11} \quad -3 \times 10^{-10} \quad 1.0000 \quad 2.0000)^T.$$

Since s is nonnegative (to an acceptable tolerance), and the duality gap $c^T x - b^T y = 7 \times 10^{-8}$ is small, we terminate. Here is a summary of the progress of the algorithm:

$\|x - x_*\|$	$\min s_i$	$c^T x - b^T y$
9×10^{-1}		
7×10^{-1}	-2×10^0	-1×10^1
8×10^{-4}	-5×10^{-2}	-3×10^{-1}
7×10^{-5}	-9×10^{-9}	6×10^{-5}
8×10^{-8}	-3×10^{-10}	7×10^{-8}

∎

The combination of affine scaling and the steepest-descent method can also be applied to the dual problem. This results in the *dual affine-scaling method*. To develop the method, we consider the dual problem

$$\begin{aligned} \text{maximize} \quad & w = b^T y \\ \text{subject to} \quad & A^T y + s = c \\ & s \geq 0. \end{aligned} \tag{D}$$

Before cluttering our notation with transformations, we discuss how we form a projected steepest-descent direction. The dual has two sets of variables: y and s. When the constraint matrix A has full rank, we can represent the problem in a more convenient form that omits the free variables y. The pair (y, s) is dual feasible if and only if $A^T y = c - s$, or equivalently, if and only if $y = (AA^T)^{-1}A(c - s)$. Substituting for y in the dual problem gives

$$\begin{aligned} \text{minimize} \quad & w = \tilde{b}^T s \\ \text{subject to} \quad & Ps = Pc \\ & s \geq 0, \end{aligned} \tag{$\tilde{\text{D}}$}$$

where

$$P = I - A^T(AA^T)^{-1}A \quad \text{and} \quad \tilde{b} = A^T(AA^T)^{-1}b.$$

The constraint matrix P is the orthogonal projection matrix corresponding to A. This matrix is usually singular. Nevertheless, it is still possible to find an orthogonal projection matrix corresponding to P; this is

$$\tilde{P} = I - P = A^T(AA^T)^{-1}A$$

(see the Exercises). The projected steepest-descent direction for $(\tilde{D})$ is

$$\Delta s = -\tilde{P}\tilde{b} = -(A^T(AA^T)^{-1}A)(A^T(AA^T)^{-1})b = -A^T(AA^T)^{-1}b.$$

This direction represents the change in the variable s. To obtain the change in y, we return to the original dual problem (D). To maintain feasibility, any step must satisfy $A^T\Delta y + \Delta s = 0$. This holds if and only if $\Delta y = -(AA^T)^{-1}A\Delta s$, or

$$\Delta y = (AA^T)^{-1}b.$$

The direction $(\Delta y, \Delta s)$ is the *dual projected steepest-descent direction*.

We now apply affine scaling to the dual problem (D). Our goal, as in the primal method, is to obtain a substantial improvement with the steepest-descent direction. To achieve this the current point should be away from the boundary. We perform an affine scaling that transforms the variables s so that s_k is sent to e. (It is not necessary to transform y, since these variables are not restricted.) The scaling matrix is now $S = \text{diag}(s_k)$, the diagonal matrix whose entries are the elements of s_k. The affine transformation is given by

$$\bar{s} = S^{-1}s \quad \text{or equivalently} \quad s = S\bar{s}$$

and the transformed problem can be written as

$$
\begin{aligned}
\text{maximize} \quad & w = b^T y \\
\text{subject to} \quad & \bar{A}^T y + \bar{s} = \bar{c} \\
& \bar{s} \geq 0,
\end{aligned}
$$

where

$$\bar{A} = AS^{-1} \quad \text{and} \quad \bar{c} = S^{-1}c.$$

We take a step in the transformed space, along the projected steepest-descent direction. The movement in y is given by

$$\Delta y = (\bar{A}\bar{A}^T)^{-1}b.$$

We now return to the original variables. Since y was not transformed, Δy is unchanged. We obtain

$$
\begin{aligned}
\Delta y &= (AS^{-2}A^T)^{-1}b \\
\Delta s &= -A^T\Delta y = -A^T(AS^{-2}A^T)^{-1}b.
\end{aligned}
$$

The direction $(\Delta y, \Delta s)$ is the *dual affine direction*.

In summary, the dual affine algorithm sets

$$y_{k+1} = y_k + \alpha \Delta y, \quad s_{k+1} = s_k + \alpha \Delta s,$$

where Δy and Δs are as given above. The step size α is typically a fraction of the maximum step to the boundary α_{max}, where

$$\alpha_{max} = \min_{\Delta s_i < 0} (-(s_k)_i / \Delta s_i)$$

is found by a ratio test. (If $\Delta s \geq 0$ and $\Delta s \neq 0$, then the dual problem is unbounded and the primal problem is infeasible.)

Termination of the dual algorithm is as in the primal algorithm. A commonly used rule is to terminate if

$$\frac{b^T y_{k+1} - b^T y_k}{\max \left\{ 1, |b^T y_k| \right\}} \leq \epsilon.$$

It is also possible to compute estimates of the primal variables:

$$x_k = -S^{-2} \Delta s.$$

If $x_k \geq 0$, it is primal feasible, since it already satisfies $A x_k = b$. Under nondegeneracy assumptions, x_k converges to the optimal primal solution.

In both the primal affine and the dual affine methods, the major effort is the computation of the projection. More specifically, both algorithms require the solution of a system of equations with respect to a matrix $A D A^T$, where D is a diagonal matrix: in the primal method $D = X^2$ and in the dual method $D = S^{-2}$. The amount of computation per iteration is similar. However, there are some differences. In the primal version, if Δx is not computed precisely, then the equation $A \Delta x = 0$ will no longer be satisfied, and feasibility will be lost. In contrast, in the dual version, even if Δy is not computed precisely, the computation $\Delta s = A^T \Delta y$ (an "easy" calculation) guarantees that dual feasibility is maintained. On the other hand, the primal affine method has the advantage that it always terminates with a primal feasible solution.

Despite their simplicity, the primal and dual affine methods are competitive with, if perhaps slightly inferior to, the best interior-point methods. The algorithms are globally convergent even for degenerate problems. They are not known to be polynomial algorithms; however, and in fact there is suggestive theoretical evidence that they are not.

In contrast, it is possible to define a *primal-dual affine-scaling method* that has polynomial complexity. In this method, the current point x_k is transformed into $(XS)^{1/2}e$, and the vector of dual slack variables s_k is transformed into $(XS)^{-1/2}e$, so that the complementarity vector XSe is transformed into e. The expressions for the search directions are similar to those for the primal affine and dual affine methods, except that the matrices X^2 and S^2 are replaced by the matrices D and D^{-1}, respectively, where $D = (XS^{-1})$. This gives the following search directions:

$$\Delta x = -(D - D A^T (A D A^T)^{-1} A D) c$$
$$\Delta y = (A D A^T)^{-1} b$$
$$\Delta s = A^T \Delta y = -A^T (A D A^T)^{-1} b.$$

As usual, the step size is determined by a ratio test. The primal-dual affine method has a complexity bound of $O(nL^2)$ iterations. In practice, however, it has not been as successful as its primal and dual counterparts. The algorithm has also been adapted to solve strictly convex quadratic programs in polynomial time.

Exercises

5.1. Consider the problem

$$\begin{array}{ll} \text{minimize} & z = 2x_1 + x_2 \\ \text{subject to} & x_1 + 2x_2 = 4 \\ & x_1, x_2 \geq 0. \end{array}$$

Plot the feasible region. Suppose the current point is $x = (2, 1)^T$. Define the corresponding affine scaling, and the transformed linear program, and plot its feasible region. Repeat the above for $x = (3, \frac{1}{2})^T$.

5.2. Use the primal affine-scaling method to solve the problem

$$\begin{array}{ll} \text{minimize} & z = 2x_1 + x_2 + 3x_3 \\ \text{subject to} & x_1 + 2x_2 - x_3 = 1 \\ & x_1 + x_2 + x_3 = 1 \\ & x_1, x_2, x_3 \geq 0. \end{array}$$

Start from the point $x_0 = (0.25, 0.5, 0.25)^T$. At each iteration use a step length that is 0.99 of the maximum step to the boundary. Terminate when the relative improvement in the objective is below 10^{-5}.

5.3. Use the dual affine-scaling method to solve the previous problem.

5.4. Prove that the direction obtained by scaling the steepest-descent direction by its norm is the solution to

$$\begin{array}{ll} \underset{p}{\text{minimize}} & z = c^T p \\ \text{subject to} & Ap = 0 \\ & p^T p = 1. \end{array}$$

Note: You can solve this problem by using the optimality conditions for constrained optimization of Chapter 14; alternatively you can solve it by using the formula $x^T y = \|x\| \cdot \|y\| \cos(\theta)$, where θ is the angle between x and y.

5.5. Prove that the projected steepest-descent direction for the primal problem is the solution to

$$\begin{array}{ll} \underset{p}{\text{minimize}} & f(p) = \|p + c\| \\ \text{subject to} & Ap = 0. \end{array}$$

5.6. Prove that in the affine-scaling algorithm, $c^T x_k - c^T x_{k+1} = \alpha \|\Delta \bar{x}\|^2 > 0$; hence the algorithm is a descent method.

5.7. Let Δx be the search direction in the primal affine method. Prove that if $\Delta x \geq 0$ and $\Delta x \neq 0$, then the problem is unbounded.

5.8. Let $y = (AX^2A^T)^{-1}AX^2c$, and let $s = c - A^Ty$. Prove that the search direction in the primal affine method is $\Delta x = -X^2s$. Prove that if $\Delta x \leq 0$, then (y, s) is dual feasible. Indicate how you would use this result to obtain a lower bound on the objective in the primal affine method.

5.9. Let A be a matrix with full row rank m, and let $P = I - A^T(AA^T)^{-1}A$. Let $\tilde{P} = I - P$. Prove that $\tilde{P}$ is symmetric, is idempotent (that is, $\tilde{P}^2 = \tilde{P}$), that it satisfies $P\tilde{P} = 0$, and that it has rank m. Conclude from this that $\tilde{P}$ is an orthogonal projection matrix for P.

5.10. In the dual affine method let $x = -S^{-2}\Delta s$. Prove that if $\Delta s \leq 0$, then x is feasible to the primal problem.

10.6   Path-Following Methods

Path-following methods are algorithms that restrict the iterates to being "close" to the *central path*. For a problem in standard form, the central path is a trajectory defined by the set of points $x(\mu)$, $y(\mu)$, and $s(\mu)$ that satisfy

$$
\begin{aligned}
Ax &= b \\
A^Ty + s &= c \\
x_js_j &= \mu, \quad j = 1, \ldots, n \\
x, s &\geq 0
\end{aligned}
$$

for some $\mu > 0$.

One motivation for these equations is that they try to maintain primal feasibility, dual feasibility, and a perturbed complementary slackness condition. This was the viewpoint taken in Section 10.2.2, where we initially developed the primal-dual path-following algorithm.

We can also motivate these equations from the perspective of the class of methods known as *logarithmic barrier methods*. This viewpoint can lead to further insight regarding the central path and also gives rise to a variety of new algorithms for solving the primal problem, and the dual problem, or the primal and dual problems simultaneously.

Barrier methods handle inequality constraints $g_i(x) \geq 0$ in an optimization problem by removing them as explicit constraints, and instead, incorporating them into the objective as a "barrier term" that prevents the iterates from reaching the boundary of these constraints. For the logarithmic barrier method the barrier term is $-\sum \log(x_i)$. A nonnegative parameter μ controls the weight of the barrier and is gradually decreased to zero as the iterates approach the solution.

Given a linear program in standard form (P), the logarithmic barrier method solves a sequence of problems of the form

$$
\begin{aligned}
\underset{x}{\text{minimize}} \quad & \beta_\mu(x) = c^Tx - \mu\sum_{j=1}^{n}\log(x_j) \\
\text{subject to} \quad & Ax = b
\end{aligned}
\qquad (P_\mu)
$$

for a decreasing sequence of positive barrier parameters μ that approach zero. The equalities $Ax = b$ can be handled explicitly by moving in the null space of the constraint matrix.

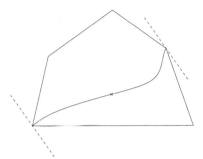

Figure 10.2. *Central path.*

The objective, $\beta_\mu(x) = c^T x - \mu \sum \log(x_j)$, will be referred to as the primal logarithmic barrier function. Its gradient and Hessian are

$$\nabla \beta_\mu(x) = c - \mu \begin{pmatrix} 1/x_1 \\ \vdots \\ 1/x_n \end{pmatrix}, \quad \nabla^2 \beta_\mu(x) = \mu \begin{pmatrix} 1/x_1^2 & & \\ & \ddots & \\ & & 1/x_n^2 \end{pmatrix}.$$

If we denote $X = \text{diag}(x)$, then

$$\nabla \beta_\mu(x) = c - \mu X^{-1} e, \quad \nabla^2 \beta_\mu(x) = \mu X^{-2}.$$

Let y denote the vector of Lagrange multipliers associated with the equality constraints of P_μ. Then the first-order optimality conditions are

$$c - \mu X^{-1} e - A^T y = 0$$
$$Ax = b.$$

It can be shown that if both the primal problem and the dual problem have strictly feasible solutions, then the barrier subproblems have a unique solution $x = x(\mu)$ for every $\mu > 0$, and that $x(\mu)$ converges to a solution of the primal problem as μ approaches zero. The central path or *barrier trajectory* for the primal problem is the set of points $\{ x(\mu) : \mu > 0 \}$. This is illustrated in Figure 10.2.

Let us take a closer look at the barrier trajectory. If we define $s = c - A^T y$, then the first equation implies that $s - \mu X^{-1} e = 0$, or equivalently, $Xs = \mu e$. Thus, for each component we have $x_j s_j = \mu$. Since x_j is positive, s_j is also positive, and $s > 0$. Thus, the multipliers y correspond to a dual feasible solution (y, s). Furthermore, when μ is small, the complementary slackness conditions are close to being satisfied—they are perturbed by μ. Denoting $S = \text{diag}(s)$, the conditions can be written in an equivalent form that describes a primal-dual central trajectory:

$$Ax = b$$
$$A^T y + s = c$$
$$XSe = \mu.$$

For a given μ, the system may be regarded as a perturbation of the optimality conditions for a linear program: primal feasibility, dual feasibility, and perturbed complementarity slackness. If we apply Newton's method to this nonlinear system of equations, then we obtain the primal-dual algorithm introduced in Section 10.2. For completeness, we present again the search directions (in a slightly different form):

$$\Delta x = -(D - DA^T(ADA^T)^{-1}AD)(c - \mu X^{-1}e)$$
$$\Delta y = (ADA^T)^{-1}(b - \mu AS^{-1}e)$$
$$\Delta s = -A^T(ADA^T)^{-1}(b - \mu AS^{-1}e),$$

where $D = S^{-1}X$. The primal-dual algorithm is therefore a path-following method.

The primal variant of the method applies Newton's method in a slightly different setting. It operates in a primal framework, updating and maintaining feasibility of the primal variables. The search direction at a point x is the projected Newton direction—the feasible direction Δx that minimizes the quadratic approximation to β_μ at x. Denoting $X = \text{diag}(x)$, the projected Newton direction solves the quadratic program

$$\text{minimize} \quad f(\Delta x) = \tfrac{1}{2}\Delta x^T \mu X^{-2} \Delta x + (c - \mu X^{-1}e)^T \Delta x$$
$$\text{subject to} \quad A\Delta x = 0.$$

Letting y be the vector of Lagrange multipliers for this problem, the first-order necessary conditions for optimality are

$$\mu X^{-2}\Delta x + (c - \mu X^{-1}e) - A^T y = 0$$
$$A\Delta x = 0.$$

Multiplying the first equation on the left by $(1/\mu)X^2$, we obtain

$$\Delta x = -\frac{1}{\mu}X^2(c - \mu X^{-1}e - A^T y).$$

Multiplying on the left by A, and recalling that $A\Delta x = 0$, we obtain an expression for y:

$$y = (AX^2A^T)^{-1}AX^2(c - \mu X^{-1}e).$$

Defining $s = c - A^T y$, the projected Newton direction is

$$\Delta x = -\frac{1}{\mu}X^2(s - \mu X^{-1}e) = x - \frac{1}{\mu}X^2 s.$$

Later we shall show that, as x converges to x_*, the vector y converges to the optimal dual solution, and the vector $s = c - A^T y$ converges to the optimal dual slack vector. For this reason we refer to (y, s) as the "dual estimates" at x.

Example 10.4 (Primal Path-Following Method). Consider again the problem in Example 10.1. We will compute the search direction generated by the primal path-following method at the initial point $x = (0.5, 0.5, 2.5, 6.5, 2.5)^T$. We use $\mu = 10$. Then

$$c - \mu X^{-1}e = \begin{pmatrix} -1 \\ -2 \\ 0 \\ 0 \\ 0 \end{pmatrix} - 10 \begin{pmatrix} 2 & 0 & 0 & 0 & 0 \\ 0 & 2 & 0 & 0 & 0 \\ 0 & 0 & 0.4 & 0 & 0 \\ 0 & 0 & 0 & 0.1539 & 0 \\ 0 & 0 & 0 & 0 & 0.4 \end{pmatrix} \begin{pmatrix} 1 \\ 1 \\ 1 \\ 1 \\ 1 \end{pmatrix} = \begin{pmatrix} -21.000 \\ -22.000 \\ -4.000 \\ -1.539 \\ -4.000 \end{pmatrix}.$$

Since

$$
AX = \begin{pmatrix} -1 & 0.5 & 2.5 & 0 & 0 \\ -0.5 & 1 & 0 & 6.5 & 0 \\ 0.5 & 0 & 0 & 0 & 2.5 \end{pmatrix} \quad \text{and} \quad AX^2A^T = \begin{pmatrix} 7.50 & 1.00 & -0.50 \\ 1.00 & 43.50 & -0.25 \\ -0.50 & -0.25 & 6.50 \end{pmatrix},
$$

we obtain

$$
y = (AX^2A^T)^{-1}AX^2(c - \mu X^{-1}e) = \begin{pmatrix} -2.7832 \\ -1.5908 \\ -4.9291 \end{pmatrix}, \quad s = c - A^T y = \begin{pmatrix} -3.2280 \\ 3.9647 \\ 2.7832 \\ 1.5908 \\ 4.9291 \end{pmatrix},
$$

and

$$
\Delta x = X\Delta \bar{x} = \begin{pmatrix} 0.5807 \\ 0.4009 \\ 0.7605 \\ -0.2211 \\ -0.5807 \end{pmatrix}. \quad \blacksquare
$$

We now re-examine the search direction Δx. We can write it as

$$
\Delta x = -\frac{1}{\mu} X^2 (c - A^T y - \mu X^{-1} e)
$$

$$
= -\frac{1}{\mu} \left(X^2 - X^2 A^T (AX^2A^T)^{-1}AX^2 \right)(c - \mu X^{-1}e)
$$

$$
= -\frac{1}{\mu} X\bar{P}X(c - \mu X^{-1}e) = -\frac{1}{\mu} X\bar{P}Xc + X\bar{P}e,
$$

where $\bar{P} = I - XA^T(AX^2A^T)^{-1}AX$ is the orthogonal projection matrix corresponding to AX. This final expression is a sum of two directions: the first is a multiple of the primal affine-scaling direction; the second, $X\bar{P}e$, is called a *centering direction*, for reasons we shall explain. Almost all primal interior-point methods move in a direction that is some combination of these two directions. (Similar observations can be made for dual methods as well as primal-dual methods.) As μ goes to zero, the contribution of the affine-scaling direction becomes increasingly dominant. Thus the limiting direction when μ approaches zero is the affine scaling direction.

The centering direction is the Newton direction for the problem

$$
\text{minimize} \quad f(x) = -\sum_{j=1}^{n} \log x_j
$$

$$
\text{subject to} \quad Ax = b.
$$

This problem finds the "analytic center" of the feasible region, the point "farthest away" from all boundaries, in the sense that the product $\prod x_j$ is maximal.

We now briefly describe the polynomial algorithms that use these search directions. In general, path-following algorithms attempt to move along the barrier trajectory. The strategy is to choose an initial parameter $\mu = \mu_0$, find an approximate solution to the appropriate subproblem (P_μ in the primal case), reduce μ (usually by some specified fraction), and repeat. The best complexity bound on the number of iterations is currently $O(\sqrt{n}L)$ iterations, for "short-step" algorithms. In such methods the iterates are confined to a small region around the barrier trajectory, so that one iteration of Newton's method will typically suffice to obtain a "near solution" to P_μ. To prevent the iterates from straying too far from the trajectory, the barrier parameter μ can be reduced only by a small amount, typically by $(1 - \gamma/\sqrt{n})$, for some appropriate constant $\gamma > 0$. "Long-step" methods allow their iterates to lie in a wider neighborhood of the barrier trajectory and reduce the barrier parameter more rapidly. They may require more than one Newton iteration to obtain a "near solution" to P_μ. For such methods, the best complexity bound on the number of iterations is currently $O(nL)$ iterations.

To conclude this section, we present a variant of the short-step primal logarithmic barrier algorithm and prove that under appropriate assumptions it requires at most $O(\sqrt{n}L)$ iterations. A similar proof can be developed for the primal-dual version of the algorithm. The algorithm is a "theoretical algorithm"; details such as the step length and the strategy for reducing the barrier parameter must be modified to obtain a computationally efficient algorithm. The algorithm proceeds as follows.

Given a strictly feasible point x, a positive barrier parameter μ, and some tolerance $\epsilon > 0$, do the following:

1. Compute
$$y = (AX^2A^T)^{-1}AX^2(c - \mu X^{-1}e), \quad s = c - A^Ty.$$

2. If $s \geq 0$ and $x^Ts \leq \epsilon$, stop—a solution has been found.

3. Let
$$\Delta x = x - \frac{1}{\mu}X^2s.$$

Set $x_+ = x + \Delta x$. Set $\mu' = \theta\mu$, where $0 < \theta < 1$, and repeat.

To establish the polynomial complexity of the algorithm, we define a measure of proximity to the central trajectory. We show that, if a point is "sufficiently close" to the trajectory (using our measure of proximity), then the Newton step will lead to a feasible point with a guaranteed reduction in the duality gap. We also show that, if μ decreases by a factor of $(1 - \gamma/\sqrt{n})$ for an appropriate value of γ (here we use $\gamma = 1/6$), then the new point will also be "sufficiently close" to the trajectory.

What is a good measure of proximity to the barrier trajectory? For a given value of μ, x is "near" the trajectory if, for some dual slack vector s, the components of x_js_j are approximately equal to μ. Here "approximately equal" is defined relative to μ. This leads to the following measure of proximity at a point x:

$$\delta(x, \mu) = \min_s \left\{ \left\| \frac{Xs}{\mu} - e \right\| : A^Ty + s = c, \, y \in \Re^m \right\}.$$

The choice of norm in this definition is crucial. If the 2-norm is used, it is possible to obtain a complexity of $O(\sqrt{n}L)$ steps. However if a "looser" ∞-norm is used, then typically the algorithm has a complexity of $O(nL)$ iterations. (Despite this inferior complexity, however, algorithms that are based on the ∞-norm can have better performance.)

In the following we use the stricter 2-norm for our measure of proximity. The measure can then be written more conveniently as

$$\delta(x, \mu) = \left\| X^{-1} \Delta x \right\|$$

since

$$
\begin{aligned}
\delta(x, \mu) &= \min \left\{ \left\| \frac{Xs}{\mu} - e \right\| : A^T y + s = c, \, y \in \mathfrak{R}^m \right\} \\
&= \mu^{-1} \min \left\{ \|Xs - \mu e, \| : A^T y + s = c, \, y \in \mathfrak{R}^m \right\} \\
&= \mu^{-1} \min \left\{ \left\| X(c - A^T y) - \mu e \right\| : y \in \mathfrak{R}^m \right\} \\
&= \mu^{-1} \min \left\{ \left\| (Xc - \mu e) - XA^T y \right\| : y \in \mathfrak{R}^m \right\} \\
&= \mu^{-1} \left\| \bar{P}(Xc - \mu e) \right\| = \left\| X^{-1} \Delta x \right\|.
\end{aligned}
$$

The vectors y and s that solve the least-squares problem are the dual estimates at x obtained from computing Δx (see the Exercises).

We now prove that under appropriate assumptions the path-following algorithm described here has polynomial complexity. We shall analyze the properties of the algorithm in a sequence of lemmas and then combine them to prove the final complexity theorem.

Within the proof, we define a "scaled complementarity" vector

$$t = \frac{Xs}{\mu},$$

whose components are $t_j = x_j s_j / \mu$. Using this notation, the proximity measure can be expressed as

$$\delta(x, \mu) = \|t - e\| = \left[\sum_{j=1}^{n} (t_j - 1)^2 \right]^{1/2}.$$

We start by showing that if $\delta(x, \mu) < 1$, the points generated by applying Newton's method to solve P_μ are feasible, and their proximity measure decreases quadratically.

Lemma 10.5 (Quadratic Reduction in Proximity Measure). *Let x be strictly feasible for the primal problem, and let $x_+ = x + \Delta x$. If $\delta(x, \mu) < 1$, then x_+ is strictly feasible and $\delta(x_+, \mu) \leq \delta(x, \mu)^2$.*

Proof. First we prove that x_+ is feasible. Since $Ax = b$ and $A\Delta x = 0$, it follows that $Ax_+ = b$, so we need only show that $x_+ > 0$. To prove this we write $x_+ = x + \Delta x = X(e + X^{-1}\Delta x)$. Now by assumption $\delta(x, \mu) = \|X^{-1}\Delta x\| < 1$, and hence each component of $X^{-1}\Delta x$ is less than 1 in absolute value. It follows that $e + X^{-1}\Delta x > 0$, and consequently $x_+ > 0$.

To prove that $\delta(x_+, \mu) \leq \delta(x, \mu)^2$ we first note that

$$\delta(x_+, \mu) = \min_{\bar{s}} \left\{ \left\| \frac{X_+ \bar{s}}{\mu} - e \right\| : A^T \bar{y} + \bar{s} = c \right\} \leq \left\| \frac{X_+ s}{\mu} - e \right\|,$$

where $X_+ = \text{diag}(x_+)$ and s is the dual slack estimate at x. Using the relation

$$x_+ = x + \Delta x = x + x - \frac{1}{\mu}X^2 s = 2x - \frac{1}{\mu}X^2 s,$$

and the relation $Xs = Sx$ where $S = \text{diag}(s)$, we obtain

$$\frac{X_+ s}{\mu} = \frac{Sx_+}{\mu} = \frac{2Sx}{\mu} - \frac{SX^2 s}{\mu^2} = \frac{2Xs}{\mu} - \frac{(XS)(Xs)}{\mu^2}.$$

Let $t = Xs/\mu$ and $T = \text{diag}(t) = XS/\mu$. Then

$$\frac{X_+ s}{\mu} - e = 2t - T^2 e - e.$$

Therefore

$$\delta(x_+, \mu)^2 \le \sum_{j=1}^{n}(2t_j - t_j^2 - 1)^2 = \sum_{j=1}^{n}(t_j - 1)^4 \le \left[\sum_{j=1}^{n}(t_j - 1)^2\right]^2 = \delta(x, \mu)^4,$$

and consequently $\delta(x_+, \mu) < \delta(x, \mu)^2$. □

Thus, if $\delta(x, \mu) < 1$, then the point x is "close" to the minimizer of P_μ, and each iteration of Newton's method will decrease the proximity measure at least quadratically. The next lemma gives a bound on the proximity measure of x with respect to a new (reduced) value of the barrier parameter μ'.

Lemma 10.6 (Proximity of x for Reduced μ). *Let $\mu' = \theta\mu$ with $0 < \theta \le 1$. Then*

$$\delta(x, \mu') \le \frac{1}{\theta}\left(\delta(x, \mu) + (1 - \theta)\sqrt{n}\right).$$

Proof. From the definition of the proximity measure we have

$$\delta(x, \mu') = \min_{\bar{s}}\left\{\left\|\frac{X\bar{s}}{\mu'} - e\right\| : A^T\bar{y} + \bar{s} = c\right\} \le \left\|\frac{Xs}{\mu'} - e\right\| = \left\|\frac{t}{\theta} - e\right\|,$$

where s is the dual slack estimate at x for barrier parameter μ, and $t = Xs/\mu$. Applying the triangle inequality, we obtain

$$\delta(x, \mu') \le \frac{1}{\theta}\|(t - e) + (1 - \theta)e\| \le \frac{1}{\theta}\|t - e\| + (1 - \theta)\|e\|$$

$$\le \frac{1}{\theta}\left(\delta(x, \mu) + (1 - \theta)\sqrt{n}\right). □$$

We now combine the results of the previous two lemmas to obtain a bound on the proximity measure of a point obtained by a Newton step and for a new value of the barrier parameter. In particular, if $\delta(x, \mu) \le \frac{1}{2}$, and the reduction in μ is sufficiently conservative (that is, θ is sufficiently close to 1), then $\delta(x_+, \mu') \le \frac{1}{2}$ also.

Lemma 10.7 (Bounded Proximity Measure). *Let x be strictly feasible for the primal problem, and suppose that $\delta(x, \mu) \leq \frac{1}{2}$. Let $x_+ = x + \Delta x$, and suppose that $\mu' = \theta\mu$. If $\theta \geq (1 - 1/(6\sqrt{n}))$, then $\delta(x_+, \mu') \leq \frac{1}{2}$.*

Proof. Applying Lemmas 10.5 and 10.6 successively, we obtain

$$\delta(x_+, \mu') \leq \frac{1}{\theta}\left(\delta(x_+, \mu) + (1 - \theta)\sqrt{n}\right)$$

$$\leq \frac{1}{\theta}\left(\delta(x, \mu)^2 + (1 - \theta)\sqrt{n}\right)$$

$$\leq \frac{1}{\theta}\left(\frac{1}{4} + \frac{1}{6}\right) = \frac{5}{12\theta} \leq \frac{1}{2}.$$

The last inequality follows since $\theta \geq 1 - 1/6\sqrt{n} \geq \frac{5}{6}$.   □

From the preceding lemmas we can conclude that if we have a strictly feasible point x_0 and a barrier parameter μ_0 satisfying $\delta(x_0, \mu_0) \leq \frac{1}{2}$, then the sequence of iterates (x_k, μ_k) obtained by repeatedly taking a single Newton step, and reducing μ by a factor of $(1 - 1/(6\sqrt{n}))$, is strictly feasible and maintains a proximity measure $\delta(x_k, \mu_k) \leq \frac{1}{2}$. Will this sequence converge to an optimal solution as μ goes to zero? The next lemma will help answer this question. It provides bounds on the duality gap at a point x in terms of μ and $\delta(x, \mu)$.

Lemma 10.8 (Bounded Duality Gap). *Let x be strictly feasible for the primal problem, and let (y, s) be the dual estimates at x with respect to μ. If $\delta(x, \mu) \leq \delta \leq 1$, then (y, s) is dual feasible, and*

$$\mu(n - \delta\sqrt{n}) \leq c^T x - b^T y \leq \mu(n + \delta\sqrt{n}).$$

Proof. By definition, $A^T y + s = c$, and hence we need only show that $s \geq 0$. Now by assumption,

$$\delta(x, \mu) = \left\|\frac{Xs}{\mu} - e\right\| \leq 1,$$

and hence each component of $\frac{Xs}{\mu} - e$ is at most 1 in absolute value. This implies that $x_j s_j \geq 0$ for all j, and since $x > 0$, it follows that $s \geq 0$.

Because x is feasible to the primal and (y, s) is feasible to the dual, $c^T x - b^T y = x^T s$. Now

$$\Delta x = x - \frac{1}{\mu}X^2 s = Xe - \frac{1}{\mu}X^2 s,$$

and hence

$$s = \mu X^{-1}(e - X^{-1}\Delta x).$$

Consequently, the duality gap is

$$x^T s = \mu x^T X^{-1}(e - X^{-1}\Delta x) = \mu e^T(e - X^{-1}\Delta x) = \mu(n - e^T X^{-1}\Delta x),$$

so that

$$\mu\left(n - \|e\|\,\|X^{-1}\Delta x\|\right) \leq x^T s \leq \mu\left(n + \|e\|\,\|X^{-1}\Delta x\|\right).$$

Since $\left\| X^{-1} \Delta x \right\| = \delta(x, \mu) \leq \delta$, we obtain the desired result:

$$\mu(n - \delta\sqrt{n}) \leq x^T s \leq \mu(n + \delta\sqrt{n}). \qquad \Box$$

It follows from the lemma that as the barrier parameter goes to zero, the iterates converge to the optimal solution. The final theorem gives an upper bound on the number of iterations required by the algorithm.

Theorem 10.9 (Complexity of the Short-Step Algorithm). *Assume that the short-step algorithm is initialized with x_0 and $\mu_0 > 0$, so that $\delta(x_0, \mu_0) < \frac{1}{2}$. Assume that at each iteration a single Newton step is taken and that the barrier parameter is updated as $\mu_{k+1} = \theta\mu_k$, where $\theta = (1 - 1/(6\sqrt{n}))$. Then the number of iterations required to find a solution with a duality gap of at most ϵ is bounded above by $6\sqrt{n}M$, where $M = \log(1.5n\mu_0/\epsilon)$.*

Proof. After the kth iteration we will have $\mu_k = \theta^k \mu_0$. Let x be the point obtained and y the corresponding dual estimate. The previous lemmas imply that x is primal feasible, y is dual feasible, and

$$c^T x - b^T y \leq \mu_k \left(n + \delta(x, \mu_k)\sqrt{n} \right) \leq \mu_k 1.5n = 1.5n\mu_0 \left(1 - 1/(6\sqrt{n}) \right)^k.$$

Thus, the algorithm will have terminated if

$$1.5n\mu_0 \left(1 - 1/(6\sqrt{n}) \right)^k \leq \epsilon,$$

or equivalently, if

$$k \log \left(1 - 1/(6\sqrt{n}) \right) \leq -\log(1.5n\mu_0/\epsilon).$$

This condition implies that

$$-k \log \left(1 - 1/(6\sqrt{n}) \right) \geq M.$$

Since $-\log(1 - \alpha) \geq \alpha$ for all $0 < \alpha < 1$ (see the Exercises), this inequality will certainly hold if $k/6\sqrt{n} \geq M$, that is, if $k \geq 6\sqrt{n}M$. Thus we have the required bound on the number of iterations. $\quad \Box$

It can be shown (see Papadimitriou and Steiglitz (1982, reprinted 1998)) that if the objective values of a primal feasible solution and a dual feasible solution differ by less than 2^{-2L}, then the objective values must be equal, so the solutions must be optimal to the primal and dual problems, respectively. Consequently, if $s_k^T x_k < 2^{-2L}$, then the optimal solution has been found. Assuming that $\mu_0 = 2^{O(L)}$, we obtain that $M = O(L)$, and hence the number of iterations required for termination is at most $O(\sqrt{n}L)$. (For this choice of μ_0, initialization procedures for the algorithm exist.)

Exercises

6.1. Write down the first-order optimality conditions for the dual logarithmic barrier problem. Prove that they are equivalent to the perturbed optimality conditions solved by the primal-dual algorithm.

6.2. Prove that the centering direction is the solution to the problem

$$\text{minimize} \quad f(x) = -\sum_{j=1}^{n} \log x_j$$

$$\text{subject to} \quad Ax = b.$$

6.3. Note that it is also possible to define a dual logarithmic barrier method—a barrier method that operates on the dual problem. The method solves a sequence of problems

$$\underset{y,s}{\text{maximize}} \quad f(y, s) = b^T y + \mu \sum_{j=1}^{n} \log(s_j)$$

$$\text{subject to} \quad A^T y + s = c$$

for a decreasing sequence of barrier parameters. Do the following:

   (i) Derive the first-order optimality conditions for this problem and prove that the points satisfying these conditions are on the primal-dual central path.

   (ii) The dual method moves in the projected Newton direction for this problem. Prove that this direction is given by

$$\Delta y = \frac{1}{\mu}(AS^{-2}A^T)^{-1}b - (AS^{-2}A^T)^{-1}AS^{-1}e,$$

   where $\Delta s = -A^T \Delta y$.

6.4. The analytic center for the dual problem is the point that solves

$$\text{minimize } f(y) = -\sum_{j=1}^{n} \log(c - A^T y)_j.$$

Find an expression for the Newton direction at a point y for the problem of finding the analytic center of the dual. This direction is called the *dual centering* direction. Prove that the projected Newton direction for the dual logarithmic barrier method is a combination of the dual affine direction and the dual centering direction.

6.5. Prove that $-\log(1 - \alpha) \geq \alpha$ for all $0 < \alpha < 1$.

10.7   Notes

Interior-Point Methods—The book by Fiacco and McCormick (1968, reprinted 1990) is a "classical" reference to interior-point methods for nonlinear optimization. See also our Chapter 16. The books by den Hertog (1994), Roos et al. (2005), Vanderbei (2007), Wright (1997), and Ye (1997) provide comprehensive overviews of interior-point methods for linear programming. The developments in interior-point methods for linear programming have been extended to wider classes of convex programming. See Sections 16.7 and 16.8.

Karmarkar's Method—In his original paper Karmarkar considered a primal problem in standard form with an additional normalizing constraint $e^T x = 1$. He showed that all linear programs can be transformed into this specific form, but the proposed transformation results in a much larger linear program. Variants of the method that are suitable for problems in standard form were proposed by Anstreicher (1986), Gay (1987), and Ye and Kojima (1987). A description of Karmarkar's method is available on the Web page for this book at http://www.siam.org/books/ot108.

Path-Following Methods—The first path-following method, which was also the first $O(\sqrt{n}L)$ algorithm for linear programming, was developed by Renegar (1988). The development of *primal-dual* path-following methods was motivated by the 1986 paper by Megiddo. The first polynomial primal-dual path-following algorithms were proposed by Kojima, Mizuno, and Yoshise (1989) and Monteiro and Adler (1989). Kojima, Mizuno, and Yoshise used an ∞-norm proximity measure to obtain an $O(nL)$ algorithm, while Monteiro and Adler used a 2-norm measure to obtain an $O(\sqrt{n}L)$ algorithm. The particular primal-dual method presented in Section 10.2 is based on the paper by Monteiro and Adler. The proof of polynomiality for the primal path-following method presented in Section 10.6 follows the article by Roos and Vial (1992).

The predictor-corrector method is described in the paper of Mehrotra (1992). The convergence behavior of the method in degenerate cases is discussed in the paper by Güler and Ye (1993). Many software packages for linear programming include an enhancement of the method proposed by Gondzio that allows efficient use of higher-order predictor and corrector terms.

A polynomial predictor-corrector primal-dual path-following algorithm is given in the paper by Mizuno, Todd, and Ye (1993). If proximity is measured using a 2-norm, the algorithm has a complexity bound of $O(\sqrt{n}L)$ iterations. A predictor-corrector algorithm based on an ∞-norm is given in the paper by Anstreicher and Bosch (1995). The algorithm requires at most $O(L)$ "predictor" steps, and each of those requires at most $O(n)$ "corrector" or centering steps, so that the algorithm requires at most $O(nL)$ steps.

The paper by Zhang and Tapia (1992) shows that the centering parameter and the step length in a primal-dual path-following method can be chosen so that both polynomiality and superlinear convergence are achieved. Further, if the solution is nondegenerate, the rate of convergence is quadratic. (When referring to "convergence rates" we assume that the method takes an infinite number of steps to converge.) The algorithm asymptotically uses the affine-scaling direction (the centering parameter—the coefficient of the centering direction—tends to zero) and allows the iterates to be close to the boundary. These features have also been observed to give good practical performance.

Computational Issues—A discussion of computational issues for interior-point methods can be found in the papers by Lustig, Marsten, and Shanno (1992, 1994a). Further developments are in the 1996 paper by Andersen et al.

Self-Dual Formulations—The concept of a self-dual linear program was introduced by Tucker (1956). The idea of embedding a linear program in a self-dual problem was proposed in the paper by Ye, Todd, and Mizuno (1994) and, in a simplified form, in the paper by Xu, Hung, and Ye (1993).

Affine-Scaling Methods—The primal affine-scaling method was proposed by Barnes (1986) and Vanderbei, Meketon, and Freedman (1986). Subsequently it was found that the method had already been proposed 12 years earlier by Dikin (1974), a student of Kantorovich.

Part III

Unconstrained Optimization

Chapter 11

Basics of Unconstrained Optimization

11.1   Introduction

In this chapter we begin studying the problem

$$\text{minimize } f(x),$$

where no constraints are placed on the variables $x = (x_1, \ldots, x_n)^T$. Unconstrained problems arise, for example, in data fitting (see Section 1.5), where the objective function measures the difference between the model and the data. Methods for unconstrained problems are of more general value, though, since they form the foundation for methods used to solve constrained optimization problems.

We will derive several optimality conditions for the unconstrained optimization problem. One of these conditions, the "first-order necessary condition," consists of a system of nonlinear equations. Applying Newton's method to this system of equations will be our fundamental technique for solving unconstrained optimization problems.

When started "close" to a solution, Newton's method converges rapidly. At an arbitrary starting point, however, Newton's method is not guaranteed to converge to a minimizer of the function f and must be refined before an acceptable algorithm can be obtained. Such refinements are described in the latter part of the chapter. These refinements can be used to ensure that Newton's method as well as other optimization methods converge from any starting point.

11.2   Optimality Conditions

We will derive conditions that are satisfied by solutions to the problem

$$\text{minimize } f(x).$$

The conditions for the problem

$$\text{maximize } f(x)$$

are analogous and will be mentioned in passing.

Let x_* denote a candidate solution to the minimization problem. In Chapter 2 we defined global solutions to optimization problems. The definition of a global optimum does not have much computational utility since it requires information about the function at every point, whereas the algorithms in common use will only have information about the function at a finite set of points. Even if the global minimizer x_* were given to us, it would be difficult or impossible to confirm that it was indeed the global minimizer. (See the Notes in Chapter 2.)

It is easier to look for local minimizers. A *local minimizer* is a point x_* that satisfies the condition

$$f(x_*) \le f(x) \quad \text{for all } x \text{ such that } \|x - x_*\| < \epsilon,$$

where ϵ is some (typically small) positive number whose value may depend on x_*. Similarly defined is a *strict local minimizer*:

$$f(x_*) < f(x) \quad \text{for all } x \text{ such that } 0 < \|x - x_*\| < \epsilon.$$

It is possible for a function to have a local minimizer and yet have no global minimizer. It is also possible to have neither global nor local minimizers, to have both global and local minimizers, to have multiple global minimizers, and various other combinations. (See the Exercises.)

In this form, these conditions are no more practical than those for a global minimizer, since they too require information about the function at an infinite number of points, and the algorithms will only have information at a finite number of points. However, with additional assumptions on the function f, practical optimality conditions can be obtained.

To obtain more practical conditions, we assume that the function f is differentiable and that its first and second derivatives are continuous in a neighborhood of the point x_*. Not all the conditions that we derive will require this many derivatives, but it will simplify the discussion if the assumptions do not change from condition to condition. (A more precise discussion can be found in the book by Ortega and Rheinboldt (1970, reprinted 2000).) All of these conditions will be derived using Taylor series expansions of f about the point x_*.

Suppose that x_* is a local minimizer of f. Consider the Taylor series with remainder term (see Section 2.6)

$$f(x_* + p) = f(x_*) + \nabla f(x_*)^T p + \tfrac{1}{2} p^T \nabla^2 f(\xi) p,$$

where p is a nonzero vector and ξ is a point between x and x_*. We will show that $\nabla f(x_*) = 0$. If x_* is a local minimizer, there can be no feasible descent directions at x_* (see Section 2.2). Hence

$$\nabla f(x_*)^T p \ge 0 \quad \text{for all feasible directions } p.$$

For an unconstrained problem, all directions p are feasible, and so the gradient at x_* must be zero; see the Exercises. Thus, if x_* is a local minimizer of f, then

$$\nabla f(x_*) = 0.$$

A point satisfying this condition is a *stationary point* of the function f.

In the one-dimensional case, there is a geometric interpretation for this condition. If f is increasing at a point x, then $f'(x) > 0$. Similarly if f is decreasing, then $f'(x) < 0$.

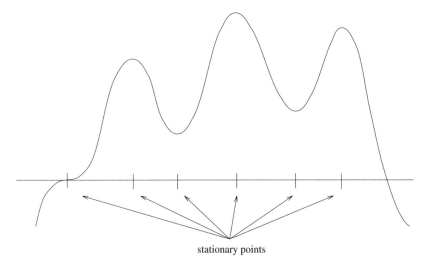

stationary points

Figure 11.1. *Stationary points.*

A point where f is increasing or decreasing cannot correspond to a minimizer. At a minimizer the function will be flat or stationary, and hence $f'(x_*) = 0$. This is illustrated in Figure 11.1.

The condition $\nabla f(x_*) = 0$ is referred to as the *first-order necessary condition* for a minimizer. The term "first-order" refers to the presence of the first derivatives of f (or to the use of the first-order term in the Taylor series to derive this condition). It is a "necessary" condition since if x_* is a local minimizer, then it "necessarily" satisfies this condition. The condition is not "sufficient" to determine a local minimizer since a point satisfying $\nabla f(x_*) = 0$ could be a local minimizer, a local maximizer, or a saddle point (a stationary point that is neither a minimizer nor a maximizer).

Local minimizers can be distinguished from other stationary points by examining second derivatives. Consider again the Taylor series expansion at $x = x_* + p$, but now using the result that $\nabla f(x_*) = 0$:

$$f(x) = f(x_* + p) = f(x_*) + \tfrac{1}{2}p^T\nabla^2 f(\xi)p.$$

We will show that $\nabla^2 f(x_*)$ must be positive semidefinite. If not, then $v^T\nabla^2 f(x_*)v < 0$ for some v. Then it is also true that $v^T\nabla^2 f(\xi)v < 0$ if $\|\xi - x_*\|$ is small. This is because $\nabla^2 f$ is assumed to be continuous at x_*. If p is chosen as some sufficiently small multiple of v, then the point ξ will be close enough to x_* to guarantee (via the Taylor series) that $f(x) < f(x_*)$, a contradiction. Hence if x_* is a local minimizer, then $\nabla^2 f(x_*)$ is positive semidefinite. This is referred to as the *second-order necessary condition* for a minimizer, with the "second-order" referring to the use of second derivatives or the second-order term in the Taylor series.

There is also a *second-order sufficient condition*, "sufficient" to guarantee that x_* is a local minimizer: If

$$\nabla f(x_*) = 0 \quad \text{and} \quad \nabla^2 f(x_*) \text{ is positive definite,}$$

then x_* is a *strict* local minimizer of f. If this condition is satisfied, then it is easy to modify the above argument to show that $f(x) = f(x_* + p) > f(x_*)$ for all $0 < \|p\| < \epsilon$ for some $\epsilon > 0$ as follows: We write down the Taylor series expansion about the point x_*, taking into account that $\nabla f(x_*) = 0$:

$$f(x) = f(x_*) + \tfrac{1}{2} p^T \nabla^2 f(\xi) p.$$

If $\nabla^2 f(x_*)$ is positive definite and $\nabla^2 f$ is continuous, then $\nabla^2 f(\xi)$ will also be positive definite if $\|\xi - x_*\|$ is sufficiently small. Since $\|\xi - x_*\| \le \|p\|$ we can choose ϵ small enough to guarantee this. Hence the second term in the Taylor series will be positive and so $f(x) > f(x_*)$, as desired.

So far we have discussed only minimization problems. There is no fundamental difference between minimization and maximization problems because

$$\max f(x) = -\min -f(x).$$

As a result, the optimality conditions for a maximizer are analogous to those for a minimizer. The necessary conditions state that if x_* is a local maximizer, then

$$\nabla f(x_*) = 0 \quad \text{and} \quad \nabla^2 f(x_*) \text{ is negative semidefinite.}$$

The sufficient conditions state that if

$$\nabla f(x_*) = 0 \quad \text{and} \quad \nabla^2 f(x_*) \text{ is negative definite,}$$

then x_* is a strict local maximizer. These optimality conditions are derived in the Exercises.

Example 11.1 (Optimality Conditions). Consider the function

$$f(x_1, x_2) = \tfrac{1}{3} x_1^3 + \tfrac{1}{2} x_1^2 + 2 x_1 x_2 + \tfrac{1}{2} x_2^2 - x_2 + 9.$$

The condition for a stationary point is

$$\nabla f(x) = \begin{pmatrix} x_1^2 + x_1 + 2x_2 \\ 2x_1 + x_2 - 1 \end{pmatrix} = 0.$$

The second component of this condition shows that $x_2 = 1 - 2x_1$, and if this is substituted into the first component, we obtain

$$x_1^2 - 3x_1 + 2 = 0 \quad \text{or} \quad (x_1 - 1)(x_1 - 2) = 0.$$

Hence there are two stationary points:

$$x_a = \begin{pmatrix} 1 \\ -1 \end{pmatrix} \quad \text{and} \quad x_b = \begin{pmatrix} 2 \\ -3 \end{pmatrix}.$$

The Hessian matrix for the function is

$$\nabla^2 f(x) = \begin{pmatrix} 2x_1 + 1 & 2 \\ 2 & 1 \end{pmatrix},$$

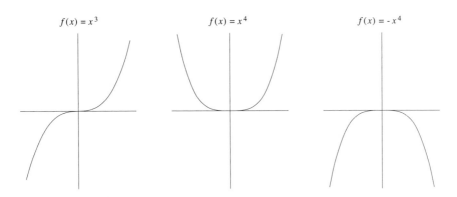

Figure 11.2. *Limitations of optimality conditions.*

so

$$\nabla^2 f(x_a) = \begin{pmatrix} 3 & 2 \\ 2 & 1 \end{pmatrix} \quad \text{and} \quad \nabla^2 f(x_b) = \begin{pmatrix} 5 & 2 \\ 2 & 1 \end{pmatrix}.$$

$\nabla^2 f(x_b)$ is positive definite, so x_b is a local minimizer. However, $\nabla^2 f(x_a)$ is indefinite, and x_a is neither a minimizer nor a maximizer of f. This function has neither a global minimizer nor a global maximizer, since f is unbounded as $x_1 \to \pm\infty$. ∎

There is a slight gap between the necessary and sufficient conditions for a minimizer, the case where $\nabla f(x_*) = 0$ and $\nabla^2 f(x_*)$ is positive *semidefinite*. This gap represents a limitation of these conditions, as can be seen by considering the one-dimensional functions $f_1(x) = x^3$, $f_2(x) = x^4$, and $f_3(x) = -x^4$. All three functions satisfy $f'(0) = f''(0) = 0$, and so $x_* = 0$ is a candidate for a local minimizer. However, while f_2 has a local minimum at $x_* = 0$, f_1 has only an inflection point, and f_3 has a local maximum. This is illustrated in Figure 11.2. More complicated conditions involving higher derivatives are required to fill this gap between the necessary and sufficient conditions.

The conditions given here require that $\nabla f(x_*)$ be exactly zero. For computer calculations this will almost never be true, and so these conditions must be adapted in a computer algorithm. This topic is discussed in Section 12.5.

Exercises

2.1. Consider the following function

$$f(x) = 15 - 12x - 25x^2 + 2x^3.$$

(i) Use the first and second derivatives to find the local maxima and local minima of f.

(ii) Show that f has neither a global maximum nor a global minimum.

2.2. Consider the function

$$f(x) = 3x^3 + 7x^2 - 15x - 3.$$

Find all stationary points of this function and determine whether they are local minimizers and maximizers. Does this function have a global minimizer or a global maximizer?

2.3. Consider the function

$$f(x_1, x_2) = 8x_1^2 + 3x_1x_2 + 7x_2^2 - 25x_1 + 31x_2 - 29.$$

Find all stationary points of this function and determine whether they are local minimizers and maximizers. Does this function have a global minimizer or a global maximizer?

2.4. Find the global minimizer of the function

$$f(x_1, x_2) = x_1^2 + x_1x_2 + 1.5x_2^2 - 2\log x_1 - \log x_2.$$

2.5. Determine the minimizers/maximizers of the following functions:

(i) $f(x_1, x_2) = x_1^4 + x_2^4 - 4x_1x_2$.
(ii) $f(x_1, x_2) = x_1^2 - 2x_1x_2^2 + x_2^4 - x_2^5$.
(iii) $f(x_1, x_2, x_3) = x_1^2 + 2x_2^2 + 5x_3^2 - 2x_1x_2 - 4x_2x_3 - 2x_3$.

2.6. Find all the values of the parameter a such that $(1, 0)^T$ is the minimizer or maximizer of the function

$$f(x_1, x_2) = a^3 x_1 e^{x_2} + 2a^2 \log(x_1 + x_2) - (a+2)x_1 + 8ax_2 + 16x_1x_2.$$

2.7. Consider the problem

$$\text{minimize } f(x_1, x_2) = (x_2 - x_1^2)(x_2 - 2x_1^2).$$

(i) Show that the first- and second-order necessary conditions for optimality are satisfied at $(0, 0)^T$.
(ii) Show that the origin is a local minimizer of f along any line passing through the origin (that is, $x_2 = mx_1$).
(iii) Show that the origin is not a local minimizer of f (consider, for example, curves of the form $x_2 = kx_1^2$). What conclusions can you draw from this?

2.8. Consider the problem

$$\text{minimize } f(x) = (x_1 - 2x_2)^2 + x_1^4.$$

Find the minimizer of f. Determine that the second-order necessary condition for a local minimizer is satisfied at this point. Is the second-order sufficient condition satisfied? Is this point a strict local minimizer? Is it a global minimizer?

2.9. Let
$$f(x) = 2x_1^2 + x_2^2 - 2x_1x_2 + 2x_1^3 + x_1^4.$$
Determine the minimizers/maximizers of f and indicate what kind of minima or maxima (local, global, strict, etc.) they are.

2.10. Let
$$f(x) = cx_1^2 + x_2^2 - 2x_1x_2 - 2x_2,$$
where c is some scalar.

(i) Determine the stationary points of f for each value of c.

(ii) For what values of c can f have a minimizer? For what values of c can f have a maximizer? Determine the minimizers/maximizers corresponding to such values of c and indicate what kind of minima or maxima (local, global, strict, etc.) they are.

2.11. Consider the following unconstrained problem:
$$\text{minimize } f(x) = x_1^2 - x_1x_2 + 2x_2^2 - 2x_1 + e^{x_1+x_2}.$$

(i) Write down the first-order necessary conditions for optimality.

(ii) Is $x = (0, 0)^T$ a local optimum? If not, find a direction p along which the function decreases.

(iii) Attempt to minimize the function starting from $x = (0, 0)^T$ along the direction p that you have chosen in part (ii). [*Hint*: Consider $F(\alpha) = f(x + \alpha p)$.]

2.12. Consider the following problem:
$$\text{minimize } f(x) = (x_1 - 2)^2 + (x_2 - 3)^2 + 1.$$

Solve this problem. Consider now the problems below: Do they all have the same optimal point? If not, explain why not.

(i) minimize $f(x) = \sqrt{(x_1 - 2)^2 + (x_2 - 3)^2 + 1}$.

(ii) minimize $f(x) = (x_1 - 2)^2 + (x_2 - 3)^2$.

(iii) minimize $f(x) = \sqrt{(x_1 - 2)^2 + (x_2 - 3)^2}$.

2.13. Consider the quadratic function
$$f(x) = \tfrac{1}{2}x^T Q x - c^T x.$$

(i) Write the first-order necessary condition. When does a stationary point exist?

(ii) Under what conditions on Q does a local minimizer exist?

(iii) Under what conditions on Q does f have a stationary point, but no local minima nor maxima?

2.14. Consider the problem
$$\text{minimize } f(x) = \|Ax - b\|_2^2,$$
where A is an $m \times n$ matrix with $m \geq n$, and b is a vector of length m. Assume that the rank of A is equal to n.

      (i) Write down the first-order necessary condition for optimality. Is this also a sufficient condition?

     (ii) Write down the optimal solution in closed form.

2.15. Give examples of functions that have the following properties:

      (i) f has a local minimizer but no global minimizer.

     (ii) f has neither global nor local minimizers.

    (iii) f has both global and local minimizers.

    (iv) f has multiple global minimizers.

2.16. Give an example of a differentiable function on $\Re^2$ which has infinitely many minimizers but not a single maximizer.

2.17. Give an example of a differentiable function on $\Re^2$ which has just one stationary point: the local but not the global minimizer.

2.18. Define the terms *global maximizer*, *strict global maximizer*, *local maximizer*, and *strict local maximizer* in analogy with the corresponding terms for minimizers.

2.19. State and prove the first-order necessary condition for a local maximizer of a function.

2.20. State and prove the second-order necessary condition for a local maximizer of a function.

2.21. State and prove the second-order sufficient condition for a local maximizer of a function.

2.22. Prove that, if f is convex, then any stationary point is also a global minimizer.

2.23. If x_* is a local minimizer of a function f, then

$$\nabla f(x_*)^T p \geq 0 \quad \text{for all feasible directions } p.$$

Prove that, for an unconstrained problem, the only way that this condition can be satisfied is if the gradient at x_* is zero.

11.3   Newton's Method for Minimization

In this section we present Newton's method in its most basic or "classical" form. In later sections we will show how the method can be adjusted to guarantee that the search directions are descent directions, to guarantee convergence, and to lower the costs of the method.

    As presented in Chapter 2, Newton's method is an algorithm for finding a zero of a nonlinear function. To use Newton's method for optimization, we apply it to the first-order necessary condition for a local minimizer:

$$\nabla f(x) = 0.$$

Since the Jacobian of $\nabla f(x)$ is $\nabla^2 f(x)$, this leads to the formula

$$x_{k+1} = x_k - [\nabla^2 f(x_k)]^{-1} \nabla f(x_k).$$

Note: $\alpha_k = 1$, this is the classical version.

Newton's method is often written as $x_{k+1} = x_k + p_k$, where p_k is the solution to the *Newton equations*:

$$[\nabla^2 f(x_k)]p = -\nabla f(x_k).$$

This emphasizes that the step p_k is usually obtained by solving a linear system of equations rather than by computing the inverse of the Hessian.

Newton's method was derived in Chapter 2 by finding a linear approximation to a nonlinear function via the Taylor series. The formula for Newton's method represents a step to a zero of this linear approximation. For the nonlinear equation $\nabla f(x) = 0$ this linear approximation is

$$\nabla f(x_k + p) \approx \nabla f(x_k) + \nabla^2 f(x_k)p.$$

The linear approximation is the gradient of the quadratic function

$$q_k(p) \equiv f(x_k) + \nabla f(x_k)^T p + \tfrac{1}{2} p^T \nabla^2 f(x_k) p.$$

$q_k(p)$ corresponds to the first three terms of a Taylor series expansion for f about the point x_k.

The quadratic function q_k provides a new interpretation of Newton's method for minimizing f. At every iteration Newton's method approximates $f(x)$ by $q_k(p)$, the first three terms of its Taylor series about the point x_k; minimizes q_k as a function of p; and then sets $x_{k+1} = x_k + p$. Hence at each iteration we are approximating the nonlinear function by a quadratic model. It is this point of view that we shall prefer.

As might be expected, Newton's method has a quadratic rate of convergence except in "degenerate" cases; it can sometimes diverge or fail. If Newton's method converges, it will converge to a stationary point. In the form that we have presented it (the "classical" Newton formula) there is nothing in the algorithm to bias it towards finding a minimum, although that topic will be discussed in Section 11.4.

Newton's method is rarely used in its classical form. The method is altered in two general ways: to make it more reliable and to make it less expensive. We have already seen that Newton's method can diverge or fail, and even if it does converge, it might not converge to a minimizer. By embedding Newton's method inside some sort of auxiliary strategy it will be possible to guarantee that the method will converge to a stationary point and possibly a local minimizer, if one exists. One approach is to use the Newton direction within our general optimization algorithm (see Section 2.4), so that the new point is defined as

$$x_{k+1} = x_k + \alpha_k p_k, \qquad \textit{(The choice of α_k can force descent)}$$

where α_k is a scalar chosen so that $f(x_{k+1}) < f(x_k)$. (In the classical Newton's method, $\alpha_k = 1$ at every iteration, and there is no guarantee that the function value is decreased.)

There are three types of costs associated with using Newton's method: derivatives, calculations, and storage. In its classical form, Newton's method requires second derivatives, the solution of a linear system, and the storage of a matrix. For an n-variable problem, there are $O(n^2)$ entries in the Hessian matrix, meaning that $O(n^2)$ expressions must be programmed to compute these derivatives. Many people find it tedious to derive and program these formulas; it is easy to make errors that can cause the optimization algorithm to perform poorly or even fail. Once the Hessian matrix has been found, it costs $O(n^3)$ arithmetic operations to solve the linear system in the Newton formula. Also, normally the Hessian matrix will have to be stored at a cost of $O(n^2)$ storage locations. As n increases, these costs grow rapidly.

Some of these concerns can be ameliorated. For example, it is possible to automate the derivative calculations (see Section 12.4). Also, many large problems have sparse Hessian matrices, and the use of sparse matrix techniques can reduce the storage and computational costs of using Newton's method (see Appendix A.6).

Alternatively, it is possible to reduce these costs by using algorithms that compromise on Newton's method. Virtually all of these algorithms get by with only first derivative calculations. Most of these algorithms avoid solving a linear system and reduce the cost of using the Newton formula to $O(n^2)$ or less. The methods designed for solving large problems reduce the storage requirements to $O(n)$. Some of these compromises will be discussed in Chapters 12 and 13.

These compromises do not come without penalties. The resulting algorithms have slower rates of convergence and tend to use more, but cheaper, iterations to solve problems.

Since Newton's method is almost never used in its classical form, why is this classical form presented here with such prominence? The reason is that Newton's method represents an "ideal" method for solving minimization problems. It may sometimes fail, and it may be too expensive to use routinely. Other algorithms strive to overcome its deficiencies while retaining its good properties, in particular, while retaining as rapid a rate of convergence as possible. It can be confusing to study all the various methods that have been proposed for solving unconstrained minimization problems. However, if it is remembered that virtually all of them are compromises on Newton's method, then the relationships among the methods become clearer, and their relative merits become easier to understand.

If Newton's method is used, the theorem below shows that under appropriate conditions the convergence rate will be quadratic.

Theorem 11.2 (Quadratic Convergence of Newton's Method). *Let f be a real-valued function of n variables defined on an <u>open convex set</u> S. Assume that $\nabla^2 f$ is Lipschitz continuous on S, that is,*

$$\left\| \nabla^2 f(x) - \nabla^2 f(y) \right\| \le L \left\| x - y \right\|$$

for all $x, y \in S$ and for some constant $L < \infty$. Consider the sequence $\{ x_k \}$ generated by

$$x_{k+1} = x_k - [\nabla^2 f(x_k)]^{-1} \nabla f(x_k).$$

Let x_* be a minimizer of $f(x)$ in S and assume that $\nabla^2 f(x_*)$ is positive definite. If $\|x_0 - x_*\|$ is sufficiently small, then $\{ x_k \}$ converges quadratically to x_*.

Proof. See the Exercises.   □

If a compromise to Newton's method is used, we cannot normally expect to achieve such a rapid rate of convergence. It is still possible, however, to achieve superlinear convergence. This is the topic of the next theorem. The theorem shows that, to achieve superlinear convergence, the search direction must approach the Newton direction in the limit as the solution is approached.

The theorem implicitly assumes that the Newton direction is defined at every iteration, that is, $\nabla^2 f(x_k)$ is nonsingular for every k. This is not an essential assumption. The conclusion of the theorem is only of interest in the limit as x_* is approached. Since $\nabla^2 f(x_*)$ is assumed to be positive definite, the continuity of $\nabla^2 f$ guarantees that $\nabla^2 f(x_k)$ will be positive definite for all sufficiently large values of k.

Theorem 11.3 (Superlinear Convergence). *Let f be a real-valued function of n variables defined on an open convex set S. Assume that $\nabla^2 f$ is Lipschitz continuous on S, that is,*

$$\left\| \nabla^2 f(x) - \nabla^2 f(y) \right\| \le L \left\| x - y \right\|$$

for all $x, y \in S$ and for some constant $L < \infty$. Consider the sequence $\{x_k\}$ generated by

$$x_{k+1} = x_k + p_k.$$

Suppose that $\{x_k\} \subset S$,

$$\lim_{k \to \infty} x_k = x_* \in S,$$

and that $x_k \ne x_*$ for all k. Also suppose that $\nabla^2 f(x_*)$ is positive definite. Then $\{x_k\}$ converges to x_* superlinearly and $\nabla f(x_*) = 0$ if and only if

$$\lim_{k \to \infty} \frac{\left\| p_k - (p_N)_k \right\|}{\left\| p_k \right\|} = 0,$$

where $(p_N)_k$ is the Newton direction at x_k.

Proof. We give here an outline of the proof. Some details are left to the Exercises.

We first prove the "if" part of the theorem, assuming that

$$\lim_{k \to \infty} \frac{\left\| p_k - (p_N)_k \right\|}{\left\| p_k \right\|} = 0.$$

This is done in two stages, first showing that $\nabla f(x_*) = 0$, and then showing that $\{x_k\}$ converges superlinearly. $(p_N)_k = -\nabla^2 f_k^{-1} \nabla f_k$

① $\nabla f(x_*) = 0$: Since $-\nabla f(x_k) = \nabla^2 f(x_k)(p_N)_k$ and $x_{k+1} - x_k = p_k$,

$$\nabla f(x_{k+1}) = [\nabla f(x_{k+1}) - \nabla f(x_k) - \nabla^2 f(x_k)(x_{k+1} - x_k)] \\ + \nabla^2 f(x_k)[p_k - (p_N)_k].$$

Thus

$$\frac{\left\| \nabla f(x_{k+1}) \right\|}{\left\| p_k \right\|} \le \frac{\left\| \nabla f(x_{k+1}) - \nabla f(x_k) - \nabla^2 f(x_k)(x_{k+1} - x_k) \right\|}{\left\| p_k \right\|} \\ + \left\| \nabla^2 f(x_k) \right\| \frac{\left\| p_k - (p_N)_k \right\|}{\left\| p_k \right\|}.$$

From Theorem B.6 in Appendix B, it follows that

$$\left\| \nabla f(x_{k+1}) - \nabla f(x_k) - \nabla^2 f(x_k)(x_{k+1} - x_k) \right\| = O(\left\| p_k \right\|^2).$$

Hence, for some positive constant γ,

$$\lim_{k\to\infty} \frac{\|\nabla f(x_{k+1})\|}{\|p_k\|} \leq \lim_{k\to\infty} \frac{\gamma \|p_k\|^2}{\|p_k\|} + \lim_{k\to\infty} \|\nabla^2 f(x_k)\| \frac{\|p_k - (p_N)_k\|}{\|p_k\|} = 0.$$

(Note that $\|\nabla^2 f(x_k)\|$ has a finite limit because of the continuity assumptions in the theorem.) Since $\lim_{k\to\infty} \|p_k\| = 0$, then

$$\nabla f(x_*) = \lim_{k\to\infty} \nabla f(x_k) = 0.$$

(ii) $\{x_k\}$ converges to x_* superlinearly: From the assumptions on f and its derivatives, there exists an $\alpha > 0$ such that

$$\|\nabla f(x_{k+1})\| = \|\nabla f(x_{k+1}) - \nabla f(x_*)\| \geq \alpha \|x_{k+1} - x_*\|$$

for all sufficiently large values of k (see the Exercises). Hence

$$\begin{aligned}
\frac{\|\nabla f(x_{k+1})\|}{\|p_k\|} &\geq \frac{\alpha \|x_{k+1} - x_*\|}{\|p_k\|} \\
&\geq \frac{\alpha \|x_{k+1} - x_*\|}{\|x_{k+1} - x_*\| + \|x_k - x_*\|} \\
&= \frac{\alpha \|x_{k+1} - x_*\| / \|x_k - x_*\|}{\|x_{k+1} - x_*\| / \|x_k - x_*\| + 1}.
\end{aligned}$$

But since

$$\lim_{k\to\infty} \frac{\|\nabla f(x_{k+1})\|}{\|p_k\|} = 0,$$

it follows that

$$\lim_{k\to\infty} \frac{\|x_{k+1} - x_*\|}{\|x_k - x_*\|} = 0,$$

and hence $\{x_k\}$ converges superlinearly.

This completes the first half of the proof. The second half, the "only if" part, more or less reverses the arguments used in the first half of the proof.

We assume that $\{x_k\}$ converges superlinearly and that $\nabla f(x_*) = 0$. Now there exists a constant $\beta > 0$ such that

$$\|\nabla f(x_{k+1})\| = \|\nabla f(x_{k+1}) - \nabla f(x_*)\| \leq \beta \|x_{k+1} - x_*\|$$

for all sufficiently large k (see the Exercises). The superlinear convergence of $\{x_k\}$ implies that

$$\begin{aligned}
0 = \lim_{k\to\infty} \frac{\|x_{k+1} - x_*\|}{\|x_k - x_*\|} \\
\geq \lim_{k\to\infty} \frac{1}{\beta} \frac{\|\nabla f(x_{k+1})\|}{\|x_k - x_*\|} \\
= \lim_{k\to\infty} \frac{1}{\beta} \frac{\|\nabla f(x_{k+1})\|}{\|p_k\|} \frac{\|x_{k+1} - x_k\|}{\|x_k - x_*\|}.
\end{aligned}$$

Since

$$\lim_{k \to \infty} \frac{\|x_{k+1} - x_k\|}{\|x_k - x_*\|} = 1$$

(see the Exercises), we obtain that

$$\lim_{k \to \infty} \frac{\|\nabla f(x_{k+1})\|}{\|p_k\|} = 0.$$

Now, by an argument similar to that used in step (i) of the first half of the proof,

$$\frac{\left\| \nabla^2 f(x_k)[p_k - (p_N)_k] \right\|}{\|p_k\|} \leq \frac{\left\| \nabla f(x_{k+1}) - \nabla f(x_k) - \nabla^2 f(x_k)(x_{k+1} - x_k) \right\|}{\|p_k\|}$$
$$+ \frac{\|\nabla f(x_{k+1})\|}{\|p_k\|}.$$

Since the limit of the right-hand side is zero, we obtain that

$$\lim_{k \to \infty} \frac{\left\| \nabla^2 f(x_k)[p_k - (p_N)_k] \right\|}{\|p_k\|} = 0.$$

Since $\nabla^2 f(x_*)$ is positive definite, $\nabla^2 f(x_k)$ will be positive definite for large values of k, and hence

$$\lim_{k \to \infty} \frac{\|p_k - (p_N)_k\|}{\|p_k\|} = 0.$$

This completes the proof.   ☐

Exercises

3.1. Let
$$f(x_1, x_2) = 2x_1^2 + x_2^2 - 2x_1 x_2 + 2x_1^3 + x_1^4.$$

What is the Newton direction at the point $x_0 = (0, 1)^T$? Use a Cholesky decomposition of the Hessian to solve the Newton equations.

3.2. Use Newton's method to solve

$$\text{minimize } f(x) = 5x^5 + 2x^3 - 4x^2 - 3x + 2.$$

Look for a solution in the interval $-2 \leq x \leq 2$. Make sure that you have found a minimum and not a maximum. You may want to experiment with different initial guesses of the solution.

3.3. Use Newton's method to solve

$$\text{minimize } f(x_1, x_2) = 5x_1^4 + 6x_2^4 - 6x_1^2 + 2x_1 x_2 + 5x_2^2 + 15x_1 - 7x_2 + 13.$$

Use the initial guess $(1, 1)^T$. Make sure that you have found a minimum and not a maximum.

3.4. Consider the problem

$$\text{minimize } f(x) = x^4 - 1.$$

Solve this problem using Newton's method. Start from $x_0 = 4$ and perform three iterations. Prove that the iterates converge to the solution. What is the rate of convergence? Can you explain this?

3.5. For a one-variable problem, suppose that $|x - x_*| = \epsilon$ where x_* is a local minimizer. Using a Taylor series expansion, find bounds on $|f(x) - f(x_*)|$ and $|f'(x) - f'(x_*)|$.

3.6. Consider the problem

$$\text{minimize } f(x) = \tfrac{1}{2}x^T Q x - c^T x,$$

where Q is a positive-definite matrix. Prove that Newton's method will determine the minimizer of f in one iteration, regardless of the starting point.

3.7. The purpose of this exercise is to prove Theorem 11.2. Assume that the assumptions of the theorem are satisfied.

(i) Prove that

$$x_{k+1} - x_* = \nabla^2 f(x_k)^{-1} \left[\nabla^2 f(x_k)(x_k - x_*) - (\nabla f(x_k) - \nabla f(x_*)) \right].$$

(ii) Prove that

$$\|x_{k+1} - x_*\| \le (L/2) \left\| \nabla^2 f(x_k)^{-1} \right\| \|x_k - x_*\|^2 .$$

Hint: Use Theorem B.6 in Appendix B.

(iii) Prove that for all large enough k,

$$\|x_{k+1} - x_*\| \le L \left\| \nabla^2 f(x_*)^{-1} \right\| \|x_k - x_*\|^2 ,$$

and from here prove the results of the theorem.

3.8. Let $\{x_k\}$ be a sequence that converges superlinearly to x_*. Prove that

$$\lim_{k \to \infty} \frac{\|x_{k+1} - x_k\|}{\|x_k - x_*\|} = 1.$$

3.9. Let f be a real-valued function of n variables and assume that f, ∇f, and $\nabla^2 f$ are continuous. Suppose that $\nabla^2 f(\bar{x})$ is nonsingular for some point $\bar{x}$. Prove that there exist constants $\epsilon > 0$ and $\beta > \alpha > 0$ such that

$$\alpha \|x - \bar{x}\| \le \|\nabla f(x) - \nabla f(\bar{x})\| \le \beta \|x - \bar{x}\|$$

for all x satisfying $\|x - \bar{x}\| \le \epsilon$.

3.10. Use the previous two problems to complete the proof of Theorem 11.3.

3.11. Assume that the conditions of Theorem 11.3 are satisfied, and that $\nabla^2 f(x_k)$ is positive definite for all k. Also assume that p_k is computed as the solution of

$$B_k p_k = -\nabla f(x_k),$$

where B_k is a positive-definite matrix, and where $\|B_k\| \leq M$ for all k, with M being some constant. Let $(p_N)_k$ be the Newton direction at the kth iteration. Prove that

$$\lim_{k \to \infty} \frac{\|[B_k - \nabla^2 f(x_k)]p_k\|}{\|p_k\|} = 0$$

if and only if

$$\lim_{k \to \infty} \frac{\|p_k - (p_N)_k\|}{\|p_k\|} = 0.$$

3.12. Consider the minimization problem in Exercise 3.1. Suppose that a change of variables $\hat{x} \equiv Ax + b$ is performed with

$$A = \begin{pmatrix} 3 & 1 \\ 4 & 1 \end{pmatrix} \quad \text{and} \quad b = \begin{pmatrix} -1 \\ -2 \end{pmatrix}.$$

Show that the Newton direction for the original problem is the same as the Newton direction for the transformed problem (when both are written using the same coordinate system).

3.13. Prove that the Newton direction remains unchanged if a change of variables $\hat{x} \equiv Ax + b$ is performed, where A is an invertible matrix.

11.4 Guaranteeing Descent

Our general optimization algorithm (see Section 2.4) determines the new estimate of the solution in the form

$$x + \alpha p,$$

where $\alpha > 0$ and $f(x + \alpha p) < f(x)$. This is possible if the search direction p is a descent direction, that is, if

$$p^T \nabla f(x) < 0.$$

In this section we show how to use a "modified matrix factorization" to guarantee this for Newton's method. (Additional requirements on p and α are needed to guarantee convergence of the overall algorithm; see Section 11.5.)

In the classical Newton method the search direction is defined by

$$p = -[\nabla^2 f(x)]^{-1} \nabla f(x).$$

If p is to be a descent direction at the point x, it must satisfy

$$p^T \nabla f(x) = -\nabla f(x)^T [\nabla^2 f(x)]^{-1} \nabla f(x) < 0$$

or

$$\nabla f(x)^T [\nabla^2 f(x)]^{-1} \nabla f(x) > 0.$$

This condition will be satisfied if $[\nabla^2 f(x)]^{-1}$ (or equivalently $\nabla^2 f(x)$) is positive definite.

Requiring that $\nabla^2 f(x)$ be positive definite is a stronger condition than $p^T \nabla f(x) < 0$. To motivate this, recall that Newton's method can be interpreted as approximating $f(x + p)$ by a quadratic:

$$f(x + p) \approx f(x) + p^T \nabla f(x) + \tfrac{1}{2} p^T \nabla^2 f(x) p.$$

The formula for Newton's method is obtained by setting the derivative of the quadratic function equal to zero. An alternative view is to *minimize* the quadratic as a function of p. If $\nabla^2 f(x)$ is positive definite, then the minimum is obtained by setting the derivative equal to zero, as before, and the two points of view are equivalent. If $\nabla^2 f(x)$ is indefinite, however, then the quadratic function does not have a finite minimum.

If the Hessian matrix is indefinite, then one possible strategy is to replace the Hessian by some related positive-definite matrix in the formula for the Newton direction. This guarantees that the search direction is a descent direction. It also implies that the search direction corresponds to the minimization of a quadratic approximation to the objective function f, a quadratic approximation obtained from the Taylor series by replacing $\nabla^2 f(x)$ with the "related positive-definite matrix."

This might seem arbitrary, but there are several justifications for it. First, if it is done appropriately, then the resulting algorithm can be shown to converge when used inside a line search method. Second, at the solution to the optimization problem ∇^2, $f(x_*)$ will usually be positive definite (it is always positive semidefinite), so that the Hessian will normally only be replaced at points distant from the solution. Third, the related positive-definite matrix can be found as a side effect of trying to use the classical Newton formula, with little additional computation required. This third point is discussed further below.

Computing the search direction involves solving the linear system

$$\nabla^2 f(x) p = -\nabla f(x).$$

If $\nabla^2 f(x)$ is positive definite, then the factorization

$$\nabla^2 f(x) = LDL^T$$

can be used, where the diagonal matrix D has positive diagonal entries (see Appendix A.7.2). If $\nabla^2 f(x)$ is not positive definite, then at some point during the computation of the factorization some diagonal entry of D will satisfy

$$d_{ii} \leq 0.$$

If this happens, then d_{ii} should be replaced by some positive entry, perhaps $|d_{ii}|$ or some small positive number.

It can be shown (via the formulas for the matrix factorization) that modifying the entries of D is equivalent to replacing $\nabla^2 f(x)$ by

$$\nabla^2 f(x) \rightarrow \nabla^2 f(x) + E,$$

where E is a diagonal matrix, and then factoring this matrix,

$$\nabla^2 f(x) + E = LDL^T,$$

and so the modified Hessian matrix is positive definite. This factorization is then used to compute the search direction:

$$(LDL^T) p = -\nabla f(x),$$

and hence the overall technique corresponds to replacing $\nabla^2 f(x)$ by the related positive-definite matrix $\nabla^2 f(x) + E$. Even if $\nabla^2 f(x)$ were always positive definite, this matrix would still be factored to compute the search direction from the Newton formula, and so this "modified" matrix factorization is obtained with little effort—just the effort of changing any negative (or zero) d_{ii} to a suitable positive number.

Example 11.4 (Modified Matrix Factorization). Suppose that

$$\nabla^2 f(x) = \begin{pmatrix} -1 & 2 & 4 \\ 2 & -3 & 6 \\ 4 & 6 & 22 \end{pmatrix}.$$

This matrix is symmetric but not positive definite. At the first stage of the factorization, $d_{1,1} = -1$; we will replace this number by 4. (In this example the entries in E have been chosen to simplify the calculations.) Then $d_{1,1} = 4$, $e_{1,1} = 5$, $\ell_{1,1} = \ell_{3,1} = 1$, and $\ell_{2,1} = \frac{1}{2}$.

At the next stage, $d_{2,2} = -4$; we will replace this number by 8. Hence $d_{2,2} = 8$, $e_{2,2} = 12$, $\ell_{2,2} = 1$, and $\ell_{3,2} = \frac{1}{2}$.

At the final stage, $d_{3,3} = 16$, so no modification is necessary. The overall factorization is

$$\nabla^2 f(x) + E = \begin{pmatrix} -1 & 2 & 4 \\ 2 & -3 & 6 \\ 4 & 6 & 22 \end{pmatrix} + \begin{pmatrix} 5 & 0 & 0 \\ 0 & 12 & 0 \\ 0 & 0 & 0 \end{pmatrix}$$

$$= \begin{pmatrix} 1 & 0 & 0 \\ \frac{1}{2} & 1 & 0 \\ 1 & \frac{1}{2} & 1 \end{pmatrix} \begin{pmatrix} 4 & 0 & 0 \\ 0 & 8 & 0 \\ 0 & 0 & 16 \end{pmatrix} \begin{pmatrix} 1 & \frac{1}{2} & 1 \\ 0 & 1 & \frac{1}{2} \\ 0 & 0 & 1 \end{pmatrix} = LDL^T.$$

This final factorization would be used to compute a search direction. ∎

There is a great deal of flexibility in choosing how to modify D in the case where $\nabla^2 f(x)$ is not positive definite. Of course, D must be chosen so that the resulting modified matrix is positive definite. To satisfy the assumptions of the convergence theorem for a line search method (see Section 11.5) the modified matrix must not be "arbitrarily" close to being singular; that is, the smallest eigenvalue of the modified matrix must be larger than some positive tolerance. In addition, the norm of the modified matrix must remain bounded. (See the Exercises in Section 11.5.) These conditions place limits on how small and large the elements of D can be. Within this range, however, any choice of D would be acceptable, at least theoretically.

We conclude by presenting a practical version of Newton's method, one that is guaranteed to converge and that does not assume that $\nabla^2 f(x_k)$ is positive definite for all values of k. Some steps in the method are left vague. It is assumed that these steps are carried out in a way that is consistent with Theorem 11.7 of Section 11.5, or some other convergence theorem for a line search method. The convergence test in this algorithm is simplified; a more complete discussion of convergence tests can be found in Section 12.5.

ALGORITHM 11.1.
Modified Newton Algorithm with Line Search

1. Specify some initial guess of the solution x_0, and specify a convergence tolerance ϵ.
2. For $k = 0, 1, \ldots$
    ⓘ If $\|\nabla f(x_k)\| < \epsilon$, stop.

(ii) Compute a modified factorization of the Hessian:

$$\nabla^2 f(x_k) + E = LDL^T$$

and solve

$$(LDL^T)p = -\nabla f(x_k)$$

for the search direction p_k. (E will be zero if $\nabla^2 f(x_k)$ is "sufficiently" positive definite.)

(iii) Perform a line search to determine

$$x_{k+1} = x_k + \alpha_k p_k,$$

the new estimate of the solution.

Exercises

4.1. Find a diagonal matrix E so that $A + E = LDL^T$, where

$$A = \begin{pmatrix} 1 & 4 & 3 \\ 4 & 2 & 5 \\ 3 & 5 & 3 \end{pmatrix}.$$

4.2. Suppose that $\nabla f(x) = 0$ and that $\nabla^2 f(x)$ is indefinite. Show how the modified matrix factorization

$$\nabla^2 f(x) + E = LDL^T$$

can be used to compute a direction along which f decreases.

4.3. Apply the result of the previous problem to the matrix

$$A = \begin{pmatrix} 1 & 4 & 3 \\ 4 & 2 & 5 \\ 3 & 5 & 3 \end{pmatrix}.$$

4.4. Let M be a positive-definite matrix and let

$$p = -M^{-1}\nabla f(x_k).$$

Prove that p is a descent direction for f at x_k.

4.5. Consider the matrix

$$A = \begin{pmatrix} \epsilon & 1 \\ 1 & 1 \end{pmatrix},$$

where ϵ is some small positive number. Consider two ways of modifying A to make it positive definite, the first where only $A_{2,2}$ is changed, and the second where both $A_{1,1}$ and $A_{2,2}$ are changed. Show that in the first case the norm of the modification is $O(\epsilon^{-1})$, whereas in the second case the modification can be chosen so that its norm is $O(1)$.

4.6. A vector d is a *direction of negative curvature* for the function f at the point x if $d^T \nabla^2 f(x)d < 0$. Prove that such a direction exists if and only if at least one of the eigenvalues of $\nabla^2 f(x)$ is negative. Also prove that, if a direction of negative curvature exists, then there also exists a direction of negative curvature that is also a descent direction.

11.5 Guaranteeing Convergence: Line Search Methods

The auxiliary techniques that are used to guarantee convergence attempt to rein in the optimization method when it is in danger of getting out of control, and they also try to avoid intervening when the optimization method is performing effectively. Far from the solution, when the Taylor series is a poor approximation to the function near the optimum, these "globalization strategies" are an active part of the algorithm, preventing movement away from the solution, or even divergence. Near the solution these strategies will remain in the background as safeguards; they are available if required, but normally they will not be invoked.

The term "globalization strategy" is used to distinguish the method used for selecting the new estimate of the solution from the method for computing the search direction. In most algorithms, the formula for the search direction is derived from the Taylor series, and the Taylor series is a "local" approximation to the function. The method for choosing the new estimate of the solution is designed to guarantee "global convergence," that is, convergence from any starting point. Note that this is convergence to a stationary point.

If the underlying optimization method produces good search directions, as is often the case with Newton's method on well-conditioned problems, then the globalization strategies will act merely as a safety net protecting against the occasional bad step. For a method that produces less effective search directions, such as a nonlinear conjugate-gradient method (see Section 13.4), they can be a major contributor to the practical success of a method.

We discuss two major types of globalization strategy. Line search methods are the topic of this section, and trust-region methods are the topic of Section 11.6. In later chapters we often assume that one of these strategies has been incorporated into the algorithms we discuss. Typically we refer to using a line search, although in many cases a trust-region strategy could also be used.

Line search methods are the oldest and most widely used of the globalization strategies. To describe them, let x_k be the current estimate of a minimizer of f, and let p_k be the search direction at the point x_k. Then the new estimate of the solution is defined by the formula

$$x_{k+1} = x_k + \alpha_k p_k,$$

where the step length α_k is some scalar chosen so that

$$f(x_{k+1}) < f(x_k).$$

Since the function value at the new point is smaller than the function value at the current point, progress has been made toward the minimum. (This is not the whole truth. Exceptions and details are discussed below.)

Example 11.5 (Line Search). Consider the problem

$$\text{minimize } f(x_1, x_2) = 5x_1^2 + 7x_2^2 - 3x_1x_2.$$

Let $x_k = (2, 3)^T$ and $p_k = (-5, -7)^T$, so that $f(x_k) = 65$. If $\alpha_k = 1$, then

$$f(x_k + \alpha_k p_k) = f(-3, -4) = 121 > f(x_k),$$

so this is not an acceptable step length. If $\alpha_k = \frac{1}{2}$, then

$$f(x_k + \alpha_k p_k) = f(-\tfrac{1}{2}, -\tfrac{1}{2}) = \tfrac{9}{4},$$

and so this step length produces a decrease in the function value, as desired. ■

Let us look more closely at the line search formula. We will assume that p_k is a descent direction at x_k; that is, p_k must satisfy

$$p_k^T \nabla f(x_k) < 0. \qquad \left(\substack{\text{so we know that an } \alpha_k \text{ can be} \\ \text{chosen to produce descent}} \right)$$

This should be guaranteed by the algorithm used to compute the search direction. For Newton's method this is discussed in Section 11.4. If p_k is a descent direction, then $f(x_k + \alpha p_k) < f(x_k)$ at least for small positive values of α. Because of this property we assume that the step length satisfies $\alpha_k > 0$.

The technique is called a "line search" or "linear search" because a search for a new point x_{k+1} is carried out along the line $y(\alpha) = x_k + \alpha p_k$. Intuitively we would like to choose α_k as the solution to

$$\underset{\alpha > 0}{\text{minimize }} F(\alpha) \equiv f(x_k + \alpha p_k).$$

That is, α_k would be the result of a one-dimensional minimization problem. It is usually too expensive to solve this one-dimensional problem exactly, so in practice an *approximate* minimizer is accepted instead. In its crudest form, this approximate minimizer merely reduces the value of the function f, as was indicated above. However, a little more than this is required to guarantee convergence, as the example below indicates.

✱ **Example 11.6** (A Naive Line Search). Consider the minimization problem

$$\text{minimize } f(x) = x^2$$

with initial guess $x_0 = -3$. At each iteration we use the search direction $p_k = 1$ with step length $\alpha_k = 2^{-k}$. Hence

$$x_{k+1} = x_k + 2^{-k}. \qquad x_n = -3 + \sum_{h=0}^{n-1} 2^{-k}$$

The sequence of approximate solutions will be

$$-3, -2, -\tfrac{3}{2}, -\tfrac{5}{4}, -\tfrac{9}{8}, \ldots \qquad \therefore x_n \to -3 + \frac{1}{1-\frac{1}{2}} = -1$$

with $x_k = -(1 + 2^{1-k})$. Each search direction is a descent direction since

$$p_k^T \nabla f(x_k) = 1 \times 2x_k = -2(1 + 2^{1-k}) < 0.$$

It is easy to check that $f(x_{k+1}) < f(x_k)$ as well.

Even though this simple algorithm produces a reduction in the function value at each iteration, it does not converge to a stationary point:

$$\lim_{k \to \infty} x_k = -1$$

and $f'(-1) = -2 \neq 0$. The solution is $x_* = 0$. Clearly more is required of a line search than just reduction in the value of f. ■

One way to guarantee convergence is to make additional assumptions, two on the search direction p_k and two on the step length α_k. The assumptions on the search direction p_k are that ⓐ it produces "sufficient descent," and ⓑ it is "gradient related." The assumptions on the step length α_k are that ⓐ it produces a "sufficient decrease" in the function f, and ⓑ it is not "too small."

Let us first discuss "sufficient descent." The search direction must first of all be a descent direction, that is, $p_k^T \nabla f(x_k) < 0$. A danger is that p_k might become arbitrarily close to being orthogonal to $\nabla f(x_k)$ while still remaining a descent direction, and thus the algorithm would make little progress toward a solution. To ensure against this we assume that

$$-\frac{p_k^T \nabla f(x_k)}{\|p_k\| \cdot \|\nabla f(x_k)\|} \geq \epsilon > 0 \qquad \text{(angle condition : sufficient descent)}$$

for all k, where $\epsilon > 0$ is some specified tolerance. This condition can also be written as

$$\cos \theta \geq \epsilon > 0,$$

where θ is the angle between the search direction p_k and the negative gradient $-\nabla f(x_k)$. For this reason, it can be referred to as the *angle condition*. If p_k and $\nabla f(x_k)$ are orthogonal, then $\cos \theta = 0$.

The search directions are said to be *gradient related* if

$$\|p_k\| \geq m \|\nabla f(x_k)\|$$

for all k, where $m > 0$ is some constant. This condition states that the norm of the search direction cannot become too much smaller than that of the gradient.

These conditions can normally be guaranteed by making slight modifications to the method used to compute the search direction. We will assume that the method used to compute the search direction has been adjusted, if necessary, to guarantee that the sufficient descent and gradient-relatedness conditions are satisfied. Techniques for doing this are discussed in the context of specific methods.

The sufficient decrease condition on α_k ensures that some nontrivial reduction in the function value is obtained at each iteration. "Nontrivial" is measured in terms of the Taylor series. A linear approximation to $f(x_k + \alpha p_k)$ is obtained from

$$f(x_k + \alpha p_k) \approx f(x_k) + \alpha p_k^T \nabla f(x_k).$$

In the line search we will demand that the step length α_k produce a decrease in the function value that is at least some fraction of the decrease predicted by the above linear approximation. More specifically, we will require that

$$f(x_k + \alpha_k p_k) \leq f(x_k) + \mu \alpha_k p_k^T \nabla f(x_k),$$

where μ is some scalar satisfying $0 < \mu < 1$. When μ is near zero this condition is easier to satisfy since only a small decrease in the function value is required. The condition is illustrated in Figure 11.3. It is sometimes referred to as an *Armijo* condition.

If α is small, the linear approximation will be good, and the sufficient decrease condition will be satisfied. If α is large, the decrease predicted by the linear approximation may

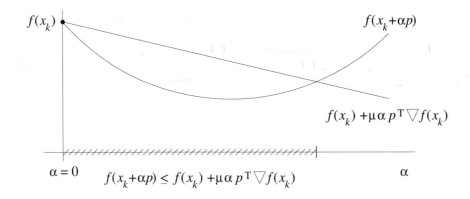

Figure 11.3. *The sufficient decrease condition.*

differ greatly from the actual decrease in f, and the condition can be violated. In this sense, the sufficient decrease condition prevents α from being "too large."

We discuss two ways of satisfying the other condition on α_k—that it not be "too small." The first, a simple line search algorithm, will be used to prove a convergence result but is not recommended for practical computations. The second, called a *Wolfe condition*, leads to better, but more complicated, algorithms; it is discussed in Section 11.5.1. The Wolfe condition is found in many widely used software packages.

The simple line search algorithm we will analyze uses backtracking: Let p_k be a search direction satisfying the sufficient descent condition. Define α_k to be the first element of the sequence

$$1, \tfrac{1}{2}, \tfrac{1}{4}, \tfrac{1}{8}, \ldots, 2^{-i}, \ldots$$

that satisfies the sufficient decrease condition. Such an α_k always exists. Because a "large" step $\alpha = 1$ is tried first and then reduced, the step lengths $\{\alpha_k\}$ that are generated by this algorithm will not be "too small."

This algorithm is easy to program on a computer. First, test if $\alpha = 1$ satisfies the sufficient decrease condition. If it does not, try $\alpha = \tfrac{1}{2}, \alpha = \tfrac{1}{4}$, etc., until an acceptable α is found. The step $\alpha = 1$ is tried first (rather than $\alpha = 5$, say) because in the classical Newton method a step of one is always used, and near the solution we would expect that a step of one would be acceptable and lead to a quadratic convergence rate.

The theorem below makes several assumptions in addition to those mentioned above. It assumes that the *level set*

$$\{ x : f(x) \le f(x_0) \}$$

is bounded. This ensures that the function takes on its minimum value at a finite point. It rules out functions such as $f(x) = e^x$ that are bounded below (in this case by zero) but only approach this bound in the limit. There is also a technical assumption that the search directions are bounded. This can usually be guaranteed by careful programming of the optimization algorithm.

In summary, the theorem requires that the function have a bounded level set, and that the gradient of the function be Lipschitz continuous. All the remaining assumptions are assumptions about the method and can be satisfied by careful design of the method. The assumptions on the optimization problem are minimal.

The conclusion to the theorem does not state that the sequence $\{x_k\}$ converges to a local minimizer of f. It only states that $\nabla f(x_k) \to 0$. To prove the stronger result using a line search algorithm, stronger assumptions must be made.

Theorem 11.7. *Let f be a real-valued function of n variables. Let x_0 be a given initial point and define $\{x_k\}$ by $x_{k+1} = x_k + \alpha_k p_k$, where p_k is a vector of dimension n and $\alpha_k \geq 0$ is a scalar. Assume that*

(i) *the set $S = \{x : f(x) \leq f(x_0)\}$ is bounded;*

(ii) *∇f is Lipschitz continuous for all x, that is,*

$$\|\nabla f(x) - \nabla f(y)\| \leq L \|x - y\|$$

for some constant $0 < L < \infty$;

(iii) *the vectors p_k satisfy a sufficient descent condition*   (angle condition)

$$-\frac{p_k^T \nabla f(x_k)}{\|p_k\| \cdot \|\nabla f(x_k)\|} \geq \epsilon > 0;$$

(iv) *the search directions are gradient related:*

$$\|p_k\| \geq m \|\nabla f(x_k)\| \quad \text{for all } k \text{ (with } m > 0\text{)},$$

and bounded in norm:
$$\|p_k\| \leq M \quad \text{for all } k;$$

(v) *the scalar α_k is chosen as the first element of the sequence $\left\{ 1, \frac{1}{2}, \frac{1}{4}, \ldots \right\}$ to satisfy a sufficient decrease condition*

$$f(x_k + \alpha p_k) \leq f(x_k) + \mu \alpha_k p_k^T \nabla f(x_k), \quad \text{(armijo, } s = \tfrac{1}{2}\text{)}$$

where $0 < \mu < 1$.

Then

$$\lim_{k \to \infty} \|\nabla f(x_k)\| = 0.$$

Proof. There are five steps in the proof. First we show that f is bounded from below on S. Second we show that $\lim f(x_k)$ exists. Third we show that

$$\lim_{k \to \infty} \alpha_k \|\nabla f(x_k)\|^2 = 0.$$

Fourth we show that if $\alpha_k < 1$, then

$$\alpha_k \geq \gamma \|\nabla f(x_k)\|^2$$

for an appropriate constant $\gamma > 0$. Finally, we show that

$$\lim \|\nabla f(x_k)\| = 0.$$

1. f is bounded from below on S: Because f is continuous, the set

$$S = \{ x : f(x) \le f(x_0) \}$$

is closed. Furthermore, by assumption (i) in the theorem, it is bounded. A continuous function on a closed and bounded set takes on its minimum value at some point in that set (see Appendix B.8). This shows that f is bounded from below on the set S, that is, $f(x) \ge C$ for some number C.

2. $\lim f(x_k)$ exists: The sufficient decrease condition ensures that $f(x_{k+1}) < f(x_k) \le f(x_0)$ so that $x_k \in S$ for all k. The sequence $\{ f(x_k) \}$ is monotone decreasing and bounded from below (by C), so it has a limit $\bar{f}$.

3. $\lim_{k\to\infty} \alpha_k \|\nabla f(x_k)\|^2 = 0$: This follows from

$$f(x_0) - \bar{f} = [f(x_0) - f(x_1)] + [f(x_1) - f(x_2)] + [f(x_2) - f(x_3)] + \cdots$$

$$= \sum_{k=0}^{\infty} [f(x_k) - f(x_{k+1})]$$

$$\ge -\sum_{k=0}^{\infty} \mu \alpha_k p_k^T \nabla f(x_k) \qquad \left(f_k - f_{k+1} \ge -\mu \alpha_k p_k^T \nabla f_k \right)$$
(from the sufficient decrease condition)

$$\ge \sum_{k=0}^{\infty} \mu \alpha_k \epsilon \|p_k\| \cdot \|\nabla f(x_k)\| \qquad \left(p_k^T \nabla f_k \ge -\epsilon \|p_k\| \|\nabla f_k\| \right)$$
(from the sufficient descent condition)

$$\ge \sum_{k=0}^{\infty} \mu \alpha_k \epsilon m \|\nabla f(x_k)\|^2 \qquad \left(\|p_k\| \ge m \|\nabla f_k\| \right)$$
(from the gradient-relatedness condition).

Since $f(x_0) - \bar{f} \le f(x_0) - C < \infty$ this final summation converges, and so the terms in the summation go to zero:

$$\lim_{k\to\infty} \mu \alpha_k \epsilon m \|\nabla f(x_k)\|^2 = 0.$$

The result now follows because m, μ, and ϵ are fixed nonzero constants.

4. If $\alpha_k < 1$, then $\alpha_k \ge \gamma \|\nabla f(x_k)\|^2$ for an appropriate constant $\gamma > 0$: This step of the proof is based on the backtracking line search. If $\alpha_k < 1$, then the sufficient decrease condition was *violated* when the step length $2\alpha_k$ was tried:

$$f(x_k + 2\alpha_k p_k) - f(x_k) > 2\mu \alpha_k p_k^T \nabla f(x_k).$$

Because ∇f is Lipschitz continuous, by Theorem B.6 of Appendix B we can conclude that

$$f(x_k + 2\alpha_k p_k) - f(x_k) - 2\alpha_k p_k^T \nabla f(x_k) \le \tfrac{1}{2} L \|2\alpha_k p_k\|^2 .$$

This can be rearranged as

$$f(x_k) - f(x_k + 2\alpha_k p_k) \geq -2\alpha_k p_k^T \nabla f(x_k) - 2L \left\| \alpha_k p_k \right\|^2 .$$

Adding this to the first inequality above and simplifying gives

$$\alpha_k L \left\| p_k \right\|^2 \geq -(1 - \mu) p_k^T \nabla f(x_k).$$

The sufficient descent and gradient-relatedness conditions then give

$$\alpha_k L \left\| p_k \right\|^2 \geq (1 - \mu)\epsilon \left\| p_k \right\| \cdot \left\| \nabla f(x_k) \right\| \geq (1 - \mu)\epsilon m \left\| \nabla f(x_k) \right\|^2 .$$

Since $\left\| p_k \right\| \leq M$, we have that $\alpha_k \geq \gamma \left\| \nabla f(x_k) \right\|^2$ with

$$\gamma = \frac{(1 - \mu)\epsilon m}{M^2 L} > 0$$

as desired.

5. $\lim_{k \to \infty} \left\| \nabla f(x_k) \right\| = 0$: Either $\alpha_k = 1$ or $\alpha_k \geq \gamma \left\| \nabla f(x_k) \right\|^2$. Hence

$$\alpha_k \geq \min \left\{ 1, \gamma \left\| \nabla f(x_k) \right\|^2 \right\}$$

and

$$\alpha_k \left\| \nabla f(x_k) \right\|^2 \geq \left[\min \left\{ 1, \gamma \left\| \nabla f(x_k) \right\|^2 \right\} \right] \left\| \nabla f(x_k) \right\|^2 \geq 0.$$

From step 3 we already know that $\lim \alpha_k \left\| \nabla f(x_k) \right\|^2 = 0$. Since $\gamma > 0$, this implies that $\lim \left\| \nabla f(x_k) \right\| = 0$ also.

The proof is completed. □

11.5.1 Other Line Searches

The backtracking line search is not the only way of guaranteeing that the step length α_k is not "too small." This is also guaranteed by conditions derived from the one-dimensional problem

$$\underset{\alpha > 0}{\text{minimize}} \ F(\alpha) \equiv f(x_k + \alpha p_k).$$

A decrease in the function value corresponds to the condition

$$f(x_k + \alpha p) < f(x_k).$$

This is equivalent to the condition

$$F(\alpha) < F(0).$$

Instead of just asking for a decrease in the function value, we could ask that α_k approximately minimize F, or that $F'(\alpha_k) \approx 0$. This condition is normally written as

$$|F'(\alpha_k)| \leq \eta |F'(0)|,$$

where η is a constant satisfying $0 \leq \eta < 1$.

An *exact* line search corresponds to choosing an $\alpha_k \geq 0$ that is a local minimizer of $F(\alpha)$. In this case the above condition is satisfied with $\eta = 0$. If f is a quadratic function, there is a simple formula to do this (see the Exercises). On general problems an exact line search is usually too expensive to be a practical technique. Exact line searches are frequently encountered in theoretical results because it can be easier to prove the convergence of an algorithm that uses an exact line search.

The term $F'(\alpha)$ is a *directional derivative* of the function f at the point $x_k + \alpha p_k$. Its formula can be derived from

$$
\begin{aligned}
F'(\alpha) &= \lim_{h \to 0} \frac{F(\alpha + h) - F(\alpha)}{h} \\
&= \lim_{h \to 0} \frac{f(x_k + \alpha p_k + h p_k) - f(x_k + \alpha p_k)}{h} \\
&= \lim_{h \to 0} \frac{f(x_k + \alpha p_k) + h p_k^T \nabla f(x_k + \alpha p_k) + \frac{1}{2} h^2 p_k^T \nabla^2 f(\xi) p_k}{h} \\
&\quad - \frac{f(x_k + \alpha p_k)}{h} \quad \text{(using a Taylor series expansion)} \\
&= \lim_{h \to 0} p_k^T \nabla f(x_k + \alpha p_k) + \frac{1}{2} h p_k^T \nabla^2 f(\xi) p_k \\
&= p_k^T \nabla f(x_k + \alpha p_k).
\end{aligned}
$$

In a similar manner it can be shown that $F''(\alpha) = p_k^T \nabla^2 f(x_k + \alpha p_k) p_k$; see the Exercises. The value $\alpha = 0$ corresponds to the point x_k. For this value,

$$
F(0) = f(x_k),
$$

the current function value, and

$$
F'(0) = p_k^T \nabla f(x_k),
$$

the directional derivative at the point x_k.

Example 11.8 (One-Dimensional Problem). Consider the function of two variables

$$
f(x) = \tfrac{1}{2} x_1^2 + x_2^2 - \log(x_1 + x_2).
$$

Let $x_k = (1, 1)^T$ and $p = (2, -1)^T$. Then

$$
F(\alpha) = f(x_k + \alpha p) = \tfrac{1}{2}(1 + 2\alpha)^2 + (1 - \alpha)^2 - \log(2 + \alpha).
$$

Note that $F(0) = \tfrac{3}{2} - \log 2 = f(x_k)$. Also,

$$
F'(\alpha) = 2(1 + 2\alpha) - 2(1 - \alpha) - \frac{1}{2 + \alpha}.
$$

We can verify that $F'(\alpha) = p^T \nabla f(x_k + \alpha p)$. First, notice that

$$
\nabla f(x) = \begin{pmatrix} x_1 - 1/(x_1 + x_2) \\ 2x_2 - 1/(x_1 + x_2) \end{pmatrix}.
$$

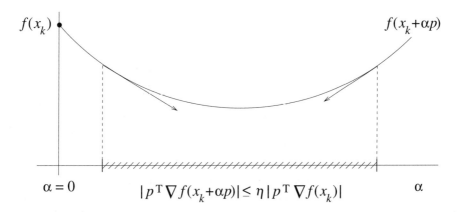

Figure 11.4. *The Wolfe condition.*

Then

$$p^T \nabla f(x_k + \alpha p) = (2 \quad -1) \begin{pmatrix} (1+2\alpha) - 1/(2+\alpha) \\ 2(1-\alpha) - 1/(2+\alpha) \end{pmatrix}$$

$$= 2(1+2\alpha) - \frac{2}{2+\alpha} - 2(1-\alpha) + \frac{1}{2+\alpha}$$

$$= 2(1+2\alpha) - 2(1-\alpha) - \frac{1}{2+\alpha} = F'(\alpha). \quad \blacksquare$$

Using the formula for the directional derivative, the condition

$$|F'(\alpha_k)| \le \eta |F'(0)|$$

becomes

$$|p_k^T \nabla f(x_k + \alpha_k p_k)| \le \eta |p_k^T \nabla f(x_k)|.$$

This is sometimes called the Wolfe condition. It is illustrated in Figure 11.4.

The Wolfe condition only finds an approximate stationary point of the function F. A local maximum of this function would satisfy the condition, so by itself it does not guarantee a decrease in the function value. For this reason, algorithms insist that the step length α also satisfy a sufficient decrease condition, with $\mu < \eta$; often the constant μ is chosen to be very small so that almost any decrease in the function is enough to be acceptable. An elegant convergence result can be derived using the combination of the Wolfe and sufficient decrease conditions; it is outlined in Exercise 5.15.

Many widely used line search algorithms are based on the Wolfe condition. These algorithms are often much more effective than the backtracking line search described earlier. However, implementing an inexact line search based on approximately minimizing F is a complicated task, requiring great attention to detail to ensure that an acceptable step length α_k that satisfies both the Wolfe and the sufficient decrease conditions can always be found.

In a line search based on the Wolfe condition, some form of one-dimensional minimization technique is used to determine a step length. A common approach is to bracket a minimizer of F, that is, to find an interval $[\underline{\alpha}, \overline{\alpha}]$ that contains a local minimizer of $F(\alpha)$. Then this interval is refined via a sequence of polynomial approximations to $F(\alpha)$.

Let us first look at bracketing. We search for an interval $[\underline{\alpha}, \overline{\alpha}]$ with $F'(\underline{\alpha}) < 0$ and $F'(\overline{\alpha}) > 0$. At some point in the interval there must be an α satisfying $F'(\alpha) = 0$. Since $F'(0) = p_k^T \nabla f(x_k) < 0$, the value $\alpha = 0$ provides an initial lower bound on the step length α_k. To obtain an upper bound, an increasing sequence of values of α are examined until one is found that satisfies $F'(\overline{\alpha}) > 0$. For example, we might try $\alpha = 1$, then $\alpha = 2$, $\alpha = 4$, etc. Then the interval $[\underline{\alpha}, \overline{\alpha}]$ brackets a minimizer of $F(\alpha)$, where $\underline{\alpha}$ is the largest trial value of α for which $F'(\alpha) < 0$.

If during the bracketing step a trial value of α that satisfies the Wolfe condition is found, the line search is terminated with that trial value as the step length α_k. On the other hand, if no upper bound $\overline{\alpha}$ is found, then the one-dimensional function F as well as the objective function f may both be unbounded below or may have no finite minimizer.

We now assume that an interval $[\underline{\alpha}, \overline{\alpha}]$ has been determined that brackets a minimizer of F, with $F'(\underline{\alpha}) < 0$ and $F'(\overline{\alpha}) > 0$. A polynomial approximation to the function F will be used to reduce the size of this interval. If cubic approximations are used, then the unique cubic polynomial $P_3(\alpha)$ satisfying

$$P_3(\underline{\alpha}) = F(\underline{\alpha}) \quad P_3'(\underline{\alpha}) = F'(\underline{\alpha})$$
$$P_3(\overline{\alpha}) = F(\overline{\alpha}) \quad P_3'(\overline{\alpha}) = F'(\overline{\alpha})$$

is computed. (In general, a cubic interpolant is uniquely determined by four independent data values.)

This cubic polynomial must have a local minimizer $\hat{\alpha}$ within the interval $[\underline{\alpha}, \overline{\alpha}]$. The point $\hat{\alpha}$ is the next estimate of the step length. If this point satisfies the Wolfe condition, then $\alpha_k = \hat{\alpha}$ is accepted as the step length. Otherwise, one of $\underline{\alpha}$ or $\overline{\alpha}$ is replaced by $\hat{\alpha}$ (depending on the sign of $F'(\hat{\alpha})$), and the process repeats.

Example 11.9 (Line Search with Wolfe Condition). Suppose that the one-dimensional function is

$$F(\alpha) = 5 - \alpha - \log(4.5 - \alpha).$$

At the initial value $\alpha = 0$,

$$F(0) = 3.4959 \quad \text{and} \quad F'(0) = -0.7778 < 0.$$

We use a Wolfe condition with $\eta = 0.1$, so that the step length α_k must satisfy

$$|F'(\alpha_k)| \leq 0.07778.$$

We first attempt to bracket the step length by trying a sequence of increasing values of α until one is found that satisfies $F'(\alpha) > 0$:

$$\alpha = 1 : F(1) = 2.7472 \quad \text{and} \quad F'(1) = -0.7143$$
$$\alpha = 2 : F(2) = 2.0837 \quad \text{and} \quad F'(2) = -0.6000$$
$$\alpha = 4 : F(4) = 1.6931 \quad \text{and} \quad F'(4) = 1.0000.$$

Thus $\overline{\alpha} = 4$, and the interval that brackets the step length is $[\underline{\alpha}, \overline{\alpha}] = [2, 4]$. None of these trial α values satisfies the Wolfe condition.

We now refine the interval using cubic polynomial approximations. Using the formulas derived in the Exercises, we determine that the cubic

$$P_3(\alpha) = 0.9309 + 2.5434\alpha - 1.3788\alpha^2 + 0.1976\alpha^3$$

matches the values of F and F' at $\underline{\alpha}$ and $\overline{\alpha}$. It has a local minimizer at $\hat{\alpha} = 3.3826$, where

$$F(\hat{\alpha}) = 1.5064 \quad \text{and} \quad F'(\hat{\alpha}) = -0.1050.$$

This point does not satisfy the Wolfe condition.

Since $F'(\hat{\alpha}) < 0$ the new interval is $[3.3826, 4]$. The new cubic is

$$P_3(\alpha) = -25.4041 + 24.7208\alpha - 7.5297\alpha^2 + 0.7608\alpha^3,$$

with local minimizer $\hat{\alpha} = 3.5294$. At this point,

$$F(\hat{\alpha}) = 1.5004 \quad \text{and} \quad F'(\hat{\alpha}) = 0.0303,$$

so this point satisfies the Wolfe condition, and the step length is $\alpha_k = 3.5294$. (This value of α also satisfies the sufficient decrease condition for $\mu = 0.1$, say.)

The exact minimizer of $F(\alpha)$ is $\alpha_* = 3.5$. This would be the step length if an exact line search were stipulated. ∎

Further care is required to transform this description of a line search algorithm into a piece of software. For example, we have made reference only to the Wolfe condition and have ignored the requirement that the step length simultaneously satisfy the sufficient decrease condition. We have also ignored the effects of computer arithmetic. These topics are discussed in the references cited in the Notes.

Exercises

5.1. Consider the problem

$$\text{minimize } f(x_1, x_2) = (x_1 - 2x_2)^2 + x_1^4.$$

(i) Suppose a Newton's method with a line search is used to minimize the function, starting from the point $x = (2, 1)^T$. What is the Newton search direction at this point? Use a Cholesky decomposition of the Hessian matrix to solve the Newton equations.

(ii) Suppose a backtracking line search is used. Does the trial step $\alpha = 1$ satisfy the sufficient decrease condition for $\mu = 0.2$? For what values of μ does $\alpha = 1$ satisfy the sufficient decrease condition?

5.2. Let

$$f(x_1, x_2) = 2x_1^2 + x_2^2 - 2x_1x_2 + 2x_1^3 + x_1^4.$$

(i) Suppose that the function is minimized starting from $x_0 = (0, -2)^T$. Verify that $p_0 = (0, 1)^T$ is a direction of descent.

(ii) Suppose that a line search is used to minimize the function $F(\alpha) = f(x_0 + \alpha p_0)$, and that a backtracking line search is used to find the optimal step length α. Does $\alpha = 1$ satisfy the sufficient decrease condition for $\mu = 0.5$? For what values of μ does $\alpha = 1$ satisfy the sufficient decrease condition?

5.3. Consider the quadratic function

$$f(x) = \tfrac{1}{2}x^T Q x - c^T x,$$

where Q is a positive-definite matrix. Let p be a direction of descent for f at the point x. Prove that the solution of the exact line search problem

$$\underset{\alpha > 0}{\text{minimize}} \ \ f(x + \alpha p)$$

is

$$\alpha = -\frac{p^T \nabla f(x)}{p^T Q p}.$$

5.4. Let f be the quadratic function in the previous problem, and assume that f is being minimized with an optimization algorithm that uses an exact line search. Prove that the current search direction p_k is orthogonal to the gradient at the new point x_{k+1}.

5.5. Let f be a differentiable function that is being minimized with an optimization algorithm that uses an exact line search. Prove that the current search direction p_k is orthogonal to the gradient at the new point x_{k+1}.

5.6. Why does the sufficient descent condition use the scaled formula

$$-\frac{p_k^T \nabla f(x_k)}{\|p_k\| \cdot \|\nabla f(x_k)\|} \geq \epsilon > 0$$

and not the simpler formula

$$-p^T \nabla f(x_k) \geq \epsilon > 0$$

as a test for descent?

5.7. Prove that $F''(\alpha) = p^T \nabla^2 f(x_k + \alpha p) p$ by using the definition

$$F''(\alpha) = \lim_{h \to 0}[F'(\alpha + h) - F'(\alpha)]/h$$

together with the formula for $F'(\alpha)$ given earlier.

5.8. Consider the objective function from the PET image reconstruction problem described in Section 1.7.5:

$$f_{ML} = -q^T x + \sum_j y_j \log \left(C^T x\right)_j.$$

Let p be the search direction at a feasible point x_k, and let $w = C^T p$. Show how you can use w to calculate the directional derivatives for a sequence of trial values of α, and show that this computation can also be used as part of the forward projection computation at the new point x_{k+1}. Thus one can test the Wolfe condition without the full expense of calculating the derivatives at the trial points.

5.9. Suppose that in a line search procedure the trial step $\hat{\alpha}$ does not satisfy the sufficient decrease condition. One strategy for selecting a new trial step is to approximate $F(\alpha) = f(x_k + \alpha p_k)$ by the one-dimensional quadratic function $q(\alpha)$ that satisfies $q(0) = F(0), q'(0) = F'(0)$, and $q(\hat{\alpha}) = F(\hat{\alpha})$. Determine the coefficients of $q(\alpha)$. Let $\bar{\alpha}$ be the minimizer of $q(\alpha)$. Prove that

$$\bar{\alpha} = -\frac{\hat{\alpha}^2 F'(0)}{2[F(\hat{\alpha}) - F(0) - \hat{\alpha} F'(0)]}.$$

Then $\bar{\alpha}$ can be used as the new trial step in the line search procedure, if $\bar{\alpha}$ is not too small. Prove also that

$$\bar{\alpha} < \frac{\hat{\alpha}}{2(1 - \mu)},$$

where μ is the constant from the sufficient decrease condition.

5.10. Prove that Theorem 11.7 is still true if the backtracking algorithm in the line search chooses α_k as the first element of the sequence

$$1, 1/\kappa, 1/\kappa^2, \ldots, 1/\kappa^i, \ldots$$

to satisfy the sufficient decrease condition, where $\kappa > 1$.

5.11. Show how to determine the cubic polynomial P_3 satisfying

$$P_3(\underline{\alpha}) = F(\underline{\alpha}) \qquad P_3'(\underline{\alpha}) = F'(\underline{\alpha})$$
$$P_3(\bar{\alpha}) = F(\bar{\alpha}) \qquad P_3'(\bar{\alpha}) = F'(\bar{\alpha}),$$

where $\underline{\alpha} < \bar{\alpha}$, $F'(\underline{\alpha}) < 0$, and $F'(\bar{\alpha}) > 0$. Why is this polynomial unique? What is the formula for the unique local minimizer of P_3 in the interval $[\underline{\alpha}, \bar{\alpha}]$? You may wish to write the polynomial in the form

$$P_3(\alpha) = c_1 + c_2(\alpha - \underline{\alpha}) + c_3(\alpha - \underline{\alpha})^2 + c_4(\alpha - \underline{\alpha})^2(\alpha - \bar{\alpha}).$$

Verify that the polynomials obtained in Example 11.9 are correct.

5.12. A line search using the Wolfe condition can also be designed using *quadratic* polynomials. In this case a quadratic polynomial P_2 is computed that satisfies

$$P_2(\underline{\alpha}) = F(\underline{\alpha}), \qquad P_2'(\underline{\alpha}) = F'(\underline{\alpha}), \quad \text{and} \quad P_2(\bar{\alpha}) = F(\bar{\alpha}).$$

Determine the coefficients of P_2, assuming that $\underline{\alpha} < \bar{\alpha}$, $F'(\underline{\alpha}) < 0$, and $F'(\bar{\alpha}) > 0$. Also determine the formula for the unique local minimizer of P_2 in the interval $[\underline{\alpha}, \bar{\alpha}]$. You may wish to write the polynomial in the form

$$P_2(\alpha) = c_1 + c_2(\alpha - \underline{\alpha}) + c_3(\alpha - \underline{\alpha})^2.$$

Apply your line search to the function in Example 11.9.

5.13. Suppose that the modified Newton direction p is computed using

$$\nabla^2 f(x) + E = LDL^T,$$

where $\|E\| \leq C \left\|\nabla^2 f(x)\right\|$ for some constant C, and where the smallest eigenvalue of $\nabla^2 f(x) + E$ is greater than or equal to $\gamma > 0$. For appropriate constants ϵ and $m > 0$, prove that the direction p is a sufficient descent direction:

$$-\frac{p^T \nabla f(x)}{\|p\| \cdot \|\nabla f(x)\|} \geq \epsilon > 0$$

and is gradient related:

$$\|p_k\| \geq m \|\nabla f(x_k)\|.$$

5.14. (The goal of this problem is to prove that <u>if Newton's method converges when an exact line search is used, then it converges quadratically</u>.) Suppose that Newton's method is used to solve

$$\text{minimize } f(x)$$

and that an exact line search is used at every iteration. Assume that the classical Newton method (that is, Newton's method with no line search) converges quadratically to x_*, and that $\nabla^2 f(x)$ is Lipschitz continuous on $\Re^n$.

√ (i) Let α_k be the step length from the exact line search. Prove that

$$\alpha_k = 1 + O(\|\nabla f(x_k)\|)$$

for all sufficiently large k. *Hint*: Let $F(\alpha) = f(x_k + \alpha p_k)$ and expand the condition $F'(\alpha) = 0$ in a Taylor series.

(ii) Prove that

$$\|x_{k+1} - x_*\| = O(\|\nabla f(x_k)\|^2) + O(\|x_k - x_*\|^2).$$

(iii) Prove that

$$\|x_{k+1} - x_*\| = O(\|x_k - x_*\|^2).$$

5.15. (The goal of this problem is to prove a convergence theorem for a line search algorithm based on the Wolfe condition.) We will assume that an algorithm is available to compute a step length α_k satisfying

$$f(x_k + \alpha_k p_k) \leq f(x_k) + \mu \alpha_k p_k^T \nabla f(x_k)$$
$$p_k^T \nabla f(x_k + \alpha_k p_k) \geq \eta p_k^T \nabla f(x_k),$$

with $0 < \mu < \eta < 1$. This is a less stringent form of the Wolfe condition presented earlier and is known as the *weak Wolfe condition*. We will also assume that the search directions satisfy a sufficient descent condition

$$-\frac{p_k^T \nabla f(x_k)}{\|p_k\| \cdot \|\nabla f(x_k)\|} \geq \epsilon > 0.$$

Let f be a real-valued function of n variables, and let x_0 be some given initial point. Assume that (i) the set $S = \{ x : f(x) \leq f(x_0) \}$ is bounded, and (ii) ∇f is Lipschitz continuous in S, that is, there exists a constant $L > 0$ such that

$$\|\nabla f(x) - \nabla f(\bar{x})\| \leq L \|x - \bar{x}\|$$

for all $x, \bar{x} \in S$.

(i) Prove that

$$(\eta - 1) p_k^T \nabla f(x_k) \leq p_k^T (\nabla f(x_{k+1}) - \nabla f(x_k)) \leq \alpha_k L \| p_k \|^2 .$$

(ii) Prove that

$$f(x_{k+1}) \leq f(x_k) - c \cos^2 \theta_k \| \nabla f(x_k) \|^2 ,$$

where $c = \mu(1 - \eta)/L$ and θ_k is the angle between p_k and $-\nabla f(x_k)$.

(iii) Prove that

$$\sum_{k=0}^{\infty} \cos^2 \theta_k \| \nabla f(x_k) \|^2 < \infty.$$

(iv) Prove that

$$\lim_{k \to \infty} \| \nabla f(x_k) \| = 0.$$

5.16. In this problem we examine conditions that guarantee the existence of a step size that satisfies both the sufficient decrease and the weak Wolfe conditions. Let f be a differentiable function, and let p be a descent direction for f at the point x_k. Let $F(\alpha) = f(x_k + \alpha p)$, and assume that F is bounded below for all positive α. Let $0 < \mu < 1$ denote the parameter in the sufficient decrease condition, and let $0 < \eta < 1$ denote the parameter of the weak Wolfe condition.

(i) Prove that $F(\alpha) < F(0) + \mu \alpha F'(0)$ for all sufficiently small $\alpha > 0$, and that $F(\alpha) > F(0) + \mu \alpha F'(0)$ for all sufficiently large α. Use this to prove that there exists some α such that

$$F(\alpha) = F(0) + \mu \alpha F'(0).$$

Let $\bar{\alpha}$ be the smallest step size that satisfies this equation. Prove that the sufficient descent condition is satisfied for any $0 < \alpha < \bar{\alpha}$.

(ii) Prove that there exists a scalar $0 < \hat{\alpha} < \bar{\alpha}$ for which

$$F'(\hat{\alpha}) = \nabla f(x_k + \hat{\alpha} p_k)^T p = \mu F'(0).$$

(iii) Suppose that $\mu < \eta$. Prove that the weak Wolfe condition and the sufficient decrease condition are satisfied for any positive α in a neighborhood of $\hat{\alpha}$.

5.17. This problem shows that, in an algorithm where the search directions approach the Newton direction, the step length $\alpha_k = 1$ satisfies the weak Wolfe condition and the sufficient decrease condition for all large k. We assume here that $0 \leq \mu \leq \frac{1}{2}$, and that $\mu \leq \eta \leq 1$, where μ and η are the parameters of the sufficient decrease condition and the Wolfe condition, respectively. We also assume that the function f has two continuous derivatives, and that the iterates are generated using $x_{k+1} = x_k + \alpha p_k$, where $p_k^T \nabla f(x_k) < 0$. Suppose that x_k converges to a point x_* at which $\nabla^2 f(x_*)$ is positive definite. Suppose also that

$$\lim_{k \to \infty} \frac{\left\| \nabla^2 f(x_k) p_k + \nabla f(x_k) \right\|}{\| p_k \|} = 0.$$

(i) Prove that there exists a $\gamma > 0$ such that

$$-p_k^T \nabla f(x_k) \geq \gamma \, \|p_k\|^2 \, .$$

Hint: Use the fact that

$$-p_k^T \nabla f(x_k) = p_k^T \nabla^2 f(x_k) p_k - p_k^T [\nabla^2 f(x_k) p_k + \nabla f(x_k)].$$

(ii) Prove that there exists a point ξ_k on the line segment between x_k and $x_k + p_k$ such that

$$f(x_k + p_k) - f(x_k) - \tfrac{1}{2} p_k^T \nabla f(x_k) = \tfrac{1}{2} p_k^T [\nabla^2 f(\xi_k) p_k + \nabla f(x_k)].$$

Use this to prove that

$$f(x_k + p_k) - f(x_k) - \tfrac{1}{2} p_k^T \nabla f(x_k) \leq (\tfrac{1}{2} - \mu) \eta \, \|p_k\|^2$$

for all large k, and hence a step length of 1 satisfies the sufficient decrease condition for all large k.

(iii) Prove that there exists a ζ_k on the line segment between x_k and $x_k + p_k$ such that

$$p_k^T \nabla f(x_k + p_k) = p_k^T [\nabla^2 f(\zeta_k) p_k + \nabla f(x_k)],$$

and conclude from this that

$$p_k^T \nabla f(x_k + p_k) \leq \gamma \eta \, \|p_k\|^2 \leq -\eta p_k^T \nabla f(x_k)$$

for all large k, and hence a step length of 1 satisfies the weak Wolfe conditions for all large k.

5.18. Write a computer program for minimizing a multivariate function using a modified Newton algorithm. If, in the Cholesky factorization of the Hessian, the diagonal entry $d_{i,i} \leq 0$, replace it by $\max \{ |d_{i,i}|, 10^{-2} \}$. Include the following:

(i) Use a backtracking line search as described in this section.

(ii) Accept x as a solution if $\|\nabla f(x)\| / (1 + |f(x)|) \leq \epsilon$, or if the number of iterations exceeds ITMAX. Use $\epsilon = 10^{-8}$ and ITMAX $= 1000$.

(iii) Print out the initial point, and then at each iteration print the search direction, the step length α, and the new estimate of the solution x_{k+1}. (If a great many iterations are required, provide this output only for the first 10 iterations and the final 5 iterations.) Indicate if no solution has been found after ITMAX iterations.

(iv) Test your algorithm on the test problems listed here:

$$f_{(1)}(x) = x_1^2 + x_2^2 + x_3^2, \qquad x_0 = (1, 1, 1)^T$$
$$f_{(2)}(x) = x_1^2 + 2x_2^2 - 2x_1 x_2 - 2x_2, \qquad x_0 = (0, 0)^T$$
$$f_{(3)}(x) = 100(x_2 - x_1^2)^2 + (1 - x_1)^2, \qquad x_0 = (-1.2, 1)^T$$
$$f_{(4)}(x) = (x_1 + x_2)^4 + x_2^2, \qquad x_0 = (2, -2)^T$$
$$f_{(5)}(x) = (x_1 - 1)^2 + (x_2 - 1)^2 + c(x_1^2 + x_2^2 - 0.25)^2, \qquad x_0 = (1, -1)^T.$$

For the final function, test three different settings of the parameter c: $c = 1$, $c = 10$, and $c = 100$. The condition number of the Hessian matrix at the solution becomes larger as c increases. Comment on how this affects the performance of the algorithm.

(v) Are your results consistent with the theory of Newton's method?

11.6 Guaranteeing Convergence: Trust-Region Methods

Trust-region methods offer an alternative framework for guaranteeing convergence. They were first used to solve nonlinear least-squares problems, but have since been adapted to more general optimization problems.

Trust-region methods make explicit reference to a "model" of the objective function. For Newton's method this model is a quadratic model derived from the Taylor series for f about the point x_k:

$$q_k(p) = f(x_k) + \nabla f(x_k)^T p + \tfrac{1}{2} p^T \nabla^2 f(x_k) p.$$

The method will only "trust" this model within a limited neighborhood of the point x_k, defined by the constraint

$$\|p\| \leq \Delta_k.$$

This will serve to limit the size of the step taken from x_k to x_{k+1}. The value of Δ_k is adjusted based on the agreement between the model $q_k(p)$ and the objective function $f(x_k + p)$. If the agreement is good, then the model can be trusted and Δ_k increased. If not, then Δ_k will be decreased. (In the discussion here we assume that $\|\cdot\| = \|\cdot\|_2$, that is, we use the Euclidean norm. Other trust-region algorithms sometimes use different norms for reasons associated with the computation of the step p_k.)

At the early iterations of the method when x_k may be far from the solution x_*, the values of Δ_k may be small and may prevent a full Newton step from being taken. However, at later iterations when x_k is closer to x_*, it is hoped that there will be greater trust in the model. Then Δ_k can be made sufficiently large so that it does not impede Newton's method, and a quadratic convergence rate is achievable.

At iteration k of a trust-region method, the following subproblem is solved to determine the step:

$$\underset{p}{\text{minimize}} \quad q_k(p) = f(x_k) + \nabla f(x_k)^T p + \tfrac{1}{2} p^T \nabla^2 f(x_k) p$$

$$\text{subject to} \quad \|p\| \leq \Delta_k.$$

This is a constrained optimization problem. The optimality conditions for this subproblem (see Section 14.5.1) show that p_k will be the solution of the linear system

$$(\nabla^2 f(x_k) + \lambda I) p_k = -\nabla f(x_k),$$

where $\lambda \geq 0$ is a scalar (called the Lagrange multiplier for the constraint), $(\nabla^2 f(x_k) + \lambda I)$ is positive semidefinite, and

$$\lambda(\Delta_k - \|p_k\|) = 0.$$

We will not derive this result here.

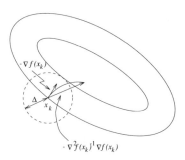

Figure 11.5. *Piecewise linear approximation to trust-region curve.*

If $\nabla^2 f(x_k)$ is positive definite and Δ_k is sufficiently large, then the solution of the subproblem is the solution to

$$\nabla^2 f(x_k)p = -\nabla f(x_k),$$

the Newton equations. Without these assumptions, the method guarantees that

$$\Delta_k \geq \|p_k\| = \left\|(\nabla^2 f(x_k) + \lambda I)^{-1}\nabla f(x_k)\right\|,$$

and so if $\Delta_k \to 0$, then $\lambda \to \infty$ and

$$p_k \approx -\frac{1}{\lambda}\nabla f(x_k).$$

(For a more rigorous demonstration of this, see the Exercises.) Hence p_k is a function of λ and indirectly a function of Δ_k. As λ varies between 0 and ∞, it can be shown that $p_k = p_k(\lambda)$ varies continuously between the Newton direction (in the positive-definite case) and a multiple of $-\nabla f(x_k)$. This is illustrated in Figure 11.5. In the figure, the arc shows the values of $p_k(\lambda)$. As $\lambda \to \infty$, $p_k(\lambda)$ points in the direction of the negative gradient. For $\lambda = 0$, $p_k(0)$ is the Newton direction.

This approach is in sharp contrast to a line search method, where the search direction is chosen (perhaps using the Newton equations) but then left fixed while the step length is computed. In a trust-region method the choice of the bound Δ_k affects both the length and the direction of the step p_k (but the step length is always one or zero).

We now specify the steps in a simple trust-region algorithm based on Newton's method.

ALGORITHM 11.2.
Trust-Region Algorithm

1. Specify some initial guess of the solution x_0. Select the initial trust-region bound $\Delta_0 > 0$. Specify the constants $0 < \mu < \eta < 1$ (perhaps $\mu = \frac{1}{4}$ and $\eta = \frac{3}{4}$).
2. For $k = 0, 1, \ldots$

    (i) If x_k is optimal, stop.

(ii) Solve

$$\underset{p}{\text{minimize}} \quad q_k(p) = f(x_k) + \nabla f(x_k)^T p + \tfrac{1}{2} p^T \nabla^2 f(x_k) p$$
$$\text{subject to} \quad \|p\| \le \Delta_k$$

for the trial step p_k.

(iii) Compute

$$\rho_k = \frac{f(x_k) - f(x_k + p_k)}{f(x_k) - q_k(p_k)} = \frac{\text{actual reduction}}{\text{predicted reduction}}.$$

(iv) If $\rho_k \le \mu$, then $x_{k+1} = x_k$ (unsuccessful step), else $x_{k+1} = x_k + p_k$ (successful step).

(v) Update Δ_k:

$$\rho_k \le \mu \implies \Delta_{k+1} = \tfrac{1}{2}\Delta_k$$
$$\mu < \rho_k < \eta \implies \Delta_{k+1} = \Delta_k$$
$$\rho_k \ge \eta \implies \Delta_{k+1} = 2\Delta_k.$$

The value of ρ_k indicates how well the model predicts the reduction in the function value. If ρ_k is small (that is, $\rho_k \le \mu$), then the actual reduction in the function value is much smaller than that predicted by $q_k(p_k)$, indicating that the model cannot be trusted for a bound as large as Δ_k; in this case the step p_k will be rejected and Δ_k will be reduced. If ρ_k is large (that is, $\rho_k \ge \eta$), then the model is adequately predicting the reduction in the function value, suggesting that the model can be trusted over an even wider region; in this case the bound Δ_k will be increased.

Example 11.10 (Trust-Region Method). Consider the unconstrained minimization problem

$$\text{minimize } f(x_1, x_2) = (x_1^4 + 2x_1^3 + 24x_1^2) + (x_2^4 + 12x_2^2)$$

with initial guess $x_0 = (2, 1)^T$ and initial trust-region bound $\Delta_0 = 1$. At the initial point,

$$f(x_0) = 141, \quad \nabla f(x_0) = \begin{pmatrix} 152 \\ 28 \end{pmatrix}, \quad \text{and} \quad \nabla^2 f(x_0) = \begin{pmatrix} 120 & 0 \\ 0 & 36 \end{pmatrix}.$$

The Newton direction is

$$p_N = -\nabla^2 f(x_0)^{-1} \nabla f(x_0) = \begin{pmatrix} -\frac{152}{120} \\ -\frac{28}{36} \end{pmatrix} \approx \begin{pmatrix} -1.2667 \\ -0.7778 \end{pmatrix}.$$

Since $\|p_N\| = 1.4864 > \Delta_0 = 1$, the Newton step cannot be used.

The trust-region step can be obtained by finding a scalar λ such that $\|p\| = 1$, where p is the solution to

$$\begin{pmatrix} 120 + \lambda & 0 \\ 0 & 36 + \lambda \end{pmatrix} \begin{pmatrix} p_1 \\ p_2 \end{pmatrix} = -\begin{pmatrix} 152 \\ 28 \end{pmatrix}.$$

A simple calculation shows that λ must satisfy

$$\left(\frac{152}{120 + \lambda} \right)^2 + \left(\frac{28}{36 + \lambda} \right)^2 = 1.$$

This equation can be solved numerically to obtain $\lambda \approx 42.655$. (Software for a trust-region method would typically find only an approximate solution to this nonlinear equation.) Hence the trust-region step is

$$p_0 = \begin{pmatrix} -0.9345 \\ -0.3560 \end{pmatrix}.$$

It is easy to verify that $\|p_0\| = 1$.

For this step the trust-region model has the value

$$q_0(p_0) = f(x_0) + \nabla f(x_0)^T p_0 + \tfrac{1}{2} p_0^T \nabla^2 f(x_0) p_0 = 43.6680.$$

The function value at $x_0 + p_0$ is $f(x_0 + p_0) = 39.8420$. Hence the ratio of actual to predicted reduction is

$$\rho_0 = \frac{f(x_0) - f(x_0 + p_0)}{f(x_0) - q_0(p_0)} = \frac{141 - 39.8420}{141 - 43.6680} = 1.0393.$$

If we use the constants $\mu = \tfrac{1}{4}$ and $\eta = \tfrac{3}{4}$ in the trust-region algorithm, then $\rho_0 > \eta$, the step is successful, $x_1 = x_0 + p_0 = (1.0655, 0.6440)^T$, and $\Delta_1 = 2\Delta_0 = 2$. ∎

The solution of the trust-region subproblem

$$\underset{p}{\text{minimize}} \quad q_k(p) = f(x_k) + \nabla f(x_k)^T p + \tfrac{1}{2} p^T \nabla^2 f(x_k) p$$

$$\text{subject to} \quad \|p\| \le \Delta_k$$

is difficult. For example, if the Newton direction does not satisfy the constraint, then it is necessary to determine λ so that

$$\|p_k\| = \left\| (\nabla^2 f(x_k) + \lambda I)^{-1} \nabla f(x_k) \right\| = \Delta_k.$$

This is a nonlinear equation in λ. In addition, it is necessary to choose λ so that $(\nabla^2 f(x_k) + \lambda I)$ is positive definite, adding a further complication.

Practical trust-region methods find only an approximate solution to the trust-region subproblem. One approach is to find an approximate solution to this nonlinear equation. Newton's method could be applied, but more efficient special methods have been derived. Even these special methods are computationally expensive, however. For each new estimate of λ, a linear system involving the matrix $(\nabla^2 f(x_k) + \lambda I)$ must be solved.

A second approach to solving the subproblem is to approximate $p_k(\lambda)$, the curve of steps defined as λ varies between 0 and ∞, by a simpler curve. It is common to use a piecewise linear curve, that is, a sequence of line segments. For $\lambda \approx 0$ the line segment should coincide with the Newton step (in the positive-definite case), and as $\lambda \to \infty$ it should coincide with $-\nabla f(x_k)$. It is easy to determine which point on this sequence of line segments solves the subproblem. The sequence of line segments is sometimes called a "dogleg" by analogy with the game of golf. For further information on both these approaches, see the book by Dennis and Schnabel (1983, reprinted 1996).

Both line search and trust-region methods have been used as the basis for optimization software. Experiments have shown them to be comparably efficient on average, although on individual problems their performance can differ. There are sometimes subtle reasons for preferring one approach over the other. Some of these reasons are discussed in the paper by Dennis and Schnabel (1989).

We now state a convergence theorem that is analogous to the convergence theorem for line search methods. The conclusion is the same, and the assumptions on the problem are similar. Additional theoretical results, including a discussion of second-order optimality conditions, can be found in the paper by Moré (1983). See the Notes.

Theorem 11.11. *Let f be a real-valued function of n variables. Let x_0 be some given initial point, and let $\{x_k\}$ be defined by the trust-region algorithm above. Assume that*

(i) *the set $S = \{x : f(x) \leq f(x_0)\}$ is bounded, and*

(ii) *f, ∇f, and $\nabla^2 f$ are continuous for all $x \in S$.*

Then

$$\lim_{k \to \infty} \|\nabla f(x_k)\| = 0.$$

Proof. The proof will be in two parts. In the first part, we will prove that a *subsequence* of $\{\|\nabla f(x_k)\|\}$ converges to zero. The proof is by contradiction. If no such subsequence converges to zero, then for all sufficiently large values of k, $\|\nabla f(x_k)\| \geq \epsilon > 0$ where ϵ is some constant. Since we are interested only in the asymptotic behavior of the algorithm, we can ignore the early iterations, so we may as well assume that $\|\nabla f(x_k)\| \geq \epsilon$ for all k. This simplifies the argument.

The first part of the proof has five major steps. The first two steps establish relationships among the quantities $f(x_k) - f(x_{k+1})$, $q_k(p_k)$, Δ_k, and $\|\nabla f(x_k)\|$. These relationships are valid in general. The remaining steps use the assumption that $\|\nabla f(x_k)\| \geq \epsilon$ to obtain a contradiction. Step 3 shows that $\lim \Delta_k = 0$. If Δ_k is small, then so is $\|p_k\|$, and the quadratic model must be a good prediction of the actual reduction in the function value (that is, $\lim \rho_k = 1$). If this is true, then the algorithm will not reduce Δ_k (that is, $\lim \Delta_k \neq 0$), thus contradicting the result of step 3 and proving the overall result.

Throughout the proof we denote $\nabla f_k = \nabla f(x_k)$ and $\nabla^2 f_k = \nabla^2 f(x_k)$. In addition, let M be a constant satisfying $\|\nabla^2 f_k\| \leq M$ for all k. The upper bound M exists because $\nabla^2 f$ is continuous and the set S is closed and bounded.

1. A bound on the predicted reduction: We prove that

$$f(x_k) - q_k(p_k) \geq \tfrac{1}{2} \|\nabla f_k\| \cdot \min\left\{\Delta_k, \frac{\|\nabla f_k\|}{M}\right\}$$

by examining how small q_k could be if p_k were a multiple of $-\nabla f_k$. To do this we define the function

$$\phi(\alpha) \equiv q_k\left(-\alpha \frac{\nabla f_k}{\|\nabla f_k\|}\right) - f(x_k)$$

$$= -\alpha \frac{\nabla f_k^T \nabla f_k}{\|\nabla f_k\|} + \tfrac{1}{2}\alpha^2 \frac{\nabla f_k^T (\nabla^2 f_k) \nabla f_k}{\|\nabla f_k\|^2}$$

$$= -\alpha \|\nabla f_k\| + \frac{1}{2}\alpha^2 M_k,$$

where $M_k = \nabla f_k^T (\nabla^2 f_k) \nabla f_k / \|\nabla f_k\|^2 \leq \|\nabla^2 f_k\| \leq M$. Let α_* be the minimizer of ϕ on the interval $[0, \Delta_k]$. Note that $\alpha_* > 0$. If $0 < \alpha_* < \Delta_k$, then α_* can be determined by setting $\phi'(\alpha) = 0$, showing that $\alpha_* = \|\nabla f_k\| / M_k$ and

$$\phi(\alpha_*) = -\tfrac{1}{2} \|\nabla f_k\|^2 / M_k \le -\tfrac{1}{2} \|\nabla f_k\|^2 / M.$$

On the other hand, suppose that $\alpha_* = \Delta_k$. It follows that $M_k \Delta_k \le \|\nabla f_k\|$ (if $M_k \le 0$, then this is trivially satisfied; otherwise, this is a consequence of setting $\phi'(\alpha) = 0$, since the solution of this equation must be $\ge \Delta_k$). Thus

$$\phi(\alpha_*) = \phi(\Delta_k) = -\Delta_k \|\nabla f_k\| + \tfrac{1}{2}\Delta_k^2 M_k \le -\tfrac{1}{2}\Delta_k \|\nabla f_k\|.$$

Finally, the desired result is obtained by noting that $q_k(p_k) - f(x_k) \le \phi(\alpha_*)$.

2. A bound on $f(x_k) - f(x_{k+1})$: If a successful step is taken, then

$$\mu \le \rho_k = \frac{f(x_k) - f(x_{k+1})}{f(x_k) - q_k(p_k)},$$

where μ is the constant used to test ρ_k in the algorithm. Hence

$$f(x_k) - f(x_{k+1}) \ge (f(x_k) - q_k(p_k))\mu \ge \frac{1}{2}\mu \|\nabla f_k\| \cdot \min\left\{ \Delta_k, \frac{\|\nabla f_k\|}{M} \right\}$$

using the result of step 1.

3. $\lim \Delta_k = 0$: First, note that $\lim f(x_k)$ exists and is finite (f is bounded below on S, and the algorithm ensures that f cannot increase at any iteration). If, as in our contrary assumption, $\|\nabla f_k\| \ge \epsilon > 0$, and if k is a successful step, then step 2 shows that

$$f(x_k) - f(x_{k+1}) \ge \frac{1}{2}\mu\epsilon \cdot \min\left\{ \Delta_k, \frac{\epsilon}{M} \right\}.$$

The limit of the left-hand side is zero, so $\lim \Delta_{k_i} = 0$, where $\{k_i\}$ are the indices of the iterations where successful steps are taken. At successful steps the trust-region bound is either kept constant or doubled, and at unsuccessful steps the bound is reduced. So between successful steps,

$$2\Delta_{k_i} \ge \Delta_{k_i+1} \ge \Delta_{k_i+2} \ge \cdots \ge \Delta_{k_{i+1}}.$$

Thus $\lim \Delta_k = 0$ also.

4. $\lim \rho_k = 1$: Using the remainder form of the Taylor series for $f(x_k + p_k)$, we obtain

$$\begin{aligned}
|f(x_k & + p_k) - q_k(p_k)| \\
&= |f(x_k) + \nabla f_k^T p_k + \tfrac{1}{2} p_k^T \nabla^2 f(x_k + \xi_k p_k) p_k - q_k(p_k)| \\
&= |-\tfrac{1}{2} p_k^T (\nabla^2 f_k) p_k + \tfrac{1}{2} p_k^T \nabla^2 f(x_k + \xi_k p_k) p_k| \\
&\le \tfrac{1}{2} M \|p_k\|^2 + \tfrac{1}{2} M \|p_k\|^2 \\
&= M \|p_k\|^2 \le M \Delta_k^2.
\end{aligned}$$

Using the bound from step 1 and the result of step 3, we obtain

$$f(x_k) - q_k(p_k) \ge \tfrac{1}{2}\epsilon \Delta_k$$

for large values of k. Hence

$$
\begin{aligned}
|\rho_k - 1| &= \left| \frac{f(x_k) - f(x_k + p_k)}{f(x_k) - q_k(p_k)} - 1 \right| \\
&= \frac{|f(x_k + p_k) - q_k(p_k)|}{|f(x_k) - q_k(p_k)|} \\
&\leq \frac{M \Delta_k^2}{\frac{1}{2} \epsilon \Delta_k} = \frac{2M}{\epsilon} \Delta_k \to 0.
\end{aligned}
$$

This is the desired result.

5. $\lim \Delta_k \neq 0$: If $\lim \rho_k = 1$, then for large values of k the algorithm will not decrease Δ_k. Hence Δ_k will be bounded away from zero.

This is the desired contradiction establishing that a subsequence of the sequence $\{\, \|\nabla f(x_k)\| \,\}$ converges to zero. This completes the first part of the proof.

The second part of the proof shows that $\lim \|\nabla f_k\| = 0$. This also is proved by contradiction. If this result is not true, then $\left\| \nabla f_{k_i} \right\| \geq \epsilon > 0$ for some subset $\{k_i\}$ of the iterations of the algorithm. (This may be a different ϵ than used above.) However, since a subsequence of $\{\, \|\nabla f(x_k)\| \,\}$ converges to zero, there must exist a set of indices $\{\ell_i\}$ such that

$$
\|\nabla f_k\| \geq \tfrac{1}{4} \epsilon \quad \text{for } k_i \leq k < \ell_i
$$
$$
\left\| \nabla f_{\ell_i} \right\| < \tfrac{1}{4} \epsilon.
$$

If $k_i \leq k < \ell_i$ and iteration k is successful, then step 2 above shows that

$$
f(x_k) - f(x_{k+1}) \geq \frac{1}{2} \mu \left(\frac{1}{4} \epsilon \right) \cdot \min \left\{ \Delta_k, \frac{\frac{1}{4} \epsilon}{M} \right\}.
$$

The left-hand side of this inequality goes to zero, so that

$$
f(x_k) - f(x_{k+1}) \geq \epsilon_1 \|x_{k+1} - x_k\|,
$$

where $\epsilon_1 = \frac{1}{8} \mu \epsilon$. Because $\|x_{k+1} - x_k\| = 0$ for an unsuccessful step, this result is valid for $k_i \leq k < \ell_i$. Using this result repeatedly, we obtain

$$
\begin{aligned}
\epsilon_1 & \left\| x_{k_i} - x_{\ell_i} \right\| \\
&\leq \epsilon_1 (\left\| x_{k_i} - x_{k_i+1} \right\| + \left\| x_{k_i+1} - x_{k_i+2} \right\| + \cdots + \left\| x_{\ell_i-1} - x_{\ell_i} \right\|) \\
&\leq f(x_{k_i}) - f(x_{k_i+1}) + f(x_{k_i+1} - f(x_{k_i+2}) + \cdots + f(x_{\ell_i-1}) - f(x_{\ell_i}) \\
&= f(x_{k_i}) - f(x_{\ell_i}).
\end{aligned}
$$

Since the right-hand side of this result goes to zero, the left-hand side can be made arbitrarily small. Because $\nabla f(x)$ is continuous on the set S, and S is closed and bounded, by choosing i large enough it is possible to guarantee that

$$
\left\| \nabla f_{k_i} - \nabla f_{\ell_i} \right\| \leq \tfrac{1}{4} \epsilon.
$$

We are now ready to obtain the desired contradiction:

$$\epsilon \leq \left\| \nabla f_{k_i} \right\| = \left\| (\nabla f_{k_i} - \nabla f_{\ell_i}) + \nabla f_{\ell_i} \right\|$$
$$\leq \left\| \nabla f_{k_i} - \nabla f_{\ell_i} \right\| + \left\| \nabla f_{\ell_i} \right\| \leq \tfrac{1}{4}\epsilon + \tfrac{1}{4}\epsilon = \tfrac{1}{2}\epsilon < \epsilon.$$

Hence $\lim \left\| \nabla f(x_k) \right\| = 0$. $\square$

Exercises

6.1. Perform an additional iteration of the trust-region method in Example 11.10.

6.2. Suppose that, in a trust-region method, $p_k = \alpha v$ where v is some nonzero vector. Show how to determine α so that p_k solves the trust-region subproblem

$$\underset{p}{\text{minimize}} \quad q_k(p) = f(x_k) + \nabla f(x_k)^T p + \tfrac{1}{2} p^T \nabla^2 f(x_k) p$$
$$\text{subject to} \quad \|p\| \leq \Delta_k.$$

6.3. Extend the result of the previous problem to the case where p_k is constrained to lie on a piecewise linear path. That is,

$$p_k = \begin{cases} \alpha v_1 & \text{if } \|p_k\| \leq \delta_1; \\ \alpha_1 v_1 + \alpha v_2 & \text{if } \delta_1 \leq \|p_k\| \leq \delta_2; \\ \alpha_1 v_1 + \alpha_2 v_2 + \alpha v_3 & \text{if } \delta_2 \leq \|p_k\| \leq \delta_3; \\ \text{etc.} \end{cases}$$

6.4. Theorem 11.11 assumes that $\nabla^2 f$ is continuous for all $x \in S$. Prove that the theorem is still true even if $\nabla^2 f$ is only bounded for all $x \in S$.

6.5. Define $p(\lambda)$ by

$$(\nabla^2 f(x) + \lambda I) p(\lambda) = -g(x).$$

(i) If $\nabla^2 f(x)$ is nonsingular, prove that

$$\lim_{\lambda \to 0} p(\lambda) = -\nabla^2 f(x)^{-1} \nabla f(x).$$

(ii) Prove that

$$\lim_{\lambda \to +\infty} \lambda p(\lambda) = -\nabla f(x),$$

and hence prove $p(\lambda) \approx -(1/\lambda) \nabla f(x)$ for sufficiently large λ.

(iii) Use (ii) to prove that

$$\lim_{\lambda \to +\infty} \|p(\lambda)\| = 0.$$

(iv) If $(\nabla^2 f(x) + \lambda I)$ is nonsingular, prove that

$$\frac{d}{d\lambda} \|p(\lambda)\| = -\frac{p(\lambda)^T (\nabla^2 f(x) + \lambda I)^{-1} p(\lambda)}{\|p(\lambda)\|}.$$

Use this result to prove that, if $(\nabla^2 f(x) + \lambda I)$ is positive definite, then $\|p(\lambda)\|$ is a monotone decreasing function of λ.

(v) Let d be a nonzero vector satisfying $d^T \nabla f(x) = 0$. Prove that

$$\lim_{\lambda \to +\infty} \frac{d^T p(\lambda)}{\|p(\lambda)\|} = 0.$$

6.6. We show in Chapter 14 that the solution to the trust-region subproblem satisfies

$$(\nabla^2 f(x_k) + \lambda I)p_k = -\nabla f(x_k)$$
$$\lambda(\Delta_k - \|p_k\|) = 0$$
$$(\nabla^2 f(x_k) + \lambda I) \text{ is positive semidefinite}$$

for some $\lambda \geq 0$. Use these conditions to answer the following questions.

(i) If $\lambda \neq 0$, prove that p_k solves

$$\begin{aligned} \text{minimize} \quad & q_k(p) \\ \text{subject to} \quad & \|p\| = \Delta_k. \end{aligned}$$

Hint: Prove that, for any p, $q_k(p_k) \leq q_k(p) + \frac{1}{2}\lambda(p^T p - p_k^T p_k)$.

(ii) If $\lambda = 0$ and $\nabla^2 f(x_k)$ is positive definite, prove that p_k solves the trust-region subproblem.

(iii) If the trust-region subproblem has no solution such that $\|p_k\| = \Delta_k$, prove that $\nabla^2 f(x_k)$ is positive definite and

$$\left\| \nabla^2 f(x_k)^{-1} \nabla f(x_k) \right\| < \Delta_k.$$

6.7. Assume that $\nabla^2 f(x)$ is positive definite, and let p_N be the Newton direction at x. Let p_c be the solution to

$$\begin{aligned} \underset{\alpha}{\text{minimize}} \quad & q(-\alpha \nabla f(x)) \\ \text{subject to} \quad & \|\alpha \nabla f(x)\| \leq \Delta. \end{aligned}$$

(i) Find a formula for $\bar{\alpha}$ satisfying $p_c = -\bar{\alpha} \nabla f(x)$.

(ii) Define

$$p(\alpha) = p_c + \alpha(p_N - pc)$$

for $0 \leq \alpha \leq 1$. Prove that $\|p(\alpha)\|$ is strictly monotone increasing as a function of α. *Hint*: Consider the derivative with respect to α of $\|p(\alpha)\|^2$.

(iii) Prove that $q(p(\alpha))$ is strictly monotone decreasing as a function of α.

(iv) Prove that there is a unique α_* satisfying

$$\|p(\alpha_*)\| = \Delta.$$

11.7   Notes

Superlinear Convergence—Theorem 11.3 is adapted from the paper by Dennis and Moré (1974).

   Guaranteeing Descent—The derivation given above of the modification E to the Hessian ignores a few details. To ensure the convergence of the descent method in Theorem 11.7

(see Section 11.5), we make a number of assumptions about the search directions. These assumptions are guaranteed to be satisfied if both $\nabla^2 f(x) + E$ and $(\nabla^2 f(x) + E)^{-1}$ are bounded. However, if the modification E is chosen carelessly, this will not be true. Safe ways of choosing E are discussed in the papers by Gill and Murray (1974a) and Schnabel and Eskow (1990). These papers also show how a search direction can be chosen in the case where $\nabla f(x) = 0$ and $\nabla^2 f(x)$ is indefinite (that is, x is a stationary point but not a local minimizer of f).

The techniques used in this section are not the only way to salvage Newton's method in the indefinite case. The trust-region approach discussed in Section 11.6 is another, where the Hessian is replaced by a matrix of the form $(\nabla^2 f(x_k) + \lambda I)$ for some $\lambda \geq 0$. Still other ideas are discussed in the book by Gill, Murray, and Wright (1981).

Line Search Methods—An extensive discussion of convergence theory for line search methods can be found in the book by Ortega and Rheinboldt (1970, reprinted 2000). Some practical line search algorithms are described in the papers by Gill and Murray (1974b) and Moré and Thuente (1994). The Wolfe condition and other conditions that can be used to design a line search are discussed in the paper by Wolfe (1969). A summary can be found in the paper by Nocedal (1992).

Trust-Region Methods—The idea of a trust region was first proposed by Levenberg (1944) and Marquardt (1963) as a technique for solving nonlinear least-squares problems. Methods for computing the search direction within a trust-region method are described in the papers by Gay (1981), Sorensen (1982), and Moré and Sorensen (1983). Convergence theory is discussed in the paper by Moré (1983). An extensive overview of trust-region methods can be found in the book by Conn, Gould, and Toint (1987).

The proof of the convergence theorem is originally due to Powell (1975) and Thomas (1975). The assumptions we make are more stringent than necessary. For example, we assume that the trust-region subproblem is solved exactly. This is not necessary. In step 1 of the first part of the proof, we examine how small q_k could be if the step p_k were a multiple of $-\nabla f_k$. As long as an iteration of the method reduces the function value by some nontrivial fraction of this amount, then the conclusion is still true. (In fact, it is not difficult to modify the proof to show this.) A great many practical methods, including the "dogleg" approach of Powell (1970), are capable of achieving this. (See also the papers of Steihaug (1983) and Toint (1981) for results applicable to large problems.)

The assumptions of the theorem are sufficient to prove a stronger result, namely that there is some limit point x_* of the sequence $\{ x_k \}$ for which $\nabla^2 f(x_*)$ is positive semidefinite. (See the papers by Moré and Sorensen mentioned above.) Thus x_* satisfies the second-order necessary conditions for a local minimizer. This stronger result may not hold, however, if the trust-region subproblem is solved approximately using dogleg and related approaches.

Chapter 12

Methods for Unconstrained Optimization

12.1 Introduction

A principal advantage of Newton's method is that it converges rapidly when the current estimate of the variables is close to the solution. It also has disadvantages, and overcoming these disadvantages has led to many ingenious techniques.

In particular, Newton's method can fail to converge, or it can converge to a point that is not a minimum. This is its most serious failing, but one which can be overcome by using strategies that guarantee progress towards the solution at every iteration, such as the line search and trust-region strategies discussed in Chapter 11.

The costs of Newton's method can also be a concern. It requires the derivation, computation, and storage of the second derivative matrix, and the solution of a system of linear equations.

Obtaining the second derivative matrix can be tedious and can be prone to error. An alternative is to automate the calculation of second derivatives, or to use a method that reduces the requirement to compute derivative values. We will consider both of these ideas.

The other costs of Newton's method are the computational costs of applying the method. If there are n variables, and if the problem is not sparse, calculating the Hessian matrix involves calculating and storing about n^2 entries, and solving a linear system requires about n^3 arithmetic operations. If n is small, these costs might be acceptable. For n of moderate size (say, $n < 200$) storing this matrix might be acceptable, but solving a linear system might not be. Also for large n (say, $n > 1000$), even storing this matrix might be undesirable.

Luckily, large problems are frequently sparse, and taking advantage of sparsity can greatly reduce the computational costs of Newton's method and make it a practical tool in such cases. Techniques for exploiting sparsity are mentioned in Chapter 13 and Appendix A.

A major topic of this chapter will be methods for solving unconstrained problems that are compromises to Newton's method and that reduce one or more of these costs. In exchange, these other methods generally have slower rates of convergence. A trade-off is made between the cost per iteration and the number of iterations.

We present two compromises on Newton's method: quasi-Newton methods and the steepest-descent method. Quasi-Newton methods are currently among the most widely used

Newton-type methods for problems of moderate size, where matrices can be stored. The steepest-descent method is an old and widely known method whose costs are low but whose performance is usually atrocious. It illustrates the dangers of compromising too much when using Newton's method.

These methods are based on Newton's method but use a different formula to compute the search direction. They are based on approximating the Hessian matrix in a way that lowers the costs of the algorithm. These methods also have slower convergence rates than Newton's method, and so there is a trade-off between the cost per iteration (higher for Newton's method) and the number of iterations (higher for the other methods). Additional methods are discussed in the next chapter that are suitable for problems with many variables.

Many of these methods can be interpreted as computing the search direction by solving a linear system of equations

$$B_k p = -\nabla f(x_k), \qquad \left(so \ p = -B_k^{-1} \nabla f_k \right)$$

where B_k is a positive-definite matrix. In the case of Newton's method, $B_k = \nabla^2 f(x_k)$, assuming that the Hessian matrix is positive definite. Intuitively, B_k should be some approximation to $\nabla^2 f(x_k)$. This interpretation emphasizes that these methods are compromises on Newton's method, where the degree of compromise reflects the degree to which B_k approximates the Hessian $\nabla^2 f(x_k)$.

In some situations it is desirable to use a method that does not require derivatives. For example, if you wish to optimize a function which does not have derivatives at all points, then Newton's method and related methods cannot be used. Effective derivative-free methods have been developed that do not require the user to compute derivatives and that do not make use of derivative values to find a solution. What is surprising is that these methods have guarantees of convergence comparable to quasi-Newton methods. It is perhaps also surprising that such methods are suitable for parallel computing.

The chapter includes some practical information about the design and use of software for unconstrained optimization, for Newton-like methods, as well as for derivative-free methods. It concludes with a summary of the historical background for these methods.

12.2   Steepest-Descent Method

The steepest-descent method is the simplest Newton-type method for nonlinear optimization. The price for this simplicity is that the method is hopelessly inefficient at solving most problems. The method has theoretical uses, though, in proving the convergence of other methods, and in providing lower bounds on the performance of better algorithms. It is well known, and has been widely used and discussed, and so it is worthwhile to be familiar with it if only to know not to use it on general problems.

The steepest-descent method is old, but not as old as Newton's method. It was invented in the nineteenth century by Cauchy, about two hundred years later than Newton's method. It is much simpler than Newton's method. It does not require the computation of second derivatives, it does not require that a system of linear equations be solved to compute the search direction, and it does not require matrix storage. So in every way it reduces the costs of Newton's method—at least, the costs per iteration. On the negative side it has a slower rate of convergence than Newton's method; it converges only at a linear rate, with a constant that is usually close to one. Hence it often converges slowly—sometimes so slowly that $x_{k+1} - x_k$ is below the precision of computer arithmetic and the method fails. As a result, even though the costs per iteration are low, the overall costs of solving an optimization problem are high.

The steepest-descent method computes the search direction from

$$p_k = -\nabla f(x_k)$$

and then uses a line search to determine $x_{k+1} = x_k + \alpha_k p_k$. Hence the cost of computing the search direction is just the cost of computing the gradient. Since the gradient must be computed to determine if the solution has been found, it is reasonable to say that the search direction is available for free. The search direction is a descent direction if $\nabla f(x_k) \neq 0$; that is, it is a descent direction unless x_k is a stationary point of the function f.

The formula for the search direction can be derived in two ways, both of which have connections with Newton's method. The first derivation is based on a crude approximation to the Hessian. If the formula for Newton's method is used $((\nabla^2 f)p = -\nabla f)$ but with the Hessian approximated by the identity matrix $(\nabla^2 f \approx I)$, then the formula for the steepest-descent method is obtained. This approach, where an approximation to the Hessian is used in the Newton formula, is the basis of the quasi-Newton methods discussed in Section 12.3.

The second, and more traditional, derivation is based on the Taylor series and explains the name "steepest descent." In our derivation of Newton's method, the function value $f(x_k + p)$ was approximated by the first three terms of the Taylor series, and the search direction was obtained by minimizing this approximation. Here we use only the first two terms of the Taylor series:

$$f(x_k + p) \approx f(x_k) + p^T \nabla f(x_k).$$

The intuitive idea is to minimize this approximation to obtain the search direction; however, this approximation does not have a finite minimum in general. Instead, the search direction is computed by minimizing a scaled version of this approximation:

$$\underset{p \neq 0}{\text{minimize}} \ \frac{p^T \nabla f(x_k)}{\|p\| \cdot \|\nabla f(x_k)\|}.$$

The solution is $p_k = -\nabla f(x_k)$ (see the Exercises).

To explain the name "steepest descent," we recall that a descent direction satisfies the condition $p^T \nabla f(x_k) < 0$. Choosing p to minimize $p^T \nabla f(x_k)$ gives the direction that provides the "most" descent possible. In the line search we also required that the search direction satisfy the sufficient descent condition

$$-\frac{p^T \nabla f(x_k)}{\|p_k\| \cdot \|\nabla f(x_k)\|} \geq \epsilon > 0.$$

For steepest descent this condition simplifies to

$$-\frac{-\nabla f(x_k)^T \nabla f(x_k)}{\nabla f(x_k)^T \nabla f(x_k)} = 1 > 0$$

and so is clearly satisfied. The gradient-relatedness condition

$$\|p_k\| \geq m \|\nabla f(x_k)\|$$

is also satisfied, with $m = 1$.

Example 12.1 (The Steepest-Descent Method). We apply the steepest-descent method to a three-dimensional quadratic problem

$$\text{minimize } f(x) = \tfrac{1}{2}x^T Q x - c^T x$$

with

$$Q = \begin{pmatrix} 1 & 0 & 0 \\ 0 & 5 & 0 \\ 0 & 0 & 25 \end{pmatrix} \quad \text{and} \quad c = \begin{pmatrix} -1 \\ -1 \\ -1 \end{pmatrix}.$$

The steepest-descent direction is

$$p_k = -\nabla f(x_k) = -(Qx_k - c).$$

An exact line search is used so that $x_{k+1} = x_k + \alpha_k p_k$ with

$$\alpha_k = -\frac{\nabla f(x_k)^T p_k}{p_k^T Q p_k}$$

(see Exercise 5.3 of Chapter 11). The solution of the minimization problem is

$$x_* = Q^{-1}c = \begin{pmatrix} -1 \\ -\frac{1}{5} \\ -\frac{1}{25} \end{pmatrix}.$$

If the initial guess of the solution is $x_0 = (0, 0, 0)^T$, then

$$f(x_0) = 0, \quad \nabla f(x_0) = \begin{pmatrix} 1 \\ 1 \\ 1 \end{pmatrix}, \quad \|\nabla f(x_0)\| = 1.7321.$$

This implies that the step length is $\alpha_0 = 0.0968$ leading to the following new estimate of the solution:

$$x_1 = \begin{pmatrix} -0.0968 \\ -0.0968 \\ -0.0968 \end{pmatrix}.$$

At this point,

$$f(x_1) = -0.1452, \quad \nabla f(x_1) = \begin{pmatrix} 0.9032 \\ 0.5161 \\ -1.4194 \end{pmatrix}, \quad \|\nabla f(x_1)\| = 1.7598.$$

The next step length is $\alpha_1 = 0.0590$ and the new estimate of the solution is

$$x_2 = \begin{pmatrix} -0.1500 \\ -0.1272 \\ -0.0131 \end{pmatrix}.$$

At the point x_2,

$$f(x_2) = -0.2365, \quad \nabla f(x_2) = \begin{pmatrix} 0.8500 \\ 0.3639 \\ 0.6732 \end{pmatrix}, \quad \|\nabla f(x_2)\| = 1.1437.$$

It takes 216 iterations before the norm of the gradient is less than 10^{-8}. ∎

Now we determine the rate of convergence for the steepest-descent method. In view of Theorem 11.3, superlinear convergence cannot in general be expected. In fact, we show that linear convergence is all that can be guaranteed. Much of our analysis is for the case of a quadratic function:

$$\text{minimize } f(x) = \tfrac{1}{2}x^T Q x - c^T x,$$

where Q is positive definite. Results for the more general nonlinear case are mentioned afterward.

The convergence rate is analyzed using $f(x_k) - f(x_*)$ instead of $\|x_k - x_*\|$ because the analysis is simpler. It can be shown that the two quantities converge at the same rate, using an argument similar to that used in Section 2.7 (see the Exercises). The argument used here is adapted from the book by Luenberger (2003).

Since $x_* = Q^{-1}c$ (or equivalently, $c = Qx_*$) we obtain

$$\begin{aligned}
f(x_k) - f(x_*) &= (\tfrac{1}{2}x_k^T Q x_k - c^T x_k) - (\tfrac{1}{2}x_*^T Q x_* - c^T x_*) \\
&= \tfrac{1}{2}x_k^T Q x_k - (Qx_*)^T x_k - (\tfrac{1}{2}x_*^T Q x_* - (Qx_*)^T x_*) \\
&= \tfrac{1}{2}x_k^T Q x_k - x_*^T Q x_k - (\tfrac{1}{2}x_*^T Q x_* - x_*^T Q x_*) \\
&= \tfrac{1}{2}x_k^T Q x_k - x_*^T Q x_k + \tfrac{1}{2}x_*^T Q x_* \\
&= \tfrac{1}{2}(x_k - x_*)^T Q (x_k - x_*).
\end{aligned}$$

We define $E(x) = \tfrac{1}{2}(x - x_*)^T Q(x - x_*)$. The convergence result is proved using the function $E(x)$ and is based on the following two lemmas.

Lemma 12.2. *Assume that $\{x_k\}$ is the sequence of approximate solutions obtained when the steepest-descent method is applied to the quadratic function $f(x) = \tfrac{1}{2}x^T Q x - c^T x$, and where an exact line search is used. Then*

$$E(x_{k+1}) = \left[1 - \frac{(\nabla f(x_k)^T \nabla f(x_k))^2}{(\nabla f(x_k)^T Q \nabla f(x_k))(\nabla f(x_k)^T Q^{-1} \nabla f(x_k))} \right] E(x_k).$$

Proof. See the Exercises. ◻

Lemma 12.3. *Let Q be a positive-definite matrix. For any vector $y \neq 0$,*

$$\frac{(y^T y)^2}{(y^T Q y)(y^T Q^{-1} y)} \geq 1 - \left[\frac{\text{cond}(Q) - 1}{\text{cond}(Q) + 1} \right]^2,$$

where $\text{cond}(Q)$ *is the condition number of Q (see Appendix A.8).*

Proof. See the book by Luenberger (2003).   □

Lemma 12.4. *Assume that* $\{x_k\}$ *is the sequence of approximate solutions obtained when the steepest-descent method is applied to the quadratic function* $f(x) = \frac{1}{2}x^T Q x - c^T x$ *and where an exact line search is used. Then for any* x_0 *the method converges to the unique minimizer* x_* *of* f, *and furthermore,*

$$f(x_{k+1}) - f(x_*) \leq \left[\frac{\text{cond}(Q) - 1}{\text{cond}(Q) + 1} \right]^2 (f(x_k) - f(x_*)),$$

that is, the method converges linearly.

Proof. This is a direct consequence of the two previous lemmas. Since the rate constant is strictly less than one, the method converges from any starting point.   □

Example 12.5 (Convergence of the Steepest-Descent Method). This convergence theory can be applied to the problem in the previous example. In this case the condition number is $\text{cond}(Q) = 25$ and the corresponding bound on the rate constant is 0.8521. Table 12.1 lists the values of

$$\frac{f(x_{k+1}) - f(x_*)}{f(x_k) - f(x_*)} = \frac{E(x_{k+1})}{E(x_k)}$$

and compares them to this rate constant. For this example, the *observed* rate constant is about 0.84, which is close to but less than the bound given in the theorem. This is typical for the steepest-descent method.   ■

The theorem provides only an upper bound on the rate constant, but in many examples this bound is close to the observed rate constant (see the Exercises). Table 12.2 compares the

Table 12.1. *Observed rate constant.*

$f(x_k)$	$E(x_{k+1})/E(x_k)$	Bound
0	—	0.8521
−0.1452	0.7659	0.8521
−0.2365	0.8077	0.8521
−0.3038	0.8246	0.8521
−0.3560	0.8348	0.8521
−0.3988	0.8379	0.8521
−0.4343	0.8397	0.8521
−0.4640	0.8401	0.8521
−0.4889	0.8404	0.8521
−0.5098	0.8405	0.8521
−0.5274	0.8405	0.8521

Table 12.2. *Rate constants for the steepest-descent method.*

cond(Q)	Constant
1	0
10	0.669421
100	0.960788
1000	0.996008
10000	0.999600
100000	0.999960
1000000	0.999996

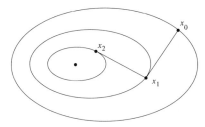

Figure 12.1. *Steepest-descent in two dimensions (part 1).*

bound on the rate constant to various values of the condition number cond(Q). Notice that, even for moderate values of cond(Q), the bound is close to one. Only when the condition number is less than about 50 does this method converge fast enough to be of practical value. For example, if cond(Q) = 100, then the steepest-descent algorithm is guaranteed to improve the solution by only about 4% per iteration.

The convergence theorem given here applies only to quadratic functions. For general nonlinear functions it is possible to show that the steepest-descent method (with an exact line search) also converges linearly, with a rate constant that is bounded by

$$\left[\frac{\text{cond}(Q) - 1}{\text{cond}(Q) + 1} \right]^2,$$

where $Q = \nabla^2 f(x_*)$, the Hessian at the solution. Hence the method behaves much the same way on general functions as it does on quadratic functions.

This poor behavior of the steepest-descent method may be surprising. In Figure 12.1 the method is applied to a two-dimensional quadratic function. The method works well in this case. For this problem, cond(Q) $\approx$ 1 and the rate constant is near 0, so the good performance is confirmed by the theory.

In cases where cond(Q) $\not\approx$ 1, the picture looks more like Figure 12.2. In this case the steepest-descent directions are almost at right angles to the direction of the minimizer and the method performs poorly, as we would expect. It is this particular figure that should be remembered.

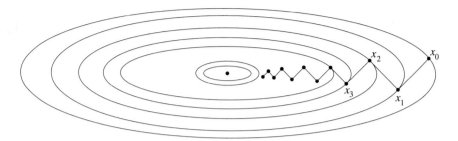

Figure 12.2. *Steepest-descent in two dimensions (part 2).*

Exercises

2.1. Use the steepest-descent method to solve

$$\text{minimize } f(x_1, x_2) = 4x_1^2 + 2x_2^2 + 4x_1x_2 - 3x_1,$$

starting from the point $(2, 2)^T$. Perform three iterations.

2.2. Apply the steepest-descent method, with an exact line search, to the three-dimensional quadratic function $f(x) = \frac{1}{2}x^TQx - c^Tx$ with

$$Q = \begin{pmatrix} 1 & 0 & 0 \\ 0 & \gamma & 0 \\ 0 & 0 & \gamma^2 \end{pmatrix} \quad \text{and} \quad c = \begin{pmatrix} 1 \\ 1 \\ 1 \end{pmatrix}.$$

Here γ is a parameter that can be varied. Try $\gamma = 1, 10, 100, 1000$. How do your results compare with the convergence theory developed above? (If you do this by hand, perform four iterations; if you are using a computer, then it is feasible to perform more iterations.)

2.3. Consider the problem

$$\text{minimize } f(x_1, x_2) = x_1^2 + 2x_2^2.$$

(i) If the starting point is $x_0 = (2, 1)^T$, show that the sequence of points generated by the steepest-descent algorithm is given by

$$x_k = \left(\tfrac{1}{3}\right)^k \begin{pmatrix} 2 \\ (-1)^k \end{pmatrix}$$

if an exact line search is used.

(ii) Show that $f(x_{k+1}) = f(x_k)/9$.

(iii) Compare the results in (ii) to the bounds on the convergence rate of the steepest-descent method when minimizing a quadratic function. What conclusions can you draw regarding this method?

2.4. The method of steepest descent applied to the problem

$$\text{minimize } f(x_1, x_2) = 4x_1^2 + x_2^2$$

generates a sequence of points $\{ x_k \}$.

(i) If $x_0 = (1, 4)^T$, show that

$$x_k = (0.6)^k \begin{pmatrix} (-1)^k \\ 4 \end{pmatrix}.$$

(ii) What is the minimizer x_* of f? What is $f(x_*)$? What is the rate of convergence of the sequence $\{ f(x_k) - f(x_*) \}$?

2.5. Suppose that the steepest-descent method (with an exact line search) is used to minimize the quadratic function

$$f(x) = \tfrac{1}{2} x^T Q x - c^T x,$$

where Q is a positive-definite matrix. Prove that

$$\nabla f_{k+1} = \nabla f_k - \frac{\nabla f_k^T \nabla f_k}{\nabla f_k^T Q \nabla f_k} Q \nabla f_k,$$

where $\nabla f_k = \nabla f(x_k)$ and $\nabla f_{k+1} = \nabla f(x_{k+1})$.

2.6. In this problem we will derive the subproblem used to determine the steepest-descent direction from the Taylor series.

(i) Prove that the subproblem

$$\underset{p}{\text{minimize }} p^T \nabla f(x_k)$$

does not have a finite solution unless $\nabla f(x_k) = 0$.

(ii) If we normalize the vectors p and $\nabla f(x_k)$ by dividing each of them by their norms, we obtain the subproblem

$$\underset{p \neq 0}{\text{minimize }} \frac{p^T \nabla f(x_k)}{\| p \| \cdot \| \nabla f(x_k) \|}.$$

Solve this problem using the formula $p^T \nabla f(x_k) = \| p \| \cdot \| \nabla f(x_k) \| \cos \theta$, where θ is the angle between p and $\nabla f(x_k)$.

2.7. Are there starting points for which the steepest-descent algorithm terminates in one iteration? This problem will partly address this question for the case of a strictly convex quadratic function. Consider the problem

$$\text{minimize } f(x) = \tfrac{1}{2} x^T Q x - c^T x,$$

where Q is a positive-definite matrix. Let x_* be the minimizer of this function. Let v be an eigenvector of Q, and let λ be the associated eigenvalue. Suppose now that the starting point for the steepest-descent algorithm is $x_0 = x_* + v$.

    (i) Prove that the gradient at x_0 is $\nabla f(x_0) = \lambda v$.

  (ii) Prove that if the steepest-descent direction is taken, then the step length which minimizes f in this direction is $\alpha_0 = 1/\lambda$.

 (iii) Prove that the steepest-descent direction with an accurate step length will lead to the minimum of the function f in one iteration.

 (iv) Confirm this result for the function

$$f(x) = 3x_1^2 - 2x_1 x_2 + 3x_2^2 + 2x_1 - 6x_2.$$

Suppose that the starting point is $x_0 = (1, 2)^T$; compute the point obtained by one iteration of the steepest-descent algorithm. Prove that the point obtained is the unique minimum x_*. Verify that $x_0 - x_*$ is an eigenvector of the Hessian matrix.

2.8. Prove Lemma 12.2.

2.9. Suppose that $\lim x_k = x_*$, where x_* is a local minimizer of the nonlinear function f (the sequence $\{x_k\}$ need not come from the steepest-descent method). Assume that $\nabla^2 f(x_*)$ is positive definite. Prove that the sequence $\{f(x_k) - f(x_*)\}$ converges linearly if and only if $\{\|x_k - x_*\|\}$ converges linearly. Prove that the two sequences converge at the same rate, regardless of what this rate is. What is the relationship between the rate constants for the two sequences? [*Hint*: See Section 2.7, but note that in that section the problem is to find a solution to $f(x) = 0$ and not minimize $f(x)$.]

2.10. Write a computer program for minimizing a multivariate function using the steepest-descent algorithm. Include the following details:

    (i) Use a backtracking line search as described in Section 11.5.

  (ii) Accept x as a solution if $\|\nabla f(x)\| / (1 + |f(x)|) \le \epsilon$, or if the number of iterations exceeds ITMAX. Use $\epsilon = 10^{-5}$ and ITMAX $= 1000$.

 (iii) Print out the initial point, and then at each iteration print the search direction, the step length α, and the new estimate of the solution x_{k+1}. (If a great many iterations are required, provide this output only for the first 10 iterations and the final 5 iterations.) Indicate if no solution has been found after ITMAX iterations.

 (iv) Test your algorithm on the test problems listed here:

$$\begin{aligned}
f_{(1)}(x) &= x_1^2 + x_2^2 + x_3^2, & x_0 &= (1, 1, 1)^T \\
f_{(2)}(x) &= x_1^2 + 2x_2^2 - 2x_1 x_2 - 2x_2, & x_0 &= (0, 0)^T \\
f_{(3)}(x) &= 100(x_2 - x_1^2)^2 + (1 - x_1)^2, & x_0 &= (-1.2, 1)^T \\
f_{(4)}(x) &= (x_1 + x_2)^4 + x_2^2, & x_0 &= (2, -2)^T \\
f_{(5)}(x) &= (x_1 - 1)^2 + (x_2 - 1)^2 + c(x_1^2 + x_2^2 - 0.25)^2, & x_0 &= (1, -1)^T.
\end{aligned}$$

For the final function, test the following three different settings of the parameter c: $c = 1$, $c = 10$, and $c = 100$. The condition number of the Hessian matrix at the solution becomes larger as c increases. Comment on how this affects the performance of the algorithm.

  (v) Are your computational results consistent with the theory of the steepest-descent method?

2.11. Consider the minimization problem in Exercise 3.1 of Chapter 11. Suppose that a change of variables $\hat{x} \equiv Ax + b$ is performed with

$$A = \begin{pmatrix} 3 & 1 \\ 4 & 1 \end{pmatrix} \quad \text{and} \quad b = \begin{pmatrix} -1 \\ -2 \end{pmatrix}.$$

Show that the steepest-descent direction for the original problem is not the same as the steepest-descent direction for the transformed problem (when both are written using the same coordinate system).

2.12. Prove that the steepest-descent direction is changed if a change of variables $\hat{x} \equiv Ax + b$ is performed, where A is an invertible matrix, unless A is an orthogonal matrix.

12.3 Quasi-Newton Methods

Quasi-Newton methods are among the most widely used methods for nonlinear optimization. They are incorporated in many software libraries, and they are effective in solving a wide variety of small to mid-size problems, in particular when the Hessian is hard to compute. In cases when the number of variables is large, other methods may be preferred, but even in this case they are the basis for *limited-memory quasi-Newton methods*, an effective method for solving large problems (see Chapter 13).

There are many different quasi-Newton methods, but they are all based on approximating the Hessian $\nabla^2 f(x_k)$ by another matrix B_k that is available at lower cost. Then the search direction is obtained by solving

$$B_k p = -\nabla f(x_k),$$

that is, from the Newton equations but with the Hessian replaced by B_k. If the matrix B_k is positive definite, then this is equivalent to minimizing the quadratic model

$$\text{minimize } q(p) = f(x_k) + \nabla f(x_k)^T p + \tfrac{1}{2} p^T B_k p.$$

The various quasi-Newton methods differ in the choice of B_k.

There are several advantages to this approach. First, an approximation B_k can be found using only first-derivative information. Second, the search direction can be computed using only $O(n^2)$ operations (versus $O(n^3)$ for Newton's method in the nonsparse case). There are also disadvantages, but they are minor. The methods do not converge quadratically, but they can converge superlinearly. At the precision of computer arithmetic, there is not much practical difference between these two rates of convergence. Also, quasi-Newton methods still require matrix storage, so they are not normally used to solve large problems. Modifications to quasi-Newton methods that do not use matrix storage are available, though; see Section 13.5.

Quasi-Newton methods are generalizations of a method for one-dimensional problems called the secant method. The secant method uses the approximation

$$f''(x_k) \approx \frac{f'(x_k) - f'(x_{k-1})}{x_k - x_{k-1}}$$

in the formula for Newton's method $x_{k+1} = x_k - f'(x_k)/f''(x_k)$. This results in the formula

$$x_{k+1} = x_k - \frac{(x_k - x_{k-1})}{f'(x_k) - f'(x_{k-1})} f'(x_k).$$

It is illustrated in the example below.

Example 12.6 (The Secant Method). We apply the secant method to

$$\text{minimize } f(x) = \sin x.$$

The secant method requires that two initial points be specified, x_0 and x_1. We use $x_0 = 0$ and $x_1 = -1$. Then

$$x_2 = x_1 - \frac{(x_1 - x_0)}{f'(x_1) - f'(x_0)} f'(x_1)$$

$$= -1 - \frac{(-1 - 0)}{\cos(-1) - \cos(0)} \cos(-1)$$

$$= -1 - \frac{(-1 - 0)}{0.5403 - 1} 0.5403 = -2.1753.$$

The next few iterates are $x_3 = -1.5728$, $x_4 = -1.5707$, and $x_5 = -1.5708 \approx -\pi/2$, and so the sequence converges to a solution of the problem.   ∎

Under appropriate assumptions, the secant method can be proved to converge super-linearly with rate $r = \frac{1}{2}(1 + \sqrt{5}) \approx 1.618$ (the "golden ratio"). See, for example, the book by Conte and de Boor (1980).

Quasi-Newton methods are based on generalizations of the formula

$$f''(x_k) \approx \frac{f'(x_k) - f'(x_{k-1})}{x_k - x_{k-1}}.$$

This formula cannot be used in the multidimensional case because it would involve division by a vector, an undefined operation. This condition is rewritten in the form

$$\nabla^2 f(x_k)(x_k - x_{k-1}) \approx \nabla f(x_k) - \nabla f(x_{k-1}).$$

From this we obtain the condition used to define the quasi-Newton approximations B_k:

$$B_k(x_k - x_{k-1}) = \nabla f(x_k) - \nabla f(x_{k-1}).$$

We will call this the *secant condition*. For an n-dimensional problem this condition rep-resents a set of n equations that must be satisfied by B_k. However, the matrix B_k has n^2 entries, and so this condition by itself is insufficient to define B_k uniquely (unless $n = 1$). Additional conditions must be imposed to specify a particular quasi-Newton method.

The secant condition has extra significance when f is a quadratic function, $f(x) = \frac{1}{2}x^T Q x - c^T x$. In this case

$$Q(x_k - x_{k-1}) = (Qx_k - c) - (Qx_{k-1} - c) = \nabla f(x_k) - \nabla f(x_{k-1})$$

so that the Hessian matrix Q satisfies the secant condition. Intuitively we are asking that the approximation B_k mimic the behavior of the Hessian matrix when it multiplies $x_k - x_{k-1}$. Although this interpretation is precise only for quadratic functions, it holds in an approximate way for general nonlinear functions. We are asking that the approximation B_k imitate the effect of the Hessian matrix along a particular direction.

Before going further it is useful to define two vectors that will appear repeatedly in the discussion of quasi-Newton methods:

$$s_k = x_{k+1} - x_k \quad \text{and} \quad y_k = \nabla f(x_{k+1}) - \nabla f(x_k).$$

This notation is used throughout the literature on quasi-Newton methods. The secant condition then becomes $B_k s_{k-1} = y_{k-1}$ or, as will be more convenient to us,

$$B_{k+1} s_k = y_k.$$

When a line search is used, $x_{k+1} = x_k + \alpha_k p_k$ where α_k is the step length and p_k is the search direction. In this case $s_k = \alpha_k p_k$.

An example of a quasi-Newton approximation is given by the formula

$$B_{k+1} = B_k + \frac{(y_k - B_k s_k)(y_k - B_k s_k)^T}{(y_k - B_k s_k)^T s_k}.$$

The numerator of the second term is the outer product of two vectors and is an $n \times n$ matrix. This approximation is illustrated in the following example.

Example 12.7 (A Quasi-Newton Approximation). We will look at a three-dimensional example. Let $k = 1$, and define

$$B_0 = I = \begin{pmatrix} 1 & 0 & 0 \\ 0 & 1 & 0 \\ 0 & 0 & 1 \end{pmatrix}, \quad s_0 = \begin{pmatrix} 2 \\ 3 \\ 4 \end{pmatrix}, \quad y_0 = \begin{pmatrix} 5 \\ 6 \\ 7 \end{pmatrix}.$$

Then $(y_0 - B_0 s_0) = (3, 3, 3)^T$. We compute

$$B_1 = B_0 + \frac{(y_0 - B_0 s_0)(y_0 - B_0 s_0)^T}{(y_0 - B_0 s_0)^T s_0} = I + \frac{\begin{pmatrix} 3 \\ 3 \\ 3 \end{pmatrix}(3 \quad 3 \quad 3)}{(3 \quad 3 \quad 3)\begin{pmatrix} 2 \\ 3 \\ 4 \end{pmatrix}}$$

$$= I + \frac{1}{27}\begin{pmatrix} 9 & 9 & 9 \\ 9 & 9 & 9 \\ 9 & 9 & 9 \end{pmatrix} = \begin{pmatrix} \frac{4}{3} & \frac{1}{3} & \frac{1}{3} \\ \frac{1}{3} & \frac{4}{3} & \frac{1}{3} \\ \frac{1}{3} & \frac{1}{3} & \frac{4}{3} \end{pmatrix}.$$

It is easy to check that

$$B_1 s_0 = \begin{pmatrix} \frac{4}{3} & \frac{1}{3} & \frac{1}{3} \\ \frac{1}{3} & \frac{4}{3} & \frac{1}{3} \\ \frac{1}{3} & \frac{1}{3} & \frac{4}{3} \end{pmatrix}\begin{pmatrix} 2 \\ 3 \\ 4 \end{pmatrix} = \begin{pmatrix} 5 \\ 6 \\ 7 \end{pmatrix} = y_0,$$

so the secant condition is satisfied. ∎

This simple formula for B_{k+1} displays many of the general properties of quasi-Newton methods.

- The secant condition will be satisfied regardless of how B_k is chosen:

$$B_{k+1}s_k = B_k s_k + \frac{(y_k - B_k s_k)(y_k - B_k s_k)^T}{(y_k - B_k s_k)^T s_k} s_k$$

$$= B_k s_k + \frac{(y_k - B_k s_k)((y_k - B_k s_k)^T s_k)}{(y_k - B_k s_k)^T s_k}$$

$$= B_k s_k + (y_k - B_k s_k) = y_k.$$

- The new approximation B_{k+1}, is obtained by modifying the old approximation B_k. To start a quasi-Newton method some initial approximation B_0 must be specified. Often $B_0 = I$ is used, but it is reasonable and often advantageous to supply a better initial approximation if one can be obtained with little effort.

- The new approximation B_{k+1} can be obtained from B_k using $O(n^2)$ arithmetic operations since the difference $B_{k+1} - B_k$ only involves products of vectors. More surprisingly the search direction can also be computed using $O(n^2)$ arithmetic operations. Normally the computational cost of solving a system of linear equations is $O(n^3)$, so this represents a significant saving. The costs are lower in this case because it is possible to derive formulas that update a Cholesky factorization of B_k rather than B_k itself. With a factorization available, the search direction can be computed via backsubstitution.

All the quasi-Newton methods we consider have the form

$$B_{k+1} = B_k + [\text{something}].$$

The "something" represents an "update" to the old approximation B_k, and so a formula for a quasi-Newton approximation is often referred to as an *update formula*.

A variety of quasi-Newton methods are obtained by imposing conditions on the approximation B_k. These conditions are usually properties of the Hessian matrix that we would like the approximation to share. For example, since the Hessian matrix is symmetric, perhaps the approximation B_k should be symmetric as well. The quasi-Newton formula in Example 12.7,

$$B_{k+1} = B_k + \frac{(y_k - B_k s_k)(y_k - B_k s_k)^T}{(y_k - B_k s_k)^T s_k},$$

preserves symmetry because B_{k+1} is symmetric if B_k is. It is called the *symmetric rank-one* update formula. The "rank-one" is in its name because the update term is a matrix of rank one. This is the only rank-one update formula that preserves symmetry, as the lemma below shows.

Lemma 12.8. *Let B_k be a symmetric matrix. Let $B_{k+1} = B_k + C$ where $C \neq 0$ is a matrix of rank one. Assume that B_{k+1} is symmetric, $B_{k+1}s_k = y_k$, and $(y_k - B_k s_k)^T s_k \neq 0$. Then*

$$C = \frac{(y_k - B_k s_k)(y_k - B_k s_k)^T}{(y_k - B_k s_k)^T s_k}.$$

Proof. If $B_{k+1} = B_k + C$ and both B_k and B_{k+1} are symmetric, then C must be symmetric also. Since C is also of rank one, C must have the form

$$C = \gamma w w^T,$$

where γ is a scalar and w is a vector of norm one. (See the Exercises.) Now we use the secant condition

$$y_k = B_{k+1} s_k = (B_k + C) s_k = (B_k + \gamma w w^T) s_k = B_k s_k + \gamma w (w^T s_k).$$

This can be rewritten as

$$\gamma (w^T s_k) w = y_k - B_k s_k.$$

If $w^T s_k = 0$, then $B_k s_k = y_k$, or in other words, B_k already satisfies the secant condition, so there is no reason to perform any update. Since the theorem assumed that $C \neq 0$, we can rule this out, and so $w^T s_k \neq 0$. Hence we can write

$$w = \theta (y_k - B_k s_k),$$

where $\theta = 1 / \| y_k - B_k s_k \|$. This shows that w is a multiple of $y_k - B_k s_k$. (Since $C = \gamma w w^T$, the sign of θ is irrelevant.)

It only remains to determine the value of γ in terms of B_k, s_k, and y_k. To do this we use $\gamma (w^T s_k) w = y_k - B_k s_k$:

$$\begin{aligned} y_k - B_k s_k &= \gamma (w^T s_k) w \\ &= \frac{\gamma}{\| y_k - B_k s_k \|^2} [(y_k - B_k s_k)^T s_k](y_k - B_k s_k), \end{aligned}$$

so

$$\gamma = \frac{\| y_k - B_k s_k \|^2}{(y_k - B_k s_k)^T s_k}.$$

If we now substitute the formulas for γ and w into $C = \gamma w w^T$, the result follows. $\square$

If a quasi-Newton method is used with a line search, then the algorithm takes the following form.

ALGORITHM 12.1.
Quasi-Newton Algorithm

1. Specify some initial guess of the solution x_0 and some initial Hessian approximation B_0 (perhaps $B_0 = I$).

2. For $k = 0, 1, \ldots$

    (i) If x_k is optimal, stop.

    (ii) Solve $B_k p = -\nabla f(x_k)$ for p_k.

    (iii) Use a line search to determine $x_{k+1} = x_k + \alpha_k p_k$.

    (iv) Compute

$$\begin{aligned} s_k &= x_{k+1} - x_k \\ y_k &= \nabla f(x_{k+1}) - \nabla f(x_k). \end{aligned}$$

(v) Compute $B_{k+1} = B_k + \cdots$ using some update formula.

This is illustrated below using the symmetric rank-one formula on a quadratic problem.

Example 12.9 (The Symmetric Rank-One Formula). We will look at a three-dimensional quadratic problem $f(x) = \frac{1}{2}x^TQx - c^Tx$ with

$$
Q = \begin{pmatrix} 2 & 0 & 0 \\ 0 & 3 & 0 \\ 0 & 0 & 4 \end{pmatrix} \quad \text{and} \quad c = \begin{pmatrix} -8 \\ -9 \\ -8 \end{pmatrix},
$$

whose solution is $x_* = (-4, -3, -2)^T$. An exact line search will be used (see Exercise 5.3 of Section 11.5). The initial guesses are $B_0 = I$ and $x_0 = (0, 0, 0)^T$. At the initial point, $\|\nabla f(x_0)\| = \|-c\| = 14.4568$, so this point is not optimal. The first search direction is

$$
p_0 = \begin{pmatrix} -8 \\ -9 \\ -8 \end{pmatrix}
$$

and the line search formula gives $\alpha_0 = 0.3333$. The new estimate of the solution, the update vectors, and the new Hessian approximation are

$$
x_1 = \begin{pmatrix} -2.6667 \\ -3.0000 \\ -2.6667 \end{pmatrix}, \ \nabla f_1 = \begin{pmatrix} 2.6667 \\ 0 \\ -2.6667 \end{pmatrix}, \ s_0 = \begin{pmatrix} -2.6667 \\ -3.0000 \\ -2.6667 \end{pmatrix}, \ y_0 = \begin{pmatrix} -5.3333 \\ -9.0000 \\ -10.6667 \end{pmatrix},
$$

and

$$
B_1 = I + \frac{(y_0 - Is_0)(y_0 - Is_0)^T}{(y_0 - Is_0)^Ts_0} = \begin{pmatrix} 1.1531 & 0.3445 & 0.4593 \\ 0.3445 & 1.7751 & 1.0335 \\ 0.4593 & 1.0335 & 2.3780 \end{pmatrix}.
$$

At this new point $\|\nabla f(x_1)\| = 3.7712$ so we keep going, obtaining the search direction

$$
p_1 = \begin{pmatrix} -2.9137 \\ -0.5557 \\ 1.9257 \end{pmatrix}
$$

and the step length $\alpha_1 = 0.3942$. This gives the new estimates

$$
x_2 = \begin{pmatrix} -3.8152 \\ -3.2191 \\ -1.9076 \end{pmatrix}, \ \nabla f_2 = \begin{pmatrix} 0.3697 \\ -0.6572 \\ 0.3697 \end{pmatrix}, \ s_1 = \begin{pmatrix} -1.1485 \\ -0.2191 \\ 0.7591 \end{pmatrix}, \ y_1 = \begin{pmatrix} -2.2970 \\ -0.6572 \\ 3.0363 \end{pmatrix},
$$

and

$$
B_2 = \begin{pmatrix} 1.6568 & 0.6102 & -0.3432 \\ 0.6102 & 1.9153 & 0.6102 \\ -0.3432 & 0.6102 & 3.6568 \end{pmatrix}.
$$

At the point x_2, $\|\nabla f(x_2)\| = 0.8397$ so we keep going, with

$$p_2 = \begin{pmatrix} -0.4851 \\ 0.5749 \\ -0.2426 \end{pmatrix}$$

and $\alpha = 0.3810$. This gives

$$x_3 = \begin{pmatrix} -4 \\ -3 \\ -2 \end{pmatrix}, \quad \nabla f_3 = \begin{pmatrix} 0 \\ 0 \\ 0 \end{pmatrix}, \quad s_2 = \begin{pmatrix} -0.1848 \\ 0.2191 \\ -0.0924 \end{pmatrix}, \quad y_2 = \begin{pmatrix} -0.3697 \\ 0.6572 \\ -0.3697 \end{pmatrix},$$

and $B_3 = Q$. Now $\|\nabla f(x_3)\| = 0$, so we stop. The final approximation matrix B_3 is equal to Q, the Hessian matrix. In exact arithmetic, this will always happen within n iterations when the symmetric rank-one formula is applied to a quadratic problem, but it is not guaranteed on more general problems. ∎

Symmetry is not the only property that can be imposed. Since the Hessian matrix at the solution x_* will normally be positive definite (it will always be positive semidefinite), it is reasonable to ask that the matrices B_k be positive definite as well. This will also guarantee that the quasi-Newton method corresponds to minimizing a quadratic model of the nonlinear function f, and that the search direction is a descent direction. (See the remarks in Section 11.4.)

There is no rank-one update formula that maintains both symmetry and positive definiteness of the Hessian approximations. However, there are infinitely many rank-two formulas that do this. The most widely used formula, and the one considered to be most effective, is the *BFGS* update formula

$$B_{k+1} = B_k - \frac{(B_k s_k)(B_k s_k)^T}{s_k^T B_k s_k} + \frac{y_k y_k^T}{y_k^T s_k}.$$

The BFGS formula gets its name from the four people who developed it: Broyden, Fletcher, Goldfarb, and Shanno. It is easy to check that $B_{k+1} s_k = y_k$. It is not as easy to check that it has the property that we want.

Lemma 12.10. *Let B_k be a symmetric positive-definite matrix, and assume that B_{k+1} is obtained from B_k using the BFGS update formula. Then B_{k+1} is positive definite if and only if $y_k^T s_k > 0$.*

Proof. If B_k is positive definite, then it can be factored as $B_k = LL^T$ where L is a nonsingular matrix. (This is just the Cholesky factorization of B_k.) If this factorization is substituted into the BFGS formula for B_{k+1}, then

$$B_{k+1} = LWL^T,$$

where

$$W = I - \frac{\hat{s}\hat{s}^T}{\hat{s}^T\hat{s}} + \frac{\hat{y}\hat{y}^T}{\hat{y}^T\hat{s}}, \quad \hat{s} = L^T s_k, \quad \text{and} \quad \hat{y} = L^{-1} y_k.$$

B_{k+1} will be positive definite if and only if W is. To test if W is positive definite, we test if $v^T W v > 0$ for all $v \neq 0$. Let θ_1 be the angle between v and $\hat{s}$, θ_2 the angle between v

and $\hat{y}$, and θ_3 the angle between $\hat{s}$ and $\hat{y}$. Then

$$
\begin{aligned}
v^T W v &= v^T v - \frac{(v^T \hat{s})^2}{\hat{s}^T \hat{s}} + \frac{(v^T \hat{y})^2}{\hat{y}^T \hat{s}} \\
&= \|v\|^2 - \frac{\|v\|^2 \|\hat{s}\|^2 \cos^2 \theta_1}{\|\hat{s}\|^2} - \frac{\|v\|^2 \|\hat{y}\|^2 \cos^2 \theta_2}{\|\hat{y}\| \cdot \|\hat{s}\| \cos \theta_3} \\
&= \|v\|^2 \left[1 - \cos^2 \theta_1 + \frac{\|\hat{y}\| \cos^2 \theta_2}{\|\hat{s}\| \cos \theta_3} \right] \\
&= \|v\|^2 \left[\sin^2 \theta_1 + \frac{\|\hat{y}\| \cos^2 \theta_2}{\|\hat{s}\| \cos \theta_3} \right].
\end{aligned}
$$

If $y_k^T s_k > 0$, then $\hat{y}^T \hat{s} > 0$ and $\cos \theta_3 > 0$; hence $v^T W v > 0$ and W is positive definite. If $y_k^T s_k < 0$, then $\cos \theta_3 < 0$; in this case, v can be chosen so that $v^T W v < 0$ and so W is not positive definite. This completes the proof.   □

The new matrix B_{k+1} will be positive definite only if $y_k^T s_k > 0$. This property can be guaranteed by performing an appropriate line search and so is not a serious limitation (see the Exercises). The BFGS formula is illustrated below.

Example 12.11 (The BFGS Formula). We will apply the BFGS formula to the same three-dimensional example that was used for the symmetric rank-one formula. Again we will choose $B_0 = I$ and $x_0 = (0, 0, 0)^T$. At iteration 0, $\|\nabla f(x_0)\| = 14.4568$, so this point is not optimal. The search direction is

$$
p_0 = \begin{pmatrix} -8 \\ -9 \\ -8 \end{pmatrix}
$$

and $\alpha_0 = 0.3333$. The new estimate of the solution and the new Hessian approximation are

$$
x_1 = \begin{pmatrix} -2.6667 \\ -3.0000 \\ -2.6667 \end{pmatrix} \quad \text{and} \quad B_1 = \begin{pmatrix} 1.1021 & 0.3445 & 0.5104 \\ 0.3445 & 1.7751 & 1.0335 \\ 0.5104 & 1.0335 & 2.3270 \end{pmatrix}.
$$

At iteration 1, $\|\nabla f(x_1)\| = 3.7712$, so we continue. The next search direction is

$$
p_1 = \begin{pmatrix} -3.2111 \\ -0.6124 \\ 2.1223 \end{pmatrix}
$$

and $\alpha_1 = 0.3577$. This gives the estimates

$$
x_2 = \begin{pmatrix} -3.8152 \\ -3.2191 \\ -1.9076 \end{pmatrix} \quad \text{and} \quad B_2 = \begin{pmatrix} 1.6393 & 0.6412 & -0.3607 \\ 0.6412 & 1.8600 & 0.6412 \\ -0.3607 & 0.6412 & 3.6393 \end{pmatrix}.
$$

At iteration 2, $\|\nabla f(x_2)\| = 0.8397$, so we continue, computing

$$
p_2 = \begin{pmatrix} -0.5289 \\ 0.6268 \\ -0.2644 \end{pmatrix}
$$

and $\alpha_2 = 0.3495$. This gives

$$x_3 = \begin{pmatrix} -4 \\ -3 \\ -2 \end{pmatrix} \quad \text{and} \quad B_3 = \begin{pmatrix} 2 & 0 & 0 \\ 0 & 3 & 0 \\ 0 & 0 & 4 \end{pmatrix}.$$

Now $\|\nabla f(x_3)\| = 0$, so we stop.

You may have noticed that the values of $\{x_k\}$ were the same here as in the previous example. This does not happen in general, but it can be shown to be a consequence of using an exact line search and solving a quadratic problem. ∎

There is a class of update formulas that preserve positive-definiteness, given by the formula

$$B_{k+1} = B_k - \frac{(B_k s_k)(B_k s_k)^T}{s_k^T B_k s_k} + \frac{y_k y_k^T}{y_k^T s_k} + \phi(s_k^T B_k s_k)v_k v_k^T,$$

where ϕ is a scalar and

$$v_k = \frac{y_k}{y_k^T s_k} - \frac{B_k s_k}{s_k^T B_k s_k}.$$

The BFGS update formula is obtained by setting $\phi = 0$. As with the BFGS update, positive-definiteness is preserved if and only if $y_k^T s_k > 0$. When $\phi = 1$ the update is called the DFP formula, which is named for its developers, Davidon, Fletcher, and Powell. The class of update formulas is sometimes referred to as the Broyden class.

We conclude this section by mentioning a convergence result for quasi-Newton methods. It applies to a subset of the Broyden class of update formulas, when the parameter satisfies $0 \le \phi < 1$. It excludes the DFP formula. In addition, the theorem assumes that $\nabla^2 f(x)$ is always positive definite, that is, the objective function is strictly convex.

Theorem 12.12. *Let f be a real-valued function of n variables. Let x_0 be some given initial point and let $\{x_k\}$ be defined by $x_{k+1} = x_k + \alpha_k p_k$, where p_k is a vector of dimension n and $\alpha_k \ge 0$ is a scalar. Assume that*

(i) *the set $S = \{x : f(x) \le f(x_0)\}$ is bounded;*

(ii) *f, ∇f, and $\nabla^2 f$ are continuous for all $x \in S$;*

(iii) *$\nabla^2 f(x)$ is positive definite for all x;*

(iv) *the search directions $\{p_k\}$ are computed using*

$$B_k p_k = -\nabla f(x_k),$$

where $B_0 = I$, and the matrices $\{B_k\}$ are updated using a formula from the Broyden class with parameter $0 \le \phi < 1$;

(v) *the step lengths $\{\alpha_k\}$ satisfy*

$$f(x_k + \alpha_k p_k) \le f(x_k) + \mu \alpha_k p_k^T \nabla f(x_k)$$
$$p_k^T \nabla f(x_k + \alpha_k p_k) \ge \eta p_k^T \nabla f(x_k),$$

with $0 < \mu < \eta < 1$, and the line search algorithm uses the step length $\alpha_k = 1$ whenever possible.

Then

$$\lim_{k \to \infty} x_k = x_*,$$

where x_* is the unique global minimizer of f on S, and the rate of convergence of $\{x_k\}$ is superlinear.

Proof. See the paper by Byrd, Nocedal, and Yuan (1987).   □

Exercises

3.1. Apply the symmetric rank-one quasi-Newton method to solve

$$\text{minimize } f(x) = \tfrac{1}{2} x^T Q x - c^T x$$

with

$$Q = \begin{pmatrix} 5 & 2 & 1 \\ 2 & 7 & 3 \\ 1 & 3 & 9 \end{pmatrix} \quad \text{and} \quad c = \begin{pmatrix} -9 \\ 0 \\ -8 \end{pmatrix}.$$

Initialize the method with $x_0 = (0,0,0)^T$ and $B_0 = I$. Use an exact line search.

3.2. Apply the BFGS quasi-Newton method to solve

$$\text{minimize } f(x) = \tfrac{1}{2} x^T Q x - c^T x$$

with

$$Q = \begin{pmatrix} 5 & 2 & 1 \\ 2 & 7 & 3 \\ 1 & 3 & 9 \end{pmatrix} \quad \text{and} \quad c = \begin{pmatrix} -9 \\ 0 \\ -8 \end{pmatrix}.$$

Initialize the method with $x_0 = (0,0,0)^T$ and $B_0 = I$. Use an exact line search.

3.3. Let f be a strictly convex quadratic function of one variable. Prove that the secant method for minimization will terminate in exactly one iteration for any initial starting points x_0 and x_1.

3.4. Let C be a symmetric matrix of rank one. Prove that C must have the form $C = \gamma w w^T$, where γ is a scalar and w is a vector of norm one.

3.5. In the proof of Lemma 12.10, show that, if $y_k^T s_k < 0$, then v can be chosen so that $v^T W v < 0$.

3.6. Let B_{k+1} be obtained from B_k using the symmetric rank-one update formula. Assume that the associated quasi-Newton method is applied to an n-dimensional, strictly convex, quadratic function, and that the vectors $s_0, \ldots, s_{n-1}$ are linearly independent. Also assume that $(y_i - B_i s_i)^T s_i \neq 0$ for all i. Prove that $B_{k+1} s_i = y_i$ for $i = 0, 1, \ldots, k$, and that the method terminates in at most $n + 1$ iterations. Use this to prove that B_n is equal to the Hessian of the quadratic function. (This exercise makes no assumptions about the line search.)

3.7. Let B_{k+1} be obtained from B_k using the update formula

$$B_{k+1} = B_k + \frac{(y_k - B_k s_k)v^T}{v^T s_k},$$

where v is a vector such that $v^T s_k \neq 0$. Prove that $B_{k+1} s_k = y_k$.

3.8. Let B_{k+1} be obtained from B_k using the BFGS update formula. Prove that $B_{k+1} s_k = y_k$.

3.9. Let B_{k+1} be obtained from B_k using the BFGS update formula. B_{k+1} is guaranteed to be positive definite only if $y_k^T s_k > 0$. Prove that if the Wolfe condition

$$|p^T \nabla f(x_k + \alpha p)| \leq \eta |p^T \nabla f(x_k)|$$

is used to terminate the line search, and η is sufficiently small, then $y_k^T s_k > 0$. Hence, if an appropriate line search is used, then B_{k+1} will be positive definite.

3.10. Consider the class of positive-definite updates depending on the parameter ϕ. What is the rank of the update formula? Prove that these updates preserve positive-definiteness if and only if $y_k^T s_k > 0$.

3.11. Write a computer program for minimizing a multivariate function using the BFGS quasi-Newton algorithm. Use $B_0 = I$ as the initial Hessian approximation. Include the following details:

(i) Use a backtracking line search as described in Section 11.5. Before updating B_k, check if $y_k^T s_k > 0$; if this condition is not satisfied, then do not update B_k at that iteration of the algorithm.

(ii) Accept x as a solution to the optimization problem if $\|\nabla f(x)\| / (1 + |f(x)|) \leq \epsilon$, or if the number of iterations exceeds ITMAX. Use $\epsilon = 10^{-8}$ and ITMAX = 1000.

(iii) Print out the initial point, and then at each iteration print the search direction, the step length α, and the new estimate of the solution x_{k+1}. (If a great many iterations are required, provide this output for only the first 10 iterations and the final 5 iterations.) Indicate if no solution has been found after ITMAX iterations.

(iv) Test your algorithm on the test problems listed here:

$$\begin{aligned}
f_{(1)}(x) &= x_1^2 + x_2^2 + x_3^2, & x_0 &= (1, 1, 1)^T \\
f_{(2)}(x) &= x_1^2 + 2x_2^2 - 2x_1 x_2 - 2x_2, & x_0 &= (0, 0)^T \\
f_{(3)}(x) &= 100(x_2 - x_1^2)^2 + (1 - x_1)^2, & x_0 &= (-1.2, 1)^T \\
f_{(4)}(x) &= (x_1 + x_2)^4 + x_2^2, & x_0 &= (2, -2)^T \\
f_{(5)}(x) &= (x_1 - 1)^2 + (x_2 - 1)^2 + c(x_1^2 + x_2^2 - 0.25)^2, & x_0 &= (1, -1)^T.
\end{aligned}$$

For the final function, test three different settings of the parameter c: $c = 1$, $c = 10$, and $c = 100$. The condition number of the Hessian matrix at the solution becomes larger as c increases. Comment on how this affects the performance of the algorithm.

(v) Are your computational results consistent with the theory of quasi-Newton methods?

12.4 Automating Derivative Calculations

One of the disadvantages of Newton's method is that it requires the computation of both first and second derivatives. This can be a disadvantage in two ways: (i) having to derive and program the formulas for these derivatives, and (ii) having to use these derivatives at all. The avoidance of second derivative calculations was discussed in the sections on the steepest-descent and quasi-Newton methods. Now we show how to avoid even calculating first derivatives. In this section, we describe two ways to automate derivative calculations. The first and most widely used technique is to use differences of function values to estimate derivatives. Next we discuss techniques that analyze formulas for the function value and derive exact formulas for the derivatives.

Another way to automate derivative calculations is to use a modeling language to describe the optimization problem. (Modeling languages are discussed in Appendix C.) A modeling language will typically build a symbolic representation for the derivatives based on the description of the optimization problem. This useful feature of modeling languages reduces the time needed for a problem's description and simplifies the modeling process overall.

There are many important applications where it is not appropriate to calculate or estimate derivative values. In some cases the derivatives of the objective may not always exist. This can happen with "noisy" functions that are subject to random errors, perhaps because of inaccuracies that arise when computing or approximating their values. In other cases, the objective function may be computed using sophisticated algorithmic techniques that lead to discontinuities or lack of smoothness. This can occur even if the underlying mathematical formulas are differentiable. See also Section 12.5.1.

In cases such as these the Newton and quasi-Newton methods can fail, or may not even be defined. It is necessary to use methods that do not rely on derivative values, either explicitly in their usage requirements, or implicitly in the derivations of the methods. Such methods are the topic of the next section.

12.4.1 Finite-Difference Derivative Estimates

Finite differencing refers to the estimation of $f'(x)$ using values of $f(x)$. The simplest formulas just use the difference of two function values which gives the technique its name. Finite differencing can also be applied to the calculation of $\nabla f(x)$ for multidimensional problems, as well as to the computation of $f''(x)$ and the Hessian matrix $\nabla^2 f(x)$. For a problem with n variables, computing $\nabla f(x)$ will be about n times as expensive as computing $f(x)$, and computing $\nabla^2 f(x)$ will be about n^2 times as expensive as $f(x)$. Hence, even though this technique relieves the burden of deriving and programming derivative formulas, it is expensive computationally. In addition, finite differencing only produces derivative *estimates*, not exact values. In contrast, the automatic differentiation techniques discussed in Section 12.4.2 have low computational costs and produce exact answers, but unfortunately they too have deficiencies.

Finite-difference estimates can be derived from the Taylor series. In one dimension,

$$f(x + h) = f(x) + hf'(x) + \tfrac{1}{2}h^2 f''(\xi).$$

A simple rearrangement gives

$$f'(x) = \frac{f(x+h) - f(x)}{h} - \frac{1}{2}hf''(\xi),$$

leading to the approximation

$$f'(x) \approx \frac{f(x+h) - f(x)}{h}.$$

This is the most commonly used finite-difference formula. It is sometimes called the *forward difference* formula because $x+h$ is a shift "forward" from the point x. This formula could also have been derived from the definition of the derivative as a limit,

$$f'(x) = \lim_{h \to 0} \frac{f(x+h) - f(x)}{h},$$

but this would not have provided an estimate of the error in the formula.

Example 12.13 (Finite Differencing). Consider the function

$$f(x) = \sin(x)$$

with derivative $f'(x) = \cos(x)$. The results of using the finite-difference formula

$$f'(x) \approx \frac{\sin(x+h) - \sin(x)}{h}$$

for $x = 2$ and for various values of h are given in Table 12.3.

The derivation of the finite-difference formula indicates that the error will be equal to $|\frac{1}{2}hf''(\xi)|$. Since ξ is between x and $x+h$,

$$\text{error} \approx |\tfrac{1}{2}hf''(x)| = |\tfrac{1}{2}h(-\sin(x))| = |\tfrac{1}{2}h(-\sin(2))| \approx |\tfrac{1}{2}h(-0.91)| = 0.455h.$$

This corresponds to the results in the table for h between 10^0 and 10^{-8}, but after that the error starts to increase, until eventually the finite-difference calculation estimates that the derivative is equal to zero. This phenomenon will be explained below by examining the errors that result when finite differencing is used. ∎

We now estimate the error in finite differencing when the calculations are performed on a computer. Part of the error is due to the inaccuracies in the formula itself; this is called the *truncation error*:

$$\text{truncation error} = \tfrac{1}{2}h|f''(\xi)|.$$

In addition there are rounding errors from the evaluation of the formula $(f(x+h) - f(x))/h$ on a computer that depend on ϵ_{mach}, the precision of the computer calculations (see Appendix B.2). There are rounding errors from the evaluations of the function f in the numerator:

$$(\text{rounding error})_1 \approx |f(x)|\epsilon_{\text{mach}}$$

Table 12.3. *Finite differencing.*

h	$f'(x)$	Estimate	Error
10^0	-0.4161468365	-0.7681774187	4×10^{-1}
10^{-1}	-0.4161468365	-0.4608806017	4×10^{-2}
10^{-2}	-0.4161468365	-0.4206863500	4×10^{-3}
10^{-3}	-0.4161468365	-0.4166014158	4×10^{-4}
10^{-4}	-0.4161468365	-0.4161923007	4×10^{-5}
10^{-5}	-0.4161468365	-0.4161513830	4×10^{-6}
10^{-6}	-0.4161468365	-0.4161472912	4×10^{-7}
10^{-7}	-0.4161468365	-0.4161468813	4×10^{-8}
10^{-8}	-0.4161468365	-0.4161468392	3×10^{-9}
10^{-9}	-0.4161468365	-0.4161468947	6×10^{-8}
10^{-10}	-0.4161468365	-0.4161471167	3×10^{-7}
10^{-11}	-0.4161468365	-0.4161448963	2×10^{-6}
10^{-12}	-0.4161468365	-0.4162226119	8×10^{-5}
10^{-13}	-0.4161468365	-0.4163336342	2×10^{-4}
10^{-14}	-0.4161468365	-0.4218847493	6×10^{-3}
10^{-15}	-0.4161468365	-0.3330669073	8×10^{-2}
10^{-16}	-0.4161468365	0	4×10^{-1}

which are then magnified and augmented by the division by h:

$$(\text{rounding error})_2 \approx \frac{|f(x)|\epsilon_{\text{mach}}}{h} + |f'(x)|\epsilon_{\text{mach}}$$

(the first rounding error is magnified by $1/h$ and then there is an additional rounding error from the division that is proportional to the result $f'(x)$). Under typical circumstances, when h is small and $f'(x)$ is not overly large, the first term will dominate, leading to the estimate

$$\text{rounding error} \approx \frac{|f(x)|\epsilon_{\text{mach}}}{h}.$$

The total error is the combination of the truncation error and the rounding error

$$\text{error} \approx \frac{1}{2}h|f''(\xi)| + \frac{|f(x)|\epsilon_{\text{mach}}}{h}.$$

For fixed x and for almost fixed ξ (ξ is between x and $x + h$, and h will be small), this formula can be analyzed as a function of h alone.

To determine the "best" value of h we minimize the estimate of the error as a function of h. Differentiating with respect to h and setting the derivative to zero gives

$$\frac{1}{2}|f''(\xi)| - \frac{|f(x)|\epsilon_{\text{mach}}}{h^2} = 0,$$

which can be rearranged to give

$$h = \sqrt{\frac{2|f(x)|\epsilon_{\text{mach}}}{|f''(\xi)|}}.$$

In cases where $f(x)$ and $f''(\xi)$ are neither especially large nor small, the simpler approximation

$$h \approx \sqrt{\epsilon_{\text{mach}}}$$

can be used. If the more elaborate formula for h is substituted into the approximate formula for the error, then the result can be simplified to

$$\text{error} \approx \sqrt{2\epsilon_{\text{mach}}|f(x) \cdot f''(\xi)|},$$

or more concisely to the result that the error is $O(\sqrt{\epsilon_{\text{mach}}})$.

In the example above, $\epsilon_{\text{mach}} \approx 10^{-16}$ and the simplified formula for h yields $h \approx \sqrt{\epsilon_{\text{mach}}} \approx 10^{-8}$. This value of h gives the most accurate derivative estimate in the example. The more elaborate formula for h yields $h \approx 2.1 \times 10^{-8}$, almost the same value. The error with this value of h is about 1.4×10^{-8}, slightly worse than the value given by the simpler formula. This does not indicate that the derivation is invalid; rather it only emphasizes that the terms used in the derivation are estimates of the various errors. As expected, the errors in this example are approximately equal to $\sqrt{\epsilon_{\text{mach}}}$.

In practical settings the value of $f''(\xi)$ will be unknown (even the value of $f''(x)$ will be unknown) and so the more elaborate formula for h cannot be used. Some software packages just use $h = \sqrt{\epsilon_{\text{mach}}}$ or some simple modification of this formula (for example, taking into account $|x|$ or $|f(x)|$). An alternative is to perform extra calculations for one value of x, perhaps the initial guess for the optimization algorithm, to obtain an estimate for $f''(\xi)$, and then use this to obtain a better value for h that will be used for subsequent finite-difference calculations.

An additional complication can arise if $|x|$ is large. If $h < \epsilon_{\text{mach}}|x|$, then the computed value of $x + h$ will be equal to x and the finite-difference estimate will be zero. Thus, in the general case the choice of h will depend on ϵ_{mach}, $|x|$, and the values of $|f''|$. For further information, see the references cited in the Notes.

If higher accuracy in the derivative estimates is required, then there are two things that can be done. One choice is to use higher-precision arithmetic (arithmetic with a smaller value of ϵ_{mach}). This might just mean switching from single to double precision, a change that can sometimes be made with an instruction to the compiler without any changes to the program. If the program is already in double precision, then on some computers it is possible to use quadruple precision, but quadruple precision arithmetic can be much slower than double precision since the instructions for it are not normally built into the computer hardware.

The other choice is to use a more accurate finite-difference formula. The simplest of these is the *central-difference* formula

$$f'(x) = \frac{f(x+h) - f(x-h)}{2h} - \frac{1}{12}h^2[f'''(\xi_1) + f'''(\xi_2)].$$

It can be derived using the Taylor series for $f(x+h)$ and $f(x-h)$ about the point h. (See the Exercises.)

Higher derivatives can also be obtained by finite differencing. For example, the formula

$$f''(x) = \frac{f(x+h) - 2f(x) + f(x-h)}{h^2} - \frac{1}{24}h^2[f^{(4)}(\xi_1) + f^{(4)}(\xi_2)]$$

can be derived from the Taylor series for $f(x+h)$ and $f(x-h)$ about the point x. (See the Exercises.)

The derivatives of multidimensional functions can be estimated by applying the finite-difference formulas to each component of the gradient or Hessian matrix. If we define the vector

$$e_j = (0 \quad \cdots \quad 0 \quad 1 \quad 0 \quad \cdots \quad 0)^T$$

having a one in the jth component and zeroes elsewhere, then

$$[\nabla f(x)]_j \approx \frac{f(x + he_j) - f(x)}{h}.$$

If the gradient is known, then the Hessian can be approximated via

$$[\nabla^2 f(x)]_{jk} = \frac{\partial^2 f(x)}{\partial x_j \partial x_k} \approx \frac{[\nabla f(x + he_k) - \nabla f(x)]_j}{h}.$$

For further details, see the Exercises.

If it is feasible to use complex arithmetic to evaluate $f(x)$, then an alternative way to estimate $f'(x)$ is to use

$$f'(x) \approx \Im[f(x + ih)]/h,$$

where $i = \sqrt{-1}$ and $\Im[f]$ is the imaginary part of the function f. This formula is capable of producing more accurate estimates of the derivative (sometimes up to full machine accuracy) with only one additional function evaluation, for a broad range of values of h.

12.4.2 Automatic Differentiation

The goal of automatic differentiation is to use software to analyze the formulas used to evaluate $f(x)$ and produce formulas to evaluate $\nabla f(x)$. The user might provide a computer program that evaluates $f(x)$, and then the automatic differentiation software would take this program and produce a new program that evaluates both the function and its gradient. The technique uses the chain rule to analyze every step in the evaluation of the function, with the results being organized in such a way that the gradient is evaluated efficiently, that is, almost as efficiently as the function itself is evaluated. The resulting software for evaluating the gradient will have accuracy comparable to the software for evaluating the function. Hence, this technique not only automates the evaluation of the gradient, but does it in a way that is in general more efficient and more accurate than finite differencing.

To explain automatic differentiation we first assume that the function evaluation has been decomposed into a sequence of simple calculations, each of which involves only one or two variables. The "variables" may represent intermediate results and need not correspond

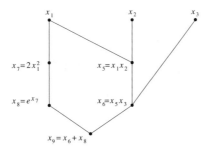

Figure 12.3. *Evaluation graph.*

to variables in the original problem. These simple calculations might be, for example, of the form

$$x_{10} = x_1 + x_2$$
$$x_{12} = x_3 x_4$$
$$x_{15} = 1/x_5$$
$$x_{21} = \sin x_7$$

and so forth. If the function evaluation is expressed in this way, it is easy to differentiate each step in the evaluation. The user need not program the function evaluation in this simple form, since the automatic differentiation software can perform this step itself. We use this representation to simplify our description of automatic differentiation.

Example 12.14 (Function Evaluation). Consider the function

$$f(x_1, x_2, x_3) = x_1 x_2 x_3 + e^{2x_1^2}.$$

It can be evaluated as follows:

$$x_5 = x_1 x_2$$
$$x_6 = x_5 x_3$$
$$x_7 = 2x_1^2$$
$$x_8 = e^{x_7}$$
$$x_9 = x_6 + x_8$$

and then $f(x) = x_9$. Each of the steps involves at most two variables.   ∎

This sequence of evaluation steps can be represented by a graph. Evaluation of the function corresponds to moving through the graph from top to bottom. This is illustrated in Figure 12.3 for the function in Example 12.14.

The graph, and the sequence of evaluation steps, can also be used to evaluate the gradient. For our example,

$$f(x) = x_9 = x_6 + x_8.$$

Hence

$$\frac{\partial f}{\partial x_9} = 1.$$

This is the initialization step for the gradient evaluation. Then

$$\frac{\partial f}{\partial x_6} = \frac{\partial f}{\partial x_9}\frac{\partial x_9}{\partial x_6} \quad \text{and} \quad \frac{\partial f}{\partial x_8} = \frac{\partial f}{\partial x_9}\frac{\partial x_9}{\partial x_8}.$$

These formulas determine the partial derivatives of f with respect to x_6 and x_8. These can in turn be used to determine the partial derivatives of f with respect to x_5 and x_7, and thus recursively the gradient of f. This process only requires calculating derivatives for each of the simple steps in the evaluation of f, which is easy to do. The entire process is illustrated in the next example.

Example 12.15 (Gradient Evaluation). To evaluate the gradient we first set

$$\frac{\partial f}{\partial x_9} = 1.$$

At the next stage

$$\frac{\partial f}{\partial x_6} = \frac{\partial f}{\partial x_9}\frac{\partial x_9}{\partial x_6} = 1 \times 1 = 1$$

$$\frac{\partial f}{\partial x_8} = \frac{\partial f}{\partial x_9}\frac{\partial x_9}{\partial x_8} = 1 \times 1 = 1.$$

In turn we can calculate

$$\frac{\partial f}{\partial x_5} = \frac{\partial f}{\partial x_6}\frac{\partial x_6}{\partial x_5} = 1 \times x_3 = x_3$$

$$\frac{\partial f}{\partial x_7} = \frac{\partial f}{\partial x_8}\frac{\partial x_8}{\partial x_7} = 1 \times e^{x_7} = e^{x_7}$$

$$\frac{\partial f}{\partial x_3} = \frac{\partial f}{\partial x_6}\frac{\partial x_6}{\partial x_3} = 1 \times x_5 = x_5$$

$$\frac{\partial f}{\partial x_1} = \frac{\partial f}{\partial x_5}\frac{\partial x_5}{\partial x_1} + \frac{\partial f}{\partial x_7}\frac{\partial x_7}{\partial x_1} = x_2 x_3 + 4x_1 e^{x_7}$$

$$\frac{\partial f}{\partial x_2} = \frac{\partial f}{\partial x_5}\frac{\partial x_5}{\partial x_2} = x_1 x_3.$$

The final three formulas determine the gradient. They include the intermediate variables from the evaluation of f, and so this approach assumes that both $f(x)$ and $\nabla f(x)$ are calculated together. For efficiency, the formulas would be left in this form, but it is possible to derive gradient formulas that involve only the original variables for the problem. In this case they are

$$\nabla f(x) = \begin{pmatrix} x_2 x_3 + 4x_1 e^{2x_1^2} \\ x_1 x_3 \\ x_1 x_2 \end{pmatrix}. \qquad \blacksquare$$

Evaluation of the gradient can be interpreted in terms of the evaluation graph. Whereas evaluating the function traverses the graph from top to bottom, evaluating the gradient traverses the graph from bottom to top. To initialize the process, the partial derivative value at the bottom node is set equal to one. (In our example, this is at the node corresponding to x_9.) Then the chain rule is used to move upward through the graph. By beginning at the bottom of the graph and moving up one level at a time, the gradient is evaluated through a sequence of calculations. This is called the *reverse mode* of automatic differentiation.

Each step in the evaluation of $f(x)$ is simple, involving at most two variables. As a result, each step in the evaluation of $\nabla f(x)$ is also simple. For example, if a step in the function evaluation involves the addition of two variables,

$$x_9 = x_6 + x_8,$$

then the step in the gradient evaluation involves two multiplications of derivative values,

$$\frac{\partial f}{\partial x_9} \frac{\partial x_9}{\partial x_6} \quad \text{and} \quad \frac{\partial f}{\partial x_9} \frac{\partial x_9}{\partial x_8}.$$

This analysis can be extended to show that the number of operations required to evaluate the gradient is proportional to the number of operations required to evaluate the function.

It would also be possible to evaluate the gradient by starting at the top of the graph and moving downward (*forward mode*). This is the traditional way of deriving the formulas for the gradient. If this is done, then in general evaluating the gradient can require about n times as many arithmetic operations as evaluating the function. The efficiency of automatic differentiation depends on evaluating the gradient starting at the bottom of the graph. Even so, software for automatic differentiation exploits both modes, since both have practical advantages.

Unfortunately, automatic differentiation is not a perfect technique. To be able to evaluate the gradient efficiently it may be necessary to store all the intermediate results in the evaluation of the function f. If the evaluation of $f(x)$ involves a large number of operations, the storage requirements for automatic differentiation can potentially be large. Modern implementations of automatic differentiation make a trade-off between efficiency and storage requirements, making it feasible to apply automatic differentiation to large classes of problems.

Exercises

4.1. Apply the forward-difference formula to the function $f(x) = \sin(100x)$ at $x = 1.0$ with various values of h. Determine the value of h that produces the best estimate of the derivative and compare it with the value predicted by the theory. How accurate is the theoretical estimate of the error for this function? How well do the "simple" estimates of h and the error perform?

4.2. Repeat the previous problem using the central-difference formula.

4.3. Repeat the previous problem using the difference formula for the second derivative.

4.4. Apply the forward-difference formula to estimate the gradient of the function

$$f(x_1, x_2) = \exp(10x_1 + 2x_2^2)$$

at $(x_1, x_2) = (-1, 1)$ with various values of h. Determine the value of h that produces the best estimate of the derivative. How well do the "simple" estimates of h and the error perform?

4.5. Estimate the Hessian of the function in the previous problem using finite differencing. First do this by taking differences of gradient values, and then repeat the calculations using differences of function values.

4.6. Derive the central-difference formula in the one-dimensional case together with the formulas for the best value of h and for the error.

4.7. Derive the one-dimensional formula for the second derivative

$$f''(x) \approx \frac{f'(x + h) - f'(x)}{h}$$

together with the formulas for the best value of h and the value of the error.

4.8. Derive the forward-difference formula for the gradient

$$[\nabla f(x)]_i \approx \frac{f(x + he_i) - f(x)}{h}$$

together with the formulas for the best value of h and the value of the error. These formulas vary from component to component. What would be an appropriate "compromise" value of h that could be used for all components?

4.9. Derive the forward-difference formula for the Hessian

$$[\nabla^2 f(x)]_{ij} = \frac{\partial^2 f(x)}{\partial x_i \partial x_j} \approx \frac{[\nabla f(x + he_j) - \nabla f(x)]_i}{h}$$

together with the formulas for the best value of h and the value of the error. These formulas vary from component to component. What would be an appropriate "compromise" value of h that could be used for all components?

4.10. Use a Taylor series approximation to show that

$$f'(x) \approx \Im[f(x + ih)]/h,$$

where $i = \sqrt{-1}$ and $\Im[f]$ is the imaginary part of the function f. Derive a formula for the error in this approximation. Repeat Exercises 4.1 and 4.4 using this derivative estimate.

4.11. Derive the evaluation graph for the function

$$f(x_1, x_2) = x_1^2 + 3x_1x_2 + 7x_2^2.$$

Use the graph to derive an evaluation technique for $\nabla f(x)$. Apply your technique at the point $x = (4, -5)^T$. Use the reverse mode of automatic differentiation.

4.12. Derive the evaluation graph for the function

$$f(x_1, x_2, x_3) = \frac{1}{\sqrt{x_1 + x_2^2 + x_3^3}} + \sin(x_1 x_2 + x_1 x_3 + x_2 x_3).$$

Use the graph to derive an evaluation technique for $\nabla f(x)$. Apply your technique at the point $x = (3, 6, 10)^T$. Use the reverse mode of automatical differentiation.

4.13. Assume that the evaluation of $f(x)$ involves only the operations of addition, subtraction, multiplication, and division. Prove that, if the automatic differentiation technique described in this section is used, then the number of arithmetic operations required to evaluate $\nabla f(x)$ is proportional to the number of arithmetic operations required to evaluate $f(x)$.

4.14. Consider the function

$$f(x_1, \ldots, x_n) = x_1 x_2 \cdots x_n.$$

Show that the number of arithmetic operations required to evaluate $\nabla f(x)$ using automatic differentiation is $O(n)$, whereas the number of arithmetic operations required to naively evaluate the gradient is $O(n^2)$. Evaluate the gradient by moving through the evaluation graph from top to bottom and show that this corresponds to using the "naive" formulas.

12.5 Methods That Do Not Require Derivatives

It is sometimes inconvenient, difficult, or impossible to calculate the derivatives of a function. For example, the function might not be differentiable at every point, so the derivative might not exist. Or the function evaluation might be carried out in terms of a long calculation involving auxiliary software—perhaps the solving of a differential equation or the running of a simulation—making it difficult to derive derivative formulas even if they exist. In these latter cases the function values might only be accurate to a few digits, so finite differencing would not be effective either.

In such cases it is desirable to have available optimization methods that do not require the calculation of derivative values. This requirement puts a severe handicap on a method, since even the optimality tests are based on derivative values. Even if such a method were given the solution to the optimization problem, it might require considerable computational effort before the method could identify such a point as the solution.

It might seem that such methods would, by necessity, fall outside the framework that we developed in Chapter 11. It is true that these methods are not directly related to Newton's method, since they cannot compute a gradient, and hence cannot compute a descent direction based on the gradient; also they cannot assess a step length using a sufficient descent condition, since that too depends on the gradient; and so forth.

What is surprising is that a large collection of derivative-free methods are guaranteed to converge, with a convergence theorem that is similar to the convergence theorem for line search methods discussed in Chapter 11. The assumptions on the objective function are the same for both convergence theorems, and the conditions on the methods are similar. This is remarkable since the conditions involve the gradient of the objective function, and the derivative-free methods do not compute gradients.

Derivative-free methods are easy to use. If it is possible to evaluate the objective function, then it is possible to use the methods. In addition, derivative-free methods are suitable for parallel computers, more so than Newton-type methods. It is gratifying that the methods have these good properties, along with guarantees of convergence.

One deficiency of derivative-free methods is their slow convergence. Another is that they are ill suited for large problems. Use of high-performance computers can ameliorate these deficiencies.

12.5.1   Simulation-Based Optimization

The most important setting for derivative-free optimization is the case where the evaluation of the objective function is based on auxiliary software, that is, the function values are obtained from a *simulation*. Suppose you are trying to design a car and would like the surface of the car to have low air resistance. You might develop a computer simulation that calculates the air flow and air resistance for a particular design. This itself could be challenging, since it involves the simulation of fluid flow at high speeds, with the potential for turbulence and wind vortices. Once this model had been perfected, the next goal would be to determine an optimal design, or at least an improved design. Presumably the preliminary design sketches would determine the approximate shape of the car, and the optimization would be used to refine this design, perhaps by adjusting the values of a relatively small number of design variables. For example, the design variables might represent the angle of the windshield, the wheel base, etc.

Given the complexity of the formulas for air flow and air resistance, it might be difficult to derive formulas for the derivative values required by a quasi-Newton method. In addition to the challenges of deriving the necessary formulas, there would be challenges in implementing the formulas in software. This would be compounded because the underlying simulation software for computing these function values would almost certainly have been prepared by others, perhaps another company, with the source code unavailable.

This might seem like a good setting for using finite-difference approximations to the gradient. In this case there are a small number of variables, the function values can be computed by the auxiliary software, and there is some hope that the function is differentiable. But this is not the case.

One reason is that the solution of the air-flow model might not be computed to high accuracy. Even with sophisticated machining equipment, the settings of the equipment would only be accurate to perhaps 3–4 digits. It would not make sense to solve the air-flow simulation to high accuracy given the limitations of machining. This would mean that the function values would be accurate to about 4 digits, and the resulting finite-difference gradient values would be accurate to only about 2 digits. (The errors in the gradient values are proportional to the square root of the error in the function values; see Section 12.4.1.) With gradients of this low accuracy, a quasi-Newton method would likely perform poorly.

There is another reason why it is inappropriate to use a finite-difference gradient, a reason that is more serious and that becomes more prominent as the auxiliary software becomes more sophisticated. Consider a simplified one-variable example with an objective function defined in terms of an integral:

$$f(x) = \int_a^b g(x, y)\, dy.$$

We will assume that we know the formula for the function $g(x, y)$, but that a formula for the integral is not available, since such a formula may not even exist. Simple algorithms approximate an integral by a weighted sum of function values:

$$\hat{f}(x) = \sum_{i=1}^k \alpha_k g(x, y_k),$$

where α_k are specified constants and y_k are specified values of y in the interval $[a, b]$.

If $g(x, y)$ is a differentiable function, then so are f and $\hat{f}$:

$$f'(x) = \int_a^b \frac{\partial}{\partial x} g(x, y)\, dy$$

$$\hat{f}'(x) = \sum_{i=1}^k \alpha_k \frac{\partial}{\partial x} g(x, y_k).$$

This is a simplistic way of estimating integrals, since the same formula is used regardless of the values of a, b, and x, and regardless of the behavior of the function g. More sophisticated software for estimating integrals does not use a single formula, but rather uses an *adaptive algorithm* that changes the calculations based on the behavior of the underlying function. For example, if the function $g(x, y)$ is changing rapidly, then a more elaborate calculation will be used. This approach makes it possible to compute accurate approximations to the integral with minimal computational effort.

Suppose we want to solve

$$\text{minimize } f(x) = \int_a^b g(x, y)\, dy$$

and that we use sophisticated software to estimate the integrals that define $f(x)$. Then it is possible, indeed likely, that as x changes, the formula used to estimate the integral also changes. Every change in the formula is likely to introduce nonsmoothness or even a discontinuity in the computed objective function. The effects on the objective function may be small, but they can have a large effect on the computation of a finite-difference gradient because of the division by the small finite-difference parameter h.

This is illustrated in Figure 12.4. The graph on the left was obtained by using an adaptive algorithm to approximate an integral. On the right is a blowup of one portion of the graph. It is clear that the approximation is not differentiable.

In summary, by using sophisticated software to estimate a smooth function $f(x)$, we can create an optimization problem involving a nonsmooth function.

These difficulties are common when using auxiliary software. Adaptive algorithms are widely used to solve differential equations, for example.

If the objective function, *as computed*, is not differentiable, then it is not possible to use finite-difference approximations. The adaptivity of the algorithm may also make it impossible to use automatic differentiation. Use of a Newton-type method would require you to derive and program analytic formulas for the required derivatives.

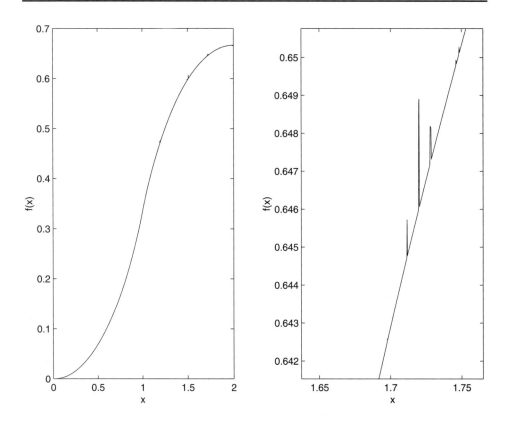

Figure 12.4. *Side effects of adaptive algorithms.*

In cases like these, derivative-free methods can be an attractive choice. If it is possible to evaluate the objective function, then it is possible to apply a derivative-free method. Whether the function is evaluated naively, or via sophisticated techniques, the derivative-free methods are an available tool.

12.5.2   Compass Search: A Derivative-Free Method

A great many derivative-free methods are based on the following basic template: Start with some trial point x. Evaluate the objective function at the point x and at a set of points in a specified pattern about x. If a better point is found (that is, one with a smaller function value), then this point becomes the new trial point and the process repeats. Otherwise, the size of the pattern is reduced and the current trial point retained. This continues until the size of the pattern is reduced below some tolerance, and then the current trial point is returned as the estimate of the solution.

If the pattern is appropriately designed, and if the function being minimized is continuously differentiable, then it is possible to prove that

$$\lim_{k \to \infty} \left\| \nabla f(x_{(k)}) \right\| = 0,$$

where the points $x_{(k)}$ are the trial points corresponding to the iterations where the pattern is reduced in size. Since the pattern is reduced only when no better point is found, these are referred to as the "unsuccessful" iterations. At "successful" iterations a better point is found, and the size of the pattern is unchanged.

Let us look more closely at one of these methods. We will do this for a two-dimensional minimization problem so the results can be displayed graphically. Later we will mention how to extend the method to problems with more than two variables.

The method we will consider is called *compass search* because the pattern corresponds to the points of a compass. We will evaluate the objective function at the trial point $x = (x_1, x_2)^T$, as well as at the points

$$\text{North} : x + \Delta \begin{pmatrix} 0 \\ 1 \end{pmatrix}$$

$$\text{South} : x + \Delta \begin{pmatrix} 0 \\ -1 \end{pmatrix}$$

$$\text{East} : x + \Delta \begin{pmatrix} 1 \\ 0 \end{pmatrix}$$

$$\text{West} : x + \Delta \begin{pmatrix} -1 \\ 0 \end{pmatrix}$$

for some scalar value Δ that determines the size of the pattern.

Example 12.16 (Compass Search Method). We apply the compass search method to the problem

$$\text{minimize } f(x_1, x_2) = x_1^2 + 3x_2^2 + 2x_1x_2 + x_1 + 3$$

whose solution is $x_* = (0.25, -0.75)^T$. We start with the initial values

$$x_0 = (2 \quad 0.75)^T$$
$$\Delta_0 = 2.$$

At the first iteration we compute

$$f(x_0) = 13.6875$$
$$f(x_0 + \Delta_0(1, 0)^T) = 30.6875$$
$$f(x_0 - \Delta_0(1, 0)^T) = 4.6875$$
$$f(x_0 + \Delta_0(0, 1)^T) = 42.6875$$
$$f(x_0 - \Delta_0(0, 1)^T) = 8.6875.$$

Since the smallest value is obtained at the new point $x_0 - \Delta_0(1, 0)^T = (0, 0.75)^T$, we consider this to be a successful iteration and set $x_1 = (0, 0.75)^T$ and $\Delta_1 = \Delta_0 = 2$. This completes the first iteration. It would also have been valid to accept $x_0 - \Delta_0(0, 1)^T = (2, -1.25)^T$ as the new point, since this also improves the value of the objective function.

The first few iterations of the compass search method are illustrated in Figure 12.5. The figure shows the contours of the objective function with the minimizer marked with a +. At each successful iteration, the point with the smallest function value is chosen as the new point. This point on the compass pattern is marked with a ∗.

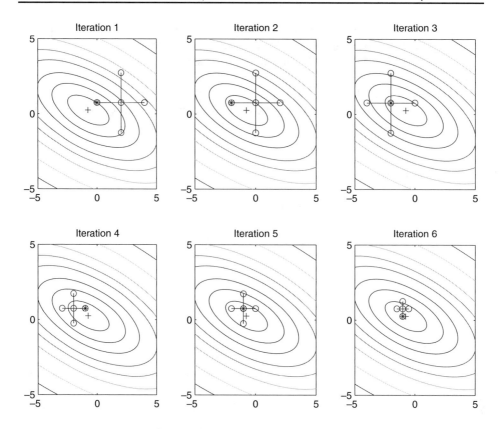

Figure 12.5. *Compass search method.*

The first two iterations are successful, giving $x_1 = (0, 0.75)^T$ and $x_2 = (-2, 0.75)^T$. The third iteration is unsuccessful, so $x_3 = x_2$ and $\Delta_3 = \Delta_2/2 = 1$. The fourth iteration is successful, giving $x_4 = (-1, 0.75)^T$. The fifth iteration is unsuccessful, so $x_5 = x_4$ and $\Delta_5 = \Delta_4/2 = 0.5$.

If the iteration is continued, the algorithm converges to the solution $x_* = (0.25, -0.75)^T$. ∎

It is easy to extend the compass search method to n dimensions. In general, the objective function is evaluated at the trial point x and at the points

$$x + \Delta d_i,$$

where d_i is either a coordinate direction or its negative. For example,

$$
\begin{aligned}
d_1 &= (1, 0, 0, \ldots, 0)^T \\
d_2 &= -(1, 0, 0, \ldots, 0)^T \\
d_3 &= (0, 1, 0, \ldots, 0)^T \\
d_4 &= -(0, 1, 0, \ldots, 0)^T
\end{aligned}
$$

$$\vdots$$

Except for the increase in the number of trial points, the method remains unchanged. Here is the algorithm.

ALGORITHM 12.2.
Compass Search Algorithm

1. Specify some initial guess of the solution x_0, an initial pattern size $\Delta_0 > 0$, and a convergence tolerance $\Delta_{\text{tol}} > 0$.
2. For $k = 0, 1, \ldots$

   (i) If $\Delta_k < \Delta_{\text{tol}}$, stop.
   (ii) Evaluate f at the points x_k and $x_k + \Delta_k d_i$, for $i = 1, \ldots, 2n$.
   (iii) If $f(x_k + \Delta_k d_i) < f(x_k)$ for some i, then set $x_{k+1} = x_k + \Delta_k d_i$ and $\Delta_{k+1} = \Delta_k$.
   (iv) Otherwise, set $\Delta_{k+1} = \Delta_k/2$.

The efficiency of the compass search method depends critically on the choice of the initial pattern size Δ_0. If Δ_0 is chosen too small and if x_0 is far from the solution, it may take many iterations to get close to the solution. If Δ_0 is chosen too large, it may take many iterations to reduce Δ_k below the tolerance Δ_{tol}.

The compass search algorithm allows some choice in the selection of x_{k_1}, since it is possible to choose any i for which $f(x_k + \Delta_k d_i) < f(x_k)$. It is possible to exploit this flexibility when using the algorithm, for example, by terminating an iteration as soon as a better point is found.

More importantly, it is possible to evaluate multiple function values $\{ f(x + \Delta d_i) \}_{i=1}^{2n}$ simultaneously on the various processors of a parallel computer. In cases where the function evaluations are expensive, such as when they involve some auxiliary simulation, this can dramatically accelerate the compass search algorithm. If sufficient processors are available, it is possible to expand the set of directions $\{ d_i \}$ by including additional directions beyond just the coordinate directions and their negatives. This can further accelerate the algorithm.

Compass search is a slow but sure method. It is guaranteed to make progress towards a solution, but might do so slowly. For this reason, the convergence tolerance Δ_{tol} is commonly set, so a coarse estimate of the solution is accepted.

12.5.3 Convergence of Compass Search

It is possible to prove a convergence theorem for compass search that is remarkably similar to the convergence theorem for line search methods (see Section 11.5). It will be necessary to make some assumptions about the objective function, and indeed we will make the same assumptions as before. Namely, we will assume that (i) the set $S = \{ x : f(x) \leq f(x_0) \}$ is bounded, and (ii) ∇f is Lipschitz continuous for all x, that is,

$$\|\nabla f(x) - \nabla f(y)\| \leq L \|x - y\|$$

for some constant $0 < L < \infty$. The first assumption assures that the method will not continue indefinitely evaluating points in an unbounded region. Some variant of the second assumption is necessary to guarantee that the algorithm will converge.

Let us look more closely at an iteration of compass search, and in particular at what happens at an unsuccessful iteration where a better point cannot be found.

Suppose that the current trial point is not a stationary point. Then $\nabla f(x_k) \neq 0$. In the two-dimensional case it is obvious that, no matter what the value of the gradient, the steepest-descent direction $-\nabla f(x_k)$ is within 45° of one of the compass directions d_i. Let d be a compass direction that satisfies this condition. Then

$$-\frac{d^T \nabla f(x_k)}{\|d\| \cdot \|\nabla f(x_k)\|} \geq \cos(45°) = \frac{1}{\sqrt{2}} > 0.$$

This is condition (iii) from the line search theorem

$$-\frac{p_k^T \nabla f(x_k)}{\|p_k\| \cdot \|\nabla f(x_k)\|} \geq \epsilon > 0$$

with $\epsilon = 1/\sqrt{2}$. In n dimensions it is possible to show that

$$-\frac{d^T \nabla f(x_k)}{\|d\| \cdot \|\nabla f(x_k)\|} \geq \frac{1}{\sqrt{n}} > 0$$

for at least one of the compass directions d. (See the Exercises.) Thus there is always a direction d that satisfies a sufficient descent condition. This condition can be rearranged as

$$\frac{1}{\sqrt{n}} \|d\| \cdot \|\nabla f(x_k)\| \leq -d^T \nabla f(x_k).$$

At an unsuccessful iteration, $f(x_k + \Delta_k d) \geq f(x_k)$. We can use the mean-value theorem to conclude that

$$0 \leq f(x_k + \Delta_k d) - f(x_k) = \Delta_k d^T \nabla f(x_k + \eta \Delta_k d)$$

for some η between 0 and 1. (See Section 2.6.) If we divide by Δ_k and then subtract $d^T \nabla f(x_k)$ from both sides, we obtain

$$-d^T \nabla f(x_k) \leq d^T[\nabla f(x_k + \eta \Delta_k d) - \nabla f(x_k)].$$

Now substitute the rearranged sufficient descent condition into this formula:

$$\frac{1}{\sqrt{n}} \|d\| \cdot \|\nabla f(x_k)\| \leq d^T[\nabla f(x_k + \eta \Delta_k d) - \nabla f(x_k)]$$

$$\leq \|d\| \cdot \|\nabla f(x_k + \eta \Delta_k d) - \nabla f(x_k)\|.$$

Earlier we assumed that the gradient of the objective function $\nabla f(x)$ was Lipschitz continuous with constant L. Applying this assumption to the right-hand side gives

$$\frac{1}{\sqrt{n}} \|d\| \cdot \|\nabla f(x_k)\| \leq \|d\| \cdot L \|\eta \Delta_k d\| \leq L \Delta_k \|d\|^2.$$

Since the compass directions all satisfy $\|d\| = 1$, this simplifies to

$$\|\nabla f(x_k)\| \leq \sqrt{n} L \Delta_k.$$

We have been able to derive a bound on $\|\nabla f(x_k)\|$ at an unsuccessful iteration of the compass search method, even though the method does not use any gradient information.

This is enough to provide us with our convergence result. If we only consider the unsuccessful iterations $x_{(k)}$, then

$$\lim_{k \to \infty} \left\| \nabla f(x_{(k)}) \right\| \leq \sqrt{n} L \lim_{k \to \infty} \Delta_{(k)}.$$

At every unsuccessful iteration, $\Delta_{k+1} = \Delta_k/2$, and Δ_k is never increased. Since the level sets are bounded, for each value of Δ there can only be a finite number of successful iterations. (See the Exercises.) Thus the limit of the right-hand side is zero and

$$\lim_{k \to \infty} \left\| \nabla f(x_{(k)}) \right\| = 0.$$

If the assumptions about the objective function are satisfied, the compass search method is guaranteed to converge. A formal statement of the theorem is given below.

Theorem 12.17 (Convergence of Compass Search). *Let f be a real-valued function of n variables. Let x_0 be a given initial point and determine $\{x_k\}$ using the compass search algorithm with initial pattern size Δ_0. Assume that*

(i) *the set $S = \{x : f(x) \leq f(x_0)\}$ is bounded;*

(ii) *∇f is Lipschitz continuous for all x, that is,*

$$\|\nabla f(x) - \nabla f(y)\| \leq L \|x - y\|,$$

for some constant $0 < L < \infty$.

Let $\{x_{(k)}\}$ be the set of unsuccessful iterations, i.e., the iterations where a better point cannot be found and Δ_k is reduced. Then

$$\lim_{k \to \infty} \left\| \nabla f(x_{(k)}) \right\| = 0.$$

It is also possible to determine the rate of convergence for the compass search algorithm. If additional assumptions are made, then it can be shown that at the unsuccessful iterations,

$$\left\| x_{(k)} - x_* \right\| \leq c \Delta_{(k)}$$

for some constant c that does not depend on k. This is similar to, but not the same as, a linear rate of convergence. There are several differences between this result and linear convergence.

First, in this result we consider only unsuccessful iterations and ignore successful iterations. Second, the result shows only that $\{x_{(k)} - x_*\}$ is *bounded* by a linearly convergent series. At each unsuccessful iteration we divide $\Delta_{(k)}$ by 2, so the sequence $\{\Delta_{(k)}\}$ converges linearly to zero. This does not guarantee that $\{x_{(k)} - x_*\}$ converges linearly. (See the Exercises.) Nevertheless it is similar to linear convergence, and this property is sometimes referred to as "r-linear" convergence.

In contrast, the steepest-descent method from Section 12.2 converges linearly, i.e., at a faster rate than the compass search method.

Exercises

5.1. Fill in the details of the iterations of the compass search method for Example 12.16.

5.2. Apply the compass search method to the problem

$$\text{minimize } f(x_1, x_2) = (x_1 + x_2 + 4)^2 + (x_1 - x_2 + 3)^2$$

using the initial point $x_0 = (1, 5)^T$ and $\Delta_0 = 6$. Perform at least four iterations of the method.

5.3. Apply the compass search method to the problem

$$\text{minimize } f(x_1, x_2, x_3) = x_1^2 + 2x_2^2 + 3x_3^2$$

using the initial point $x_0 = (1, 2, 1)^T$ and $\Delta_0 = 3$. Perform at least five iterations of the method.

5.4. Variants of compass search can be developed using different sets of search directions. For example, the directions could be based on the corners of a box or cube centered about the origin. In two dimensions this gives the directions $d_i = (\pm 1, \pm 1)^T$:

$$
\begin{aligned}
d_1 &= (\,1 \quad\ \ 1\,)^T \\
d_2 &= (\,1 \quad -1\,)^T \\
d_3 &= (-1 \quad\ \ 1\,)^T \\
d_4 &= (-1 \quad -1\,)^T.
\end{aligned}
$$

In three dimensions this gives the directions $d_i = (\pm 1, \pm 1, \pm 1)^T$. Repeat the previous two problems using this variant of compass search. This is sometimes called *box* search.

5.5. If compass search is used to solve an n-dimensional problem, how many directions d_i are needed? How many directions are needed for box search in this case? Which method is preferable as n increases?

5.6. Program the compass search algorithm, and apply it to the problem in Example 12.16 with $\Delta_{\text{tol}} = 10^{-5}$. In your program, choose the best value of $x_k + \Delta_k d_i$ at each iteration. At which iteration is $x_k = x_*$? How many additional iterations are required before the algorithm terminates?

5.7. Repeat the previous problem, but choose the *first* value of $x_k + \Delta_k d_i$ that yields a better function value. How do the two versions of the algorithm compare? Be sure to take into account the number of function evaluations required to find a solution, and not just the number of iterations.

5.8. The compass search algorithm terminates when $\Delta_k < \Delta_{\text{tol}}$. Apply the compass search algorithm to the minimization problems in the previous problems and compare

Δ_k and $\|\nabla f(x_k)\|$ at the unsuccessful iterations. (You should use a small value of Δ_{tol}.) Is the value of Δ_k an effective substitute for the value of $\|\nabla f(x_k)\|$ in the convergence test? (It would be desirable to use $\|\nabla f(x_k)\|$ in the convergence test, but compass search is often used when it is impossible to compute $\nabla f(x_k)$.)

5.9. Repeat the previous exercise, but compare Δ_k and $\|\nabla f(x_k)\|$ at all iterations, both successful and unsuccessful. Can you identify a relationship between these two sets of values?

5.10. Consider the two-variable problem

$$\text{minimize } f(x) = \tfrac{1}{2} \max \left\{ \|x - c\|^2, \|x - d\|^2 \right\},$$

where $c = (1, -1)^T$ and $d = -c$.

   (i) Prove that the function is continuous and strictly convex, but that its gradient is discontinuous at points of the form $\hat{x} = (a, a)^T$.

   (ii) Prove that the minimizer of $f(x)$ is $x_* = (0, 0)^T$.

   (iii) Let $x_0 = (a, a)^T$ for some $a \neq 0$. Prove that any positive value of Δ_0 will result in an unsuccessful iteration of the compass search algorithm. Use this result to prove that the compass search algorithm will terminate at $(a, a)^T$ and make no progress towards the solution.

   (iv) Program the compass search algorithm and run it on this example. Use initial guesses of the form $x_0 = (a, a)^T$ as well as points that do not lie on this line. Is compass search able to find the minimizer of this function?

5.11. Give an example of a sequence $\{ y_k \}$ that satisfies

$$|y_k| \leq 1/2^k,$$

but where the sequence does not converge linearly. This shows that a sequence that converges "r-linearly" need not converge linearly.

5.12. In the compass search method, prove that for each value of Δ there can only be a finite number of successful iterations if the level sets are bounded.

5.13. Let $f(x)$ be a function of n variables and assume that $\|\nabla f(x_k)\| > 0$. Prove that there is at least one compass direction d that satisfies

$$-\frac{d^T \nabla f(x_k)}{\|d\| \cdot \|\nabla f(x_k)\|} \geq \frac{1}{\sqrt{n}} > 0.$$

12.6 Termination Rules

Ideally an algorithm for unconstrained optimization would terminate at an estimate of the solution x_k that satisfied

$$\nabla f(x_k) = 0 \quad \text{and} \quad \nabla^2 f(x_k) \text{ positive semidefinite.}$$

There are two reasons why this is not realistic. First, it is unlikely that the calculated value of the gradient would ever be exactly zero because of rounding errors in computer calculations. Second, even if there are no rounding errors, no algorithm is guaranteed to find such a point in a finite amount of time. Another difficulty, although not as serious, is that many commonly used algorithms for unconstrained optimization do not have available to them the Hessian matrix $\nabla^2 f(x_k)$, and so cannot directly verify if this matrix is positive semidefinite.

As an alternative, we might consider replacing the above conditions by the test

$$\|\nabla f(x_k)\| \le \epsilon$$

for some small number ϵ. This immediately raises the question of how small to make ϵ. If (say) 16-digit arithmetic were used, then perhaps $\epsilon \approx 10^{-16}$ would be appropriate. In Section 2.7, we have shown that if the error in the function value $|f(x_k) - f(x_*)|$ is $O(h)$, then the norm of the gradient $\|\nabla f(x_k)\|$ and the errors in the variables $\|x_k - x_*\|$ will both be $O(\sqrt{h})$, suggesting that a value

$$\epsilon \approx \sqrt{10^{-16}} = 10^{-8}$$

would be more appropriate.

Suppose now that the objective function were changed by changing the units in which it was measured. For example, suppose that instead of measuring the objective in terms of kilometers it were now measured in terms of millimeters. This would cause the objective function to be multiplied by 10^6, and hence would cause $\|\nabla f(x_k)\|$ to be multiplied by 10^6. This minor change to the objective function would make the convergence test much more difficult to satisfy, even though the underlying optimization problem was essentially unchanged.

To alleviate this difficulty, the convergence test could be modified to

$$\|\nabla f(x_k)\| \le \epsilon |f(x_k)|,$$

where $\epsilon \approx \sqrt{\epsilon_{\text{mach}}}$, in this case $\epsilon \approx 10^{-8}$. This test is also flawed. If the optimal value of the objective function is zero (which is not uncommon when least-squares problems are solved), then this convergence test will be unnecessarily stringent and may very well be impossible to satisfy. We are thus led to consider a convergence test of the form

$$\|\nabla f(x_k)\| \le \epsilon(1 + |f(x_k)|)$$

which attempts to cope with all the above-mentioned difficulties. When $|f(x_k)|$ is close to zero, this test resembles our original test, whereas when $|f(x_k)|$ is large, it resembles our second test.

When Newton's method is being used, it is also appropriate to ask that $\nabla^2 f(x_k)$ be positive semidefinite. Due to the properties of rounded arithmetic, however, we weaken this requirement and only demand that

$$\nabla^2 f(x_k) + \epsilon I$$

be positive semidefinite. (The choice of ϵ is discussed below.) Since a factorization $\nabla^2 f(x_k)$ is normally computed to determine the search direction in Newton's method, this test can be performed at low cost as a by-product of the method. (See Section 11.4.)

Since it is not possible to design a perfect convergence test for terminating an algorithm, it is common to insist that additional tests be satisfied before a point x_k is accepted as
an approximate minimizer of the function f. For example, the algorithm might attempt to
ensure that the sequences $\{ f(x_k) \}$ and $\{ x_k \}$ are both converging. The combined test might
take the form

$$\| \nabla f(x_k) \| \leq \epsilon_1 (1 + |f(x_k)|)$$
$$f(x_{k-1}) - f(x_k) \leq \epsilon_2 (1 + |f(x_k)|)$$
$$\| x_{k-1} - x_k \| \leq \epsilon_3 (1 + \| x_k \|)$$
$$\nabla^2 f(x_k) + \epsilon_4 I \text{ is positive semidefinite.}$$

The fourth test on $\nabla^2 f(x_k)$ would be performed only if the Hessian matrix were available.

The tolerances ϵ_i should be selected based on the accuracy of the computer calculations. The tolerance ϵ_2 should represent the accuracy with which the function and derivative
values can be calculated. If they can be calculated to full machine accuracy, then it is appropriate to take $\epsilon_2 = \epsilon_{mach}$, or perhaps some multiple of ϵ_{mach} that depends on the number
of variables in the problem. If the function and derivative values are only accurate to, say,
8 digits, then $\epsilon_2 = 10^{-8}$ is a more appropriate choice. (Such a choice would be necessary, for example, if finite-difference derivative estimates were used; see Section 12.4.1.)
Under many circumstances, it will be appropriate to choose $\epsilon_4 = \epsilon_2 \| \nabla^2 f(x_k) \|$, unless
the accuracy of the Hessian calculations is different from that of the function and gradient
values.

Since x_k and $\nabla f(x_k)$ can only be expected to have half the precision of $f(x_k)$ (see
Section 2.7), we are led to choose $\epsilon_1 = \epsilon_3 = \sqrt{\epsilon_2}$. The book by Gill, Murray, and Wright
(1981) recommends the larger value $\epsilon_1 = \epsilon_3 = \sqrt[3]{\epsilon_2}$, based on computational tests that the
authors have performed.

The tests on $\nabla f(x_k)$ and x_k both are based on norms. If, for example, $\nabla f(x_k) = (\gamma, \ldots, \gamma)^T$, then

$$\| \nabla f(x_k) \|_2 = \sqrt{n} |\gamma| \quad \text{and} \quad \| \nabla f(x_k) \|_\infty = |\gamma|,$$

where n is the number of variables. If n is large, then the 2-norm of the gradient can be
large even if γ is small. This can distort the convergence tests and so it is wise to use the
infinity norm when large problems are solved.

The use of norms in the convergence tests can have other side effects. Suppose that

$$x_{k-1} = (\, 1.44453 \quad 0.00093 \quad 0.0000079 \,)^T$$
$$x_k = (\, 1.44441 \quad 0.00012 \quad 0.0000011 \,)^T.$$

If we had chosen $\epsilon_3 = 10^{-3}$, then

$$\| x_{k-1} - x_k \|_\infty = \big\| (\, 0.00012 \quad 0.00081 \quad 0.0000068 \,)^T \big\|$$
$$= 0.00081 \leq \epsilon_3 \| x_k \| = 1.44441 \times 10^{-3}$$

and so x_k would pass this test. If, for our application, it was important that the *significant*
digits of all the variables be accurate, then x_k would not be satisfactory since its second

and third components are still changing in their first significant digit. The use of norms emphasizes the larger components in a vector, and so the smaller components may have poor *relative* accuracy.

This effect can be ameliorated by scaling the variables. Suppose that we knew that

$$x_* \approx (1 \quad 10^{-4} \quad 10^{-6})^T.$$

(We might know this from practical experience, or from a previous attempt at computing x_*.) Then we could perform a change of variables $x \to \hat{x}$ so that

$$\hat{x}_* \approx (1 \quad 1 \quad 1)^T.$$

At the optimum, all the variables in the transformed problem would have approximately the same relative accuracy. In this case the change of variables would be

$$x = (x_1 \quad x_2 \quad x_3)^T \to (x_1 \quad 10^4 x_2 \quad 10^6 x_3)^T = \hat{x}.$$

Note that the formulas for f and ∇f would have to be adjusted accordingly.

Example 12.18 (Scaling). Consider the quadratic function

$$f(x) = \tfrac{1}{2}x^T Q x - c^T x,$$

where

$$Q = \begin{pmatrix} 8 & 3 \times 10^4 & 0 \\ 3 \times 10^4 & 4 \times 10^8 & 1 \times 10^{10} \\ 0 & 1 \times 10^{10} & 6 \times 10^{12} \end{pmatrix} \quad \text{and} \quad c = \begin{pmatrix} 11 \\ 8 \times 10^4 \\ 7 \times 10^6 \end{pmatrix}.$$

This function has a minimum at

$$x_* = Q^{-1}c = (1 \quad 10^{-4} \quad 10^{-6})^T.$$

We make the change of variable

$$\hat{x} = Dx,$$

where

$$D = \begin{pmatrix} 1 & 0 & 0 \\ 0 & 10^4 & 0 \\ 0 & 0 & 10^6 \end{pmatrix},$$

and let $\hat{f}$ be the transformed function. Then $x = D^{-1}\hat{x}$, or

$$x_1 = \hat{x}_1, \quad x_2 = 10^{-4}\hat{x}_2, \quad \text{and} \quad x_3 = 10^{-6}\hat{x}_3.$$

Substituting into the formulas for f and ∇f gives

$$\hat{f}(\hat{x}) = \tfrac{1}{2}\hat{x}^T(D^{-1}QD^{-1})\hat{x} - (D^{-1}c)^T\hat{x}$$

and

$$\nabla \hat{f}(\hat{x}) = (D^{-1}QD^{-1})\hat{x} - (D^{-1}c),$$

where

$$D^{-1}QD^{-1} = \begin{pmatrix} 8 & 3 & 0 \\ 3 & 4 & 1 \\ 0 & 1 & 6 \end{pmatrix} \quad \text{and} \quad D^{-1}c = \begin{pmatrix} 11 \\ 8 \\ 7 \end{pmatrix}.$$

The minimizer of the scaled function is

$$\hat{x}_* = (D^{-1}QD^{-1})^{-1}(D^{-1}c) = (1 \quad 1 \quad 1)^T.$$

Notice that

$$\nabla \hat{f}(\hat{x}) = D^{-1}\nabla f(x).$$

This result can be derived directly using the chain rule (see below). ■

In general suppose we perform a change of variable of the form

$$x \to Dx = \hat{x},$$

where D is an invertible scaling matrix. Then by the chain rule,

$$\hat{f}(\hat{x}) = f(D^{-1}\hat{x}) = f(x) \quad \text{and} \quad \nabla \hat{f}(\hat{x}) = D^{-1}\nabla f(D^{-1}\hat{x}) = D^{-1}\nabla f(x).$$

Once the minimization has been performed with respect to the variables $\hat{x}$, the solution to the original problem is obtained from $x_* = D^{-1}\hat{x}_*$.

Exercises

6.1. Suppose that a change of variables of the form

$$x \to Dx = \hat{x}$$

has been performed, where D is an invertible matrix. What is the formula for $\nabla^2 \hat{f}(\hat{x})$?

6.2. Suppose that a change of variables of the form

$$x \to Dx = \hat{x}$$

has been performed, where D is an invertible matrix. Prove that the Newton direction (in terms of the original variables) is unchanged. Under what conditions on D are the steepest-descent and the BFGS quasi-Newton directions unchanged?

6.3. Suppose that a change of variables of the form

$$x \to Dx = \hat{x} + v$$

has been performed, where D is an invertible matrix and v is a nonzero vector. Determine the formulas for $\hat{f}(\hat{x})$, $\nabla \hat{f}(\hat{x})$, and $\nabla^2 \hat{f}(\hat{x})$. Apply your results to the function in Example 12.18 using

$$D = \begin{pmatrix} 1 & 3 & 0 \\ 0 & 10^3 & 5 \\ 0 & 0 & 10^5 \end{pmatrix} \quad \text{and} \quad v = (4 \quad 7 \quad 2).$$

6.4. Use Newton's method (with or without a line search) to solve

$$\text{minimize } f(x_1, x_2) = 5x_1^4 + 6x_2^4 - 6x_1^2 + 2x_1x_2 + 5x_2^2 + 15x_1 - 7x_2 + 13.$$

Use the initial guess $(1, 1)^T$, and use the termination rules derived in this section. Experiment with the choices of ϵ_1–ϵ_4, and determine how small you can make them before you reach the limits of the precision on your computer.

6.5. Repeat the previous problem using the steepest-descent method. You need not test if the Hessian at x_k is (approximately) positive semidefinite.

12.7 Historical Background

Many of the ideas in this chapter owe a great debt to Isaac Newton, and in particular to his manuscripts of the 1660s.[11] In these manuscripts Newton examines the roots of polynomial equations from both a numerical and an analytic point of view and develops a technique (now called Newton's method) for approximating these roots to any desired accuracy.

Newton's own derivation is different from ours, and we present it here, using the example that Newton himself employed:

$$x^3 - 2x - 5 = 0.$$

The first step is to find an integer adjacent to a root. A few simple calculations indicate that there is a root between 2 and 3, and so 2 is an acceptable initial guess. The root can be written as

$$x_* = 2 + p$$

for some p. Substituting this into the cubic polynomial gives

$$p^3 + 6p^2 + 10p - 1 = 0.$$

Since $|p| < 1$, the terms $p^3 + 6p^2$ "are neglected on account of their smallness" leaving us with the approximate equation

$$10p - 1 = 0,$$

or $p \approx 0.1$, "very nearly true." We now repeat this process, applying it to p instead of x. The exact value of p will satisfy

$$p = 0.1 + q$$

for some q, and this expression can be substituted into the cubic equation for p, giving

$$q^3 + 6.3q^2 + 11.23q + 0.061 = 0.$$

---

[11]Newton's manuscripts are available in an edition due to D. T. Whiteside that includes the Latin original, an English translation, and extensive annotations.

If the terms $q^3 + 6.3q^2$ are ignored, we solve for q in

$$11.23q + 0.061 = 0,$$

obtaining $q \approx -0.0054$. Repeating this procedure one more time using $q = -0.0054 + r$ gives (after ignoring nonlinear terms) the equation

$$0.000541708 + 11.16196r = 0$$

or $r \approx -0.00004853$. Combining these results together gives

$$x_* \approx 2.09455147.$$

The root is $x_* = 2.0945514815\ldots$. Newton goes on, a few pages later, to show how to use his technique to obtain analytic series approximations to roots of polynomial equations. He also points out that a quadratic equation could be solved to obtain the update to the solution (for example, p could be obtained by solving $6p^2 + 10p - 1 = 0$) and thus improve the convergence rate of the method. His comments indicate that he is aware that his basic method converges at a quadratic rate, and that using a quadratic equation would cause it to converge at a cubic rate.

The approach used by Newton solves a nonlinear equation by guessing at a solution, approximating the nonlinear equation near that guess by a linear equation (in fact the equation for the tangent line), and using that linear approximation to obtain a new guess of the solution. This is the same approach used in Chapter 2 to derive Newton's method, but the organization of the calculations used there is different. The algorithm in Chapter 2 was used by Joseph Raphson in 1690, although Raphson did not refer to derivatives. Raphson was aware of Newton's work, but he did not realize that the two methods were equivalent.

From a practical, pencil-and-paper point of view the methods *are different*. Newton's original formulation requires that a new polynomial be derived every time the estimate of the solution is refined, whereas Raphson's version always works with the original polynomial. It is also more obvious how to generalize Raphson's formula to more general nonlinear functions. For these reasons, the more modern formulas are often referred to collectively as the Newton–Raphson method. (Further information about Raphson can be found in the paper by Bićanić and Johnson (1979).)

Many other important enhancements, such as the interpretation of the method using derivatives, are due to Thomas Simpson (1740). A more detailed history of Newton's method can be found in the paper by Ypma (1995).

This was not the first approach that Newton had used to solve polynomial equations. In late 1664[12] Newton examined existing algorithms due to Viète and Oughtred; these methods converge linearly, finding the root one digit at a time. In early 1665 Newton discovered the secant method. His derivation appears to be based on geometric reasoning, although he does not include a diagram. Other relevant results that can be found in Newton's manuscripts include the Taylor series and Horner's rule for evaluating a polynomial. Newton was averse to publishing many of his results. However, some of his manuscripts were circulated during and after his lifetime, inspiring publications by other British scientists such as Raphson,

---

[12]Many of Newton's manuscripts cannot be precisely dated; the dates given here are those proposed by Whiteside.

Taylor, Horner, Halley, Maclaurin, Simpson, and others, and it is those publications that give these results their modern names.

It would be pleasing to report that Newton was the first to discover Newton's method, but this is not the case. For example, according to Whiteside, a "primitive form of it" was used to solve $x^n - a = 0$ by the "fifteenth-century Arabic mathematician al-Kāšī." It appears, however, that Newton was unaware of these earlier developments.

The optimality conditions for one-dimensional optimization problems were known to Newton, and were developed as he developed his calculus. The corresponding conditions for multidimensional problems were known to Simpson, although there is indirect evidence that Newton may have known these as well. The multidimensional version of Newton's method is also due to Simpson. Multidimensional versions of the secant method (i.e., quasi-Newton methods) were developed by Davidon in 1959, although his report was not published until 1991; the first published account of these techniques was by Fletcher and Powell in (1963) and was based on Davidon's technical report.

The steepest-descent method was developed by Cauchy in 1847. He discussed it as a technique for solving a nonlinear equation of the form

$$f(x_1, \ldots, x_n) = 0,$$

where f is a real-valued function that cannot take on negative values. This is the same as solving

$$\text{minimize } f(x_1, \ldots, x_n),$$

and the fact that in Cauchy's case the optimal value of f is zero is but a minor technicality. Cauchy, like Gauss, was motivated by the study of planets and comets, and his specific motivation was presented in a paper he published earlier the same year.

12.8 Notes

Quasi-Newton Methods—Although the BFGS formula is the most widely used of the quasi-Newton formulas for optimization, the symmetric rank-one formula also has desirable properties. The relative merits of the two methods have been studied in several papers (see, for example, Nocedal (1991)), and these studies may lead to modifications in the way quasi-Newton methods are programmed. The symmetric rank-one formula must be used with some care, since the denominator in the update can be zero or negative, leading to computational difficulties for the optimization method.

The description of the quasi-Newton methods given above uses the formula $B_k p = -\nabla f(x_k)$ to compute the search direction. This would seem to require the solution of a linear system at a cost of $O(n^3)$ operations. Some authors recommend using a more sophisticated technique. The Cholesky factorization of the initial matrix $B_0 = L_0 L_0^T$ is computed in the usual way. (If $B_0 = I$, then $L_0 = I$ as well, making this especially easy.) At later iterations the update formula is applied to L_{k-1} rather than B_{k-1} so that at every iteration the Cholesky factorization of B_k is automatically available. In this way, computing the search direction requires only two applications of backsubstitution at a cost of only $O(n^2)$ operations. This is possible because the update formula is of low rank (for example, the BFGS update is of rank two). Details of this approach can be found in the paper by Gill and Murray (1972).

This approach also can be used to ensure that the sufficient descent and gradient-relatedness conditions on the search direction are satisfied.

The first developers of quasi-Newton methods used a different set of update formulas. Instead of computing a set of approximations $B_k \approx \nabla^2 f(x_k)$ they computed $H_k \approx [\nabla^2 f(x_k)]^{-1}$. For each of the update formulas given above there is a corresponding update formula for H_k. Using these approximations, the search direction is computed from $p = -H_k \nabla f(x_k)$ without having to solve a system of linear equations. This makes the algorithms easy to program. However, there is a price associated with this simplicity. If the BFGS formula for H_k is programmed on a computer, then the matrices $\{H_k\}$ ought to be positive definite. But in rounded computer arithmetic there is a danger that this will not always be true, and this has led people to abandon this approach. If, however, the formula for B_k is used and a factorization of this matrix is updated, then it is possible to monitor the effects of computer arithmetic and ensure that the matrices remain positive definite. (An alternative is to update H_k and monitor the conditions specified in the line search convergence theorem. If these conditions are violated, then the method can be "restarted" by setting $H_k = I$ and using the current value of x_k as a new "initial guess" of the solution. Both of these approaches can be used to guarantee global convergence, but both may also interfere with the asymptotic properties of the method and prevent superlinear convergence rates.)

An extensive discussion of quasi-Newton methods can be found in the paper of Dennis and Moré (1977). The convergence result that we mention is but one of many results that have been proved for quasi-Newton methods. For further information, see for example the paper by Byrd, Nocedal, and Yuan (1987).

Finite Differencing—Further discussion of finite differencing can be found in the books by Dennis and Schnabel (1983, reprinted 1996) and Gill, Murray, and Wright (1981). Special finite-differencing techniques have been derived to approximate sparse Hessian matrices; for details see the papers by Curtis, Powell, and Reid (1974) and Powell and Toint (1979).

Finite differencing using complex numbers was first described in the paper by Lyness and Moler (1967); see also the paper by Squire and Trapp (1998).

In our discussion we assumed that the rounding errors were proportional to machine precision ϵ_{mach}. More generally, the rounding errors will be proportional to the accuracy ϵ_f with which the function f can be computed, which may be larger than machine precision. In this case, the formulas for h will involve ϵ_f as well as ϵ_{mach}.

Automatic Differentiation—A thorough discussion of automatic differentiation can be found in the book by Griewank (2000). Early descriptions of automatic differentiation can be found in the papers by Wengert (1964) and Ostrovskii, Wolin, and Borisov (1971). A number of papers on the topic are contained in the book edited by Griewank and Corliss (1991). We have discussed using automatic differentiation just to compute the gradient. It can also be used effectively for computing the Hessian and for other derivative calculations, such as directional derivatives. (This latter technique can be useful within a truncated-Newton method; see Section 13.3.) Although the number of operations required to calculate the gradient is comparable to the number of operations required to calculate the function, computation of the Hessian is roughly n times as expensive for a function with n variables; this cost is inherent, and not just a deficiency of current techniques.

Derivative-Free Methods—One of the most widely used derivative-free methods is due to Nelder and Mead (1965). For this particular method the gradient need not go to zero in the limit, although a number of other methods have convergence properties similar to the one described in this section. Many types of derivative-free methods have been developed and studied, including methods suitable for parallel computing, and methods that can solve constrained problems. A survey of such methods can be found in the paper by Kolda, Lewis, and Torczon (2003).

Chapter 13

Low-Storage Methods for Unconstrained Problems

13.1   Introduction

The methods discussed in this chapter are suitable for solving large problems. They are suitable for two reasons. One is that their storage needs are low (just a few vectors are needed, unlike the matrix storage needed for the quasi-Newton methods of the last chapter). The other is that the work per iteration (the work required to compute a search direction) is also low, usually proportional to the number of variables n. For quasi-Newton methods the work per iteration is proportional to n^2. The search directions computed by these low-storage methods are generally of lower quality than those computed by Newton's method, in the sense that the rate of convergence is slower, and hence more iterations will often be required to find a solution. The hope is that the increase in the number of iterations will be more than compensated by the savings in the costs per iteration—savings that grow ever more dramatic as n increases.

Large problems can arise in several ways. In a model designed to optimize the operations of a corporation, adding variables may allow a more detailed examination of the corporate structure—for example, it might allow budget decisions to be examined at the level of individual departments rather than in terms of larger divisions within the company. In an engineering design problem, such as the design of a bridge or an airplane, the number of variables might be proportional to the number of components and so would be large if there were a large number of parts. In optimal control problems the solution is a function that might be approximated by determining its values at a finite number of points; if the solution varied rapidly, a large number of points would be needed to approximate the solution accurately.

The chapter begins with a discussion of the linear conjugate-gradient method, a method for solving systems of linear equations. This may seem an odd introduction to this topic, but two of the three groups of methods discussed here adapt the linear conjugate-gradient method to the solution of nonlinear optimization problems. Truncated-Newton methods use the linear conjugate-gradient method to find a search direction. Nonlinear conjugate-gradient methods generalize its formulas to create a new method. Limited-memory quasi-Newton methods are not as easy to categorize. They use the formulas of quasi-Newton methods but are often considered as extensions of nonlinear conjugate-gradient methods. The chapter concludes with a discussion of preconditioning, a powerful tool for accelerating the convergence of these algorithms.

In many cases, large problems are also sparse. By exploiting sparsity, it is possible to extend Newton's method to solve large problems. Although we do not discuss this topic in detail, various approaches to exploiting sparsity are described in the Notes for this chapter and for Appendix A.

Of these methods, nonlinear conjugate-gradient methods typically have the lowest *per iteration* computational costs, followed by limited-memory quasi-Newton methods, truncated-Newton methods, and sparse versions of Newton's method. Nonlinear conjugate-gradient methods usually require many iterations to find a solution and may have difficulty in finding a high-accuracy solution to an optimization problem. However, such methods can be successful in finding a low-accuracy solution efficiently. In contrast, Newton's method is likely to require far fewer iterations and is often able to compute a solution to high accuracy.

13.2 The Conjugate-Gradient Method for Solving Linear Equations

The most commonly used technique for solving systems of linear equations is Gaussian elimination. It is referred to as a "direct" method because it determines the solution in a fixed number of arithmetic operations that can be predicted in advance. "Iterative" methods, on the other hand, do not have fixed costs since the solution is obtained from a sequence of approximate solutions, and the algorithm is terminated when some measure of the error has been made adequately small.

Iterative methods are a valuable tool for solving large systems of linear equations. They have several potential advantages over direct methods in this case. First, since the coefficient matrix need not be factored, there is no fill-in and loss of sparsity. Second, storage requirements are often lower for iterative methods than for direct methods. In some cases, it may not be necessary to store the coefficient matrix at all. Third, if a good approximation to the coefficient matrix is available, and this approximation can be inverted at low cost, then an iterative method can take advantage of this information to obtain the solution more rapidly. This is not normally possible with a direct method.

A great many iterative methods have been invented, but we will only consider one of them: the conjugate-gradient method. (Many of the other iterative methods are applied primarily in the solution of differential equations.) The conjugate-gradient method is designed to solve

$$Ax = b$$

in the case where the matrix A is symmetric and positive definite. It can be considered as a technique for solving the equivalent problem

$$\text{minimize } f(x) = \tfrac{1}{2}x^T A x - b^T x.$$

To see that the problems are equivalent, set the gradient of f to zero:

$$\nabla f(x) = Ax - b = 0.$$

This finds the minimum since the Hessian matrix $\nabla^2 f(x) = A$ is positive definite, and the sufficient conditions for a minimum are satisfied.

The conjugate-gradient method gets its name from the fact that it generates a set of vectors $\{ p_i \}$ that are *conjugate* with respect to the coefficient matrix A; that is,

$$p_i^T A p_j = 0 \quad \text{if} \quad i \neq j.$$

In the special case where $A = I$, conjugate vectors are just orthogonal vectors. In general, any set of nonzero conjugate vectors will be linearly independent (see the Exercises).

To see the significance of the conjugacy property, assume that the vectors $\{ p_i \}$ are known. Consider a trial point y that is a linear combination of $m + 1$ such vectors

$$y = \sum_{i=0}^{m} \alpha_i p_i$$

and evaluate $f(y)$:

$$f(y) = f\left(\sum_{i=0}^{m} \alpha_i p_i\right) = \frac{1}{2}\left(\sum_{i=0}^{m} \alpha_i p_i\right)^T A \left(\sum_{j=0}^{m} \alpha_j p_j\right) - b^T \left(\sum_{i=0}^{m} \alpha_i p_i\right)$$

$$= \frac{1}{2}\sum_{i=0}^{m}\sum_{j=0}^{m} \alpha_i \alpha_j p_i^T A p_j - \sum_{i=0}^{m} \alpha_i b^T p_i$$

$$= \frac{1}{2}\sum_{i=0}^{m} \alpha_i^2 p_i^T A p_i - \sum_{i=0}^{m} \alpha_i b^T p_i \qquad \text{(from conjugacy)}$$

$$= \sum_{i=0}^{m} \left(\frac{1}{2}\alpha_i^2 p_i^T A p_i - \alpha_i b^T p_i\right).$$

It is then easy to minimize the function over all vectors y of this form:

$$\min_{y} f(y) = \min_{\{\alpha_i\}} f\left(\sum_{i=0}^{m} \alpha_i p_i\right)$$

$$= \min_{\{\alpha_i\}} \sum_{i=0}^{m} \left(\frac{1}{2}\alpha_i^2 p_i^T A p_i - \alpha_i b^T p_i\right)$$

$$= \sum_{i=0}^{m} \min_{\alpha_i} \left(\frac{1}{2}\alpha_i^2 p_i^T A p_i - \alpha_i b^T p_i\right).$$

The original problem has been reduced to the sum of one-dimensional minimization problems. Each of these one-dimensional problems can be solved by setting the derivative with respect to α_i equal to zero:

$$\alpha_i (p_i^T A p_i) - b^T p_i = 0 \qquad \text{or} \qquad \alpha_i = \frac{b^T p_i}{p_i^T A p_i}.$$

(The formula for α_i corresponds to an exact line search along the direction p_i for the function f; see the Exercises.) Thus, if we can represent the solution as a linear combination of conjugate vectors, the solution can be found easily.

The conjugate-gradient method iteratively determines a set of conjugate vectors $\{\, p_i \,\}$ and their coefficients $\{\, \alpha_i \,\}$. Here is one version of the method. The vector r_i is equal to the residual $b - Ax_i$, and the scalar β_i is used to determine the vector p_i.

ALGORITHM 13.1.
Conjugate-Gradient Method

1. Set $x_0 = 0$, $r_0 = b = b - Ax_0$, $p_{-1} = 0$, $\beta_0 = 0$, and specify the convergence tolerance ϵ.
2. For $i = 0, 1, \ldots$

      (i) If $\| r_i \| < \epsilon$, stop.
      (ii) If $i > 0$, set $\beta_i = r_i^T r_i / r_{i-1}^T r_{i-1}$.
      (iii) Set $p_i = r_i + \beta_i p_{i-1}$.
      (iv) Set $\alpha_i = r_i^T r_i / p_i^T A p_i$.
      (v) Set $x_{i+1} = x_i + \alpha_i p_i$.
      (vi) Set $r_{i+1} = r_i - \alpha_i A p_i$.

In this algorithm the initial guess is specified as $x_0 = 0$. It may be that a better initial guess is known. In this case the method can still be used, but with $r_0 = b - Ax_0$ instead of $r_0 = b$; no other changes are necessary (see the Exercises).

The formulas for α_i and r_i in the conjugate-gradient method are different from the formulas given earlier in this section, but the different formulas can be shown to be equivalent; see the Exercises. One reason for using different formulas is to reduce computational effort. For example, the formula for α_i reuses the value $r_i^T r_i$ from the computation of β_i. More importantly, the formula for r_i reuses $A p_i$ from the computation of α_i. The matrix A only appears as part of $A p_i$, and as long as this matrix-vector product can be computed the algorithm can be used, even if A is not available explicitly. The formation of $A p_i$ is discussed further below.

In this algorithm $x_{i+1} = x_i + \alpha_i p_i$, and hence

$$x_{i+1} = \alpha_i p_i + \alpha_{i-1} p_{i-1} + \cdots + \alpha_0 p_0$$

so the approximate solutions from the conjugate-gradient method have the same form as the point y used earlier. This shows that x_{i+1} is in the subspace spanned by $\{p_j\}_{j=0}^i$. In fact, x_{i+1} minimizes the quadratic function $f(x) = \frac{1}{2} x^T A x - b^T x$ over this subspace.

It will be shown later in this section that, if exact arithmetic is used, this algorithm produces a set of conjugate vectors $\{\, p_i \,\}$ and a set of orthogonal vectors $\{\, r_i \,\}$.

Although this algorithm has been written with subscripts on all the variables, only the current values of the variables need be saved. For example, the new value x_{i+1} can be overwritten on the old one x_i. Only the inner product $r_{i-1}^T r_{i-1}$ from the previous iteration need be saved.

Example 13.1 (The Conjugate-Gradient Method). We will apply the conjugate-gradient method to the problem $Ax = b$ with

$$A = \begin{pmatrix} 1 & 0 & 0 \\ 0 & 2 & 0 \\ 0 & 0 & 3 \end{pmatrix} \quad \text{and} \quad b = \begin{pmatrix} 1 \\ 1 \\ 1 \end{pmatrix}.$$

The method is initialized with

$$x_0 = \begin{pmatrix} 0 \\ 0 \\ 0 \end{pmatrix}, \quad r_0 = \begin{pmatrix} 1 \\ 1 \\ 1 \end{pmatrix}, \quad p_{-1} = \begin{pmatrix} 0 \\ 0 \\ 0 \end{pmatrix}, \quad \beta_0 = 0, \quad \text{and} \quad \epsilon = 10^{-12}.$$

At iteration 0, $\|r_0\| = 1.7321 > \epsilon$, so we do not stop. Then

$$p_0 = \begin{pmatrix} 1 \\ 1 \\ 1 \end{pmatrix}$$

and so $\alpha_0 = 0.5000$,

$$x_1 = \begin{pmatrix} 0.5000 \\ 0.5000 \\ 0.5000 \end{pmatrix}, \quad \text{and} \quad r_1 = \begin{pmatrix} 0.5000 \\ 0 \\ -0.5000 \end{pmatrix}.$$

At iteration 1, $\|r_1\| = 0.7071 > \epsilon$, so we continue and obtain $\beta_1 = 0.1667$ and

$$p_1 = \begin{pmatrix} 0.6667 \\ 0.1667 \\ -0.3333 \end{pmatrix}.$$

This gives $\alpha_1 = 0.6000$,

$$x_2 = \begin{pmatrix} 0.9000 \\ 0.6000 \\ 0.3000 \end{pmatrix}, \quad \text{and} \quad r_2 = \begin{pmatrix} 0.1000 \\ -0.2000 \\ 0.1000 \end{pmatrix}.$$

At iteration 2, $\|r_2\| = 0.2449 > \epsilon$, so we continue, getting $\beta_2 = 0.1200$ and

$$p_2 = \begin{pmatrix} 0.1800 \\ -0.1800 \\ 0.0600 \end{pmatrix}.$$

Then $\alpha_2 = 0.5556$,

$$x_3 = \begin{pmatrix} 1.0000 \\ 0.5000 \\ 0.3333 \end{pmatrix}, \quad \text{and} \quad r_3 = 10^{-16} \times \begin{pmatrix} 0.1388 \\ 0.2776 \\ 0.1388 \end{pmatrix}.$$

(We used 16-digit arithmetic for these calculations.) At iteration 3, $\|r_3\| = 3.4 \times 10^{-17} < \epsilon$, so we stop. ∎

The conjugate-gradient method does not require the matrix A explicitly. It requires only the computation of matrix-vector products of the form Ap for an arbitrary vector p. If the matrix A were stored explicitly, then of course this matrix-vector product could be computed using the traditional formulas. However, the traditional formulas are not often used. Large problems are often sparse, or have other special structure that makes it possible to compute Ap efficiently. In such cases software can be written to compute Ap that does not use any matrix storage.

Example 13.2 (Efficient Matrix-Vector Product). Consider the $n \times n$ matrix

$$A = \begin{pmatrix} 4 & 1 & & & & \\ 1 & 4 & 1 & & 0 & \\ & 1 & \ddots & \ddots & & \\ & & \ddots & 4 & 1 & \\ 0 & & & 1 & 4 & 1 \\ & & & & 1 & 4 \end{pmatrix} + ww^T,$$

where w is a vector. If all the entries of w are nonzero, then A is a dense matrix. Nevertheless the matrix-vector product $y = Ap$ can be computed efficiently via the following algorithm.

ALGORITHM 13.2.
Efficient Matrix-Vector Product

    1. For $i = 1, 2, \ldots, n$

        (i) $y_i = 4p_i$
        (ii) If $i > 1$, then $y_i \leftarrow y_i + p_{i-1}$
        (iii) If $i < n$, then $y_i \leftarrow y_i + p_{i+1}$

    2. $y \leftarrow y + (w^T p)w$.

The cost of this algorithm is about $6n$ arithmetic operations, far less than the $2n^2$ operations of the traditional matrix-vector product. Notice that the matrix A is not stored. ∎

With the possible exception of the matrix-vector product, all of the operations in the conjugate-gradient method are standard vector operations (inner product, multiplication of a vector by a scalar, and sums of vectors). High-performance computers often have special hardware that performs these operations rapidly; hence this algorithm is well suited to such computers if the matrix-vector product can be programmed efficiently.

The conjugacy and orthogonality properties of the conjugate-gradient method are proved in the theorem below. The proof given here uses only elementary arguments and is based on mathematical induction. Other derivations of the conjugate-gradient method are possible; some can be found in the paper by Hestenes and Stiefel (1952).

Theorem 13.3. *Assume that the vectors $\{\, p_i \,\}$ and $\{\, r_i \,\}$ are defined by the formulas for the conjugate-gradient method. Then*

$$r_i^T r_j = 0, \quad r_i^T p_j = 0, \quad \text{and} \quad p_i^T A p_j = 0$$

for $i > j$. That is, the residuals $\{\, r_i \,\}$ are orthogonal, the residuals are orthogonal to the directions $\{\, p_j,\ j < i \,\}$, and the directions $\{\, p_i \,\}$ are conjugate.

Proof. The proof is by mathematical induction, with the induction hypothesis consisting of all three of the equations in the statement of the theorem.

For $i = 0$ we need only show that $r_0^T p_{-1} = 0$ and $p_0^T A p_{-1} = 0$ since no other cases are possible. These results follow immediately because $p_{-1} = 0$.

Assume now that the three results are true for i; we wish to prove them for $i + 1$. This will be done in five steps. Each of these steps makes regular use of the formulas for the conjugate-gradient method.

1. $r_{i+1}^T r_j = 0$ for $j < i$:

$$\begin{aligned}
r_{i+1}^T r_j &= (r_i - \alpha_i A p_i)^T r_j = r_i^T r_j - \alpha_i r_j^T A p_i \\
&= 0 - \alpha_i (p_j - \beta_j p_{j-1})^T A p_i \\
&= -\alpha_i p_j^T A p_i + \alpha_i \beta_j p_{j-1}^T A p_i = 0.
\end{aligned}$$

Each of the individual terms in the second-to-last line vanishes because of the induction hypothesis.

2. $r_{i+1}^T r_i = 0$:

$$\begin{aligned}
r_{i+1}^T r_i &= r_i^T r_i - \alpha_i r_i^T A p_i = r_i^T r_i - \alpha_i (p_i - \beta_i p_{i-1})^T A p_i \\
&= r_i^T r_i - \alpha_i p_i^T A p_i + \alpha_i \beta_i p_{i-1}^T A p_i \\
&= r_i^T r_i - \frac{r_i^T r_i}{p_i^T A p_i} p_i^T A p_i + 0 = 0.
\end{aligned}$$

3. $r_{i+1}^T p_j = 0$ for $j < i + 1$:

$$\begin{aligned}
r_{i+1}^T p_j &= r_{i+1}^T r_j + \beta_j r_{i+1}^T p_{j-1} \\
&= 0 + \beta_j (r_i - \alpha_i A p_i)^T p_{j-1} \\
&= \beta_j (r_i^T p_{j-1} - \alpha_i p_i^T A p_{j-1}) = 0.
\end{aligned}$$

Notice that $r_{i+1}^T r_j = 0$ because of steps 1 and 2 in this proof.

4. $p_{i+1}^T A p_j = 0$ for $j < i$:

$$\begin{aligned}
p_{i+1}^T A p_j &= (r_{i+1} + \beta_{i+1} p_i)^T A p_j = r_{i+1}^T A p_j + \beta_{i+1} p_i^T A p_j \\
&= r_{i+1}^T [\alpha_j^{-1} (r_j - r_{j+1})] + 0 \\
&= \alpha_j^{-1} (r_{i+1}^T r_j - r_{i+1}^T r_{j+1}) = 0
\end{aligned}$$

because of steps 1 and 2.

5. $p_{i+1}^T A p_i = 0$:

$$
\begin{aligned}
p_{i+1}^T A p_i &= (r_{i+1} + \beta_{i+1} p_i)^T [\alpha_i^{-1}(r_i - r_{i+1})] \\
&= \alpha_i^{-1}(r_{i+1}^T r_i + \beta_{i+1} p_i^T r_i - r_{i+1}^T r_{i+1} - \beta_{i+1} p_i^T r_{i+1}) \\
&= \alpha_i^{-1}(0 + \beta_{i+1} p_i^T r_i - r_{i+1}^T r_{i+1} - 0) \\
&= \alpha_i^{-1}[\beta_{i+1}(r_i + \beta_i p_{i-1})^T r_i - r_{i+1}^T r_{i+1}] \\
&= \alpha_i^{-1}[\beta_{i+1} r_i^T r_i + \beta_{i+1} \beta_i p_{i-1}^T r_i - r_{i+1}^T r_{i+1}] \\
&= \alpha_i^{-1}[\beta_{i+1} r_i^T r_i + 0 - r_{i+1}^T r_{i+1}] = 0.
\end{aligned}
$$

The final line follows from the definition of $\beta_{i+1} = r_{i+1}^T r_{i+1} / r_i^T r_i$.

These five steps together show that the induction hypothesis is true for $i + 1$, and hence this completes the proof.   ☐

The approximate solution x_i minimizes the function $f(x) = \frac{1}{2} x^T A x - b^T x$ over all linear combinations of vectors of the form

$$
x_i = c_1 b + c_2 (Ab) + c_3 (A^2 b) + \cdots + c_i (A^{i-1} b)
$$

(see the Exercises). The set of vectors of this form is called a *Krylov subspace*. This result can be used to show that, in exact arithmetic, the conjugate-gradient method converges in a finite number of iterations equal to the number of distinct eigenvalues of the matrix A. The number of distinct eigenvalues is, at most, the number of variables.

Most iterative methods do not converge in a finite number of iterations. Instead they generate an infinite sequence of approximate solutions that approach the exact solution in the limit. For this reason it may seem odd to consider the conjugate-gradient method as an iterative method. When the number of variables is large, however, it may be too time consuming to use a number of iterations equal to the number of variables. Also, in rounded arithmetic the finite termination property may not be satisfied, and the number of iterations can be much larger than the number of variables. In a truncated-Newton method (see Section 13.3) the matrix-vector products Ap may only be approximated, and this can lead to a loss of conjugacy and orthogonality more pronounced than that caused by rounded arithmetic. For these reasons the conjugate-gradient method is usually treated as an iterative method like any other.

When considered as an iterative method, the linear conjugate-gradient method can be shown to converge linearly with

$$
\frac{\| x_{i+1} - x_* \|_A}{\| x_i - x_* \|_A} \le \frac{\sqrt{\text{cond}(A)} - 1}{\sqrt{\text{cond}(A)} + 1}.
$$

The error is measured in the norm defined by

$$
\| y \|_A^2 \equiv \tfrac{1}{2} y^T A y.
$$

If $\text{cond}(A) = 1$, then the rate constant above is zero and the conjugate-gradient method converges in one iteration. If $\text{cond}(A) = 100$, then the rate constant is about 0.82, and if $\text{cond}(A) = 1{,}000{,}000$, then the rate constant is about 0.998.

The conjugate-gradient method is often used with a "preconditioning" matrix, that is, a matrix $M \approx A$. The matrix M usually corresponds to a related linear system that is easier to solve. For example, the matrix M might be diagonal and represent a scaling of the variables. If such a matrix is available, then the conjugate-gradient method can be adjusted to take advantage of this additional information, and this can lead to much faster convergence of the method. This is discussed in Section 13.6.

Exercises

2.1. Apply the conjugate-gradient method to the problem $Ax = b$, where

$$
A = \begin{pmatrix} 2 & 1 & 0 \\ 1 & 2 & 1 \\ 0 & 1 & 2 \end{pmatrix} \quad \text{and} \quad b = \begin{pmatrix} 1 \\ 1 \\ 1 \end{pmatrix}.
$$

Verify that the vectors $\{ p_i \}$ are conjugate and the vectors $\{ r_i \}$ are orthogonal.

2.2. The conjugate-gradient method is applied to the minimization of a function of three variables. Initially $r_0 = (1, -1, 2)^T$. At iteration 1, the first two components of r_1 are 2 and 2, respectively. What is the direction of search prescribed by the conjugate-gradient method at the second iteration?

2.3. The conjugate-gradient method was applied to the minimization of a function. At some iteration the following data were given: $r_i = (5, 3, -1)^T$ and $p_i = (4, -2, 1)^T$. Why cannot these data be correct?

2.4. The choice of the starting direction $p_0 = r_0$ is important in the conjugate-gradient method. To see this consider the problem

$$
\begin{pmatrix} 2 & 0 \\ 0 & 1 \end{pmatrix} \begin{pmatrix} x_1 \\ x_2 \end{pmatrix} = \begin{pmatrix} 3 \\ 4 \end{pmatrix}.
$$

Suppose that $x_0 = (1, 1)^T$, and instead of $p_0 = r_0$ take $p_0 = (-1, 0)^T$. Perform two iterations of the conjugate-gradient method. List all properties of the "regular" method that are not satisfied in these two iterations.

2.5. Prove that the two formulas for α_i given in this section are equivalent, i.e.,

$$
\alpha_i = \frac{b^T p_i}{p_i^T A p_i} = \frac{r_i^T r_i}{p_i^T A p_i}.
$$

2.6. Show that the conjugate-gradient method can be adjusted to take advantage of some initial guess $x_0 \neq 0$ by taking $r_0 = b - Ax_0$. *Hint*: Show that this is equivalent to applying the method to a related linear system with initial guess equal to zero.

2.7. Prove by induction that the vector r_i in the conjugate-gradient method is equal to the residual $b - Ax_i$.

2.8. Prove that the conjugate-gradient method converges in one iteration if $A = \kappa I$, where κ is some positive number.

2.9. Prove that if the nonzero vectors $\{\,v_i\,\}$ are mutually conjugate with respect to a positive-definite matrix A, then they are also linearly independent.

2.10. In the conjugate-gradient method prove that p_{i-1} can be written as a linear combination of the set of vectors $\{\,b,\ Ab,\ A^2b,\ \dots,\ A^{i-1}b\,\}$. Also prove that x_i minimizes the quadratic function $f(x) = \frac{1}{2}x^TAx - b^Tx$ over all linear combinations of vectors from this set.

2.11. Use the result of the previous problem to prove that, if exact arithmetic is used, the conjugate-gradient method converges in a number of iterations equal to the number of distinct eigenvalues of the matrix A.

2.12. The goal of this problem is to prove that, when the conjugate-gradient method is used to solve $Ax = b$, the norms of the errors decrease monotonically.

  (i) Prove that the conjugate directions satisfy $p_i^Tp_j \geq 0$ for all i and j.

  (ii) Prove that the estimates of the solution satisfy $\|x_i\| \geq \|x_{i-1}\|$ for all i. (Here the 2-norm is used.)

  (iii) Let x_* solve $Ax = b$. Prove that $\|x_i - x_*\| \leq \|x_{i-1} - x_*\|$ for all i.

2.13. Write a computer program that uses the conjugate-gradient method to minimize an n-dimensional quadratic function

$$\operatorname*{minimize}_{x}\ f(x) = \tfrac{1}{2}x^TAx - b^Tx,$$

where A is a positive-definite matrix. You may assume that the starting point x_0 is the zero vector. Test your algorithm on the quadratic function with data

$$A = \begin{pmatrix} 4 & 0 & 0 & 1 & 0 & 0 \\ 0 & 4 & 0 & 0 & 1 & 0 \\ 0 & 0 & 5 & 0 & 0 & 1 \\ 1 & 0 & 0 & 5 & 0 & 0 \\ 0 & 1 & 0 & 0 & 6 & 0 \\ 0 & 0 & 1 & 0 & 0 & 6 \end{pmatrix} \quad \text{and} \quad b = \begin{pmatrix} 4 \\ -8 \\ 16 \\ 1 \\ -2 \\ 9 \end{pmatrix}.$$

Your program should be written so that it only uses matrix-vector products of the form Ap, instead of explicitly using the matrix A. This means that no matrices are used in the course of the algorithm.

13.3 Truncated-Newton Methods

Truncated-Newton methods are used to solve the problem

$$\operatorname{minimize}\ f(x).$$

They are a compromise on Newton's method, and they compute a search direction by finding an approximate solution to the Newton equations

$$\nabla^2 f(x_k)p \approx -\nabla f(x_k)$$

using some iterative method, usually the conjugate-gradient method. Notice that the Newton equations are a linear system of the form $Ax = b$ with $A = \nabla^2 f(x_k)$ and $b = -\nabla f(x_k)$. The iterative method is stopped ("truncated") before the exact solution to the Newton equations has been found, giving the method its name.

The overall optimization method now consists of nested iterations. There is an outer iteration that corresponds to our general optimization method (see Section 2.4). At each outer iteration we compute a search direction and perform a line search. The computation of the search direction uses an inner iteration corresponding to the iterative method used to solve the Newton equations. (The line search also represents an inner iteration of the optimization method.)

This appears to be a minor adjustment to Newton's method, but it is enough to allow Newton's method to be extended to solve large problems. The line search only requires the storage of a few vectors: the current point x_k, the gradient at x_k, the search direction p, plus perhaps trial values of $x_k + \alpha p$ together with their corresponding gradients. The conjugate-gradient method to compute the search direction also only requires the storage of a few vectors. The only possible obstacle to using this approach might be the computation of matrix-vector products of the form $Av = \nabla^2 f(x_k)v$ for arbitrary vectors v. Of course, if explicit second-derivative information were available and easy to compute, then these matrix-vector products would be computed in the traditional way.

It is possible to approximate the matrix-vector products using values of the gradient in such a way that the Hessian matrix need not be computed or stored. By rearranging the Taylor series

$$\nabla f(x_k + hv) = \nabla f(x_k) + h\nabla^2 f(x_k)v + O(h^2),$$

we obtain that

$$\nabla^2 f(x_k)v = \lim_{h \to 0} \frac{\nabla f(x_k + hv) - \nabla f(x_k)}{h}.$$

Hence the matrix-vector product $\nabla^2 f(x_k)v$ can be approximated using

$$\nabla^2 f(x_k)v \approx \frac{\nabla f(x_k + hv) - \nabla f(x_k)}{h}$$

for some small value of h. (The selection of h is discussed in Section 12.4.1, although in this setting there is an additional complication since the choice of h depends on v as well as x and f.) Since $\nabla f(x_k)$ is the right-hand side of the Newton equations and so is already available, this shows that a matrix-vector product can be approximated using one gradient evaluation.

It would also be possible to compute the matrix-vector products using automatic differentiation (see Section 12.4.2). With this approach, two additional gradient evaluations are required for each matrix-vector product, instead of just one for finite differencing. Automatic differentiation would typically produce a more accurate result, however.

With either of these techniques for computing the matrix-vector products, the implementation of a truncated-Newton method requires only that the function $f(x)$ and the gradient $\nabla f(x)$ be calculated. Vector storage is needed for the conjugate-gradient method and for the line search (these can be the same vectors), but no matrix storage. Many other optimization methods designed for solving large problems have the same requirements for function information and storage, although the exact amount of vector storage may vary slightly from method to method.

If the costs of the matrix-vector product are ignored, then each inner iteration of such a truncated-Newton method requires $O(n)$ arithmetic operations, where n is the number of variables. The cost of each outer iteration, however, is variable since it depends on the number of inner iterations performed. If more inner iterations are performed, then we would expect that a "better" search direction would be obtained, in the sense that we would expect the search direction to be a better approximation to the Newton direction. Hence, there is a trade-off between the arithmetic costs of each outer iteration and the quality of each search direction.

This trade-off can be made explicit, since there is a relationship between the accuracy with which the Newton equations are solved and the convergence rate of the truncated-Newton method. Let us assume that the iterative method is stopped when the norm of the residual satisfies

$$\left\| \nabla^2 f(x_k)p + \nabla f(x_k) \right\| \le \phi_k \left\| \nabla f(x_k) \right\|$$

for some tolerance $\phi_k \ge 0$. In this test, the index k refers to the outer iteration and corresponds to the current approximation x_k to the solution x_*. This condition involves the term

$$\nabla^2 f(x_k)p + \nabla f(x_k),$$

which is the residual for the Newton equations, and which measures how close p is to the Newton direction. This residual is compared to $\|\nabla f(x_k)\|$, the norm of the right-hand side of the Newton equations. This test measures the *relative* size of the residual. If the objective function f is multiplied by a nonzero constant (if the units used to measure the objective function are changed), this test is unaffected.

The rate of convergence of the truncated-Newton method can be directly related to the choice of the tolerances $\{\phi_k\}$. If

$$\phi_k = c < 1,$$

then the truncated-Newton method will converge linearly with rate constant c. If

$$\phi_k \to 0,$$

then the truncated-Newton method will converge superlinearly. (This is a consequence of Theorem 11.3; see the Exercises.) If

$$\phi_k \le \|\nabla f(x_k)\|,$$

then the truncated-Newton method will converge quadratically. If $\phi_k = 0$, then Newton's method is obtained.

The more rapid rates of convergence carry a price. The smaller ϕ_k is, the more inner iterations will be required to compute a search direction. On the other hand, the higher the rate of convergence, the fewer outer iterations will normally be required to solve the optimization problem. Many implementations of truncated-Newton methods choose $\{\phi_k\}$ to obtain linear or superlinear rates of convergence. One such implementation is illustrated in the example below.

Example 13.4 (Truncated-Newton Method). We will demonstrate the usage of a simple truncated-Newton algorithm. The overall algorithm is of the form described in Theorem 11.7, that is, it will use a backtracking line search. (We will use the parameter value $\mu = 0.1$ in the sufficient decrease condition.) The conjugate-gradient method from Section 13.2 will be used to compute the search direction; the search direction p will be accepted when

$$\frac{\|\nabla^2 f(x_k)p + \nabla f(x_k)\|}{\|\nabla f(x_k)\|} \le 0.001.$$

In this example, the matrix-vector products will be computed in the traditional way, using the Hessian matrix $\nabla^2 f(x_k)$.

We apply the truncated-Newton method to the n-dimensional problem

$$\text{minimize } f(x) = \tfrac{1}{10}(x - e)^T D(x - e) + (x^T x - \tfrac{1}{4})^2,$$

where

$$e = (1 \quad \cdots \quad 1)^T$$

and D is a diagonal matrix with diagonal entries $1, 2, \ldots, n$. The gradient and Hessian are given by the formulas

$$\nabla f(x) = \tfrac{1}{5}D(x - e) + 4(x^T x - \tfrac{1}{4})x$$
$$\nabla^2 f(x) = \tfrac{1}{5}D + 4(x^T x - \tfrac{1}{4})I + 8xx^T.$$

The matrix-vector product in the conjugate-gradient method can be computed using the formula

$$\nabla^2 f(x)v = \tfrac{1}{5}Dv + 4(x^T x - \tfrac{1}{4})v + 8(x^T v)x;$$

this requires only $O(n)$ arithmetic operations instead of the $O(n^2)$ operations required for a traditional matrix-vector product.

We give detailed results for the first iteration for the case $n = 4$ and summaries of the complete results for $n = 4$ and $n = 100$. The initial guess is

$$x_0 = (1 \quad -1 \quad 1 \quad -1)^T,$$

and at this point

$$f(x_0) = 16.462$$
$$\nabla f(x_0) = (15.0 \quad -15.8 \quad 15.0 \quad -16.6)^T$$
$$\nabla^2 f(x_0) = \begin{pmatrix} 23.2 & -8 & 8 & -8 \\ -8 & 23.4 & -8 & 8 \\ 8 & -8 & 23.6 & -8 \\ -8 & 8 & -8 & 23.8 \end{pmatrix}.$$

Table 13.1. *Truncated-Newton method ($n = 4$ and $n = 100$).*

k	ls	cg	$\|\nabla f(x_k)\|$	k	ls	cg	$\|\nabla f(x_k)\|$
0	1	0	2×10^1	0	1	0	4×10^2
1	2	2	5×10^0	1	2	2	1×10^2
2	3	5	2×10^0	2	3	5	5×10^1
3	4	8	7×10^{-1}	3	4	8	2×10^1
4	5	11	5×10^{-1}	4	5	12	7×10^0
5	6	15	1×10^{-1}	5	6	16	2×10^0
6	7	19	1×10^{-2}	6	7	20	1×10^{-1}
7	8	23	2×10^{-4}	7	8	24	4×10^{-4}
8	9	27	5×10^{-8}	8	9	28	8×10^{-7}
				9	10	32	1×10^{-10}

For the conjugate-gradient iteration, the norms of the scaled residuals at the first three iterations are

$$1, \quad 2.6 \times 10^{-2}, \quad 2.8 \times 10^{-4}.$$

The final value is smaller than the tolerance, and the corresponding approximate solution to the Newton equations is accepted as the search direction:

$$p = (-0.29602 \quad 0.34340 \quad -0.28794 \quad 0.38588)^T.$$

In the line search,

$$\alpha = 1$$
$$f(x_0 + \alpha p) = 4.061$$
$$f(x_0) + \mu \alpha p^T \nabla f(x_0) = 14.403$$

so the step length $\alpha = 1$ is accepted. The new estimate of the solution is

$$x_1 = (0.70398 \quad -0.65661 \quad 0.71206 \quad -0.61412)^T.$$

This completes the first iteration of the truncated-Newton method.

The complete results for $n = 4$ and $n = 100$ are given in Table 13.1. The outer iteration number is indicated by k. The costs of the method are given by "ls" (the number of gradient evaluations used in the line search) and "cg" (the number of gradient evaluations used inside the conjugate-gradient method). (The numbers in the table record the cumulative costs. The norm used is the infinity norm.) Note that, even when $n = 100$, only 2–4 conjugate-gradient iterations are used to compute the search directions for this example. ∎

Truncated-Newton methods are a flexible class of methods. They can be implemented in a general way to solve a wide class of optimization problems. They can also be adapted to specific settings. For example, a special-purpose matrix-vector product could be used, or a carefully chosen preconditioning strategy (see Section 13.6). The tolerances $\{\phi_k\}$ can be adjusted. A special iterative method could be used so that the method would be suitable for a parallel computer. There are many opportunities for a knowledgeable user to enhance the method. Further information can be found in the Notes section.

Exercises

3.1. Verify the results obtained in Example 13.4 for $n = 4$.

3.2. Apply the truncated-Newton method described in Example 13.4 to the problem

$$\text{minimize } f(x) = \sum_{i=1}^{n-1} (x_i - 2x_{i+1}^2)^2.$$

Use the initial guess $x_0 = (1, \ldots, 1)^T$. Solve this problem for $n = 4$ and $n = 10$.

3.3. How accurate is the approximation

$$\nabla^2 f(x_k)v \approx \frac{\nabla f(x_k + hv) - \nabla f(x_k)}{h}$$

for "small" values of h?

3.4. Consider a truncated-Newton method where the inner conjugate-gradient iteration is terminated after one iteration. Prove that the resulting search direction is a multiple of the steepest-descent direction.

3.5. Suppose that the truncated-Newton method from the previous problem is applied to a quadratic function. Prove that using a step length of $\alpha = 1$ corresponds to an exact line search.

3.6. Suppose that this same truncated-Newton method is applied to a nonlinear function f that has four continuous derivatives. Define $F(\alpha) = f(x_k) + \alpha p$. Examine the Taylor series of $F'(\alpha)$ for $\alpha = 1$ for the search direction produced by the truncated-Newton method. How many terms vanish?

3.7. Suppose that the truncated-Newton method of Example 13.4 is applied to a strictly convex nonlinear function f that has two continuous derivatives. Prove that the search direction is always a descent direction, regardless of how many inner iterations of the conjugate-gradient method are performed.

3.8. Suppose that the truncated-Newton method of Example 13.4 is applied to a general nonlinear function f, and that finite differencing is used to approximate the required matrix-vector products. What are the vector storage requirements for the method? You can ignore any storage that might be required to evaluate the function f and its gradient.

3.9. Suppose that the inner iteration of a truncated-Newton method uses a set of parameters $\{\phi_k\}$ satisfying $\phi_k \to 0$. Assume that $x_k \to x_*$ and that $\nabla^2 f(x_*)$ is positive definite. Use Theorem 11.3 to prove that the truncated-Newton method converges at a superlinear rate.

3.10. Write a program that minimizes an unconstrained function using a truncated-Newton method. Take the basic details of the method and the test functions from Exercise 5.18 in Chapter 11. Use the following approach.

(i) Write a program to solve a linear system of equations $Ay = b$ using the conjugate-gradient method. Assume that A is positive definite. See Exercise 2.13 of this chapter. You may find it helpful to denote the search direction in the conjugate-gradient method as v, rather than p, to avoid confusion between the search direction in the inner iteration and the outer iteration.

(ii) Modify your program from (i) so that instead of explicitly using the matrix A, it uses only matrix-vector products of the form Av. This means that no matrices are used in the course of the algorithm.

(iii) Write a program for minimizing a convex nonlinear function via a truncated-Newton method. Terminate the inner iteration when the norm of the scaled residual is less than 0.1. Do not form the Hessian. The only quantities required are Hessian-vector products of the form $\nabla^2 f(x)v$, and these may be approximated by $(\nabla f(x + hv) - \nabla f(x))/h$. Use $h = \sqrt{\epsilon_{mach}}$, where ϵ_{mach} is machine epsilon. Run your program on test functions 1, 2, 4, and 5.

(iv) Modify your code to handle nonconvex problem as follows: Exit the inner iteration if $v^T \nabla^2 f(x)v \le \theta$. If this occurs in the first iteration of the inner iteration, set $p = -\nabla f(x)$. Otherwise, set p to be the solution estimate presently available from the inner iteration. Use $\theta = 10^{-6}$. Run your program on test function 3.

13.4   Nonlinear Conjugate-Gradient Methods

Nonlinear conjugate-gradient methods adapt the formulas of the linear conjugate-gradient method so that nonlinear problems can be solved. The linear conjugate-gradient method solves the problem

$$\text{minimize } f(x) = \tfrac{1}{2}x^T A x - b^T x.$$

At each iteration it computes a new conjugate direction p_i using the residual $r_i = b - Ax_i$ and the old direction p_{i-1}. Then it computes a step length α that minimizes $f(x_i + \alpha p_i)$ as a function of α. The residual can also be written as

$$r_i = b - Ax_i = -\nabla f(x_i).$$

If we do this and replace the computation of α by a line search, then the algorithm has the following form.

ALGORITHM 13.3.
Nonlinear Conjugate-Gradient Method

1. Set $p_{-1} = 0$, $\beta_0 = 0$, and set the convergence tolerance ϵ.
2. For $i = 0, 1, \ldots$

(i) If $\|\nabla f(x_i)\| < \epsilon$, stop.

(ii) If $i > 0$, set

$$\beta_i = \nabla f(x_i)^T \nabla f(x_i) / \nabla f(x_{i-1})^T \nabla f(x_{i-1}).$$

(iii) Set $p_i = -\nabla f(x_i) + \beta_i p_{i-1}$.

(iv) Use a line search to determine $x_{i+1} = x_i + \alpha_i p_i$.

This algorithm can be applied to general unconstrained problems:

$$\text{minimize } f(x).$$

It requires the computation of the gradient $\nabla f(x)$, but no second-derivative calculations. Its storage requirements are low—just the three vectors x_i, p_i, and $\nabla f(x_i)$, plus whatever temporary storage is required by the line search. At each iteration the algorithm requires only a small number of operations on vectors, plus the computation of f and ∇f for various values of x. Hence it is suitable for large problems.

Example 13.5 (Nonlinear Conjugate-Gradient Method). We apply the above nonlinear conjugate-gradient method to the minimization problem from Example 13.4. As in Example 13.4, we use a backtracking line search with parameter $\mu = 0.1$. We give detailed results for the first two iterations, as well as a summary of the complete results for $n = 4$. The results for $n = 100$ are similar.

The initial point is

$$x_0 = (1 \quad -1 \quad 1 \quad -1)^T.$$

At this point,

$$f(x_0) = 16.462$$
$$\nabla f(x_0) = (15.0 \quad -15.8 \quad 15.0 \quad -16.6)^T.$$

At the first iteration the steepest-descent direction is used:

$$p = -\nabla f(x_0) = (-15.0 \quad 15.8 \quad -15.0 \quad 16.6)^T.$$

In the line search, the first few trial values of α are rejected:

$$\alpha = 1$$
$$f(x_0 + \alpha p) = 7.3 \times 10^5$$
$$f(x_0) + \mu \alpha p^T \nabla f(x_0) = -81.058$$
$$\vdots$$
$$\alpha = 0.0625$$
$$f(x_0 + \alpha p) = 0.0985$$
$$f(x_0) + \mu \alpha p^T \nabla f(x_0) = 10.367$$

but the step length $\alpha = 0.0625$ is accepted.

The new point is

$$x_1 = (0.0625 \quad -0.0125 \quad 0.0625 \quad 0.0375)^T$$

with

$$f(x_1) = 0.98506$$
$$\nabla f(x_1) = (\,-0.24766 \quad -0.39297 \quad -0.62266 \quad -0.80609\,)^T.$$

At this iteration

$$\beta_1 = 1.2851 \times 10^{-3}$$

and the search direction is

$$p = (\,0.22838 \quad 0.41327 \quad 0.60338 \quad 0.82743\,)^T.$$

In the line search

$$\alpha = 1$$
$$f(x_1 + \alpha p) = 1.5712$$
$$f(x_1) + \mu \alpha p^T \nabla f(x_1) = 0.85889$$

$$\alpha = 0.5$$
$$f(x_1 + \alpha p) = 0.46348$$
$$f(x_1) + \mu \alpha p^T \nabla f(x_1) = 0.92197$$

and the step length $\alpha = 0.5$ is accepted. The new point is

$$x_2 = (\,0.17669 \quad 0.19414 \quad 0.36419 \quad 0.45121\,)^T.$$

This concludes the second iteration.

The complete results for $n = 4$ are given in Table 13.2. The iteration number is indicated by k. The costs of the method are given by "ls" (the number of gradient evaluations

Table 13.2. *Nonlinear conjugate-gradient method ($n = 4$).*

k	ls	$\|\nabla f(x_k)\|$	k	ls	$\|\nabla f(x_k)\|$
0	1	2×10^1	13	33	6×10^{-5}
1	6	8×10^{-1}	14	35	5×10^{-5}
2	8	2×10^{-1}	15	37	2×10^{-5}
3	11	2×10^{-1}	16	38	2×10^{-5}
4	13	1×10^{-1}	17	41	8×10^{-6}
5	16	6×10^{-2}	18	43	7×10^{-6}
6	18	3×10^{-2}	19	46	2×10^{-6}
7	20	1×10^{-2}	20	48	9×10^{-7}
8	23	1×10^{-3}	21	50	5×10^{-7}
9	24	1×10^{-3}	22	52	1×10^{-7}
10	27	7×10^{-4}	23	54	1×10^{-7}
11	29	5×10^{-4}	24	57	7×10^{-8}
12	32	2×10^{-4}			

used in the line search). The numbers in Table 13.2 record the cumulative costs. The norm used is the infinity norm.

Notice that the nonlinear conjugate-gradient method required more iterations to solve this problem than the truncated-Newton method, but that the iterations for the truncated-Newton method required more gradient evaluations. Overall, however, the truncated-Newton method solved the problem more efficiently. This performance is typical for these methods. ∎

There are several versions of the nonlinear conjugate-gradient method that differ in the formula for β_i. When applied to a quadratic function these versions are equivalent, but they behave differently on general nonlinear functions. The three best known formulas for β_i are

$$\beta_i^{(1)} = \nabla f(x_i)^T \nabla f(x_i) / \nabla f(x_{i-1})^T \nabla f(x_{i-1})$$
$$\beta_i^{(2)} = y_{i-1}^T \nabla f(x_i) / \nabla f(x_{i-1})^T \nabla f(x_{i-1})$$
$$\beta_i^{(3)} = y_{i-1}^T \nabla f(x_i) / y_{i-1}^T p_{i-1},$$

where $y_{i-1} = \nabla f(x_i) - \nabla f(x_{i-1})$. The first formula for β_i is the one used in the algorithm above and is called the "Fletcher–Reeves" formula because of the paper by Fletcher and Reeves (1964) that discussed the method. If this formula is used, then the method is guaranteed to converge under appropriate assumptions.

The second formula is known as the "Polak–Ribiére" formula, and the third as the "Hestenes–Stiefel" formula. Computational experiments have suggested that they generally perform better than the Fletcher–Reeves formula, even though examples have been constructed which cause them to display especially poor performance, bordering on failure.

All of these methods have low storage requirements: 3–5 vectors of length n. The line search in a nonlinear conjugate-gradient method plays a more critical role than in, say, Newton's method. The tolerances in the line search may be more stringent or, depending on the formula used to compute β_i, an additional condition must be imposed on the step length α_i to ensure that the search direction at the next iteration is a descent direction. (See the Exercises.)

At one time nonlinear conjugate-gradient methods were the only effective technique for solving large unconstrained optimization problems. There are still circumstances where they are recommended, for example, if storage is severely limited. They are usually less effective than truncated-Newton methods or limited-memory quasi-Newton methods (see Section 13.5), methods whose storage requirements are only slightly greater but which tend to be more reliable and more efficient. The relative merits of these methods are discussed in the paper by Nash and Nocedal (1991).

Exercises

4.1. Verify the results obtained in Example 13.5 for $n = 4$.

4.2. Apply the nonlinear conjugate-gradient method described in Example 13.5 to

$$\text{minimize } f(x) = \sum_{i=1}^{n-1} (x_i - 2x_{i+1}^2)^2.$$

Use the initial guess $x_0 = (1, \ldots, 1)^T$. Solve this problem for $n = 4$ and $n = 10$.

4.3. Prove that the three formulas for computing β_i are all equivalent when f is a quadratic function and the linear conjugate-gradient method is used.

4.4. Examine the three formulas for β_i and determine under what conditions it is possible to guarantee that p_i is a descent direction.

4.5. Suppose that the nonlinear conjugate-gradient method of Example 13.5 is applied to a general nonlinear function f. What are the vector storage requirements for the method? You can ignore any storage that might be required to evaluate the function f and its gradient. Do the vector storage requirements change if one of the other formulas for β_i is used?

4.6. Consider a nonlinear conjugate-gradient method that uses the formula $\beta_i^{(1)}$. Under what circumstances, if any, will it be possible to guarantee that the conditions of the line search convergence theorem, Theorem 11.7, are satisfied?

13.5   Limited-Memory Quasi-Newton Methods

Limited-memory quasi-Newton methods compromise on quasi-Newton methods in an attempt to achieve some semblance of their performance, but with much lower storage requirements and much lower arithmetic costs per iteration. Limited-memory quasi-Newton methods use the quasi-Newton update formulas to describe an approximation to the inverse of the Hessian matrix, but do not form explicitly the matrix corresponding to this approximation. Instead the formulas are used directly, and only the information required to evaluate the formulas is stored. Because no matrices are stored, the techniques can be applied to problems with large numbers of variables. We will derive these methods from the point of view of quasi-Newton methods. However, in their simplest versions they are similar to nonlinear conjugate-gradient methods, and they share many practical properties with these methods.

We begin with a brief review of quasi-Newton methods (see also Section 12.3). Let $\{x_k\}$ be the sequence of approximate solutions to the optimization problem. Then we define

$$s_k \equiv x_{k+1} - x_k \quad \text{and} \quad y_k \equiv \nabla f(x_{k+1}) - \nabla f(x_k).$$

If the BFGS quasi-Newton method is used, the Hessian approximations $B_k \approx \nabla^2 f(x_k)$ are defined by

$$B_{k+1} = B_k - \frac{(B_k s_k)(B_k s_k)^T}{s_k^T B_k s_k} + \frac{y_k y_k^T}{y_k^T s_k}$$

with (say) $B_0 = I$. At each iteration the search direction for the optimization method is obtained by solving

$$B_k p = -\nabla f(x_k)$$

and then a line search can be used to determine x_{k+1}.

Every update formula for B_k has a companion formula for updating $H_k \equiv B_k^{-1}$. It can be derived using the Sherman–Morrison formula (see Appendix A.9). For the BFGS formula,

$$H_{k+1} = H_k - \frac{s_k(H_k y_k)^T + (H_k y_k)s_k^T}{y_k^T s_k}$$

$$+ \frac{y_k^T s_k + y_k^T H_k y_k}{(y_k^T s_k)^2}(s_k s_k^T)$$

$$= \left[I - \frac{s_k y_k^T}{y_k^T s_k}\right] H_k \left[I - \frac{y_k s_k^T}{y_k^T s_k}\right] + \frac{s_k s_k^T}{y_k^T s_k}.$$

If the inverse matrices are updated, then the search direction is

$$p_k = -H_k \nabla f(x_k).$$

Limited-memory quasi-Newton methods are based on these inverse formulas.

Assume that y_k, s_k, and H_k are known. Then $p_{k+1} = -H_{k+1}\nabla f(x_{k+1})$ can be computed using the update formula for H_{k+1}:

$$p_{k+1} = -H_{k+1}\nabla f(x_{k+1})$$

$$= -\left[I - \frac{s_k y_k^T}{y_k^T s_k}\right] H_k \left[I - \frac{y_k s_k^T}{y_k^T s_k}\right] \nabla f(x_{k+1}) - \frac{s_k s_k^T}{y_k^T s_k}\nabla f(x_{k+1}).$$

This final formula requires no matrix storage (other than the storage used to represent H_k) and uses only vector operations. It is one version of a limited-memory quasi-Newton method.

More elaborate formulas can be developed by using additional pairs

$$(y_{k-1}, s_{k-1}), \ (y_{k-2}, s_{k-2}), \ (y_{k-3}, s_{k-3}), \ \ldots.$$

Then instead of using a specified H_k, H_k would be defined using the update formula in terms of y_{k-1}, s_{k-1}, and H_{k-1}. Then H_{k-1} could be defined in terms of s_{k-2}, y_{k-2}, and H_{k-2}. Each succeeding term requires additional vector storage and additional arithmetic to perform the updates. Practical experience has shown that 3–5 updates are appropriate for many problems. If r updates are used to define H_{k+1}, then H_{k+1-r} must be initialized in some manner. A simple approach is to choose $H_{k+1-r} = I$; more sophisticated options are described in the papers cited in the Notes. There is a recursive algorithm that can be used to efficiently compute the search direction using r updates (see the Exercises).

Example 13.6 (Limited-Memory Quasi-Newton Method). We apply a limited-memory quasi-Newton method to the minimization problem from Example 13.4. We use a single BFGS update initialized with $H_k = I$ to compute the search direction. As in Example 13.4, we use a backtracking line search with parameter $\mu = 0.1$. We give detailed results for the first two iterations, as well as a summary of the complete results, for $n = 4$. The results for $n = 100$ are similar.

The first iteration of the limited-memory quasi-Newton method is the same as the first iteration of the nonlinear conjugate-gradient iteration (at the first iteration of both methods, the steepest-descent direction is used):

$$x_0 = (1 \quad -1 \quad 1 \quad -1)^T$$
$$f(x_0) = 16.462$$
$$\nabla f(x_0) = (15.0 \quad -15.8 \quad 15.0 \quad -16.6)^T$$
$$p_0 = (-15.0 \quad 15.8 \quad -15.0 \quad 16.6)^T$$
$$\alpha = 0.0625$$
$$x_1 = (0.0625 \quad -0.0125 \quad 0.0625 \quad 0.0375)^T$$
$$f(x_1) = 0.98506$$
$$\nabla f(x_1) = (-0.24766 \quad -0.39297 \quad -0.62266 \quad -0.80609)^T.$$

At the second iteration

$$s_0 = (-0.93750 \quad 0.98750 \quad -0.93750 \quad 1.0375)^T$$
$$y_0 = (-15.248 \quad 15.407 \quad -15.623 \quad 15.794)^T.$$

For computational purposes, we expand the update formula for p_1 in terms of vector operations. It is convenient to use the following intermediate quantities:

$$t_0 = y_0^T s_0 = 60.542$$
$$t_1 = s_0^T \nabla f(x_1) = -0.40846$$
$$t_2 = t_1/t_0 = -0.0067468.$$

Then we form the vector u:

$$u = \nabla f(x_1) - t_2 y_0 = (-0.35053 \quad -0.28902 \quad -0.72806 \quad -0.69954)^T.$$

In the general case we would then compute $u \leftarrow H_0 u$. Here we are using $H_0 = I$ so no computation is necessary. We can then compute the search direction p_1 via

$$t_3 = y_0^T u = 1.2176$$
$$t_4 = t_3/t_0 = 0.20112$$

and finally

$$p_1 = -u + (t_4 - t_2)s_0$$
$$= (0.32535 \quad 0.31554 \quad 0.70288 \quad 0.72740)^T.$$

In the line search

$$\alpha = 1$$
$$f(x_1 + \alpha p_1) = 1.5261$$
$$f(x_1) + \mu \alpha p_1^T \nabla f(x_1) = 0.86220$$

$$\alpha = 0.5$$
$$f(x_1 + \alpha p_1) = 0.47636$$
$$f(x_1) + \mu \alpha p_1^T \nabla f(x_1) = 0.92363$$

Table 13.3. *Limited-memory quasi-Newton method ($n = 4$).*

k	ls	$\|\nabla f(x_k)\|$	k	ls	$\|\nabla f(x_k)\|$
0	1	2×10^{1}	13	30	3×10^{-5}
1	6	8×10^{-1}	14	32	9×10^{-6}
2	8	3×10^{-1}	15	34	8×10^{-6}
3	10	2×10^{-1}	16	36	4×10^{-6}
4	11	9×10^{-2}	17	37	3×10^{-6}
5	13	9×10^{-2}	18	39	3×10^{-6}
6	16	2×10^{-2}	19	41	7×10^{-7}
7	18	6×10^{-3}	20	43	6×10^{-7}
8	21	2×10^{-3}	21	45	4×10^{-7}
9	22	4×10^{-4}	22	46	3×10^{-7}
10	24	3×10^{-4}	23	48	3×10^{-7}
11	27	6×10^{-5}	24	50	1×10^{-7}
12	28	3×10^{-5}	25	52	8×10^{-8}

so the step length $\alpha = 0.5$ is accepted. The new point is

$$x_2 = (\,0.22517 \quad 0.14527 \quad 0.41394 \quad 0.40120\,)^T.$$

This concludes the second iteration.

The complete results for $n = 4$ are given in Table 13.3. The iteration number is indicated by k. The number of gradient evaluations used in the line search is given by "ls." (The numbers in Table 13.3 record the cumulative costs. The norm used is the infinity norm.)   ■

Exercises

5.1. Verify the results obtained in Example 13.6 for $n = 4$.

5.2. Apply the limited-memory quasi-Newton method in Example 13.6 to

$$\text{minimize } f(x) = \sum_{i=1}^{n-1} (x_i - 2x_{i+1}^2)^2.$$

Use the initial guess $x_0 = (1, \ldots, 1)^T$. Solve this problem for $n = 4$ and $n = 10$.

5.3. Suppose that the limited-memory quasi-Newton method of Example 13.6 is applied to a general nonlinear function f. What are the vector storage requirements for the method? You can ignore any storage that might be required to evaluate the function f and its gradient.

5.4. Use the Sherman–Morrison formula to derive the BFGS update formula for H_k.

5.5. Use the Sherman–Morrison formula to derive the update formula for H_k based on the symmetric rank-one update

$$B_{k+1} = B_k + \frac{(y_k - B_k s_k)(y_k - B_k s_k)^T}{(y_k - B_k s_k)^T s_k}.$$

5.6. Show how to compute the search direction when two updates are used. That is, $H_{k-2} = I$ and the pairs (y_{k-1}, s_{k-1}) and (y_{k-2}, s_{k-2}) are given. What are the formulas when three updates are used? How much additional computation is required for each additional update?

5.7. Suppose that a limited-memory quasi-Newton method based on the BFGS formula is used to minimize a nonlinear function. Prove that the search directions are descent directions if all the pairs (y_i, s_i) satisfy the condition $y_i^T s_i > 0$.

5.8. Consider a limited-memory quasi-Newton method based on r BFGS updates, with the formula initialized with $H_{k+1-r} = I$. Prove that the following algorithm computes the corresponding search direction p_{k+1}. Within the algorithm, $\bar{p}$ records the current estimate of $H_{k+1} \nabla f(x_{k+1})$.

> If $(k + 1) < r$ set $J = 0$, $K = k + 1$; else set $J = k + 1 - r$, $K = r$
> Set $\bar{p} = \nabla f(x_{k+1})$.
> For $i = (K - 1), \ldots, 0$
> $j = i + J$
> $\rho_j = 1/(s_j^T y_j)$
> $\gamma_i = \rho_j s_j^T \bar{p}$
> $\bar{p} \leftarrow \bar{p} - \gamma_i y_j$
> For $i = 0, 1, \ldots, (K - 1)$
> $j = i + J$
> $\phi_i = \rho_j y_j^T \bar{p}$
> $\bar{p} \leftarrow \bar{p} + s_j(\gamma_i - \phi_i)$
> Set $p_{k+1} = -\bar{p}$.

13.6   Preconditioning

The idea behind preconditioning is to take advantage of auxiliary information about a problem to accelerate the convergence of an algorithm. We discuss this idea in the context of the linear conjugate-gradient method where the technique is well developed, but it is a general idea that can be applied to many algorithms.

A preconditioner for the conjugate-gradient method is a positive-definite matrix M. Instead of solving

$$Ax = b$$

we will solve the equivalent problem

$$M^{-1}Ax = M^{-1}b.$$

The hope is that this transformed system will be easier to solve than the original.

Example 13.7 (Preconditioning). Consider the linear system

$$Ax = \begin{pmatrix} 2000 & 1000 \\ 1 & 2 \end{pmatrix} \begin{pmatrix} x_1 \\ x_2 \end{pmatrix} = \begin{pmatrix} 1 \\ 1 \end{pmatrix} = b.$$

The coefficient matrix has condition number $\text{cond}(A) \approx 1700$. If we use the preconditioning matrix

$$M = \begin{pmatrix} 1000 & 0 \\ 0 & 1 \end{pmatrix},$$

then we obtain the transformed system

$$M^{-1}Ax = \begin{pmatrix} 2 & 1 \\ 1 & 2 \end{pmatrix} \begin{pmatrix} x_1 \\ x_2 \end{pmatrix} = \begin{pmatrix} 0.001 \\ 1 \end{pmatrix} = M^{-1}b.$$

The transformed matrix has condition number $\text{cond}(M^{-1}A) = 3$, a considerable improvement over the original matrix. It can be verified that the original and the transformed systems have the same solution $x \approx (-0.3227, 0.6663)^T$. ∎

If the conjugate-gradient method is applied to the transformed system, then, after a change of variables, the formulas for the "preconditioned conjugate-gradient method" are obtained (for details, see the Exercises).

ALGORITHM 13.4.
Preconditioned Conjugate-Gradient Method

1. Set $x_0 = 0$, $r_0 = b$, $p_{-1} = 0$, $\beta_0 = 0$, and specify the convergence tolerance ϵ.
2. For $i = 0, 1, \ldots$

    (i) If $\|r_i\| < \epsilon$, stop.
    (ii) Set $z_i = M^{-1}r_i$.
    (iii) If $i > 0$, set $\beta_i = r_i^T z_i / r_{i-1}^T z_{i-1}$.
    (iv) Set $p_i = z_i + \beta_i p_{i-1}$.
    (v) Set $\alpha_i = r_i^T z_i / p_i^T A p_i$.
    (vi) Set $x_{i+1} = x_i + \alpha_i p_i$.
    (vii) Set $r_{i+1} = r_i - \alpha_i A p_i$.

This algorithm requires one more vector than the original conjugate-gradient method (z_i) as well as the computation $z_i = M^{-1}r_i$. This means that M^{-1} must be "easy" to compute, or that linear systems of the form $Mz = r$ be "easy" to solve. For example, M might be a diagonal matrix. The matrix M is called the *preconditioning matrix*.

The convergence of the preconditioned algorithm depends on the matrix $M^{-1}A$. If $M = A$, then $M^{-1}A = I$ and the preconditioned method converges in one iteration, but normally this choice is impractical. We would like $M^{-1}A \approx I$ or $M \approx A$. More precisely, in exact arithmetic the preconditioned method converges in a number of iterations equal to the number of distinct eigenvalues of the matrix $M^{-1}A$. Also, from iteration to iteration the

Table 13.4. *Conjugate-gradient iterations.*

Preconditioned			Unpreconditioned		
i	$\|r_i\|$	$\|x_i - x_*\|$	i	$\|r_i\|$	$\|x_i - x_*\|$
0	2×10^0	4×10^{-1}	0	2×10^0	4×10^{-1}
1	1×10^0	4×10^{-1}	1	9×10^{-1}	3×10^{-1}
2	1×10^0	2×10^{-1}	2	4×10^{-1}	2×10^{-1}
3	4×10^{-1}	4×10^{-2}	3	2×10^{-1}	1×10^{-1}
4	3×10^{-2}	3×10^{-3}	4	2×10^{-1}	9×10^{-2}
5	4×10^{-3}	3×10^{-4}	5	1×10^{-1}	5×10^{-2}
6	3×10^{-4}	3×10^{-5}	6	9×10^{-2}	2×10^{-2}
7	1×10^{-5}	2×10^{-6}	7	3×10^{-2}	7×10^{-3}
8	1×10^{-6}	8×10^{-8}	8	2×10^{-2}	3×10^{-3}
9	2×10^{-16}	8×10^{-17}	9	5×10^{-3}	1×10^{-3}
			10	1×10^{-0}	6×10^{-4}
			11	9×10^{-4}	2×10^{-4}
			12	4×10^{-4}	5×10^{-5}
			13	1×10^{-4}	1×10^{-5}
			14	6×10^{-6}	7×10^{-7}
			15	7×10^{-18}	8×10^{-17}

method converges linearly with a rate constant that depends on $\text{cond}(M^{-1}A)$. Normally M is chosen to reduce the condition number, or to reduce the number of distinct eigenvalues, or both.

Example 13.8 (Preconditioned Conjugate-Gradient Method). Consider the system $Ax = b$ where A is a 15×15 diagonal matrix with $A_{i,i} = i$ and $b = (1, \ldots, 1)^T$. We use a diagonal preconditioning matrix M with diagonal entries

$$1, 2, 3, 4, 5, 6, 7, 1, \ldots, 1.$$

Then $M^{-1}A$ has nine distinct eigenvalues (1, 8, 9, 10, 11, 12, 13, 14, 15). As can be seen from Table 13.4, the preconditioned conjugate-gradient method converges in nine iterations, which is consistent with the theory. If the method is used with $M = I$ (no preconditioning), then the method requires 15 iterations, again as expected. ■

The choice of the preconditioner is sometimes specific to the application. For example, if the linear system $Ax = b$ represents an analysis of variance problem with missing data, then M might correspond to the same problem with complete data, a much easier problem to solve. There are also some general choices available. If A is a sparse matrix, then M might represent a factorization of A where all the fill-in is ignored, called an "incomplete Cholesky" factorization. Just as with the matrix-vector product, it is not necessary to store M explicitly.

Preconditioning (sometimes called "scaling") can also be applied to other algorithms. In a limited-memory quasi-Newton method a preconditioning matrix M can be used as the initial approximation to the Hessian. In a truncated-Newton method the inner conjugate-gradient iteration can be preconditioned. In general, if it is known that the variables in a problem will be of vastly different magnitudes (for example, if x_1 is measured in light years and x_2 in millimeters), then it may be useful to scale the variables so that they all have approximately the same magnitude (see Section 12.5).

The conjugacy and orthogonality properties of the preconditioned conjugate-gradient method are stated in the theorem below, an analog of Theorem 13.3.

Theorem 13.9. *Assume that the vectors $\{\, p_i \,\}$ and $\{\, r_i \,\}$ are defined by the formulas for the preconditioned conjugate-gradient method. Then*

$$r_i^T M^{-1} r_j = 0, \quad r_i^T p_j = 0, \quad \text{and} \quad p_i^T M^{1/2} A M^{1/2} p_j = 0$$

for $i > j$.

Proof. See the Exercises. $\square$

Exercises

6.1. Verify the results in Example 13.8.

6.2. Repeat Example 13.8, but with varying choices of the preconditioning matrix M. Let M be a diagonal matrix with diagonal entries

$$1, 2, \ldots, j, 1, \ldots, 1,$$

where j varies between 1 and 15. How many iterations are required to find the solution? Is this result consistent with the theory?

6.3. Apply the preconditioned conjugate-gradient method to the 10×10 linear system $Ax = b$, where $A = D + vv^T$, D is a diagonal matrix with entries $d_{i,i} = i$, $v = (1, \ldots, 1)^T$, and $b = (1, \ldots, 1)^T$. Use the preconditioner $M = D$. How many iterations are required? Is this result consistent with the theory?

6.4. Suppose that the preconditioned conjugate-gradient method is applied to a linear system $Ax = b$ with preconditioner $M = A$. Prove directly that the method terminates in one iteration.

6.5. The purpose of this problem is to derive the formulas for the preconditioned conjugate-gradient method. Assume that A and M are both symmetric positive-definite matrices. You may use the fact that a positive-definite matrix M can be written as $M^{1/2} M^{1/2}$, where $M^{1/2}$ is also symmetric and positive definite.

   (i) Show that the linear system $M^{-1} Ax = M^{-1} b$ is equivalent to $\hat{A}\hat{x} = \hat{b}$ where

$$\hat{A} = M^{-1/2} A M^{-1/2}, \quad \hat{x} = M^{1/2} x, \quad \hat{b} = M^{-1/2} b.$$

   (ii) Write out the traditional linear conjugate-gradient method for $\hat{A}\hat{x} = \hat{b}$.

(iii) Show that your result from (ii) is equivalent to the preconditioned conjugate-gradient method given above.

6.6. Prove Theorem 13.9. While it is possible to do this by mimicking the proof of Theorem 13.3, the results can be obtained more easily by using the approach described in Exercise 6.5.

13.7   Notes

The Conjugate-Gradient Method—The conjugate-gradient method was first described by Hestenes and Stiefel (1952). This paper contains a wealth of information about the method.

In rounded arithmetic, the conjugacy and orthogonality properties of the method are lost. For some problems such as Example 13.1 the loss may be negligible, but it is not hard to find examples where almost complete loss of conjugacy occurs. Some authors do not see this as a serious deficiency, but just treat the conjugate-gradient method as another iterative method that keeps iterating until the residual is small enough (see, for example, the book by Cullum and Willoughby (1985, reprinted 2002)). At the other extreme it is possible to save all the vectors $\{\, p_i \,\}$ and $\{\, r_i \,\}$ and do additional calculations to guarantee orthogonality and conjugacy. This can be expensive. In between these extremes there is a technique called "selective" orthogonalization that performs the addition calculations only as necessary and monitors the algorithm for loss of orthogonality (see the book by Parlett (1980, reprinted 1998)).

The conjugate-gradient method has been generalized to problems where the matrix is symmetric, but not positive definite, as well as to nonsymmetric systems; see, for example, the paper by Saad (1981). Many of these generalizations are based on the Lanczos algorithm, an algorithm developed by Cornelius Lanczos in 1950. The Lanczos algorithm is normally used to find eigenvalues of symmetric matrices but can be shown to be equivalent to the conjugate-gradient method. For more information see the paper by Paige and Saunders (1975).

Truncated-Newton Methods—A survey of truncated-Newton methods can be found in the paper by Nash (2000). The convergence theory for truncated-Newton methods is described in the paper by Dembo, Eisenstat, and Steihaug (1982). The rules given here are based on the residual of the Newton equations. Other rules are possible and perhaps even preferable; see the paper by Nash and Sofer (1990).

The conjugate-gradient method assumes that the coefficient matrix is positive definite, suggesting that truncated-Newton methods can only be applied to convex problems. This is not true. Truncated-Newton methods can be extended to general problems in much the same way as Newton's method (see the paper by Nash (1984)), as well as by exploiting the properties of the conjugate-gradient method within a trust-region strategy (see the papers by Steihaug (1983) and Toint (1981)). Software incorporating these ideas is available from NETLIB (www.netlib.org).

A truncated-Newton algorithm suitable for parallel computers is described in the paper by Nash and Sofer (1991). Software is also available from NETLIB and is described in the paper by Nash and Sofer (1992).

Nonlinear Conjugate-Gradient Methods— An extensive discussion of nonlinear conjugate-gradient methods can be found in the paper by Nocedal (1992). The convergence of such a method with the Fletcher–Reeves formula was proved in the paper by Al-Baali (1985). It should be noted that these methods may require a different line search from that described in Theorem 11.7, since it may not be possible to guarantee convergence, or even to guarantee that the search direction is a descent direction, unless additional conditions are imposed on the step length α. There have been attempts to produce a hybrid method that performs as well as the Polak–Ribiére formula with the guaranteed convergence of the Fletcher–Reeves formula, but the results have only been partially successful. For further details, see the paper by Gilbert and Nocedal (1992).

Implementations of nonlinear conjugate-gradient methods often include "restart" procedures, where β_i is set to zero intermittently, and the search direction is thus reset to the steepest-descent direction. This is motivated theoretically by the conjugacy properties of the linear conjugate-gradient method (the method terminates after n iterations in exact arithmetic) as well as the practical observation that these methods can "stall" before finding the solution. One approach, described in the paper by Powell (1977), proposes that the method be restarted if

$$|\nabla f(x_i)^T \nabla f(x_{i-1})| > \nu \|\nabla f(x_{i-1})\|^2 ,$$

where ν is some small positive number. For a quadratic function, successive gradients (residuals) will be orthogonal; Powell's test monitors the deterioration in this property.

Limited-Memory Quasi-Newton Methods—Limited-memory quasi-Newton methods were first described in the papers by Perry (1977) and Shanno (1978). More recently developed techniques can be found in the papers by Liu and Nocedal (1989) and Byrd et al. (1995).

Another way of adapting quasi-Newton methods for large problems is based on a property called "partial separability." This assumes that the objective function $f(x)$ can be written as the sum of simpler functions, each depending only on a small number of variables. Many large problems can be expressed in this way. For such functions, the Hessian matrix can be decomposed into smaller submatrices, and the usual quasi-Newton formulas can be used to approximate each of these. For further information on this topic, see the paper by Griewank and Toint (1982).

Preconditioning—A general discussion of preconditioning can be found in the book by Chen (2005). Preconditioning within the conjugate-gradient method is hinted at in the paper of Hestenes and Stiefel (1952), although the idea was not fully developed until the paper of Concus, Golub, and O'Leary (1976). The choice of a preconditioner may depend on the particular problem being solved, although there are some general techniques available. If the coefficient matrix is sparse, then an incomplete Cholesky factorization may be used as a preconditioner; see the paper by Meijerink and Van Der Vorst (1977). Automatic preconditioning strategies for the conjugate-gradient method within a truncated-Newton methods are discussed in the paper by Nash (1985). The scaling of limited-memory quasi-Newton methods is discussed in the paper by Gilbert and Lemaréchal (1989).

Part IV

Nonlinear Optimization

Chapter 14

Optimality Conditions for Constrained Problems

14.1 Introduction

In this part we study techniques for solving nonlinear optimization problems. We concentrate on problems that can be written in the general form

$$
\begin{array}{ll}
\text{minimize} & f(x) \\
\text{subject to} & g_i(x) = 0, \quad i \in \mathcal{E} \\
& g_i(x) \geq 0, \quad i \in \mathcal{I}.
\end{array}
$$

Here $\mathcal{E}$ is an index set for the equality constraints and $\mathcal{I}$ is an index set for the inequality constraints. We assume that the objective function f and the constraint functions g_i are twice continuously differentiable.

In this chapter we study the conditions satisfied by solutions to the constrained optimization problem. We shall focus only on local solutions, for the same reasons as in the unconstrained case. In the case of convex problems, that is, when the feasible region is convex and f is a convex function, any local solution is also a global solution.

In the unconstrained case the optimality conditions were derived by using a Taylor series approximation to examine the behavior of the objective function f about a local minimizer x_*. In particular, at points "near" x_* the value of f does not decrease.

A similar approach is used in the constrained case. Taylor series approximations are used to analyze the behavior of the objective f and the constraints g_i about a local constrained minimizer x_*. In this case, at *feasible* points "near" x_* the value of f does not decrease.

The optimality conditions will be derived in stages, first for problems with linear constraints, and then for problems with nonlinear constraints. The intuition in both cases is similar, but is easier to comprehend when the constraints are linear. In the nonlinear case the details are more complicated and can disguise the basic ideas involved.

If all the constraints are linear, feasible movements are completely characterized by feasible directions. (See Section 3.1.) At a local minimizer there can be no feasible directions of descent for f, hence

$$p^T \nabla f(x_*) \geq 0 \quad \text{for all feasible directions } p \text{ at } x_*. \tag{14.1}$$

The first-order optimality condition is a direct result of this statement.

483

If the problem has nonlinear constraints, it may no longer be possible to move to nearby points along feasible directions. Instead, movements will be made along feasible *curves*. Analyzing movement along curves is more complicated than along directions, and more complicated situations can arise. Even so, the basic idea is that the objective value will not decrease at feasible points near x_*.

Some new concepts arise in the constrained case, in particular, the Lagrange multipliers and the Lagrangian function. The Lagrange multipliers are analogous to the dual variables in linear optimization. The Lagrangian is a single function that combines the objective and constraint functions; it plays a central role in the theory and algorithms of constrained optimization. We shall also develop a theory of duality, a generalization of the duality theory for linear optimization.

14.2   Optimality Conditions for Linear Equality Constraints

In this section we discuss the optimality conditions for nonlinear problems where all constraints are linear equalities:

$$\begin{aligned} \text{minimize} \quad & f(x) \\ \text{subject to} \quad & Ax = b, \end{aligned}$$

where A is an $m \times n$ matrix. We assume that f is twice continuously differentiable over the feasible region. We also assume that the rows of A are linearly independent, that is, A has full row rank. This is not an unduly restrictive assumption since in theory, if a problem is consistent, we can discard any redundant constraints.

The main idea is to transform this constrained problem into an equivalent unconstrained problem. The theory and methods for unconstrained optimization can then be applied to the new problem.

To demonstrate the approach consider the problem

$$\begin{aligned} \text{minimize} \quad & f(x) = x_1^2 - 2x_1 + x_2^2 - x_3^2 + 4x_3 \\ \text{subject to} \quad & x_1 - x_2 + 2x_3 = 2. \end{aligned}$$

At any feasible point, the variable x_1 can be expressed in terms of x_2 and x_3 using $x_1 = 2 + x_2 - 2x_3$. Substituting this into the formula for $f(x)$, we obtain the equivalent unconstrained problem

$$\text{minimize} \quad 2x_2^2 + 3x_3^2 - 4x_2x_3 + 2x_2.$$

(The number of variables has been reduced from three to two.) It is easy to verify that a strict local minimizer to the unconstrained problem is $x_2 = -1.5$, $x_3 = -1$. The solution to the original problem is $x_* = (2.5, -1.5, -1)^T$ with an optimal objective value of $f(x_*) = -1.5$.

Any problem with linear equality constraints $Ax = b$ can be recast as an equivalent unconstrained problem. Suppose we have a feasible point $\bar{x}$, that is, $A\bar{x} = b$. Then any other feasible point can be expressed as $x = \bar{x} + p$, where p is a feasible direction. Any feasible direction must lie in the null space of A, the set of vectors p satisfying $Ap = 0$. Denoting this null space by $\mathcal{N}(A)$, the feasible region can be described by $\{ x : x = \bar{x} + p, \ p \in \mathcal{N}(A) \}$. Let Z be an $n \times r$ null-space matrix for A (with $r \geq n - m$). Then the feasible region is

Note: $x = \bar{x} + Zv$
$Ax = A\bar{x} + AZv$
$\quad = b$

given by $\{x : x = \bar{x} + Zv, \text{ where } v \in \Re^r\}$. Consequently, our constrained problem in x is equivalent to the unconstrained problem

$$\underset{v \in \Re^r}{\text{minimize}} \; \phi(v) = f(\bar{x} + Zv).$$

The function ϕ is the restriction of f onto the feasible region; we shall refer to it as the *reduced function*.

If Z is a *basis* matrix for the null space of A, then ϕ will be a function of $n - m$ variables. Not only has the constrained problem been transformed into an unconstrained problem, but also the number of variables has been reduced as well.

Example 14.1 (Reduced Function). Consider again the problem

$$\text{minimize} \quad f(x) = x_1^2 - 2x_1 + x_2^2 - x_3^2 + 4x_3$$
$$\text{subject to} \quad x_1 - x_2 + 2x_3 = 2.$$

Select

$$Z = \begin{pmatrix} 1 & -2 \\ 1 & 0 \\ 0 & 1 \end{pmatrix}$$

$A = [1 \; -1 \; 2]$

$p \in Null(A)$ if $Ap = 0$
if $P_1 - P_2 + 2P_3 = 0$
so, $P_1 = P_1 \quad P_3 = \frac{1}{2}P_2 - \frac{1}{2}P_1$
$P_2 = P_2$

as a null-space matrix for the constraint matrix $A = (1, -1, 2)$. Using the (arbitrary) feasible point $\bar{x} = (2, 0, 0)^T$, any feasible point can be written as

$$x = \bar{x} + Zv = \begin{pmatrix} 2 \\ 0 \\ 0 \end{pmatrix} + \begin{pmatrix} 1 & -2 \\ 1 & 0 \\ 0 & 1 \end{pmatrix} v$$

Obtain two LI vectors
e.g $\begin{bmatrix} 1 \\ 1 \\ 0 \end{bmatrix}, \begin{bmatrix} -2 \\ 0 \\ 1 \end{bmatrix}$

This is a basis for the null space, which has $n - m = 2$ L.I vectors.

for some $v = (v_1, v_2)^T$. Substituting into f, we obtain the reduced function $\phi(v) = 2v_1^2 + 3v_2^2 - 4v_1v_2 + 2v_1$. This is the same reduced function as before, except that now the variables are called v_1 and v_2 rather than x_2 and x_3. ∎

The optimality conditions involve the derivatives of the reduced function. If $x = \bar{x} + Zv$, then by the chain rule (see Appendix B.7),

$$\nabla\phi(v) = Z^T\nabla f(\bar{x} + Zv) = Z^T\nabla f(x)$$

where $x = \bar{x} + Zv$

and

$$\nabla^2\phi(v) = Z^T\nabla^2 f(\bar{x} + Zv)Z = Z^T\nabla^2 f(x)Z.$$

(See the Exercises.) The vector $\nabla\phi(v) = Z^T\nabla f(x)$ is called the *reduced gradient* of f at x. If Z is an orthogonal projection matrix, it is sometimes called the *projected gradient*. Similarly the matrix $\nabla^2\phi(v) = Z^T\nabla^2 f(x)Z$ is called the *reduced Hessian matrix*, or sometimes the *projected Hessian matrix*. The reduced gradient and Hessian matrix are the gradient and Hessian of the restriction of f onto the feasible region, evaluated at x.

If x_* is a local solution of the constrained problem, then $x_* = \bar{x} + Zv_*$ for some v_*, and v_* is a local minimizer of ϕ. Hence

$$\nabla\phi(v_*) = 0 \quad \text{and} \quad \nabla^2\phi(v_*) \text{ is positive semidefinite.}$$

Using the formulas for the reduced gradient and Hessian matrix, we obtain the first- and second-order necessary conditions for a local minimizer. They are summarized in the following lemma.

Lemma 14.2 (Necessary Conditions, Linear Equality Constraints). *If x_* is a local minimizer of f over $\{x : Ax = b\}$ and Z is a null-space matrix for A, then*

$\nabla \phi(v^*) = 0$

$\nabla^2 \phi(v^*) \geqslant 0$

- $Z^T \nabla f(x_*) = 0$,   *and*
- $Z^T \nabla^2 f(x_*) Z$ *is positive semidefinite;*

where $x^* = \bar{x} + Zv^*$

that is, the reduced gradient is zero and the reduced Hessian matrix is positive semidefinite.

A point at which the reduced gradient is zero is a *stationary point*. Such a point may be a local minimizer of f, or a local maximizer, or neither, in which case it is a *saddle point*. Second derivative information is used to distinguish local minimizers from other stationary points.

The second-order condition is equivalent to the condition

$$v^T Z^T \nabla^2 f(x_*) Z v \geq 0 \quad \text{for all } v.$$

Observing that $p = Zv$ is a null-space vector, this can be rewritten as

$$p^T \nabla^2 f(x_*) p \geq 0 \quad \text{for all } p \in \mathcal{N}(A);$$

that is, the Hessian matrix at x_* must be positive semidefinite on the null space of A.

This condition does not require that the Hessian matrix itself be positive semidefinite. It is a less stringent requirement. If the Hessian matrix at x_* is positive semidefinite, however, then of course the second-order condition will be satisfied.

The second-order sufficiency conditions are also analogous to the unconstrained case. We will assume that Z is a basis matrix for the null space of A, so that the columns of Z are linearly independent. The corresponding second-order sufficiency conditions are given in the lemma below. Results for other null-space matrices are given in the Exercises.

Lemma 14.3 (Sufficient Conditions, Linear Equality Constraints). *If x_* satisfies*

- $Ax_* = b$,
- $Z^T \nabla f(x_*) = 0$,   *and*
- $Z^T \nabla^2 f(x_*) Z$ *is positive definite,*

where Z is a basis matrix for the null space of A, then x_* is a strict local minimizer of f over $\{x : Ax = b\}$.

The following example illustrates the optimality conditions.

Example 14.4 (Necessary Conditions for Optimality). We examine again the problem

$$\text{minimize} \quad f(x) = x_1^2 - 2x_1 + x_2^2 - x_3^2 + 4x_3$$
$$\text{subject to} \quad x_1 - x_2 + 2x_3 = 2.$$

Since $\nabla f(x) = (2x_1 - 2, 2x_2, -2x_3 + 4)^T$, then at the feasible point $x_* = (2.5, -1.5, -1)^T$ the gradient of f is $(3, -3, 6)^T$. Selecting

$$Z = \begin{pmatrix} 1 & -2 \\ 1 & 0 \\ 0 & 1 \end{pmatrix}$$

as the null-space matrix of $A = (1, -1, 2)$, it is easily verified that $Z^T \nabla f(x_*) = (0, 0)^T$. Thus, the reduced gradient vanishes at x_*, and the first-order necessary condition for a local minimum is satisfied at this point. Checking the reduced Hessian matrix, we find that

$$Z^T \nabla^2 f(x_*) Z = \begin{pmatrix} 1 & 1 & 0 \\ -2 & 0 & 1 \end{pmatrix} \begin{pmatrix} 2 & 0 & 0 \\ 0 & 2 & 0 \\ 0 & 0 & -2 \end{pmatrix} \begin{pmatrix} 1 & -2 \\ 1 & 0 \\ 0 & 1 \end{pmatrix} = \begin{pmatrix} 4 & -4 \\ -4 & 6 \end{pmatrix}.$$

The reduced Hessian matrix is positive definite at x_*. Hence the second-order sufficiency conditions are satisfied, and x_* is a strict local minimizer of f. Notice that $\nabla^2 f(x_*)$ itself is not positive definite.

Let us choose some other feasible point, say $x = (2, 0, 0)^T$. The reduced gradient at this point is

$$Z^T \nabla f(x) = \begin{pmatrix} 2 \\ 0 \end{pmatrix} \neq \begin{pmatrix} 0 \\ 0 \end{pmatrix};$$

hence this point is not a local minimizer. To move to a better point we should use a descent direction. Any vector $v = (v_1, v_2)^T$ such that $v^T(Z^T \nabla f(x)) = 2v_1 < 0$ will be a descent direction for the reduced function at this point. The corresponding direction $p = Zv$ will be a feasible descent direction for f. ∎

Let us take another look at the first-order necessary condition. Let x_* be a local minimum, and let Z be any $n \times r$ null-space matrix for A. Breaking $\nabla f(x_*)$ into its null-space and range-space components gives

$$\nabla f(x_*) = Zv_* + A^T \lambda_*,$$

(ie $\nabla f(1_*) = q + p$,
where $q \in \mathcal{Null}(A)$, $p \in R(A^T)$

see. p. 82)

where v_* is in $\mathfrak{R}^r$ and λ_* is in $\mathfrak{R}^m$. Premultiplying by Z^T and recalling that the reduced gradient vanishes at x_*, we find that $Z^T Z v_* = 0$. This can occur only if $Zv_* = 0$, that is, if the null-space component of the gradient is zero. Therefore, if x_* is a local minimizer,

$$\nabla f(x_*) = A^T \lambda_* \tag{14.2}$$

for some m-vector λ_*. Thus, at a local minimum the gradient of the objective is a linear combination of the gradients of the constraints. The vector λ_* gives the coefficients of this linear combination. It is known as the vector of *Lagrange multipliers*. Its ith component is the Lagrange multiplier for the ith constraint.

The optimality conditions are demonstrated in Figure 14.1. This problem involves a single linear constraint $a^T x = b$. At the minimizer x_* the gradient is parallel to the vector a. Therefore there exists some number λ_* such that $\nabla f(x_*) = a\lambda_*$. On the other hand, at the point $\bar{x}$ the gradient is not parallel to the vector a; thus there is no λ that satisfies $\nabla f(\bar{x}) = a\lambda$, and the point is not optimal.

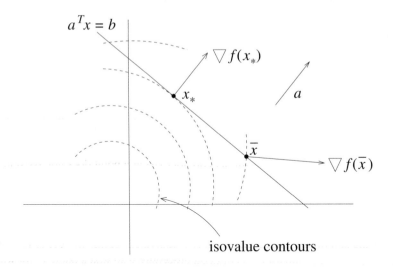

Figure 14.1. *Existence of Lagrange multipliers.*

Example 14.5 (Necessary Conditions for Optimality—Lagrange Multipliers). Consider again the problem in Example 14.1. The first-order necessary condition is

$$\begin{pmatrix} 2x_1 - 2 \\ 2x_2 \\ -2x_3 + 4 \end{pmatrix} = \begin{pmatrix} 1 \\ -1 \\ 2 \end{pmatrix} \lambda.$$

This implies that $x_1 = 1 + \lambda/2$, $x_2 = -\lambda/2$, and $x_3 = 2 - \lambda$. Since the solution must be feasible, we substitute these values into the constraint $x_1 - x_2 + 2x_3 = 2$ to obtain $\lambda_* = 3$ as the only solution. This indicates that $x_* = (2.5, -1.5, -1)^T$ is the unique stationary point. Since we have seen that the second-order sufficiency conditions are satisfied at x_*, this is the unique local solution. ∎

In Example 14.5 we used condition (14.2) to obtain a local solution. In most cases, however, these equations will not have a closed-form solution. This is demonstrated in Example 14.6.

Example 14.6 (Intractability of the Optimality Conditions). Consider the problem

$$\text{minimize} \quad f(x) = x_1^4 x_2^2 + x_1^2 x_3^4 + \tfrac{1}{2} x_1^2 + x_1 x_2 + x_3$$
$$\text{subject to} \quad x_1 + x_2 + x_3 = 1.$$

The first-order necessary condition implies that, at a local minimum,

$$\begin{pmatrix} 4x_1^3 x_2^2 + 2x_1 x_3^4 + x_1 + x_2 \\ 2x_1^4 x_2 + x_1 \\ 4x_1^2 x_3^3 + 1 \end{pmatrix} = \begin{pmatrix} 1 \\ 1 \\ 1 \end{pmatrix} \lambda$$

for some number λ. These three equations together with the constraint $x_1 + x_2 + x_3 = 1$ give four equations in the four unknowns x_1, x_2, x_3, and λ. These equations are not easy to solve, however. If we try to solve them numerically, there is no guarantee that the solution will be a local minimizer; it may be a saddle point or even a local maximizer. ■

We have shown that if the reduced gradient is zero, then there exists a vector of Lagrange multipliers λ_* that satisfies the optimality condition (14.2). The reverse is also true; that is, (14.2) implies that the reduced gradient vanishes (see the Exercises). Thus, the two versions of the first-order optimality condition are equivalent. From a practical point of view there is a difference, however. If the reduced gradient at a given point is nonzero, it can be used to find a descent direction for the reduced function, and in turn for f. In contrast, the fact that Lagrange multipliers do not exist at a point does not assist in finding a better estimate of a solution.

Then why do we care about Lagrange multipliers? The Lagrange multipliers provide important information in sensitivity analysis (see Section 14.3). Furthermore, for problems with inequality constraints, estimates of the multipliers can indicate how to improve an estimate of the solution. Consequently, the two equivalent optimality conditions are used together in optimization software. A common procedure is to find a point x_* for which the reduced gradient is zero; at x_* condition (14.2) is consistent and the corresponding Lagrange multipliers can be computed.

Our derivation assumes that the matrix A has full row rank, that is, its rows are linearly independent. This assumption is called a *regularity assumption*. The results in this section can be extended to the case where the rows of A are linearly dependent, but then the vector of Lagrange multipliers will not generally be unique. For problems with nonlinear constraints, some assumption, such as a regularity assumption on the gradients of the constraints at the local minimum, is needed to state the optimality conditions.

Exercises

2.1. Consider the problem

$$\text{minimize} \quad f(x) = x_1^2 + x_1^2 x_3^2 + 2x_1 x_2 + x_2^4 + 8x_2$$
$$\text{subject to} \quad 2x_1 + 5x_2 + x_3 = 3.$$

(i) Determine which of the following points are stationary points: (i) $(0, 0, 2)^T$; (ii) $(0, 0, 3)^T$; (iii) $(1, 0, 1)^T$.

(ii) Determine whether each stationary point is a local minimizer, a local maximizer, or a saddle point.

2.2. Determine the minimizers/maximizers of the following functions subject to the given constraints.

(i) $f(x_1, x_2) = x_1 x_2^3$ subject to $2x_1 + 3x_2 = 4$.

(ii) $f(x_1, x_2) = 2x_1 - 3x_2$ subject to $x_1^2 + x_2^2 = 25$.

(iii) $f(x_1, x_2) = x_1^2 + 2x_1 x_2 + x_2^2$ subject to $3x_1^2 + x_2^2 = 9$.

(iv) $f(x_1, x_2) = 3x_1^3 + 2x_2^3$ subject to $x_1^2 + x_2^2 = 4$.

(v) $f(x_1, x_2) = x_2$ subject to $x_1^3 + x_2^3 - 3x_1 x_2 = 0$.

(vi) $f(x_1, x_2) = x_1^3 + x_2^3$ subject to $2x_1 + x_2 = 1$.

(vii) $f(x_1, x_2) = \frac{1}{3}x_1^3 + x_2$ subject to $x_1^2 + x_2^2 = 1$.

2.3. Find all the values of the parameters a and b such that $(0, 0)^T$ minimizes or maximizes the following function subject to the given constraint:

$$f(x_1, x_2) = (a + 2)x_1 - 2x_2 \quad \text{subject to} \quad a(x_1 + e^{x_1}) + b(x_2 + e^{x_2}) = 1.$$

2.4. Consider the problem of finding the minimum distance from a point r to a set $\{x : a^T x = b\}$. The problem can be written as

$$\begin{array}{ll} \text{minimize} & f(x) = \frac{1}{2}(x - r)^T(x - r) \\ \text{subject to} & a^T x = b. \end{array}$$

Prove that the solution is given by

$$x_* = r + \frac{b - a^T r}{a^T a}a.$$

2.5. Solve the problem

$$\begin{array}{ll} \text{maximize} & f(x) = x_1 x_2 x_3 \\ \text{subject to} & \dfrac{x_1}{a_1} + \dfrac{x_2}{a_2} + \dfrac{x_3}{a_3} = 1 \quad (a_1, a_2, a_3 > 0). \end{array}$$

2.6. Solve the problem

$$\begin{array}{ll} \text{maximize} & f(x) = x_1 x_2 \cdots x_n \\ \text{subject to} & \dfrac{x_1}{a_1} + \dfrac{x_2}{a_2} + \cdots + \dfrac{x_n}{a_n} = 1 \quad (a_1, a_2, \ldots, a_n > 0). \end{array}$$

2.7. Let A be a matrix of full row rank. Find the point in the set $Ax = b$ which minimizes $f(x) = \frac{1}{2}x^T x$.

2.8. *Heron's problem.* The two-dimensional points A and B lie in the same half plane with respect to the line l. If C minimizes the sum of the distances AC and BC, prove that the angle between AC and l is equal to the angle between BC and l.

2.9. *Euclid's problem.* In a given triangle ABC, inscribe a parallelogram $ADEF$ with side DE parallel to AC, and FE parallel to AB. Determine the parallelogram of this form with the largest area.

2.10. Consider the linear program

$$\begin{array}{ll} \text{minimize} & f(x) = c^T x \\ \text{subject to} & Ax = b. \end{array}$$

Prove that if the problem has a feasible solution, then either the problem is unbounded, or all feasible points are optimal.

2.11. (From Luenberger (2003).) Consider the quadratic program

$$\begin{array}{ll} \text{minimize} & f(x) = \frac{1}{2}x^T Q x - c^T x \\ \text{subject to} & Ax = b. \end{array}$$

Prove that x_* is a local minimum point if and only if it is a global minimum point (no convexity is assumed).

2.12. Derive the formulas for the reduced gradient and Hessian, that is,

$$\nabla\phi(v) = Z^T\nabla f(\bar{x} + Zv) = Z^T\nabla f(x)$$
$$\nabla^2\phi(v) = Z^T\nabla^2 f(\bar{x} + Zv)Z = Z^T\nabla^2 f(x)Z,$$

where $x = \bar{x} + Zv$.

2.13. Prove Lemma 14.3.

2.14. Let Z be a null-space matrix for the matrix A. Prove that if $\nabla f(x_*) = A^T\lambda$ for some λ, then $Z^T\nabla f(x_*) = 0$.

2.15. Consider the problem of minimizing a twice continuously differentiable function f subject to the linear constraints $Ax = b$. Let x_* be a feasible point for the constraints. Let Z be an $n \times r$ null-space matrix for A which is not a basis, that is, some of its columns are linearly dependent (and $r > n - m$).

  (i) Prove that the matrix $Z^T\nabla^2 f(x_*)Z$ cannot be positive definite.

  (ii) Prove that if

$$Z^T\nabla f(x_*) = 0 \quad \text{and} \quad p^T\nabla^2 f(x_*)p > 0 \quad \text{for all } p \in \mathcal{N}(A),\ p \neq 0,$$

  then x_* is a strict local minimizer of f over the set $Ax = b$. This is an alternative form of the second-order sufficiency conditions.

2.16. Consider the problem

$$\text{minimize} \quad f(x) = x_1^2 + x_2^2$$
$$\text{subject to} \quad x_1 + x_2 = 2.$$

Let

$$Z = \begin{pmatrix} \frac{1}{2} & -\frac{1}{2} \\ -\frac{1}{2} & \frac{1}{2} \end{pmatrix}$$

be a null-space matrix for the constraint set. Show that the first-order necessary condition is satisfied at the point $x_* = (1, 1)^T$, but that $Z^T\nabla^2 f(x_*)Z$ is not positive definite. Show also that the second-order conditions given in the previous problem are satisfied, and hence x_* is a strict local minimum point.

2.17. Consider the problem

$$\text{maximize} \quad f(x)$$
$$\text{subject to} \quad Ax = b,$$

where f is twice continuously differentiable, and A is a matrix of full row rank.

  (i) State and prove the first-order necessary condition for a local solution.

  (ii) State and prove the second-order necessary conditions for a local solution.

  (iii) State and prove the second-order sufficiency conditions for a local solution.

14.3 The Lagrange Multipliers and the Lagrangian Function

The Lagrange multipliers express the gradient at the optimum as a linear combination of the rows of the constraint matrix A. These multipliers have a significance which goes beyond this purely mathematical interpretation. In this section we shall see that they indicate the sensitivity of the optimal objective value to changes in the data. We also present the Lagrangian function and show how it can be used to express the optimality conditions in a concise way.

In most applications, only approximate data are available. Measurement errors, fluctuations in data, and unavailability of information are some of the factors that contribute to imprecision in the optimization model. In the absence of precise data, there may be no choice but to solve the problem using the best available estimates. Once a solution is obtained, the next step is to assess the quality of the resulting solution. A key question is, how sensitive is the solution to variations in the data?

Here we address this question for the particular case where small variations are made in the right-hand side of the constraints and investigate their effect on the optimal objective value. Our presentation will be informal. A more formal proof is somewhat more complex.

We start with the problem

$$\text{minimize} \quad f(x)$$
$$\text{subject to} \quad Ax = b.$$

We assume that f is twice continuously differentiable, and that A is an $m \times n$ matrix of full row rank. We also assume that a local minimizer x_* has been found, with corresponding optimal objective value $f(x_*)$. Suppose now that the right-hand side b is perturbed to $b + \delta$, where δ is a vector of "small" perturbations. We shall investigate how the optimal objective value changes as a result of these perturbations. If the perturbations are sufficiently small, it is reasonable to assume that the new problem has an optimum that is close to x_*. In fact this can be shown to be true, provided that the second-order sufficiency conditions are satisfied at x_*. For $\bar{x}$ close to x_* with $A\bar{x} = b + \delta$, we can use a Taylor series approximation to obtain

$$\begin{aligned}
f(\bar{x}) &\approx f(x_*) + (\bar{x} - x_*)^T \nabla f(x_*) \\
&= f(x_*) + (\bar{x} - x_*)^T A^T \lambda_* \\
&= f(x_*) + \delta^T \lambda_* \\
&= f(x_*) + \sum_{i=1}^{m} \delta_i \lambda_{*i}.
\end{aligned}$$

In particular, this is valid if $\bar{x}$ is the minimizer of the perturbed problem. If the right-hand side of the ith constraint changes by δ_i, then the optimal objective value changes by approximately $\delta_i \lambda_{*i}$. Hence λ_{*i} represents the change in the optimal objective per unit change in the ith right-hand side. For this reason, the Lagrange multipliers are also called *shadow prices* or *dual variables*.

Example 14.7 (Solution of a Perturbed Problem). Consider again the problem

$$\text{minimize} \quad f(x) = x_1^2 - 2x_1 + x_2^2 - x_3^2 + 4x_3$$
$$\text{subject to} \quad x_1 - x_2 + 2x_3 = 2.$$

In Example 14.5 we determined that $x_* = (2.5, -1.5, -1)^T$, with $f(x_*) = -1.5$ and $\lambda_* = 3$.

Consider now the perturbed problem

$$\text{minimize} \quad f(x) = x_1^2 - 2x_1 + x_2^2 - x_3^2 + 4x_3$$
$$\text{subject to} \quad x_1 - x_2 + 2x_3 = 2 + \delta,$$

and denote its minimum value by $f_*(\delta)$. The interpretation of the Lagrange multipliers as shadow prices indicates that a first-order estimate of this minimum value is

$$f_*(\delta) \approx -1.5 + 3\delta.$$

For example, if $\delta = 0.5$, the approximate optimal objective value is zero.

The precise solution to the perturbed problem is $x_1 = 2.5 - \delta/2$, $x_2 = -1.5 + \delta/2$, and $x_3 = -1 + \delta$, with an objective value of

$$f_*(\delta) = -1.5 + 3\delta - 0.5\delta^2.$$

If $\delta = 0.5$, the true value of the optimal objective is -0.125. ∎

Let us now take another look at the optimality conditions (14.2). Since any solution must be feasible, a local optimum is the solution to the system of $n + m$ equations in the $n + m$ unknowns x and λ:

$$\nabla f(x) - A^T\lambda = 0$$
$$Ax = b.$$

This is another representation of the first-order optimality conditions.

These conditions were used by Lagrange, although his work was done in a more general setting (see Section 14.9). Following Lagrange's approach we can construct a function of x and λ:

$$\mathcal{L}(x, \lambda) = f(x) - \sum_{i=1}^{m} \lambda_i (a_i^T x - b_i) = f(x) - \lambda^T(Ax - b),$$

where a_i^T denotes the ith row of A. This function is called the *Lagrangian function*. The gradient of the Lagrangian with respect to x is $\nabla_x \mathcal{L}(x, \lambda) = \nabla f(x) - A^T\lambda$, and the gradient with respect to λ is $\nabla_\lambda \mathcal{L}(x, \lambda) = b - Ax$. Hence, the first-order optimality conditions can simply be stated as

$$\nabla \mathcal{L}(x_*, \lambda_*) = 0.$$

Thus a local minimizer is a stationary point of the Lagrangian function.

Exercises

3.1. Consider the problem

$$\begin{aligned} \text{minimize} \quad & f(x) = 3x_1^2 - \tfrac{1}{2}x_2^2 - \tfrac{1}{2}x_3^2 + x_1x_2 - x_1x_3 + 2x_2x_3 \\ \text{subject to} \quad & 2x_1 - x_2 + x_3 = 2. \end{aligned}$$

Solve this problem. Use the Lagrange multiplier to estimate the minimum value of f under the perturbed constraint $2x_1 - x_2 + x_3 = 2 + \delta$. Compare the estimated minimum objective value to the actual minimum for $\delta = 0.25$.

3.2. For the previous problem, determine the exact solution to the perturbed problem as a function of δ. Are there any limits on how large δ can be before this result ceases to be valid?

3.3. The following problem demonstrates that if the second-order sufficiency conditions are not satisfied at a local minimum x_*, the perturbed problem may not have an optimum. Consider the problem

$$\text{minimize} \quad f(x) = x_1^2 - x_1 x_2 - x_1$$
$$\text{subject to} \quad x_1 - x_2 = 1.$$

Show that f has a constant value zero at all feasible points, and thus that all feasible points are local optima. Show that the second-order sufficiency conditions are not satisfied anywhere. Consider now the problem of minimizing f subject to the perturbed constraint $x_1 - x_2 = 1 + \delta$. Show that the perturbed problem has no local minimum for any nonzero value of δ.

3.4. In Example 14.5, suppose that the constraint is perturbed to

$$x_1 - x_2 + (2 + \delta)x_3 = 2.$$

What is the solution to the perturbed problem? Can this solution be approximated using Lagrange multipliers? Can you approximate the solution to the problem with constraint

$$(1 + \delta_1)x_1 - (1 + \delta_2)x_2 + (2 + \delta_3)x_3 = 2 + \delta_4,$$

where all the coefficients are independently perturbed?

14.4 Optimality Conditions for Linear Inequality Constraints

We now turn our attention to problems where the constraints are linear inequalities. We start with an example:

$$\text{minimize} \quad f(x) = \tfrac{1}{2}x_1^2 + \tfrac{1}{2}x_2^2$$
$$\begin{aligned}\text{subject to} \quad & x_1 + 2x_2 \geq 2 \\ & x_1 - x_2 \geq -1 \\ & -x_1 \geq -3.\end{aligned}$$

The objective of this problem is to find the feasible point closest in Euclidean norm to the origin. This is depicted in Figure 14.2.

The figure suggests that the solution x_* is on the line $x_1 + 2x_2 = 2$. Let us assume for the moment that this is the case. Then x_* is a minimizer of f along this line and will solve the equality-constrained problem

$$\text{minimize} \quad f(x) = \tfrac{1}{2}x_1^2 + \tfrac{1}{2}x_2^2$$
$$\text{subject to} \quad x_1 + 2x_2 = 2.$$

The unique solution of the first-order optimality conditions for this problem is $x_* = (\tfrac{2}{5}, \tfrac{4}{5})^T$ with objective value $f(x_*) = \tfrac{2}{5}$ and Lagrange multiplier $\lambda_* = \tfrac{2}{5}$. Since the

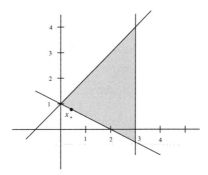

Figure 14.2. *Graphical solution.*

objective function is strictly convex, this point is the global minimizer of f along the line $x_1 + 2x_2 = 2$.

To determine if x_* solves the original inequality-constrained problem, we must verify that the objective value does not decrease as we make a small movement away from x_* into the interior of the feasible region, where $x_1 + 2x_2 > 2$. For small positive values δ, the minimum objective value along $x_1 + 2x_2 = 2 + \delta$ will change by approximately $\lambda_* \delta$. If λ_* were negative, the objective value would decrease, indicating that the point x_* would not be a local minimizer. In our example $\lambda_* = \frac{2}{5}$, and thus as we move from the boundary into the interior where $x_1 + 2x_2 = 2 + \delta$, the objective will increase by roughly $\frac{2}{5}\delta$. Thus, the point x_* has a better objective value than all its neighboring feasible points and consequently is a local minimizer.

In this example a graph helped us identify the constraints that are binding at the solution. We then minimized f with all binding constraints treated as equalities. In most problems we cannot determine the optimal active set (the set of binding constraints) so easily. We may need to examine many trial active sets and rule out those that are not optimal. How can we verify whether a guess of the active set is indeed optimal?

Suppose, for example, that we erroneously guessed that only the third constraint was active at the solution. This would mean that the solution to the problem also solved the equality constrained problem

$$
\begin{aligned}
\text{minimize} \quad & f(x) = \tfrac{1}{2}x_1^2 + \tfrac{1}{2}x_2^2 \\
\text{subject to} \quad & -x_1 = -3.
\end{aligned}
$$

This problem has the solution $x_1 = 3$, $x_2 = 0$, with associated multiplier $\lambda = -3$. Since $\lambda < 0$, moving into the interior of the constraint will yield points with better objective values. Hence $(3, 0)^T$ is not a local minimizer.

Our example highlights two major points in the treatment of inequality-constrained problems. First, any solution of the problem is also a solution of the equality-constrained problem obtained by requiring that the active constraints at that point be satisfied exactly. Second, the associated Lagrange multipliers at this point cannot be negative; a negative multiplier indicates that the objective can be improved by making a small move into the interior of the corresponding constraint.

We now develop the optimality conditions in a more rigorous manner. We are interested in solving

$$\text{minimize} \quad f(x)$$

$$\text{subject to} \quad Ax \geq b,$$

where A is an $m \times n$ matrix whose rows are the vectors a_i^T. We shall assume that the system has a feasible solution.

Let x_* be a local solution. Since the constraints that are inactive at x_* have no impact on the local optimality of x_*, they can be omitted from the discussion. Let $\hat{A}$ be the matrix whose rows are the vectors a_i^T of the active constraints, and let $\hat{b}$ be the corresponding vector of right-hand-side coefficients. Then $\hat{A}x_* = \hat{b}$. Any other feasible point may be reached from x_* by moving along some feasible direction p. Note that p is a feasible direction at x_* if and only if $\hat{A}p \geq 0$. (See Section 3.1.) To simplify matters, we shall make the regularity assumption that the rows of $\hat{A}$ are linearly independent; however, the final results below will hold even without this assumption.

Since x_* is a local solution for the inequality-constrained problem it is also a local solution for the equality-constrained problem

$$\text{minimize} \quad f(x)$$

$$\text{subject to} \quad \hat{A}x = \hat{b}.$$

Let Z be a null-space matrix for $\hat{A}$. The first-order necessary condition for the equality-constrained problem implies that

$$Z^T \nabla f(x_*) = 0 \quad \text{or equivalently} \quad \nabla f(x_*) = \hat{A}^T \hat{\lambda}_*,$$

where $\hat{\lambda}_*$ is the vector of Lagrange multipliers (if there are t active constraints, then $\hat{\lambda}$ is of length t). The second-order necessary conditions imply that $Z^T \nabla^2 f(x_*)Z$ must be positive semidefinite.

We now show that $\hat{\lambda}_* \geq 0$. If not, then some component of $\hat{\lambda}_*$, say $\hat{\lambda}_{*1}$, is negative. Let e_1 be a t-dimensional vector whose first component is 1 and all other components are 0. The rows of $\hat{A}$ are linearly independent, and so we can find a vector p such that $\hat{A}p = e_1$ (see the Exercises). Since $\hat{A}p \geq 0$, p is a feasible direction for the constraints. But

$$p^T \nabla f(x_*) = p^T \hat{A}^T \hat{\lambda}_* = e_1^T \hat{\lambda}_* = \hat{\lambda}_{*1} < 0.$$

Thus p is a feasible direction of descent at x_*, contradicting the fact that x_* is a local minimizer. Therefore, at a local solution the vector of multipliers $\hat{\lambda}_*$ must be nonnegative.

We can look at the first-order conditions in another way by defining the multiplier of an inactive constraint to be zero. Then we may form an m-vector λ_* of the multipliers associated with *all* the constraints. The conditions $\nabla f(x_*) = \hat{A}^T \hat{\lambda}_*$ and $\hat{\lambda}_* \geq 0$ are now equivalent to the conditions $\nabla f(x_*) = A^T \lambda_*$ and $\lambda_* \geq 0$. (Each column of A corresponding to an inactive constraint is multiplied by a zero Lagrange multiplier.) The requirement that any inactive constraint have a zero Lagrange multiplier can be expressed as

$$\lambda_{*i}(a_i^T x_* - b_i) = 0, \quad i = 1, \ldots, m.$$

These are known as the *complementary slackness conditions*. They state that either a constraint is active ($a_i^T x_* - b_i = 0$) or its associated Lagrange multiplier is zero ($\lambda_{*i} = 0$).

At least one of the two must hold. The situation where *exactly* one of the two holds—either $a_i^T x_* - b_i = 0$ or $\lambda_{*i} = 0$, but not both—is called *strict complementarity*. In this case, the Lagrange multiplier for an active constraint will be positive (that is, $\hat{\lambda}_* > 0$). If strict complementarity does not hold, then some active constraint will have a Lagrange multiplier equal to zero. Such a constraint is called *degenerate*. Degeneracy is an undesirable feature, since it may cause algorithms to progress very slowly, or even to fail.

We summarize the first- and second-order necessary conditions in the following lemma.

Lemma 14.8 (Necessary Condition, Linear Inequality Constraints). *If x_* is a local minimizer of f over the set $\{ x : Ax \geq b \}$, then for some vector λ_* of Lagrange multipliers,*

- $\nabla f(x_*) = A^T \lambda_*$, *or equivalently,* $Z^T \nabla f(x_*) = 0$,
- $\lambda_* \geq 0$,
- $\lambda_*^T (Ax_* - b) = 0$, *and*
- $Z^T \nabla^2 f(x_*) Z$ *is positive semidefinite,*

where Z is a null-space matrix for the matrix of active constraints at x_*.

We also develop sufficiency conditions that guarantee that a stationary point of f is indeed a local minimizer. For the sake of clarity, we consider only the case when Z is a basis matrix for the null space of $\hat{A}$.

One might anticipate that the sufficiency conditions would require $Z^T \nabla^2 f(x_*) Z$ to be positive definite, as is the case for equality-constrained problems. Unfortunately, the situation is not so straightforward, as is demonstrated by the following example.

Example 14.9 (Anomaly in the Sufficiency Conditions). Consider the problem

$$\text{minimize} \quad f(x) = x_1^3 + x_2^2$$
$$\text{subject to} \quad -1 \leq x_1 \leq 0.$$

At the point $\bar{x} = (0, 0)^T$ the active set consists of the upper bound constraint on x_1. Writing this constraint as $-x_1 \geq 0$ we find that $\hat{A} = (-1, 0)$ is the matrix of the active constraints. Since $\nabla f(\bar{x}) = (0, 0)^T$, then $\nabla f(\bar{x}) = \hat{A}^T \hat{\lambda}$ for $\hat{\lambda} = 0$, and the first-order necessary condition is satisfied at this point. We now examine the second-order necessary conditions, using $Z = (0, 1)^T$ as a basis matrix for the null space of $\hat{A}$. Then

$$Z^T \nabla^2 f(\bar{x}) Z = (0, 1) \begin{pmatrix} 0 & 0 \\ 0 & 2 \end{pmatrix} \begin{pmatrix} 0 \\ 1 \end{pmatrix} = 2 > 0,$$

and so the reduced Hessian matrix is positive definite at $\bar{x}$. The point $\bar{x}$ is not optimal, however, since any nearby point of the form $(-\epsilon, 0)^T$ (with ϵ small and positive) has a lower objective value. ■

The culprit in this example is a degenerate constraint, namely an active constraint having Lagrange multiplier equal to zero. If there are no degenerate constraints (if strict complementarity holds), we can extend the second-order sufficiency conditions in a straightforward manner.

Lemma 14.10 (Sufficient Conditions, Linear Inequality Constraints I). *If x_* satisfies*

- $Ax_* \geq b$,
- $\nabla f(x_*) = A^T \lambda_*$,
- $\lambda_* \geq 0$,
- *strict complementarity holds, and*
- $Z^T \nabla^2 f(x_*) Z$ *is positive definite,*

then x_* is a strict local minimizer for the inequality-constrained problem.

Proof. We show that along any feasible direction at x_*, the function f is increasing. Notice first that x_* is a strict local minimizer of f on the set $\{x : \hat{A}x = \hat{b}\}$ defined by the active constraints at x_*. Therefore, along any direction p such that $\hat{A}p = 0$, the function f is increasing. Consider now a direction p such that $\hat{A}p \geq 0$, where some components of $\hat{A}p$ are strictly positive. The direction p points into the interior of the feasible region. Since $\nabla f(x_*) = A^T \lambda_* = \hat{A}^T \hat{\lambda}_*$, then

$$ p^T \nabla f(x_*) = p^T \hat{A}^T \hat{\lambda}_* > 0, $$

and p is a direction of ascent. Thus, along any feasible direction at x_*, the function f is increasing, and therefore x_* is a strict local minimizer.   □

As we have seen, if the Lagrange multiplier associated with an active constraint is positive, than a small movement into the interior of the constraint will cause an increase in the objective value. If the Lagrange multiplier is zero, however, it is impossible to predict from first-order information whether a small move into the interior of the constraint will cause an increase or a decrease in the objective value. Thus, in the case of a degenerate constraint, more stringent conditions on the Hessian are required to guarantee that a point is a local minimum. We just state the result as a lemma.

Lemma 14.11 (Sufficient Conditions, Linear Inequality Constraints II). *Let $\hat{A}_+$ be the submatrix of $\hat{A}$ corresponding to the nondegenerate active constraints at x_*, that is, those constraints whose associated multipliers are positive. Let Z_+ be a basis matrix for the null space of $\hat{A}_+$. If x_* satisfies*

- $Ax_* \geq b$,
- $\nabla f(x_*) = A^T \lambda_*$,
- $\lambda_* \geq 0$,
- $\lambda_*^T(Ax_* - b) = 0$,   *and*
- $Z_+^T \nabla^2 f(x_*) Z_+$ *is positive definite,*

then x_* is a strict local minimizer for the inequality-constrained problem.

Example 14.12 (Sufficiency Conditions in the Presence of Degenerate Constraints). For the problem in Example 14.9, $\hat{A}_+$ is an empty matrix. Hence $Z_+ = I$, and

$$ Z_+^T \nabla^2 f(\bar{x}) Z_+ = \begin{pmatrix} 0 & 0 \\ 0 & 2 \end{pmatrix}. $$

Since this matrix is not positive definite, the second-order sufficiency conditions are not satisfied at $\bar{x}$. ■

We conclude this section with an example involving a small inequality-constrained problem.

Example 14.13 (A Small Inequality-Constrained Problem). Consider the problem

$$
\begin{array}{ll}
\text{minimize} & f(x) = x_1^3 - x_2^3 - 2x_1^2 - x_1 + x_2 \\
\text{subject to} & -x_1 - 2x_2 \geq -2 \\
& x_1 \geq 0 \\
& x_2 \geq 0.
\end{array}
$$

Let $\lambda = (\lambda_1, \lambda_2, \lambda_3)^T$ be the vector of Lagrange multipliers associated with the three constraints. Then the necessary conditions for a local minimum are

$$
-x_1 - 2x_2 \geq -2, \quad x_1 \geq 0, \quad x_2 \geq 0,
$$

$$
\begin{pmatrix} 3x_1^2 - 4x_1 - 1 \\ -3x_2^2 + 1 \end{pmatrix} = \begin{pmatrix} -1 \\ -2 \end{pmatrix} \lambda_1 + \begin{pmatrix} 1 \\ 0 \end{pmatrix} \lambda_2 + \begin{pmatrix} 0 \\ 1 \end{pmatrix} \lambda_3,
$$

$$
\lambda_1(2 - x_1 - 2x_2) = 0, \quad \lambda_2 x_1 = 0, \quad \lambda_3 x_2 = 0, \quad \lambda_1 \geq 0, \lambda_2 \geq 0, \lambda_3 \geq 0.
$$

We must consider all possible combinations in the complementary slackness conditions, that is, either a constraint is active, or its Lagrange multiplier is zero.

If all three constraints are active, there are no feasible points.

If the first two constraints are active and $\lambda_3 = 0$, then

$$
\begin{array}{l}
-x_1 - 2x_2 = -2 \\
x_1 = 0
\end{array}
\implies x = \begin{pmatrix} 0 \\ 1 \end{pmatrix}
$$

so that

$$
\begin{pmatrix} -1 \\ -2 \end{pmatrix} = \begin{pmatrix} -1 \\ -2 \end{pmatrix} \lambda_1 + \begin{pmatrix} 1 \\ 0 \end{pmatrix} \lambda_2 \implies \begin{array}{l} \lambda_1 = 1 \\ \lambda_2 = 0, \end{array}
$$

and hence this point is a stationary point, where the second constraint is degenerate.

If the first and third constraints are active and $\lambda_2 = 0$, then

$$
\begin{array}{l}
-x_1 - 2x_2 = -2 \\
x_2 = 0
\end{array}
\implies x = \begin{pmatrix} 2 \\ 0 \end{pmatrix}
$$

so that

$$
\begin{pmatrix} 3 \\ 1 \end{pmatrix} = \begin{pmatrix} -1 \\ -2 \end{pmatrix} \lambda_1 + \begin{pmatrix} 0 \\ 1 \end{pmatrix} \lambda_3 \implies \begin{array}{l} \lambda_1 = -3 \\ \lambda_3 = -5, \end{array}
$$

and hence this point is not optimal.

If the second and third constraints are active and $\lambda_1 = 0$, then

$$\begin{array}{c} x_1 = 0 \\ x_2 = 0 \end{array} \quad \Longrightarrow \quad x = \begin{pmatrix} 0 \\ 0 \end{pmatrix}$$

so that

$$\begin{pmatrix} -1 \\ 1 \end{pmatrix} = \begin{pmatrix} 1 \\ 0 \end{pmatrix} \lambda_2 + \begin{pmatrix} 0 \\ 1 \end{pmatrix} \lambda_3 \quad \Longrightarrow \quad \begin{array}{ccc} \lambda_2 & = & -1 \\ \lambda_3 & = & 1, \end{array}$$

and hence this point is not optimal.

If the first constraint is active and $\lambda_2 = \lambda_3 = 0$, then $x_1 = 2 - 2x_2$ and

$$\begin{pmatrix} 12x_2^2 - 16x_2 + 3 \\ -3x_2^2 + 1 \end{pmatrix} = \begin{pmatrix} -1 \\ -2 \end{pmatrix} \lambda_1.$$

This leads to two solutions. The first is $x_2 = 1$ which gives $x_1 = 0$, which leads to the first point above. The second is $x_2 = 0.1852$ which in turn gives $x_1 = 1.6297$ and $\lambda_1 = -0.4485$, and hence this cannot be a solution.

If the second constraint is active and $\lambda_1 = \lambda_3 = 0$, then $x_1 = 0$ and

$$\begin{pmatrix} -1 \\ -3x_2^2 + 1 \end{pmatrix} = \begin{pmatrix} 1 \\ 0 \end{pmatrix} \lambda_2,$$

which implies that $\lambda_2 = -1$ and hence is not optimal.

If the third constraint is active and $\lambda_1 = 0$ and $\lambda_2 = 0$, then $x_2 = 0$ and

$$\begin{pmatrix} 3x_1^2 - 4x_1 - 1 \\ 1 \end{pmatrix} = \begin{pmatrix} 0 \\ 1 \end{pmatrix} \lambda_3,$$

which gives two solutions $x = (1.5486, 0)^T$ and $x = (-0.2153, 0)^T$. The first has $\lambda_3 = 1$, and hence it is a stationary point; the second solution is infeasible.

Finally, it is easy to see that if no constraints are active, then

$$\begin{pmatrix} 3x_1^2 - 4x_1 - 1 \\ -3x_2^2 + 1 \end{pmatrix} = \begin{pmatrix} 0 \\ 0 \end{pmatrix},$$

which gives only infeasible solutions.

It remains to verify which stationary points satisfy the second-order necessary conditions, and whether any of those satisfies the sufficiency conditions. This is left as an exercise. Since the feasible region is bounded and the objective is continuous, the problem must have a minimum. ∎

This example demonstrates the major difficulty in solving problems with inequality constraints: the combinatorial issue of determining the correct active set. Even for a small problem the number of possibilities may be large. To avoid a prohibitive amount of computation, we must devise methods that do not require the consideration of every active set. In Chapter 15 we discuss active-set methods, in which we usually need only consider a fraction of the possible active sets. Another approach, one that avoids the combinatorial difficulty entirely by moving through the interior of the feasible region, is described in Chapter 16.

Exercises

4.1. Solve the problem

$$
\begin{array}{ll}
\text{minimize} & f(x) = \tfrac{1}{2}x_1^2 + x_2^2 \\
\text{subject to} & 2x_1 + x_2 \geq 2 \\
& x_1 - x_2 \leq 1 \\
& x_1 \geq 0.
\end{array}
$$

4.2. Solve the problem

$$
\begin{array}{ll}
\text{minimize} & f(x) = -x_1^2 + x_2^2 - x_1 x_2 \\
\text{subject to} & 2x_1 - x_2 \geq 2 \\
& x_1 + x_2 \leq 4 \\
& x_1 \geq 0.
\end{array}
$$

4.3. Determine which stationary points in Example 14.13 are local minimizers.

4.4. Consider the linear program

$$
\begin{array}{ll}
\text{minimize} & f(x) = c^T x \\
\text{subject to} & Ax \geq b.
\end{array}
$$

(i) Write the first- and second-order necessary conditions for a local solution.

(ii) Show that the second-order sufficiency conditions do not hold anywhere, but that any point x_* satisfying the first-order necessary conditions is a global minimizer. (*Hint*: Show that there are no feasible directions of descent at x_*, and that this implies that x_* is a global minimizer.)

4.5. Consider the quadratic problem

$$
\begin{array}{ll}
\text{minimize} & f(x) = \tfrac{1}{2}x^T Q x - c^T x \\
\text{subject to} & Ax \geq b,
\end{array}
$$

where Q is a symmetric matrix.

(i) Write the first- and second-order necessary optimality conditions. State all assumptions that you are making.

(ii) Is it true that any local minimum to the problem is also a global minimum?

4.6. Let A be an $m \times n$ matrix whose rows are linearly independent. Prove that there exists a vector p such that $Ap = e_1$, where $e_1 = (1, 0, \ldots, 0)^T$.

4.7. Prove Lemma 14.11.

4.8. (From Avriel (1976, reprinted 2003).) Prove the following result, first proved by Gibbs in 1876. Let x_* be a solution to the problem

$$
\begin{array}{ll}
\text{minimize} & f(x) = \displaystyle\sum_{j=1}^{n} f_j(x_j) \\[2ex]
\text{subject to} & \displaystyle\sum_{j=1}^{n} x_j = 1 \\[2ex]
& x_j \geq 0, \quad j = 1, \ldots, n,
\end{array}
$$

where each f_j is differentiable. Then there exists a number η such that

$$f'_j(x_{*j}) = \eta \quad \text{if} \quad x_{*j} > 0$$
$$f'_j(x_{*j}) \geq \eta \quad \text{if} \quad x_{*j} = 0.$$

4.9. (From Avriel (1976, reprinted 2003).) Solve the problem

$$\text{minimize} \quad f(x) = \sum_{j=1}^{n} \frac{c_j}{x_j}$$

$$\text{subject to} \quad \sum_{j=1}^{n} a_j x_j = 1$$

$$x_j \geq 0, \quad j = 1, \ldots, n,$$

where $\{ a_j \}$ and $\{ c_j \}$ are positive constants.

4.10. Consider the bound-constrained problem

$$\text{minimize} \quad f(x)$$

$$\text{subject to} \quad l \leq x \leq u,$$

where l and u are vectors of lower and upper bounds, such that $l < u$. Let x_* be a local minimizer. Show that

$$\text{if } x_{*i} = l_i \quad \text{then} \quad \frac{\partial f(x_*)}{\partial x_i} \geq 0,$$

$$\text{if } x_{*i} = u_i \quad \text{then} \quad \frac{\partial f(x_*)}{\partial x_i} \leq 0,$$

$$\text{if } l_i < x_{*i} < u_i \quad \text{then} \quad \frac{\partial f(x_*)}{\partial x_i} = 0.$$

Note that the Lagrange multipliers can be obtained at no additional cost in a bound-constrained problem.

4.11. Consider the problem

$$\text{maximize} \quad f(x)$$

$$\text{subject to} \quad Ax \geq b,$$

where A is an $m \times n$ matrix, b is an m-vector, and f is a twice continuously differentiable function. Assume that the system has a feasible solution.

(i) State and prove the first-order necessary conditions for a local solution.

(ii) State and prove the second-order necessary conditions for a local solution.

(iii) State and prove the second-order sufficient conditions for a local solution. (Assume that strict complementarity holds.)

14.5   Optimality Conditions for Nonlinear Constraints

The optimality conditions for problems with nonlinear constraints are similar in form to those for problems with linear constraints. Their derivation, however, is more complicated, even though it is based on related principles. The intuition behind both derivations is the same, but in the case of nonlinear constraints different technical tools are required to give substance to this intuition. In addition, nonlinear constraints can give rise to situations that are impossible in the case of linear constraints.

The optimality conditions for nonlinearly constrained problems form the basis for algorithms for solving such problems, and so are of great importance. However, not all readers may be interested in studying the derivation of these conditions. For this reason, we state the optimality conditions first, together with some examples. We then discuss the use of these optimality conditions within optimization algorithms. Only then do we present the derivation of the optimality conditions.

14.5.1 Statement of Optimality Conditions

We present the optimality conditions separately for problems with equality and inequality constraints. It is straightforward to combine these results into a more general optimality condition (see the Exercises).

The problem with equality constraints is written in the general form

$$\text{minimize} \quad f(x)$$
$$\text{subject to} \quad g_i(x) = 0, \quad i = 1, \dots, m.$$

The problem with inequality constraints is

$$\text{minimize} \quad f(x)$$
$$\text{subject to} \quad g_i(x) \geq 0, \quad i = 1, \dots, m.$$

We assume that all the functions are twice continuously differentiable.

Some additional assumption must be made to ensure the validity of the optimality conditions. We have chosen to assume that a solution x_* to the optimization problem is a "regular" point. In the case of equality constraints this means that the gradients of the constraints $\{ \nabla g_i(x_*) \}$ are linearly independent. In the case of inequality constraints this means that the gradients of the *active* constraints at x_*, $\{ \nabla g_i(x_*) : g_i(x_*) = 0 \}$, are linearly independent.

Example 14.14 (Regularity). Consider an equality-constrained problem with the two constraints

$$g_1(x) = x_1^2 + x_2^2 + x_3^2 - 3 = 0$$
$$g_2(x) = 2x_1 - 4x_2 + x_3^2 + 1 = 0$$

at the feasible point $x_* = (1, 1, 1)^T$. The gradients of the constraints at x_* are

$$\nabla g_1(x_*) = (2 \quad 2 \quad 2)^T$$
$$\nabla g_2(x_*) = (2 \quad -4 \quad 2)^T.$$

These two gradients are linearly independent, and so x_* is a regular point.

Now consider an inequality-constrained problem with the single constraint

$$g_1(x) = (\tfrac{1}{2}x_1^2 + \tfrac{1}{2}x_2^2 - 1)^3 \geq 0$$

at the feasible point $x_* = (1, 1)^T$. The constraint is binding at this point, and its gradient is

$$\nabla g_1(x_*) = (0 \quad 0)^T.$$

Hence x_* is not a regular point. ∎

The optimality conditions are expressed in terms of the *Lagrangian function*

$$\mathcal{L}(x, \lambda) = f(x) - \sum_{i=1}^{m} \lambda_i g_i(x) = f(x) - \lambda^T g(x),$$

where λ is a vector of Lagrange multipliers, and g is the vector of constraint functions $\{\, g_i \,\}$. We state these conditions below. They are derived in Section 14.7.

Theorem 14.15 (Necessary Conditions, Equality Constraints). *Let x_* be a local minimizer of f subject to the constraints $g(x) = 0$. Let $Z(x_*)$ be a null-space matrix for the Jacobian matrix $\nabla g(x_*)^T$. If x_* is a regular point of the constraints, then there exists a vector of Lagrange multipliers λ_* such that*

- $\nabla_x \mathcal{L}(x_*, \lambda_*) = 0$, *or equivalently* $Z(x_*)^T \nabla f(x_*) = 0$, *and*
- $Z(x_*)^T \nabla^2_{xx} \mathcal{L}(x_*, \lambda_*) Z(x_*)$ *is positive semidefinite.*

Theorem 14.16 (Sufficiency Conditions, Equality Constraints). *Let x_* be a point satisfying $g(x_*) = 0$. Let $Z(x_*)$ be a basis for the null space of $\nabla g(x_*)^T$. Suppose there exists a vector λ_* such that*

- $\nabla_x \mathcal{L}(x_*, \lambda_*) = 0$, *and*
- $Z(x_*)^T \nabla^2_{xx} \mathcal{L}(x_*, \lambda_*) Z(x_*)$ *is positive definite.*

Then x_* is a strict local minimizer of f in the set $\{\, x : g(x) = 0 \,\}$. (If the reduced Hessian is negative definite, then x_* is a local maximizer of f.)

The theorems involve the Jacobian matrix $\nabla g(x_*)^T$, which is the matrix of gradients of the constraint functions (see Appendix B.4). For a system of linear equality constraints $Ax = b$, the Jacobian would be equal to A, and so the condition

$$Z(x_*)^T \nabla f(x_*) = 0,$$

or equivalently

$$\nabla_x \mathcal{L}(x_*, \lambda_*) = \nabla f(x_*) - \nabla g(x_*) \lambda_* = 0,$$

is analogous to the condition

$$Z^T \nabla f(x_*) = 0, \quad \text{or equivalently} \quad \nabla f(x_*) = A^T \lambda_*,$$

for linear equality constraints.

The second-order conditions are based on the reduced Hessian

$$Z(x_*)^T \nabla_{xx}^2 \mathcal{L}(x_*, \lambda_*) Z(x_*).$$

These conditions involve the Hessian of the Lagrangian $\mathcal{L}$, while in the case of linear constraints they involve the Hessian of the objective f. The second derivatives of linear constraints are zero, however, and so

$$\nabla_{xx}^2 \mathcal{L}(x_*, \lambda_*) = \nabla^2 f(x_*)$$

in this case. Thus, the second-order conditions for linearly constrained problems are a special case of the conditions above.

These optimality conditions are illustrated in the following example.

Example 14.17 (Optimality Conditions, Equality Constraints). Consider the problem

$$\begin{aligned} \text{minimize} \quad & f(x) = x_1^2 - x_2^2 \\ \text{subject to} \quad & x_1^2 + 2x_2^2 = 4. \end{aligned}$$

Here we have a single constraint $g(x) = x_1^2 + 2x_2^2 - 4 = 0$. The Lagrangian function is $\mathcal{L}(x, \lambda) = x_1^2 - x_2^2 - \lambda(x_1^2 + 2x_2^2 - 4)$. An optimal point must therefore satisfy

$$\begin{aligned} 2x_1 - 2\lambda x_1 &= 0 \\ -2x_2 - 4\lambda x_2 &= 0, \end{aligned}$$

together with the feasibility requirement. The first equation has two possible solutions: $x_1 = 0$ and $\lambda = 1$. If $x_1 = 0$, then from feasibility $x_2 = \pm\sqrt{2}$. In either case, the second equation implies that $\lambda = -\frac{1}{2}$. If on the other hand $\lambda = 1$, then from the second equation we get $x_2 = 0$, and from feasibility $x_1 = \pm 2$. There are four possible solutions:

$$\begin{aligned} x &= (0, \quad \sqrt{2})^T, & \lambda &= -\tfrac{1}{2}; \\ x &= (0, -\sqrt{2})^T, & \lambda &= -\tfrac{1}{2}; \\ x &= (2, \quad 0)^T, & \lambda &= 1; \\ x &= (-2, \quad 0)^T, & \lambda &= 1. \end{aligned}$$

These are all stationary points of f. We can determine which are minimizers by examining the Hessian matrix

$$\nabla_{xx}^2 \mathcal{L}(x, \lambda) = \begin{pmatrix} 2 & 0 \\ 0 & -2 \end{pmatrix} - \lambda \begin{pmatrix} 2 & 0 \\ 0 & 4 \end{pmatrix} = \begin{pmatrix} 2(1 - \lambda) & 0 \\ 0 & -2(1 + 2\lambda) \end{pmatrix}.$$

Consider the solution $x = (0, \sqrt{2})^T$ with Lagrange multiplier $\lambda = -\frac{1}{2}$. Since $\nabla g(x) = (2x_1, 4x_2)^T = (0, 4\sqrt{2})^T$ we can choose the null-space matrix $Z = Z(x) = (1, 0)^T$. Taking $\lambda = -\frac{1}{2}$ we obtain

$$Z^T \nabla_{xx}^2 \mathcal{L}(x, \lambda) Z = 3 > 0,$$

and hence the reduced Hessian is positive definite and the point is a strict local minimizer of f. Similarly, the solution $x = (0, -\sqrt{2})^T$ is also a strict local minimizer.

If we take the solution $x = (2, 0)^T$, $\lambda = 1$, then $\nabla g(x) = (2x_1, 4x_2)^T = (4, 0)^T$, and we can choose the null-space matrix $Z = (0, 1)^T$. The reduced Hessian is $Z^T \nabla^2_{xx} \mathcal{L}(x, \lambda) Z = -6 < 0$, and hence the point is a local maximizer of f. A similar conclusion holds for the point $x = (-1, 0)^T$. For this problem, all feasible points are regular points. ∎

The next theorem gives the necessary conditions for problems with inequality constraints. These conditions are sometimes called the Karush–Kuhn–Tucker conditions, the KKT conditions, or the Kuhn–Tucker conditions; see Section 14.9.

Theorem 14.18 (Necessary Conditions, Inequality Constraints). *Let x_* be a local minimum point of f subject to the constraints $g(x) \geq 0$. Let the columns of $Z(x_*)$ form a basis for the null space of the Jacobian of the active constraints at x_*. If x_* is a regular point for the constraints, then there exists a vector of Lagrange multipliers λ_* such that*

- $\nabla_x \mathcal{L}(x_*, \lambda_*) = 0$, *or equivalently* $Z(x_*)^T \nabla f(x_*) = 0$,
- $\lambda_* \geq 0$,
- $\lambda_*^T g(x_*) = 0$, *and*
- $Z(x_*)^T \nabla^2_{xx} \mathcal{L}(x_*, \lambda_*) Z(x_*)$ *is positive semidefinite.*

The condition $\lambda_*^T g(x_*) = 0$ is the *complementary slackness condition*. Since the vectors λ_* and $g(x_*)$ are both nonnegative, it implies that $\lambda_{*i} g_i(x_*) = 0$ for each i. This means that either a constraint is active, or its associated Lagrange multiplier is zero. In particular, any inactive constraint has a Lagrange multiplier of zero. If the multipliers corresponding to the active constraints are all positive, then we have *strict complementarity*; otherwise, if a Lagrange multiplier corresponding to an active constraint is zero, the constraint is said to be *degenerate*.

The second-order sufficiency conditions for a local minimum point are stated below.

Theorem 14.19 (Sufficiency Conditions, Inequality Constraints). *Let x_* be a point satisfying $g(x_*) \geq 0$. Suppose there exists a vector λ_* such that*

- $\nabla_x \mathcal{L}(x_*, \lambda_*) = 0$,
- $\lambda_* \geq 0$,
- $\lambda_*^T g(x_*) = 0$, *and*
- $Z_+(x_*)^T \nabla^2_{xx} \mathcal{L}(x_*, \lambda_*) Z_+(x_*)$ *is positive definite,*

where Z_+ is a basis for the null space of the Jacobian matrix of the nondegenerate constraints (the active constraints with positive Lagrange multipliers) at x_*. Then x_* is a strict local minimizer of f in the set $\{ x : g(x) \geq 0 \}$.

These optimality conditions are illustrated in the following example.

Example 14.20 (Optimality Conditions, Inequality Constraints). Consider the problem

$$\text{minimize} \quad f(x) = x_1$$
$$\text{subject to} \quad (x_1 + 1)^2 + x_2^2 \geq 1$$
$$x_1^2 + x_2^2 \leq 2.$$

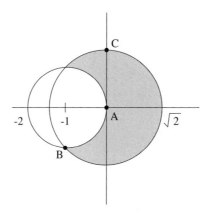

Figure 14.3. *Problem with nonlinear inequalities.*

We shall test whether the points $A = (0, 0)^T$, $B = (-1, -1)^T$, and $C = (0, \sqrt{2})^T$ are optimal (see Figure 14.3).

Rearranging the constraints to the "$\geq$" form we obtain

$$\mathcal{L}(x, \lambda) = x_1 - \lambda_1((x_1 + 1)^2 + x_2^2 - 1) + \lambda_2(x_1^2 + x_2^2 - 2).$$

Therefore

$$\nabla_x \mathcal{L}(x, \lambda) = \begin{pmatrix} 1 - 2\lambda_1(x_1 + 1) + 2\lambda_2 x_1 \\ -2\lambda_1 x_2 + 2\lambda_2 x_2 \end{pmatrix}$$

and

$$\nabla_{xx}^2 \mathcal{L}(x, \lambda) = \begin{pmatrix} 2(\lambda_2 - \lambda_1) & 0 \\ 0 & 2(\lambda_2 - \lambda_1) \end{pmatrix}.$$

At the point A, only the first constraint is active, and hence $\lambda_2 = 0$. Solving for λ_1 we obtain

$$\begin{matrix} 1 - 2\lambda_1 = 0 \\ 0 = 0 \end{matrix} \quad \Longrightarrow \quad \lambda_1 = \tfrac{1}{2}.$$

Therefore, this is a candidate for a local minimizer. Taking $Z = (0, 1)^T$ as a basis matrix for the null space of the Jacobian matrix $(2, 0)$, we get

$$Z^T \nabla_{xx}^2 \mathcal{L}(x, \lambda) Z = (0, 1) \begin{pmatrix} -1 & 0 \\ 0 & -1 \end{pmatrix} \begin{pmatrix} 0 \\ 1 \end{pmatrix} = -1,$$

and hence the reduced Hessian matrix is negative definite, and the sufficiency conditions are not satisfied. This point is not a local maximizer since $\lambda_1 > 0$ (see Exercise 5.9).

At the point B both constraints are active. Solving for the Lagrange multipliers we obtain

$$\begin{matrix} 1 - 2\lambda_2 = 0 \\ 2\lambda_1 - 2\lambda_2 = 0 \end{matrix} \quad \Longrightarrow \lambda_1 = \lambda_2 = \tfrac{1}{2}.$$

Therefore the point satisfies the first-order necessary condition for optimality. Moving to the sufficiency conditions, we note that the null-space matrix for the Jacobian is empty.

Therefore the sufficiency conditions are trivially satisfied, and the point is a strict local minimizer.

At the point $C = (0, \sqrt{2})^T$, only the second constraint is active, and hence $\lambda_1 = 0$. Solving for λ_2 we obtain

$$1 + 2\lambda_2(0) = 0$$
$$2\lambda_2\sqrt{2} = 0.$$

This system is inconsistent. Hence the first-order necessary condition is not satisfied and the point is not optimal.   ∎

As in the linearly constrained case (see Section 14.3), the Lagrange multipliers provide a measure of the sensitivity of the optimal objective value to changes in the constraints. This shows up in the optimality conditions which include the requirement that $\lambda_* \geq 0$ for the inequality-constrained problem. The magnitude of the multipliers also has meaning, with a large multiplier indicating a constraint more sensitive to changes in its right-hand side.

The following example shows that if the regularity condition is not satisfied at a local minimizer, the first-order necessary condition for optimality may not hold.

Example 14.21 (Regularity Condition Not Satisfied). Consider the problem

$$\begin{aligned} \text{minimize} \quad & f(x) = 3x_1 + 4x_2 \\ \text{subject to} \quad & (x_1 + 1)^2 + x_2^2 = 1 \\ & (x_1 - 1)^2 + x_2^2 = 1. \end{aligned}$$

The solution to this problem is $x_* = (0, 0)^T$, which is also the only feasible point. The gradients of the constraints at x_* are $(2, 0)^T$ and $(-2, 0)^T$, and thus are linearly dependent. Setting the gradient of the Lagrangian with respect to x equal to zero yields

$$3 - 2\lambda_1 + 2\lambda_2 = 0$$
$$4 = 0.$$

This is an inconsistent system. Hence there are no multipliers λ_1 and λ_2 for which the gradient of the Lagrangian is zero, even though the point is optimal.   ∎

Exercises

5.1. Consider the constraint $g_1(x) = (\frac{1}{2}x_1^2 + \frac{1}{2}x_2^2 - 1)^3 \geq 0$ of Example 14.14.

   (i) Determine which feasible points are regular points and which are not.

   (ii) Consider the constraint $\hat{g}(x) = \frac{1}{2}x_1^2 + \frac{1}{2}x_2^2 - 1 \geq 0$. Show that the constraint $\hat{g}(x) \geq 0$ is equivalent to the constraint $g_1(x) \geq 0$. Show also that every feasible point for $\hat{g}(x) \geq 0$ is also a regular point of this constraint. Thus regularity is not a result of the geometry of the feasible region, but a result of its algebraic representation.

5.2. Solve the problem
$$\text{minimize} \quad f(x) = c^T x$$
$$\text{subject to} \quad \sum_{i=1}^{n} x_i = 0$$
$$\sum_{i=1}^{n} x_i^2 = 1.$$

5.3. Solve the problem
$$\text{minimize} \quad f(x) = x_1 + x_2$$
$$\text{subject to} \quad \log x_1 + 4 \log x_2 \geq 1.$$

5.4. Determine the minimizers/maximizers of the following functions subject to the given constraints. If helpful, graph the feasible set to guess minimizers or maximizers. However, you must verify your guess using the optimality conditions.

  (i) $f(x_1, x_2) = x_2$ subject to $x_1^2 + x_2^2 \leq 1$, $-x_1 + x_2^2 \leq 0$, and $x_1 + x_2 \geq 0$.
  (ii) $f(x_1, x_2) = x_1^2 + 2x_2^2$ subject to $x_1^3 + x_2^3 \leq 1$ and $x_1^2 + x_2^2 \geq 1$.

5.5. Consider the problem
$$\text{maximize} \quad f(x) = c^T x$$
$$\text{subject to} \quad x^T Q x \leq 1,$$
where Q is a positive-definite symmetric matrix.

  (i) Solve the problem. What is the optimal objective value?
  (ii) What is the solution when the objective function is to be minimized?

5.6. Let Q be an $n \times n$ symmetric matrix.

  (i) Find all stationary points of the problem
$$\text{maximize} \quad f(x) = x^T Q x$$
$$\text{subject to} \quad x^T x = 1.$$

  (ii) Determine which of the stationary points are global maximizers.
  (iii) How do your results in part (i) change if the constraint is replaced by
$$x^T A x \leq 1,$$

  where A is positive definite?

5.7. Use the optimality conditions to find all local solutions to the problem
$$\text{minimize} \quad f(x) = x_1 + x_2$$
$$\text{subject to} \quad (x_1 - 1)^2 + x_2^2 \leq 2$$
$$(x_1 + 1)^2 + x_2^2 \geq 2.$$

5.8. For the problem in Example 14.20, perturb the right-hand side of the first constraint:
$$(x_1 + 1)^2 + x_2^2 \geq 1 + \delta.$$

Solve the perturbed problem (assume that δ is "small"). Compare the optimal objective value to the value predicted by λ_* for the original problem.

5.9. Let x_* be a local maximizer of f subject to the constraints $g_i(x) \geq 0$.

    (i) Use the results of Theorem 14.18 to prove that if x_* is a regular point for the constraints, then there exists a vector of Lagrange multipliers $\lambda_* \leq 0$ such that $\nabla_x \mathcal{L}(x_*, \lambda_*) = 0$ and $\lambda_*^T g(x_*) = 0$.

    (ii) Let $Z(x_*)$ be a basis for the null space of the active constraints at x_*. Prove that $Z(x_*)^T \nabla_{xx}^2 \mathcal{L}(x_*, \lambda_*) Z(x_*)$ must be negative semidefinite.

5.10. Consider the problem

$$\begin{aligned} \text{minimize} \quad & f(x) \\ \text{subject to} \quad & g_i(x) = 0, \quad i = 1, \ldots, l \\ & g_i(x) \geq 0, \quad i = l+1, \ldots, m. \end{aligned}$$

    (i) Derive a definition for a regular point of the constraints for this problem.

    (ii) State the first- and second-order necessary conditions for optimality.

    (iii) State the second-order sufficiency conditions for optimality at a feasible point.

5.11. Consider the problem

$$\text{minimize} \quad \max_{1 \leq i \leq m} f_i(x).$$

This problem is called a *minimax* problem.

    (i) Formulate the minimax problem as a constrained optimization problem.

    (ii) Use the optimality conditions for the constrained optimization problem to derive the optimality conditions for the minimax problem.

14.6   Preview of Methods

In the unconstrained case, we derived methods for solving

$$\text{minimize} \quad f(x)$$

by applying Newton's method to the optimality condition

$$\nabla f(x) = 0.$$

Convergence was guaranteed by insisting that at each iteration the value of the objective function improve:

$$f(x_{k+1}) < f(x_k).$$

Methods for constrained problems can be derived in a similar way, although guaranteeing convergence in the constrained case is more difficult. The goal in this section is to indicate some of the difficulties that arise. The next two chapters will provide considerably more detail.

We focus our attention here on the equality-constrained problem

$$\begin{aligned} \text{minimize} \quad & f(x) \\ \text{subject to} \quad & g(x) = 0. \end{aligned}$$

Here g is a vector of m functions g_i.

The Lagrangian function in this case is

$$\mathcal{L}(x, \lambda) = f(x) - \sum_{i=1}^{m} \lambda_i g_i(x) = f(x) - \lambda^T g(x).$$

The optimality conditions can be expressed as conditions on the derivatives of the Lagrangian:

$$\nabla_x \mathcal{L}(x_*, \lambda_*) = \nabla f(x_*) - \nabla g(x_*)\lambda_* = 0$$
$$\nabla_\lambda \mathcal{L}(x_*, \lambda_*) = -g(x_*) = 0,$$

where $\nabla g(x_*)$ is the matrix with columns $\nabla g_i(x_*)$ (i.e., the transpose of the Jacobian matrix at x_*). We can summarize these conditions compactly as $\nabla \mathcal{L}(x_*, \lambda_*) = 0$. Thus a local minimizer is a stationary point of the Lagrangian function. *(for equality Constrained problems)*

One way to determine a solution to the optimization problem would be to apply Newton's method to the system of nonlinear equations

$$\nabla \mathcal{L}(x, \lambda) = 0,$$

where the gradient is taken with respect to both x and λ. In fact, the formulas for many methods can be interpreted as doing this, or some variant of this.

The formula for Newton's method is

$$\begin{pmatrix} x_{k+1} \\ \lambda_{k+1} \end{pmatrix} = \begin{pmatrix} x_k \\ \lambda_k \end{pmatrix} + \begin{pmatrix} p_k \\ v_k \end{pmatrix},$$

where p_k and v_k are obtained as the solution to the linear system

$$\nabla^2 \mathcal{L}(x_k, \lambda_k) \begin{pmatrix} p_k \\ v_k \end{pmatrix} = -\nabla \mathcal{L}(x_k, \lambda_k).$$

This linear system has the form

$$\begin{pmatrix} \nabla_{xx}^2 \mathcal{L}(x_k, \lambda_k) & -\nabla g(x_k) \\ -\nabla g(x_k)^T & 0 \end{pmatrix} \begin{pmatrix} p_k \\ v_k \end{pmatrix} = \begin{pmatrix} -\nabla_x \mathcal{L}(x_k, \lambda_k) \\ g(x_k) \end{pmatrix}.$$

As in the unconstrained case, if a method is to converge at a superlinear rate, the search direction for the method must approach the solution of this linear system in the limit as the solution to the optimization problem is approached.

This linear system plays a fundamental role in the "local" behavior of the methods, that is, their behavior near the solution. On its own, however, it is inadequate to define a method since its "global" behavior is inadequate. As we saw in Chapter 11, there is no guarantee that convergence to a local minimum will be obtained. The first-order conditions define a stationary point, and this need not be a constrained minimizer. In addition, there is the danger that the iterates could diverge. Some globalization strategy is required to guarantee convergence to a local solution.

It might be hoped that techniques from the unconstrained case could be applied directly. For example, we might insist that

$$\mathcal{L}(x_{k+1}, \lambda_{k+1}) < \mathcal{L}(x_k, \lambda_k).$$

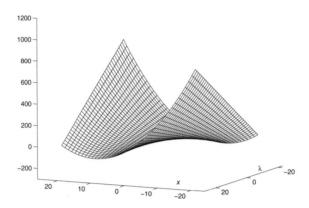

Figure 14.4. *Lagrangian function.*

This would only be satisfactory if the solution (x_*, λ_*) were a local minimizer of the Lagrangian function with respect to both x and λ. Unfortunately (x_*, λ_*) is in general a saddle point of the Lagrangian—not a minimizer—as the following example shows.

Example 14.22 (Stationary Point of the Lagrangian). Consider the one-dimensional problem

$$\text{minimize} \quad f(x) = x^2$$
$$\text{subject to} \quad x = 1.$$

The solution to this problem is $x_* = 1$, with Lagrange multiplier $\lambda_* = 2$. The Lagrangian function is

$$\mathcal{L}(x, \lambda) = x^2 - \lambda(x - 1).$$

Since $\nabla \mathcal{L}(x, \lambda) = (2x - \lambda, -(x - 1))^T$, then $\nabla \mathcal{L}(1, 2) = (0, 0)^T$, and indeed (x_*, λ_*) is a stationary point of the Lagrangian. The Hessian matrix of the Lagrangian is

$$\nabla^2 \mathcal{L}(x_*, \lambda_*) = \begin{pmatrix} 2 & -1 \\ -1 & 0 \end{pmatrix}.$$

This is an indefinite matrix. Thus (x_*, λ_*) is a saddle point of the Lagrangian function. This is illustrated in Figure 14.4. ∎

We will describe two general approaches that are used to guarantee convergence. The first attempts to ensure that x_k is feasible at every iteration. Then convergence can be guaranteed, as in the unconstrained case, by insisting that

$$\mathcal{L}(x_{k+1}, \lambda_{k+1}) = f(x_{k+1}) < f(x_k) = \mathcal{L}(x_k, \lambda_k).$$

If only feasible points are considered, then the solution x_* will be a local minimizer of the Lagrangian. When the constraints are linear it is not difficult to maintain feasibility at every iteration, but when nonlinear constraints are present this idea must be interpreted more freely. This approach is the topic of Chapter 15.

The second approach constructs a new function, related to the Lagrangian, that (ideally) has a minimum at (x_*, λ_*). This new function can be considered as "distorting" the Lagrangian at infeasible points so as to create a minimum at (x_*, λ_*). Unconstrained minimization techniques can then be applied to the new function. This approach can make it easier to guarantee convergence to a local solution, but there is the danger that the local convergence properties of the method can be damaged. The "distortion" of the Lagrangian function can lead to a "distortion" in the Newton equations for the method. Hence the behavior of the method near the solution may be poor unless care is taken. This approach is the topic of Chapter 16.

These two general approaches, along with all the possible methods within each approach, lead to many different algorithms for solving constrained optimization problems. It is difficult to identify an ideal method, in part because there is no clear way to determine whether the new point x_{k+1} is better than the current point x_k.

The difficulty arises because two criteria are used to compare two successive iterates: the value of the objective function, and the infeasibility in the constraints. If all the iterates are feasible points, then the latter criterion is irrelevant and the objective function can be used to measure progress towards the solution. Without that, it is necessary to balance the two criteria to compare two estimates of the solution, and there is no simple way to do this.

One approach is to use auxiliary *merit functions* as an indirect way to measure progress and guarantee convergence. Consider an optimization problem of the form

$$\text{minimize} \quad f(x)$$
$$\text{subject to} \quad g(x) = 0.$$

An example of a merit function for this problem is

$$\mathcal{M}(x) = f(x) + \rho \|g(x)\|, \quad \rho > 0.$$

(Here, the norm could be any of the norms discussed in Appendix A.3.) This merit function reflects the combined decrease of the two criteria involved in minimization: the objective function and the infeasibility.

The parameter $\rho > 0$ specifies the relative importance of one criterion with respect to the other. For example, if ρ is large, then the merit function emphasizes the importance of infeasibility when comparing two different iterates. The best choice of ρ is typically unknown. Practical algorithms usually adjust ρ in the course of solving an optimization problem.

It can be difficult to choose an appropriate merit function and, once one is chosen, to determine an appropriate value of the parameter ρ. If the choices are inappropriate, the merit function may reject good choices of $x_k + \alpha p$ and even prevent the algorithm from achieving its ideal rate of convergence. That is, a poor choice of merit function might reject the iterates generated by Newton's method, interfere with the quadratic convergence rate, and result in a method that converges only at a slow linear rate.

For these reasons, methods have been developed that replace the merit function with a device called a *filter*. Filter methods assess iterates by explicitly using the objective function and the infeasibility as separate criteria, without combining them into a single value. The filter keeps track of a *set* of iterates x_k that are the best found so far by the algorithm. Because the iterates are compared using two separate criteria, there may be no "best" iterate. For example, suppose that we have two iterates x_1 and x_5, where x_1 has an objective value of

5.2 and the norm of the infeasibility is 2.7, and x_5 has objective value of 3.1 and infeasibility 8.2. Then x_5 is better in terms of objective value but worse in terms of infeasibility. In this case, the filter would keep track of both iterates. Filter methods ameliorate many of the difficulties associated with merit functions, but they require a more complicated approach to assess the successive estimates of the solution.

Merit functions, filters, and related techniques are discussed in greater detail in the next two chapters.

There is a further difficulty, and that is the difficulty of finding a point that satisfies the constraints. If the constraints are linear, then it is possible to find a feasible point by solving a linear program (see Section 5.5). In the nonlinear case, finding a feasible point requires solving a system of nonlinear equations and inequalities, a problem that is as difficult in general as the original optimization problem. Methods have been developed whose purpose is to find a feasible point, but they are not guaranteed to succeed. Although unattractive, it may be necessary to try a method repeatedly with varying starting points to obtain a feasible point.

Nonlinear constraints can define complicated feasible regions. A constraint might define a single point

$$(x_1 - 5)^2 + (x_2 - 3)^2 = 0,$$

a curve of points

$$(x_1 - 5)^2 + (x_2 - 3)^2 = 1,$$

a discrete set of points

$$\sin(\pi x_1) = 0,$$

or no points

$$x_1^2 + 1 = 0.$$

Small changes in the coefficients in a nonlinear constraint can lead to qualitative differences in the shape of the feasible region. This further exacerbates the difficulty of solving nonlinear optimization problems.[13]

Exercises

6.1. Consider the equality-constrained problem

$$\begin{aligned} \text{minimize} \quad & f(x) \\ \text{subject to} \quad & g(x) = 0. \end{aligned}$$

Let (x_*, λ_*) be the optimal solution vector and the associated Lagrange multipliers. Assume that the regularity conditions and the second-order sufficiency conditions hold at x_*. Prove that $\mathcal{L}(x_*, \lambda_*)$ is a minimizer of $\mathcal{L}(x, \lambda)$ in the direction $(p^T, v^T)^T$, when p is a vector in the null space of $\nabla g(x_*)$ and $v = 0$. Prove also that $\mathcal{L}(x_*, \lambda_*)$ is a maximizer of $\mathcal{L}(x, \lambda)$ in the direction $(p^T, v^T)^T$, when $p = 0$ and v is in the range space of $\nabla g(x_*)^T$. Hence (x_*, λ_*) is a saddle point of the Lagrangian.

---

[13]It is possible to omit reading the rest of this chapter without loss of continuity.

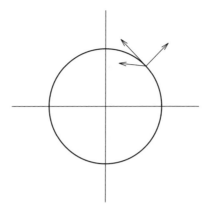

Figure 14.5. *No feasible directions.*

14.7 Derivation of Optimality Conditions for Nonlinear Constraints

We now examine the derivation of the optimality conditions for problems with nonlinear constraints. We again begin with a problem that has equality constraints only:

$$\text{minimize} \quad f(x)$$
$$\text{subject to} \quad g_i(x) = 0, \quad i = 1, \ldots, m.$$

Each of the functions f and g_i is assumed to be twice continuously differentiable. If we define $g(x)$ as the vector of constraint functions $\{ g_i(x) \}$, then the problem is to minimize $f(x)$ subject to $g(x) = 0$. The set of points x such that $g(x) = 0$ is called a *surface*.

We derive first- and second-order optimality conditions for this problem. The main difficulty is the characterization of small movements that maintain feasibility. In the linear case, such movements can be completely represented in terms of feasible directions (see Chapter 3). In the nonlinear case, this may not be possible. For example, consider the nonlinear equality constraint $x_1^2 + x_2^2 = 2$, and let x be any feasible point, say $x = (1, 1)^T$. Any small step taken from x along any direction will result in the loss of feasibility (see Figure 14.5). Thus there are no feasible directions at this point, or at any other feasible point. To define small movements that maintain feasibility, we will use *feasible curves*.

A curve is a set of points $\{ x = x(t) : t_0 \le t \le t_1 \}$. It is termed a *feasible curve* with respect to the surface $g(x) = 0$ if $g(x(t)) = 0$ for all such t. A curve passing through the point x_* satisfies $x(t_*) = x_*$ for some $t_0 \le t_* \le t_1$ (by a shift of the parameter t we can always assume that $t_* = 0$).[14] The first derivative of $x(t)$ is the vector $x'(t) = d(x(t))/dt$; this is the tangent to the curve at $x(t)$. The second derivative is the vector $x''(t) = d^2(x(t))/dt^2$. In the following discussion we shall consider only curves that have two continuous derivatives.

---

[14]In this section, we assume that any feasible curve $x(t)$ through x_* satisfies $x(0) = x_*$.

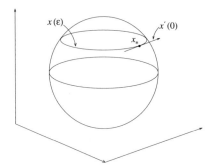

Figure 14.6. *Feasible curve.*

Example 14.23 (Feasible Curve). Consider the constraint $g(x) = x_1^2 + x_2^2 + x_3^2 - 3 = 0$, and consider the curve

$$x(t) = \begin{pmatrix} \sqrt{2}\cos(t + \pi/4) \\ \sqrt{2}\sin(t + \pi/4) \\ 1 \end{pmatrix}, \quad -\pi \le t \le \pi.$$

Then $x(0) = (1, 1, 1)^T$ and $g(x(t)) = 2\cos^2(t + \pi/4) + 2\sin^2(t + \pi/4) + 1 - 3 = 0$. Hence $x(t)$ is a feasible curve passing through the point $(1, 1, 1)^T$. The tangent to the curve at $(1, 1, 1)^T$ is (see also Figure 14.6)

$$x'(0) = \frac{d}{dt}x(t)\Big|_{t=0} = \begin{pmatrix} -\sqrt{2}\sin(t + \pi/4) \\ \sqrt{2}\cos(t + \pi/4) \\ 0 \end{pmatrix}_{t=0} = \begin{pmatrix} -1 \\ 1 \\ 0 \end{pmatrix}. \quad \blacksquare$$

Suppose that x_* is a local solution of the optimization problem. Then x_* is a local minimizer of f along any feasible curve passing through x_*. Let $x(t)$ be any such curve with $x(0) = x_*$. Then $t = 0$ is a local minimizer of the one-dimensional function $f(x(t))$, and the derivative of $f(x(t))$ with respect to t must vanish at $t = 0$. Using the chain rule we obtain

$$\frac{d}{dt}f(x(t))\Big|_{t=0} = x'(t)^T\nabla f(x(t))\Big|_{t=0} = x'(0)^T\nabla f(x_*) = 0.$$

Thus, if x_* is a local minimizer of f, then

$$x'(0)^T\nabla f(x_*) = 0 \quad \text{for all feasible curves } x(t) \text{ through } x_*. \tag{14.3}$$

Define

$$T(x_*) = \big\{\, p : p = x'(0) \text{ for some feasible curve } x(t) \text{ through } x_* \,\big\}.$$

This is the set of all tangents to feasible curves through x_*. Assume for convenience that $0 \in T(x_*)$. The set has the property that if $p \in T(x_*)$, then $\alpha p \in T(x_*)$ for any nonnegative scalar α (see the Exercises). A set with this property is called a *cone*, and for this reason

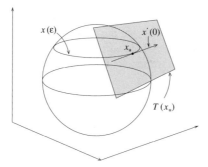

Figure 14.7. *Tangent cone.*

$T(x_*)$ is sometimes called the *tangent cone* at x_*. The tangent cone at the point $(1, 1, 1)^T$ in Example 14.23 is shown in Figure 14.7. It is parallel to the tangent plane at $(1, 1, 1)^T$ but passes through the origin.

From equation (14.3) we obtain a condition for optimality of a feasible point x_*:

$$p^T \nabla f(x_*) = 0 \quad \text{for all } p \in T(x_*).$$

In this form, the optimality condition is not yet practical, since it is not always easy to represent the set of all feasible curves explicitly. We shall develop an alternative characterization of the tangent cone. To this end, we notice that $g_i(x(t))$ is a constant function of t (it is zero for all t), and hence its derivative with respect to t vanishes everywhere, i.e., $\frac{d}{dt} g_i(x(t)) = 0$. Using the chain rule we obtain

$$x'(t)^T \nabla g_i(x(t)) = 0.$$

In particular, at $t = 0$ we obtain $x'(0)^T \nabla g_i(x_*) = 0$. Since this is true for all feasible arcs through x_*, we obtain

$$p^T \nabla g_i(x_*) = 0 \quad \text{for all } p \in T(x_*).$$

The equation above holds for each constraint i. It will be useful to define $A(x_*)$ as the $m \times n$ matrix whose ith row is $\nabla g_i(x_*)^T$. This is the Jacobian matrix of g at x_*. The equation above can be written as $A(x_*)p = 0$, so that any vector in the tangent cone at x_* also lies in the null space of the Jacobian matrix at x_*:

$$p \in T(x_*) \Rightarrow p \in \mathcal{N}(A(x_*)).$$

Hence the tangent cone at a point is contained in the null space of the Jacobian matrix at the point.

Example 14.24 (Null Space of the Jacobian). Consider the problem in Example 14.23. At $x_* = (1, 1, 1)^T$ we have $\nabla g(x_*)^T = (2, 2, 2)^T$. Thus any vector p in the tangent cone must satisfy $p^T \nabla g(x_*) = 2p_1 + 2p_2 + 2p_3 = 0$, that is, $p_1 + p_2 + p_3 = 0$. In this example, the tangent cone and the null space of the Jacobian are both equal to the set $\{p : p_1 + p_2 + p_3 = 0\}$. ∎

In the previous example, the tangent cone and the null space of the Jacobian were equal. It can be difficult to characterize the tangent cone, but it is easy to compute the Jacobian matrix and generate its associated null space. Hence it would be useful if these two sets were always equal. Unfortunately, this is not always the case, as the next example shows.

Example 14.25 $(T(x_*) \neq \mathcal{N}(A(x_*)))$. Consider the constraint $g(x) = (\frac{1}{2}x_1^2 + \frac{1}{2}x_2^2 - 1)^2 = 0$. The feasible set is a circle of radius $\sqrt{2}$. The tangent cone at the point $x_* = (1, 1)^T$ is the set $T(x_*) = \{ p : p_1 + p_2 = 0 \}$. Since $\nabla g(x) = (2(\frac{1}{2}x_1^2 + \frac{1}{2}x_2^2 - 1)x_1, 2(\frac{1}{2}x_1^2 + \frac{1}{2}x_2^2 - 1)x_2)^T$, the Jacobian matrix at x_* is $A(x_*) = (0, 0)$. Therefore the null space of the Jacobian is $\mathcal{N}(A(x_*)) = \Re^2$, and $T(x_*) \neq \mathcal{N}(A(x_*))$. ∎

Luckily, examples such as the above are uncommon. In the majority of problems the tangent cone at a feasible point is indeed equal to the null space of the Jacobian matrix at the point. One condition that guarantees this is regularity, that is, the assumption that the gradient vectors $\nabla g_i(x_*)$, $i = 1, \ldots, m$, are linearly independent (or equivalently, that their Jacobian matrix has full row rank). In the next lemma we prove that if x_* is a regular point, then $T(x_*) = \mathcal{N}(A(x_*))$.

Lemma 14.26. *If x_* is a regular point of the constraints, then $T(x_*) = \mathcal{N}(A(x_*))$.*

Proof. We need only show that $p \in \mathcal{N}(A(x_*))$ implies that $p \in T(x_*)$; that is, there exists some feasible curve $x(t)$ through x_* satisfying $x'(0) = p$.

To prove the existence of a feasible curve we shall use the implicit function theorem (see Appendix B.9). Let y be an m-dimensional vector, and consider the following system of nonlinear equations in y and t:

$$g(x_* + tp + \nabla g(x_*)y) = 0.$$

The system has a solution at $(\hat{y}, \hat{t}) = (0, 0)$. Its Jacobian with respect to y at this point is

$$\nabla g(x_*)^T \nabla g(x_* + tp + \nabla g(x_*)y)|_{(y,t)=(0,0)} = \nabla g(x_*)^T \nabla g(x_*),$$

which by the regularity assumption is nonsingular. Therefore, by the implicit function theorem, there exists a continuously differentiable function $y = y(t)$ in a neighborhood of $t = 0$ satisfying

$$g(x_* + tp + \nabla g(x_*)y(t)) = 0.$$

Letting $x(t) = x_* + tp + \nabla g(x_*)y(t)$, we obtain that $x(t)$ is a feasible curve through x_* with $x(0) = x_*$. It remains only to show that $x'(0) = p$. From the formula for $x(t)$ we have that $x'(0) = p + \nabla g(x_*)y'(0)$, so we need to show that the second term is zero. Since $x(t)$ is a feasible curve it satisfies $\nabla g(x_*)^T x'(0) = 0$. Hence

$$\nabla g(x_*)^T p + \nabla g(x_*)^T \nabla g(x_*) y'(0) = 0.$$

The first term above is zero because $p \in \mathcal{N}(A(x_*))$. The lemma now follows because of the regularity assumption. ☐

If we assume that a local minimizer is a regular point, we can obtain a more useful optimality condition. Let x_* be a local solution that satisfies the regularity condition. Then any vector $p \in \mathcal{N}(A(x_*))$ is also in $T(x_*)$. It follows from (14.7) that

$$p^T \nabla f(x_*) = 0 \quad \text{for all } p \in \mathcal{N}(A(x_*)).$$

If $Z(x_*)$ is a null-space matrix for $A(x_*)$, then

$$Z(x_*)^T \nabla f(x_*) = 0.$$

This is the first-order necessary condition for optimality. It states that the reduced gradient at a local minimum must be zero. We caution here that the same condition is also satisfied at a local maximum point. The reduced gradient may also be zero at a point that is neither a local maximum nor a local minimum point, that is, at a saddle point.

As in the linear case, we can show that the reduced gradient is zero if and only if there exists an m-dimensional vector λ_* such that

$$\nabla f(x_*) = A(x_*)^T \lambda_* = \sum_{i=1}^{m} \lambda_{*i} \nabla g_i(x_*).$$

This is an equivalent statement of the first-order necessary condition for optimality. The coefficients $\{\lambda_{*i}\}$ are the Lagrange multipliers.

We now derive the second-order conditions for optimality. Recall that if x_* is a local minimizer, then x_* is a local minimizer along any feasible curve passing through x_*. Let $x(t)$ be any such curve with $x(0) = x_*$. Then since $t = 0$ is a local minimizer of the function $f(x(t))$, the second derivative of $f(x(t))$ with respect to t must be nonnegative at $t = 0$. Using the chain rule we obtain

$$\frac{d^2}{dt^2} f(x(t)) = \frac{d}{dt}\left[x'(t)^T \nabla f(x(t))\right] = x'(t)^T \nabla^2 f(x(t))x'(t) + \nabla f(x(t))^T x''(t).$$

Hence

$$\frac{d^2}{dt^2} f(x(0)) = p^T \nabla^2 f(x_*)p + \nabla f(x_*)^T x''(0) \geq 0,$$

where $p = x'(0)$ is the tangent to the curve at x_*. In the expression above, the term $\nabla f(x_*)^T x''(0)$ does not necessarily vanish. Therefore the second derivative along an arc depends not only on the Hessian of the objective, but also on the curvature of the constraints (that is, on the term $x''(0)$).

To transform this into a more useful condition, it will be convenient to get rid of the term involving $x''(0)$. To do this, we notice that $g_i(x(t))$ is constant, so its second derivative with respect to t must vanish for all t, in particular at $t = 0$. Using the chain rule we obtain

$$p^T \nabla^2 g_i(x_*)p + \nabla g_i(x_*)^T x''(0) = 0.$$

We can multiply the last equality by λ_{*i} and sum over all i. If we subtract the result from the previous inequality, then, because $\nabla_x \mathcal{L}(x_*, \lambda_*) = 0$, the term involving $x''(0)$ will be eliminated. The final result is that

$$p^T \left[\nabla^2 f(x_*) - \sum_{i=1}^{m} \lambda_{*i} \nabla^2 g_i(x_*) \right] p \geq 0$$

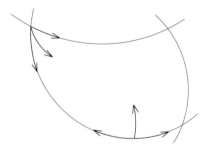

Figure 14.8. *Feasible arcs.*

for all tangent vectors $p \in T(x_*)$. The term in brackets is the Hessian of $\mathcal{L}$ with respect to x at the point (x, λ). Therefore

$$p^T \left[\nabla_{xx}^2 \mathcal{L}(x_*, \lambda_*) \right] p \geq 0$$

for all tangent vectors $p \in T(x_*)$. Under the regularity assumption, this inequality will hold for any p in $\mathcal{N}(A(x_*))$. Consequently, the reduced Hessian $Z(x_*)^T \nabla_{xx}^2 \mathcal{L}(x_*, \lambda_*) Z(x_*)$ must be positive semidefinite. This is the second-order necessary condition for optimality.

The proof of the sufficiency conditions uses similar techniques.

Finally, we consider a problem with nonlinear inequality constraints:

$$\begin{aligned} \text{minimize} \quad & f(x) \\ \text{subject to} \quad & g_i(x) \geq 0, \quad i = 1, \dots, m. \end{aligned}$$

Optimality conditions can be derived by combining the ideas developed for problems with nonlinear equalities with those for problems with linear inequalities. There are a few issues which are unique to problems with nonlinear inequalities, however. We discuss them briefly.

Let x_* be a feasible solution to the inequality-constrained problem. Whereas in the case of equality constraints we can maintain feasibility by moving in either direction along a feasible curve through x_*, here it is often possible to move in only one direction; we shall call this "movement along a feasible arc." More formally, we define an *arc* emanating from x_* as a directed curve $x(t)$ parameterized by the variable t in an interval $[0, T]$ for which $x(0) = x_*$. An arc is feasible if $g(x(t)) \geq 0$ for t in $[0, T]$. Some examples of feasible arcs are illustrated in Figure 14.8. The optimality conditions are a result of the requirement that if a small movement is made along a feasible arc, the objective value will not decrease.

The constraints that are inactive at x_* can be ignored, since they do not influence the local optimality conditions. With the regularity assumption, it is possible to derive the first- and second-order conditions for optimality. The proofs are left to the Exercises.

Exercises

    7.1. Consider the constraint $g(x) = (x_2 - x_1^2)(x_1^2 - x_2^2) = 0$. Find the tangent cone and the null space of the Jacobian matrix at $x_* = (0, 0)^T$.

7.2. Consider the problem

$$\begin{aligned} \text{minimize} \quad & f(x) = x_2 \\ \text{subject to} \quad & x_2 = 0 \\ & x_2 - x_1^3 = 0. \end{aligned}$$

Show that the solution is $x_* = (0, 0)^T$. Show that there exists a vector of multipliers λ_* such that $\nabla \mathcal{L}(x_*, \lambda_*) = 0$, even though $T(x) \neq \mathcal{N}(A(x))$.

7.3. Let x_* be a feasible point to an equality-constrained problem, and let $T(x_*)$ be the tangent cone at x_*. Prove that if $0 \in T(x_*)$, then $T(x_*)$ is a cone. (*Hint*: Prove that if $x(t)$ is a feasible arc through x_*, then so is $x(\alpha t)$.)

7.4. Let x_* be a feasible point to the inequality-constrained problem. Prove that if $x(t)$ is a feasible arc emanating from x_*, then $x'(0)^T \nabla g_i(x_*) \geq 0$ for all binding constraints i.

7.5. Assume that x_* is a regular point for the inequality-constrained problem, and that there exists some λ such that $\nabla_x \mathcal{L}(x_*, \lambda) = 0$ and $\lambda^T g(x_*) = 0$. Suppose that $\lambda_1 < 0$. Prove that there exists some feasible arc $x(t)$ emanating from x_* such that

$$\left. \frac{df(x(t))}{dt} \right|_{t=0} = \lambda_1.$$

Hence, x_* cannot be a local minimizer.

7.6. Use the results of the previous two problems to prove Theorem 14.18.

7.7. Prove Theorem 14.19.

7.8. Consider the problem with a single constraint:

$$\begin{aligned} \text{minimize} \quad & f(x) \\ \text{subject to} \quad & g(x) \geq 0. \end{aligned}$$

It is possible to transform the inequality constraint into an equality constraint by subtracting a squared variable, say y^2, from g. We obtain a problem with $n + 1$ variables and a single equality constraint. Write the first- and second-order optimality conditions for this problem, and compare them to the optimality conditions for the original inequality-constrained problem. Also, determine the conditions satisfied by a regular point for this problem and compare them to the conditions satisfied by a regular point of the original problem.

7.9. Consider the problem

$$\begin{aligned} \text{minimize} \quad & f(x) \\ \text{subject to} \quad & g(x) = 0, \end{aligned}$$

where $x \in \mathfrak{R}^n$ and g is a vector function. Let x_* be a regular point of the constraints. Consider now the equivalent problem

$$\begin{aligned} \text{minimize} \quad & f(x) \\ \text{subject to} \quad & g(x) \geq 0 \\ & -g(x) \geq 0. \end{aligned}$$

Prove that x_* is no longer a regular point of the constraints for this new problem, but that there exists a vector of Lagrange multipliers that satisfies Theorem 14.18. Is the vector of multipliers unique?

7.10. It is possible to derive optimality conditions that do not require that $T(x_*) = \mathcal{N}(A(x_*))$. This was done by Fritz John (1948) who defined a "weak Lagrangian" that includes a multiplier for the objective. Consider the problem

$$\text{minimize} \quad f(x)$$
$$\text{subject to} \quad g(x) \geq 0,$$

where $x \in \mathfrak{R}^n$, g is vector function, and all functions are continuously differentiable. John showed that if x_* is a local minimizer of the problem, then there exist multipliers θ_*, λ_* not all zero so that $\theta_* \nabla f(x_*) - \lambda_*^T \nabla g(x_*) = 0$, $\lambda_*^T g(x_*) = 0$, and $\lambda_* \geq 0$.

(i) Use this result to prove that if x_* is a regular point of the constraints, then $\theta_* \neq 0$.

(ii) Consider the problem of minimizing $f(x) = -x_1$ subject to the constraint $(1 - x_1)^3 - x_2 \geq 0$ and $x_1 x_2 \geq 0$. The local minimizer is $x_* = (1, 0)^T$ but it is not a regular point. Show that there are no multipliers that satisfy our regular optimality condition, but the Fritz John conditions are satisfied.

(iii) Consider now a problem with equality constraints $g(x) = 0$ that are split into two inequalities as in Exercise 7.9. Prove that the Fritz John conditions are satisfied by every feasible point. This illustrates one major weakness of these conditions.

14.8   Duality

We have already encountered the concept of duality in the chapters on linear programming (Chapters 4–10), in which we saw that every linear programming problem has an associated *dual* linear programming problem that is intimately related to it. There are several connections between a linear problem (the *primal* problem) and its dual: (i) if the primal is a minimization problem, the dual problem is a maximization problem, and vice versa; (ii) the dual of the dual problem is the primal problem; (iii) the objective value for any feasible solution to the maximization problem is a lower bound on the objective value for any feasible solution to the minimization problem (the weak duality theorem); (iv) if one problem has an optimal solution, then so does the other, and the optimal objective values of the two problems are equal (the strong duality theorem).

The relationship between primal and dual linear programs has immense value from both theoretical and computational points of view. First, the dual problem may be easier to solve, and if the optimal solution to the dual problem is known, then (in nondegenerate cases) the optimal solution to the primal problem can easily be computed. Second, a good estimate of the optimal dual solution may assist in obtaining a good estimate of the optimal primal solution. In addition, the dual variables have an important economic significance and can be used for sensitivity analysis (see Section 6.4).

The first-order optimality conditions for nonlinear optimization are stated not only in terms of the vector of variables x_* but also in terms of a vector of optimal Lagrange multipliers λ_*. The Lagrange multipliers are the dual variables. If the optimal Lagrange multipliers were known a priori, the optimization problem would often be easier to solve. Thus it would be useful if we could easily compute λ_*. Hence the question, can we define a new nonlinear optimization problem, where the unknown variables are the Lagrange multipliers, and the solution is λ_*? Furthermore, under what conditions will the solution to this new problem also provide us with a solution to the original problem? Duality theory examines these questions.

In this section we develop a duality theory for nonlinear optimization. Specifically, we define for every nonlinear optimization problem a related dual problem. Ideally we would like the relationships between a pair of primal and dual nonlinear problems to replicate those of primal and dual linear programs. As we shall see, this is not always possible.

Various possible formulations have been proposed for the dual of a nonlinear problem. The fundamental idea underlying most approaches is to represent a constrained minimization problem in a form that will be termed a "min-max" problem. Its dual problem will be represented in a form termed a "max-min" problem. These two problems can be viewed as strategy problems for two players in a game. We describe this idea next.

14.8.1 Games and Min-Max Duality

Let us consider a game between two players, Peter and Harriet. In this game, Peter has a set of possible strategies X, while Harriet has a set of possible strategies Y. The game proceeds as follows: Peter chooses a strategy $x \in X$, and Harriet chooses a strategy $y \in Y$. The choices of strategies are then revealed simultaneously. As a result, Peter pays an amount $\mathcal{F}(x, y)$ to Harriet (the amount paid can be negative, in which case, Peter actually gains). The game is called a *zero-sum game* because whatever one player wins, the other player loses.

We now make the assumption that both Peter and Harriet act rationally to maximize their rewards. Suppose in fact that both players would like to take a course of action that will guarantee their largest gain, regardless of what their opponent does. Then both players would attempt to optimize their worst-case scenario. Based on these premises, we now analyze the players' optimal strategies.

Consider first Peter, who is worried about making a large payoff to Harriet. If Peter chooses strategy $x \in X$, then in the worst case (that is, if Harriet is either very clever or very lucky) his payoff to Harriet will be

$$\mathcal{F}^*(x) = \max_{y \in Y} \mathcal{F}(x, y).$$

(In strict mathematical terms, we should use the "supremum" operator[15] rather than the "maximum" operator, because there may be no $y \in Y$ for which a maximum value is achieved; here for simplicity we ignore this possibility.) To minimize this worst-case payoff, Peter has to choose the strategy that solves the optimization problem

$$\underset{x \in X}{\text{minimize}} \ \mathcal{F}^*(x).$$

This problem is referred to as a min-max problem, since it seeks the value

$$\min_{x \in X} \max_{y \in Y} \mathcal{F}(x, y).$$

---

[15]The supremum of a set is its least upper bound; the infimum of the set is its greatest lower bound. If the supremum or infimum exists, it need not belong to the set.

Harriet, on the other hand, is worried about receiving a payoff that is too small. If she chooses a strategy $y \in Y$, then in the worst case, the payoff that she will receive will only be

$$\mathcal{F}_*(y) = \min_{x \in X} \mathcal{F}(x, y).$$

(For simplicity we use the "minimum" operator instead of the more precise "infimum" operator.) Harriet's optimal strategy is to maximize the worst-case payoff that she will receive. Hence she must solve the problem

$$\underset{y \in Y}{\text{maximize }} \mathcal{F}_*(y).$$

This problem is referred to as a max-min problem, since it seeks the value

$$\max_{y \in Y} \min_{x \in X} \mathcal{F}(x, y).$$

The two problems are said to be *dual* to each other. We refer to the min-max problem faced by Peter as the *primal* problem. The objective to be minimized, $\mathcal{F}^*(x)$, is referred to as the *primal function*. We refer to the max-min problem faced by Harriet as the *dual* problem. The objective to be maximized, $\mathcal{F}_*(y)$, is referred to as the *dual function*.

For any $x \in X$ and $y \in Y$,

$$\mathcal{F}_*(y) = \min_{x \in X} \mathcal{F}(x, y) \le \mathcal{F}(x, y) \le \max_{y \in Y} \mathcal{F}(x, y) = \mathcal{F}^*(x), \qquad (14.4)$$

and thus in particular

$$\mathcal{F}_*(y) \le \mathcal{F}^*(x).$$

This is the statement of *weak duality*.

A consequence of the weak duality statement is that the optimal solution of the max-min problem is bounded above by the optimal value of the min-max problem. This result is summarized in the following lemma.

Lemma 14.27.

$$\max_{y \in Y} \min_{x \in X} \mathcal{F}(x, y) \le \min_{x \in X} \max_{y \in Y} \mathcal{F}(x, y).$$

Proof. From (14.4) it follows that

$$\min_{x \in X} \mathcal{F}(x, y) \le \max_{y \in Y} \mathcal{F}(x, y).$$

This holds for any $x \in X$ and $y \in Y$, so it holds for the y that maximizes the left-hand term and for the x that minimizes the right-hand term. Thus

$$\max_{y \in Y} \min_{x \in X} \mathcal{F}(x, y) \le \min_{x \in X} \max_{y \in Y} \mathcal{F}(x, y). \qquad \square$$

Example 14.28 (Two-Person Zero-Sum Game). Suppose that Peter has a set of two possible strategies, denoted $X = \{1, 2\}$, and that Harriet also has a set of two possible strategies, denoted $Y = \{1, 2\}$. If Peter chooses strategy $i \in X$ and Harriet chooses strategy $j \in Y$, Peter's payoff will be $F(i, j) = a_{i,j}$, where $A = (a_{i,j})$ is given as

$$A = \begin{pmatrix} -1 & 2 \\ 4 & 3 \end{pmatrix}.$$

Then Peter's min-max problem is to choose the strategy that solves

$$\text{minimize} \max_j a_{i,j}.$$

Now

$$\min_i \max_j a_{i,j} = \min \left\{ \max_j a_{1,j}, \max_j a_{2,j} \right\} = \min\{2, 4\} = 2,$$

and hence Peter's optimal strategy would be to choose row $i = 1$, thereby guaranteeing that his payoff to Harriet will be at most 2.

Harriet's max-min problem is to choose the strategy that solves

$$\text{maximize} \min_i a_{i,j}.$$

Now

$$\max_j \min_i a_{i,j} = \max \left\{ \min_i a_{i,1}, \min_i a_{i,2} \right\} = \max\{-1, 2\} = 2,$$

and hence Harriet's optimal strategy would be to choose column $j = 2$, thereby guaranteeing that her reward will be at least 2. In this case we see that

$$\text{max-min} = 2 = \text{min-max}.$$

The game can be considered to be in equilibrium, in the sense that neither player can gain from a change in strategy as long as the opponent's strategy remains fixed.

It is not always true that max-min = min-max. If the matrix

$$\bar{A} = \begin{pmatrix} -1 & 2 \\ 4 & 1 \end{pmatrix}$$

were used instead of A, then we would obtain that

$$\text{max-min} = 1 \leq 2 = \text{min-max}. \quad \blacksquare$$

In this example, element a_{12} of A is largest in its row and smallest in its column. If such an element exists, a game of this type is said to have a saddle point. The matrix $\bar{A}$ does not have a saddle point. As we will see, the optimal primal and dual objectives are equal if and only if the game has a saddle point.

We first give a more general definition. A point (x_*, y_*) with $x_* \in X$ and $y_* \in Y$ is said to satisfy the *saddle-point condition* for $\mathcal{F}$ if

$$\mathcal{F}(x_*, y) \leq \mathcal{F}(x_*, y_*) \leq \mathcal{F}(x, y_*)$$

for all $x \in X$ and $y \in Y$. Thus, x_* is a minimizer of $\mathcal{F}$ when y is fixed at y_*, and y_* is a maximizer of $\mathcal{F}$ when x is fixed at x_*.

The following lemma states under what conditions the optimal values of the two problems are equal.

Lemma 14.29 (Strong Duality). *The condition*

$$\max_{y \in Y} \min_{x \in X} \mathcal{F}(x, y) = \min_{x \in X} \max_{y \in Y} \mathcal{F}(x, y) \tag{14.5}$$

holds if and only if there exists a pair (x_*, y_*) *that satisfies the saddle-point condition for* $\mathcal{F}$.

Proof. Suppose that (x_*, y_*) satisfies the saddle-point condition. Then

$$\max_{y \in Y} \mathcal{F}(x_*, y) \leq \mathcal{F}(x_*, y_*) \leq \min_{x \in X} \mathcal{F}(x, y_*).$$

Now the left-hand term in the inequality above is bounded below by

$$\min_{x \in X} \max_{y \in Y} \mathcal{F}(x, y) \leq \max_{y \in Y} \mathcal{F}(x_*, y),$$

and the right-hand term in the inequality above is bounded above by

$$\min_{x \in X} \mathcal{F}(x, y_*) \leq \max_{y \in Y} \min_{x \in X} \mathcal{F}(x, y).$$

Combining the last three inequalities, we obtain that

$$\min_{x \in X} \max_{y \in Y} \mathcal{F}(x, y) \leq \mathcal{F}(x_*, y_*) \leq \max_{y \in Y} \min_{x \in X} \mathcal{F}(x, y),$$

or in short, min-max $\leq$ max-min. But in view of Lemma 14.27, this can be true only if min-max = max-min, and therefore

$$\max_{y \in Y} \min_{x \in X} \mathcal{F}(x, y) = \mathcal{F}(x_*, y_*) = \min_{x \in X} \max_{y \in Y} \mathcal{F}(x, y).$$

Suppose now that (14.5) holds. Then

$$\max_{y \in Y} \min_{x \in X} \mathcal{F}(x, y) = \mathcal{F}(x_*, y_*) = \min_{x \in X} \max_{y \in Y} \mathcal{F}(x, y)$$

for some (x_*, y_*). Therefore for any $x \in X$ and $y \in Y$,

$$\mathcal{F}(x_*, y) \leq \max_{y \in Y} \mathcal{F}(x_*, y) = \mathcal{F}(x_*, y_*) = \min_{x \in X} \mathcal{F}(x, y_*) \leq \mathcal{F}(x, y_*),$$

and therefore (x_*, y_*) satisfies the saddle-point condition for $\mathcal{F}$. $\square$

14.8.2   Lagrangian Duality

Min-max duality serves as the basis for developing a dual problem to a nonlinear problem. The main idea is to define some game with payoff function $\mathcal{F}$ so that the solution to the min-max problem with respect to $\mathcal{F}$ is also the solution to the nonlinear minimization problem. The resulting max-min problem is then the dual problem. There are various ways in which this can be done. The approaches differ in the way the sets X and Y and the function $\mathcal{F}$ are defined. Here we describe an approach sometimes termed *Lagrangian duality* that is useful from a computational point of view. An alternative approach termed *conjugate duality* is more intricate, yet it is more powerful from a theoretical point of view.

We assume that the nonlinear problem has the form

$$\begin{aligned} \text{minimize} \quad & f(x) \\ \text{subject to} \quad & g(x) \geq 0 \\ & x \in X. \end{aligned} \tag{14.6}$$

Here g is a vector of m functions g_i. The *set constraint* $x \in X$ is used to impose additional requirements that we may wish to handle separately. These may be additional explicit functional constraints, such as nonnegativity constraints on the variables. Or, if this were a discrete problem, these may be the integrality requirements, and X would be the set of points in $\Re^n$ with integer components (this option is useful in developing an approach for solving integer programs). When $X = \Re^n$, there are no "special" requirements and, we can omit this constraint from the statement of the problem.

Consider now the Lagrangian function

$$\mathcal{L}(x, \lambda) = f(x) - \lambda^T g(x)$$

for $x \in X \subset \Re^n$ and $\lambda \in \Re^m, \lambda \geq 0$; here g is the vector of constraint functions g_i. We will show that the nonlinear optimization problem can be represented as a min-max problem. The Lagrangian $\mathcal{L}(x, \lambda)$ will assume the role of $\mathcal{F}$ in this representation, and the set $\{\lambda \in \Re^m, \lambda \geq 0\}$ will assume the role of the set Y.

We first define

$$\mathcal{L}^*(x) = \max_{\lambda \geq 0} \mathcal{L}(x, \lambda)$$

for any $x \in X$. This is the *primal function*. Let us take a closer look at this function for a fixed x:

$$\mathcal{L}^*(x) = \max_{\lambda \geq 0} [f(x) - \lambda^T g(x)].$$

If $g(x) \geq 0$, then $\lambda^T g(x)$ will be nonnegative. Thus the Lagrangian will be maximized when the second term is zero. This is achieved, for example, if $\lambda = 0$, and we then obtain that $\mathcal{L}^*(x) = f(x)$. If on the contrary $g_i(x) < 0$ for some constraint, then by letting the multiplier λ_i increase without limit and keeping all other multipliers equal to zero, the Lagrangian will increase without limit. Therefore

$$\mathcal{L}^*(x) = \begin{cases} f(x) & \text{if } g(x) \geq 0 \\ \infty & \text{otherwise.} \end{cases}$$

Now the min-max primal problem is

$$\underset{x \in X}{\text{minimize}} \; \mathcal{L}^*(x).$$

If we ignore the regions where $\mathcal{L}^*$ is infinite, then this problem becomes our original constrained problem

$$\begin{aligned} \text{minimize} \quad & f(x) \\ \text{subject to} \quad & g(x) \geq 0 \\ & x \in X. \end{aligned}$$

Thus the original constrained nonlinear problem is represented as a min-max problem.

We can now utilize min-max duality to formulate a dual problem. For any $\lambda \geq 0$ the *dual function* is defined as

$$\mathcal{L}_*(\lambda) = \min_{x \in X} \mathcal{L}(x, \lambda).$$

The resulting max-min dual problem becomes

$$\underset{\lambda \geq 0}{\text{maximize}} \ \mathcal{L}_*(\lambda).$$

This is the nonlinear dual problem, with dual variables λ. More explicitly, the dual problem can be written in the form

$$\underset{\lambda \geq 0}{\text{maximize}} \ \min_{x \in X} [f(x) - \lambda^T g(x)].$$

Throughout this discussion we have assumed that in the original problem, all the constraints involving $\{\,g_i\,\}$ are inequalities. There is no conceptual difficulty if some constraints are equalities, however. If some g_i is required to be exactly zero, then in the definition of the dual function its associated multiplier λ_i will be unrestricted in sign.

We now give an example. Unless stated otherwise, in the examples in this section we assume that the set X is the entire space $\Re^n$.

Example 14.30 (Dual Problem). Consider the problem

$$\text{minimize} \quad f(x) = x^2$$
$$\text{subject to} \quad x \geq 1.$$

The solution to this problem is $x_* = 1$. The Lagrangian function is

$$\mathcal{L}(x, \lambda) = x^2 - \lambda(x - 1).$$

Therefore the dual function is

$$\mathcal{L}_*(\lambda) = \min_x \mathcal{L}(x, \lambda) = \min_x [x^2 - \lambda(x - 1)]$$

for $\lambda \geq 0$. The function on the right is a quadratic in x whose minimizer is $x = \lambda/2$. Thus the dual function takes the form

$$\mathcal{L}_*(\lambda) = \lambda - \tfrac{1}{4}\lambda^2$$

for $\lambda \geq 0$. The dual problem is therefore

$$\underset{\lambda \geq 0}{\text{maximize}} \ \lambda - \tfrac{1}{4}\lambda^2.$$

It is easy to see that the solution is $\lambda_* = 2$, which is indeed the Lagrange multiplier corresponding to x_*. The optimal dual objective value is $\mathcal{L}_*(\lambda_*) = 1$, and it is equal to the optimal objective value $f(x_*)$. ∎

In this example we succeeded in obtaining an explicit expression for the dual function. Unfortunately, this will not always be possible, as is demonstrated in the following example.

Example 14.31 (Dual Function Not Available Explicitly). Consider the problem

$$\text{minimize} \quad f(x) = e^x$$
$$\text{subject to} \quad 1 - x^2 \geq 0.$$

The dual function is

$$\mathcal{L}_*(\lambda) = \min_x \mathcal{L}(x, \lambda) = \min_x e^x - \lambda(1 - x^2).$$

For any fixed $\lambda \geq 0$, the function $e^x - \lambda(1 - x^2)$ is convex, and hence any local minimizer will also be a global minimizer. This minimizer is the solution $x = x(\lambda)$ of the nonlinear equation

$$e^x + 2x\lambda = 0.$$

For a given value λ, the solution can be found numerically, but this solution cannot be expressed as an explicit function of λ. The dual problem is thus

$$\underset{\lambda \geq 0}{\text{maximize}} \ e^x - \lambda(1 - x^2),$$

where x solves

$$e^x + 2x\lambda = 0.$$

Equivalently, we can write this problem in the form

$$\underset{x, \lambda}{\text{maximize}} \quad e^x - \lambda(1 - x^2)$$
$$\text{subject to} \quad e^x + 2x\lambda = 0$$
$$\lambda \geq 0.$$

If we represent the primal problem in the form

$$\text{minimize} \quad f(x) = e^x$$
$$\text{subject to} \quad -1 \leq x \leq 1,$$

then the dual can be written explicitly. Thus the form of the dual depends on the particular way in which the primal is written. ∎

Examples of problems where the dual function cannot be given explicitly are common. The statement of the dual problem may include the original variables as well as the dual variables. Thus we define a point $(\bar{x}, \bar{\lambda})$ to be *dual feasible* if $\bar{x} \in X, \bar{\lambda} \geq 0$, and $\bar{x}$ minimizes $\mathcal{L}(x, \bar{\lambda})$. In the previous example the point $(\bar{x}, \bar{\lambda}) = (-1, \frac{1}{2}e^{-1})$ is a dual feasible solution.

Let us now examine the relationship between a nonlinear problem and its dual. If x is feasible to the primal problem, then $\mathcal{L}^*(x) = f(x)$, and if $(\bar{x}, \bar{\lambda})$ is feasible to the dual problem, then $\mathcal{L}_*(\bar{\lambda}) = \mathcal{L}(\bar{x}, \bar{\lambda})$. From weak duality, we know that $\mathcal{L}_*(\bar{\lambda}) \leq \mathcal{L}^*(x)$. We therefore obtain the following theorem.

Theorem 14.32 (Weak Duality). *Let x be a feasible solution to the primal problem* (14.6), *and let $(\bar{x}, \bar{\lambda})$ be a feasible solution to its dual problem. Then*

$$f(\bar{x}) - \bar{\lambda}^T g(\bar{x}) \leq f(x).$$

The following is a consequence of the weak duality theorem.

Corollary 14.33.

$$\max_{\lambda \geq 0} \mathcal{L}_*(\lambda) \leq \min_{x \in X} \{ f(x) : g(x) \geq 0 \}.$$

If we define the dual problem to be infeasible when $\mathcal{L}_*(\lambda) = -\infty$ for all $\lambda \geq 0$, then we have the following corollary.

Corollary 14.34. *If a (primal) nonlinear optimization problem is unbounded, then its dual is infeasible. If the dual is unbounded, then the primal is infeasible.*

Corollary 14.33 states that the optimal objective value of the primal problem is greater than or equal to the optimal value of the dual problem. In Example 14.30 the optimal objective values of the primal and dual problems were equal. This may lead us to believe that, just as in linear programming, this result will always be true whenever an optimal solution to the problems exists. This is not the case, however. There are problems for which a *duality gap* exists, where the optimal value of the primal is strictly greater than the optimal value of the dual. This is illustrated in the next example.

Example 14.35 (Duality Gap). Consider the problem

$$\begin{aligned} \text{minimize} \quad & f(x) = -x^2 \\ \text{subject to} \quad & x = 1 \\ & x \in X, \end{aligned}$$

where $X = \{ x : 0 \leq x \leq 2 \}$. The solution is clearly $x_* = 1$ with optimal objective value -1.

Denote by λ the Lagrange multiplier associated with the constraint $x = 1$. Because the constraint is an equality, λ is no longer restricted to being nonnegative. The Lagrangian function is

$$\mathcal{L}(x, \lambda) = -x^2 - \lambda(x - 1), \quad x \in X, \quad -\infty \leq \lambda \leq \infty.$$

The dual function is

$$\mathcal{L}_*(\lambda) = \min_{x \in X} \mathcal{L}(x, \lambda) = \min_{0 \leq x \leq 2} -x^2 - \lambda(x - 1).$$

The function $-x^2 - \lambda(x - 1)$ has no local minimizer, and hence it will attain its minimum in X either at $x = 0$ or at $x = 2$. Comparing function values, we obtain

$$\mathcal{L}_*(\lambda) = \begin{cases} \lambda & \text{if } \lambda < -2 \\ -4 - \lambda & \text{if } \lambda \geq -2. \end{cases}$$

The dual function is depicted in Figure 14.9. The maximum of the dual function is at $\lambda_* = -2$. At this point the optimal dual objective is $\mathcal{L}_*(\lambda_*) = -2$. Since the optimal primal objective is -1, the two optimal objective values are not equal. The duality gap is the difference $f(x_*) - \mathcal{L}_*(\lambda_*) = 1$. ∎

It is possible also to construct examples with an infinite duality gap, as the next example shows.

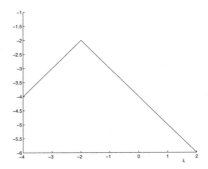

Figure 14.9. *Dual function.*

Example 14.36 (Infinite Duality Gap). Consider the problem

$$\text{minimize} \quad f(x) = -x^2$$
$$\text{subject to} \quad 0 \le x \le 1.$$

The solution is $x_* = 1$ with corresponding Lagrange multiplier $\lambda_{*1} = 0$ for the lower bound constraint and $\lambda_{*2} = 2$ for the upper bound constraint. The Lagrangian function is

$$\mathcal{L}(x, \lambda) = -x^2 - \lambda_1 x - \lambda_2 (1 - x),$$

and the dual function is

$$\mathcal{L}_*(\lambda_1, \lambda_2) = \min_x \mathcal{L}(x, \lambda) = \min_x [-x^2 - \lambda_1 x - \lambda_2 (1 - x)] = -\infty$$

for all $\lambda_1 \ge 0$ and $\lambda_2 \ge 0$. Therefore the dual problem

$$\underset{\lambda}{\text{maximize}} \ \mathcal{L}_*(\lambda_1, \lambda_2)$$

has an optimal objective value of $-\infty$. ∎

Under what conditions are the optimal primal and dual objectives guaranteed to be equal? The strong duality theorem of the previous section indicates that this will be true if and only if there is some point (x_*, λ_*) that satisfies the saddle-point condition

$$\mathcal{L}(x_*, \lambda) \le \mathcal{L}(x_*, \lambda_*) \le \mathcal{L}(x, \lambda_*)$$

for all $x \in X$ and $\lambda \ge 0$. The last two examples show that it is possible that no (x_*, λ_*) satisfies the saddle-point condition, even if the problem has an optimal solution.

There is one important class of problems for which, under mild conditions, the saddle-point condition is guaranteed to be satisfied. These are problems of the form

$$\text{minimize} \quad f(x)$$
$$\text{subject to} \quad g_i(x) \ge 0, \quad i = 1, \dots, m,$$

(14.7)

where f is a convex function and each g_i is a concave function. Such problems are convex optimization problems. (The result can be extended to include linear equalities; see Example 14.38.)

Theorem 14.37 (Convex Duality). *Consider the convex optimization problem* (14.7) *in which all functions are assumed to be continuously differentiable. Let x_* be a solution to the problem, and assume that x_* is a regular point of the constraints. Let λ_* be the vector of Lagrange multipliers corresponding to x_*. Then (x_*, λ_*) is dual feasible, λ_* solves the dual problem, and the optimal primal and dual function values are equal.*

Proof. It is sufficient to show that (x_*, λ_*) satisfies the saddle-point condition for $\mathcal{L}$.
Since f is convex, g_i are concave, and $\lambda_* \geq 0$, the function

$$\mathcal{L}(x, \lambda_*) = f(x) - \lambda^T g(x)$$

is a convex function of x. Consequently, the first-order optimality condition

$$\nabla_x \mathcal{L}(x_*, \lambda_*) = 0$$

implies that x_* minimizes $\mathcal{L}(x, \lambda_*)$, and hence

$$\mathcal{L}(x_*, \lambda_*) \leq \mathcal{L}(x, \lambda_*)$$

for all x. In addition, the complementary slackness conditions at x_* imply that $\lambda_*{}^T g(x_*) = 0$. Thus $\mathcal{L}(x_*, \lambda_*) = f(x_*)$, and for any $\lambda \geq 0$ we have

$$\mathcal{L}(x_*, \lambda) = f(x_*) - \lambda^T g(x_*) \leq f(x_*) = \mathcal{L}(x_*, \lambda_*).$$

Therefore

$$\mathcal{L}(x_*, \lambda) \leq \mathcal{L}(x_*, \lambda_*) \leq \mathcal{L}(x, \lambda_*)$$

and (x_*, λ_*) satisfies the saddle-point condition.   □

The problem of Example 14.30 is convex, and indeed the optimal primal and dual objectives are equal. In Examples 14.35 and 14.36, where we observed a duality gap, the problems are not convex. It is also possible to construct examples of problems which are not convex, but where the duality gap is zero.

14.8.3   Wolfe Duality

When all problem functions are continuously differentiable, the dual of a convex problem can be represented in a more convenient form. For any fixed λ, the point $\bar{x}$ minimizes $\mathcal{L}(x, \lambda)$ if and only if

$$\nabla_x \mathcal{L}(x, \lambda)|_{x=\bar{x}} = 0.$$

Because of this we can state the dual in the form

$$\begin{array}{ll} \underset{x, \lambda \geq 0}{\text{maximize}} & \mathcal{L}(x, \lambda) \\ \text{subject to} & \nabla_x \mathcal{L}(x, \lambda) = 0. \end{array}$$

This representation is sometimes called the *Wolfe dual*.

Example 14.38 (Dual of a Linear Program). Consider the linear program

$$\begin{array}{ll} \text{minimize} & f(x) = c^T x \\ \text{subject to} & Ax = b \\ & x \geq 0. \end{array}$$

Denoting by y the vector of Lagrange multipliers for the linear equalities, and by $\lambda \geq 0$ the vector of Lagrange multipliers for the nonnegativity constraints, we obtain the Lagrangian

$$\mathcal{L}(x, y, \lambda) = c^T x - y^T (Ax - b) - \lambda^T x.$$

The dual problem is

$$\begin{array}{ll} \underset{\lambda \geq 0}{\text{maximize}} & \mathcal{L}(x, y, \lambda) = c^T x - y^T (Ax - b) - \lambda^T x \\ \text{subject to} & c - A^T y - \lambda = 0. \end{array}$$

The equality constraint implies that $c^T x - y^T Ax - \lambda^T x = 0$. Substituting this into the dual objective, the dual problem may be rewritten in the form

$$\begin{array}{ll} \text{maximize} & b^T y \\ \text{subject to} & A^T y \leq c \end{array}$$

which is equivalent to the dual of a linear program in standard form. ∎

The dual of a linear program is a linear program. The next example shows that the dual of a strictly convex quadratic program is a quadratic program.

Example 14.39 (Dual of a Quadratic Program). Consider the quadratic program

$$\begin{array}{ll} \text{minimize} & f(x) = \frac{1}{2} x^T Q x + c^T x \\ \text{subject to} & Ax \geq b, \end{array}$$

where Q is a positive-definite matrix. This is a convex problem, and hence the Wolfe dual problem exists:

$$\begin{array}{ll} \underset{x, \lambda}{\text{maximize}} & \frac{1}{2} x^T Q x + c^T x - \lambda^T (Ax - b) \\ \text{subject to} & Qx + c - A^T \lambda = 0 \\ & \lambda \geq 0. \end{array}$$

It follows from the equality constraint that $x = Q^{-1}(A^T \lambda - c)$. When this is used to eliminate x, the resulting dual problem involves only λ:

$$\begin{array}{ll} \underset{\lambda}{\text{maximize}} & -\frac{1}{2} \lambda^T (A Q^{-1} A^T) \lambda + (A Q^{-1} c + b)^T \lambda - \frac{1}{2} c^T Q^{-1} c \\ \text{subject to} & \lambda \geq 0. \end{array}$$

This is also a quadratic program. ∎

14.8.4 More on the Dual Function

The correspondence between the solutions of the primal and dual problems has useful implications. First, the dual problem could be used to obtain estimates of the optimal Lagrange multipliers. Good multiplier estimates could in turn provide good solution estimates for the primal. Alternatively, if deemed more convenient, the dual instead of the primal problem could be solved directly.

For convex problems (14.7), there is a correspondence between the primal and dual solutions. As we have already seen, for nonconvex problems such a correspondence may not hold. Nevertheless, it is possible in certain cases to define a *local duality* theory. We use the name "local duality" because we restrict the problem to a neighborhood of a local solution. To obtain the primal/dual correspondence we assume that the Lagrangian at the local solution is strictly convex. In this section we develop the local duality theory. We also study properties of the dual function that are useful in algorithms.

First, we show that the dual function is concave. Surprisingly, this property holds even if the primal problem is not convex. To prove this result, let λ_1 and λ_2 be any two nonnegative multipliers, and let α be a scalar such that $0 \leq \alpha \leq 1$. Then

$$
\begin{aligned}
\mathcal{L}_*(\alpha\lambda_1 &+ (1-\alpha)\lambda_2) \\
&= \min_{x \in X} \left\{ f(x) - (\alpha\lambda_1 + (1-\alpha)\lambda_2)^T g(x) \right\} \\
&= \min_{x \in X} \left\{ \alpha[f(x) - \lambda_1^T g(x)] + (1-\alpha)[f(x) - \lambda_2^T g(x)] \right\} \\
&\geq \alpha \min_{x \in X} \left\{ f(x) - \lambda_1^T g(x) \right\} + (1-\alpha) \min_{x \in X} \left\{ f(x) - \lambda_2^T g(x) \right\} \\
&= \alpha\mathcal{L}_*(\lambda_1) + (1-\alpha)\mathcal{L}_*(\lambda_2),
\end{aligned}
$$

and so the dual function is concave.

The concavity of the dual function is an appealing property, since it implies that every local maximizer of the dual function is also a global maximizer. It suggests also that an algorithm for solving the dual may not encounter some of the difficulties associated with maximizing a nonconcave function (these difficulties are similar to those incurred when minimizing a nonconvex function). Note that solving the dual only gives bounds on the primal objective unless the duality gap is known to be zero. The latter will be true for a convex optimization problem under the assumptions stated in Lemma 14.10. It will also be true under the convexity assumptions of the local duality theory presented below.

To develop the local duality theory we consider the problem

$$
\begin{aligned}
&\text{minimize}   &&f(x) \\
&\text{subject to} &&g_i(x) \geq 0, \quad i = 1, \ldots, m,
\end{aligned}
\tag{14.8}
$$

where f and all g_i are twice continuously differentiable. Let x_* be a local solution to this problem, and assume that it is a regular point. Let λ_* be the corresponding vector of Lagrange multipliers, so that

$$
\nabla\mathcal{L}(x_*, \lambda_*) = \nabla f(x_*) - \nabla g(x_*)\lambda_* = 0.
$$

We will now require the assumption that the Lagrangian Hessian

$$
\nabla^2_{xx}\mathcal{L}(x_*, \lambda_*)
$$

is positive definite. This assumption is stronger (that is, more restrictive) than the assumption made in the second-order sufficiency conditions that the *reduced* Lagrangian Hessian is positive definite.

Consider the system of equations

$$\nabla \mathcal{L}(x, \lambda) = \nabla f(x) - \nabla g(x)\lambda = 0.$$

The system has a solution (x_*, λ_*), and its Jacobian with respect to x at this point, the matrix $\nabla^2_{xx}\mathcal{L}(x_*, \lambda_*)$, is positive definite. It follows from the implicit function theorem (see Appendix B.9) that there exist neighborhoods of x_* and λ_* such that every λ in the neighborhood of λ_* has a unique corresponding point $x = x(\lambda)$ in the neighborhood of x_* that solves this system of equations, and furthermore, $\nabla^2_{xx}\mathcal{L}(x, \lambda)$ is positive definite. Let us denote such a neighborhood of x_* by X. Then every λ sufficiently close to λ_* has a unique corresponding x that is the global minimizer of $\mathcal{L}(x, \lambda)$ in X. Suppose now that we restrict our attention to the solution of (14.8) in the neighborhood X of x_*. Then the primal problem can be written locally as

$$\begin{aligned} &\underset{x \in X}{\text{minimize}} \quad f(x) \\ &\text{subject to} \quad g_i(x) \geq 0. \end{aligned}$$

The dual function for this local problem takes the form

$$\mathcal{L}_*(\lambda) = \min_{x \in X} \mathcal{L}(x, \lambda).$$

When λ is sufficiently close to λ_*,

$$\mathcal{L}_*(\lambda) = \mathcal{L}(x(\lambda), \lambda) = f(x(\lambda)) - \lambda^T g(x(\lambda)), \tag{14.9}$$

where $x(\lambda)$ is the unique point near x_* such that

$$\nabla f(x(\lambda)) - \nabla g(x(\lambda))\lambda = 0.$$

The problem

$$\underset{\lambda \geq 0}{\text{maximize}} \ \mathcal{L}_*(\lambda)$$

is termed the *local dual problem*. The equality of the local optimal primal and dual objective values is established in the following theorem.

Theorem 14.40 (Local Duality). *Consider the problem* (14.8) *in which all functions are assumed to be twice continuously differentiable. Let x_* be a local solution to the problem that is a regular point of the constraints. Let λ_* be the corresponding vector of Lagrange multipliers, and assume that $\nabla^2_{xx}\mathcal{L}(x_*, \lambda_*)$ is positive definite. Then λ_* is a solution of the local dual problem, (x_*, λ_*) is dual feasible, and the optimal primal and dual function values are equal.*

Proof. We prove that (x_*, λ_*) satisfies the saddle-point condition for the Lagrangian. The results then follow from the strong duality theorem.

Because $\nabla_x \mathcal{L}(x_*, \lambda_*) = 0$ and $\nabla^2_{xx}\mathcal{L}(x_*, \lambda_*)$ is positive definite, x_* minimizes $\mathcal{L}(x, \lambda_*)$, and

$$\mathcal{L}(x_*, \lambda_*) \le \mathcal{L}(x, \lambda_*)$$

for all $x \in X$. Also, because $\lambda \ge 0$ and $g(x_*) \ge 0$, and because of the complementary slackness at x_*,

$$\mathcal{L}(x_*, \lambda) = f(x_*) - \lambda^T g(x_*) \le f(x_*) = \mathcal{L}(x_*, \lambda_*)$$

for all $\lambda \ge 0$. Therefore (x_*, λ_*) satisfies the saddle-point condition.   □

As already mentioned, for λ sufficiently close to λ_*, the dual function takes the form (14.9), where $x(\lambda)$ is the minimizer of the Lagrangian close to x_*. It is not necessary to specify a set X explicitly to compute the point $x(\lambda)$ and the corresponding value of the dual function $\mathcal{L}_*(\lambda)$.

Example 14.41 (Local Duality). Consider the problem

$$\text{minimize} \quad f(x) = -3x$$
$$\text{subject to} \quad 1 - x^3 \ge 0.$$

The optimal point is clearly $x_* = 1$ and the corresponding objective value is -3. The Lagrangian function for the problem is

$$\mathcal{L}(x, \lambda) = -3x - \lambda(1 - x^3),$$

and the first-order necessary condition

$$\nabla_x \mathcal{L}(x, \lambda) = -3 + 3\lambda x^2 = 0$$

at x_* yields $\lambda_* = 1$.

The dual function is

$$\mathcal{L}_*(\lambda) = \min_x \mathcal{L}(x, \lambda) = \min_x\{-3x - \lambda(1 - x^3)\} = -\infty$$

for all $\lambda \ge 0$, and hence the dual problem is infeasible. Notice, however, that

$$\nabla^2_{xx}\mathcal{L}(x, \lambda) = 6\lambda x \quad \Longrightarrow \quad \nabla^2_{xx}\mathcal{L}(x_*, \lambda_*) = 6 > 0$$

and as a result, the local duality theorem applies. We thus focus on the problem restricted to an appropriate neighborhood of x_*. To find a local minimizer of the Lagrangian for λ close to $\lambda_* = 1$, note that

$$\nabla_x \mathcal{L}(x, \lambda) = -3 + 3\lambda x^2 = 0 \quad \Longrightarrow \quad x^2 = 1/\lambda.$$

Since $\nabla^2_{xx}\mathcal{L}(x, \lambda)$ is positive definite when both x and λ are positive, then $x = x(\lambda) = 1/\sqrt{\lambda}$ is the minimizer of $\mathcal{L}(x, \lambda)$ for $\lambda > 0$. It follows that the local dual function has the form

$$\mathcal{L}_*(\lambda) = -2/\sqrt{\lambda} - \lambda$$

for $\lambda > 0$. Since

$$\nabla \mathcal{L}_*(\lambda_*) = (\lambda_*)^{-1.5} - 1 = 0 \quad \text{and} \quad \nabla^2 \mathcal{L}_*(\lambda_*) = -1.5(\lambda_*)^{-2.5} = -1.5 < 0,$$

then $\lambda_* = 1$ is indeed the maximizer of $\mathcal{L}_*(\lambda)$. Furthermore, $x(\lambda_*) = x_* = 1$ and $\mathcal{L}_*(\lambda_*) = -3 = f(x_*)$.   ∎

The equality of the primal and dual objectives, combined with the concavity of the dual function, suggest that solving the dual problem could be a sensible alternative to solving the primal. However, a potential difficulty in this approach is that the dual function is not usually available in explicit form. The evaluation of $\mathcal{L}_*$ at a point λ may require the solution of the optimization problem $\min_x \mathcal{L}(x, \lambda)$, and so the dual function can be expensive to compute. Despite this, given the value of the dual function at some point λ close to λ_*, its first and second derivatives at that point can be computed readily. The assumption that is needed again is positive-definiteness of $\nabla_{xx}^2 \mathcal{L}(x_*, \lambda_*)$. It follows then from the implicit function theorem that $x(\lambda)$ is twice continuously differentiable for λ close to λ_*. This result is used below.

To compute the derivatives of the dual function, we note that in a neighborhood of λ_* it can be written in the form

$$\mathcal{L}_*(\lambda) = \mathcal{L}(x(\lambda), \lambda) = f(x(\lambda)) - \lambda^T g(x(\lambda)).$$

To obtain the gradient of this function we use the chain rule:

$$\nabla \mathcal{L}_*(\lambda) = \nabla x(\lambda) \nabla_x \mathcal{L}(x(\lambda), \lambda) + \nabla_\lambda \mathcal{L}(x(\lambda), \lambda),$$

where $\nabla x(\lambda)$ is an $m \times n$ matrix whose jth column is $\nabla_\lambda x_j(\lambda)$. Now $\nabla_x \mathcal{L}(x(\lambda), \lambda) = 0$ because $x(\lambda)$ is the minimizer of the Lagrangian. Also, $\nabla_\lambda \mathcal{L}(x, \lambda) = -g(x)$. Therefore we obtain

$$\nabla \mathcal{L}_*(\lambda) = -g(x(\lambda)),$$

or in short

$$\nabla \mathcal{L}_* = -g.$$

To obtain the second derivative, we differentiate $\nabla \mathcal{L}_*(\lambda)$. Using the previous result and the chain rule, we obtain

$$\nabla^2 \mathcal{L}_*(\lambda) = -\nabla x(\lambda) \nabla g(x(\lambda)).$$

Although we do not have an explicit expression for $x(\lambda)$, we can obtain an expression for its gradient. Since

$$\nabla_x \mathcal{L}(x(\lambda), \lambda) = 0$$

for all λ, we can differentiate this equality with respect to λ to obtain

$$\nabla x(\lambda) \nabla_{xx}^2 \mathcal{L}(x(\lambda), \lambda) + \nabla_{\lambda x} \mathcal{L}(x(\lambda), \lambda)^T$$
$$= \nabla x(\lambda) \nabla_{xx}^2 \mathcal{L}(x(\lambda), \lambda) - \nabla g(x(\lambda))^T = 0.$$

This implies that

$$\nabla x(\lambda) = \nabla g(x(\lambda))^T [\nabla_{xx}^2 \mathcal{L}(x(\lambda), \lambda)]^{-1},$$

and hence

$$\nabla^2 \mathcal{L}_*(\lambda) = -\nabla g(x(\lambda))^T [\nabla_{xx}^2 \mathcal{L}(x(\lambda), \lambda)]^{-1} \nabla g(x(\lambda)).$$

By omitting the function arguments, we can write this compactly as

$$\nabla^2 \mathcal{L}_* = -\nabla g^T [\nabla_{xx}^2 \mathcal{L}]^{-1} \nabla g.$$

Example 14.42 (Derivatives of the Dual Function). Consider the quadratic programming problem of Example 14.39. In this case $x(\lambda)$ is available explicitly: $x(\lambda) = Q^{-1}(A^T\lambda - c)$. Using the above formulas with $g(x) = Ax - b$ we obtain

$$\nabla \mathcal{L}_*(\lambda) = -g(x(\lambda)) = -(A(x(\lambda)) - b) = -AQ^{-1}A^T\lambda + AQ^{-1}c + b.$$

Also, since $\nabla g(x) = A^T$ and $\nabla^2 \mathcal{L}(x, \lambda) = Q$, we obtain that

$$\nabla^2 \mathcal{L}_* = -\nabla g^T[\nabla_{xx}\mathcal{L}]^{-1}\nabla g = -AQ^{-1}A^T.$$

The same results could be obtained directly by differentiating the quadratic objective of the dual problem. ∎

14.8.5 Duality in Support Vector Machines

Duality plays an important role in pattern classification via support vector machines, a problem that was presented in Section 1.7.2. Suppose that we have a set of m training points $x_i \in \Re^n$ that are classified by a scalar y_i, where $y_i = 1$ if x_i has a specified characteristic, and $y_i = -1$ if it does not. We wish to find a separating hyperplane of the form $w^Tx + b$, so that all points with $y_i = 1$ lie on one side of the hyperplane, all points with $y_i = -1$ lie on the other side, and with the margin of separation as large as possible. This separating hyperplane will serve as a "machine" that predicts the classification of any new point $\bar{x}$ based on the sign of $w^T\bar{x} + b$.

We will assume that our training data might not be separable, and we denote the violation of the margin by ξ. Our problem is to find an n-dimensional vector w, an n-dimensional vector ξ, and a scalar b that solve the quadratic program

$$\text{minimize} \quad \tfrac{1}{2}w^Tw + C\sum_{i=1}^{m}\xi_i$$
$$\text{subject to} \quad y_i(x_i^Tw + b) \geq 1 - \xi_i, \quad i = 1, \dots, m,$$
$$\xi_i \geq 0.$$

Here C is a penalty coefficient representing the trade-off between misclassification and a wide separation margin.

To define the dual of this problem it will be convenient to represent the problem constraints in matrix-vector form. Denote by X the $n \times m$ matrix whose columns are the training vectors x_i. Let $Y = \text{diag}(y)$ be a diagonal matrix whose ith diagonal term is y_i, and let $e = (1 \quad \cdots \quad 1)^T$ be a vector of length m whose entries are all equal to one. Then the problem becomes

$$\text{minimize} \quad \tfrac{1}{2}w^Tw + Ce^T\xi$$
$$\text{subject to} \quad YX^Tw + yb \geq e - \xi$$
$$\xi \geq 0.$$

Let α and η be the m-dimensional vectors of Lagrange multipliers corresponding to the hyperplane constraints and the constraints $\xi \geq 0$, respectively. Setting the gradient of the

Lagrangian to zero implies that

$$w = XY\alpha = \sum_{i=1}^{m} \alpha_i y_i x_i.$$

The dual problem is therefore given by

$$\text{maximize} \quad \sum_{i=1}^{m} \alpha_i - \tfrac{1}{2}\alpha^T (YX^TXY)\alpha$$

$$\text{subject to} \quad \sum_{i=1}^{m} y_i \alpha_i = 0$$
$$\alpha_i + \eta_i = C$$
$$\alpha, \eta \geq 0$$

(see the Exercises). The problem can be written just in terms of α as

$$\text{maximize} \quad \sum_{i=1}^{m} \alpha_i - \tfrac{1}{2}\alpha^T YX^TXY\alpha$$

$$\text{subject to} \quad \sum_{i=1}^{m} y_i \alpha_i = 0$$
$$0 \leq \alpha_i \leq C.$$

The dual, like the primal, is a quadratic problem. However, it is usually easier to solve because, with the exception of one equality, all constraints are simple upper and lower bounds.

From complementary slackness, any α_i that is positive corresponds to a binding hyperplane constraint, and hence the corresponding point x_i is a support vector. We will denote the set of support vectors by SV. The coefficients of the hyperplane w can then be computed as

$$w = \sum_{i \in SV} \alpha_i y_i x_i.$$

Complementary slackness also implies that if $\alpha_j < C$, then $\xi_j = 0$. Thus any point x_j for which $0 < \alpha_j < C$ satisfies $y_j(w^T x_j - b) = 1$. Any such point can be used to compute the value of b:

$$b = w^T x_j - y_j = \sum_{i \in SV} \alpha_i y_i x_i^T x_j - y_j.$$

If there are several such points, the average computed value of b is commonly taken to ensure the highest accuracy.

The significance of the dual formulation, however, is not just its computational ease. It also allows us to expand the power of support vector machines to data that are not linearly separable.

Consider, for example, the points shown on the left-hand side of Figure 14.10. They are clearly not separable by a linear equation, and it seems more plausible to separate them by a quadratic equation, say an ellipse centered at the origin, as is shown in the figure on the right. For every point $x = (x_1, x_2) \in \Re^2$, the quadratic equation would involve some linear combination of the terms x_1^2, $x_1 x_2$, and x_2^2. If we transformed each two-dimensional data vector x into the higher-dimensional "feature" vector $\Phi(x) = (x_1^2 \quad x_1 x_2 \quad x_2^2)^T$, we

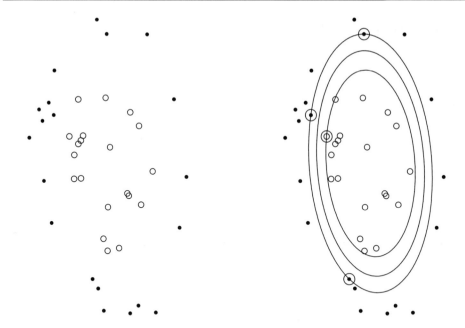

Figure 14.10. *Nonlinear support vector machines.*

could consider a three-dimensional hyperplane of the form $\Phi(x)^T w + b$ that separates the feature vectors corresponding to the data. The coefficients of the hyperplane would in turn provide the coefficient of the best ellipse that separates the data. We could also generalize this idea to define other separating nonlinear functions such as higher-order polynomials, which would involve even more "features."

While this idea is appealing, the dimensionality of the problem could easily explode. The problem of separating an n-dimensional set of data by an ellipsoid centered at the origin is transformed into a separating hyperplane problem of dimension $n(n + 1)/2$. Separation by higher-order polynomials makes the problem substantially larger. For many applied problems, such as voice recognition or face recognition, the dimension of the data is so large as to make this prohibitive. Luckily the dual formulation offers an approach for efficient computation.

Suppose that we mapped the lower-dimensional points x to points $\Phi(x)$ in some higher-dimensional space (called "feature space") and suppose that we used a separating hyperplane for the points $\Phi(x)$ in feature space. The dual to the resulting problem would be

$$\text{maximize} \quad \sum_{i=1}^{m} \alpha_i - \tfrac{1}{2}\alpha^T Y(\Phi(X)^T\Phi(X))Y\alpha$$
$$\text{subject to} \quad \sum_{i=1}^{m} y_i\alpha_i = 0$$
$$0 \le \alpha_i \le C,$$

where $\Phi(X)$ is the matrix whose columns are $\Phi(x_i)$.

We now make the observation that the only way the data appear in the dual problem is through the matrix $\Phi(X)^T\Phi(X)$ whose elements are $K(x_i, x_j) = \Phi(x_i)^T\Phi(x_j)$. It would be advantageous to have an efficient way to compute inner products of the form

$$K(x, z) = \Phi(x)^T\Phi(z)$$

for any two points x and z without going through the trouble of forming $\Phi(x)$ and $\Phi(z)$.

Consider the above problem of separating the data by an ellipse. With $\Phi(x) = (x_1^2 \quad x_1 x_2 \quad x_2^2)^T$ there are no special savings to be gained in computing the inner product of transformed points. Suppose instead that we defined the transformation to be $\Phi(x) = (x_1^2 \quad \sqrt{2}x_1 x_2 \quad x_2^2)^T$. Then we would still transform the data space to the same feature space, but now, the computation of $K(x, z)$ is easy, since $K(x, z) = \Phi(x)^T\Phi(z) = (x^Tz)^2$. Notice that the inner product is computed in the lower-dimensional input space rather than the higher-dimensional feature space, and it does not require explicit construction of the Φ vectors. In fact, as long as our function $K(x, z)$ corresponds to inner products of Φ vectors, we do not even need to know the exact form of Φ to compute $K(x, y)$, nor do we need to formulate and solve the dual problem. Furthermore, the classification of any new test point $\bar{x}$ does not require explicit construction of Φ vectors either. The decision is determined by the sign of $w^T\Phi(\bar{x}) + b$. From the formula $w = \sum_{i \in SV} \alpha_i y_i \Phi(x_i)$ we find that

$$w^T\Phi(\bar{x}) + b = \sum_{i \in SV} \alpha_i y_i \Phi(x_i)^T\Phi(\bar{x}) + b = \sum_{i \in SV} \alpha_i y_i K(x_i, \bar{x}) + b.$$

The computation requires only the products $K(x_i, \bar{x})$ and does not require that the Φ's (or the vector w) be constructed explicitly. (Note that the computation of b requires only products of the form $K(x_i, x_j)$; see the Exercises.)

The function $K(x, z) = \Phi(x)^T\Phi(z)$ is called the *kernel* of the machine. Our interest is in kernel functions that can be computed efficiently without constructing Φ. In fact we will not even be concerned with the specific form of Φ. The question is, under what conditions is a given function $K(x, z)$ an admissible kernel? The answer is given in a theorem due to Mercer. Simplified, it states that the matrix of all inner products of any number of points in the data space must be positive semidefinite. Examples of kernels that meet the Mercer condition are the homogeneous polynomial kernel $K(x, z) = (x^Tz)^p$ and the nonhomogeneous polynomial kernel $K(x, z) = (x^Tz + 1)^p$, both for any positive integer p. If, for example, we wish to separate a set of data by a cubic polynomial, we would use $p = 3$ without worrying about the precise form of the underlying Φ vector. Another admissible kernel is the widely used Gaussian radial-based kernel

$$K(x, z) = e^{-\|x-z\|^2/2}.$$

The associated Φ vector would be hard to construct since it corresponds to an infinite-dimensional feature space. Another example is the sigmoidal kernel $K(x, z) = \tanh(\gamma x^Tz + \mu)$, which is admissible for any positive scalars γ and μ.

Support vector machines have been successfully used in a wide range of applications. The search for appropriate kernels is a crucial part of effective classification. It is the dual viewpoint that makes this so successful.

Exercises

8.1. Consider the problem

$$\text{minimize} \quad f(x) = e^x$$

$$\text{subject to} \quad x \geq 1.$$

What is the solution to this problem? Formulate and solve the dual problem. Formulate the dual to the dual problem and show that it is equivalent to the primal problem.

8.2. Consider the problem of Example 14.30. Formulate the dual to the dual problem and show that it is equivalent to the primal problem.

8.3. Suppose that in the problem of Example 14.35 the constraints $0 \leq x \leq 2$ are treated explicitly. What would be the dual problem?

8.4. (Fletcher (2000)) Consider the problem

$$\text{minimize} \quad f(x) = \tfrac{1}{2}\eta x_1^2 + \tfrac{1}{2}x_2^2 + x_1$$

$$\text{subject to} \quad x_1 \geq 0.$$

Determine the solution to this problem for the cases $\eta = 1$ and $\eta = -1$. For each of the two cases, formulate the dual and determine whether its local solution gives the Lagrange multipliers at the optimal primal solution.

8.5. Consider the problem

$$\underset{x \in X}{\text{minimize}} \quad f(x) = \sum_{i=1}^{n} x_i \log(x_i/c_i)$$

$$\text{subject to} \quad Ax = b,$$

where the constants $\{c_i\}$ are positive, A is a matrix of full row rank, and $X = \{x : x > 0\}$. What is the dual to this problem? Determine expressions for the first and second derivatives of the dual function.

8.6. Consider the problem

$$\text{minimize} \quad f(x) = -3x$$

$$\text{subject to} \quad 1 - x^3 \geq 0$$

presented in Example 14.41. Formulate and solve the dual to the problem using each of the following sets. In each case compare the solution of the dual to the solution of the local dual in Example 14.41.

  (i) $X = \{x : 0.5 \leq x \leq 1.5\}$;
  (ii) $X = \{x : 0 < x\}$;
  (iii) $X = \{x : -1 \leq x\}$.

8.7. Formulate the dual to the following problem:

$$\text{minimize} \quad f(x) = \sum_{i=1}^{n} |x_i - a_i|$$

$$\text{subject to} \quad \sum_{i=1}^{n} x_i = 0.$$

Solve the dual problem, then determine the solution to the primal problem.

8.8. Let x_* be an optimal solution to problem (14.6), and assume that it is a regular point of the constraints. Let λ_* be an optimal solution to the dual problem, and suppose that $f(x_*) = \mathcal{L}_*(\lambda_*)$. Prove that λ_* is the vector of optimal Lagrange multipliers corresponding to x_*.

8.9. Verify the formulas for the first and second derivatives of the dual function.

8.10. Consider the least-squares problem

$$\text{minimize} \quad f(x) = \sum_{i=1}^{n} \tfrac{1}{2}(x_i - \bar{x}_i)^2$$

$$\text{subject to} \quad x_1 \geq x_2 \geq \cdots \geq x_n,$$

where $\bar{x}_i$ are given constants. Find an explicit form for the dual problem for the case $n = 4$. Generalize this result for all n.

8.11. Consider the problem

$$\text{minimize} \quad f(x)$$

$$\text{subject to} \quad g(x) \geq 0,$$

where f is a convex function and g is a vector of concave functions. Let $(\bar{x}, \bar{\lambda})$ be a dual feasible solution for this problem. Determine a lower bound on the optimal objective value $f(x_*)$.

8.12. Verify the formulation of the dual to the separating hyperplane problem.

8.13. Suppose you wish to separate a set of n-dimensional training points by a polynomial of degree 4. What would be the size of the feature space?

8.14. Let $K(x, z)$ be the kernel of a nonlinear support vector machine, and suppose an optimal solution α to the dual problem is found. Describe how the coefficient b can be computed without explicitly computing w.

14.9 Historical Background

The foundations of nonlinear optimization were developed in the eighteenth and nineteenth centuries in the study of the *calculus of variations*. The calculus of variations solves optimization problems whose parameters are not simple variables, but rather functions. For example, how should the shape of an automobile hood be chosen so as to minimize air resistance? Or, what path does a ray of light follow in an irregular medium? The calculus of variations is closely related to optimal control theory, where a set of "controls" is used to achieve a certain goal in an optimal way. For example, the pilot of an aircraft might wish to use the throttle and flaps to achieve a particular cruising altitude and velocity in a minimum amount of time or using a minimum amount of fuel. We are surrounded by devices designed using optimal control—in cars, elevators, heating systems, audio systems, etc.

The calculus of variations was inspired by problems in mechanics, especially the study of three-dimensional motion. It was used to derive many important laws of physics. This was done using the principle of least action. Action is defined to be the integral of the product of mass, velocity, and distance. The principle of least action asserts that nature acts so as to minimize this integral. To apply the principle, the formula for the action integral would be specialized to the setting under study, and then the calculus of variations would be used to optimize the integral. This general approach was used to derive important equations in mechanics, fluid dynamics, and other fields.

The most famous problem in the calculus of variations was posed in 1696 by John Bernoulli. It is called the brachistochrone ("least time") problem, and it asks that we find the shape of the curve down which a bead will slip from one point to another in the least time when accelerated only by gravity. The solution to the brachistochrone problem can be found by solving

$$\underset{x(t)}{\text{minimize}} \quad \frac{1}{\sqrt{2g}} \int_{t_1}^{t_2} \sqrt{\frac{1 + x'(t)^2}{x(t)}} dt,$$

where g is the gravitational constant. If this were a finite-dimensional problem, then it could be solved by setting the derivative of the objective function equal to zero, but seventeenth-century mathematics did not know how to take a derivative with respect to a function.

The brachistochrone problem was solved at the time by Newton and others, but the general techniques that inspired the name "calculus of variations" were not developed until several decades later. The first major results were obtained by Euler in the 1740s. He considered various problems of the general form

$$\underset{x(t)}{\text{minimize}} \quad \int_{t_1}^{t_2} f(t, x(t), x'(t))dt.$$

The brachistochrone problem is of this form. Euler solved these problems by discretizing the solution $x(t)$—approximating the solution by its values at finitely many points. This gave a finite-dimensional problem that could be solved using the techniques of calculus. Euler then took the limit of the approximate solutions as the number of discretization points tended to infinity. This approach was effortful and restrictive because it had to be adapted to the specifics of the problem being solved, and because there were restrictions on the types of problems for which it was successful.

Far more influential was the approach of Lagrange. He suggested that the solution be perturbed or "varied" from $x(t)$ to $x(t) + \epsilon y(t)$, where ϵ is a small number and $y(t)$ is some arbitrary function that satisfies $y(t_1) = y(t_2) = 0$. For the brachistochrone problem this latter condition ensures that the perturbed function still represents a path between the two points.

If $x(t)$ is a solution to the problem

$$\underset{x(t)}{\text{minimize}} \quad \int_{t_1}^{t_2} f(t, x(t), x'(t))dt,$$

then $\epsilon = 0$ will be a solution to

$$\underset{\epsilon}{\text{minimize}} \quad \int_{t_1}^{t_2} f(t, \, x(t) + \epsilon y(t), \, x'(t) + \epsilon y'(t)) \, dt.$$

This observation allowed Lagrange to convert the original infinite-dimensional problem to a one-dimensional problem that could be analyzed using ordinary calculus. Setting the derivative of the integral with respect to ϵ equal to zero at the point $\epsilon = 0$ leads to the equation

$$\frac{d}{dt}\frac{\partial f}{\partial x'} - \frac{\partial f}{\partial x} = 0.$$

This final condition is a first-order optimality condition for an unconstrained calculus-of-variations problem. It was first discovered by Euler, but the derivation here is due to Lagrange.

The name "calculus of variations" was chosen by Euler and was inspired by Lagrange's approach in "varying" the function $x(t)$. The optimality condition is stated as "the first variation must equal zero" by analogy with the condition $f'(x) = 0$ for a one-variable optimization problem. Euler was so impressed with Lagrange's work that he held back his own papers so that Lagrange could publish first, a magnanimous gesture by the renowned Euler towards the then young and unknown Lagrange.

There are additional first-order optimality conditions for calculus-of-variations problems. The theory is more complicated than for finite-dimensional optimization, and the necessary and sufficient conditions for an optimal solution were not fully understood until the 1870s, when Weierstrass studied this topic. A discussion of this theory can be found in the book by Gregory and Lin (1992, reprinted 2007).

Constraints can be added to problems in the calculus of variations just as in other optimization problems. A constraint might represent the principle of conservation of energy, or perhaps that the motion was restricted in some way, for example, that a planet was traveling in a particular orbit around the sun.

Both Euler and Lagrange considered problems of this type, and both were led to the concept of a multiplier. In the calculus of variations the multiplier might be a scalar (as it is in finite-dimensional problems) or, depending on the form of the constraint, it might be a function of the independent variable t. They have come to be called "Lagrange multipliers" but, as with the optimality condition, Euler discovered them first.

In his book *Mécanique Analytique* (1778), Lagrange includes an interpretation of the multiplier terms. He writes that they can be considered as representing the moments of forces acting on the moving particle, and serve to keep the constraints satisfied. This point of view is the basis for duality theory, although Lagrange does not seem to have followed up on this idea.

Euler and Lagrange considered only problems with equality constraints, but later authors allowed inequality constraints as well. When specialized to finite-dimensional problems, the optimality condition is referred to as the Karush–Kuhn–Tucker condition. Kuhn and Tucker derived this result in a 1951 paper. It was later discovered that Karush had proven the same result in his Master's thesis (1939) at the University of Chicago in the department headed by Bliss. We mention two aspects of his result: its treatment of inequality constraints, and the assumption or "constraint qualification" that was used to prove it. The first idea can be traced to Weierstrass and the second to Mayer, and both are outgrowths of the calculus of variations. (There are actually earlier examples of these results, due to Cauchy (1847a) and Gauss (1850–51), but these were isolated results and do not seem to have been influential.)

In the 1870s Weierstrass studied the calculus of variations and presented lectures on the results of his investigations. Weierstrass did not publish his work, and it only became widely known years later through the writings of those in attendance. According to the book by Bolza (1904, reprinted 2005), Weierstrass converted the inequality constraint

$$g(x) \leq 0$$

into an equivalent equality constraint

$$g(x) + s^2 = 0$$

using a squared "slack" variable s. He then applied Lagrange's results to obtain the optimality conditions (see the Exercises in Section 14.7). This technique is described in many sources dating from 1900 onward. Bolza later became a professor at the University of Chicago, establishing a connection between Weierstrass, Bliss, and Karush. Karush used this technique in his thesis.

The constraint qualification used by Karush, Kuhn, and Tucker relates feasible arcs (paths of feasible points leading to the solution) to the gradients of the constraints at the solution. This same condition was used by Mayer in 1886, although applied to a calculus-of-variations problem with equality constraints, and then in a chain of papers by various authors (including Bliss) leading to Karush's thesis. In these papers it is called a "normality" condition, and it is equivalent to requiring that the matrix of constraint gradients at the solution be of full rank. The implicit function theorem (see Appendix B.9) can be used to relate this to the condition on feasible arcs, an observation that is explicit in Mayer's work.

Duality theory did not become fully developed until the early 20th century, with many of the important steps coming from the calculus of variations. At first there were only isolated examples of duality. That is, someone would notice that a pair of problems—one a maximization problem, one a minimization problem—would have optimal solutions that were related to each other. An early example of this type was published in 1755 and is described in the paper by Kuhn (1991). In the 19th century various other examples were noticed, such as the relationship between currents and voltages in an electrical circuit. Gradually it was understood that duality was not an accidental phenomenon peculiar to these examples, but rather a general principle that applied to wide classes of optimization problems. By the 1920s, techniques had been developed for obtaining upper and lower bounds on the solutions to optimization problems by finding approximate solutions to the primal and dual problems. Duality as a general idea is described in the book by Courant and Hilbert (1953, reprinted 1989); the same material can be found in the 1931 German edition of this book; however, it is not present in the 1924 edition, suggesting a date of origin in the late 1920s.

14.10   Notes

Optimality Conditions—In our previous development we saw that an optimal point for a nonlinearly constrained problem must satisfy the first-order optimality conditions, provided that this point is regular. Furthermore, an optimal point that is not regular may not satisfy the first-order optimality conditions. A condition such as regularity that guarantees that the first-order optimality conditions will be satisfied at an optimal point is called a *constraint qualification*. For example, the condition that $T(x) = \mathcal{N}(A(x))$ for all feasible x is a constraint qualification for the equality-constrained problem. Whenever it holds, an optimal point must satisfy the first-order optimality conditions. The regularity condition is also a constraint qualification, but stronger (that is, more restrictive) than the condition $T(x) = \mathcal{N}(A(x))$ in the sense that any regular point will satisfy this condition. Unfortunately, the condition $T(x) = \mathcal{N}(A(x))$ is not practical because it is not easy to verify. Regularity is easy to verify and therefore practical.

The condition $T(x) = \mathcal{N}(A(x))$ is not the weakest (that is, least restrictive) constraint qualification, in the sense that there are points that satisfy the first-order optimality conditions that do not satisfy this constraint qualification. One example is the problem

$$
\begin{aligned}
\text{minimize} \quad & f(x) = x_2 \\
\text{subject to} \quad & x_2 = 0 \\
& x_2 - x_1^3 = 0.
\end{aligned}
$$

At the solution $x_* = (0, 0)^T$ there exists a vector of multipliers λ_* such that $\nabla \mathcal{L}(x_*, \lambda_*) = 0$, even though $T(x) \neq \mathcal{N}(A(x))$. (See the Exercises.)

Our comments here focus primarily on optimization problems with equality constraints. The constraint qualifications can be extended to problems that include inequality constraints as well.

During the 1960s and early 1970s, much research was conducted to develop constraint qualifications that would guarantee the fulfillment of the optimality conditions; see, for example, Abadie (1967), Mangasarian (1969, reprinted 1994), and Gould and Tolle (1971). The constraint qualification developed by Guignard (1969) is the weakest in the sense that it is not only sufficient but also necessary for the fulfillment of the optimality conditions. Unfortunately it too is not practical for computation.

A stronger but more useful constraint qualification for inequality-constrained problems is the Slater condition. It applies to a feasible set $\{ x : g_i(x) \geq 0, i = 1, \ldots, m \}$ where the functions g_i are concave. The condition states that, if S has a nonempty interior (that is, there is some point x_0 satisfying $g_i(x_0) > 0$ for all i), then an optimal point must satisfy the first-order necessary condition.

Duality—For more information on duality, see the books by Avriel (1976, reprinted 2003) and Rockafellar (1974). The use of the dual for creating nonlinear classifiers was proposed in 1992 by Boser et al. This work was built on the ideas of Aizerman et al. (1964) who first proposed the presentation of kernels as inner products in a feature space.

Chapter 15

Feasible-Point Methods

15.1 Introduction

In this chapter we examine methods that solve constrained optimization problems by attempting to remain feasible at every iteration. If all the constraints are linear, maintaining feasibility is straightforward. We discuss this case first. When nonlinear constraints are present, then more elaborate procedures are required. We discuss two such approaches to nonlinear constraints: sequential quadratic programming (SQP) and reduced-gradient methods. Both of these approaches generalize the techniques for linear constraints. Although they strive to maintain feasibility at every iteration, they do not always achieve this.

15.2 Linear Equality Constraints

The vast majority of methods for solving problems with linear equality constraints are feasible-point methods: they start from a feasible point and move along feasible descent directions to consecutively better feasible points. There are two features that make this approach particularly attractive. First, from a practical point of view it is advantageous that all iterates are feasible. Even if the algorithm fails to solve the problem to the accuracy desired, it might still provide a feasible solution that is usable. Second, by restricting movement to feasible directions, the equality-constrained problem is transformed into an unconstrained problem in the null space of the constraints. This new problem may then be solved using unconstrained minimization techniques.

As in the previous chapter we write the problem in the form

$$\text{minimize} \quad f(x)$$
$$\text{subject to} \quad Ax = b.$$

We assume that f is twice continuously differentiable, and that A is an $m \times n$ matrix of full row rank. As in the unconstrained case, the methods we describe are only guaranteed to find a stationary point of the problem. If the objective function is convex, this point will be a global minimizer of f. In the more general case, there is no guarantee that the point will be a global minimizer. In some cases it may not even be a local minimizer.

Let $\bar{x}$ be a feasible point. Since any other feasible point can be reached from $\bar{x}$ by moving in a feasible direction, the solution to the problem can be written as $x_* = \bar{x} + p$, where p solves the problem

$$\underset{p \in \Re^n}{\text{minimize}} \quad f(\bar{x} + p)$$

$$\text{subject to} \quad Ap = 0.$$

If Z denotes an $n \times (n - m)$ basis matrix for the null space of A, then $p = Zv$ for some $(n - m)$-dimensional vector v. This problem is equivalent to the unconstrained problem

$$\underset{v \in \Re^{n-m}}{\text{minimize}} \quad \phi(v) = f(\bar{x} + Zv).$$

We have reduced the problem of finding the "best" n-dimensional vector p to the unconstrained problem of finding the "best" $(n - m)$-dimensional vector v, that is, the v that minimizes the reduced function ϕ.

Conceptually, it is possible to minimize the reduced function ϕ using any of the unconstrained methods described in Chapters 11–13. In practice it is not necessary to provide an explicit expression for $\phi(v)$. Instead, it is possible to work directly with the original variables x, using

$$\nabla \phi(v) = Z^T \nabla f(x) \quad \text{and} \quad \nabla^2 \phi(v) = Z^T \nabla^2 f(x) Z,$$

where $x = \bar{x} + Zv$. This idea will be demonstrated shortly.

The basic tool for any solution method is the Taylor series. Expanding ϕ about zero (notice that $\phi(0) = f(\bar{x})$), we obtain

$$\phi(v) = f(\bar{x} + Zv) = f(\bar{x}) + v^T Z^T \nabla f(\bar{x}) + \tfrac{1}{2} v^T Z^T \nabla^2 f(\bar{x}) Z v + \cdots.$$

Ideally we would like to use the Newton direction; it is obtained by minimizing the quadratic approximation to $\phi(v)$ obtained from the Taylor series. Setting the gradient of the quadratic approximation to zero gives the following linear system in v:

$$\left[Z^T \nabla^2 f(\bar{x}) Z \right] v = -Z^T \nabla f(\bar{x}).$$

This system is known as the *reduced Newton equation*, or the *null-space equation*. Its solution may be written explicitly as

$$v = - \left(Z^T \nabla^2 f(\bar{x}) Z \right)^{-1} Z^T \nabla f(\bar{x}).$$

This is an estimate of the "best" v; in turn it provides an estimate of the "best" p

$$p = Zv = -Z \left(Z^T \nabla^2 f(\bar{x}) Z \right)^{-1} Z^T \nabla f(\bar{x}).$$

This direction is termed the *reduced Newton direction* at $\bar{x}$.

We can now derive the equality-constrained analog of the classical Newton method. The method sets

$$x_{k+1} = x_k + p,$$

where $p = -Z(Z^T \nabla^2 f(x_k)Z)^{-1}Z^T \nabla f(x_k)$ is the reduced Newton direction at x_k. This is just the mathematical representation of the method, and in practice explicit inverses are not normally computed. The method does not require that the reduced function be formed explicitly.

Example 15.1 (Reduced Newton Direction). Consider the problem

$$\text{minimize} \quad f(x) = \tfrac{1}{2}x_1^2 - \tfrac{1}{2}x_3^2 + 4x_1x_2 + 3x_1x_3 - 2x_2x_3$$
$$\text{subject to} \quad x_1 - x_2 - x_3 = -1,$$

and consider the feasible point $\bar{x} = (1, 1, 1)^T$. Choose

$$Z = \begin{pmatrix} 1 & 1 \\ 1 & 0 \\ 0 & 1 \end{pmatrix}$$

as a basis for the null space of $A = (1, -1, -1)$. The reduced gradient at $\bar{x}$ is

$$Z^T \nabla f(\bar{x}) = \begin{pmatrix} 1 & 1 & 0 \\ 1 & 0 & 1 \end{pmatrix} \begin{pmatrix} 8 \\ 2 \\ 0 \end{pmatrix} = \begin{pmatrix} 10 \\ 8 \end{pmatrix},$$

and the reduced Hessian matrix at $\bar{x}$ is

$$Z^T \nabla^2 f(\bar{x}) Z = \begin{pmatrix} 1 & 1 & 0 \\ 1 & 0 & 1 \end{pmatrix} \begin{pmatrix} 1 & 4 & 3 \\ 4 & 0 & -2 \\ 3 & -2 & -1 \end{pmatrix} \begin{pmatrix} 1 & 1 \\ 1 & 0 \\ 0 & 1 \end{pmatrix} = \begin{pmatrix} 9 & 6 \\ 6 & 6 \end{pmatrix}.$$

The reduced Newton equation yields

$$v = \begin{pmatrix} -2/3 \\ -2/3 \end{pmatrix},$$

and hence the reduced Newton direction is

$$p = Zv = \begin{pmatrix} -4/3 \\ -2/3 \\ -2/3 \end{pmatrix}.$$

Since the objective function is a quadratic and the reduced Hessian matrix is positive definite, a step length of $\alpha = 1$ leads to the optimum $x_* = (-\tfrac{1}{3}, \tfrac{1}{3}, \tfrac{1}{3})^T$. At x_*, the reduced gradient is $Z^T \nabla f(x_*) = Z^T(2, -2, -2)^T = (0, 0)^T$ as expected. The corresponding Lagrange multiplier (satisfying $\nabla f(x_*) = A^T \lambda_*$) is $\lambda_* = 2$. ∎

The reduced Newton direction is invariant with respect to the null-space matrix Z: mathematically, any choice of basis matrix Z will yield the same search direction p (see the Exercises). Numerically, however, the choice of Z can have a dramatic effect on the computation. In fact, the condition number of $Z^T \nabla^2 f(x) Z$ may be substantially larger than the condition number of $\nabla^2 f(x)$. It can be shown that

$$\text{cond}(Z^T \nabla^2 f(x) Z) \leq \text{cond}(\nabla^2 f(x)) \times \text{cond}(Z^T Z),$$

and the upper bound on the right can sometimes be tight. Thus if the matrix $Z^T Z$ has a large condition number (this may happen, for example, if the columns of Z are "almost" linearly dependent), then the reduced Hessian matrix may have a condition number that is much larger than that of the Hessian matrix itself. For this reason it is advisable to select

Z so that the condition number of $Z^T Z$ is small. One such matrix is the orthogonal basis Z obtained from the QR factorization, which satisfies $Z^T Z = I$ (see Section 3.3.4). This choice guarantees that the reduced Hessian matrix has a condition number no larger than that of the Hessian matrix, and consequently it is a good choice for small dense problems. For large sparse problems the QR method may be too expensive or may give bases which are too dense to be practical.

The classical reduced Newton method has all the properties of the classical Newton method. In particular, if the reduced Hessian matrix at the solution is positive definite, and if the starting point is sufficiently close to the solution, then the iterates will converge quadratically. In the more general case, however, the method may diverge or fail. To make the method reliable, some globalization strategy should be used (see Sections 11.5 and 11.6). Such strategies are essentially the same as in the unconstrained case. For example, we might use a line search to obtain a sufficient decrease in the objective value f and insist that the search direction produce sufficient descent.

If x_k is not a local solution and if the reduced Hessian matrix is positive definite, then the reduced Newton direction is a descent direction, since

$$p^T \nabla f(x_k) = -\nabla f(x_k) Z \left(Z^T \nabla^2 f(x_k) Z \right)^{-1} Z^T \nabla f(x_k) < 0.$$

If the reduced Hessian matrix is not positive definite, then the search direction may not be a descent direction and—worse still—may not even be defined. In this case, the modified factorizations described in Section 11.4 can be applied to the reduced Hessian matrix to provide a descent direction. (The reduced Hessian matrix will be positive definite in a neighborhood of a local solution point whenever this point satisfies the second-order sufficiency conditions.)

Other compromises on Newton's method may be made to obtain cheaper iterations. The simplest of all methods is of course the steepest-descent method. For the reduced function, this strategy gives the direction

$$v = -Z^T \nabla f(x_k)$$

in the reduced space, which yields the *reduced steepest-descent direction*

$$p = -Z Z^T \nabla f(x_k)$$

in the original space. Here the matrix Z may be any null-space matrix for A. Note, however, that the direction will vary with the particular choice of Z, unlike the reduced Newton direction.

Example 15.2 (Reduced Steepest-Descent Direction). The reduced gradient at the initial point of Example 15.1 is $Z^T \nabla f(\bar{x}) = (10, 8)^T$, and hence the reduced steepest-descent direction is $p = -Z Z^T \nabla f(\bar{x}) = (-18, -10, -8)^T$. Had we chosen

$$\hat{Z} = \begin{pmatrix} 2 & 0 \\ 1 & 4 \\ 1 & -4 \end{pmatrix}$$

as the null-space matrix for A, the reduced gradient would be

$$\hat{Z}^T \nabla f(\bar{x}) = \hat{Z}^T (8, 2, 0)^T = (18, 8)^T,$$

and the reduced steepest-descent direction would be

$$p = -\hat{Z}\hat{Z}^T \nabla f(\bar{x}) = (-36, -50, 14)^T. \quad \blacksquare$$

The reduced steepest-descent method has the same properties as that of its unconstrained counterpart. The iterations are cheap, but convergence may be very slow (see Section 12.2).

Quasi-Newton methods are a more sophisticated compromise. A common approach is to construct an approximation B_k to the reduced Hessian matrix at x_k. Let Z be a basis for the null space of A. The search direction is computed as $p = Zv$, where v is obtained by solving

$$B_k v = -Z^T \nabla f(x_k).$$

The approximation B_k is updated in much the same way as in the unconstrained case, except that all quantities are in the reduced space. For example, the symmetric rank-one update formula becomes

$$B_{k+1} = B_k + \frac{(\bar{y}_k - B_k \bar{s}_k)(\bar{y}_k - B_k \bar{s}_k)^T}{(\bar{y}_k - B_k \bar{s}_k)^T \bar{s}_k},$$

where $\bar{y}_k = Z^T (\nabla f(x_{k+1}) - \nabla f(x_k))$ and $\bar{s}_k = Z^T(x_{k+1} - x_k)$.

For large problems, "reduced" versions of the low-storage methods in Chapter 13 may be suitable. For example, a truncated-Newton method or limited-memory quasi-Newton methods could be used. If the matrix A is sparse, a null-space matrix Z that preserves sparsity should be chosen.

In principle, it is possible to solve an equality-constrained problem using any standard technique for unconstrained optimization. In practice, the numerical difficulties encountered when solving equality-constrained problems are not quite the same as those encountered when solving unconstrained problems, and it is not always possible to solve a large equality-constrained problem by simply applying general-purpose software for unconstrained optimization. One reason is that the reduced Hessian matrix in a constrained problem is often different in structure from the Hessian matrix in an unconstrained minimization problem. This is demonstrated in the following example.

Example 15.3 (Special Structure of Matrix Destroyed by Reduction). Consider the quadratic problem

$$\text{minimize} \quad f(x) = \tfrac{1}{2}x_1^2 + x_2^2 + 2x_3^2 + 4x_4^2$$
$$\text{subject to} \quad x_1 + x_2 + x_3 + x_4 = 1.$$

Taking the matrix

$$Z = \begin{pmatrix} \frac{1}{2} & \frac{1}{2} & \frac{1}{2} \\ \frac{1}{2} & -\frac{1}{2} & -\frac{1}{2} \\ -\frac{1}{2} & \frac{1}{2} & -\frac{1}{2} \\ -\frac{1}{2} & -\frac{1}{2} & \frac{1}{2} \end{pmatrix}$$

as a basis for the null space of the constraint matrix $A = (1, 1, 1, 1)$, we obtain

$$\nabla^2 f(x) = \begin{pmatrix} 1 & 0 & 0 & 0 \\ 0 & 2 & 0 & 0 \\ 0 & 0 & 4 & 0 \\ 0 & 0 & 0 & 8 \end{pmatrix} \quad \text{and} \quad Z^T \nabla^2 f(x) Z = \begin{pmatrix} \frac{15}{4} & \frac{3}{4} & -\frac{5}{4} \\ \frac{3}{4} & \frac{15}{4} & -\frac{9}{4} \\ -\frac{5}{4} & -\frac{9}{4} & \frac{15}{4} \end{pmatrix}.$$

Thus the reduced matrix $Z^T \nabla^2 f(x) Z$ is dense, even though the matrix $\nabla^2 f(x)$ is sparse. The special diagonal structure of the Hessian matrix is destroyed by the reduction. ■

To overcome these problems, special implementations that are tailored to the equality-constrained problem may be needed. For example, if a conjugate-gradient method is used to solve the reduced Newton equation (as in a truncated-Newton method), then it is not necessary to form the reduced Hessian matrix explicitly. The reason is that the conjugate-gradient method requires only Hessian-vector products—in this case, products of the form $(Z^T \nabla^2 f(x_k) Z) y$ for some vector y. This can be done by first computing $y_1 = Zy$, then computing $y_2 = \nabla^2 f(x_k) y_1$, and finally by computing the required Hessian-vector product as $y_3 = Z^T y_2$. If $\nabla^2 f$ is sparse, and Z can be represented in sparse form, then for large problems such techniques may bring substantial computational savings.

Once an optimal solution to the equality-constrained problem is obtained, the associated vector of Lagrange multipliers is computed. There are several reasons for this. First, the Lagrange multipliers measure the sensitivity of the solution to changes in the constraints. Second, the equality-constrained problem could be one of a sequence of problems generated by an algorithm for solving a problem with inequality constraints (see Section 15.4). In this case, the Lagrange multipliers indicate how to improve the current solution.

Computation of the Lagrange multipliers should be done in a cost-effective manner. We discuss this issue in more detail in the next section.

As was mentioned in the last chapter, the optimality conditions for

$$\text{minimize} \quad f(x)$$
$$\text{subject to} \quad Ax = b$$

can be used directly to derive algorithms. The conditions are

$$\nabla f(x) - A^T \lambda = 0$$
$$b - Ax = 0,$$

where λ is the vector of Lagrange multipliers. This is a system of nonlinear equations in the variables x and λ.

If Newton's method is applied to this system, then

$$x_{k+1} = x_k + p_k$$
$$\lambda_{k+1} = \lambda_k + v_k,$$

where the updates p_k and v_k are the solution to the Newton equations

$$\begin{pmatrix} \nabla^2 f(x_k) & -A^T \\ -A & 0 \end{pmatrix} \begin{pmatrix} p \\ v \end{pmatrix} = \begin{pmatrix} A^T \lambda_k - \nabla f(x_k) \\ Ax_k - b \end{pmatrix}.$$

(See the Exercises.) Some algorithms for constrained optimization work directly with this linear system, although some care must be taken to ensure that descent directions are obtained.

This linear system is closely related to the reduced Newton equation derived earlier. If x_k is feasible, then $b - Ax_k = 0$, and p must satisfy $Ap = 0$. Hence $p = Zv$ for some $(n - m)$-dimensional vector v. If we substitute for p in the first equation of our system, then

$$\nabla^2 f(x_k) Zv - A^T v = A^T \lambda_k - \nabla f(x_k).$$

Multiplying this equation on the left by Z^T, and using the fact that $Z^T A^T = 0$, we obtain

$$[Z^T \nabla^2 f(x_k) Z] v = -Z^T \nabla f(x_k),$$

the reduced Newton equation. Thus we can interpret the reduced Newton equation as a technique for applying Newton's method to the optimality conditions for the constrained optimization problem.

There is yet another way to compute the Newton direction. The derivation is based on the Newton equations, and it assumes that $\nabla^2 f(x)$ is positive definite for all x. Suppose that x_k is feasible so that $Ax_k - b = 0$, and let $\lambda_{k+1} = \lambda_k + v$. If we write $\nabla^2 f = \nabla^2 f(x_k)$ and $\nabla f = \nabla f(x_k)$, then the Newton equations are equivalent to

$$\begin{pmatrix} \nabla^2 f & -A^T \\ -A & 0 \end{pmatrix} \begin{pmatrix} p \\ \lambda_{k+1} \end{pmatrix} = \begin{pmatrix} -\nabla f \\ 0 \end{pmatrix}.$$

From the first equation we obtain

$$p = (\nabla^2 f)^{-1} A^T \lambda_{k+1} - (\nabla^2 f)^{-1} \nabla f.$$

Since $Ap = 0$,

$$0 = Ap = A(\nabla^2 f)^{-1} A^T \lambda_{k+1} - A(\nabla^2 f)^{-1} \nabla f.$$

This equation can be used to solve for λ_{k+1}:

$$\lambda_{k+1} = [A(\nabla^2 f)^{-1} A^T]^{-1} A(\nabla^2 f)^{-1} \nabla f.$$

Substituting this into the equation for p gives

$$p = -\left((\nabla^2 f)^{-1} - (\nabla^2 f)^{-1} A^T [A(\nabla^2 f)^{-1} A^T]^{-1} A(\nabla^2 f)^{-1} \right) \nabla f.$$

This is sometimes called the *projected Newton direction*.

Exercises

2.1. Use a reduced Newton method to solve the problem

$$\text{minimize} \quad f(x) = \tfrac{1}{2} x^T Q x$$
$$\text{subject to} \quad Ax = b,$$

where

$$Q = \begin{pmatrix} 0 & -13 & -6 & -3 \\ -13 & 23 & -9 & 3 \\ -6 & -9 & -12 & 1 \\ -3 & 3 & 1 & -1 \end{pmatrix},$$

$$A = \begin{pmatrix} 2 & 1 & 2 & 1 \\ 1 & 1 & 3 & -1 \end{pmatrix}, \quad \text{and} \quad b = \begin{pmatrix} 3 \\ 2 \end{pmatrix}.$$

Initialize the method with $x_0 = (1, 1, 0, 0)^T$.

2.2. Use a reduced steepest-descent method to solve the problem above, starting from $x_0 = (1, 1, 0, 0)^T$. Perform three iterations, using an exact line search.

2.3. Prove that the reduced steepest-descent direction is a descent direction at any point that is not a stationary point.

2.4. The particular variant of the reduced steepest-descent method, in which the null-space matrix for the constraints $Ax = b$ is the orthogonal projection matrix, is called the *projected gradient method*. Derive an explicit expression for the search direction at each iteration of the projected gradient method.

2.5. Let Z_1 and Z_2 be two basis matrices for the null space of a full row rank $m \times n$ matrix A. Let H be a matrix such that $Z_1^T H Z_1$ is positive definite. Prove that the matrix $Z_2^T H Z_2$ is also positive definite, and that

$$Z_1(Z_1^T H Z_1)^{-1} Z_1^T = Z_2 (Z_2^T H Z_2)^{-1} Z_2^T.$$

2.6. Let H be the three-dimensional identity matrix, and let $A = (2, 1, 1)$. Compute a basis matrix Z for the null space of A for which (a) $\text{cond}(Z^T H Z) = 1$, and (b) $\text{cond}(Z^T H Z) > 1000$.

2.7. Derive the formula for the reduced version of the symmetric rank-one update.

2.8. Derive a reduced version of the BFGS update formula.

2.9. Apply Newton's method to the first-order optimality condition for

$$\text{minimize} \quad f(x)$$
$$\text{subject to} \quad Ax = b.$$

Show that the iteration has the form

$$x_{k+1} = x_k + p_k$$
$$\lambda_{k+1} = \lambda_k + \nu_k,$$

where the updates p_k and ν_k form the solution to

$$\begin{pmatrix} \nabla^2 f(x_k) & -A^T \\ -A & 0 \end{pmatrix} \begin{pmatrix} p \\ \nu \end{pmatrix} = \begin{pmatrix} A^T \lambda_k - \nabla f(x_k) \\ Ax_k - b \end{pmatrix}.$$

Under what conditions on $\nabla^2 f(x)$ and A will this iteration have a quadratic convergence rate? Under what conditions on $\nabla^2 f(x)$ and A will p_k be a descent direction for f at x_k?

2.10. Let Z be a null-space matrix, and assume that $\nabla^2 f(x)$ is positive definite. Prove that

$$\text{cond}(Z^T \nabla^2 f(x)Z) \le \text{cond}(\nabla^2 f(x)) \times \text{cond}(Z^T Z).$$

15.3   Computing the Lagrange Multipliers

Consider the linear equality-constrained problem

$$\text{minimize} \quad f(x)$$
$$\text{subject to} \quad Ax = b.$$

Assume that the regularity condition holds, that is, that the rows of A are linearly independent. Consider the optimality condition

$$A^T\lambda_* = \nabla f(x_*).$$

This is a system of n equations in $m \le n$ unknowns, and so it cannot normally be expected to have a solution. At most feasible points x_*, this overdetermined system will be inconsistent, but if x_* is a local solution of the optimization problem, then the system will have a solution. How can such a solution λ_* be computed?

A useful tool is a matrix known as the *right inverse*. We define an $n \times m$ matrix A_r to be a *right inverse* for the $m \times n$ matrix A, if $AA_r = I_m$. It is easy to see that a matrix A has a right inverse only if it has full row rank. In this case, and if $m = n$, then the right inverse is unique, and $A_r = A^{-1}$. If $m < n$, the right inverse is generally not unique. For example, the matrices

$$\begin{pmatrix} \frac{3}{4} & 0 \\ -\frac{1}{4} & 0 \\ 0 & \frac{1}{2} \\ 0 & \frac{1}{2} \end{pmatrix} \quad \text{and} \quad \begin{pmatrix} 1 & 0 \\ 0 & 0 \\ 0 & 1 \\ 0 & 0 \end{pmatrix}$$

are both right inverses for the matrix

$$A = \begin{pmatrix} 1 & -1 & 0 & 0 \\ 0 & 0 & 1 & 1 \end{pmatrix}.$$

To see how right inverses are of use in solving the system $A^T\lambda_* = \nabla f(x_*)$, suppose that a solution to this system exists. If both sides of this equation are multiplied by A_r^T, then we obtain

$$\lambda_* = A_r^T \nabla f(x_*).$$

(Here A_r^T refers to $(A_r)^T$ and not $(A^T)_r$.) If the system $A^T\lambda_* = \nabla f(x_*)$ is consistent, its solution $\lambda_* = A_r^T \nabla f(x_*)$ is unique, even though the right inverse may not be unique. To verify this, note that $A^T\lambda_* = \nabla f(x_*)$ implies that $AA^T\lambda_* = A\nabla f(x_*)$, and so the unique solution is

$$\lambda_* = (AA^T)^{-1}A\nabla f(x_*).$$

(If A has full row rank, the matrix AA^T is positive definite, and hence its inverse exists.)

The linear system $A^T\lambda_* = \nabla f(x_*)$ is consistent if and only if $\nabla f(x_*)$ is a linear combination of the rows of A. Hence a vector λ_* computed via $\lambda_* = A_r^T \nabla f(x_*)$ will be a solution to the system if and only if

$$(I - A^T A_r^T)\nabla f(x_*) = 0$$

(see the Exercises).

In practice we will almost never find a point x_* that satisfies the optimality conditions precisely. Rather, we will (if successful) find some point x_k that satisfies the optimality conditions to within some specified tolerance. The point x_k will be an estimate of the optimal solution. Correspondingly, the vector $\lambda = A_r^T \nabla f(x_k)$ will be only an estimate of the vector of Lagrange multipliers at the solution. It is sometimes termed a *first-order* estimate, because, for sufficiently small ϵ, if $\|x_k - x_*\| = O(\epsilon)$, then $\|\lambda - \lambda_*\| = O(\epsilon)$ also.

In the rest of this section, we discuss methods for computing a right-inverse matrix. To avoid unnecessary work, the computation of a right-inverse matrix for a matrix A should be performed in conjunction with the computation of the null-space matrix for A. We will show that each of the methods for computing a null-space matrix for A (see Section 3.3) provides a right-inverse matrix at little or no additional cost. The discussion assumes that A is an $m \times n$ matrix of full row rank.

- *The variable reduction method (see Section 3.3.1).* In this method the variables are partitioned into m basic and $n - m$ nonbasic variables. The matrix A is partitioned into basic and nonbasic columns correspondingly. Assuming that the first m columns are basic, we have $A = (B, N)$ where B is an $m \times m$ nonsingular matrix, and the $n \times (n - m)$ matrix

$$Z = \begin{pmatrix} -B^{-1}N \\ I \end{pmatrix}$$

  is a basis matrix for the null space of A. The matrix

$$A_r = \begin{pmatrix} B^{-1} \\ 0 \end{pmatrix}$$

  is a right-inverse matrix for A that is available with no additional computation.

- *Orthogonal projection matrix (see Section 3.3.2).* Let the $n \times n$ matrix

$$P = I - A^T(AA^T)^{-1}A$$

  be the orthogonal projection matrix into the null space of A. A right inverse for A associated with the orthogonal projection is the matrix

$$A_r = A^T(AA^T)^{-1}.$$

This matrix, which we will denote by A^+, is a special right inverse. It satisfies the following four conditions:

$$AA^+A = A, \qquad (AA^+)^T = AA^+,$$
$$A^+AA^+ = A^+, \qquad (A^+A)^T = A^+A.$$

It can be shown that, for *any* $m \times n$ matrix A, there is a unique $n \times m$ matrix A^+ that satisfies these conditions. A^+ is called the *Penrose–Moore generalized inverse* of A. If A has full row rank, then $A^+ = A^T(AA^T)^{-1}$, and if A has full column rank, then $A^+ = (A^TA)^{-1}A^T$. Formulas for A^+ can also be developed when A does not have full row or column rank; see the book by Golub and Van Loan (1996).

Given a point x_k, the vector of Lagrange multiplier estimates $(A^+)^T \nabla f(x_k)$ obtained from the Penrose–Moore generalized inverse has the appealing property that it solves the problem

$$\underset{\lambda \in \Re^m}{\text{minimize}} \ \left\| A^T\lambda - \nabla f(x_k) \right\|_2$$

(see the Exercises). For this reason it is termed the *least-squares Lagrange multiplier estimate* at x_k.

Because the condition number of AA^T is the square of the condition number of A, the computation of $(AA^T)^{-1}$ is potentially unstable. The QR factorization provides a stable approach to computing this matrix that is practical for smaller problems (see below).

- *A nonorthogonal projection (see Section 3.3.3).* Let D be a positive-definite $n \times n$ matrix. Then the $n \times n$ projection matrix

$$P_D = I - DA^T(ADA^T)^{-1}A$$

is a null-space matrix for A. A right inverse for A associated with this projection is

$$A_r = DA^T(ADA^T)^{-1}.$$

- *The QR factorization (see Section 3.3.4).* The QR factorization represents A^T as a product of an orthogonal matrix Q and an upper triangular matrix R. Denoting the first m columns of Q by Q_1 and the last $n - m$ columns by Q_2, we have

$$A^T = QR = (\, Q_1 \quad Q_2 \,) \begin{pmatrix} R_1 \\ 0 \end{pmatrix},$$

where R_1 is an $m \times m$ triangular matrix. The $n \times (n - m)$ matrix

$$Z = Q_2$$

is an orthogonal basis for the null space of A. The matrix

$$A_r = Q_1 R_1^{-T}$$

is a right inverse for A available from the QR factorization at little additional cost. In fact, this matrix need not be formed explicitly: a computation of the form $\lambda = A_r^T \nabla f(x_k)$ may be done by first computing $y_1 = Q_1^T \nabla f(x_k)$ and then solving the triangular system $R_1 \lambda = y_1$. It is easy to show that $A_r = A^T(AA^T)^{-1}$, and hence this right inverse is in fact the Penrose–Moore generalized inverse of A (see the Exercises).

Just as with the "regular" matrix inverse, a right inverse is a useful notational tool, but it should rarely be formed explicitly. Instead, computations with respect to the right inverse should use the specific matrix factorizations that were employed to obtain the null-space matrix.

Example 15.4 (Right Inverses). We will construct several right inverses for

$$A = \begin{pmatrix} 1 & -1 & 0 & 0 \\ 0 & 0 & 1 & 1 \end{pmatrix}.$$

If variable reduction is used, with columns 2 and 3 of A being selected as the basic columns, then

$$B = \begin{pmatrix} -1 & 0 \\ 0 & 1 \end{pmatrix} \quad \text{and} \quad N = \begin{pmatrix} 1 & 0 \\ 0 & 1 \end{pmatrix}.$$

From these we determine that

$$Z = \begin{pmatrix} 1 & 0 \\ 1 & 0 \\ 0 & -1 \\ 0 & 1 \end{pmatrix} \quad \text{and} \quad A_r = \begin{pmatrix} 0 & 0 \\ -1 & 0 \\ 0 & 1 \\ 0 & 0 \end{pmatrix}.$$

(For all the right inverses that we compute in this example, it is straightforward to verify that $AZ = 0$ and $AA_r = I$.)

If the orthogonal projection matrix is used, then

$$P = I - A^T(AA^T)^{-1}A = \begin{pmatrix} \frac{1}{2} & \frac{1}{2} & 0 & 0 \\ \frac{1}{2} & \frac{1}{2} & 0 & 0 \\ 0 & 0 & \frac{1}{2} & -\frac{1}{2} \\ 0 & 0 & -\frac{1}{2} & \frac{1}{2} \end{pmatrix}.$$

The corresponding right inverse is

$$A_r = A^T(AA^T)^{-1} = \begin{pmatrix} \frac{1}{2} & 0 \\ -\frac{1}{2} & 0 \\ 0 & \frac{1}{2} \\ 0 & \frac{1}{2} \end{pmatrix}.$$

A nonorthogonal projection can also be used. If

$$D = \begin{pmatrix} 1 & 0 & 0 & 0 \\ 0 & 2 & 0 & 0 \\ 0 & 0 & 3 & 0 \\ 0 & 0 & 0 & 4 \end{pmatrix},$$

then

$$P_D = I - DA^T(ADA^T)^{-1}A = \begin{pmatrix} \frac{2}{3} & \frac{1}{3} & 0 & 0 \\ \frac{2}{3} & \frac{1}{3} & 0 & 0 \\ 0 & 0 & \frac{4}{7} & -\frac{3}{7} \\ 0 & 0 & -\frac{4}{7} & \frac{3}{7} \end{pmatrix}.$$

The corresponding right inverse is

$$A_r = DA^T(ADA^T)^{-1} = \begin{pmatrix} \frac{1}{3} & 0 \\ -\frac{2}{3} & 0 \\ 0 & \frac{3}{7} \\ 0 & \frac{4}{7} \end{pmatrix}.$$

If a QR factorization of A^T is used, then

$$Q = \begin{pmatrix} -\frac{1}{\sqrt{2}} & 0 & -\frac{1}{2} & -\frac{1}{2} \\ \frac{1}{\sqrt{2}} & 0 & -\frac{1}{2} & -\frac{1}{2} \\ 0 & -\frac{1}{\sqrt{2}} & \frac{1}{2} & -\frac{1}{2} \\ 0 & -\frac{1}{\sqrt{2}} & -\frac{1}{2} & \frac{1}{2} \end{pmatrix} \quad \text{and} \quad R = \begin{pmatrix} -\sqrt{2} & 0 \\ 0 & -\sqrt{2} \\ 0 & 0 \\ 0 & 0 \end{pmatrix}.$$

Z consists of the last two columns of Q:

$$Z = \begin{pmatrix} -\frac{1}{2} & -\frac{1}{2} \\ -\frac{1}{2} & -\frac{1}{2} \\ \frac{1}{2} & -\frac{1}{2} \\ -\frac{1}{2} & \frac{1}{2} \end{pmatrix}.$$

The right inverse is obtained from the formula

$$A_r = Q_1 R_1^{-T},$$

where Q_1 consists of the first two columns of Q,

$$Q_1 = \begin{pmatrix} -\frac{1}{\sqrt{2}} & 0 \\ \frac{1}{\sqrt{2}} & 0 \\ 0 & -\frac{1}{\sqrt{2}} \\ 0 & -\frac{1}{\sqrt{2}} \end{pmatrix}$$

and R_1 consists of the first two rows of R,

$$R_1 = \begin{pmatrix} -\sqrt{2} & 0 \\ 0 & -\sqrt{2} \end{pmatrix}.$$

Hence

$$A_r = \begin{pmatrix} \frac{1}{2} & 0 \\ -\frac{1}{2} & 0 \\ 0 & \frac{1}{2} \\ 0 & \frac{1}{2} \end{pmatrix}.$$

If

$$\nabla f(x) = (7 \quad -7 \quad -2 \quad -2)^T,$$

then for all of the above right inverses,

$$\lambda = A_r^T \nabla f(x) = \begin{pmatrix} 7 \\ -2 \end{pmatrix}.$$

No matter which right inverse is used, the same values of the Lagrange multipliers are obtained. ∎

Exercises

3.1. Compute four different right inverses for the matrix $A = (1, 4, -2)$. Demonstrate that each of the right inverses yields the same solution when used to solve the system $A^T \lambda = (-3, -12, 6)^T$.

3.2. Use the variable reduction method to find a right inverse for the matrix

$$A = \begin{pmatrix} 2 & -1 & 4 & 6 \\ 1 & -1 & 7 & -5 \end{pmatrix},$$

where the basis consists of the second and third variables, i.e., $x_B = (x_2, x_3)^T$. Use your computations to solve the system $A^T \lambda = (3, -1, 1, 17)^T$.

3.3. Consider the system $A^T \lambda = \nabla f(x_*)$, where A is a matrix with full row rank. Let A_r be a right inverse matrix for A. Prove that the system has a solution if and only if

$$(I - A^T A_r^T) \nabla f(x_*) = 0.$$

3.4. Consider the equality-constrained problem

$$\begin{aligned} \text{minimize} \quad & f(x) \\ \text{subject to} \quad & Ax = b. \end{aligned}$$

Let x_* be a local solution to the problem with associated Lagrange multiplier vector λ_*. Assume that A has full row rank, and let A_r be a right inverse for a matrix A. Prove that $x_* + A_r \delta$ is a feasible solution to the perturbed constraints $Ax = b + \delta$. Confirm that a first-order approximation to the objective function at this point is $f(x_*) + \delta^T \lambda_*$.

3.5. Let A be an $m \times n$ matrix of full row rank.

   (i) Prove that $A^T (AA^T)^{-1}$ satisfies the four conditions for a Penrose–Moore generalized inverse.

   (ii) Prove that if $m = n$, then $A^+ = A^{-1}$.

3.6. Let A be an $m \times n$ matrix of full column rank. Prove that $(A^T A)^{-1} A^T$ satisfies the four conditions for a Penrose–Moore generalized inverse.

3.7. Let A be an $m \times n$ matrix of full row rank, and let P be the projection matrix associated with $\mathcal{N}(A)$. Prove that $P^+ = P$.

3.8. Let A be an $m \times n$ matrix of full row rank. Prove that $\bar{\lambda} = (AA^T)^{-1} A \nabla f(x_k)$ solves the least-squares problem

$$\underset{\lambda \in \Re^m}{\text{minimize}} \ \left\| A^T \lambda - \nabla f(x_k) \right\|_2.$$

3.9. Let A be a full row rank matrix, and let A_r be a right inverse for A. Prove that the matrix $P = I - A_r A$ is a projection matrix.

3.10. Let $A_r = Q_1 R_1^{-T}$ be the right-inverse matrix for A obtained from the QR factorization of A^T. Prove that $A_r = A^T (AA^T)^{-1}$.

3.11. Write a computer program that solves

$$\begin{aligned} \text{minimize} \quad & f(x) = \tfrac{1}{2} x^T Q x - c^T x \\ \text{subject to} \quad & Ax = b, \end{aligned}$$

where Q is positive definite and A is an $m \times n$ matrix of full row rank. Use a reduced Newton direction with a step size of 1. Assume that a feasible starting point is available. Use your program to solve the problem with

$$A = \begin{pmatrix} 1 & 2 & 1 & 2 & 0 \\ 0 & 1 & 0 & 0 & -1 \\ 1 & 0 & 2 & 1 & 1 \end{pmatrix}, \quad Q = \begin{pmatrix} 2 & -1 & & & \\ -1 & 2 & & & \\ & & 1 & & \\ & & & 1 & -1 \\ & & & -1 & 1 \end{pmatrix}, \quad b = \begin{pmatrix} 5 \\ 1 \\ 3 \end{pmatrix},$$

$$c = (-7 \quad 3 \quad -3 \quad -8 \quad -2)^T, \quad \text{and} \quad x_0 = (0 \quad 1 \quad 1 \quad 1 \quad 0)^T.$$

Print out the optimal solution and the associated Lagrange multipliers.

15.4 Linear Inequality Constraints

In this section we discuss methods for solving problems with linear inequality constraints. The problem will be written in the form

$$\text{minimize} \quad f(x)$$
$$\text{subject to} \quad Ax \geq b,$$

where f is a twice continuously differentiable function. For simplicity we assume that the problem has only inequality constraints. The extension of the methods to include equality constraints is straightforward and will be mentioned at the end of the section.

Suppose that the point x_* is a local solution to this problem. Let $\hat{A}$ be a matrix whose rows are the coefficients of the active constraints at x_*, and let Z be a null-space matrix for $\hat{A}$. The first-order optimality conditions state that there exists $\lambda_* \geq 0$ such that $\nabla f(x_*) = \hat{A}^T \lambda_*$ (or equivalently, $Z^T \nabla f(x_*) = 0$).

Problems that have inequality constraints are significantly more difficult to solve than problems in which all constraints are equations. The reason is that it is not known in advance which inequality constraints are active at the solution. If we knew a priori the correct active set, then we could ignore the inactive constraints and minimize the objective function with all active constraints treated as equalities. In practice, unfortunately, we do not know what the correct active set is.

How can we resolve this combinatorial issue? A brute force approach would be to solve the equality-constrained problem for all possible selections of active constraints, and then choose the best solution. Even for a small problem, however, the number of such subproblems is enormous, and the amount of work could be prohibitive.

Active-set methods attempt to overcome this difficulty by moving sequentially from one choice of active constraints to another choice that is guaranteed to produce at least as good a solution. The hope is that only a fraction of the potential subproblems will be considered.

The most commonly used active-set methods are feasible-point methods. An initial feasible point, if none is provided, can be obtained much as in linear programming (see Section 5.5). At each iteration of the active-set method, we select a *working set* of constraints that are assumed to be active at the optimum. We attempt to minimize f with all constraints in the working set treated as *equalities*. All other constraints are considered inactive and temporarily ignored. (We should keep in mind, however, that these constraints should not be violated—more on that later.)

In general, the working set at the current point x is a subset of the constraints that are active at x, so that x is a feasible point for the working set. There may also be constraints that are active at x but that are not included in the working set; hence the working set is not necessarily equal to the active set.

Example 15.5 (Working Set). Consider the problem

$$\begin{array}{ll}
\text{minimize} & f(x) \\
\text{subject to} & x_1 + x_2 + x_3 \geq 1 \\
& x_1 \geq 0 \\
& x_2 \geq 0 \\
& x_3 \geq 0.
\end{array}$$

At $x_a = (0, \frac{1}{2}, \frac{1}{2})^T$ the first two inequalities are active. If both constraints are chosen for the working set, then we will attempt to minimize f on the set

$$\{ x : x_1 + x_2 + x_3 = 1, \ x_1 = 0 \} .$$

If only the first constraint is selected to be in the working set, then we will attempt to minimize f on the set

$$\{ x : x_1 + x_2 + x_3 = 1 \} . \quad \blacksquare$$

The problem of minimizing f subject to the constraints defined by the working set is an equality-constrained problem. Therefore we can use any of the techniques described in Section 15.2 to obtain a feasible search direction p. We could then use this search direction within a line search method and find an appropriate step length α. If this were truly an equality-constrained problem, we could use a standard line search method for unconstrained minimization and take a step considered to be "acceptable" by the line search. For example, we could test an initial step length of one, and use backtracking to find a step that satisfies a sufficient decrease condition (see Section 11.5). In a problem with inequality constraints, however, an "acceptable" step might lead to a point that is infeasible, that is, a point that violates one or more of the constraints that we have ignored. Geometrically, as we move from x along p, we may encounter the boundary of some constraint for some step length $\bar{\alpha}$. The value $\bar{\alpha}$ represents the largest possible step that may be taken without violating feasibility. The line search procedure must place an upper bound $\bar{\alpha}$ on any trial step, and the step length must never exceed $\bar{\alpha}$.

It is possible that the "best" acceptable step length that does not exceed $\bar{\alpha}$ is $\bar{\alpha}$ itself. This step leads to a point on the boundary of a constraint. The constraint encountered is now satisfied exactly at the new point, and it is added to the working set. With the step length determined, and any necessary adjustments made to the working set, the entire process is repeated.

It is also possible that the "best" acceptable step length is zero. This is an exceptional case, and it occurs when $\bar{\alpha} = 0$. If $\bar{\alpha} = 0$ occurs, no step is taken, and a constraint is added to the working set (since $\alpha = \bar{\alpha} = 0$).

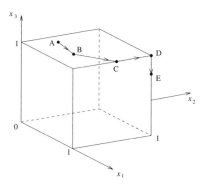

Figure 15.1. *Sequence of movements in an active-set method.*

Suppose now that we have found a point x that minimizes f on a given working set. Then the first-order optimality conditions for the equality-constrained problem are satisfied at x, and we can compute the Lagrange multipliers corresponding to the constraints in the working set. (The Lagrange multipliers for all other constraints are assumed to be zero.) If the Lagrange multipliers are all nonnegative, then x is also a solution to the original inequality-constrained problem and the problem is solved. However, if some Lagrange multiplier is negative, then x is not an optimal solution. The negative multiplier indicates that the function can be decreased if we move away from the corresponding constraint into the interior of the feasible region. Hence we can drop this constraint from the working set. We now have a new working set, and the process is repeated. (For simplicity, we allow only one constraint at a time to enter or leave the working set, although alternatives are possible.)

In summary, active-set methods have two major components. One component is the selection of the search direction. For a given working set, it is a feasible descent direction for the equality-constrained problem associated with the working set. The other component is a strategy for changing the working set. A constraint may be added to the working set if it is encountered in the course of a line search; a constraint may be dropped from the working set at an optimal point of the equality-constrained problem if its corresponding Lagrange multiplier is negative.

Example 15.6 (Sequence of Movements in an Active-Set Method). Figure 15.1 illustrates a possible sequence of movements in an active-set method for minimizing a convex function f on the box $0 \leq x_i \leq 1$, $i = 1, 2, 3$. The directions of movement in this example are arbitrary feasible descent directions and do not correspond to any specific method.

Let the point A be the starting point. The initial working set is chosen as the active constraints at A. It consists of the upper bound constraint on x_3, which is the single constraint that is active at A. An attempt is now made to minimize f subject to $x_3 = 1$, starting from A. Suppose that the first search ends at the point B. If B is still not a minimizer of f on $x_3 = 1$, then another search is made, starting from B. Suppose that this time the boundary of the constraint $x_1 \leq 1$ is encountered at the point C. Then C is accepted as the new solution, and the upper bound constraint on x_1 is added to the working set. If the point C is not a minimizer of f on the set $x_3 = 1$ and $x_1 = 1$, then a search is made on this constraint set. Suppose that a constraint boundary is encountered at the point $D = (1, 1, 1)^T$. Then the constraint $x_2 = 1$ is added to the working set. There are now three constraints in the working set: $x_3 = 1$, $x_2 = 1$, and $x_1 = 1$. Thus the point D is the only feasible solution to these constraints, and D is trivially a local minimizer of f on these constraints. The Lagrange multipliers corresponding to the three constraints are now computed. Suppose that the Lagrange multiplier corresponding to the constraint $x_3 = 1$ is negative, so f can be improved by decreasing x_3. After removing the constraint $x_3 = 1$, the working set consists of the constraints $x_2 = 1$ and $x_1 = 1$. A search along these constraints leads to, say, the point E. Assume that E is a local minimizer of f along these two constraints. If the Lagrange multipliers corresponding to these constraints are both nonnegative, then E is also a local minimizer for the original problem, and the algorithm is terminated.   ∎

We now discuss some of the above ideas in further detail. We assume that we have a point x that is feasible for the inequality-constrained problem. Denote the working set at x by $\mathcal{W}$; that is, $\mathcal{W}$ is the set of indices of the constraints in the working set. Denote the coefficient matrix for the constraints in the working set by $\bar{A}$ and the corresponding right-hand-side vector by $\bar{b}$. Let $\bar{Z}$ be a null-space matrix for $\bar{A}$. The equality-constrained problem for the working set $\mathcal{W}$ can be written in the form

$$\text{minimize} \quad f(x)$$
$$\text{subject to} \quad \bar{A}x = \bar{b}.$$

This problem is commonly solved using some feasible direction method. Thus, if p is the search direction at x, p satisfies $\bar{A}p = 0$. The step-size procedure will attempt to find an acceptable step length, while retaining feasibility with respect to *all* the constraints.

It is easy to compute the maximum feasible step length that can be taken along a direction p using a ratio test (see Section 3.1):

$$\bar{\alpha} = \max \left\{ \alpha : x + \alpha p \text{ is feasible} \right\}$$
$$= \min \left\{ (a_i^T x - b_i)/(-a_i^T p) : a_i^T p < 0, \quad i \notin \mathcal{W} \right\}.$$

We now outline a simple active-set method. Assume that a feasible starting point x_0 is given and let $\mathcal{W}$ be the index set of the active constraints at x_0. Let $\bar{A}$ be the constraint matrix for the active constraints, let $\bar{Z}$ be a null-space matrix for $\bar{A}$, and let $\bar{A}_r$ be a right inverse for $\bar{A}$. Set $k = 0$.

1. *The Optimality Test*—If $\bar{Z}^T \nabla f(x_k) = 0$, then

    (i) If no constraints are active, then the current point is a local (unconstrained) stationary point—stop.

    (ii) Else, compute Lagrange multipliers:

    $$\bar{\lambda} = \bar{A}_r^T \nabla f(x_k).$$

    (iii) If $\bar{\lambda} \geq 0$, then stop (a local stationary point has been reached). Otherwise, drop a constraint corresponding to a negative multiplier from the active set, and update $\mathcal{W}, \bar{A}, \bar{Z},$ and $\bar{A}_r$.

2. *The Search Direction*—Compute a descent direction p that is feasible with respect to the constraints in the working set.

3. *The Step*—Compute a step length satisfying $f(x_k + \alpha p) < f(x_k)$ and $\alpha \leq \bar{\alpha}$ where $\bar{\alpha}$ is the maximum feasible step along p.

4. *The Update*—Find the new solution point

$$x_{k+1} = x_k + \alpha p.$$

If a new constraint boundary is encountered ($\alpha = \bar{\alpha}$), add it to the working set and update $\mathcal{W}$, $\bar{A}$, Z, and $\bar{A}_r$ accordingly. (If more than one constraint boundary is encountered, then pick one of the constraints to enter the working set; this is a degenerate case.) Set $k = k + 1$ and return to 1.

Example 15.7 (Active-Set Method). Consider the problem

$$\begin{aligned} \text{minimize} \quad & f(x) = \tfrac{1}{2}(x_1 - 3)^2 + (x_2 - 2)^2 \\ \text{subject to} \quad & 2x_1 - x_2 \geq 0 \\ & -x_1 - x_2 \geq -4 \\ & x_2 \geq 0. \end{aligned}$$

We use an active-set method to solve this problem. The equality-constrained subproblems will be solved using a reduced Newton method. Since the objective function is quadratic, we use a step $\alpha = 1$ whenever it is feasible, that is, whenever $\bar{\alpha} \geq 1$. Otherwise, we shall take the step $\alpha = \bar{\alpha}$.

We use the variable reduction method to compute the null-space matrix for the active constraints, and for simplicity always use the "left-hand" submatrix of $\bar{A}$ as the basis matrix. This gives

$$\bar{A} = (\, B \quad N \,), \quad \bar{Z} = \begin{pmatrix} -B^{-1}N \\ I \end{pmatrix}, \quad \bar{A}_r = \begin{pmatrix} B^{-1} \\ 0 \end{pmatrix}.$$

Let $x_0 = (0, 0)^T$ be our feasible starting point. Since the first and third constraints are active, we have

$$\bar{A} = \begin{pmatrix} 2 & -1 \\ 0 & 1 \end{pmatrix}, \quad \bar{A}_r = \begin{pmatrix} 1/2 & 1/2 \\ 0 & 1 \end{pmatrix}.$$

The basis matrix $\bar{Z}$ is empty, and hence the reduced gradient $\bar{Z}^T \nabla f(x_0)$ vanishes trivially. We therefore compute the Lagrange multipliers

$$\bar{\lambda} = \begin{pmatrix} \lambda_1 \\ \lambda_3 \end{pmatrix} = \bar{A}_r^T \nabla f(x_0) = \begin{pmatrix} 1/2 & 0 \\ 1/2 & 1 \end{pmatrix} \begin{pmatrix} -3 \\ -4 \end{pmatrix} = \begin{pmatrix} -3/2 \\ -11/2 \end{pmatrix}.$$

Both multipliers are negative, and thus we should drop one of the constraints from the working set. We drop the third constraint because its multiplier is more negative. Updating the working set gives

$$\bar{A} = (\, 2 \quad -1 \,), \quad \bar{Z} = \begin{pmatrix} 1/2 \\ 1 \end{pmatrix}, \quad \bar{A}_r = \begin{pmatrix} 1/2 \\ 0 \end{pmatrix},$$

so that the new reduced gradient is $\bar{Z}^T \nabla f(x_0) = (-11/2)$. We now compute the reduced Newton search direction to obtain

$$p = -\bar{Z}(\bar{Z}^T \nabla^2 f(x_0)\bar{Z})^{-1}\bar{Z}^T \nabla f(x_0) = -\begin{pmatrix} 1/2 \\ 1 \end{pmatrix}(9/4)^{-1}(-11/2) = \begin{pmatrix} 11/9 \\ 22/9 \end{pmatrix}.$$

The maximum step to the boundary of the constraints is $\bar{\alpha} = 12/11$, and hence a step length of 1 is feasible ($\alpha = 1$). The full Newton step is taken to reach the point

$$x_1 = \begin{pmatrix} 0 \\ 0 \end{pmatrix} + 1\begin{pmatrix} 11/9 \\ 22/9 \end{pmatrix} = \begin{pmatrix} 11/9 \\ 22/9 \end{pmatrix}.$$

The next iteration begins with the optimality test:

$$\bar{Z}^T \nabla f(x_1) = (1/2 \quad 1)\begin{pmatrix} -16/9 \\ 8/9 \end{pmatrix} = (0).$$

Thus the reduced gradient vanishes at x_1, as expected. Since a local minimum of f with respect to the working set has been found, we compute the Lagrange multiplier corresponding to the active constraint:

$$\bar{\lambda} = (\lambda_1) = \bar{A}_r^T \nabla f(x_1) = (1/2 \quad 0)\begin{pmatrix} -16/9 \\ 8/9 \end{pmatrix} = (-8/9).$$

As this multiplier is negative, we drop the constraint from the active set. We are now left with no active constraints, which means that the problem is locally unconstrained. The reduced gradient is simply the gradient itself ($\bar{Z} = I$), and the search direction is simply the unconstrained Newton direction

$$p = -\nabla^2 f(x_1)^{-1} \nabla f(x_1) = -\begin{pmatrix} 1 & 0 \\ 0 & 1/2 \end{pmatrix}\begin{pmatrix} -16/9 \\ 8/9 \end{pmatrix} = \begin{pmatrix} 16/9 \\ -4/9 \end{pmatrix}.$$

Since the largest feasible step to the boundary is

$$\bar{\alpha} = \min\{1/4, 11/2\} = 1/4,$$

we use $\alpha = 1/4$, and at the new point

$$x_2 = \begin{pmatrix} 11/9 \\ 22/9 \end{pmatrix} + (1/4)\begin{pmatrix} 16/9 \\ -4/9 \end{pmatrix} = \begin{pmatrix} 5/3 \\ 7/3 \end{pmatrix},$$

the second constraint is active. We now update

$$\bar{A} = (-1 \quad -1), \quad \bar{Z} = \begin{pmatrix} -1 \\ 1 \end{pmatrix}, \quad \bar{A}_r = \begin{pmatrix} -1 \\ 0 \end{pmatrix}.$$

Again testing for optimality, we find that the reduced gradient at the new point is

$$\bar{Z}^T \nabla f(x_2) = (-1 \quad 1)\begin{pmatrix} -4/3 \\ 2/3 \end{pmatrix} = (2).$$

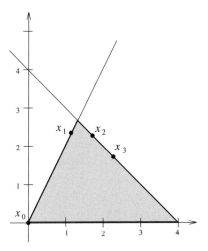

Figure 15.2. *Illustration of active-set algorithm.*

Since it is not zero, we continue with a search in the Newton direction. This gives

$$p = -\bar{Z}\left(\bar{Z}^T\nabla^2 f(x_2)\bar{Z}\right)^{-1}\bar{Z}^T\nabla f(x_2) = -\begin{pmatrix} -1 \\ 1 \end{pmatrix}(3)^{-1}(2) = \begin{pmatrix} 2/3 \\ -2/3 \end{pmatrix}.$$

The maximum step to the boundary is $\bar{\alpha} = 7/2$, and hence the step length is $\alpha = 1$. This gives

$$x_3 = \begin{pmatrix} 5/3 \\ 7/3 \end{pmatrix} + 1\begin{pmatrix} 2/3 \\ -2/3 \end{pmatrix} = \begin{pmatrix} 7/3 \\ 5/3 \end{pmatrix}.$$

At the new point the reduced gradient is

$$\bar{Z}^T\nabla f(x_3) = (-1 \quad 1)\begin{pmatrix} -2/3 \\ -2/3 \end{pmatrix} = (0).$$

Since

$$\bar{\lambda} = (\lambda_2) = \bar{A}_r^T\nabla f(x_3) = (-1 \quad 0)\begin{pmatrix} -2/3 \\ -2/3 \end{pmatrix} = (2/3) > 0,$$

the point satisfies the first-order optimality conditions and we terminate. Since the objective function f is strictly convex, the solution $x = (7/3, 5/3)^T$ is a strict global minimizer. The Lagrange multipliers corresponding to the three constraints are $\lambda_1 = 0$, $\lambda_2 = 2/3$, $\lambda_3 = 0$.

The progress of the algorithm is shown in Figure 15.2. ■

There are various modifications that can be made to the basic active-set method. Here we just give a brief overview of them.

One possible modification is to solve the equality-constrained subproblems *inexactly*, whenever there is reason to believe that the working set is not the optimal active set. The rationale is that faster progress may be made by obtaining a "better" working set than by getting a few extra digits of accuracy on an incorrect set. This idea requires care in

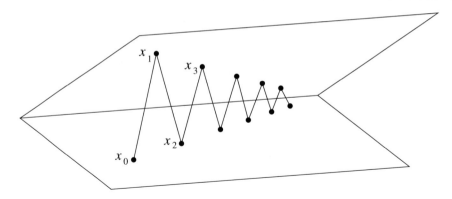

Figure 15.3. *Zigzagging.*

implementation, however. The reason is that as the solution to a problem becomes less accurate, the computed Lagrange multipliers also become less accurate. These inaccuracies can affect the sign of a computed Lagrange multiplier. Consequently, a constraint may erroneously be deleted from the working set, thereby wiping out any potential savings.

Another possible danger is zigzagging. This phenomenon can occur if the iterates cycle repeatedly between two working sets. This situation is depicted in Figure 15.3. Zigzagging cannot occur if the equality-constrained problems are solved sufficiently accurately before constraints are dropped from the working set.

To conclude, we indicate how the active-set method can be adapted to solve a problem of the form

$$\begin{array}{ll} \text{minimize} & f(x) \\ \text{subject to} & A_1 x \geq b_1 \\ & A_2 x = b_2 \end{array}$$

containing a mix of equality and inequality constraints. In this case, the equality constraints are kept permanently in the working set $\mathcal{W}$ since they must be kept satisfied at every iteration. The Lagrange multipliers for equality constraints can be positive or negative, and so do not play a role in the optimality test. The equality constraints also do not play a role in the selection of the maximum allowable step length $\bar{\alpha}$. These are the only changes that need be made to the active-set method.

15.4.1   Linear Programming

The simplex method for linear programming is a special case of an active-set method.[16] Suppose that we were trying to solve a linear program with n variables and m linearly independent equality constraints:

$$\begin{array}{ll} \text{minimize} & f(x) = c^T x \\ \text{subject to} & Ax = b \\ & x \geq 0. \end{array}$$

---

[16]This section uses the notation of Chapters 4 and 5.

Assume for simplicity that the problem is not degenerate. The matrix A represents only the general constraints and does not include the nonnegativity constraints. Let x_k be a basic feasible solution, with the variables ordered so that

$$x_k = \begin{pmatrix} x_B \\ x_N \end{pmatrix},$$

where x_B is the vector of basic variables and x_N is the (currently zero) vector of nonbasic variables. We write

$$A = (B \quad N) \quad \text{and} \quad c = \begin{pmatrix} c_B \\ c_N \end{pmatrix}.$$

The current working set corresponds to the equations

$$\bar{A}x_k \equiv \begin{pmatrix} B & N \\ 0 & I \end{pmatrix} \begin{pmatrix} x_B \\ x_N \end{pmatrix} = \begin{pmatrix} b \\ 0 \end{pmatrix}.$$

This is a system of n equations in n unknowns, and so

$$\bar{A}_r = \bar{A}^{-1} = \begin{pmatrix} B^{-1} & -B^{-1}N \\ 0 & I \end{pmatrix}.$$

The right-most portion of $\bar{A}_r$ is the same as the null-space matrix

$$\bar{Z} = \begin{pmatrix} -B^{-1}N \\ I \end{pmatrix}$$

that would be obtained if variable reduction were applied to A.

The first step in the active-set method is the optimality test. Since $\bar{A}$ is an $n \times n$ invertible matrix, the corresponding null-space matrix $\bar{Z}$ is empty, and so

$$\bar{Z}^T \nabla f(x_k) = \bar{Z}^T c = 0.$$

(This is guaranteed to occur, so this step is not an explicit part of the algorithm.) Lagrange multipliers are computed from the formula

$$\begin{aligned}
\bar{\lambda} &= \bar{A}_r^T \nabla f(x_k) \\
&= \begin{pmatrix} B^{-T} & 0 \\ -N^T B^{-T} & I \end{pmatrix} \begin{pmatrix} c_B \\ c_N \end{pmatrix} \\
&= \begin{pmatrix} B^{-T} c_B \\ c_N - N^T B^{-T} c_B \end{pmatrix}.
\end{aligned}$$

Hence $\bar{\lambda}$ consists of the vector of simplex multipliers $y = B^{-T} c_B$ together with the vector of reduced costs $\hat{c}_N = c_N - N^T B^{-T} c_B$. The simplex multipliers y are the Lagrange multipliers for the general constraints $Ax = b$, and so there is no sign restriction on them. The reduced costs are the multipliers for the constraints $x_N \geq 0$. For x_k to be optimal, the reduced costs must be nonnegative.

If the optimality conditions are not satisfied, then $\hat{c}_j < 0$ for some j, and the constraint $x_j \geq 0$ is dropped from the working set. This corresponds to deleting one of the bottom rows from $\bar{A}$. A feasible descent direction must then be determined. Such a direction p must satisfy $Ap = 0$ and so, if variable reduction is used, p can be written in the form

$$p = \begin{pmatrix} -B^{-1}N \\ I \end{pmatrix} v = \bar{Z}v$$

for some v. In addition, the other nonbasic variables must remain zero. This further restriction implies that we can choose

$$p = \bar{Z}_j,$$

where $\bar{Z}_j$ is the column of $\bar{Z}$ corresponding to $\hat{c}_j$. This is the same search direction as in the simplex method (see Section 5.3.2).

If a step of length α is taken along the search direction p, then the value of the objective function is given by

$$f(x_k + \alpha p) = c^T x_k + \alpha \hat{c}_j$$

and so $f(x_k + \alpha p) < f(x_k)$ for all $\alpha > 0$. The greatest reduction in f is obtained by taking the maximum feasible step along p. The upper bound $\bar{\alpha}$ on α is determined by a ratio test, a test that is the same in the simplex method as in the active-set method.

In the simplex method, the step length is $\alpha = \bar{\alpha}$, and a new constraint boundary is encountered. This boundary corresponds to one of the nonnegativity constraints becoming active, that is, to a variable leaving the basis. Hence a row is added to $\bar{A}$, making it again an $n \times n$ invertible matrix of the same general form as before. This completes an iteration of the active-set method.

The above remarks show that the simplex method is an active-set method in which variable reduction is used to represent the null space of the general constraints.

Exercises

4.1. Apply the simple active-set method to the problem in Example 15.5, using the initial guess $x_0 = (5, 9, 2)^T$, and with $f(x) = x_1^2 + 2x_2^2 + 3x_3^2$. Compute a search direction using a reduced Newton method.

4.2. Suppose that a quasi-Newton approximation is used to compute the search direction within the simple active-set method, and that only an approximation to the reduced Hessian is maintained.

 (i) Indicate how to adjust the quasi-Newton approximation to the reduced Hessian when a constraint is added or dropped from the working set.

 (ii) If a positive-definite quasi-Newton update formula is used, do your techniques guarantee that the approximation to the reduced Hessian remains positive definite?

4.3. Suppose that a constraint is dropped from the working set in the optimality test for the simple active-set method. If the search direction is computed using a reduced Newton method, and the Hessian is positive definite, is the search direction a feasible descent direction at the current iteration? What if some other method is used to compute the search direction?

4.4. Let $\bar{A}$ be the matrix corresponding to the working set, and assume that an orthogonal projection matrix $P = I - \bar{A}^T(\bar{A}\bar{A}^T)^{-1}\bar{A}^T$ is used as the null-space matrix. Let the constraint $a^T x \geq b$ be added to the working set, and assume that $Pa \neq 0$. Prove that the new orthogonal projection matrix $\hat{P}$ satisfies

$$\hat{P} = P - Pa(a^T P a)^{-1} a^T P.$$

What happens if $Pa = 0$?

4.5. Derive a "steepest-edge" rule to select a constraint to drop from the active set. (See Section 7.6.1.)

4.6. Consider the optimization problem in Example 15.7. At the solution x_*, the active set includes only the second constraint. For every feasible point x corresponding to this active set, compute an estimate λ of the optimal Lagrange multiplier λ_* using the right inverse in the example. What is the relationship between $\|x - x_*\|$ and $|\lambda - \lambda_*|$? For what values of x is the *sign* of the multiplier estimate correct?

4.7. Suppose that the line search within the active-set method uses a backtracking approach (see Section 11.5). Specify the algorithm for this line search. What conditions must the step length α satisfy?

4.8. Suppose that the line search within the active-set method is based on a Wolfe condition (see Section 11.5.1). What conditions must the step length α satisfy?

4.9. Consider the linear programming example from Section 5.2. Reinterpret the iterations of the simplex method as steps in an active-set method.

4.10. Write a computer program to solve the problem

$$\text{minimize} \quad f(x) = \tfrac{1}{2}x^T Q x - c^T x$$
$$\text{subject to} \quad Ax \geq b,$$

where Q is positive definite and A is an $m \times n$ matrix of full row rank. Use an active-set method based on a reduced Newton search direction. You may assume that the initial guess is a feasible point and that $\bar{\alpha}$ is always positive. Use your method to solve the problem

$$\text{minimize} \quad f(x) = x_1^2 + 2x_2^2$$
$$\text{subject to} \quad x_1 - x_2 \geq 3$$
$$2x_1 - 3x_2 \geq 6.$$

Use the initial guess $x_0 = (8, 2)^T$. Include an upper limit on the number of iterations (10 should be more than sufficient here). At each iteration print out the current point, the working set, the reduced gradient, the Lagrange multipliers (if applicable), the search direction, and the step size. Plot the sequence of iterates obtained.

15.5 Sequential Quadratic Programming

Sequential quadratic programming is a popular and successful technique for solving non-linearly constrained problems. The main idea is to obtain a search direction by solving a quadratic program, that is, a problem with a quadratic objective function and linear constraints. This approach is a generalization of Newton's method for unconstrained minimization.

As was pointed out in Section 14.6, methods for solving

$$\text{minimize} \quad f(x)$$
$$\text{subject to} \quad g(x) = 0$$

can be derived by applying Newton's method to the corresponding optimality conditions. (Here g is a vector of m functions g_i.) The Lagrangian for this problem is

$$\mathcal{L}(x, \lambda) = f(x) - \lambda^T g(x),$$

and the first-order optimality condition is

$$\nabla \mathcal{L}(x, \lambda) = 0.$$

Then the formula for Newton's method is

$$\begin{pmatrix} x_{k+1} \\ \lambda_{k+1} \end{pmatrix} = \begin{pmatrix} x_k \\ \lambda_k \end{pmatrix} + \begin{pmatrix} p_k \\ v_k \end{pmatrix},$$

where p_k and v_k are obtained as the solution to the linear system

$$\nabla^2 \mathcal{L}(x_k, \lambda_k) \begin{pmatrix} p_k \\ v_k \end{pmatrix} = -\nabla \mathcal{L}(x_k, \lambda_k).$$

This linear system has the form

$$\begin{pmatrix} \nabla_{xx}^2 \mathcal{L}(x_k, \lambda_k) & -\nabla g(x_k) \\ -\nabla g(x_k)^T & 0 \end{pmatrix} \begin{pmatrix} p_k \\ v_k \end{pmatrix} = \begin{pmatrix} -\nabla_x \mathcal{L}(x_k, \lambda_k) \\ g(x_k) \end{pmatrix}.$$

We will use these formulas to develop one of the most successful classes of methods for constrained optimization.

This system of equations represents the first-order optimality conditions for the optimization problem

$$\text{minimize} \quad q(p) = \tfrac{1}{2} p^T [\nabla_{xx}^2 \mathcal{L}(x_k, \lambda_k)] p + p^T [\nabla_x \mathcal{L}(x_k, \lambda_k)]$$
$$\text{subject to} \quad [\nabla g(x_k)]^T p + g(x_k) = 0,$$

with v_k the vector of Lagrange multipliers. This optimization problem is a *quadratic program*; that is, it is the minimization of a quadratic function subject to linear constraints. The quadratic function is a Taylor series approximation to the Lagrangian at (x_k, λ_k), and the constraints are a linear approximation to $g(x_k + p) = 0$.

In Chapter 11, where we examined unconstrained problems, the formulas for Newton's method corresponded to the minimization of a quadratic approximation to the objective function. Now in the constrained case, the formulas for Newton's method correspond to the constrained minimization of a quadratic approximation to the Lagrangian.

In a *sequential quadratic programming method*, at each iteration a quadratic program is solved to obtain (p_k, v_k). These are used to update (x_k, λ_k), and the process repeats at the new point. Each of the quadratic programs is solved using the techniques described in Section 15.2.

Example 15.8 (Sequential Quadratic Programming Method). We apply the SQP method to the problem

$$\text{minimize} \quad f(x_1, x_2) = e^{3x_1} + e^{-4x_2}$$
$$\text{subject to} \quad g(x_1, x_2) = x_1^2 + x_2^2 - 1 = 0.$$

The solution to this problem is $x_* \approx (-0.74834, 0.66332)^T$ with $\lambda_* \approx -0.21233$.

We use the initial guess $x_0 = (-1, 1)^T$ and $\lambda_0 = -1$. At this point

$$\nabla f = \begin{pmatrix} 3e^{3x_1} \\ -4e^{-4x_2} \end{pmatrix} = \begin{pmatrix} 0.14936 \\ -0.07326 \end{pmatrix}$$

$$\nabla^2 f = \begin{pmatrix} 9e^{3x_1} & 0 \\ 0 & 16e^{-4x_2} \end{pmatrix} = \begin{pmatrix} 0.44808 & 0 \\ 0 & 0.29305 \end{pmatrix}$$

$$g = x_1^2 + x_2^2 - 1 = (1)$$

$$\nabla g = \begin{pmatrix} 2x_1 \\ 2x_2 \end{pmatrix} = \begin{pmatrix} -2 \\ 2 \end{pmatrix} \quad \nabla^2 g = \begin{pmatrix} 2 & 0 \\ 0 & 2 \end{pmatrix}$$

$$\nabla_x \mathcal{L} = \nabla f - \lambda \nabla g = \begin{pmatrix} -1.85064 \\ 1.92674 \end{pmatrix}$$

$$\nabla_{xx}^2 \mathcal{L} = \nabla^2 f - \lambda \nabla^2 g = \begin{pmatrix} 2.44808 & 0 \\ 0 & 2.29305 \end{pmatrix}.$$

The corresponding quadratic program is

$$\text{minimize} \quad q(p) = \tfrac{1}{2} p^T [\nabla_{xx}^2 \mathcal{L}] p + p^T [\nabla_x \mathcal{L}]$$
$$\text{subject to} \quad [\nabla g]^T p + g = 0.$$

Its solution can be found using the first-order optimality conditions for the quadratic program:

$$\begin{pmatrix} \nabla_{xx}^2 \mathcal{L} & -\nabla g \\ -\nabla g^T & 0 \end{pmatrix} \begin{pmatrix} p \\ v \end{pmatrix} = \begin{pmatrix} -\nabla_x \mathcal{L} \\ g \end{pmatrix}$$

or

$$\begin{pmatrix} 2.44808 & 0 & 2 \\ 0 & 2.29305 & -2 \\ 2 & -2 & 0 \end{pmatrix} \begin{pmatrix} p_1 \\ p_2 \\ v \end{pmatrix} = \begin{pmatrix} 1.85064 \\ -1.92674 \\ 1 \end{pmatrix}.$$

The solution of the quadratic program is

$$p_0 = \begin{pmatrix} 0.22577 \\ -0.27423 \end{pmatrix} \quad \text{and} \quad v_0 = (-0.64896),$$

and the new estimates of the solution are

$$x_1 = x_0 + p_0 = \begin{pmatrix} -0.77423 \\ 0.72577 \end{pmatrix}$$
$$\lambda_1 = \lambda_0 + v_0 = (-0.35104).$$

The complete iteration is given in Table 15.1. Notice the rapid convergence rate, as expected for Newton's method. ∎

Table 15.1. *Sequential quadratic programming method.*

k	x_k		λ_k	$\|\nabla_x \mathcal{L}\|$	$\|g\|$
0	-1.00000	1.00000	-1.00000	3×10^0	1×10^0
1	-0.77423	0.72577	-0.35104	4×10^{-1}	1×10^{-1}
2	-0.74865	0.66614	-0.21606	1×10^{-2}	4×10^{-3}
3	-0.74834	0.66332	-0.21232	4×10^{-6}	8×10^{-6}
4	-0.74834	0.66332	-0.21233	3×10^{-11}	2×10^{-11}
5	-0.74834	0.66332	-0.21233	1×10^{-16}	0

Computing the updates p_k and ν_k by solving the quadratic program corresponds to applying Newton's method to the optimality conditions for the original problem. As a result, if the initial point is "close enough" to the solution, this method will have a quadratic convergence rate, provided that $\nabla^2 \mathcal{L}(x_*, \lambda_*)$ is nonsingular. This rapid convergence rate is observed in the example. The Hessian of the Lagrangian will be nonsingular if the regularity condition and the second-order sufficiency conditions for the original optimization problem are satisfied, that is, if $\nabla g(x_*)$ is of full rank and if $Z^T \nabla_{xx}^2 \mathcal{L}(x_*, \lambda_*) Z$ is positive definite, where Z is a basis matrix for the null space of $\nabla g(x_*)^T$; see the Exercises.

The sequential quadratic programming method outlined above is not often used in this simple form. There are two major reasons for this. First, it is not guaranteed to converge to a local solution to the optimization problem. Second, it is not always appropriate to use Newton's method because of its computational costs. These are the same reasons that the classical Newton method is not always used in the unconstrained case.

We first discuss the issue of computational costs. The method outlined above requires the Hessians of the objective function f and the constraint functions g. Once these have been obtained, a quadratic program must be solved to determine the updates p_k and ν_k (this corresponds to solving the linear system of optimality conditions for the quadratic program). As in the unconstrained case, it would be desirable to reduce these requirements for derivatives and to reduce the number of arithmetic operations required at each iteration of the algorithm. If large problems are being solved, it would also be desirable to reduce the storage requirements by using computational techniques that do not require matrix storage.

One way to reduce the expense of this algorithm is to use a quasi-Newton approximation to the Hessian of the Lagrangian (see Section 12.3). If this is done, then second derivatives need not be computed. In addition, the quasi-Newton matrix can be maintained in the form of a factorization, and in this way the arithmetic costs of the method can also be reduced.

Choosing an update formula for the quasi-Newton approximation in the constrained case can be a more complicated decision than in the unconstrained case. For example, one might choose to approximate the full Hessian of the Lagrangian, or perhaps just the reduced Hessian. This choice can affect the convergence theory for the method. (See the Notes.)

Convergence results for the sequential quadratic programming method are obtained by insisting that (x_{k+1}, λ_{k+1}) be a "better" estimate of the solution than (x_k, λ_k). In the unconstrained case, progress is measured in terms of the objective function. In the constrained case it is common to measure progress in terms of an auxiliary *merit function*. Usually, a merit function is the sum of terms that include the objective function and the amount of infeasibility of the constraints. If the new point reduces the objective function and reduces infeasibility, then the value of the merit function will decrease. In many instances, however, improvements in the objective value come at the expense of feasibility, and vice versa, so the merit function must balance these two goals. One example of a merit function is the *quadratic penalty function*

$$\mathcal{M}(x) = f(x) + \rho g(x)^T g(x) = f(x) + \rho \sum_{i=1}^{m} g_i(x)^2,$$

where ρ is some positive number. The greater the value of ρ, the greater the penalty for infeasibility. This merit function is a function of x only; other examples of merit functions may be functions of both x and λ.

Example 15.9 (Merit Function). Let us use the merit function

$$\mathcal{M}(x) = f(x) + 10g(x)^T g(x)$$

to measure the progress of the sequential quadratic programming method in the previous example. At the first iteration

$$\begin{aligned}
x_0 &= (-1.00000 \quad 1.00000)^T \\
f(x_0) &= 0.068103 \\
g(x_0) &= 1.00000 \\
\mathcal{M}(x_0) &= 10.068103
\end{aligned}$$

and

$$\begin{aligned}
x_1 &= (-0.77423 \quad 0.72577)^T \\
f(x_1) &= 0.152864 \\
g(x_1) &= 0.126174 \\
\mathcal{M}(x_1) &= 0.312063
\end{aligned}$$

so, in terms of this merit function, the point x_1 is better than the point x_0. At the next iteration,

$$\begin{aligned}
x_2 &= (-0.74865 \quad 0.66614)^T \\
f(x_2) &= 0.175457 \\
g(x_2) &= 0.004219 \\
\mathcal{M}(x_2) &= 0.175635,
\end{aligned}$$

indicating that x_2 is better than both x_1 and x_0 for this merit function.   ∎

Ideally, the merit function would be chosen so that (x_*, λ_*) would be a local minimizer of the merit function if and only if it were a local solution of the optimization problem. If this were true, then a line search with respect to the merit function could be performed:

$$\begin{pmatrix} x_{k+1} \\ \lambda_{k+1} \end{pmatrix} = \begin{pmatrix} x_k \\ \lambda_k \end{pmatrix} + \alpha \begin{pmatrix} p_k \\ v_k \end{pmatrix},$$

where α is chosen so that

$$\mathcal{M}(x_{k+1}, \lambda_{k+1}) < \mathcal{M}(x_k, \lambda_k).$$

For this to be successful, the search direction from the quadratic program would have to be a descent direction for the merit function. The convergence theorem for unconstrained minimization (Theorem 11.7) could then be adapted to prove convergence in the constrained case.

Unfortunately, it is rarely possible to guarantee that the local minimizers of the merit function and the local solutions of the optimization problem coincide. For the merit function

$$\mathcal{M}(x) = f(x) + \rho g(x)^T g(x),$$

some of the local minimizers of $\mathcal{M}$ approach local solutions of the constrained problem in the limit as $\rho \to \infty$ (see Section 16.2). In addition $\mathcal{M}$ may have local minima at points where $g(x) \neq 0$, that is, at infeasible points. Other merit functions have analogous deficiencies that can limit the applicability of convergence theorems or can complicate the development of sequential quadratic programming methods.

In the unconstrained case we assume that search directions are descent directions with respect to the objective function. In this setting, we assume that either p_k or the combined vector (p_k, v_k) is a descent direction with respect to the merit function. A common way to guarantee this is to insist that the reduced Hessian for the quadratic program be positive definite. (Recall that in the unconstrained case we insisted that the Hessian of the quadratic model be positive definite.)

If a quasi-Newton approximation to the Hessian is used to define the quadratic program, then positive-definiteness of the reduced Hessian is often guaranteed by the choice of quasi-Newton update formula. If Newton's method is used, so that the Hessian in the quadratic program is $\nabla_{xx}^2 \mathcal{L}(x_k, \lambda_k)$, then it is necessary to test if the reduced Hessian is positive definite, and to modify it if it is not. (See also Section 11.4.) In the constrained case, testing whether the reduced Hessian is positive definite is more complicated than in the unconstrained case, particularly if the quadratic program is solved via the linear system

$$\begin{pmatrix} \nabla_{xx}^2 \mathcal{L}(x_k, \lambda_k) & -\nabla g(x_k) \\ -\nabla g(x_k)^T & 0 \end{pmatrix} \begin{pmatrix} p_k \\ v_k \end{pmatrix} = \begin{pmatrix} -\nabla_x \mathcal{L}(x_k, \lambda_k) \\ g(x_k) \end{pmatrix}.$$

In this case, the reduced Hessian may not be available, and more elaborate tests for positive-definiteness must be used.

Near the solution of the constrained problem we would normally like to take a step of $\alpha = 1$ in the line search so that the quadratic convergence rate of Newton's method could be achieved. Hence the merit function should be chosen so that a step of $\alpha = 1$ is guaranteed to be accepted in the limit as the solution is approached. For certain merit functions this is not true. In such cases it is possible to give examples where a step of $\alpha = 1$ is unacceptable at every iteration, no matter how close the current point is to the solution.

We will ignore many of these difficulties and only prove that p_k is a descent direction for the quadratic penalty merit function. Even more, we will make the simplifying assumption that the full Hessian of the Lagrangian (or some approximation to it) is positive definite. If a quasi-Newton approximation to the Hessian is used, this assumption is not unreasonable, but for Newton's method it is restrictive. The lemma below (together with Theorem 11.7) can be used to show that

$$\lim_{k \to \infty} \nabla \mathcal{M}(x_k) = 0.$$

For large values of ρ, local solutions of the constrained problem are approximate local minimizers of $\mathcal{M}(x)$ to within $O(1/\rho)$, and so this argument provides the rough outline of a convergence theorem. For further results, ones with less restrictive assumptions and with more satisfying conclusions, see the references mentioned in the Notes.

Lemma 15.10 (Descent Direction for Merit Function). *Assume that (p_k, ν_k) is computed as the solution to the quadratic program*

$$\begin{array}{ll} \text{minimize} & q(p) = \frac{1}{2} p^T H p + p^T [\nabla_x \mathcal{L}(x_k, \lambda_k)] \\ \text{subject to} & [\nabla g(x_k)]^T p + g(x_k) = 0, \end{array}$$

where H is some positive-definite approximation to $\nabla_{xx}^2 \mathcal{L}(x_k, \lambda_k)$. If $p_k \neq 0$, then

$$p_k^T \nabla \mathcal{M}(x_k) < 0$$

for all sufficiently large values of ρ, where

$$\mathcal{M}(x) = f(x) + \rho g(x)^T g(x);$$

that is, p_k is a descent direction with respect to this merit function.

Proof. For simplicity, we omit the subscript k on most vectors and write $\nabla_x \mathcal{L} \equiv \nabla_x \mathcal{L}(x_k, \lambda_k)$, etc. The proof is based directly on the optimality conditions for the quadratic program

$$\begin{aligned} Hp &= -\nabla f + \nabla g(\lambda + \nu) \\ \nabla g^T p &= -g. \end{aligned}$$

The first of these conditions implies that

$$p^T \nabla f = -p^T H p + p^T \nabla g \hat{\lambda},$$

where $\hat{\lambda} \equiv \lambda + \nu$.

Using these conditions we obtain

$$\begin{aligned} p^T \nabla \mathcal{M} &= p^T \nabla f + 2\rho p^T (\nabla g) g \\ &= p^T \nabla f - 2\rho g^T g \\ &= -p^T H p + p^T (\nabla g) \hat{\lambda} - 2\rho g^T g \\ &= -p^T H p - g^T \hat{\lambda} - 2\rho g^T g. \end{aligned}$$

The first term is negative, since we have assumed that H is positive definite and $p \neq 0$. If $g = 0$, then $p^T \nabla \mathcal{M} < 0$. If $g \neq 0$, then we can ensure that $p^T \nabla \mathcal{M} < 0$ by choosing ρ large enough. To see this, we write the last two terms as

$$\sum_{i=1}^{m} (-g_i \hat{\lambda}_i - 2\rho g_i^2),$$

and so if

$$\rho > \max_i \left\{ -\frac{\hat{\lambda}_i}{2g_i} : g_i \neq 0 \right\},$$

the summation will be negative, showing that p is a descent direction.   □

So far we have discussed only problems with equality constraints. Consider now a problem of the form

$$\begin{array}{ll} \text{minimize} & f(x) \\ \text{subject to} & g_1(x) \geq 0 \\ & g_2(x) = 0, \end{array}$$

where g_1 and g_2 are vectors of constraint functions. We can develop a sequential quadratic programming method for this problem as well.

Our earlier quadratic program was based on a quadratic approximation to the Lagrangian function and a linear approximation to the constraints. If the same approximations are used here, we obtain

$$\begin{array}{ll} \text{minimize} & q(p) = \frac{1}{2}p^T [\nabla^2_{xx}\mathcal{L}(x_k, \lambda_k)]p + p^T [\nabla f(x_k)] \\ \text{subject to} & [\nabla g_1(x_k)]^T p + g_1(x_k) \geq 0 \\ & [\nabla g_2(x_k)]^T p + g_2(x_k) = 0. \end{array}$$

The solution p_k to this quadratic program provides the step to the next estimate of the solution of the original constrained problem, and the Lagrange multipliers provide an estimate of the Lagrange multipliers of the original problem. The objective for this quadratic problem includes the term $p^T[\nabla f(x_k)]$—not $p^T[\nabla_x\mathcal{L}(x_k, \lambda_k)]$ as might have been expected. Without this modification, the method is not guaranteed to converge. (See the Exercises.)

The references in the Notes discuss the case where the quadratic program has no solution. This quadratic program can be solved using the active-set method discussed in Section 15.4.

Exercises

5.1. Show that the solution to

$$\begin{array}{ll} \text{minimize} & f(x) = e^{3x_1 + 4x_2} \\ \text{subject to} & g(x_1, x_2) = x_1^2 + x_2^2 - 1 = 0 \end{array}$$

is $x_* = (-\frac{3}{5}, -\frac{4}{5})^T$ with $\lambda_* = -\frac{5}{2}e^{-5}$. Set up and solve the quadratic subproblem at $x = (\frac{3}{5}, \frac{4}{5})^T$ and $\lambda = \frac{5}{2}e^{-5}$.

5.2. Prove that the Hessian of the Lagrangian function

$$\mathcal{L}(x, \lambda) = f(x) - \lambda^T g(x)$$

will be nonsingular if the regularity condition and the second-order sufficiency conditions for the original optimization problem are satisfied, that is, if $\nabla g(x_*)$ is of full rank and if $Z^T \nabla^2_{xx}\mathcal{L}(x_*, \lambda_*)Z$ is positive definite, where Z is a basis matrix for the null space of $\nabla g(x_*)$.

5.3. Consider the problem

$$\text{minimize} \quad f(x) = x^T Q x$$
$$\text{subject to} \quad x^T x \leq 1,$$

where

$$Q = \begin{pmatrix} 2 & 0 \\ 0 & 1 \end{pmatrix}.$$

Suppose that $x_0 = (1, 1)^T$ and that λ_0 is specified. Find x_1 and λ_1 using the sequential quadratic programming method. Under what conditions is $\lambda_1 = \lambda_0$?

5.4. (This problem illustrates a pathology of sequential quadratic programming called the *Maratos effect*.) Consider the problem

$$\text{minimize} \quad f(x) = 2(x_1^2 + x_2^2 - 1) - x_1$$
$$\text{subject to} \quad x_1^2 + x_2^2 - 1 = 0.$$

(i) Prove that $x_* = (1, 0)^T$ is the minimizer with associated multiplier $\lambda_* = 3/2$.

(ii) Suppose that $x_k = (\cos\theta, \sin\theta)^T$, where $\theta \approx 0$. Verify that x_k is feasible and close to x_*. Let $\lambda_k = \lambda_*$. Set up and solve the corresponding quadratic program at this point. Show that the solution is $p_k = (\sin^2\theta, -\sin\theta\cos\theta)^T$. What is v_k? Show that if $x_{k+1} = x_k + p_k$, then $f(x_{k+1}) > f(x_k)$ and that x_{k+1} is infeasible. This shows that, even when close to the solution, a unit step along the Newton direction may increase the value of *any* merit function.

5.5. Suppose that you wish to solve an optimization problem of the form

$$\text{minimize} \quad f(x)$$
$$\text{subject to} \quad g_1(x) \geq 0$$
$$\quad\quad\quad\quad\quad g_2(x) = 0,$$

where the search directions are computed using the quadratic subproblem

$$\text{minimize} \quad q(p) = \tfrac{1}{2} p^T [\nabla_{xx}^2 \mathcal{L}(x_k, \lambda_k)] p + p^T [\nabla_x \mathcal{L}(x_k, \lambda_k)]$$
$$\text{subject to} \quad [\nabla g_1(x_k)]^T p + g_1(x_k) \geq 0$$
$$\quad\quad\quad\quad\quad [\nabla g_2(x_k)]^T p + g_2(x_k) = 0.$$

Suppose that the initial guess for the Lagrange multipliers satisfies $\lambda_0 > \lambda_*$. Prove that the estimates of the Lagrange multipliers for the inequality constraints do not converge to the optimal Lagrange multipliers.

15.6 Reduced-Gradient Methods

Reduced-gradient methods try to maintain feasibility at every iteration. This approach has several advantages. If each estimate of the solution is feasible, the algorithm can be stopped before it converges, and the approximate solutions may still be useful. Also, guaranteeing convergence is simpler because progress can be measured directly using the value of the objective function, rather than with an auxiliary merit function. The disadvantage of reduced-gradient methods is the computational expense of ensuring that nonlinear constraints remain satisfied at every iteration.

We apply a version of the reduced-gradient method to the problem

$$\begin{array}{ll} \text{minimize} & f(x) \\ \text{subject to} & g(x) = 0, \end{array}$$

where g is a vector of m functions g_i. Our derivation begins the same way as for sequential quadratic programming. The Lagrangian is

$$\mathcal{L}(x, \lambda) = f(x) - \lambda^T g(x),$$

and if Newton's method is applied to the first-order optimality condition, then

$$\begin{pmatrix} x_{k+1} \\ \lambda_{k+1} \end{pmatrix} = \begin{pmatrix} x_k \\ \lambda_k \end{pmatrix} + \begin{pmatrix} p_k \\ v_k \end{pmatrix},$$

where p_k and v_k are obtained as the solution to the linear system

$$\nabla^2 \mathcal{L}(x_k, \lambda_k) \begin{pmatrix} p_k \\ v_k \end{pmatrix} = -\nabla \mathcal{L}(x_k, \lambda_k).$$

This linear system has the form

$$\begin{pmatrix} \nabla_{xx}^2 \mathcal{L}(x_k, \lambda_k) & -\nabla g(x_k) \\ -\nabla g(x_k)^T & 0 \end{pmatrix} \begin{pmatrix} p_k \\ v_k \end{pmatrix} = \begin{pmatrix} -\nabla_x \mathcal{L}(x_k, \lambda_k) \\ g(x_k) \end{pmatrix}.$$

It is at this point that the derivations for the two methods diverge.

In the reduced-gradient method we use these formulas to derive a *portion* of the search direction, that is, the portion that lies in the null space of the constraint gradients. If Z_k is a basis matrix for the null space of $\nabla g(x_k)^T$, and Y_k is a basis matrix for the range space of $\nabla g(x_k)$, then

$$p_k = Z_k p_z + Y_k p_Y,$$

where p_z is the solution to the reduced system

$$Z_k^T [\nabla_{xx}^2 \mathcal{L}(x_k, \lambda_k)] Z_k p_z = -Z_k^T \nabla_x \mathcal{L}(x_k, \lambda_k).$$

This formula determines p_z in the reduced-gradient method.

If all the constraints are linear, then this formula is equivalent to the formula for the reduced Newton method derived in Section 15.2. In that case the matrix Z_k will be the same at every iteration, and

$$\begin{aligned} \nabla_{xx}^2 \mathcal{L}(x_k, \lambda_k) &= \nabla^2 f(x_k) \\ Z_k^T \nabla_x \mathcal{L}(x_k, \lambda_k) &= Z_k^T \nabla f(x_k). \end{aligned}$$

(This last equation is valid even in the nonlinear case; see the Exercises.) Hence the reduced-gradient method is a generalization of the reduced Newton method for linearly constrained problems.

The remaining portion of the search direction is determined from the condition that the new estimate of the solution must be feasible: $g(x_{k+1}) = 0$. Since $x_{k+1} = x_k + p_k$, this condition has the form

$$g(x_k + Z_k p_z + Y_k p_y) = 0.$$

This is a system of m nonlinear equations in the m variables p_y. (We are assuming here that $\nabla g(x_k)$ is a matrix of full rank.) If the constraints are linear, then $p_y = 0$. If the constraints are nonlinear, some auxiliary algorithm must be applied to this nonlinear system to determine p_y. For example, Newton's method could be used.

The reduced-gradient method is illustrated in the following example.

Example 15.11 (Reduced-Gradient Method). We apply the reduced-gradient method to the same problem used in Example 15.8:

$$\text{minimize} \quad f(x_1, x_2) = e^{3x_1} + e^{-4x_2}$$
$$\text{subject to} \quad g(x_1, x_2) = x_1^2 + x_2^2 - 1 = 0.$$

The solution to this problem is $x_* \approx (-0.74834, 0.66332)^T$ with $\lambda_* \approx -0.212325$.

We again use the initial guess $x_0 = (-1, 1)^T$, even though this point does not satisfy the constraints. At this point

$$\nabla f = \begin{pmatrix} 3e^{3x_1} \\ -4e^{-4x_2} \end{pmatrix} = \begin{pmatrix} 0.14936 \\ -0.07326 \end{pmatrix}$$

$$\nabla^2 f = \begin{pmatrix} 9e^{3x_1} & 0 \\ 0 & 16e^{-4x_2} \end{pmatrix} = \begin{pmatrix} 0.44808 & 0 \\ 0 & 0.29305 \end{pmatrix}$$

$$g = x_1^2 + x_2^2 - 1 = (1)$$

$$\nabla g = \begin{pmatrix} 2x_1 \\ 2x_2 \end{pmatrix} = \begin{pmatrix} -2 \\ 2 \end{pmatrix} \quad \nabla^2 g = \begin{pmatrix} 2 & 0 \\ 0 & 2 \end{pmatrix}.$$

An estimate of the Lagrange multiplier is needed to determine the gradient and Hessian of the Lagrangian. We compute a multiplier estimate λ_k by solving

$$\underset{\lambda}{\text{minimize}} \ \|\nabla f(x_k) - \lambda \nabla g(x_k)\|_2^2.$$

At the initial point x_0, the multiplier estimate is $\lambda_0 = -0.055655$. Using this value we obtain

$$\nabla_x \mathcal{L} = \nabla f - \lambda \nabla g = \begin{pmatrix} 0.03805 \\ 0.03805 \end{pmatrix}$$

$$\nabla_{xx}^2 \mathcal{L} = \nabla^2 f - \lambda \nabla^2 g = \begin{pmatrix} 0.55939 & 0 \\ 0 & 0.40436 \end{pmatrix}.$$

We use variable reduction to compute the null-space matrix Z_k and the range-space matrix Y_k, based on

$$\nabla g^T = (\, 2x_1 \quad 2x_2 \,) \equiv (\, B \quad N \,).$$

Then

$$Z_k = \begin{pmatrix} -B^{-1}N \\ I \end{pmatrix} = \begin{pmatrix} -x_2/x_1 \\ 1 \end{pmatrix} \quad \text{and} \quad Y_k = \frac{1}{\sqrt{x_1^2 + x_2^2}} \begin{pmatrix} x_1 \\ x_2 \end{pmatrix}.$$

At this iteration

$$Z_0 = \begin{pmatrix} 1 \\ 1 \end{pmatrix}.$$

The null-space portion of the search direction is obtained by solving

$$Z_k^T [\nabla_{xx}^2 \mathcal{L}] Z_k p_z = -Z_k^T \nabla_x \mathcal{L},$$

or

$$(0.963754) p_z = -0.076099,$$

so that $p_z = -0.078961$.

The remaining portion of the search direction, $Y_k p_Y$, is determined by solving

$$g(x_k + Z_k p_z + Y_k p_Y) = 0$$

using Newton's method. In this example,

$$Y_k p_Y = \begin{pmatrix} y_1 \\ y_2 \end{pmatrix} \gamma$$

for some unknown γ, where y_1 and y_2 are the components of Y_k. If we define $\hat{x}_k = x_k + Z_k p_z$, then the condition for $Y_k p_Y$ has the form

$$\phi(\gamma) \equiv (\hat{x}_1 + \gamma y_1)^2 + (\hat{x}_2 + \gamma y_2)^2 - 1 = 0.$$

Applying Newton's method to this equation gives the iteration

$$\gamma_{i+1} = \gamma_i - \frac{\phi(\gamma_i)}{\phi'(\gamma_i)} = \gamma_i - \frac{(\hat{x}_1 + \gamma_i y_1)^2 + (\hat{x}_2 + \gamma y_2)^2 - 1}{2 y_1 (\hat{x}_1 + \gamma_i y_1) + 2 y_2 (\hat{x}_2 + \gamma_i y_2)}.$$

In this example we initialize the iteration with $\gamma_0 = 0$. Then

$$
\begin{aligned}
\gamma_0 &= 0, & \phi(\gamma_0) &= 1 \times 10^0 \\
\gamma_1 &= -0.35796205114318, & \phi(\gamma_1) &= 1 \times 10^{-1} \\
\gamma_2 &= -0.41861845196074, & \phi(\gamma_2) &= 4 \times 10^{-3} \\
\gamma_3 &= -0.42046619052517, & \phi(\gamma_3) &= 3 \times 10^{-6} \\
\gamma_4 &= -0.42046790833490, & \phi(\gamma_4) &= 3 \times 10^{-12} \\
\gamma_5 &= -0.42046790833638, & \phi(\gamma_5) &= -2 \times 10^{-16}.
\end{aligned}
$$

The overall search direction is

$$p_0 = Z_0 p_z + Y_0 p_Y = \begin{pmatrix} 1 \\ 1 \end{pmatrix} (-0.078961) + \begin{pmatrix} -0.70711 \\ 0.70711 \end{pmatrix} (-0.420468) = \begin{pmatrix} 0.218355 \\ -0.376276 \end{pmatrix}$$

and

$$x_1 = x_0 + p_0 = \begin{pmatrix} -0.78164 \\ 0.62372 \end{pmatrix}.$$

The complete iteration is given in Table 15.2. The initial guess of the solution is not feasible, but all later estimates of the solution are feasible (to 15 digits). The method converges rapidly, as expected for Newton's method. ∎

Table 15.2. *Reduced-gradient method.*

k	x_k		λ_k	$\|\nabla_x \mathcal{L}\|$	$\|g\|$
0	-1.00000	1.00000	-0.055655	5×10^{-2}	1×10^{0}
1	-0.78164	0.62372	-0.215306	8×10^{-2}	8×10^{-15}
2	-0.74950	0.66201	-0.212401	3×10^{-3}	8×10^{-15}
3	-0.74834	0.66332	-0.212325	2×10^{-6}	3×10^{-15}

The reduced-gradient method corresponds to using Newton's method in the null space of the constraints, and so it can be expected to converge quadratically in nondegenerate cases. This quadratic convergence rate is observed in Example 15.11. As before, Newton's method is not guaranteed to converge, and even if it does converge, it may converge to a maximum or stationary point and not a minimum. Hence some globalization strategy must be employed to ensure convergence to a local solution of the optimization problem.

A line search can be used to guarantee convergence, as was done with the sequential quadratic programming method. If all the solution estimates x_k are feasible points, then the value of the quadratic penalty merit function is

$$\mathcal{M}(x_k) = f(x_k) + \rho g(x_k)^T g(x_k) = f(x_k),$$

and so a line search can be performed using the objective function f itself as a merit function. As long as the conditions of Theorem 11.7 are satisfied, convergence is guaranteed.

Example 15.12 (Progress Measured by Objective Function). For the problem in Example 15.11, the successive values of the objective function are

$$f(x_0) = 0.068103$$
$$f(x_1) = 0.178359$$
$$f(x_2) = 0.176349$$
$$f(x_3) = 0.176347,$$

so in this case the objective value increases at the first iteration while the reduced-gradient method finds a feasible point, but the objective value decreases at every other iteration, even though no line search is employed. ∎

The line search for a reduced-gradient method is more complicated than in the unconstrained case. For each trial value of the step length α, the trial point must satisfy

$$g(x_k + \alpha Z_k p_z + Y_k p_y) = 0.$$

Hence p_y depends on α and must be computed by solving a nonlinear system of equations. For "large" values of α there may be no p_y that satisfies the constraints, further complicating the line search algorithm. For these reasons, it is not entirely correct to say that the reduced-gradient method produces a search "direction" since in fact the method must search along an arc defined by p_z and α.

Theorem 11.7 makes several assumptions about the search direction and the step length. For example, the search direction must be a descent direction, that is, we require

$$f(x_k + \epsilon Z_k p_z + Y_k p_Y(\epsilon)) < f(x_k)$$

for "small" positive values of ϵ. (Here we have written $p_Y(\epsilon)$ to emphasize that p_Y depends on ϵ.) Conditions that guarantee descent are given in the following lemma. The remaining requirements of the convergence theorem can be derived in a similar manner.

Lemma 15.13 (Descent Direction). *Assume that the reduced-gradient method is applied to the problem*

$$\begin{aligned} \text{minimize} \quad & f(x) \\ \text{subject to} \quad & g(x) = 0. \end{aligned}$$

Let x_k be the kth estimate of the solution, with $g(x_k) = 0$. Assume that $\nabla g(x)$ is of full rank for all x in a neighborhood of x_k. Also assume that the null-space portion p_z of the search direction is computed from

$$Z_k^T[\nabla_{xx}^2 \mathcal{L}(x_k, \lambda_k)]Z_k p_z = -Z_k^T \nabla_x \mathcal{L}(x_k, \lambda_k),$$

where λ_k is an estimate of the Lagrange multipliers, and where

$$Z_k^T[\nabla_{xx}^2 \mathcal{L}(x_k, \lambda_k)]Z_k$$

is positive definite. Define $p_Y(\epsilon)$ as the solution to

$$g(x_k + \epsilon Z_k p_z + Y_k p_Y(\epsilon)) = 0.$$

If $Z_k^T \nabla_x \mathcal{L}(x_k, \lambda_k) \neq 0$, then

$$f(x_k + \epsilon Z_k p_z + Y_k p_Y(\epsilon)) < f(x_k)$$

for all sufficiently small positive values of ϵ.

Proof. To simplify the proof, we omit the subscript k on all quantities and use the notation $H = Z_k^T[\nabla_{xx}^2 \mathcal{L}(x_k, \lambda_k)]Z_k$. Using a Taylor series expansion, we obtain

$$\begin{aligned} f(x + \epsilon Z p_z + Y p_Y(\epsilon)) &= f(x) + \epsilon \nabla f(x)^T Z p_z + \nabla f(x)^T Y p_Y(\epsilon) + O(\epsilon^2) \\ &= f(x) - \epsilon \nabla f(x)^T Z H^{-1} Z^T \nabla f(x) + \nabla f(x)^T Y p_Y(\epsilon) + O(\epsilon^2). \end{aligned}$$

If the matrix H is positive definite, then the second term in this expansion will be negative. If ϵ is small enough, the final term will be negligible in comparison with the second term. We will prove that the third term is proportional to ϵ^2, and so (for small enough values of ϵ) it too will be negligible.

We first prove that

$$Y p_Y(\epsilon) = O(\epsilon),$$

and then use this result to prove that

$$Y p_Y(\epsilon) = O(\epsilon^2).$$

We will analyze p_Y using the condition

$$g(x + Zp_z + Yp_Y(\epsilon)) = 0.$$

Expanding in a Taylor series, we obtain

$$g(x) + \nabla g(x + \xi)^T(\epsilon Zp_z + Yp_Y(\epsilon)) = 0,$$

where $x + \xi$ is some unknown point between x and $x + Zp_z + Yp_Y(\epsilon)$. Since $g(x) = 0$, and since $\nabla g(x)$ is of full rank, we can rearrange this equation to obtain

$$Yp_Y(\epsilon) = -\epsilon[\nabla g(x + \xi)\nabla g(x + \xi)^T]^{-1}\nabla g(x + \xi)^T Zp_z.$$

This shows that $Yp_Y(\epsilon) = O(\epsilon)$; it follows that $\xi = O(\epsilon)$ also.

To prove that $Yp_Y(\epsilon) = O(\epsilon^2)$, we take the equation

$$\nabla g(x + \xi)^T(\epsilon Zp_z + Yp_Y(\epsilon)) = 0$$

and further expand $\nabla g(x + \xi)$ using a Taylor series. We obtain

$$[\nabla g(x) + O(\epsilon)]^T(\epsilon Zp_z + Yp_Y(\epsilon)) = 0.$$

Since $Z^T\nabla g(x) = 0$ (because Z is a null-space matrix for $\nabla g(x)$), this is equivalent to

$$\nabla g(x)^T Yp_Y(\epsilon) = -O(\epsilon^2)Zp_z.$$

Hence, because $\nabla g(x)$ is full rank,

$$Yp_Y(\epsilon) = -O(\epsilon^2)[\nabla g(x)\nabla g(x)^T]^{-1}Zp_z = O(\epsilon^2).$$

Thus,

$$f(x + \epsilon Zp_z + Yp_Y(\epsilon)) = f(x) - \epsilon\nabla f(x)^T ZH^{-1}Z^T\nabla f(x) + O(\epsilon^2),$$

and it now follows that descent is obtained for "small" positive values of ϵ. $\quad\square$

The lemma assumes that

$$Z_k^T[\nabla_{xx}^2\mathcal{L}(x_k, \lambda_k)]Z_k$$

is positive definite. This will not be true in general. If this condition is not satisfied, then the techniques in Section 11.4 can be used to guarantee that a descent direction is obtained.

The reduced-gradient method need not be based on Newton's method. A quasi-Newton approximation to the Hessian, or the steepest-descent method, or a truncated-Newton method could be used instead to compute p_z. As in the unconstrained case, the overall method would converge at a slower rate.

The main disadvantage of a reduced-gradient method is that each iteration can be expensive. The method (at least in the idealized form given here) insists that the estimates of the solution x_k remain feasible. When the constraints are nonlinear, this means that a system of nonlinear equations must be solved at each trial point to restore feasibility. If the constraints are "nearly" linear, or if there are not too many nonlinear constraints, then this will not be difficult. Otherwise, solving the nonlinear system may require a great deal of computation, and in some cases the nonlinear system will have no solution.

Variants of the reduced-gradient method have been developed that are more flexible in that they allow some violation of the constraints. Such methods might be considered a compromise between reduced-gradient and sequential quadratic programming methods; see the Notes.

Exercises

6.1. Prove that
$$Z(x)^T \nabla_x \mathcal{L}(x, \lambda) = Z(x)^T \nabla f(x).$$

15.7   Filter Methods

In this section we describe a multicriteria approach that guarantees convergence of optimization methods. This approach uses *filters* instead of merit functions to measure the progress of algorithms. For several reasons filter methods are an attractive alternative to the methods using merit functions. First, filter methods rely on primary goals— a decrease of the objective function and infeasibility—and therefore may not be sensitive to the choice of parameters or a particular merit function. Second, filter methods tend to allow more freedom in selecting the next trial point than methods based on penalty or merit functions. This can contribute to the efficiency of filter methods. Finally, filter methods demonstrate robust and efficient performance in practice.

Filter methods are a new and active area of research. They can be based on a variety of different underlying optimization methods. Filter methods are analogous to the use of a merit function, since they are used to guarantee convergence. In this section we discuss filter methods in the context of SQP algorithms.

Consider an optimization problem with equality constraints

$$\begin{aligned}
\text{minimize} \quad & f(x) \\
\text{subject to} \quad & g_i(x) = 0, \quad i = 1, \dots, m.
\end{aligned}$$

We will discuss later how to treat inequality constraints.

To guarantee convergence of the SQP method, in Section 15.5 we used a quadratic penalty function as a merit function,

$$\mathcal{M}(x) = f(x) + \rho g(x)^T g(x) = f(x) + \rho \sum_{i=1}^{m} g_i(x)^2,$$

where ρ is the positive penalty parameter. Alternatively, some other merit functions can be used, e.g.,

$$\pi(x, \rho) = f(x) + \rho \|g(x)\|,$$

where $\|g(x)\|$ is some norm of the vector function $g(x)$. (See Chapter 16.) A penalty function is the weighted sum of two terms that includes the objective function and some measure of infeasibility.

Temporarily, let us use the following penalty function $\pi(x, \rho) = f(x) + \rho \|g(x)\|$ as a merit function to control convergence of the iterates x_k. If $\pi(x_{k+1}, \rho) < \pi(x_k, \rho)$, then x_{k+1} is considered a better estimate of the solution than x_k. The penalty function measures the decrease of a combination of the objective function $f(x)$ and the infeasibility $\|g(x)\|$. The parameter $\rho > 0$ specifies the relative importance of one criterion with respect to the other. For example, if ρ is small, then the penalty function emphasizes decrease of the objective function. However, if ρ is large, then the penalty function emphasizes decrease of the infeasibility.

Numerical experiments reveal that the value of ρ can influence the performance of the algorithm. What is the best choice of ρ? This difficult question lacks a definitive answer. The best value of ρ may depend on the problem and on the distance from x_k to the solution. Usually, practical algorithms start with a certain value of ρ and adjust it as required. Various strategies are possible for adjusting ρ.

Is it possible to eliminate ρ altogether? The answer is yes, and filter methods are one way to achieve this. Instead of using penalty functions, filter methods consider decrease of the objective function and decrease of infeasibility as two criteria that must be controlled simultaneously. The main idea of filter methods is that the next trial point x_{k+1} is accepted if it improves at least one of two criteria: the value of the objective function or the infeasibility of the constraints.

Our filter method will incorporate a function $h(x) = \|g(x)\|$ that measures the infeasibility of the constraints; that is, (i) $h(x) = 0$ if and only if x is a feasible point, and (ii) the larger the value of $h(x)$, the more infeasible x is considered to be.

Suppose that we apply some optimization algorithm to the equality-constrained problem, and that it generates a sequence $\{x_k\}$ of estimates of the solution. We will use $f_k = f(x_k)$ and $h_k = h(x_k)$ to denote the corresponding objective values and infeasibilities. If $f_{k+1} < f_k$ and $h_{k+1} < h_k$, then we will say that x_{k+1} is closer than x_k to the solution of the problem. It would be ideal if this were true at every iteration, but that will not always happen. Even convergent algorithms do not guarantee that both inequalities $f_{k+1} < f_k$ and $h_{k+1} < h_k$ hold. Often outside of the feasible set, reduction of infeasibility results in an increase of the value of the objective function (see Figure 15.4). It can be better to accept x_{k+1} if there is a substantial decrease in just one of the two criteria, even if the other increases. That gives the algorithm more freedom in selecting iterates without compromising convergence properties.

In the following we formalize the idea of filters. A pair (f_q, h_q) is said to *dominate* another pair (f_r, h_r) if and only if both $f_q \leq f_r$ and $h_q \leq h_r$. A *filter* is said to be a list of pairs $\mathcal{F} = \{(f_j, h_j), \ j = 1, \ldots, s\}$ such that no pair dominates any other. A filter $\mathcal{F} = \{(f_1, h_1), (f_2, h_2), (f_3, h_3)\}$ is illustrated in Figure 15.5 (left). In the figure, one point dominates another if the first point is below and to the left of the second point.

Every point in the filter generates a block consisting of the points dominated by this point, and the union of these blocks represents the set of points that are dominated by the filter. For example, (f_2, h_2) generates the grey block shown in Figure 15.5 (left). Every point in this block, e.g., $(\bar{f}, \bar{h})$, is dominated by (f_2, h_2).

The filter is used to accept or reject the trial point x_{k+1}. The trial point x_{k+1} is accepted if and only if (f_{k+1}, h_{k+1}) is included in the filter. In turn, (f_{k+1}, h_{k+1}) is included

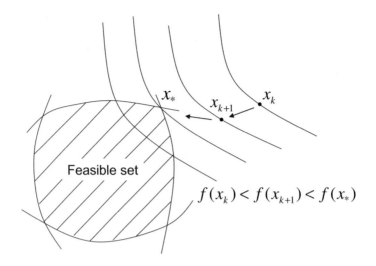

Figure 15.4. *Iterates can approach the solution outside the feasible set.*

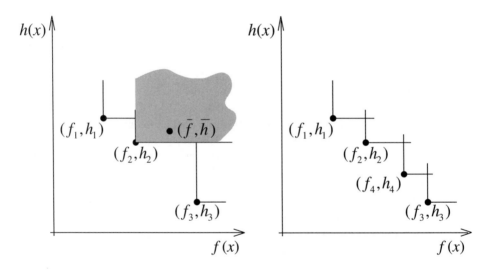

Figure 15.5. *Filter.*

in the filter if and only if (f_{k+1}, h_{k+1}) is not dominated by any point from the filter. For example, in Figure 15.5 (left), $(\bar{f}, \bar{h})$ is dominated by (f_2, h_2). Therefore $(\bar{f}, \bar{h})$ is not included in the filter. However, (f_4, h_4) in Figure 15.5 (right) is not dominated by (f_1, h_1), (f_2, h_2), or (f_3, h_3). Therefore (f_4, h_4) is included in the filter, and the updated filter is $\mathcal{F} = \{(f_1, h_1), (f_2, h_2), (f_3, h_3), (f_4, h_4)\}$.

In mathematical terms, (f_{k+1}, h_{k+1}) is included in the filter if and only if

$$\text{either} \quad f_{k+1} < f_j \quad \text{or} \quad h_{k+1} < h_j \quad \text{for all} \quad j = 1, \ldots, s.$$

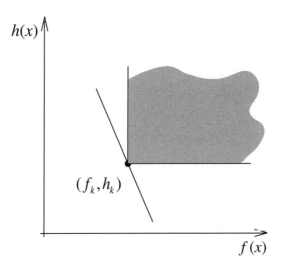

Figure 15.6. *Filter and penalty.*

After a new pair (f_{k+1}, h_{k+1}) is included in the filter, all the pairs from the filter dominated by (f_{k+1}, h_{k+1}) are eliminated from the filter. Thus, a filter can consist of as many as $k + 1$ pairs. However, at a certain iteration a new pair can dominate more than one old entry of the filter. If this happens, we can eliminate multiple entries from the filter and thus decrease the number of points in the filter. In general, we assume that a filter $\mathcal{F}_k$ consists of $s \leq k + 1$ points.

Figure 15.6 illustrates the difference between using filters and merit functions to accept trial points. For example, the contours of the merit function $\pi(x_k, \rho)$ would be straight lines with slope $-\rho$ (see Figure 15.6) and only points to the left of this line would be accepted. Filter methods would accept all the points outside the grey area. The filter method is able to consider a larger set of points when choosing the next trial point. If a trial point is accepted by the merit function for some $\rho > 0$, it is also accepted by the filter. However, the reverse is not true. (See the Exercises.) If ρ is not chosen correctly, then the merit function can reject good trial points that would be accepted by the filter, or by the merit function with a different value of ρ. Thus, filter methods tend to be less restrictive than methods controlled by merit functions when the number of entries in the filter is small.

The filter is used to accept or reject a trial point, but by itself it is not sufficient to guarantee convergence. It must be combined with either a line search or trust-region approach. (See Chapter 11.) Here we will use a trust-region approach, which was the approach used when filter methods were first developed.

We will apply the filter method in the context of an SQP algorithm. At each iteration a quadratic program is solved to obtain a search direction p. The trust-region strategy imposes the following additional constraint to control the distance between x_{k+1} and x_k:

$$\|p\| \leq \Delta_k,$$

where $\Delta_k > 0$ is the maximum distance that we will allow between trial and current points.

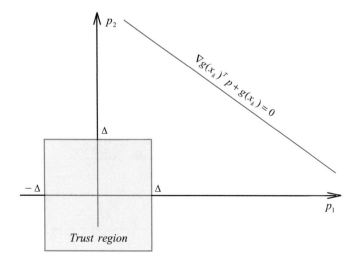

Figure 15.7. *Infeasibility.*

At each iteration the SQP-filter method solves the following problem:

$$\text{minimize} \quad q(p) = \tfrac{1}{2} p^T H_k p + p^T \nabla f(x_k)$$
$$\text{subject to} \quad \nabla g(x_k)^T p + g(x_k) = 0$$
$$\|p\| \le \Delta_k,$$

where $H_k \approx \nabla_{xx}^2 \mathcal{L}(x_k, \lambda_k)$ approximates the Hessian of the Lagrangian. If a solution to the quadratic program is found, then the trial point is

$$x_{k+1} = x_k + p_k.$$

If x_{k+1} is not accepted by the filter, then Δ_k is reduced and the quadratic program is solved again. If $p_k = 0$, then x_k is a first-order optimality point of the original problem with the Lagrange multipliers v_k from the quadratic program. (See the Exercises.)

If the infinity norm ($\|p\| = \max_{1 \le i \le n} |p_i|$) is used, then the trust-region constraint $\|p\| \le \Delta_k$ is equivalent to a set of bound constraints on the components of p. If the 2-norm $\left(\|p\| = \sqrt{\sum_{i=1}^n p_i^2}\right)$ is used, then the constraint $\|p\| \le \Delta_k$ is equivalent to one quadratic inequality. (See the Exercises.) For a discussion of the norms see Appendix A.3.

If the quadratic problem is infeasible, then two situations can occur. The first situation is when the linearized constraints

$$\nabla g(x_k)^T p + g(x_k) = 0$$

are consistent, but Δ_k is too small (see Figure 15.7). This situation can be resolved by increasing Δ_k and solving the quadratic problem again.

The second situation is when the linearized constraints are inconsistent. If this happens, the algorithm tries to restore feasibility before continuing. It ignores the value of $f(x)$ and focuses on reducing $h(x)$. For details on this topic, see the Notes. Practical algorithms may use various criteria when choosing between increasing Δ_k or restoring feasibility. The algorithm presented in this section always performs feasibility restoration followed by resetting Δ_k to some fixed initial value in the case where the quadratic problem is infeasible.

Inequality constraints can be treated by the SQP-filter method in a similar way. Consider an optimization problem of the form

$$
\begin{array}{ll}
\text{minimize} & f(x) \\
\text{subject to} & g_i(x) \geq 0, \quad i \in \mathcal{I} \\
& g_i(x) = 0, \quad i \in \mathcal{E}.
\end{array}
$$

We can use the function $\psi_i(x) = \min(g_i(x), 0)$ to measure infeasibility of the ith inequality constraint. The functions $h_{\mathcal{I}}(x) = \max_{i \in \mathcal{I}} |\psi_i(x)|$ and $h_{\mathcal{E}}(x) = \max_{i \in \mathcal{E}} |g_i(x)|$ measure the infeasibility of the inequality and equality constraints, respectively. Then $h(x) = \max(h_{\mathcal{I}}(x), h_{\mathcal{E}}(x))$ measures the overall infeasibility of the problem. There are other examples of functions that measure the infeasibility of the problem. Often the choice of $h(x)$ is related to the method used for feasibility restoration. For details on this topic, see the Notes.

Now the filter approach is applied to the general problem with objective function $f(x)$ and infeasibility measure $h(x)$. The quadratic program solved at each iteration now includes the linearized inequality constraints:

$$
\begin{array}{ll}
\underset{p}{\text{minimize}} & \frac{1}{2}p^T H_k p + p^T \nabla f(x_k) \\
\text{subject to} & \nabla g_{\mathcal{I}}(x_k)^T p + g_{\mathcal{I}}(x_k) \geq 0 \\
& \nabla g_{\mathcal{E}}(x_k)^T p + g_{\mathcal{E}}(x_k) = 0 \\
& \|p\| \leq \Delta_k,
\end{array}
$$

where $\nabla g_{\mathcal{I}}^T$ and $\nabla g_{\mathcal{E}}^T$ are the Jacobians corresponding to the inequality and equality constraints, respectively. The rest is the same as before.

We are now able to present the SQP-filter algorithm.

ALGORITHM 15.1.
SQP-Filter Algorithm

1. Select x_0, λ_0. Set $f_0 = f(x_0)$, $h_0 = h(x_0)$. Initialize the filter: $\mathcal{F}_0 = \{(f_0, h_0)\}$. Select the initial trust-region radius $\Delta_0 = \bar{\Delta}$.

2. For $k = 0, 1, \ldots$

   (i) *Solving the Quadratic Subproblem*—Compute p_k and the Lagrange multipliers v_k by solving the quadratic subproblem

$$
\begin{array}{ll}
\underset{p}{\text{minimize}} & \frac{1}{2}p^T H_k p + p^T \nabla f(x_k) \\
\text{subject to} & \nabla g_{\mathcal{I}}(x_k)^T p + g(x_k) \geq 0 \\
& \nabla g_{\mathcal{E}}(x_k)^T p + g(x_k) = 0 \\
& \|p\| \leq \Delta_k.
\end{array}
$$

   (ii) *The Optimality Test*—If $p_k = 0$, then stop. Output x_k, $\lambda_k = v_k$.

(iii) *The Restoration Test*—If the quadratic subproblem is infeasible, then perform feasibility restoration: determine x_{k+1} such that

$$h_{k+1} = h(x_{k+1}) < \min_{(f_j, h_j) \in \mathcal{F}_k} h_j.$$

Reset $\Delta_{k+1} = \bar{\Delta}$.

(iv) *The Update*—If p_k is the solution to the quadratic subproblem, then set

$$x_{k+1} = x_k + p_k.$$

If x_{k+1} is accepted, then set $\lambda_{k+1} = \nu_k$, $f_{k+1} = f(x_{k+1})$, $h_{k+1} = h(x_{k+1})$, $\Delta_{k+1} = \bar{\Delta}$; update the filter using $\mathcal{F}_{k+1} = \mathcal{F}_k \cup (f_{k+1}, h_{k+1})$, and then remove the points dominated by (f_{k+1}, h_{k+1}) from the filter. If x_{k+1} is not accepted, then set $\Delta_{k+1} = \Delta_k/2$, $x_{k+1} = x_k$, $\mathcal{F}_{k+1} = \mathcal{F}_k$.

Example 15.14 (The SQP-Filter Method). We apply the SQP-filter method to the following problem:

$$\text{minimize} \quad f(x_1, x_2) = e^{3x_1} + e^{-4x_2}$$
$$\text{subject to} \quad g(x_1, x_2) = x_1^2 + x_2^2 - 1 = 0.$$

The solution to this problem is $x_* = (-0.74834, 0.66332)^T$ with $\lambda_* = -0.21233$.

We use $\Delta_0 = 1$, the infinity norm, and the initial guess $x_0 = (0, 0)^T$. We set $H_k = \nabla_{xx}^2 \mathcal{L}(x_k, \lambda_k)$, so we need also an initial guess for the Lagrange multiplier $\lambda_0 = -1$. At this point,

$$f = e^{3x_1} + e^{-4x_2} = (2)$$
$$\nabla f = \begin{pmatrix} 3e^{3x_1} \\ -4e^{-4x_2} \end{pmatrix} = \begin{pmatrix} 3 \\ -4 \end{pmatrix}$$
$$\nabla^2 f = \begin{pmatrix} 9e^{3x_1} & 0 \\ 0 & 16e^{-4x_2} \end{pmatrix} = \begin{pmatrix} 9 & 0 \\ 0 & 16 \end{pmatrix}$$
$$g = x_1^2 + x_2^2 - 1 = (-1)$$
$$\nabla g = \begin{pmatrix} 2x_1 \\ 2x_2 \end{pmatrix} = \begin{pmatrix} 0 \\ 0 \end{pmatrix}, \quad \nabla^2 g = \begin{pmatrix} 2 & 0 \\ 0 & 2 \end{pmatrix}$$
$$\nabla_{xx}^2 \mathcal{L} = \nabla^2 f - \lambda \nabla^2 g = \begin{pmatrix} 11 & 0 \\ 0 & 18 \end{pmatrix}.$$

We initialize the filter as $\mathcal{F}_0 = \{(f_0, h_0)\} = \{(2, 1)\}$. The corresponding quadratic program is

$$\text{minimize} \quad q(p) = \tfrac{1}{2} p^T \nabla_{xx}^2 \mathcal{L} p + p^T \nabla f$$
$$\text{subject to} \quad \nabla g^T p + g = 0$$
$$-\Delta \le p_1 \le \Delta$$
$$-\Delta \le p_2 \le \Delta.$$

Since $\nabla g = (0, 0)^T$ but $g = -1$, the linear system $\nabla g^T p + g = 0$ does not have a solution. Therefore, the quadratic problem is infeasible. The SQP-filter method invokes the feasibility restoration phase, which can be done by minimizing the function $g(x_1, x_2)^2 = (x_1^2 + x_2^2 - 1)^2$ using methods of unconstrained minimization. In fact, to improve infeasibility, it is enough to move the iterate a little in any direction. For example, $x_1 = (0.5, 0.5)^T$ improves the infeasibility since $g(0.5, 0.5) = -0.5$. Therefore $x_1 = (0.5, 0.5)^T$ is accepted and the filter is updated as follows:

$$\mathcal{F}_1 = \mathcal{F}_0 \cup (f_1, h_1) = \{(2, 1)\,(4.61702, 0.5)\}.$$

Note that neither $(2, 1)$ nor $(4.61702, 0.5)$ dominates the other. At x_1,

$$f = (4.61702)$$
$$\nabla f = \begin{pmatrix} 13.44507 \\ -0.54134 \end{pmatrix}$$
$$\nabla^2 f = \begin{pmatrix} 40.33520 & 0 \\ 0 & 2.16536 \end{pmatrix}$$
$$g = (-0.5)$$
$$\nabla g = \begin{pmatrix} 1 \\ 1 \end{pmatrix}, \quad \nabla^2 g = \begin{pmatrix} 2 & 0 \\ 0 & 2 \end{pmatrix}$$
$$\nabla_{xx}^2 \mathcal{L} = \nabla^2 f - \lambda \nabla^2 g = \begin{pmatrix} 42.33520 & 0 \\ 0 & 4.16536 \end{pmatrix}.$$

This time the quadratic problem is feasible and we have the following solution:

$$p_1 = \begin{pmatrix} -0.25599 \\ 0.75599 \end{pmatrix} \quad \text{and} \quad \lambda_1 = (2.60764).$$

The candidate for the next iterate is

$$x_2 = x_1 + p_1 = \begin{pmatrix} 0.24401 \\ 1.25599 \end{pmatrix}.$$

At this point

$$f = (2.08587)$$
$$g = (0.63705).$$

The new filter entry $(f_2, h_2) = (2.08587, 0.63705)$ neither dominates nor is dominated by the other filter entries, so x_2 is accepted, and the updated filter is

$$\mathcal{F}_2 = \{(2, 1),\ (4.61702, 0.5),\ (2.08587, 0.63705)\}.$$

The next iteration produces $x_3 = (-0.20627, 1.08990)^T$ and a new filter entry $(f_3, h_3) = (0.55137, 0.23035)$. All the other elements of the filter are dominated by this new entry and therefore are excluded from the filter:

$$\mathcal{F}_3 = \{(0.55137, 0.23035)\}.$$

Several more iterations are given in Table 15.3.

The final filter

$$\mathcal{F}_8 = \{(0.17635,\ 3 \times 10^{-7})\,(0.17629,\ 3 \times 10^{-4})\,(0.17425,\ 1 \times 10^{-2})\,(0.15447,\ 1 \times 10^{-1})\}$$

has four entries. ∎

Table 15.3. *The SQP-filter method.*

k	x_k		λ_k	$\|\nabla_x \mathcal{L}\|$	$\mathcal{F}_k$
0	0	0	-1	5×10^0	$(2, 1)$
1	0.5	0.5	-1	1×10^1	$(4.61702, 5 \times 10^{-1})$
					$(2, 1)$
2	0.24401	1.25599	2.60764	8×10^0	$(2.08587, 6 \times 10^{-1})$
					$(4.61702, 5 \times 10^{-1})$
					$(2, 1)$
3	-0.20627	1.08990	0.32747	2×10^0	$(0.55137, 2 \times 10^{-1})$
4	-0.59301	0.91099	0.01350	5×10^{-1}	$(0.19495, 2 \times 10^{-1})$
5	-0.84205	0.64922	-0.11364	2×10^{-1}	$(0.15447, 1 \times 10^{-1})$
6	-0.74258	0.67771	-0.19838	3×10^{-2}	$(0.17425, 1 \times 10^{-2})$
					$(0.15447, 1 \times 10^{-1})$
7	-0.74876	0.66303	-0.21201	9×10^{-4}	$(0.17629, 3 \times 10^{-4})$
					$(0.17425, 1 \times 10^{-2})$
					$(0.15447, 1 \times 10^{-1})$
8	-0.74834	0.66332	-0.21232	5×10^{-7}	$(0.17635, 3 \times 10^{-7})$
					$(0.17629, 3 \times 10^{-4})$
					$(0.17425, 1 \times 10^{-2})$
					$(0.15447, 1 \times 10^{-1})$

To guarantee convergence, the SQP-filter methods must be modified by imposing stricter rules for acceptance of a new trial iterate:

$$\text{either} \quad f_{k+1} < f_j - \gamma h_{k+1} \quad \text{or} \quad h_{k+1} < \eta h_j \quad \text{for all} \quad j = 1, \ldots, s,$$

for some $0 < \eta < 1$, $0 < \gamma < 1$. The acceptance area is shown below the dashed line in Figure 15.8.

Another modification is enforcing the sufficient reduction condition, similar to the one used in unconstrained minimization (see Chapter 12). First, the predicted reduction

$$\Delta q_k = \tfrac{1}{2} p_k^T H_k p_k + p_k^T \nabla f(x_k)$$

is calculated. Then if $\Delta q_k < 0$, the algorithm verifies if

$$f_{k+1} < f_k + \sigma \Delta q_k.$$

In other words, the filter algorithm verifies whether a new iterate sufficiently reduces the objective function.

Convergence can be guaranteed for the filter method with the following acceptance rule. A trial point $x_{k+1} = x_k + p_k$ is accepted at iteration k if

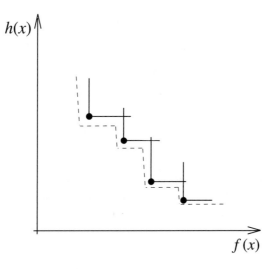

Figure 15.8. *Modified filter.*

- x_{k+1} is acceptable to the modified filter, that is,

  either $\quad f_{k+1} < f_j - \gamma h_{k+1} \quad$ or $\quad h_{k+1} < \eta h_j \quad$ for all $\quad j = 1, \ldots, s,$

  for some $0 < \eta < 1, 0 < \gamma < 1$;

- the new iterate achieves sufficient reduction of the objective function:

$$f_{k+1} < f_k + \sigma \Delta q_k \quad \text{if} \quad \Delta q_k < 0.$$

We already mentioned that filters are a general tool for controlling convergence. Filter methods can be based on a variety of different optimization techniques and not just on an SQP method, for example, with interior-point methods. The Notes contain additional information on practical filter algorithms.

Exercises

7.1. Consider the problem

$$\text{minimize} \quad f(x) = e^{-x_1} + e^{-2x_2}$$
$$\text{subject to} \quad g(x) = x_1^2 + x_2^2 - 4 = 0.$$

Starting with $x_0 = (1, 1)^T$, $\lambda_0 = 0$, and $\bar{\Delta} = 1$, perform three iterations of the SQP-filter method.

7.2. Let (p_*, λ_*) be a first-order optimality point of the following subproblem:

$$\text{minimize} \quad q(p) = \tfrac{1}{2} p^T H_k p + p^T \nabla f(x_k)$$
$$\text{subject to} \quad \nabla g_{\mathcal{I}}(x_k)^T p + g(x_k) \geq 0$$
$$\|p\| \leq \Delta.$$

Prove that if $p_* = 0$, then (x_k, λ_*) is a first-order optimality point of the following problem:

$$\begin{array}{ll} \text{minimize} & f(x) \\ \text{subject to} & g_i(x) \geq 0, \quad i \in \mathcal{I}. \end{array}$$

7.3. Suppose that a constraint $\|x\| \leq M$ based on the infinity norm is added to the problem

$$\begin{array}{ll} \text{minimize} & f(x) \\ \text{subject to} & g_i(x) \geq 0, \quad i \in \mathcal{I}. \end{array}$$

Prove that, for M sufficiently large, once $\|x^0\| \leq M$, then all the quadratic subproblems generate directions p_k such that $\|x_k\| \leq M$, $k = 1, 2, \ldots$.

7.4. Consider a filter with one entry, as illustrated in Figure 15.6. Prove that if a trial point x_{k+1} is accepted by the penalty function $\pi(x, \rho)$, i.e., if the following inequality holds:

$$\pi(x_{k+1}, \rho) < \pi(x_k, \rho)$$

for some penalty parameter $\rho > 0$, then x_{k+1} will be accepted by the filter.

7.5. Consider again a filter with one entry, as illustrated in Figure 15.6. Prove that if a trial point x_{k+1} is accepted by the filter, then there exists $\bar{\rho} > 0$ such that x_{k+1} will be accepted by the penalty function $\pi(x, \bar{\rho})$, i.e., the following inequality holds:

$$\pi(x_{k+1}, \bar{\rho}) < \pi(x_k, \bar{\rho}).$$

7.6. Prove the following.

(i) The trust-region constraint $\|p\| \leq \Delta_k$ with the infinity norm ($\|p\| = \max_{1 \leq i \leq n} |p_i|$) is equivalent to a set of linear constraints.

(ii) The trust-region constraint $\|p\| \leq \Delta_k$ with the 2-norm $\left(\|p\| = \sqrt{\sum_{i=1}^{n} p_i^2}\right)$ is equivalent to one quadratic inequality.

15.8 Notes

Active-Set Methods—One of the first practical active-set methods for nonlinear optimization was given in the paper by Rosen (1960). A detailed discussion can be found in the book by Gill, Murray, and Wright (1981).

Sequential Quadratic Programming—SQP was first suggested in the Ph.D. thesis of Wilson (1963). Since then, the approach has been examined by a great many authors. A recent survey can be found in the paper by Boggs and Tolle (1995).

If all the constraints are equalities, then algorithms can be designed that maintain feasibility at every iteration. (Hence the nonlinear objective function can be used as a merit function, a valuable simplification.) See, for example, the paper by Panier and Tits (1993).

Reduced-Gradient Methods—A reduced-gradient method for problems with nonlinear constraints was first discussed in the paper by Abadie and Carpentier (1965). More recent results can be found in the papers by Abadie (1978) and Lasdon et al. (1978). Sophisticated versions of reduced-gradient methods have been developed that allow some violation of the constraints. An example of such an algorithm is described in the paper by Drud (1992).

Filter Methods—SQP-filter methods based on a trust-region approach were developed and analyzed by Fletcher and Leyffer (2002) and Fletcher et al. (2002). Since then, line search versions have been derived, and filter methods have been applied in the context of other optimization algorithms. For further information, see the survey paper by Fletcher et al. (2007).

Chapter 16

Penalty and Barrier Methods

16.1   Introduction

In Chapter 15 we discussed the use of feasible-point methods for solving constrained optimization problems. These methods are based on minimizing the Lagrangian function while attempting to attain and maintain feasibility. When inequality constraints are present, these methods generalize the simplex method. They solve a sequence of subproblems with a changing active set (or basis) until a solution to the original constrained problem is found.

There are some major disadvantages to this approach. First, as the number of constraints increases, the number of potential subproblems increases exponentially. While the hope is that the algorithm will consider only a small proportion of these subproblems, there is no known method to guarantee that this indeed will be the case. (This was illustrated for the simplex method in Chapter 9.) Second, the idea of keeping the constraints satisfied exactly, although easily achieved in the case of linear constraints, is much more difficult to accomplish in the case of nonlinear constraints, and in some cases may not be desirable.

In this chapter we discuss a group of methods, referred to here as *penalization methods*, that remove some of these difficulties. These methods solve a constrained optimization problem by solving a sequence of unconstrained optimization problems. The hope is that in the limit, the solutions of the unconstrained problems will converge to the solution of the constrained problem. The unconstrained problems involve an auxiliary function that incorporates the objective function or the Lagrangian function, together with "penalty" terms that measure violations of the constraints (various techniques differ in the way the auxiliary function is defined). The auxiliary function also includes one or more parameters that determine the relative importance of the constraints in the auxiliary function. By changing these parameters appropriately, a sequence of problems is generated where the effect of the constraints becomes increasingly pronounced. In contrast to active-set methods, the auxiliary function takes into account *all* constraints, even when inequalities are present, and thus the combinatorial difficulties of guessing a correct active set are avoided. Further, since penalization techniques do not attempt to keep the constraints satisfied exactly, they can be more suitable for handling nonlinear constraints.

Although penalization methods ameliorate some of the difficulties associated with feasible-point methods, they introduce difficulties of their own. In particular, straightforward applications of penalization can give rise to ill-conditioning. Sophisticated modern interpretations of this idea, however, resolve some of these issues and can be effective methods for challenging optimization problems.

16.2   Classical Penalty and Barrier Methods

The general class of penalization methods includes two groups of methods: one group imposes a penalty for *violating* a constraint, and the other imposes a penalty for *reaching the boundary* of an inequality constraint. We refer to the first group as *penalty methods* and to the second group as *barrier methods*. In this section we present the "classical" version of these methods. The methods in later sections can be viewed as modifications or enhancements to the classical penalty and barrier methods.

We start with a geometrical motivation for these methods. Suppose that our constrained problem is given in the form

$$\text{minimize} \quad f(x)$$
$$\text{subject to} \quad x \in S,$$

where S is the set of feasible points. Define

$$\sigma(x) = \begin{cases} 0 & \text{if } x \in S \\ +\infty & \text{if } x \notin S. \end{cases}$$

The function σ can be considered as an infinite penalty for violating feasibility. Hence the constrained problem can be transformed into an equivalent unconstrained problem

$$\text{minimize } f(x) + \sigma(x). \tag{16.1}$$

Conceptually, if we could solve this unconstrained minimization problem, we would be done (a point x_* is a solution to the constrained problem if and only it is a solution to this unconstrained problem). Unfortunately, this is not a practical idea, since the objective function of the unconstrained minimization is not defined outside the feasible region. Even if we were to replace the "∞" by a large number, the resulting unconstrained problem would be difficult to solve because of its discontinuities.

Instead, barrier and penalty methods solve a sequence of unconstrained subproblems that are more "manageable," and that gradually approximate problem (16.1). This is achieved by replacing the "ideal" penalty σ by a continuous function that gradually approaches σ.

In barrier methods, this function (called a *barrier term*) approaches σ from the interior of the feasible region. It creates a barrier that prevents the iterates from reaching the boundary of the feasible region. In penalty methods this function (called a *penalty term*) approaches σ from the exterior of the feasible region. It serves as a penalty for being infeasible.

Barrier methods generate a sequence of strictly feasible iterates that converge to a solution of the problem from the interior of the feasible region. For this reason they are also called *interior-point methods*. Since these methods require the interior of the feasible region to be nonempty, they are not appropriate for problems with equality constraints.

In contrast, penalty methods permit the iterates to be infeasible. A gradually increasing penalty is imposed for violation of feasibility, however. Penalty methods usually generate a sequence of points that converges to a solution of the problem from the exterior of the feasible region. These methods are usually more convenient on problems with equality constraints. Although the methods can also handle inequality constraints, the resulting unconstrained problems usually do not have continuous second derivatives, and barrier methods are often preferable.

Despite their apparent differences, barrier and penalty methods have much in common. Their convergence theories are similar, and the underlying structure of their unconstrained problems is similar. Much of the theory for barrier methods can be replicated for penalty methods and vice versa. It is common to use the generic name "penalty methods" to describe both methods, with *interior penalty methods* referring to barrier methods, and *exterior penalty methods* referring to penalty methods.

16.2.1 Barrier Methods

Consider the nonlinear inequality-constrained problem

$$
\begin{aligned}
\text{minimize} \quad & f(x) \\
\text{subject to} \quad & g_i(x) \geq 0, \quad i = 1, \ldots, m.
\end{aligned}
\tag{16.2}
$$

The functions are assumed to be twice continuously differentiable.

Barrier methods are strictly feasible methods; that is, the iterates lie in the interior of the feasible region. We assume, therefore, that the feasible set has a nonempty interior; that is, there exists some point x_0 such that $g_i(x_0) > 0$ for all i. We also assume that it is possible to reach any boundary point by approaching it from the interior.

Barrier methods maintain feasibility by creating a barrier that keeps the iterates away from the boundary of the feasible region. The methods use a barrier term that approaches the infinite penalty function σ. Let $\phi(x)$ be a function that is continuous on the interior of the feasible set, and that becomes unbounded as the boundary of the set is approached from its interior:

$$
\phi(x) \to \infty \quad \text{as} \quad g_i(x) \to 0_+.
$$

Two examples of such a function are the logarithmic function

$$
\phi(x) = -\sum_{i=1}^{m} \log(g_i(x))
$$

and the inverse function

$$
\phi(x) = \sum_{i=1}^{m} \frac{1}{g_i(x)}.
$$

Now let μ be a positive scalar. Then $\mu\phi(x)$ will approach $\sigma(x)$ as μ approaches zero. This is demonstrated in Figure 16.1 for a one-dimensional problem with bound constraints.

By adding a barrier term of the form $\mu\phi(x)$ to the objective, we obtain a *barrier function*

$$
\beta_\mu(x) = f(x) + \mu\phi(x),
$$

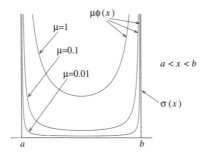

Figure 16.1. *Effect of barrier term.*

where μ is referred to as the *barrier parameter*. The best-known barrier function is the logarithmic barrier function

$$\beta_\mu(x) = f(x) - \mu \sum_{i=1}^{m} \log(g_i(x)),$$

but the inverse barrier function

$$\beta_\mu(x) = f(x) - \mu \sum_{i=1}^{m} \frac{1}{g_i(x)}$$

is also widely used.

Barrier methods solve a sequence of unconstrained minimization problems of the form

$$\underset{x}{\text{minimize}} \ \beta_{\mu_k}(x)$$

for a sequence $\{\mu_k\}$ of positive barrier parameters that decrease monotonically to zero. Because the barrier term is infinite on the boundary of the feasible region, it acts as a repelling force that drives the minimizers of the barrier function away from the boundary into the interior of the feasible region. Thus any minimizer of the barrier function will be strictly feasible. As the barrier parameter is decreased, however, the effect of the barrier term is diminished, so that the iterates can gradually approach the boundary of the feasible region.

Why solve a sequence of problems? It might seem better to solve a single unconstrained problem using a small value of μ, but this is not normally practical. The reason is that when the barrier parameter is small, the problems are difficult to solve (recall that the function $\mu\phi$ will be "close" in shape to the infinite penalty σ). For this reason we start with larger values of the barrier parameter. If μ is decreased gradually, and if the solution of one unconstrained problem is used as the starting point of the next problem, these unconstrained minimization problems tend to be much easier to solve.

Example 16.1 (Barrier Method). Consider the nonlinear optimization problem

$$\begin{aligned}
\text{minimize} \quad & f(x) = x_1 - 2x_2 \\
\text{subject to} \quad & 1 + x_1 - x_2^2 \geq 0 \\
& x_2 \geq 0.
\end{aligned}$$

Then the logarithmic barrier function gives the unconstrained problem

$$\text{minimize } \beta_\mu(x) = x_1 - 2x_2 - \mu \log(1 + x_1 - x_2^2) - \mu \log x_2$$

for a sequence of decreasing barrier parameters. For a specific parameter μ, the first-order necessary conditions for optimality are

$$1 - \frac{\mu}{1 + x_1 - x_2^2} = 0$$

$$-2 + \frac{2\mu x_2}{1 + x_1 - x_2^2} - \frac{\mu}{x_2} = 0.$$

If the constraints are strictly satisfied, the denominators are positive. We obtain an equation for x_2:

$$x_2^2 - x_2 - \tfrac{1}{2}\mu = 0.$$

We can solve this equation to determine x_2 in terms of μ:

$$x_2(\mu) = \frac{1 + \sqrt{1 + 2\mu}}{2}.$$

(Why was the positive root taken?) Since $x_1 = x_2^2 - 1 + \mu$, we obtain

$$x_1(\mu) = \frac{\sqrt{1 + 2\mu} + 3\mu - 1}{2}.$$

The unconstrained objective is strictly convex, and hence this solution is the unique local minimizer in the feasible region.

As μ approaches zero, we obtain

$$\lim_{\mu \to 0_+} x_1(\mu) = \frac{\sqrt{1 + 2(0)} + 3(0) - 1}{2} = 0$$

and

$$\lim_{\mu \to 0_+} x_2(\mu) = \frac{1 + \sqrt{1 + 2(0)}}{2} = 1,$$

and it is easy to verify that $x_* = (0, 1)^T$ is indeed the solution to this problem.

Table 16.1 shows the values of $x_1(\mu)$ and $x_2(\mu)$ for a sequence of barrier parameters. The initial barrier parameter is selected to be 1, and consecutive parameters are decreased by a factor of 10. Observe that $x(\mu) = (x_1(\mu), x_2(\mu))^T$ exhibits a linear rate of convergence to the optimal solution. ∎

Example 16.1 illustrates a number of features that typically occur in a barrier method. First, the sequence of barrier minimizers $x(\mu)$ converges to the optimal solution x_*. Indeed, it is possible to prove convergence for barrier methods under mild conditions. (See Section 16.2.3.) Further, the sequence of barrier minimizers defines a differentiable curve $x(\mu)$. This curve is known as the *barrier trajectory*. It exists when the logarithmic or inverse barrier methods are used to solve (16.2), provided that x_* is a regular point of the constraints

Table 16.1. *Barrier function minimizers.*

μ	$x_1(\mu)$	$x_2(\mu)$
10^0	1.8660254	1.3660254
10^{-1}	0.1977226	1.0477226
10^{-2}	0.0199752	1.0049752
10^{-3}	0.0019998	1.0004998
10^{-4}	0.0002000	1.0000500
10^{-5}	0.0000200	1.0000050
10^{-6}	0.0000020	1.0000005

that satisfies the second-order sufficiency conditions as well as the strict complementarity conditions. The existence of a trajectory can be used to develop algorithms such as path-following algorithms for linear programming (see Chapter 10). It can also be used to develop techniques that accelerate the convergence of a barrier method.

There is another important feature. We discuss it for the case of the logarithmic barrier function, but an analogous result holds for the inverse barrier function. Consider a point $x = x(\mu)$ that is a minimizer of the logarithmic barrier function for a specific barrier parameter μ. Setting the gradient of the barrier function to zero, we obtain

$$\nabla f(x) - \mu \sum_{i=1}^{m} \frac{\nabla g_i(x)}{g_i(x)} = 0.$$

This may be written as

$$\nabla f(x) - \sum_{i=1}^{m} \frac{\mu}{g_i(x)} \nabla g_i(x) = 0.$$

In turn, we can write the last expression in the form

$$\nabla f(x) - \sum_{i=1}^{m} \lambda_i \nabla g_i(x) = 0,$$

where $\lambda_i = \lambda_i(\mu)$ is defined by

$$\lambda_i = \frac{\mu}{g_i(x)}.$$

We therefore have a feasible point $x(\mu)$ and a vector $\lambda(\mu)$ that satisfy the following relations:

$$\nabla f(x(\mu)) - \sum_{i=1}^{m} \lambda_i(\mu) \nabla g_i(x(\mu)) = 0$$
$$\lambda_i(\mu) g_i(x(\mu)) = \mu, \quad i = 1, \ldots, m$$
$$\lambda_i(\mu) \geq 0, \quad i = 1, \ldots, m.$$

These relations resemble the first-order necessary conditions for optimality of the constrained problem. The only difference is that the complementary slackness conditions are perturbed, so that $\lambda_i(\mu)g_i(x(\mu))$ is equal to μ rather than to zero. Thus $\lambda(\mu)$ can be viewed as an estimate of the Lagrange multipliers λ_* at the optimal point x_*. Indeed, if x_* is a regular point of the constraints, then as $x(\mu)$ converges to x_*, $\lambda(\mu)$ converges to λ_*. The above results show that the points on the barrier trajectory, together with their associated Lagrange multiplier estimates, are the solutions to a perturbation of the first-order optimality conditions.

Example 16.2 (Lagrange Multiplier Estimates). Consider the problem

$$\begin{aligned} \text{minimize} \quad & f(x) = x_1^2 + x_2^2 \\ \text{subject to} \quad & x_1 - 1 \geq 0 \\ & x_2 + 1 \geq 0. \end{aligned}$$

The solution to this problem is $x_* = (1,0)^T$. The first inequality is active at x_*, and the corresponding Lagrange multiplier is $(\lambda_*)_1 = 2$. The second constraint is inactive; hence its Lagrange multiplier is $(\lambda_*)_2 = 0$.

Suppose the problem is solved via a logarithmic barrier method. Then the method solves the unconstrained minimization problem

$$\text{minimize } \beta_\mu(x) = x_1^2 + x_2^2 - \mu \log(x_1 - 1) - \mu \log(x_2 + 1)$$

for a decreasing sequence of barrier parameters μ that converge to zero. The first-order necessary conditions for optimality are

$$2x_1 - \frac{\mu}{x_1 - 1} = 0$$

$$2x_2 - \frac{\mu}{x_2 + 1} = 0,$$

yielding the unconstrained minimizers

$$x_1(\mu) = x_1 = \frac{1 + \sqrt{1 + 2\mu}}{2}, \qquad x_2(\mu) = x_2 = \frac{-1 + \sqrt{1 + 2\mu}}{2}.$$

The Lagrange multiplier estimates at this point are

$$\lambda_1(\mu) = \frac{\mu}{x_1 - 1} = \frac{2\mu}{\sqrt{1 + 2\mu} - 1} = \sqrt{1 + 2\mu} + 1$$

and

$$\lambda_2(\mu) = \frac{\mu}{x_1 + 1} = \frac{2\mu}{\sqrt{1 + 2\mu} + 1} = \sqrt{1 + 2\mu} - 1.$$

When μ approaches zero, we obtain

$$\lim_{\mu \to 0} x_1(\mu) = \frac{1 + \sqrt{1 + 2(0)}}{2} = 1, \qquad \lim_{\mu \to 0} x_2(\mu) = \frac{-1 + \sqrt{1 + 2(0)}}{2} = 0.$$

Also we obtain

$$\lim_{\mu \to 0} \lambda_1(\mu) = \sqrt{1 + 2(0)} + 1 = 2, \qquad \lim_{\mu \to 0} \lambda_2(\mu) = \sqrt{1 + 2(0)} - 1 = 0.$$

Thus $x(\mu) \to x_*$ and $\lambda(\mu) \to \lambda_*$.   ∎

Barrier methods have several attractive features. They avoid the combinatorial difficulty associated with active-set constraints. They converge under mild conditions. The barrier minimizers provide estimates of the Lagrange multipliers at the optimum. Yet another desirable property shared by both the logarithmic barrier function and the inverse barrier function is that the barrier function is convex if the constrained problem is a convex optimization problem defined in terms of a convex objective function and concave constraint functions (see the Exercises).

Despite these attractive features, barrier methods also have potential difficulties. The property for which barrier methods have drawn the most severe criticism is that the unconstrained problems become increasingly difficult to solve as the barrier parameter decreases. The reason is that (with the exception of some special cases) the condition number of the Hessian matrix of the barrier function at its minimum point becomes increasingly large, tending to infinity as the barrier parameter tends to zero. We discuss this ill-conditioning in more detail in Section 16.3. Here we demonstrate it via an example.

Example 16.3 (Ill-Conditioning of the Barrier Hessian Matrix). Consider the problem of Example 16.2. Then

$$\nabla^2_x \beta_\mu(x) = \begin{pmatrix} 2 + \dfrac{\mu}{(x_1 - 1)^2} & 0 \\ 0 & 2 + \dfrac{\mu}{(x_2 + 1)^2} \end{pmatrix}.$$

Suppose now that $x(\mu) = (x_1, x_2)^T$ is a minimizer of the barrier function for some value of μ. Recall from Example 16.2 that $\lambda_1(\mu) = \mu/(x_1 - 1)$ and that $\lambda_2(\mu) = \mu/(x_2 + 1)$. When μ is small, $\lambda_1(\mu) \approx 2$ and $\lambda_2(\mu) \approx 0$. Therefore

$$\nabla^2_x \beta_\mu(x) = \begin{pmatrix} 2 + \dfrac{\lambda_1^2(\mu)}{\mu} & 0 \\ 0 & 2 + \dfrac{\lambda_2^2(\mu)}{\mu} \end{pmatrix} \approx \begin{pmatrix} 2 + \dfrac{4}{\mu} & 0 \\ 0 & 2 \end{pmatrix}.$$

The condition number of the Hessian matrix is approximately equal to

$$\frac{2 + \frac{4}{\mu}}{2} = 1 + \frac{2}{\mu} = O\left(\frac{1}{\mu}\right);$$

hence the matrix is ill conditioned. Although the calculations were performed for a point on the barrier trajectory, the same results will hold at all points in a neighborhood of the solution.

The contours of the barrier function in this example are shown in Figure 16.2 for the parameters $\mu = 1$ and $\mu = 0.1$. We see that for the smaller barrier parameter, the contours of the barrier function are almost parallel to the line $x_1 = 1$. More precisely, the contours

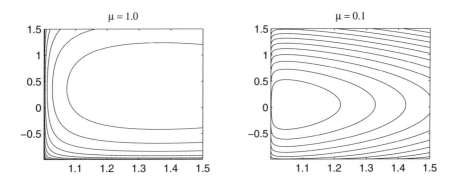

Figure 16.2. *Contours of the logarithmic barrier function.*

are almost parallel to the null space of the gradient of the active constraint at the solution. This is characteristic of barrier functions. ∎

The ill-conditioning of the Hessian matrix of the barrier function has several ramifications. It rules out the use of an unconstrained method whose convergence rate depends on the condition number of the Hessian matrix at the solution. Therefore Newton-type methods are usually the method of choice. The solution to the Newton equations is also sensitive to the ill-conditioning of the Hessian matrix, however. The numerical errors can result in a poor search direction.

The ill-conditioning of the barrier functions led to their abandonment in the early 1970s. Interest in barrier methods was renewed in 1984, with the announcement of Karmarkar's method for linear programming and the discovery, shortly thereafter, that this method is in fact a special case of a barrier method. (When applied to a linear program with a unique solution, a barrier method does not introduce ill-conditioning; see the Exercises.) Recently, special attention has been given to the development of specialized linear algebra techniques that compute a numerically stable, approximate solution to the Newton equations for a barrier function. One such approach is studied in Section 16.4.

We conclude this section by mentioning a few properties of barrier methods. First, barrier functions are singular at the boundary of the feasible region. Therefore special safeguards should be used in the (unconstrained) line search to avoid the evaluation of the barrier function at trial points that are not strictly feasible. Furthermore, because of the singularity, standard line search techniques such as backtracking or polynomial interpolation can be ineffective. Special line search techniques have been developed that model the specific singularity of the barrier function.

A consequence of the ill-conditioning of the barrier Hessian is that, if the initial barrier parameter μ_1 is "too small," the corresponding unconstrained minimization problem may be extremely difficult to solve. At the same time, there are also drawbacks to using an initial barrier parameter which is "too large." A large parameter will likely have a minimizer that is "far" from the constrained minimizer, and would require that more unconstrained minimization problems be solved. In addition, if the feasible region is unbounded, the unconstrained problem may not have a minimizer. This situation can occur only if the problem is nonconvex. It is illustrated in the following example.

Example 16.4 (Unbounded Barrier Function). Consider the one-dimensional problem

$$\text{minimize} \quad f(x) = \log(x + 1)$$
$$\text{subject to} \quad x \geq 0,$$

which has a solution at the point $x_* = 0$. The corresponding logarithmic barrier function is $\beta_\mu(x) = \log(x + 1) - \mu \log(x)$. Its derivative is

$$\frac{d}{dx}\beta_\mu(x) = \frac{1}{x + 1} - \frac{\mu}{x} = \frac{(1 - \mu)x - \mu}{x(x + 1)}.$$

When $\mu \geq 1$, the numerator in the expression above is negative for all $x \geq 0$. Therefore, for $\mu \geq 1$, the logarithmic barrier function has no minimizer—in fact it is unbounded.

If $\mu < 1$, then the problem has a solution: $x(\mu) = \mu/(1 - \mu)$. This solution approaches the optimal solution $x_* = 0$ as μ approaches zero. ∎

In most applied problems, the unconstrained minimization problem will have a solution if the barrier parameter is sufficiently small. This is not always guaranteed to be true, however. It is possible to construct examples where the barrier function is unbounded for any positive value of μ, even though the original problem has a solution (see the Exercises). Fortunately, such examples are rare in practice.

Barrier methods require that the initial guess of the solution be strictly feasible. In our examples, such an initial guess has been provided, but for general problems a strictly feasible point may not be known. It is sometimes possible to find an initial point by solving an auxiliary optimization problem. This is analogous to the use of a two-phase method in linear programming, where in the first phase a linear program is solved to determine an initial feasible point, and in the second phase, the original linear program is solved. For further details, see the Exercises.

16.2.2   Penalty Methods

In contrast to barrier methods, penalty methods solve a sequence of unconstrained optimization problems whose solution is usually infeasible to the original constrained problem. A penalty for violation of the constraints is incurred, however. As this penalty is increased, the iterates are forced towards the feasible region.

An advantage of penalty methods is that they do not require the iterates to be strictly feasible. Thus, unlike barrier methods, they are suitable for problems with equality constraints.

Consider first the equality-constrained problem

$$\text{minimize} \quad f(x)$$
$$\text{subject to} \quad g(x) = 0,$$

where $g(x)$ is an m-dimensional vector whose ith component is $g_i(x)$. We assume that all functions are twice continuously differentiable.

The penalty for constraint violation will be a continuous function ψ with the property that

$$\begin{aligned} \psi(x) &= 0 \quad \text{if } x \text{ is feasible,} \\ \psi(x) &> 0 \quad \text{otherwise.} \end{aligned} \tag{16.3}$$

The best-known such penalty is the *quadratic-loss function*

$$\psi(x) = \frac{1}{2} \sum_{i=1}^{m} g_i(x)^2 = \frac{1}{2} g(x)^T g(x).$$

Also possible is a penalty of the form

$$\psi(x) = \frac{1}{\gamma} \sum_{i=1}^{m} |g_i(x)|^{\gamma},$$

where $\gamma \geq 1$. (The case $\gamma = 1$ is discussed in Section 16.5.) The weight of the penalty is controlled by a positive *penalty parameter* ρ. As ρ increases, the function $\rho\psi$ approaches the "ideal penalty" σ. By adding the term $\rho\psi$ to f we obtain the *penalty function*

$$\pi_\rho(x) = f(x) + \rho\psi(x).$$

The penalty method consists of solving a sequence of unconstrained minimization problems of the form

$$\text{minimize } \pi_{\rho_k}(x)$$

for an increasing sequence $\{\rho_k\}$ of positive values tending to infinity. In general, the minimizers of the penalty function violate the constraints $g(x) = 0$. The growing penalty gradually forces these minimizers towards the feasible region.

Penalty methods share many of the properties of barrier methods. First, under mild conditions, it is possible to guarantee convergence. Also, under appropriate conditions, the sequence of penalty function minimizers defines a continuous trajectory. In the latter case, it is possible to get estimates of the Lagrange multipliers at the solution. For example, consider the quadratic-loss penalty function

$$\pi_\rho(x) = f(x) + \frac{1}{2}\rho \sum_{i=1}^{m} g_i(x)^2.$$

Its minimizer $x(\rho)$ satisfies

$$\nabla_x \pi_\rho(x(\rho)) = \nabla f(x(\rho)) + \rho \sum_{i=1}^{m} \nabla g_i(x(\rho)) g_i(x(\rho)) = 0.$$

Defining $\lambda_i(\rho) = -\rho g_i(x(\rho))$, we obtain that

$$\nabla f(x(\rho)) - \sum_{i=1}^{m} \lambda_i(\rho) \nabla g_i(x(\rho)) = 0.$$

If $x(\rho)$ converges to a solution x_* that is a regular point of the constraints, then $\lambda(\rho)$ converges to the Lagrange multiplier λ_* associated with x_*.

Penalty functions suffer from the same problems of ill-conditioning as do barrier functions. As the penalty parameter increases, the condition number of the Hessian matrix of $\pi_\rho(x(\rho))$ increases, tending to ∞ as $\rho \to \infty$. Therefore the unconstrained minimization problems can become increasingly difficult to solve.

Example 16.5 (Penalty Method). Consider the problem

$$\text{minimize} \quad f(x) = -x_1 x_2$$
$$\text{subject to} \quad g(x) = x_1 + 2x_2 - 4 = 0.$$

Suppose that this problem is solved via a penalty method using the quadratic-loss penalty function. Then a sequence of unconstrained minimization problems

$$\text{minimize} \quad \pi_\rho(x) = -x_1 x_2 + \tfrac{1}{2}\rho(x_1 + 2x_2 - 4)^2$$

is solved for increasing values of the penalty parameter ρ. The necessary conditions for optimality for the unconstrained problem are

$$-x_2 + \rho(x_1 + 2x_2 - 4) = 0$$
$$-x_1 + \rho(x_1 + 2x_2 - 4)(2) = 0.$$

For $\rho > 1/4$ this yields the solution

$$x_1(\rho) = x_1 = \frac{8\rho}{4\rho - 1}, \qquad x_2(\rho) = x_2 = \frac{4\rho}{4\rho - 1},$$

which is a local as well as a global minimizer. (The unconstrained problem has no minimum if $\rho \le 1/4$.) Note that $x(\rho)$ is infeasible to the original constrained problem, since

$$g(x(\rho)) = x_1 + 2x_2 - 4 = \frac{16\rho}{4\rho - 1} - 4 = \frac{4}{4\rho - 1}.$$

At any solution $x(\mu)$ we can define a Lagrange multiplier estimate as

$$\lambda = -\rho g(x(\rho)) = \frac{-4\rho}{4\rho - 1}.$$

As ρ tends to ∞ we obtain

$$\lim_{\rho \to \infty} x_1(\rho) = \lim_{\rho \to \infty} \frac{2}{1 - 1/4\rho} = 2, \qquad \lim_{\rho \to \infty} x_2(\rho) = \lim_{\rho \to \infty} \frac{1}{1 - 1/4\rho} = 1,$$

and indeed $x_* = (2, 1)^T$ is the minimizer for the constrained problem. Further,

$$\lim_{\rho \to \infty} \lambda(\rho) = \lim_{\rho \to \infty} \frac{-1}{1 - 1/4\rho} = -1,$$

and indeed $\lambda_* = -1$ is the Lagrange multiplier at x_*.

To demonstrate the ill-conditioning of the penalty function, we compute its Hessian matrix at $x(\rho)$:

$$\nabla_x^2 \pi_\rho(x(\rho)) = \begin{pmatrix} \rho & 2\rho - 1 \\ 2\rho - 1 & 4\rho \end{pmatrix}.$$

It can be shown that its condition number is approximately $25\rho/4$. When ρ is large, the Hessian matrix is ill conditioned. ∎

It is also possible to apply penalty methods to problems with inequality constraints. Consider, for example, the problem

$$\begin{aligned} \text{minimize} \quad & f(x) \\ \text{subject to} \quad & g_i(x) \geq 0, \ i = 1, \ldots, m. \end{aligned}$$

Any continuous function ψ that satisfies the conditions (16.3) can serve as a penalty. Thus, for example, the quadratic-loss penalty function in this case is

$$\psi(x) = \frac{1}{2} \sum_{i=1}^{m} [\min(g_i(x), 0)]^2.$$

This function has continuous first derivatives

$$\nabla \psi(x) = \sum_{i=1}^{m} [\min(g_i(x), 0)] \nabla g_i(x),$$

but its second derivatives can be discontinuous at points where some constraint g_i is satisfied exactly (see the Exercises). The same observation holds for other simple forms of the penalty function. Thus, one cannot safely use Newton's method to minimize the function. For this reason, straightforward penalty methods have not been widely used for solving general inequality-constrained problems.

16.2.3 Convergence[17]

In this subsection we discuss the convergence properties of penalization methods. We focus on the convergence of barrier methods when applied to the inequality-constrained problem. Convergence results for penalty methods can be developed in a similar manner.

We consider the problem

$$\begin{aligned} \text{minimize} \quad & f(x) \\ \text{subject to} \quad & g_i(x) \geq 0, \quad i = 1, \ldots, m. \end{aligned}$$

Denote the feasible set by

$$S = \{ x : g_i(x) \geq 0, \quad i = 1, \ldots, m \},$$

and its interior by

$$S^0 = \{ x : g_i(x) > 0, \quad i = 1, \ldots, m \}.$$

In developing the convergence theory, we make the following assumptions.

(i)   The functions f and $g_1, \ldots, g_m$ are continuous on $\Re^n$.

(ii) The set $\{ x : x \in S, f(x) \leq \alpha \}$ is bounded for any finite α.

(iii) The set S^0 is nonempty.

(iv) S is the closure of S^0.

---

[17]This subsection can be omitted from study without loss of continuity.

Assumptions (i) and (ii) imply that the function f has a minimum value on the set S. We denote this minimum value by f_*. Assumption (iii) is necessary to define the barrier subproblems. Assumption (iv) is necessary to avoid situations where the minimum point is isolated and does not have neighboring interior points. As an example, consider the problem of minimizing x subject to the constraints $x^2 - 1 \geq 0$ and $x + 1 \geq 0$. The feasible region consists of the points $\{x \geq 1\}$ and the isolated point $x = -1$. The point $x = -1$ is the minimizer, but because it is isolated it is not possible to approach it from the interior of the feasible region, and a barrier method could not converge to this solution.

The barrier function will be of the form $\beta_\mu(x) = f(x) + \mu\phi(x)$, where ϕ can be any function that is continuous on the interior of the feasible set, and that satisfies

$$\phi(x) \to \infty \quad \text{as} \quad g_i(x) \to 0_+.$$

We will show here that under mild conditions, the sequence of barrier minimizers has a convergent subsequence, and the limit of any such convergent subsequence is a solution to the problem. Although in practice, convergence of the entire sequence of minimizers is observed, from a theoretical point of view it is not always possible to guarantee convergence of the entire sequence, but only convergence of some subsequence. The following example illustrates this.

Example 16.6 (Nonconvergent Sequence with Convergent Subsequence). Consider the one-variable problem

$$\text{minimize} \quad f(x) = -x^2$$
$$\text{subject to} \quad 1 - x^2 \geq 0.$$

The logarithmic barrier function is $\beta_\mu(x) = -x^2 - \mu \log(1 - x^2)$. It has a single minimizer $x = 0$ if $\mu \geq 1$, and two minimizers $x = \pm\sqrt{1-\mu}$ if $\mu < 1$. (The point $x = 0$ is a local maximizer if $\mu < 1$.) Suppose that $\mu_0, \mu_1, \ldots$ is a sequence of decreasing barrier parameters less than 1. Then a possible sequence of minimizers of β is $x_k = (-1)^k \sqrt{1 - \mu_k}$. This sequence oscillates between neighborhoods of -1 and $+1$, and hence is nonconvergent. However, the subsequences $\{x_{2k}\}$ and $\{x_{2k+1}\}$ both converge to solutions of the original constrained problem. ∎

This example was contrived to illustrate the limitations of what can be said about a sequence of minimizers. In practice, because the solution of one subproblem is used as the starting point of the next subproblem, it is unlikely that a sequence of minimizers obtained in computation would jump back and forth in the manner described by $\{x_k\}$. Rather, consecutive minimizers would likely either all be in a neighborhood of $+1$ (eventually) or all be in the neighborhood of -1, and the resulting sequence would converge.

In the following, we assume that it is possible to compute global minimizers of the barrier function. This is true, for example, if the problem is convex (see the Exercises).

Theorem 16.7. *Suppose that a nonlinear inequality-constrained problem satisfies conditions (i)–(iv) above. Suppose that a sequence of unconstrained minimization problems*

$$\textit{minimize } \beta_\mu(x) = f(x) + \mu\phi(x)$$

is solved for μ taking values $\mu_1 > \mu_2 > \cdots > \mu_k > \cdots$, where $\lim_{k \to \infty} \mu_k = 0$. Suppose also that the functions $\beta_{\mu_k}(x)$ have a minimum in S^0 for each k. Let x_k denote a global minimizer of $\beta_{\mu_k}(x)$. Then

(i) $f(x_{k+1}) \leq f(x_k)$,

(ii) $\phi(x_{k+1}) \geq \phi(x_k)$,

(iii) *the sequence x_k has a convergent subsequence,*

(iv) *if $\{ x_k : k \in \mathcal{K} \}$ is any convergent subsequence of unconstrained minimizers of β, then its limit point is a global solution of the constrained problem.*

Proof. We prove the results in the order stated in the theorem. Only the last is of any difficulty.

1. Since x_k is the minimizer of $\beta_{\mu_k}(x)$, then $\beta_{\mu_k}(x_k) \leq \beta_{\mu_k}(x_{k+1})$, so

$$f(x_k) + \mu_k \phi(x_k) \leq f(x_{k+1}) + \mu_k \phi(x_{k+1}).$$

Also, since x_{k+1} is the minimizer of $\beta_{\mu_{k+1}}(x)$, then $\beta_{\mu_{k+1}}(x_{k+1}) \leq \beta_{\mu_{k+1}}(x_k)$, or

$$f(x_{k+1}) + \mu_{k+1} \phi(x_{k+1}) \leq f(x_k) + \mu_{k+1} \phi(x_k).$$

Multiplying the first inequality by μ_{k+1}, the second inequality by μ_k, adding the resulting inequalities, and reordering yields

$$(\mu_k - \mu_{k+1}) f(x_{k+1}) \leq (\mu_k - \mu_{k+1}) f(x_k).$$

Since $\mu_k > \mu_{k+1}$, we conclude that

$$f(x_{k+1}) \leq f(x_k).$$

2. As before, since x_k is the minimizer of $\beta_{\mu_k}(x)$, then

$$f(x_k) + \mu_k \phi(x_k) \leq f(x_{k+1}) + \mu_k \phi(x_{k+1}).$$

Since $f(x_k) \geq f(x_{k+1})$ this implies that

$$\phi(x_k) \leq \phi(x_{k+1}).$$

3. Consider the set $S^1 = \{ x \in S : f(x) \leq f(x_1) \}$. The continuity of f implies that S^1 is closed, and assumption (ii) implies that it is bounded. Hence S^1 is compact. Now in view of (i), $f(x_k) \leq f(x_1)$ for all k. Thus the sequence $\{ x_k \}$ lies in the compact set S^1, and therefore it has a convergent subsequence in S^1, and thus also in S.

4. Let $\{ x_k : k \in \mathcal{K} \}$ be a convergent subsequence of $\{ x_k \}$, and let $\hat{x}$ be its limit point. Since $g_i(x_k) > 0$ for all k, $g_i(\hat{x}) \geq 0$, and hence $\hat{x}$ is feasible to the constrained problem. Let f_* be the minimum value of f in the feasible region. We will show that $f(\hat{x}) = f_*$ by contradiction, assuming that

$$f(\hat{x}) > f_*. \tag{16.4}$$

It follows from (16.4) and assumption (iv) that there exists some *strictly feasible* point $y \in S^0$ such that

$$f(y) < f(\hat{x}).$$

Denote $\epsilon = f(\hat{x}) - f(y)$; then $\epsilon > 0$. Because f is continuous, it holds that

$$\lim_{\substack{k \in \mathcal{K} \\ k \to \infty}} f(x_k) = f(\hat{x}),$$

and thus for sufficiently large $k \in \mathcal{K}$ we have

$$f(y) + \tfrac{1}{2}\epsilon < f(x_k). \qquad (16.5)$$

Also, because x_k is a minimizer of $\beta_{\mu_k}()$ we have

$$f(x_k) + \mu_k\phi(x_k) \leq f(y) + \mu_k\phi(y). \qquad (16.6)$$

We consider now two cases.

(i) $\hat{x}$ is strictly feasible. Then for large enough $k \in \mathcal{K}$, x_k is strictly feasible, and therefore $\phi(x_k)$ is bounded. Also, because y is strictly feasible, $\phi(y)$ is bounded. Therefore, for k sufficiently large,

$$-\tfrac{1}{8}\epsilon \leq \mu_k\phi(x_k) \quad \text{and} \quad \mu_k\phi(y) \leq \tfrac{1}{8}\epsilon.$$

Combining this with (16.6) yields

$$f(x_k) - \tfrac{1}{8}\epsilon \leq f(y) + \tfrac{1}{8}\epsilon$$

or

$$f(x_k) \leq f(y) + \tfrac{1}{4}\epsilon.$$

But this is a contradiction to (16.5), and therefore to (16.4).

(ii) $\hat{x}$ is not strictly feasible. It follows from (16.5) that

$$f(y) < f(x_k).$$

Adding this to (16.6), rearranging, and dividing by μ_k gives

$$\phi(x_k) < \phi(y).$$

Because y is strictly feasible, the right-hand side is finite. Nevertheless, because x_k approaches the boundary, the left-hand side is unbounded above as k tends to ∞. We therefore have a contradiction to (16.4).   □

A key assumption in the theorem is that the barrier functions have a minimizer. We now state conditions that guarantee this.

Lemma 16.8. *Assume that the set S is compact. Then for any fixed positive value μ there exists a point $x(\mu) \in S^0$ that minimizes $\beta_\mu(x)$.*

Proof. Let x_0 be any point in S^0, and define the set

$$W = \left\{ x \in S^0 : \beta_\mu(x) \le \beta_\mu(x_0) = \beta_0 \right\}.$$

Because S is compact, W is bounded. We now show that W is also closed; that is, it contains all its limits points.

Let $\left\{ w_j \right\}$ be any convergent sequence in W with limit point $\hat{w}$. Since $\left\{ w_j \right\} \subset S$ and S is compact, then $\hat{w} \in S$. Now if $\hat{w}$ were on the boundary of S (that is, some $g_i(\hat{w}) = 0$), then $\lim_{j \to \infty} \beta_\mu(w_j) = \infty$, which contradicts the fact that $w_j \in W$. Thus $\hat{w} \in S^0$. Now because $w_j \in W$, it follows that $\beta_\mu(w_j) \le \beta_0$ for all j. The continuity of β in S^0 implies that $\beta_\mu(\hat{w}) \le \beta_0$ also. Thus $\hat{w} \in W$. Therefore W contains all its limit points, and hence it is closed. Because it is also bounded, W is compact.

Since β is continuous on the compact set W, it attains its global minimum $x(\mu)$ in W. But by definition, the value of β at any point in S that is not in W must be larger than β_0, and in turn, larger than the minimum value $\beta_\mu(x(\mu))$. Thus $x(\mu)$ is the global minimizer of β in S. $\quad\Box$

Another case where the barrier minimizers always exist is when the problem is convex. This is true even if the feasible region is unbounded. A proof can be found in the book by Fiacco and McCormick (1968, reprinted 1990).

Theorem 16.7 shows that a subsequence of global minimizers of the barrier function converges to a global minimum of the constrained problem. It is also possible to prove convergence to a local minimum of the constrained problem if additional assumptions are made. In practice, if started close to some local solution, it is likely that consecutive minimizers of the barrier functions will naturally satisfy these conditions, and that the sequence of barrier minimizers will therefore have a limit.

It is not true that every limit point of a sequence of local minimizers of the barrier function is a constrained minimizer. (See the Exercises.)

Exercises

2.1. Consider the problem

$$\begin{aligned}
\text{minimize} \quad & f(x) = \tfrac{1}{2}x_1^2 + \tfrac{1}{2}x_2^2 \\
\text{subject to} \quad & x_1 + x_2 \ge 1.
\end{aligned}$$

Suppose that the logarithmic barrier method is used to solve this problem.

(i) What is $x(\mu)$? What is $\lambda(\mu)$? What is x_*? What is λ_*?

(ii) Compute the Hessian matrix B of the logarithmic barrier function for $\mu = 10^{-4}$. What is the condition number of B? What is B^{-1}?

2.2. Repeat Exercise 2.1 for

$$\text{minimize} \quad f(x) = x_1 + x_2$$
$$\text{subject to} \quad x_1 \geq 0$$
$$x_2 \geq 0.$$

This example illustrates that if an n-variable problem has n constraints in the optimal active set, then the barrier method does not introduce ill-conditioning.

2.3. Repeat Exercise 2.1 for

$$\text{maximize} \quad f(x) = x_1 + x_2$$
$$\text{subject to} \quad x_1 \geq 0$$
$$x_2 \geq 0$$
$$x_1 + x_2 \leq 1.$$

2.4. (Due to Powell (1972).) Consider the one-dimensional problem

$$\text{minimize} \quad \frac{-1}{x^2 + 1}$$
$$\text{subject to} \quad x \geq 1.$$

Show that the logarithmic barrier function is unbounded below in the feasible region. Show also that the logarithmic barrier function has a local minimizer that approaches the solution $x_* = 1$ as $\mu \to 0$.

2.5. For barrier methods, we assume that it is possible to reach any boundary point of the feasible region by approaching it from the interior. Give three examples of feasible regions that violate this assumption.

2.6. (Due to Fiacco (1973) and Wright (1992).) Consider the problem

$$\text{minimize} \quad f(x) = x$$
$$\text{subject to} \quad x^2 \geq 0$$
$$x \geq -1.$$

Prove that the sequence of global minimizers of the logarithmic barrier function converges to the constrained global minimizer $x = -1$, but that the sequence of local, but nonglobal, minimizers converges to 0, which is not a constrained minimizer.

2.7. Show how to use a two-phase approach to obtain an initial feasible point for a barrier method.

2.8. Suppose that f is bounded below in the feasible region. Prove that if $\beta_{\mu_0}(x)$ is bounded from below in S^0, then so is $\beta_\mu(x)$ for all μ such that $0 < \mu < \mu_0$.

16.3   Ill-Conditioning

We now analyze the structure of the Hessian matrix of the barrier function and show that as μ decreases, the Hessian matrix becomes increasingly ill conditioned. Although our discussion will focus on the logarithmic barrier function method, the same results are true in general for other barrier and penalty methods. This ill-conditioning is the main reason that classical barrier and penalty methods were abandoned in the early 1970s. In Section 16.4

we show how to overcome this ill-conditioning by deriving numerically stable formulas for the Newton direction in a barrier method.

The logarithmic barrier method attempts to solve the constrained problem by solving a sequence of unconstrained problems of the form

$$\text{minimize } \beta_\mu(x) = f(x) - \mu \sum_{i=1}^{m} \log(g_i(x))$$

for a sequence of positive barrier parameters μ which decrease to zero.

The ill-conditioning can be observed by examining the formulas for the Hessian matrix of β. They are

$$\nabla_x \beta = \nabla f - \mu \sum_{i=1}^{m} \frac{\nabla g_i}{g_i}$$

$$\nabla_{xx}^2 \beta = \nabla^2 f - \mu \sum_{i=1}^{m} \frac{\nabla^2 g_i}{g_i} + \mu \sum_{i=1}^{m} \frac{\nabla g_i \nabla g_i^T}{g_i^2},$$

where we have written f in place of $f(x)$, etc. Let $x(\mu)$ be an unconstrained minimizer of $\beta_\mu(x)$, and let $\lambda_i = \mu/g_i(x(\mu))$ be the associated Lagrange multiplier estimate. The formulas for the gradient and Hessian matrix can be rewritten as

$$\nabla_x \beta = \nabla f - \sum_{i=1}^{m} \lambda_i \nabla g_i$$

$$\nabla_{xx}^2 \beta = \nabla^2 f - \sum_{i=1}^{m} \lambda_i \nabla^2 g_i + \frac{1}{\mu} \sum_{i=1}^{m} \lambda_i^2 \nabla g_i \nabla g_i^T.$$

Notice that the gradient is an approximation to the gradient of the Lagrangian at the optimum.

Let us inspect the expression for the Hessian matrix. The term $\nabla^2 f - \sum_{i=1}^{m} \lambda_i \nabla^2 g_i$ is an approximation to the Hessian matrix of the Lagrangian of the problem. In general, its condition number reflects the conditioning of the Lagrangian Hessian for the original constrained problem.

Ill-conditioning in the barrier function arises because of the final term in the Hessian matrix. If a constraint is binding at the solution of the constrained problem (and its corresponding Lagrange multiplier is not zero), then the ratio λ_i^2/μ tends to infinity as μ approaches zero. As a result, the Hessian matrix in general becomes increasingly ill conditioned as the solution is approached. This structural ill-conditioning occurs even when the underlying constrained problem is well conditioned.

16.4 Stabilized Penalty and Barrier Methods[18]

Despite the ill-conditioning of the Hessian matrix of the barrier function, it is still possible to compute a Newton-type direction in a numerically stable manner. We show how to do

---

[18]This section can be omitted from study without loss of continuity.

this for the logarithmic barrier function applied to the problem

$$\text{minimize} \quad f(x)$$
$$\text{subject to} \quad g(x) \geq 0.$$

The results can be extended to other penalty and barrier methods.

The ill-conditioning is avoided by using an approximate formula for the Newton direction for the barrier problem. This formula becomes more accurate as the barrier parameter goes to zero, that is, as the ill-conditioning becomes more severe.

For simplicity, we assume that all the constraints are binding at the solution, and that all of these constraints have positive Lagrange multipliers. In practice, there is likely to be a mix of binding and nonbinding constraints, and in that case the techniques that we describe would be applied only to the binding constraints.

Our approach is to separate the Newton direction into two components, one in the null space and one in the range space of the constraint gradients. We show that approximations to both these components can be computed in a stable manner.

Let $A = \nabla g(x)^T$ be the Jacobian matrix of the constraints, and assume that A has full rank. Let Z be a basis matrix for the null space of A, and let A_r be a right-inverse matrix for A. We assume that Z and A_r have been obtained from an orthogonal QR factorization (see Section 3.3.4), so that

$$(Z \quad A^T) \begin{pmatrix} Z^T \\ A_r^T \end{pmatrix} = I.$$

We also define the Lagrange multiplier estimates $\lambda_i = \mu/g_i(x)$ and the diagonal matrix D, whose ith diagonal entry is λ_i^2. Finally, let $B = \nabla_{xx}^2 \beta$ be the Hessian of the barrier function.

With these definitions, we obtain

$$B = \nabla_{xx}^2 \beta = \nabla^2 f - \sum_{i=1}^m \lambda_i \nabla^2 g_i + \frac{1}{\mu} \sum_{i=1}^m \lambda_i^2 \nabla g_i \nabla g_i{}^T = H + \frac{1}{\mu} A^T D A,$$

where

$$H = \nabla^2 f - \sum_{i=1}^m \lambda_i \nabla^2 g_i$$

is the Hessian matrix of the Lagrangian. From the identity

$$B = IBI = (Z \quad A^T) \begin{pmatrix} Z^T \\ A_r^T \end{pmatrix} B (Z \quad A_r) \begin{pmatrix} Z^T \\ A \end{pmatrix},$$

we obtain

$$B^{-1} = (Z \quad A_r) \begin{pmatrix} Z^T B Z & Z^T B A_r \\ A_r^T B Z & A_r^T B A_r \end{pmatrix}^{-1} \begin{pmatrix} Z^T \\ A_r^T \end{pmatrix}.$$

To compute the search direction, we approximate the inverse of the middle matrix on the right-hand side.

To find an explicit expression for this inverse matrix, we use the *bordered-inverse formula*: If A_1 and A_3 are symmetric matrices, then

$$
\begin{pmatrix} A_1 & A_2 \\ A_2^T & A_3 \end{pmatrix}^{-1} = \begin{pmatrix} A_1^{-1} + A_1^{-1}A_2G^{-1}A_2^TA_1^{-1} & -G^{-1}A_2^TA_1^{-1} \\ -A_1^{-1}A_2G^{-1} & G^{-1} \end{pmatrix}
$$

$$
= \begin{pmatrix} A_1^{-1} & 0 \\ 0 & 0 \end{pmatrix} + \begin{pmatrix} A_1^{-1}A_2 \\ -I \end{pmatrix} G^{-1} (A_2^TA_1^{-1} \quad -I),
$$

where

$$
G = (A_3 - A_2^TA_1^{-1}A_2).
$$

Applying this formula, and noting that $AZ = 0$, gives

$$
B^{-1} = (Z \quad A_r) \left[\begin{pmatrix} (Z^THZ)^{-1} & 0 \\ 0 & 0 \end{pmatrix} \right.
$$

$$
\left. + \begin{pmatrix} (Z^THZ)^{-1}Z^THA_r \\ -I \end{pmatrix} G^{-1} (A_r^THZ(Z^THZ)^{-1} \quad -I) \right] \begin{pmatrix} Z^T \\ A_r^T \end{pmatrix},
$$

where

$$
G = \frac{1}{\mu}D + A_r^THA_r - A_r^THZ(Z^THZ)^{-1}Z^THA_r.
$$

When μ is small (that is, as we approach the solution where the ill-conditioning becomes apparent), $G^{-1} \approx \mu D^{-1}$. Hence

$$
B^{-1} \approx Z(Z^THZ)^{-1}Z^T
$$

$$
+ \mu(Z(Z^THZ)^{-1}Z^TH - I)A_rD^{-1}A_r^T(HZ(Z^THZ)^{-1}Z^T - I).
$$

This approximation to B^{-1} determines an approximation to the Newton direction $p = B^{-1}\nabla_x\beta$. It is straightforward to show that

$$
p = B^{-1}\nabla_x\beta \approx p_1 + \mu p_2,
$$

where

$$
p_1 = -Z(Z^THZ)^{-1}Z^T\nabla_x\beta,
$$
$$
\lambda = A_r^T(Hp_1 + \nabla_x\beta),
$$
$$
p_2 = (Z(Z^THZ)^{-1}Z^TH - I)A_rD^{-1}\lambda.
$$

The main point to notice here is that to compute an approximate Newton direction, the only matrix inverse required is that of Z^THZ. This matrix has no structural ill-conditioning, that is, no ill-conditioning due to the presence of the barrier parameter.

The matrix Z^THZ appears twice in these formulas. This inverse matrix should not be formed. Instead a factorization should be computed. This factorization need only be computed once; backsubstitution can be used to compute p_1 and p_2.

As $\mu \to 0$, it can be shown that the error in the approximate search direction is $O(\mu)$; see the references in the Notes.

Example 16.9 (Stabilized Formulas). Consider the problem from Example 16.2 with the nonbinding constraint removed:

$$\text{minimize} \quad f(x) = x_1^2 + x_2^2$$
$$\text{subject to} \quad g(x) = x_1 - 1 \geq 0.$$

The corresponding barrier problem is

$$\text{minimize} \ \beta_\mu(x) = x_1^2 + x_2^2 - \mu \log(x_1 - 1).$$

If we set $\mu = 10^{-4}$ and $x = (1.001, 0.001)^T$, then the multiplier estimate is $\lambda = \mu/(x_1 - 1) = 0.1$. At this point,

$$\nabla_x \beta = \begin{pmatrix} 2x_1 - \mu/(x_1 - 1) \\ 2x_2 \end{pmatrix} = \begin{pmatrix} 1.902 \\ 0.002 \end{pmatrix}$$

$$B = \nabla_{xx}^2 \beta = \begin{pmatrix} 2 + \mu/(x_1 - 1)^2 & 0 \\ 0 & 2 \end{pmatrix} = \begin{pmatrix} 102 & 0 \\ 0 & 2 \end{pmatrix}.$$

Then $\text{cond}(B) = 50.1$. The Newton direction is

$$p = -B^{-1}b = \begin{pmatrix} -0.0186 \\ -0.0010 \end{pmatrix}.$$

We now determine the approximate Newton direction. For this problem,

$$H = \begin{pmatrix} 2 & 0 \\ 0 & 2 \end{pmatrix} \quad \text{and} \quad A = (1 \quad 0).$$

Hence

$$A_r = \begin{pmatrix} 1 \\ 0 \end{pmatrix}, \quad Z = \begin{pmatrix} 0 \\ 1 \end{pmatrix}, \quad \text{and} \quad Z^T H Z = (2).$$

Now $\text{cond}(Z^T H Z) = 1$. The diagonal matrix is $D = (\lambda^2) = (0.01)$. From these we determine

$$p_1 = -Z(Z^T H Z)^{-1} Z^T \nabla_x \beta = \begin{pmatrix} 0 \\ -0.001 \end{pmatrix}$$
$$\lambda = A_r^T(H p_1 + \nabla_x \beta) = (1.902)$$
$$p_2 = (Z(Z^T H Z)^{-1} Z^T H - I) A_r D^{-1} \lambda = \begin{pmatrix} -190.2 \\ 0 \end{pmatrix}.$$

The approximate Newton direction is

$$\bar{p} = p_1 + \mu p_2 = \begin{pmatrix} -0.0190 \\ -0.0010 \end{pmatrix}.$$

The difference between the Newton direction and the approximation Newton direction is

$$p - \bar{p} = \begin{pmatrix} 4 \times 10^{-4} \\ 0 \end{pmatrix}.$$

This is as expected, since the theory predicts that this difference will be proportional to $\mu = 10^{-4}$. ∎

If the matrix $Z^T H Z$ is positive definite at the point x, then these formulas will produce a descent direction provided that the barrier parameter μ is sufficiently small. If $Z^T H Z$ is not positive definite, it is still possible to compute a descent direction by modifying the Hessian matrix (see Section 11.4).

Exercises

4.1. For Exercise 2.1, compute $Z^T H Z$. Note that the only inverse matrix required in the stabilized formulas is that of $Z^T H Z$.

4.2. Verify the formulas for the first and second derivatives of the barrier function in Section 16.3.

4.3. Verify the expressions for B^{-1}.

4.4. Verify the bordered-inverse formula.

16.5   Exact Penalty Methods

Barrier methods and penalty methods both solve a constrained problem by solving a sequence of unconstrained problems. In general, a sequence of unconstrained optima will *approach* the solution of the constrained problem, but will not attain it exactly, even if exact arithmetic is used. Is it possible to obtain the exact solution of a constrained problem by solving a finite number of unconstrained problems? Or better still, is it possible to obtain the exact solution of a constrained problem by solving a single unconstrained problem?

In this section we present a family of methods that achieve these goals. These methods are called *exact penalty methods* because they make it possible to solve the constrained problem *exactly* for a finite positive penalty parameter.

To illustrate these methods, we consider a problem with both equality and inequality constraints:

$$\begin{aligned} \text{minimize} \quad & f(x) \\ \text{subject to} \quad & g_i(x) = 0, \quad i \in \mathcal{E} \\ & g_i(x) \geq 0, \quad i \in \mathcal{I}. \end{aligned}$$

We assume that the objective function f and the constraint functions g_i are continuous. Consider the penalty function

$$\pi_\rho(x) = f(x) + \rho \sum_{i \in \mathcal{E}} |g_i(x)| + \rho \sum_{i \in \mathcal{I}} |\min(0, g_i(x))|$$

$$= f(x) + \rho \sum_{i \in \mathcal{E}} |g_i(x)| - \rho \sum_{i \in \mathcal{I}} \min(0, g_i(x)).$$

The penalty imposed for violating a constraint is simply the absolute value of the violation. The function π is continuous, but it is not differentiable at all points. This penalty function is sometimes called the ℓ_1 *exact penalty function*, because the penalty term is the ℓ_1-norm (1-norm) of the constraint violations.

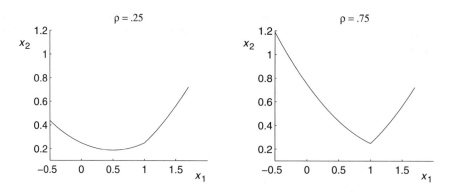

Figure 16.3. *Exact penalty function.*

Example 16.10 (Exact Penalty Method). Consider the problem

$$\text{minimize} \quad f(x) = \tfrac{1}{2}(x_1 - x_2)^2 + \tfrac{1}{2}x_2^2$$
$$\text{subject to} \quad x_1 \geq 1.$$

This problem has a unique minimizer $x_1 = 1, x_2 = \tfrac{1}{2}$, with a corresponding Lagrange multiplier $\lambda = \tfrac{1}{2}$. Let us use the exact ℓ_1 penalty function to solve the problem. The penalty function is

$$\pi_\rho(x) = \tfrac{1}{2}(x_1 - x_2)^2 + \tfrac{1}{2}x_2^2 - \rho \min(0, x_1 - 1).$$

This function is not differentiable when $x_1 = 1$. It will be convenient to write π in the form

$$\pi_\rho(x) = (\tfrac{1}{2}x_1 - x_2)^2 + (\tfrac{1}{4}x_1^2 - \rho \min(0, x_1 - 1)).$$

Then the minimum of π is obtained at a point where $x_2 = \tfrac{1}{2}x_1$, and where x_1 minimizes the function defined by

$$\begin{cases} \tfrac{1}{4}x_1^2 - \rho(x_1 - 1) & \text{if } x_1 < 1 \\ \tfrac{1}{4}x_1^2 & \text{if } x_1 \geq 1. \end{cases}$$

This function is graphed in Figure 16.3 for two values of ρ. When $\rho \leq \tfrac{1}{2}$, the left branch of this function is decreasing for $x_1 \leq 2\rho$ and then increasing for $2\rho \leq x_1 \leq 1$. At this point it meets the right branch of the function which also increases. When $\rho > \tfrac{1}{2}$, the left branch of the function is decreasing for $x_1 \leq 1$, and the right branch is increasing for $x_1 \geq 1$. Hence, for any $\rho > \tfrac{1}{2}$, $x_1 = 1$ is the minimizer of the function. In turn this yields $x_2 = \tfrac{1}{2}$. Thus, for any $\rho > \tfrac{1}{2}$, the solution x_* is also a minimizer of the penalty function. ∎

The following lemma shows that, for a large-enough penalty parameter, the minimizer of the penalty function also solves the constrained minimization problem.

Lemma 16.11. *Let (x_*, λ_*) be a solution to the constrained optimization problem. Assume that the second-order sufficiency conditions are satisfied, and that $\|\lambda_*\| \leq M$ for some constant M. There exists a $\bar{\rho}$ such that if $\rho \geq \bar{\rho}$, then x_* is also a local minimizer of $\pi_\rho(x)$.*

Proof. We prove the result in the case when only equality constraints are present; the remainder of the proof is left to the Exercises.

The proof is based on ideas from sensitivity analysis. Let h be some vector, with $\|h\|$ "small," and consider the perturbed optimization problem

$$\begin{aligned} \text{minimize} \quad & f(x) \\ \text{subject to} \quad & g(x) = h. \end{aligned}$$

Let $F(h)$ be the optimal objective value. Then

$$F(h) = F(0) + h^T \nabla F(\xi) = f(x_*) + h^T \nabla F(\xi),$$

where $\|\xi\| \leq \|h\|$.

We now examine the penalty function:

$$\begin{aligned} \min_x \pi_\rho(x) &= \min_x f(x) + \rho \sum |g_i(x)| \\ &= \min_{x,h} \left\{ f(x) + \rho \sum |h_i| : g_i(x) = h_i \right\} \\ &= \min_h F(h) + \rho \sum |h_i|. \end{aligned}$$

We now show that $F(h) + \rho \sum |h_i| > F(0)$ if $h \neq 0$ and ρ is large. Since $F(0) = \pi_\rho(x_*) = f(x_*)$, this will prove the lemma.

Using the Taylor series expansion, we obtain

$$F(h) + \rho \sum |h_i| = f(x_*) + h^T \nabla F(\xi) + \rho \sum |h_i|.$$

Let ϵ be some small positive number. Since $\nabla F(0) = \lambda$ (see, for example, the book by Luenberger (1984)) and $\|\xi\| < \|h\|$, we can guarantee that $|(\nabla F(\xi))_i| < |\lambda_i| + \epsilon$ if $\|h\|$ is sufficiently small. Hence

$$\begin{aligned} h^T \nabla F(\xi) &\geq -\left(\max_i \{ |(\nabla F(\xi))_i| \} \right) \sum |h_i| \\ &\geq -\left(\max_i \{ |\lambda_i| + \epsilon \} \right) \sum |h_i|. \end{aligned}$$

Let $|\hat{\lambda}|$ be the largest Lagrange multiplier in absolute value. Then

$$F(h) + \rho \sum |h_i| \geq f(x_*) + (\rho - |\hat{\lambda}| - \epsilon) \sum |h_i|.$$

If $\rho > |\hat{\lambda} + \epsilon|$, then this is greater than $f(x_*)$. Since this is true for all $\epsilon > 0$, the lemma is true with $\bar{\rho} = |\hat{\lambda}|$. $\quad \square$

Optimization methods such as Newton's method use the derivatives of the problem functions. They cannot be applied directly to this exact penalty function, since it is not differentiable at all points. So, even though it improves on classical penalty methods by allowing the penalty parameter to remain bounded, it has its own deficiencies.

In the next section we will derive a differentiable exact penalty function, but it too will not be free of complications.

Exercises

5.1. Apply an exact penalty method to

$$\text{minimize} \quad f(x) = \tfrac{1}{3}x^3 - x$$
$$\text{subject to} \quad x = 1.$$

Find the minimizer of the penalty function for any penalty parameter ρ. Determine the critical value of the paramter $\bar{\rho}$ that guarantees that, for any $\rho \geq \bar{\rho}$, the problem can be solved via a single subproblem.

5.2. Consider the linear programming problem

$$\text{minimize} \quad f(x) = c^T x$$
$$\text{subject to} \quad Ax = b$$
$$x \geq 0.$$

(i) Find the dual problem.

(ii) Suppose that there is no known initial feasible solution to this problem. A common technique is to add (nonnegative) artificial variables to the equality constraints and use a big-M method to solve the problem. On paper, if the problem is solved by the simplex method, one can treat "M" as a symbol. In practice, a value of M must be specified. How large must M be? Show that it is sufficient to choose $M \geq \max\{\,|y_i|\,\}$, where y is the vector of dual variables corresponding to the equality constraints. You may assume that the original linear program has a unique optimal solution.

(iii) Show also that the big-M method may be viewed as an exact penalty method, and indicate how the results above confirm the theory of exact penalty methods.

5.3. Complete the proof of Lemma 16.11 by showing that the result is also true if inequality constraints are present.

16.6   Multiplier-Based Methods

The ill-conditioning of penalty methods can be ameliorated by including multipliers explicitly in the penalty function. This complicates the algorithms, because both the variables x and the multipliers λ must be updated, but in exchange it offers the possibility of more rapid rates of convergence. Of course, multipliers appear in the context of the classical penalty method, but in that case they are a by-product of the method. For example, in the classical penalty method, the multiplier estimate is

$$\lambda = -\rho g(x),$$

where g is the vector of constraint functions. These multiplier estimates are used in the termination tests, as well as for sensitivity analysis. But multiplier estimates can be used in a more active way to drive an optimization algorithm. We examine this idea in this section.

Let us begin by examining problems of the form

$$\text{minimize} \quad f(x)$$
$$\text{subject to} \quad g(x) = 0,$$

where g is a vector-valued function. As was pointed out in Section 14.6, the solution x_* to this problem is a stationary point of the Lagrangian function, but is a *minimizer* of this function if x is restricted to be feasible. Thus x_* solves

$$\text{minimize} \quad \mathcal{L}(x, \lambda) = f(x) - \lambda^T g(x)$$
$$\text{subject to} \quad g(x) = 0.$$

If penalty terms are added to the Lagrangian, then it might be possible to make x_* an *unconstrained* minimizer of this penalized Lagrangian function. For example, it might be possible to solve

$$\underset{x}{\text{minimize}} \; \mathcal{A}_\rho(x, \lambda) = f(x) - \lambda^T g(x) + \tfrac{1}{2}\rho g(x)^T g(x)$$

to obtain x_*. This is referred to as an *augmented Lagrangian method* because it is based on minimizing an *augmented Lagrangian function*.

A simple augmented Lagrangian method has the following form: Values of x_0, λ_0, and $\rho_0 > 0$ are chosen to initialize the method. Then for $k = 0, 1, \ldots$ do the following:

1. *The Optimality Test*—If $\nabla \mathcal{L}(x_k, \lambda_k) = 0$, then stop.
2. *The Unconstrained Subproblem*—Compute x_{k+1} by solving

$$\underset{x}{\text{minimize}} \; \mathcal{A}_{\rho_k}(x, \lambda_k) = f(x) - \lambda_k^T g(x) + \tfrac{1}{2}\rho_k g(x)^T g(x)$$

   using any of our unconstrained optimization methods.
3. *The Update*—Determine λ_{k+1} and ρ_{k+1}. For example, for the multiplier update we could use

$$\lambda_{k+1} = \lambda_k - \rho_k g(x_{k+1}).$$

   The new penalty parameter should be chosen so that $\rho_{k+1} \geq \rho_k$.

The final step requires comment. If x_{k+1} minimizes $\mathcal{A}_{\rho_k}(x, \lambda_k)$, then

$$\nabla_x \mathcal{A}_{\rho_k}(x_{k+1}, \lambda_k) = 0$$

or

$$\nabla f(x_{k+1}) - \nabla g(x_{k+1})\lambda_k + \rho_k \nabla g(x_{k+1})g(x_{k+1}) = 0.$$

This can be rearranged as

$$\nabla f(x_{k+1}) - \nabla g(x_{k+1})[\lambda_k - \rho g(x_{k+1})] = 0.$$

If we define $\lambda_{k+1} = \lambda_k - \rho_k g(x_{k+1})$, then

$$\nabla_x \mathcal{L}(x_{k+1}, \lambda_{k+1}) = \nabla f(x_{k+1}) - \nabla g(x_{k+1})\lambda_{k+1} = 0;$$

that is, the optimality conditions for the constrained problem are partially satisfied. The algorithm will terminate when

$$\nabla_\lambda \mathcal{L}(x_{k+1}, \lambda_{k+1}) = -g(x_{k+1}) = 0,$$

that is, when feasibility has been achieved.

The update step also selects ρ_{k+1}. This parameter should be "large enough" so that the augmented Lagrangian function has a local minimizer in x. If ρ is too small, then the unconstrained subproblem may not have a solution. (See the Exercises.) If a failure is detected when attempting to solve this subproblem, then ρ should be increased. Software for these methods often includes elaborate rules for updating ρ.

Example 16.12 (Augmented Lagrangian Method). We apply the augmented Lagrangian method to the problem in Example 15.8:

$$\begin{aligned} \text{minimize} \quad & f(x_1, x_2) = e^{3x_1} + e^{-4x_2} \\ \text{subject to} \quad & g(x_1, x_2) = x_1^2 + x_2^2 - 1 = 0. \end{aligned}$$

The solution to this problem is $x_* = (-0.74834, 0.66332)^T$ with $\lambda_* \approx -0.21233$. The augmented Lagrangian function is

$$\mathcal{A}_\rho(x, \lambda) = e^{3x_1} + e^{-4x_2} - \lambda(x_1^2 + x_2^2 - 1) + \tfrac{1}{2}\rho(x_1^2 + x_2^2 - 1)^2.$$

We again use the initial guess $x_0 = (-1, 1)^T$, together with $\lambda_0 = -1$. The penalty parameter will be kept constant: $\rho_0 = \rho_k = 10$. At the initial point

$$\nabla f = \begin{pmatrix} 3e^{3x_1} \\ -4e^{-4x_2} \end{pmatrix} = \begin{pmatrix} 0.14936 \\ -0.07326 \end{pmatrix}$$

$$\nabla^2 f = \begin{pmatrix} 9e^{3x_1} & 0 \\ 0 & 16e^{-4x_2} \end{pmatrix} = \begin{pmatrix} 0.44808 & 0 \\ 0 & 0.29305 \end{pmatrix}$$

$$g = x_1^2 + x_2^2 - 1 = (1)$$

$$\nabla g = \begin{pmatrix} 2x_1 \\ 2x_2 \end{pmatrix} = \begin{pmatrix} -2 \\ 2 \end{pmatrix}, \quad \nabla^2 g = \begin{pmatrix} 2 & 0 \\ 0 & 2 \end{pmatrix}.$$

We use Newton's method to solve the unconstrained subproblem. For simplicity, we use the classical method, without a line search, so that

$$x \leftarrow x - (\nabla_{xx}^2 \mathcal{A})^{-1}(\nabla_x \mathcal{A}).$$

For $x = x_0$,

$$\nabla_x \mathcal{A} = \begin{pmatrix} -21.851 \\ 21.927 \end{pmatrix} \quad \text{and} \quad \nabla_{xx}^2 \mathcal{A} = \begin{pmatrix} 62.448 & -40.000 \\ -40.000 & 62.293 \end{pmatrix}$$

and

$$x \leftarrow \begin{pmatrix} -1 \\ 1 \end{pmatrix} - \begin{pmatrix} -0.21138 \\ 0.21626 \end{pmatrix} = \begin{pmatrix} -0.78862 \\ 0.78374 \end{pmatrix}.$$

Table 16.2. *Augmented Lagrangian method.*

k	x_k		λ_k	$\|\nabla_x \mathcal{L}\|$	$\|g\|$
0	-1.00000	1.00000	-1.000000	3×10^0	1×10^0
1	-0.71795	0.63937	-0.242431	6×10^{-12}	8×10^{-2}
2	-0.74719	0.66242	-0.213381	1×10^{-15}	3×10^{-3}
3	-0.74830	0.66329	-0.212362	7×10^{-10}	1×10^{-4}
4	-0.74833	0.66332	-0.212326	2×10^{-15}	4×10^{-6}
5	-0.74834	0.66332	-0.212325	2×10^{-10}	1×10^{-7}
6	-0.74834	0.66332	-0.212325	2×10^{-13}	4×10^{-9}

At this point $\|\nabla_x \mathcal{A}\| = 7.1533$. We continue using Newton's method until

$$\|\nabla_x \mathcal{A}\| < 10^{-9},$$

obtaining the successive iterates

$$\begin{pmatrix} -0.70745 \\ 0.68719 \end{pmatrix}, \quad \begin{pmatrix} -0.70887 \\ 0.65110 \end{pmatrix}, \quad \cdots \quad \begin{pmatrix} -0.71795 \\ 0.63937 \end{pmatrix}.$$

This final point is x_1. (Even though no line search is used, the iteration converges to a local minimizer of the augmented Lagrangian function in this case.)

To complete an iteration of the augmented Lagrangian method we update the Lagrange multiplier estimate:

$$\lambda_1 = \lambda_0 - \rho_0 g(x_1) = -1 - 10(-0.075757) = -0.24243.$$

Now the whole process repeats at the new point.

The complete iteration is summarized in Table 16.2. From the first iteration to the last iteration, the optimality condition is almost satisfied. This is a consequence of the update formula for the Lagrange multiplier. (The optimality condition is not exactly satisfied because the unconstrained subproblems are not solved exactly.) ∎

We now establish some basic properties of the augmented Lagrangian function. Assume that x_* is a regular point, and let λ_* be the vector of Lagrange multipliers at x_*. For any $\rho > 0$,

$$\mathcal{A}_\rho(x_*, \lambda_*) = f(x_*) - \lambda_*^T g(x_*) + \tfrac{1}{2} \rho g(x_*)^T g(x_*) = f(x_*),$$

since the solution must be a feasible point (that is, $g(x_*) = 0$). This shows that the objective function and the augmented Lagrangian function have the same value at the solution. In addition,

$$\nabla_x \mathcal{A}_\rho(x_*, \lambda_*) = \nabla f(x_*) - \nabla g(x_*)\lambda_* + \rho \nabla g(x_*)g(x_*)$$
$$= \nabla f(x_*) - \nabla g(x_*)\lambda_* = 0.$$

Hence the gradient of the augmented Lagrangian is equal to the gradient of the Lagrangian and vanishes as the solution. It is also possible to show that the Hessian matrix of the augmented Lagrangian is positive definite under reasonable conditions.

Lemma 16.13 (Hessian Matrix of Augmented Lagrangian). *Let (x_*, λ_*) be a solution to the constrained optimization problem*

$$\text{minimize} \quad f(x)$$
$$\text{subject to} \quad g(x) = 0.$$

Assume that the second-order sufficiency conditions are satisfied at (x_*, λ_*) and that x_* is a regular point of the constraints. Then $\nabla^2_{xx} \mathcal{A}_\rho(x_*, \lambda_*)$ is positive definite for all $\rho \geq \bar{\rho}$, for some value $\bar{\rho}$.

Proof. We prove that $x^T \nabla^2_{xx} \mathcal{A}_\rho(x_*, \lambda_*) x > 0$ for $x \neq 0$ and ρ sufficiently large. To simplify the proof, we denote $H = \nabla^2_{xx} \mathcal{L}(x_*, \lambda_*)$ and $A = \nabla g(x_*)^T$. Then

$$\nabla^2_{xx} \mathcal{A}_\rho(x_*, \lambda_*) = H + \rho A^T A.$$

Any x can be written as $x = Zv + q$, where Z is a basis matrix for the null space of A and q is in the range of A^T. (See Section 3.2.) If $x \neq 0$, then v and q cannot both be zero.

We now test if the Hessian matrix of the augmented Lagrangian function is positive definite (using the fact that $AZ = 0$):

$$\begin{aligned}
x^T \nabla^2_{xx} \mathcal{A}_\rho(x_*, \lambda_*) x &= (Zv + q)^T (H + \rho A^T A)(Zv + q) \\
&= v^T (Z^T H Z) v + 2v^T Z^T H q + q^T H q + \rho q^T A^T A q.
\end{aligned}$$

Let α be the smallest eigenvalue of $Z^T H Z$; because of the second-order sufficiency condition, $\alpha > 0$. Also, let $\beta = \|Z\|$, $\gamma = \|H\|$, and let δ be the smallest nonzero eigenvalue of $A^T A$. Note that $\delta > 0$. Then (see the Exercises)

$$x^T \nabla^2_{xx} \mathcal{A}_\rho(x_*, \lambda_*) x \geq \alpha \|v\|^2 - 2\beta\gamma \|v\| \|q\| + (\rho\delta - \gamma) \|q\|^2.$$

This final result will be positive if

$$\rho > \frac{1}{\delta} \left(\frac{\beta^2 \gamma^2}{\alpha} + \gamma \right)$$

(see the Exercises). This proves that the Hessian matrix is positive definite for sufficiently large ρ, and the right-hand side of this inequality determines $\bar{\rho}$. ☐

The lemma shows that, under the above conditions, the Hessian matrix of the augmented Lagrangian is positive definite for $\rho \geq \bar{\rho}$ at a local solution to the constrained optimization problem. Since the functions f and g are twice continuously differentiable, the result is also true for $x \approx x_*$ and $\lambda \approx \lambda_*$, although a larger value of $\bar{\rho}$ might have to be used.

At a local solution to the constrained problem, the gradient of the augmented Lagrangian is zero. If the Hessian matrix of the augmented Lagrangian is positive definite at this point, then this local solution is a local minimizer of the augmented Lagrangian function. Thus the solution to the constrained problem can be found using a single unconstrained minimization of the augmented Lagrangian function if $\rho \geq \bar{\rho}$ and if $\lambda = \lambda_*$. Thus the augmented Lagrangian function is an exact penalty function (see Section 16.5). Unlike a classical penalty method where the penalty parameter must tend to infinity, an augmented Lagrangian method can maintain ρ at some fixed, sufficiently large value. This ameliorates some of the ill-conditioning associated with penalty methods.

This lemma assumes that $\lambda = \lambda_*$, that is, the optimal Lagrange multipliers are known. This will not normally be true. If only an estimate of the multipliers is provided, then progress towards a solution can still be guaranteed if the penalty parameter is large enough. This is the topic of the next theorem. The theorem shows that the progress of the algorithm is controlled by the accuracy of the multiplier estimates, in the sense that the accuracy of λ at the current iteration determines the accuracy of both x and λ at the next iteration. For this reason, an augmented Lagrangian method is sometimes referred to as a *method of multipliers*.

Theorem 16.14 (Local Convergence of Augmented Lagrangian Method). *Assume that the second-order sufficiency conditions are satisfied for*

$$\text{minimize} \quad f(x)$$
$$\text{subject to} \quad g(x) = 0.$$

Let $\bar{\rho} > 0$ be large enough so that $\nabla^2_{xx}\mathcal{A}_{\bar{\rho}}(x_*, \lambda_*)$ is positive definite. Then there exist positive constants δ, ϵ, and M such that

(i) *for all $(\lambda, \rho) \in D \equiv \{\, (\lambda, \rho) : \|\lambda - \lambda_*\| < \delta\rho, \; \rho \geq \bar{\rho} \,\}$ the problem*

$$\text{minimize}_x \quad \mathcal{A}_\rho(x, \lambda)$$
$$\text{subject to} \quad \|x - x_*\| \leq \epsilon$$

has a unique solution $x(\lambda, \rho)$. The function $x(\lambda, \rho)$ is continuously differentiable in the interior of D, and for all $(\lambda, \rho) \in D$,

$$\|x(\lambda, \rho) - x_*\| \leq \frac{M}{\rho} \|\lambda - \lambda_*\| \,;$$

(ii) *for all $(\lambda, \rho) \in D$,*

$$\|\lambda_+(\lambda, \rho) - \lambda_*\| \leq \frac{M}{\rho} \|\lambda - \lambda_*\| \,,$$

where

$$\lambda_+(\lambda, \rho) = \lambda - \rho g(x(\lambda, \rho))$$

are the updated multipliers;

(iii) *for all $(\lambda, \rho) \in D$, $\nabla^2_{xx}\mathcal{A}_\rho(x(\lambda, \rho), \lambda)$ is positive definite, and $\nabla g(x(\lambda, \rho))$ has full rank.*

Proof. There are two major steps in the proof. In the first step, we examine the governing equations for an augmented Lagrangian method. We make a change of variables so that the implicit function theorem can be applied to them, and we conclude that a solution to these equations exists for $\lambda \approx \lambda_*$ and for $\rho \geq \bar{\rho}$. Most of the conclusions in the theorem follow from this step, with the exception of the bounds on $\|x(\lambda, \rho) - x_*\|$ and $\|\lambda_+(\lambda, \rho) - \lambda_*\|$.

The second step establishes these bounds by analyzing the first-order terms in Taylor series expansions of $x(\lambda, \rho)$ and $\lambda_+(\lambda, \rho)$. This step requires a number of minor technical arguments.

1. *Use of the Implicit Function Theorem*: Let $\rho \geq \bar{\rho}$, let x be a local minimizer of the augmented Lagrangian function for a given λ and ρ, and let λ_+ be the new vector of Lagrange multipliers. (The local minimizer x will exist for $\lambda \approx \lambda_*$; see the remarks after Lemma 16.13.) Then

$$\nabla f(x) - \nabla g(x)\lambda_+ = 0$$
$$\frac{1}{\rho}(\lambda - \lambda_+) - g(x) = 0.$$

(The first equation comes from the optimality of x, and the second is a rearrangement of the update formula for the multipliers.)

We make a change of variables: $\zeta \equiv (\lambda - \lambda_*)/\rho$ and $\mu = 1/\rho$. (If ρ is large and if $\lambda \approx \lambda_*$, then both ζ and μ will be near zero.) The equations become

$$\nabla f(x) - \nabla g(x)\lambda_+ = 0$$
$$\zeta + \mu(\lambda_* - \lambda_+) - g(x) = 0.$$

This is a system of $n + m$ equations in the $n + m$ unknowns x and λ_+. The system has the solution $x = x_*$ and $\lambda_+ = \lambda_*$ at $\zeta = 0$, $\mu = 0$. We differentiate with respect to x and λ_+ and evaluate the resulting Jacobian at this solution:

$$\begin{pmatrix} \nabla^2_{xx}\mathcal{L}(x_*, \lambda_*) & -\nabla g(x_*) \\ -\nabla g(x_*)^T & -\mu I \end{pmatrix}.$$

This matrix is invertible for $0 \leq \mu \leq 1/\bar{\rho}$ (see the Exercises). Hence by the implicit function theorem (see Appendix B.9) this system has a unique solution $\hat{x}(\zeta, \mu)$, $\hat{\lambda}(\zeta, \mu)$ at nearby points: There exist positive constants ϵ and δ such that

$$\nabla f(\hat{x}(\zeta, \mu)) - \nabla g(\hat{x}(\zeta, \mu))\hat{\lambda}(\zeta, \mu) = 0$$
$$\zeta + \mu(\lambda_* - \hat{\lambda}(\zeta, \mu)) - g(\hat{x}(\zeta, \mu)) = 0 \tag{16.7}$$

for $0 \leq \mu \leq 1/\bar{\rho}$ and $\|\zeta\| \leq \delta$, with

$$\sqrt{\left\| \hat{x}(\zeta, \mu) - x_* \right\|^2 + \left\| \hat{\lambda}(\zeta, \mu) - \lambda_* \right\|^2} < \epsilon.$$

In addition, $\hat{x}(\zeta, \mu)$ and $\hat{\lambda}(\zeta, \mu)$ are continuously differentiable with respect to ζ and μ. We can choose δ and ϵ sufficiently small so that $\nabla g(\hat{x}(\zeta, \mu))$ has full rank and $\nabla^2_{xx}\mathcal{A}_\rho(\hat{x}(\zeta, \mu), \hat{\lambda}(\zeta, \mu))$ is positive definite.

If we now reverse the change of variables, the same conclusions apply to the implicit solutions $x(\lambda, \rho)$ and $\lambda_+(\lambda, \rho)$. These conclusions are sufficient to prove all the results in the theorem, with the exception of the bounds on $\|x(\lambda, \rho) - x_*\|$ and $\|\lambda_+(\lambda, \rho) - \lambda_*\|$.

2. *Bounds on the Errors*: We derive Taylor series expansions for $\hat{x}(\zeta, \mu)$ and $\hat{\lambda}(\zeta, \mu)$. To do this, we require formulas for their Jacobians. To obtain these, we differentiate (16.7) with respect to ζ and μ and rearrange:

$$\nabla \begin{pmatrix} \hat{x}(\zeta, \mu) \\ \hat{\lambda}(\zeta, \mu) \end{pmatrix}^T = A(\zeta, \mu) \begin{pmatrix} 0 & 0 \\ -I & \hat{\lambda}(\zeta, \mu) - \lambda_* \end{pmatrix},$$

where

$$A(\zeta, \mu) = \begin{pmatrix} \nabla_{xx}^2 \mathcal{L}(\hat{x}(\zeta, \mu), \hat{\lambda}(\zeta, \mu)) & -\nabla g(\hat{x}(\zeta, \mu)) \\ -\nabla g(\hat{x}(\zeta, \mu))^T & -\mu I \end{pmatrix}^{-1}.$$

We expand in a Taylor series about $\zeta = 0$ and $\mu = 0$. At this point $\hat{x}(0, 0) = x_*$ and $\hat{\lambda}(0, 0) = \lambda_*$. Using the remainder form of the Taylor series for each component, we obtain

$$\begin{pmatrix} \hat{x}(\zeta, \mu) - x_* \\ \hat{\lambda}(\zeta, \mu) - \lambda_* \end{pmatrix}_j = \left[A(\eta_j \zeta, \eta_j \mu) \begin{pmatrix} 0 & 0 \\ -I & \hat{\lambda}(\eta_j \zeta, \eta_j \mu) - \lambda_* \end{pmatrix} \begin{pmatrix} \zeta \\ \mu \end{pmatrix} \right]_j, \qquad (16.8)$$

where $0 \le \eta_j \le 1$.

Bounds on the errors are obtained by taking norms and using equation (16.8). If δ is sufficiently small, and if $\|\zeta\| \le \delta$, then $\|A(\eta_j \zeta, \eta_j \mu)\| \le C$ for some constant C. Now pick δ so that $C\delta < 1$, and pick $\mu < \delta$. Then from (16.8),

$$\sqrt{\left\|\hat{x}(\zeta, \mu) - x_*\right\|^2 + \left\|\hat{\lambda}(\zeta, \mu) - \lambda_*\right\|^2} \le C\left[\|\zeta\| + \mu \max_{0 \le \eta \le 1} \left\|\hat{\lambda}(\eta\zeta, \eta\mu) - \lambda_*\right\| \right]. \quad (16.9)$$

This implies that

$$\left\|\hat{\lambda}(\zeta, \mu) - \lambda_*\right\| \le C\|\zeta\| + \mu C \max_{0 \le \eta \le 1} \left\|\hat{\lambda}(\eta\zeta, \eta\mu) - \lambda_*\right\|.$$

This condition remains satisfied if, on the left-hand side, we replace ζ by $\eta\zeta$ and μ by $\eta\mu$, so that

$$\max_{0 \le \eta \le 1} \left\|\hat{\lambda}(\eta\zeta, \eta\mu) - \lambda_*\right\| \le C\|\zeta\| + \mu C \max_{0 \le \eta \le 1} \left\|\hat{\lambda}(\eta\zeta, \eta\mu) - \lambda_*\right\|.$$

Hence

$$\max_{0 \le \eta \le 1} \left\|\hat{\lambda}(\eta\zeta, \eta\mu) - \lambda_*\right\| \le \frac{C}{1 - C\mu} \|\zeta\|.$$

Combining this with (16.9) produces

$$\sqrt{\left\|\hat{x}(\zeta, \mu) - x_*\right\|^2 + \left\|\hat{\lambda}(\zeta, \mu) - \lambda_*\right\|^2} \le \left(C + \frac{C^2 \mu}{1 - C\mu}\right) \|\zeta\| \le \frac{C}{1 - C\mu} \|\zeta\| \le 2C \|\zeta\|.$$

Let $M = 2C$. If we now reverse the change of variables made in step 1, we obtain

$$\|x(\lambda, \rho) - x_*\| \le \frac{M}{\rho} \|\lambda - \lambda_*\|$$

$$\|\lambda_+(\lambda, \rho) - \lambda_*\| \le \frac{M}{\rho} \|\lambda - \lambda_*\|$$

for $\|\lambda - \lambda_*\| \le \delta\rho$ and $\rho > \max\{\bar{\rho}, 1/\delta\}$. Because $x(\lambda, \rho)$ is continuously differentiable, we can find a constant M that produces the same result for $\bar{\rho} \le \rho \le \max\{\bar{\rho}, 1/\delta\}$. This completes the proof. $\quad\square$

This theorem can be used to prove convergence of an augmented Lagrangian method. First, notice that if $\lambda = \lambda_*$, then the algorithm converges in one iteration. More generally, if the algorithm is controlled so that the multiplier estimates λ_k remain bounded, then $\|\lambda_k - \lambda_*\|$ will also remain bounded. By choosing ρ large enough, the error bound in the theorem will guarantee that

$$\|\lambda_{k+1} - \lambda_*\| \leq \frac{M}{\rho} \|\lambda_k - \lambda_*\|$$

with $M/\rho < 1$. This shows that the multipliers converge linearly, and hence $\{ x_k \}$ converges also. If a sequence $\{ \rho_k \}$ of barrier parameters are used, with $\lim \rho_k = +\infty$, then

$$\lim_{k \to \infty} \frac{\|\lambda_{k+1} - \lambda_*\|}{\|\lambda_k - \lambda_*\|} \leq \lim_{k \to \infty} \frac{M}{\rho_k} = 0,$$

showing that the multiplier estimates $\{ \lambda_k \}$ converge superlinearly.

The "convergence rate" here is not directly comparable to the convergence rates of the other methods we have discussed. Each iteration of the augmented Lagrangian method involves the solution of an unconstrained minimization subproblem, whereas an iteration of the other methods required at most the solution of a system of linear equations. Thus the iterations here are in general much more expensive than the iterations of the other methods. In special cases (for example, when Newton's method is applied to a convex problem) it can be shown that, for points $x_k \approx x_*$, only one Newton iteration need be applied to each subproblem. Under such circumstances, the convergence rate of an augmented Lagrangian method is comparable to that of other methods.

If ρ is large, then the augmented Lagrangian method will suffer from ill-conditioning, just as the penalty method did. Thus there is a conflict between rate of convergence and conditioning of the method. This ill-conditioning can be controlled by using stabilized formulas for the search directions, analogous to the techniques described in Section 16.4.

A multiplier-based method can also be derived for problems with inequality constraints:

$$\text{minimize} \quad f(x)$$
$$\text{subject to} \quad g(x) \geq 0.$$

In this case, if a multiplier estimate $\lambda_k \approx \lambda_*$ and a barrier parameter $\mu_k > 0$ are specified, then at the kth iteration a *modified barrier function* is minimized to obtain x_{k+1}:

$$\underset{x}{\text{minimize}} \; \mathcal{S}_{\mu_k}(x, \lambda_k) = f(x) - \mu_k \sum_{i=1}^{m} (\lambda_k)_i \log(\mu_k^{-1} g_i(x) + 1).$$

Then the multipliers are updated using

$$(\lambda_{k+1})_i = \frac{(\lambda_k)_i}{\mu_k^{-1} g_i(x_{k+1}) + 1}.$$

If this update formula is used, then (as before)

$$\nabla_x \mathcal{L}(x_{k+1}, \lambda_{k+1}) = \nabla f(x_{k+1}) - \nabla g(x_{k+1})\lambda_{k+1} = 0,$$

so that this part of the optimality conditions is satisfied. The iteration is continued until complementary slackness is attained.

In the barrier method, every estimate of the solution has to be strictly feasible so that the logarithmic barrier term can be evaluated. The modified barrier function is defined as long as $\mu^{-1} g_i(x) + 1 > 0$, or

$$g_i(x) > -\mu.$$

Thus the modified barrier function permits the constraints to be violated by as much as μ.

The modified barrier function has many of the same properties as the augmented Lagrangian function. For any $\mu > 0$,

$$\mathcal{S}_\mu(x_*, \lambda_*) = f(x_*)$$

and

$$\nabla_x \mathcal{S}_\mu(x_*, \lambda_*) = \nabla f(x_*) - \nabla g(x_*)\lambda_* = 0.$$

Hence the gradient of the modified barrier function is equal to the gradient of the Lagrangian and vanishes as the solution. The Hessian matrix of the modified barrier function is positive definite under certain conditions for μ less than some critical value $\bar{\mu}$. Finally, there is an analog of our convergence theorem that shows that

$$\|x_{k+1} - x_*\| \le M\mu \|\lambda_k - \lambda_*\| \quad \text{and} \quad \|\lambda_{k+1} - \lambda_*\| \le M\mu \|\lambda_k - \lambda_*\|$$

for some constant M. (See the Exercises.)

16.6.1 Dual Interpretation

The convergence of the augmented Lagrangian method is determined by the Lagrange multiplier estimates, and in fact the entire method can be considered as an algorithm for solving the dual problem. If the constrained problem is

$$\begin{aligned} \text{minimize} \quad & f(x) \\ \text{subject to} \quad & g(x) = 0, \end{aligned}$$

then the iteration has the form

$$\lambda_{k+1} = \lambda_k - \rho_k g(x_{k+1}).$$

We can interpret $-g(x_{k+1})$ as a search direction for the dual problem, with ρ_k as the step length. In this subsection we examine this interpretation in greater detail.

Our discussion is based on the theory of local duality (see Section 14.8.3). The constrained optimization problem may not be convex, but it is equivalent to

$$\begin{aligned} \text{minimize} \quad & \pi_\rho(x) = f(x) + \tfrac{1}{2}\rho g(x)^T g(x) \\ \text{subject to} \quad & g(x) = 0. \end{aligned}$$

Lemma 16.13 shows that this problem is locally strictly convex near (x_*, λ_*) for ρ sufficiently large.

The corresponding dual problem (in a neighborhood of (x_*, λ_*)) is

$$\underset{\lambda}{\text{maximize}} \ \mathcal{L}_*(\lambda),$$

where the dual function is defined as

$$\mathcal{L}_*(\lambda) = \min_x \left\{ f(x) - \lambda^T g(x) + \tfrac{1}{2}\rho g(x)^T g(x) \right\}.$$

We could apply any unconstrained optimization method to maximize $\mathcal{L}_*$, and in fact any such method would give an update formula for λ_k. Of most relevance to us will be Newton's method and the steepest-ascent method. (The steepest-ascent method for a maximization problem is the analog of the steepest-descent method for a minimization problem.)

In Section 14.8 we derived formulas for the gradient and Hessian matrix of the dual function. When applied to this case we obtain

$$\nabla \mathcal{L}_*(\lambda) = -g(x(\lambda))$$
$$\nabla^2 \mathcal{L}_*(\lambda) = -\nabla g(x(\lambda))^T [\nabla^2_{xx} \mathcal{L}(x(\lambda), \lambda) $$
$$+ \rho \nabla g(x(\lambda)) \nabla g(x(\lambda))^T]^{-1} \nabla g(x(\lambda)),$$

where $x(\lambda)$ is the value of x obtained when evaluating $\mathcal{L}_*(\lambda)$.

The steepest-ascent method uses the search direction

$$p_k = \nabla \mathcal{L}_*(\lambda_k) = -g(x(\lambda_k))$$

and so the update formula

$$\lambda_{k+1} = \lambda_k - \rho_k g(x_{k+1})$$

for the augmented Lagrangian method corresponds to applying the steepest-ascent method to the dual problem with a fixed step length ρ_k. As was pointed out in Chapter 12, the steepest-ascent method is often slow to converge, and so it is normally advisable to use a better method, such as Newton's method. What we will show is that, in the limit as $\rho \to \infty$, this update formula becomes the classical Newton method applied to the dual problem. Thus, at least for large values of ρ, using the steepest-ascent method (even with this fixed step length) can achieve rapid convergence.

The convergence of the steepest-ascent method depends on the condition number of the Hessian matrix of $\mathcal{L}_*$ (see Section 12.2). The following lemma determines this condition number. Note that

$$-\nabla g(x(\lambda))^T [\nabla^2_{xx} \mathcal{L}(x(\lambda), \lambda)]^{-1} \nabla g(x(\lambda))$$

is the Hessian matrix of the dual function for the *original* problem.

Lemma 16.15 (Condition Number of Dual Hessian). *Let $\mathcal{L}_*(\lambda)$ be the dual function for the augmented optimization problem. Then the condition number of $\nabla^2 \mathcal{L}_*(\lambda)$ is*

$$\frac{W^{-1} + \rho}{w^{-1} + \rho},$$

where w and W are the smallest and largest eigenvalues, respectively, of

$$\nabla g(x(\lambda))^T [\nabla^2_{xx} \mathcal{L}(x(\lambda), \lambda)]^{-1} \nabla g(x(\lambda)).$$

Proof. Let $H = \nabla^2_{xx} \mathcal{L}(x(\lambda, \lambda)$ and $A = \nabla g(x(\lambda))^T$. Then the Hessian matrix of the dual function is

$$-A[H + \rho A^T A]^{-1} A^T.$$

We must relate the eigenvalues of this matrix to those of $AH^{-1}A^T$. We do this by showing that these matrices have the same eigenvectors.

The matrix version of the Sherman–Morrison formula (see Appendix A.9) can be used to show that

$$\rho A[H + \rho A^T A]^{-1} A^T = I - (I + \rho A H^{-1} A^T)^{-1}$$

(see the Exercises). Multiplying on the right by $(I + \rho A H^{-1} A^T)$ we obtain

$$\rho A[H + \rho A^T A]^{-1} A^T (I + \rho A H^{-1} A^T) = \rho A H^{-1} A^T.$$

Let y be an eigenvector of $AH^{-1}A^T$ with corresponding eigenvalue v. Then

$$\rho A[H + \rho A^T A]^{-1} A^T (1 + \rho v) y = \rho v y.$$

Hence y is also an eigenvector of $A[H + \rho A^T A]^{-1} A^T$ with eigenvalue

$$\frac{\rho v}{\rho(1 + \rho v)} = \frac{1}{v^{-1} + \rho}.$$

Letting v be the smallest and largest eigenvalues of $AH^1 A^T$ completes the proof. $\qquad \square$

To see what happens for large values of ρ, we examine

$$\lim_{\rho \to \infty} \frac{W^{-1} + \rho}{w^{-1} + \rho} = 1.$$

Hence, as ρ increases, the Hessian matrix becomes better conditioned and the steepest-ascent method performs better.

The search direction for Newton's method in the dual is

$$p_k = -[\nabla^2 \mathcal{L}_*(\lambda_k)]^{-1} \nabla \mathcal{L}_*(\lambda_k).$$

The Hessian here is

$$\nabla^2 \mathcal{L}_*(\lambda) = -\nabla g(x(\lambda))^T [\nabla^2_{xx} \mathcal{L}(x(\lambda), \lambda) \\ + \rho \nabla g(x(\lambda)) \nabla g(x(\lambda))^T]^{-1} \nabla g(x(\lambda)).$$

The Sherman–Morrison formula (see the proof of Lemma 16.15) implies that, for large values of ρ,

$$\nabla^2 \mathcal{L}_*(\lambda) \approx -\rho^{-1} I.$$

Hence the Newton direction satisfies

$$p_k \approx \rho \nabla \mathcal{L}_*(\lambda_k) = -\rho g(x(\lambda_k))$$

and the update formula

$$\lambda_{k+1} = \lambda_k - \rho_k g(x_{k+1})$$

for the augmented Lagrangian method corresponds to an approximate Newton method when ρ is large. Unfortunately, we can expect the augmented Lagrangian function to be ill conditioned in this case. This provides an additional explanation of the conflict between rapid convergence and numerical stability for this method.

Exercises

6.1. Consider the one-dimensional problem

$$\text{minimize} \quad f(x) = -\tfrac{1}{2}x^2$$
$$\text{subject to} \quad x = 1.$$

Verify that the solution is $x_* = 1, \lambda_* = -1$. Show that if $\rho < 1$, then the augmented Lagrangian function $\mathcal{A}_\rho(x, \lambda)$ is unbounded below. Show also that for $\rho > 1$ the augmented Lagrangian function is strictly convex when considered as a function of x, and thus has a unique minimizer. What happens when $\rho = 1$?

6.2. For the example in the previous problem, suppose that $x_0 = 0$ and $\lambda_0 = 0$. Prove that if ρ is fixed and $\rho > 1$, then

$$x_k = 1 - \frac{(-1)^k}{(\rho - 1)^k} \quad \text{and} \quad \lambda_k = -1 + \frac{(-1)^k}{(\rho - 1)^k}.$$

What is the limit point of (x_k, λ_k)? What is the convergence rate?

6.3. Use the matrix version of the Sherman–Morrison formula (see Appendix A.9) to prove that

$$\rho A[H + \rho A^T A]^{-1} A^T = I - (I + \rho A H^{-1} A^T)^{-1}.$$

6.4. Use the previous problem to prove that

$$\nabla^2 \mathcal{L}_*(\lambda) \approx -\rho^{-1} I.$$

6.5. (From Avriel (1976, reprinted 2003).) Consider the quadratic problem

$$\underset{x}{\text{minimize}} \quad f(x) = \tfrac{1}{2}x^T Q x + \theta b^T x$$
$$\text{subject to} \quad b^T x = 0,$$

where Q is a nonsingular symmetric matrix that is positive definite on the subspace $b^T x = 0$. The vector $b \neq 0$ and the scalar $\theta > 0$ are given.

(i) What is the optimal solution to this problem? What is the corresponding Lagrange multiplier?

(ii) Suppose the problem is solved by the augmented Lagrangian method, where the multiplier estimates are updated in the standard way. Let λ_k be the multiplier estimate at the kth iteration, and let ρ be the penalty parameter. Show that, if the augmented Lagrangian has a minimum, its minimizer is

$$x_k = \frac{(\lambda_k - \theta)Q^{-1}b}{1 + \rho b^T Q^{-1} b}.$$

(iii) If we initialize the method with $\lambda_1 = 0$, show that

$$x_k = -\frac{\theta Q^{-1} b}{(1 + \rho b^T Q^{-1} b)^k} \quad \text{and} \quad \lambda_{k+1} = -\theta \left[\frac{1}{(1 + \rho b^T Q^{-1} b)^k} - 1 \right].$$

(iv) Assume that the penalty parameter ρ is held fixed throughout the solution process. For what values of ρ is the iterative process guaranteed to converge? (Assume as in part (iii) that $\lambda_0 = 0$.)

(v) Comment on the effect of the value of the penalty parameter ρ on (i) the rate of convergence of the algorithm, and (ii) the condition number of the Hessian matrix of the augmented Lagrangian function.

6.6. Consider an optimization problem with a single constraint:

$$\text{minimize} \quad f(x)$$
$$\text{subject to} \quad g(x) \geq 0.$$

Make the constraint an equality by subtracting a squared slack variable s.

(i) What is the augmented Lagrangian function $\mathcal{A}_\rho(x, s, \lambda)$ for this problem?

(ii) Determine the update formula for the Lagrange multiplier.

(iii) Let (x_*, λ_*) be the solution to the original problem, and let $s_* = \sqrt{g(x_*)}$. Prove that $\mathcal{A}_\rho(x_*, s_*, \lambda_*) = f(x_*)$, and that $\nabla_{xs}\mathcal{A}_\rho(x_*, s_*, \lambda_*) = 0$.

6.7. Suppose that

$$\nabla^2_{xx}\mathcal{A} = H + \rho A^T A.$$

Write $x = Zv + q$, where Z is a basis matrix for the null space of A and q is in the range of A^T. Let α be the smallest eigenvalue of $Z^T H Z$ and assume that $\alpha > 0$. Also, let $\beta = \|Z\|$, $\gamma = \|H\|$, and let δ be the smallest nonzero eigenvalue of $A^T A$.

(i) Prove that

$$x^T(\nabla^2_{xx}\mathcal{A})x \geq \alpha \|v\|^2 - 2\beta\gamma \|v\| \|q\| + (\rho\delta - \gamma) \|q\|^2.$$

(ii) Prove that $x^T(\nabla^2_{xx}\mathcal{A})x > 0$ if

$$\rho > \frac{1}{\delta}\left(\frac{\beta^2\gamma^2}{\alpha} + \gamma\right).$$

6.8. In the proof of Theorem 16.14, prove that

$$\begin{pmatrix} \nabla^2_{xx}\mathcal{L}(x_*, \lambda_*) & -\nabla g(x_*) \\ -\nabla g(x_*)^T & -\mu I \end{pmatrix}$$

is invertible for $0 \leq \mu \leq 1/\bar{\rho}$.

6.9. Let (x_*, λ_*) be a solution to

$$\text{minimize} \quad f(x)$$
$$\text{subject to} \quad g(x) \geq 0.$$

Prove that, for any $\mu > 0$, the modified barrier function satisfies

$$\mathcal{S}_\mu(x_*, \lambda_*) = f(x_*)$$

and

$$\nabla_x \mathcal{S}_\mu(x_*, \lambda_*) = \nabla f(x_*) - \nabla g(x_*)\lambda_* = 0.$$

6.10. The modified-barrier function is a barrier function with *shifted* constraints, that is, it perturbs the constraint $g(x) \geq 0$ to $g(x) \geq -\mu$. Prove that the augmented Lagrangian function is equivalent to the shifted penalty function

$$f(x) + \tfrac{1}{2}\rho(g(x) - \rho^{-1}\lambda)^T(g(x) - \rho^{-1}\lambda).$$

6.11. Prove an analog of Theorem 16.14 for the modified-barrier function, showing that (for some constant M)

$$\|x_{k+1} - x_*\| \leq M\mu \|\lambda_k - \lambda_*\| \quad \text{and} \quad \|\lambda_{k+1} - \lambda_*\| \leq M\mu \|\lambda_k - \lambda_*\|.$$

Assume that strict complementarity slackness holds.

16.7   Nonlinear Primal-Dual Methods

The penalty and barrier methods that we have discussed so far assign different roles to the primal and dual variables. An unconstrained optimization problem is solved to find the primal iterate x_k; then the dual iterate λ_k is computed from x_k using a simple formula. This asymmetry between the primal and dual variables limits the efficiency of the methods.

In this section we describe a family of methods that assign an equal role to both primal and dual variables. The resulting methods are in general more efficient, with a faster convergence rate and better performance. They were originally inspired by the success of primal-dual interior-point methods for linear programming (see Chapter 10).

These methods share some characteristics of barrier methods. In particular, they solve a sequence of subproblems with a barrier parameter that is shrinking to zero. But, unlike in a traditional barrier method, for each value of the barrier parameter only a small number of Newton iterations are used, perhaps only one.

The Newton iteration is based on a system of nonlinear equations derived from the optimality conditions for the barrier problem. The solution to this system is used to update the primal and dual iterates simultaneously. Under an appropriate condition, and if started sufficiently close to the solution, the algorithm can achieve a quadratic rate of convergence, which is a most desirable property.

The overall method is called a *primal-dual interior-point method*. It is a "primal-dual" method because the two sets of variables are given equal prominence, and it is an "interior-point" method because the iterates lie in the interior of the feasible region.

Since the development of Karmarkar's method in 1984, primal-dual interior-point methods have become a major focus of theoretical research and practical experiments. For linear programming, there are polynomial-time algorithms of this type. Computational experiments have demonstrated that primal-dual interior-point methods are very efficient for linear, convex quadratic, and general nonlinear problems.

In the following subsections we derive a primal-dual interior-point method for solving general nonlinear optimization problems and present a convergence theorem.

16.7.1 Primal-Dual Interior-Point Methods

We will focus on a minimization problem with inequality constraints:

$$\text{minimize} \quad f(x)$$
$$\text{subject to} \quad g_i(x) \geq 0, \quad i = 1, \ldots, m.$$

Equality constraints can be included in the formulation, but we ignore them to simplify the presentation.

The primal-dual interior-point method will be derived from a lagarithmic barrier method

$$\text{minimize} \; \beta_{\mu_k}(x),$$

where $\beta_{\mu_k}(x) = f(x) - \mu_k \sum_{i=1}^{m} \log g_i(x)$. The minimization problem will be solved for a decreasing sequence of barrier parameters μ_k.

The solution $(x(\mu_k), \lambda(\mu_k))$ to the barrier subproblem satisfies the following *perturbed* primal-dual system of equations and inequalities (see Section 16.2):

$$g(x) \geq 0$$
$$\nabla_x L(x, \lambda) = \nabla f(x) - \sum_{i=1}^{m} \lambda_i \nabla g_i(x) = 0$$
$$\lambda \geq 0$$
$$\lambda_i g_i(x) = \mu_k, \quad i = 1, \ldots, m,$$

where $L(x, \lambda) = f(x) - \sum_{i=1}^{m} \lambda_i g_i(x)$ is the Lagrangian of the problem. The equations will serve as the foundation for the new method.

At each iteration the primal-dual interior-point method obtains a search direction by computing the Newton directions $(\Delta x_k, \Delta \lambda_k)$ for the following nonlinear system of equations:

$$\nabla_x L(x, \lambda) = \nabla f(x) - \sum_{i=1}^{m} \lambda_i \nabla g_i(x) = 0$$
$$\lambda_i g_i(x) = \mu_k, \quad i = 1, \ldots, m.$$

To derive the formulas for the Newton directions, we linearize this system at (x_k, λ_k) using Taylor's series, ignoring the terms of second and higher order, and set the linearized system to zero.

If we write $x = x_k + \Delta x_k$ and $\lambda = \lambda_k + \Delta \lambda_k$, then the Taylor series expansions are

$$\begin{aligned}
\nabla_x L(x, \lambda) &= \nabla_x L(x_k + \Delta x_k, \lambda_k + \Delta \lambda_k) \\
&\approx \nabla_x L(x_k, \lambda_k) + \nabla_{xx}^2 L(x_k, \lambda_k) \Delta x_k + \nabla_{\lambda x}^2 L(x_k, \lambda_k) \Delta \lambda_k \\
&= \nabla_x L(x_k, \lambda_k) + \nabla_{xx}^2 L(x_k, \lambda_k) \Delta x_k - \nabla g(x_k) \Delta \lambda_k
\end{aligned}$$

and

$$\begin{aligned}
\lambda_i g_i(x) &= ((\lambda_k)_i + (\Delta \lambda_k)_i) g_i(x_k + \Delta x_k) \\
&\approx (\lambda_k)_i g_i(x_k) + \lambda_i \nabla g_i(x_k)^T \Delta x_k + (\Delta \lambda_k)_i g_i(x_k), \quad i = 1, \ldots, m,
\end{aligned}$$

where $\nabla_{xx}^2 L(x, \lambda) = \nabla^2 f(x) - \sum_{i=1}^m \lambda_i \nabla^2 g_i(x)$ is the Hessian of the Lagrangian and $\nabla g(x)$ is the transpose of the Jacobian of the vector-valued function $g(x) = (g_1(x), \ldots, g_m(x))^T$. If we ignore the second- and higher-order terms, we obtain the following linear system for the primal and dual directions:

$$\begin{bmatrix} \nabla_{xx}^2 L(x_k, \lambda_k) & -\nabla g(x_k) \\ \Lambda_k \nabla g(x_k)^T & G(x_k) \end{bmatrix} \begin{bmatrix} \Delta x_k \\ \Delta \lambda_k \end{bmatrix} = \begin{bmatrix} -\nabla_x L(x_k, \lambda_k) \\ -\Lambda_k g(x_k) + \mu_k e \end{bmatrix},$$

where $G(x)$ is the diagonal matrix with elements $g_i(x)$, $i = 1, \ldots, m$; Λ is the diagonal matrix with the elements λ_i, $i = 1, \ldots, m$; and $e = (1, \ldots, 1)^T$.

The primal-dual interior-point method requires that the inequalities of the perturbed system be strictly satisfied. This means that the iterates x_k are strictly feasible ($g_i(x_k) > 0$) and that $(\lambda_k)_i > 0$, $i = 1, \ldots, m$. If the initial values (x_0, λ_0) are interior to the feasible region, then the choice of the step lengths α_P and α_D will guarantee that the subsequent iterates are also interior to the feasible region. If the feasible set does not have an interior point, then the method cannot be used.

The following ratio test can be used to keep the dual iterates λ_{k+1} positive:

$$\alpha_D = \min_{1 \le i \le m} \left\{ 1; \ -\kappa \frac{(\lambda_k)_i}{(\Delta \lambda_k)_i} : (\Delta \lambda_k)_i < 0 \right\},$$

where $0 < \kappa < 1$. To keep the primal iterates in the interior of the primal feasible, set the algorithm uses a backtracking line search. As the step length is reduced, the trial point gets closer to x_k, which satisfies $g_i(x_k) > 0$. If the constraint functions $g_i(x)$ are continuous, then for a small-enough step length we are guaranteed that the trial point must satisfy $g_i(x_{k+1}) > 0$ also.

The optimality test is based on a *merit function* that measures the violation of optimality conditions

$$v(x, \lambda) = \max\{\|\nabla_x L(x, \lambda)\|, \ \|G(x)\lambda\|\}.$$

At the solution, $v(x_*, \lambda_*) = 0$. (For additional discussion of merit functions, see Section 14.6.)

We are now able to present our algorithm.

ALGORITHM 16.1.
Primal-Dual Interior-Point Algorithm

1. Choose (x_0, λ_0) such that $g_i(x_0) > 0$ and $(\lambda_0)_i > 0$ for $i = 1, \ldots, m$, and select an initial barrier parameter $\mu_0 > 0$.

2. For $k = 0, 1, \ldots$ do the following:

   (i) *The Optimality Test*—If $v(x_k, \lambda_k) = 0$, then stop.

   (ii) *Find the Primal-Dual Directions*—Compute $(\Delta x_k, \Delta \lambda_k)$ by solving the linear system of equations

$$\begin{bmatrix} \nabla_{xx}^2 L(x_k, \lambda_k) & -\nabla g(x_k) \\ \Lambda_k \nabla g(x_k)^T & G(x_k) \end{bmatrix} \begin{bmatrix} \Delta x_k \\ \Delta \lambda_k \end{bmatrix} = \begin{bmatrix} -\nabla_x L(x_k, \lambda_k) \\ -\Lambda_k g(x_k) + \mu_k e \end{bmatrix}.$$

(iii) *Find the Primal and the Dual Step Lengths*—Compute α_D from the formula

$$\alpha_D = \min_{1 \le i \le m} \left\{ 1; \ -\kappa \frac{(\lambda_k)_i}{(\Delta \lambda_k)_i} : (\Delta \lambda_k)_i < 0 \right\}.$$

Compute α_P using a backtracking line search: choose α_P as the first element of the sequence $\{ 1, 1/2, 1/4, \ldots \}$ for which $g_i(x_k + \alpha_P \Delta x_k) > 0$ for $i = 1, \ldots, m$. The choices of the step lengths α_P and α_D guarantee that x_{k+1} is in the interior of the primal feasible regions and λ_{k+1} is positive.

(iv) *The Update*—Determine x_{k+1} and λ_{k+1} from

$$x_{k+1} = x_k + \alpha_P \Delta x_k$$
$$\lambda_{k+1} = \lambda_k + \alpha_D \Delta \lambda_k.$$

(v) *The Barrier Parameter Update*—Compute μ_{k+1} using an appropriate rule, e.g.,

$$\mu_{k+1} = \theta \mu_k, \quad 0 < \theta < 1.$$

Example 16.16 (Primal-Dual Interior-Point Method). We apply the primal-dual interior-point method to the problem from Example 16.1:

$$\begin{aligned}
\text{minimize} \quad & f(x) = x_1 - 2x_2 \\
\text{subject to} \quad & g_1(x) = 1 + x_1 - x_2^2 \ge 0 \\
& g_2(x) = x_2 \ge 0.
\end{aligned}$$

The initial guess is chosen in the interior of the primal and dual feasible sets: $x_0 = (0.5, 0.5)^T$ and $\lambda_0 = (0.5, 0.5)^T$. The initial barrier parameter is $\mu_0 = 1$. For these values, the merit function $v(x_0, \lambda_0)$ is nonzero, so the optimality condition is not satisfied.

The system for finding the Newton directions is

$$\begin{bmatrix} 0 & 0 & -1 & 0 \\ 0 & 2\lambda_1 & 2x_2 & -1 \\ \lambda_1 & -2x_2\lambda_1 & 1 + x_1 - x_2^2 & 0 \\ 0 & \lambda_2 & 0 & x_2 \end{bmatrix} \begin{bmatrix} \Delta x \\ \Delta \lambda \end{bmatrix} = \begin{bmatrix} -1 + \lambda_1 \\ 2 - 2x_2\lambda_1 + \lambda_2 \\ -\lambda_1(1 + x_1 - x_2^2) + \mu \\ -\lambda_2 x_2 + \mu \end{bmatrix}.$$

At the initial point the system is

$$\begin{bmatrix} 0 & 0 & -1.00 & 0 \\ 0 & 1.00 & 1.00 & -1.00 \\ 0.50 & -0.50 & 1.25 & 0 \\ 0 & 0.50 & 0 & 0.50 \end{bmatrix} \begin{bmatrix} \Delta x \\ \Delta \lambda \end{bmatrix} = \begin{bmatrix} 0.500 \\ -2.000 \\ -0.375 \\ -0.750 \end{bmatrix}.$$

We solve the system to obtain $\Delta x = (1.0, 1.5)^T$ and $\Delta \lambda = (0.5, 0)^T$. The primal step length α_P is calculated by backtracking: $\alpha_P = 1$ fails to keep the next iterate $x_0 + \alpha_P \Delta x$ in the interior of the primal feasible region; however, $\alpha_P = 0.5$ produces an acceptable primal iterate $x_1 = x_0 + 0.5\Delta x = (1.00, 1.25)^T$.

The dual step length $\alpha_D = 1$ is calculated using the ratio test with $\kappa = 0.9$, giving $\lambda_1 = \lambda_0 + \Delta \lambda = (1.0, 0.5)^T$.

Table 16.3. *Interior-point method.*

k	μ	α_P	x_k	α_D	λ_k
1	1	0.5	$(1.00000, 1.25000)^T$	1	$(1.00000, 0.50000)^T$
2	10^{-1}	1	$(0.22500, 1.07500)^T$	1	$(1.00000, 0.15000)^T$
3	10^{-2}	1	$(0.02424, 1.00924)^T$	1	$(1.00000, 0.01847)^T$
4	10^{-3}	1	$(0.00207, 1.00057)^T$	0.96	$(1.00000, 0.00185)^T$
5	10^{-4}	1	$(0.00020, 1.00005)^T$	0.95	$(1.00000, 0.00018)^T$
6	10^{-5}	1	$(0.00002, 1.00000)^T$	0.95	$(1.00000, 0.00001)^T$
7	10^{-6}	1	$(0.00000, 1.00000)^T$	0.95	$(1.00000, 0.00000)^T$

If we update the barrier parameter using $\mu_{k+1} = 0.1\mu_k$, then $\mu_1 = 0.1$.

At the next iteration, the optimality test gives $\nu(x_1, \lambda_1) \neq 0$, so the algorithm executes the second step. The results of the next few iterations are listed in Table 16.3. The table shows the barrier parameter, the primal-dual iterates x_k and λ_k, and their step lengths α_P and α_D. In this example, the algorithm converges linearly to the primal-dual solution $x_* = (0, 1)^T$ and $\lambda_* = (1, 0)^T$.   ∎

Primal-dual interior-point methods are well defined if the matrix

$$M(x, \lambda) = \begin{bmatrix} \nabla^2_{xx} L(x, \lambda) & -\nabla g(x) \\ \Lambda \nabla g(x)^T & G(x) \end{bmatrix}$$

from the system for finding the search directions is nonsingular. The second-order optimality, strict complementarity, and regularity conditions guarantee the nonsingularity of $M(x, \lambda)$ in some neighborhood of the primal-dual solution. Further from the solution, the algorithm may *regularize* the Hessian $\nabla^2_{xx} L(x, y)$, if necessary, to guarantee that the matrix $M(x, \lambda)$ is nonsingular:

$$R_t(x, y) = \nabla^2_{xx} L(x, y) + tI, \quad t \geq 0,$$

where I is the identity matrix and $t \geq 0$ is the regularization parameter. It can be shown that if $t \geq 0$ is large enough, then replacing the Hessian with the regularized Hessian guarantees that the matrix M is nonsingular (see the Exercises).

The efficiency of the algorithm depends critically on the strategy for updating the barrier parameter μ at each iteration. This strategy affects the rate of convergence of the interior-point method and therefore the number of steps required to find the solution to a specified accuracy. We will discuss the rate of convergence in Section 16.7.2.

Another factor that affects the performance of the primal-dual interior-point algorithm is the efficiency of solving the Newton system for the search directions at each step. An effective implementation must exploit any sparsity or other structure in the matrix.

The system can be modified by a simple division of its rows such that the system matrix becomes symmetric (see the Exercises). Also by eliminating $\Delta\lambda$, the system can be reduced so the system matrix becomes symmetric and positive definite (see the Exercises). Therefore a direct method using an LDL^T or Cholesky factorization can be an efficient technique, especially if the matrix is sparse.

In the next subsection we discuss convergence of the primal-dual interior-point method.

16.7.2 Convergence of the Primal-Dual Interior-Point Method

In primal-dual interior-point methods, the primal and dual variables play equally important roles in convergence. This symmetry enables interior-point methods to have a better convergence rate than either barrier or penalty methods. In this section, we present an interior-point method that has a quadratic convergence rate in the neighborhood of the solution. The interior-point method is applied to an equivalent problem that includes slack variables. This leads to a somewhat more complicated presentation, but the resulting algorithm has better theoretical and practical properties.

The original problem can be reformulated using slack variables s_i, $i = 1, \ldots, m$:

$$\begin{aligned} \text{minimize} \quad & f(x) \\ \text{subject to} \quad & g_i(x) - s_i = 0, \quad s_i \geq 0, \quad i = 1, \ldots, m. \end{aligned}$$

Although this problem has more variables and constraints, the inequality constraints now have a simple form: $s_i \geq 0, i = 1, \ldots, m$.

Starting from an initial guess $z^0 = (x^0, s^0, \lambda^0)$, a primal-dual interior-point method for this problem generates a sequence of primal-dual iterates $z_k = (x_k, s_k, \lambda_k)$, $k = 1, 2, 3, \ldots$. If the iterate $z^k = (x^k, s^k, \lambda^k)$ of the algorithm is close to the solution $z_* = (x_*, s_*, \lambda_*)$, i.e., z^k is in the neighborhood

$$\begin{aligned} U_\varepsilon(x_*, s_*, \lambda_*) = \{(x, s, \lambda) : \\ \|x - x_*\| &\leq \varepsilon, \\ \|s - s_*\| &\leq \varepsilon, \\ \|\lambda - \lambda_*\| &\leq \varepsilon, \\ x \in \Re^n, \ s \in \Re^m_+, \ \lambda \in \Re^m_+\}, \end{aligned}$$

for some $\varepsilon > 0$, then the primal-dual interior-point method converges to the solution with a quadratic rate, under the conditions of the theorem below.

This version of the interior-point method does not guarantee that the equality constraints $g_i(x) - s_i = 0, i = 1, \ldots, m$, are satisfied until the problem is solved, and therefore does not require all the functions $g_i(x), i = 1, \ldots, m$, to be positive at an initial guess. In fact, the algorithm we describe can choose an arbitrary initial guess x_0. What matters is that the slacks $(s_0)_i$ and the dual variables $(\lambda_0)_i$ are positive and remain so until the algorithm finds a solution. The following ratio tests are used to keep s_k and λ_k positive:

$$\alpha_P = \min_{1 \leq i \leq m} \left\{ 1; -\kappa \frac{(s_k)_i}{(\Delta s_k)_i} : (\Delta s_k)_i < 0 \right\}$$

$$\alpha_D = \min_{1 \leq i \leq m} \left\{ 1; -\kappa \frac{(\lambda_k)_i}{(\Delta \lambda_k)_i} : (\Delta \lambda_k)_i < 0 \right\},$$

where $0 < \kappa < 1$.

The first-order optimality conditions for the problem with slacks have the following form:

$$
\begin{aligned}
\nabla f(x_*) - \nabla g(x_*)^T \lambda_* &= 0 \\
S_* \Lambda_* e &= 0 \\
g(x_*) - s_* &= 0 \\
s_* &\geq 0 \\
\lambda_* &\geq 0,
\end{aligned}
$$

where S is the diagonal matrix with the elements s_i, $i = 1, \ldots, m$ (see the Exercises). Since the algorithm keeps $(s_k)_i$ and $(\lambda_k)_i$ positive, the merit function

$$
v(x, s, \lambda) = \max \left\{ \|\nabla f(x) - \nabla g(x)\lambda\|, \ \|S\Lambda e\|, \ \|g(x) - s\| \right\}
$$

measures the violation of the optimality conditions.

The rate of convergence of the primal-dual interior-point method depends on the rule used to update the barrier parameter. If we use

$$
\mu_k = \theta \mu_{k-1}
$$

or

$$
\mu_k = \theta \, v(x_k, s_k, \lambda_k)
$$

for some $0 < \theta < 1$, the interior-point algorithm converges only at a linear rate. However, the following rule for changing the barrier parameter,

$$
\mu_k = \min \left\{ \theta v(x_k, s_k, \lambda_k), \ v(x_k, s_k, \lambda_k)^2 \right\},
$$

makes it possible to achieve a quadratic rate of convergence in a neighborhood $U_\varepsilon(x_*, s_*, \lambda_*)$ of the solution.

To achieve this fast convergence we must also vary the parameter κ in the ratio tests:

$$
\kappa = \max \left\{ \bar{\kappa}, 1 - v(x_k, s_k, \lambda_k) \right\},
$$

where $0 < \bar{\kappa} < 1$. The following theorem describes the local convergence properties of this primal-dual interior-point method.

Theorem 16.17. *Assume that the second-order sufficiency, regularity, and strict complementarity conditions are satisfied for the problem with inequality constraints, and that the Hessians $\nabla^2 f(x)$ and $\nabla^2 g_i(x)$, $i = 1, \ldots, m$, are Lipschitz continuous with constant $L > 0$:*

$$
\begin{aligned}
\|\nabla^2 f(x) - \nabla^2 f(y)\| &\leq L \|x - y\| \\
\|\nabla^2 g_i(x) - \nabla^2 g_i(y)\| &\leq L \|x - y\|, \quad i = 1, \ldots, m,
\end{aligned}
$$

in some neighborhood $z^* = (x^*, s^*, \lambda^*)$ of the solution, i.e., $(x, s, \lambda) \in U_{\varepsilon_0}(x^*, s^*, \lambda^*)$ for some $\varepsilon_0 > 0$. Also assume that the primal-dual interior-point method updates the barrier and step length parameters using the formulas

$$
\mu_k = \min \left\{ \theta v(x_k, s_k, \lambda_k), \ v(x_k, s_k, \lambda_k)^2 \right\}, \quad 0 < \theta < 1
$$

and

$$\kappa = \max \{\bar{\kappa}, 1 - \nu(x_k, s_k, \lambda_k)\}, \quad 0 < \bar{\kappa} < 1.$$

Then there exists $0 < \varepsilon \leq \varepsilon_0$ such that for any $z_k = (x_k, s_k, \lambda_k) \in U_\varepsilon(z^*)$, the new approximation $z_{k+1} = (x_{k+1}, s_{k+1}, \lambda_{k+1})$ satisfies

$$\|z_{k+1} - z^*\| \leq c\|z_k - z^*\|^2,$$

where the constant $0 < c < \infty$ depends only on the data in the optimization model.

The convergence theorem can be used to estimate the number of steps required to solve the problem to a given accuracy if a primal-dual iterate z^k is in the neighborhood $U_{\varepsilon_0}(z^*)$ of the solution. However, outside this neighborhood, further modifications are needed to guarantee convergence of the algorithm. (See, for example, Section 14.6.)

Exercises

7.1. Consider the problem[19]

$$\begin{aligned}
\text{minimize} \quad & f(x) = x^2 + 1 \\
\text{subject to} \quad & g(x) = x - 2 \geq 0.
\end{aligned}$$

  (i) Find the primal-dual solution (x_*, λ_*) to this problem.

 (ii) Write the perturbed primal-dual system corresponding to the interior-point method.

(iii) Find the solution $(x(\mu), \lambda(\mu))$.

(iv) Verify that $\lim_{\mu \to 0} x(\mu) = x_*$ and $\lim_{\mu \to 0} \lambda(\mu) = \lambda_*$.

 (v) Derive the matrix

$$M(x, \lambda) = \begin{bmatrix} \nabla_{xx}^2 L(x, \lambda) & -\nabla g(x) \\ \Lambda \nabla g(x)^T & G(x) \end{bmatrix}.$$

(vi) Prove that the inverse of the matrix $M(x, \lambda)$ exists for any x and λ in the interior of the primal and dual feasible sets.

(vii) Starting with $x_0 = 2$, $\lambda_0 = \frac{1}{2}$, and $\mu_0 = 0.1$, perform two iterations of the primal-dual interior-point method with the rule $\mu_{k+1} = 0.1\mu_k$ for updating the barrier parameter. Show that (x_1, λ_1) is closer to (x_*, λ_*) than (x_0, λ_0) is, and that (x_2, λ_2), in turn, is closer to (x_*, λ_*) than (x_1, λ_1) is. Use $\kappa = 0.9$ for the ratio test.

7.2. Consider the problem

$$\begin{aligned}
\text{minimize} \quad & f(x) = \frac{1}{3}x^3 \\
\text{subject to} \quad & x \geq 1 \\
& x \leq 3.
\end{aligned}$$

---

[19]The earlier exercises are based on the ideas of subsection 16.7.1. Starting with Exercise 7.8, the exercises are based on the ideas of subsection 16.7.2.

(i) What is the solution (x_*, λ_*) to this problem?

(ii) Starting with $x_0 = 1$, $\lambda_0 = (1, 1)^T$, and $\mu_0 = 0.1$, perform two iterations of the primal-dual interior-point method with the rule $\mu_{k+1} = 0.1\mu_k$ for updating the barrier parameter. Use $\kappa = 0.9$ for the ratio test. Show that (x_1, λ_1) is closer to (x_*, λ_*) than (x_0, λ_0) is, and that (x_2, λ_2), in turn, is closer to (x_*, λ_*) than (x_1, λ_1) is.

7.3. Consider the problem

$$\text{minimize} \quad f(x) = x_1 + x_2$$
$$\text{subject to} \quad x_1^2 + x_2^2 \leq 2.$$

(i) What is the solution (x_*, λ_*) to this problem?

(ii) Starting with $x_0 = (1, 0)^T$, $\lambda_0 = 1$, and $\mu_0 = 0.1$, perform two iterations of the primal-dual interior-point method with the rule $\mu_{k+1} = 0.1\mu_k$ for updating the barrier parameter. Use $\kappa = 0.9$ for the ratio test. Show that (x_1, λ_1) is closer to (x_*, λ_*) than (x_0, λ_0) is, and that (x_2, λ_2), in turn, is closer to (x_*, λ_*) than (x_1, λ_1) is.

7.4. Prove that if the second-order sufficiency, regularity, and strict complementarity conditions are satisfied at the solution, then the matrix $M(x_*, \lambda_*)$ is nonsingular.

7.5. Consider a problem with inequality constraints. Assuming that the feasible set is bounded, prove that for any $\mu > 0$ there exists a solution to the barrier subproblem $(x(\mu), \lambda(\mu))$.

7.6. Prove that for any primal and dual vectors x and λ in the interior of the primal and dual feasible sets, if the regularized Hessian

$$R_t(x, y) = \nabla_{xx}^2 L(x, y) + tI, \quad t \geq 0,$$

where I is the identity matrix, is used instead of $\nabla_{xx}^2 L(x, \lambda)$, then the matrix

$$M_t(x, \lambda) = \begin{bmatrix} R_t(x, \lambda) & -\nabla g(x) \\ \Lambda \nabla g(x)^T & G(x) \end{bmatrix}$$

is nonsingular for sufficiently large $t \geq 0$.

7.7. In the primal-dual interior-point methods, the matrix of the system for the search directions is not symmetric. Derive an equivalent symmetric system of equations of the same size. Derive an equivalent symmetric positive semidefinite system of equations of reduced size n, where n is a number of variables.

7.8. Derive the first-order optimality conditions for the problem formulation with slacks.

7.9. Derive the perturbed system for the problem formulation with slacks.

7.10. For the interior-point method that uses slacks, derive the linear system for finding the Newton directions. What is the coefficient matrix $M(x, s, \lambda)$?

7.11. Consider the problem

$$\text{minimize} \quad f(x) = \tfrac{1}{2}x^2 + 1$$
$$\text{subject to} \quad x \geq 1.$$

(i) Reformulate the problem using a slack variable s.

(ii) Verify that the solution is $x_* = 1$, $s_* = 0$, $\lambda_* = 1$.

(iii) Write the primal-dual system corresponding to the interior-point method.

(iv) Find the solution $(x(\mu), s(\mu), \lambda(\mu))$.

(v) Verify that $\lim_{\mu \to 0} x(\mu) = x_*$, $\lim_{\mu \to 0} s(\mu) = s_*$, and $\lim_{\mu \to 0} \lambda(\mu) = \lambda_*$.

(vi) Derive the matrix $M(x, s, \lambda)$.

(vii) Prove that the solution of the primal-dual system for finding the search directions exists if and only if $s + \lambda > 0$.

7.12. Derive the second-order sufficiency conditions for the formulation with slack variables.

7.13. Prove that the second-order sufficiency conditions for the formulation with slack variables are equivalent to the second-order sufficiency conditions for the original formulation.

7.14. Prove that the regularity conditions for the formulation with slack variables are equivalent to the regularity conditions for the original formulation.

7.15. Prove that if the second-order sufficiency, regularity, and strict complementarity conditions are satisfied at the solution, then the matrix $M(x_*, s_*, \lambda_*)$ is nonsingular.

7.16. Prove that for any primal-dual vector $z = (x, s, \lambda)$ such that $s > 0$, $\lambda > 0$ and for large enough $t \geq 0$, the regularized matrix

$$
M_t(z) = \begin{bmatrix} R_t(x, \lambda) & 0 & -\nabla g(x) \\ 0 & \Lambda & S \\ \nabla g(x)^T & -I & 0 \end{bmatrix}
$$

is nonsingular. ($R_t(x, \lambda)$ is the regularized Hessian. See Exercise 7.6.)

16.8 Semidefinite Programming

In this section we consider another area of optimization: *semidefinite programming (SDP)*, or the optimization of a linear function on a cone of symmetric positive semidefinite matrices.

This problem is a special case of *convex cone programming*, for which important complexity results are established. Convex cone programming is the area of optimization that studies problems of finding a minimum of a convex, real-valued function defined on a convex cone. Convex cone problems include linear programming problems as a special case and can be solved in polynomial time using interior-point or ellipsoid methods. A general cone can be defined on a linear space of vectors or matrices. The formal definitions of a linear space and a general convex cone can be found in elementary books on linear algebra. We do not define a linear space here since the main goal of this section is to present a particular case of convex cone programming, that is, SDP, which is important for a number of reasons.

First, many known classes of problems, such as linear programs and convex quadratic problems with quadratic constraints, can be formulated as SDP problems, and hence consequently as convex cone problems. Therefore all these classes of problems can be solved in polynomial time.

Second, SDP problems can address a wide range of applications including circuit design, sensor network localization, structural design, antennae design, quantum chemistry, control theory, and matrix completion problems, in addition to the applications of linear and convex quadratic programming.

Third, solutions to many problems of integer optimization and nonconvex quadratic programming with many local minima can be approximated by solving semidefinite problems. Therefore many applications of integer and nonconvex quadratic programming can be addressed using SDP. These applications include, e.g., design of very large integrated circuits, where one needs to minimize cross-layer connections in a circuit design subject to layout constraints; as well as selecting traveling routes subject to logistics constraints.

Fourth, since for the matrix X to be semidefinite it is necessary and sufficient that all the eigenvalues of X be nonnegative, SDP is closely related to eigenvalue optimization problems. The latter can address some hard problems in graph partitioning and stability theory.

Fifth, some algorithms for linear programming, e.g., interior-point methods, can be naturally extended to SDP. SDP problems, in principle, can be formulated as nonlinear optimization problems and solved directly using nonlinear solvers. However, the specific structure of SDP problems allows putting them in a separate class closely related to linear programming. Some results form the duality theory developed for linear programming can be extended in a similar way to SDP.

Finally, SDP problems can be solved efficiently in practice. There are many efficient software packages based on various methods. We mention some reports on SDP software in the Notes.

We consider a general SDP problem in the *primal standard form* together with its *dual* companion. The SDP problem is an optimization problem, where the "variable" is a positive semidefinite matrix X of size $n \times n$:

$$\begin{aligned}
\text{minimize} \quad & C \bullet X \\
\text{subject to} \quad & A_i \bullet X = b_i, \quad i = 1, \ldots, m \\
& X \succeq 0.
\end{aligned}$$

Here C and A_i are symmetric matrices of size $n \times n$, b_i are numbers, and $X \bullet Y$ denotes the trace of the matrix $Z = XY$. The trace of an $n \times n$ matrix $Z = (Z_{ij})$ is the sum of its diagonal elements: $\text{tr}(Z) = \sum_{k=1}^{n} z_{kk}$. The notation $X \succeq 0$ means that X is positive semidefinite (see Appendix A.7.1). This constraint is analogous to the nonnegativity of a vector $x \geq 0$ in linear programming.

The dual problem is derived using principles similar to those for linear programming:

$$\begin{aligned}
\text{maximize} \quad & b^T y \\
\text{subject to} \quad & \sum_{i=1}^{m} y_i A_i + S = C \\
& S \succeq 0.
\end{aligned}$$

This dual SDP problem sometimes is written in a slightly different equivalent form:

$$\begin{aligned}
\text{maximize} \quad & b^T y \\
\text{subject to} \quad & C - \sum_{i=1}^{m} y_i A_i \succeq 0.
\end{aligned}$$

The constraint

$$C - \sum_{i=1}^{m} y_i A_i \succeq 0$$

is often called a *linear matrix inequality*.

For simplicity, we assume that the matrices A_i, $i = 1, \ldots, m$, are linearly independent. If they are not, we can always express dependent matrices using linearly independent matrices, and therefore can eliminate dependent matrices altogether.

SDP has a wide range of applications. Most of them require an introduction of many additional concepts, which are beyond the scope of this book. Here, we mention just a few of them. The first example is minimizing the maximum eigenvalue of a matrix. This problem is useful, e.g., for stabilizing a system of differential equations. Suppose that we have a symmetric matrix $A(z)$ that depends linearly on a vector z:

$$A(z) = A_0 + z_1 A_1 + \cdots + z_m A_m.$$

We would like to minimize the maximum eigenvalue of $A(z)$,

$$\underset{z}{\text{minimize}} \quad \lambda_{\max}(A(z)),$$

which is equivalent to

$$\text{minimize} \quad t$$
$$\text{subject to} \quad \lambda_{\max}(A(z)) \leq t.$$

It can be shown that $\lambda_{\max}(A(z)) \leq t$ is equivalent to $\lambda_{\max}(A(z) - tI) \leq 0$, which, in turn, is equivalent to $\lambda_{\min}(tI - A(z)) \geq 0$. Therefore, we have the following semidefinite problem:

$$\text{maximize} \quad -t$$
$$\text{subject to} \quad tI - A_0 - z_1 A_1 - \cdots - z_m A_m \succeq 0.$$

This problem is in the form of the dual semidefinite problem described above.

A related problem that can be solved using SDP is the minimization of the k largest eigenvalues of a matrix $A(z)$. This problem can be formulated as an SDP problem:

$$\text{minimize} \quad kt + I \bullet X$$
$$\text{subject to} \quad tI + X - A_0 - z_1 A_1 - \cdots - z_m A_m \succeq 0$$
$$X \succeq 0.$$

This SDP problem can be reformulated in the form of the dual SDP (see the Exercises).

Our next example shows that a so-called *semidefinite relaxation* can be used to solve integer programming problems.

One way to approach an integer programming problem is to first solve an easier *relaxation* of the problem, and then to systematically add new constraints that refine the solution so it satisfies the constraints of the original problem. The new constraints, called *cutting planes*, cut off fractional solutions and "strengthen" the relaxation.

For example, consider the following integer programming problem:

$$\text{maximize} \quad b^T x$$
$$\text{subject to} \quad Ax \leq c$$
$$x_i \in \{0, 1\}, \quad i = 1, \ldots, n.$$

One way to define a relaxation to this problem is to replace the integrality constraints $x_i \in \{0, 1\}$, $i = 1, \ldots, n$, with the relaxations $0 \leq x_i \leq 1$, $i = 1, \ldots, n$. The resulting linear programming problem can be solved efficiently, but the solution to the relaxed problem can have many noninteger components that are strictly between 0 and 1. Therefore, we may need many additional cutting planes to find an integer solution. In cases where the solution to the relaxation is not close to the solution of the original problem, we say that the relaxation is not strong enough. While using a linear programming relaxation can be successful for some combinatorial problems, in general such relaxations are poor approximations to integer problems.

Therefore we need other approaches that produce stronger relaxations. At the same time we still need to be able to solve the relaxed problems efficiently. Applying SDP for constructing new types of relaxations has turned out to be successful.

For example, the problem of finding a maximum cut on a graph is a difficult integer problem. The solution to this problem is used for designing electric circuits. The relaxation of the max-cut problem provides an upper bound for the solution to the max-cut problem. The difference between the solution to the relaxation and that of the original max-cut problem measures the strength of the relaxation. It is possible to construct an SDP relaxation that has an optimal value at most 14% greater than that of the max-cut problem, while for some graphs the linear programming relaxation can produce an upper bound almost 100% greater, i.e., almost twice as large as the optimal value of the max-cut problem. Therefore, using SDP relaxations, it is possible to construct new approximation algorithms with good performance for difficult integer problems.

The details of these algorithms go beyond the scope of this book; however, we consider one method of how semidefinite relaxation can be introduced based on our example.

Let us first define the following matrix:

$$Y = \begin{pmatrix} x \\ 1 \end{pmatrix} (x^T \ 1) = \begin{pmatrix} xx^T & x \\ x^T & 1 \end{pmatrix}.$$

This matrix is semidefinite (see the Exercises). We will add the additional constraints $Y_{ii} = Y_{i,n+1} = Y_{n+1,i}$, which are equivalent to $x_i^2 = x_i$. It is easy to see that $x_i^2 = x_i$ if and only if $x_i \in \{0, 1\}$, $i = 1, \ldots, n$.

The relaxation of the integral constraints $x_i \in \{0, 1\}$, $i = 1, \ldots, n$, is obtained by replacing the matrix Y with

$$Z = \begin{pmatrix} X & x \\ x^T & 1 \end{pmatrix},$$

where the matrix X is symmetric and $x = \operatorname{diag}(X)$, i.e., $x_i = X_{ii}$. The latter condition is equivalent to $Z_{ii} = Z_{i,n+1} = Z_{n+1,i}$, $i = 1, \ldots, n$.

The class of matrices Z is wider than that of Y. We need to impose an additional constraint

$$Z = \begin{pmatrix} X & x \\ x^T & 1 \end{pmatrix} \succeq 0$$

since, unlike Y, the matrix Z is not guaranteed to be semidefinite.

The linear constraints $Ax \leq c$ can be reformulated as

$$\operatorname{diag}(c - Ax) \succeq 0.$$

Therefore, we have the following problem:

$$\text{maximize} \quad b^T x$$

$$\text{subject to} \quad \begin{pmatrix} \text{diag}(c - Ax) & 0 & 0 \\ 0 & X & x \\ 0 & x^T & 1 \end{pmatrix} \succeq 0, \quad x = \text{diag}(X),$$

where the variables are the vector x and the matrix X. This problem can be reformulated in the form of the dual semidefinite problem described above (see the Exercises).

Next, we show how SDP can be used in linear control systems. Suppose that a trajectory $x(t)$ satisfies the following system of differential equations:

$$\dot{x}(t) = Ax(t), \quad x(0) = x_0,$$

where A is an $n \times n$ matrix. We want to know whether $x(t)$ remains bounded.

Results from control theory state that $x(t)$ remains bounded if and only if there exists some positive-definite matrix $P \succ 0$ such that a Lyapunov function $V(t) = x(t)^T P x(t)$ remains uniformly bounded; i.e., the inequality $V(t) \leq M$ holds for all $t \geq 0$ and some $M > 0$.

To guarantee the uniform boundedness of $V(t)$ it is enough to prove that $V(t)$ is nonincreasing for $t \geq 0$, which holds if and only if

$$\tfrac{d}{dt} V(t) = \dot{x}^T P x + x^T P \dot{x} = x^T A^T P x + x^T P A x = x^T (A^T P + P A) x \leq 0$$

for all $t \geq 0$. Therefore if the matrix

$$-A^T P - P A$$

is positive semidefinite for some $P \succ 0$, then $x(t)$ is bounded.

How can we find such a positive-definite matrix P and not just positive semidefinite? For any positive-definite matrix P and any $\alpha > 0$, the matrix αP is also positive definite. Therefore, we can use the condition

$$P \succeq I \quad \Longleftrightarrow \quad P - I \succeq 0$$

to determine P; here I is the identity matrix. If we seek a matrix with a minimum condition number $\text{cond}(P)$, then we obtain the following SDP problem to find matrix P:

$$\text{minimize} \quad c$$

$$\text{subject to} \quad -A^T P - P A \succeq 0$$
$$I \preceq P \preceq cI,$$

which can be reformulated in the form of the dual SDP problem.

There are many other applications of SDP. For further information, see the Notes.

Semidefinite and linear programming problems have many common features. Here, we briefly formulate some duality theorems for SDP similar to those proven for linear programming.

Theorem 16.18 (Weak Duality). *If X is primal feasible and* (y, S) *is dual feasible, then there is a nonnegative duality gap:*

$$C \bullet X - b^T y = X \bullet S \geq 0.$$

Proof. See the Exercises.   ☐

Theorem 16.19 (Complementarity). *If X and S are positive semidefinite matrices, then* $X \bullet S = 0$ *implies* $XS = 0$.

Proof. See the Exercises.   ☐

To guarantee the existence of optimal solutions of primal and dual problems with zero duality gap, it is enough to assume the existence of positive-definite matrices $\bar{X} \succ 0$ and $\bar{S} \succ 0$, which satisfy the linear constraints.

Theorem 16.20 (Strong Duality). *If* $\bar{X} \succ 0$ *and* $\bar{S} \succ 0$ *are primal and dual feasible positive-definite matrices, then there exist optimal primal and dual solutions* X^* *and* (y^*, S^*), *and there is no duality gap:*

$$C \bullet X^* - b^T y^* = X^* \bullet S^* = 0.$$

Proof. See the Notes.   ☐

The conditions of the strong duality theorem for SDP are stricter than similar conditions for linear programming. In linear programming we need only assume that *feasible* primal and dual points exist to guarantee that primal and dual *optimal* solutions exist.

A detailed discussion of the theory and algorithms of SDP is beyond the scope of this book. We mention only that interior-point methods can be generalized to solve SDP problems. It is possible to obtain polynomial complexity results using interior-point algorithms and the theory of self-concordance. See the Notes for details.

Exercises

8.1. Reformulate the following problem

$$\text{maximize} \quad b^T x$$

$$\text{subject to} \quad \begin{pmatrix} \text{diag}\,(c - Ax) & 0 & 0 \\ 0 & X & x \\ 0 & x^T & 1 \end{pmatrix} \succeq 0, \quad x = \text{diag}\,(X),$$

where the vector x and the symmetric matrix X are variables, in the form of a dual SDP problem:

$$\text{maximize} \quad b^T y$$

$$\text{subject to} \quad C - \sum_{i=1}^{m} y_i A_i \succeq 0.$$

8.2. Prove that the 2×2 matrix

$$A = \begin{pmatrix} a & b \\ b & c \end{pmatrix}$$

is positive semidefinite if and only if the following three inequalities are satisfied:

$$a \geq 0, \quad c \geq 0, \quad b^2 \leq ac.$$

8.3. Using the fact that any symmetric matrix A can be represented as $A = V \Lambda V^T$, where Λ is a diagonal matrix with the eigenvalues on the diagonal and V is an orthogonal matrix (see Appendix A.2), prove that a matrix A is positive semidefinite if and only if there exists a matrix B such that $A = B^T B$.

8.4. Prove that a convex quadratic optimization problem with quadratic constraints

$$
\begin{aligned}
\text{minimize} \quad & f(x) = x^T Q_0 x + p_0^T x + r_0 \\
\text{subject to} \quad & x^T Q_1 x + p_1^T x + r_1 \leq 0 \\
& \qquad \vdots \\
& x^T Q_m x + p_m^T x + r_m \leq 0
\end{aligned}
$$

with positive semidefinite matrices Q_i, $i = 0, 1, \ldots m$, can be formulated as an SDP problem. (*Hint*: Use the fact that any positive semidefinite matrix A can be represented as $A = B^T B$.)

8.5. Consider the following problem:

$$
\begin{aligned}
\underset{x}{\text{minimize}} \quad & z = c^T x \\
\text{subject to} \quad & Ax = b \\
& \|D_i x + e_i\|_2 \leq f_i^T x + g_i, \quad i = 1, \ldots, m.
\end{aligned}
$$

Here x, c, and f_i are vectors of length n; b is a vector of length r; e_i are vectors of length n_i; A is an $r \times n$ matrix; D_i are $n_i \times n$ matrices; and g_i are numbers. This problem is called a *second-order cone* programming problem, because the feasible set is a rotated second-order Lorenz cone (see Figure 16.4).

(i) Prove that the second-order cone programming problem is convex.

(ii) Prove that the second-order cone programming problem can be reformulated as an SDP problem.

8.6. Prove that $A \bullet B = B \bullet A$.

8.7. Using the fact that any symmetric matrix A can be represented as $A = V \Lambda V^T$, where Λ is a diagonal matrix with the eigenvalues on the diagonal and V is an orthogonal matrix (see Appendix A.2), prove that any positive semidefinite matrix A can be represented as $A = RR$ with a positive semidefinite matrix R.

8.8. Show that the problem

$$
\begin{aligned}
\text{minimize} \quad & kt + I \bullet X \\
\text{subject to} \quad & tI + X - A_0 - z_1 A_1 - \cdots - z_m A_m \succeq 0 \\
& X \succeq 0
\end{aligned}
$$

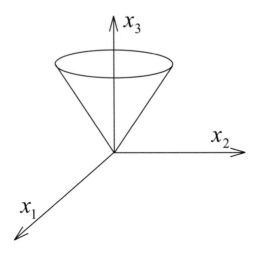

Figure 16.4. *Second-order cone.*

can be represented as the standard semidefinite problem in the dual form

$$\text{maximize} \quad b^T y$$
$$\text{subject to} \quad C - \sum_{i=1}^{m} y_i A_i \succeq 0.$$

8.9. Prove that minimization of the k largest eigenvalues of a matrix $A_0 + z_1 A_1 + \cdots + z_m A_m$ can be formulated as

$$\text{minimize} \quad kt + I \bullet X$$
$$\text{subject to} \quad tI + X - A_0 - z_1 A_1 - \cdots - z_m A_m \succeq 0$$
$$X \succeq 0.$$

8.10. Prove that if X is primal feasible and (y, S) is dual feasible, then there is a nonnegative duality gap

$$C \bullet X - b^T y = X \bullet S \geq 0.$$

8.11. Prove that if $X \succeq 0$, $S \succeq 0$, and $X \bullet S = 0$, then $XS = 0$. Is the converse true?

16.9   Notes

Penalty and Barrier Methods—The most important reference on penalty and barrier methods is the book by Fiacco and McCormick (1968, reprinted 1990). It includes an extensive survey on the methods. A more recent discussion can be found in the paper by Wright (1992). The quadratic penalty function was first described by Courant (1943). Frisch (1955) is often cited as the originator of the logarithmic barrier method; even though the formulas he uses in solving linear programming problems are related to the gradient of the barrier function,

Frisch does not mention the barrier function, nor does he use a barrier method. The history of these methods is more complex, and it is discussed in the paper by Fiacco (1967).

Ill-Conditioning—A detailed discussion of the ill-conditioning of penalty and barrier methods can be found in the paper by Murray (1971).

Stabilized Methods—The results in Section 16.4 are adapted from the paper by Nash and Sofer (1993). A related result can be found in the paper by Murray (1971). A different approach to controlling ill-conditioning is described in the paper by Wright (1994).

In the large-scale case it is often advantageous to use a truncated-Newton method. In this case, a linear conjugate gradient (or some other iterative method) is used to solve the systems of equations associated with $(Z^T H Z)^{-1}$. Consequently, these formulas require two runs of the conjugate-gradient method. A related set of formulas can be derived that requires only one use of the conjugate-gradient method and that is more suitable in this case; see the paper by Nash and Sofer (1993).

Exact Penalty Methods—A nondifferentiable exact penalty function was first suggested by Ablow and Brigham (1955). A more general discussion can be found in the paper by Pietrzykowski (1969). Special algorithms designed to cope with the nondifferentiability are described in the papers by Zangwill (1967), Conn (1973), Lemaréchal (1975), Coleman (1979), and Mayne and Maratos (1979), for example.

Augmented Lagrangian Methods—Augmented Lagrangian methods were suggested independently by Hestenes (1969) and Powell (1969). A general discussion can be found in the book by Bertsekas (1982, reprinted 1996); many ideas in the section are adapted from this book. Computational issues associated with these methods are discussed in the book by Conn, Gould, and Toint (1992). Modified-barrier methods were developed by Polyak (1992).

Interior-Point Methods—The book by Fiacco and McCormick (1968, reprinted 1990) is a "classical" reference to interior-point methods for nonlinear programming. The book by den Hertog (1994) provides a comprehensive introduction to modern interior-point methods.

A discussion of computational issues for interior-point methods can be found in the papers by Lustig, Marsten, and Shanno (1992) and (1994a). The proof of a quadratic rate of convergence in the neighborhood of the solution can be found in the papers of Yamashita and Yabe (1996, 2005). Practical interior-point algorithms for nonlinear programming are described in the papers of Vanderbei and Shanno (1999); Byrd, Hribar, and Nocedal (1999); and Wächter and Biegler (2006).

Semidefinite Programming (SDP)—Although SDP is a relatively new topic, the idea of constraining a matrix to be positive semidefinite is an old one. In the 1880s Lyapunov used a matrix inequality constraint to characterize stability of the solution of a system of differential equations. In the 1960s Bellman and Fan introduced primal and dual SDP problems and developed duality theorems. In the 1970s it was recognized that some difficult graph partitioning problems can be addressed using SDP. Since then the number of applications of SDP has grown substantially. Another wave of interest in SDP was stimulated by the application of SDP to approximate the solutions to some difficult integer and nonconvex quadratic programming problems. Also, SDP is a special case of convex cone programming (see below). Surveys of SDP can be found in Todd (2001) and Wolkowicz et al. (2000).

SDP is a special case of convex cone programming. Nesterov and Nemirovski (1994) applied self-concordance theory in this unifying framework to show that convex cone problems can be solved in polynomial time. For results based on the primal-dual interior-point method for cone programming, see Nesterov and Todd (1998).

There are many software packages developed for SDP. Information about them can be found in the sources mentioned in Appendix C.

Part V

Appendices

Appendix A

Topics from Linear Algebra

A.1 Introduction

We give a brief overview of various topics in numerical linear algebra. For further information, consult the books by Golub and Van Loan (1996) and Horn and Johnson (1991, reprinted 1994).

An n-dimensional *vector* x is an array of n scalars $x_1, x_2, \ldots, x_n$. The notation x represents the array when its elements are arranged in a column, while the notation x^T represents the array when its elements are arranged in a row. The collection of all real vectors of dimension n is denoted by $\Re^n$. Vectors are usually denoted by lowercase Roman letters.

A *matrix* is a rectangular array of scalars. If the matrix has m rows and n columns it is referred to as an $m \times n$ matrix; if $m = n$, it is called a *square* matrix. Matrices are usually denoted by uppercase Roman letters, while their elements are denoted by lowercase letters. For example, the element in row i and column j of the matrix A is denoted by $a_{i,j}$. The transpose of an $m \times n$ matrix A is the $n \times m$ matrix A^T obtained by interchanging the rows of A with its columns. A square matrix is said to be *symmetric* if $A = A^T$.

A.2 Eigenvalues

Let A be an $n \times n$ matrix. If a scalar λ and a nonzero vector v satisfy

$$Av = \lambda v,$$

then λ is an *eigenvalue* and v is an *eigenvector* of the matrix A. The eigenvalues are the solutions to the equation

$$\det(A - \lambda I) = 0,$$

where "det" denotes the determinant. The left-hand side of this equation is a polynomial of degree n called the *characteristic equation*, and the eigenvalues are the roots of this polynomial.

Example A.1 (Eigenvalues and Eigenvectors). Let

$$A = \begin{pmatrix} 2 & 1 \\ 1 & 2 \end{pmatrix}.$$

The equation $\det(A - \lambda I) = (2 - \lambda)^2 - 1 = 0$ yields the eigenvalues $\lambda_1 = 1$ and $\lambda_2 = 3$. To compute the eigenvectors we solve the system of equations $(A - \lambda I)v = 0$ for the computed values of λ. We obtain that $v_1 = (1 \quad -1)^T$ is an eigenvector corresponding to $\lambda_1 = 1$, and $v_2 = (1 \quad 1)^T$ is an eigenvector corresponding to $\lambda_2 = 3$. The eigenvectors v_1 and v_2 are not unique. Any nonzero multiple of an eigenvector is also an eigenvector. ∎

If A is symmetric, then its n eigenvalues are all real, and it has n linearly independent eigenvectors corresponding to these eigenvalues. Labeling these eigenvectors $v_1, \ldots, v_n$, it is possible to choose them to be orthogonal to each other (that is, $v_i^T v_j = 0$) for $i \neq j$ and to normalize them so that $v_i^T v_i = 1$. Denoting the corresponding eigenvalues by $\lambda_1, \ldots, \lambda_n$, we can define the matrices

$$V = (v_1 \quad \cdots \quad v_n) \quad \text{and} \quad \Lambda = \text{diag}\{\lambda_1, \ldots, \lambda_n\}.$$

The matrix V is orthogonal because of the way the eigenvectors were selected. It is easy to verify that $AV = V\Lambda$, or that

$$A = V\Lambda V^T.$$

This representation of A is called the *spectral decomposition* of A.

A.3 Vector and Matrix Norms

Norms give a rough measure of the magnitude of the entries in vectors and matrices. They generalize the notion of absolute value for real numbers.

In general, a norm $\|\cdot\|$ is defined as a real-valued function with the properties that

- $\|x\| \geq 0$ for all x
- $\|x\| = 0$ if and only if $x = 0$
- $\|\alpha x\| = |\alpha| \, \|x\|$ for all real numbers α
- $\|x + y\| \leq \|x\| + \|y\|$ for all x and y.

This definition applies to both vector and matrix norms (in fact, to norms of all types). It is useful but not essential to ask that matrix norms have the additional property

- $\|xy\| \leq \|x\| \cdot \|y\|$ for all x and y whose product xy exists.

This last property is satisfied by most commonly used matrix norms.

Let us first consider vector norms for vectors of the form

$$x = (x_1 \quad \cdots \quad x_n)^T.$$

The most commonly used is the Euclidean, or 2-norm,

$$\|x\|_2 \equiv \sqrt{x_1^2 + \cdots + x_n^2}.$$

Also widely used are the 1-norm

$$\|x\|_1 \equiv \sum_{i=1}^{n} |x_i|$$

and the infinity norm

$$\|x\|_\infty \equiv \max_{1 \le i \le n} |x_i|.$$

These are special cases of the norms

$$\|x\|_p \equiv \left(\sum_{i=1}^{n} |x_i|^p \right)^{\frac{1}{p}}$$

defined for $p \ge 1$.

Example A.2 (Vector Norms). Let $x = (1, 2, 3)^T$ then

$$\|x\|_1 = 1 + 2 + 3 = 6$$
$$\|x\|_2 = \sqrt{1 + 4 + 9} = \sqrt{14} \approx 3.74$$
$$\|x\|_\infty = \max\{1, 2, 3\} = 3. \quad \blacksquare$$

In many cases it does not make much difference which norm is used, and so it is sometimes convenient to use the notation $\|x\|$ without specifying a specific norm. In software for large problems, it may be preferable to use $\|x\|_\infty$. If $x = (1, \ldots, 1)^T$, then

$$\|x\|_1 = n, \quad \|x\|_2 = \sqrt{n}, \quad \text{and} \quad \|x\|_\infty = 1.$$

If n is large, then both $\|x\|_1$ and $\|x\|_2$ will be large even though the entries in x are all equal to one. The infinity norm does not share this property.

The most widely used matrix norms are defined in terms of vector norms. If A is a matrix and $\|x\|$ is a vector norm, then the *induced* matrix norm $\|A\|$ is defined by

$$\|A\| = \max_{\|x\|=1} \|Ax\|.$$

Every induced matrix norm satisfies

$$\|Ax\| \le \|A\| \cdot \|x\|$$

for all vectors x. It is also true that $\|AB\| \le \|A\| \cdot \|B\|$ for matrices A and B.

The matrix norms corresponding to the vector norms above are

$$\|A\|_1 = \max_{1 \le j \le n} \sum_{i=1}^{n} |A_{i,j}|$$
$$\|A\|_2 = \sqrt{\lambda_{max}(A^T A)}$$
$$\|A\|_\infty = \max_{1 \le i \le n} \sum_{j=1}^{n} |A_{i,j}|,$$

where $\lambda_{max}(A^T A)$ is the largest eigenvalue of $A^T A$. In the special case where A is a square symmetric matrix, we have

$$\|A\|_2 = |\lambda_{max}(A)|,$$

where $\lambda_{max}(A)$ is the eigenvalue of largest magnitude of A. In general, $\|A\|_2$ is generally much more expensive to compute than either $\|A\|_1$ or $\|A\|_\infty$.

Example A.3 (Matrix Norms). Let

$$A = \begin{pmatrix} 1 & 3 & -2 & 4 \\ -5 & 7 & 9 & -3 \\ 2 & -1 & 6 & 8 \end{pmatrix}.$$

Then

$$\|A\|_1 = \max\{1 + 5 + 2, 3 + 7 + 1, 2 + 9 + 6, 4 + 3 + 8\} = \max\{8, 11, 17, 15\} = 17$$

and

$$\|A\|_\infty = \max\{1 + 3 + 2 + 4, 5 + 7 + 9 + 3, 2 + 1 + 6 + 8\} = \max\{10, 24, 17\} = 24.$$

The eigenvalues of $A^T A$ are $\{0, 23.4771, 108.0513, 167.4716\}$, so

$$\|A\|_2 = \sqrt{167.4716} = 12.9411. \qquad \blacksquare$$

The *condition number* of a nonsingular matrix is defined as

$$\mathrm{cond}(A) = \|A\| \cdot \|A^{-1}\|.$$

Any matrix norm can be used here. From the definition that $\mathrm{cond}(I) = 1$, and since $\|AA^{-1}\| \le \|A\| \|A^{-1}\|$, it follows that

$$\mathrm{cond}(A) \ge 1$$

for all matrices A. If A is symmetric, and the 2-norm is used, then

$$\mathrm{cond}(A) = \frac{\lambda_{max}}{\lambda_{min}},$$

where λ_{max} is the largest eigenvalue of A in absolute value, and λ_{min} is the smallest. The condition number of a matrix A is an indication of the sensitivity to perturbation of a linear system involving A. See Appendix A.8.

A.4 Systems of Linear Equations

To illustrate the techniques for solving systems of linear equations, we use the example

$$\begin{aligned} x_1 + 2x_2 + 4x_3 &= 1 \\ x_1 + 3x_2 + 9x_3 &= 6 \\ x_1 + 5x_2 + 25x_3 &= 4. \end{aligned}$$

In matrix form the system looks like

$$\begin{pmatrix} 1 & 2 & 4 \\ 1 & 3 & 9 \\ 1 & 5 & 25 \end{pmatrix} \begin{pmatrix} x_1 \\ x_2 \\ x_3 \end{pmatrix} = \begin{pmatrix} 1 \\ 6 \\ 4 \end{pmatrix}.$$

Denote this system as $Ax = b$.

This system is square, having the same number of equations as variables. The techniques discussed here applies only to square systems.

If a system of equations has a unique solution, then the system as well as the matrix A are said to be *nonsingular*. For the matrix, this is the same as saying that A has an inverse, or that A^{-1} exists. If A^{-1} does not exist, then the matrix and the system are *singular*.

The matrix A will be *singular* if its rows (or equivalently, its columns) are linearly dependent; that is, if one of the rows is a linear combination of the others. The matrix

$$A = \begin{pmatrix} 1 & 2 & 3 \\ 4 & 5 & 6 \\ 7 & 8 & 9 \end{pmatrix}$$

is singular because the last row is equal to twice the second row minus the first row. A system of equations with a singular matrix A has either no solution or an infinite family of solutions, depending on the components of the right-hand-side vector b. For this particular matrix A, there is no solution if $b = (10, 10, 20)^T$. To see this, subtract twice the second equation minus the first equation from the last equation to obtain $0 = 10$, an equation with no solution. On the other hand, if $b = (10, 10, 10)^T$, then the last equation is redundant and can be deleted from the system. This leaves a system of two equations in three unknowns:

$$1x_1 + 2x_2 + 3x_3 = 10$$
$$4x_1 + 5x_2 + 6x_3 = 10.$$

We can solve for any two of the variables in terms of the third. For example, if we solve for x_1 and x_2 in terms of x_3, we obtain the infinite family of solutions

$$x_1 = x_3 - 10 \quad \text{and} \quad x_2 = 10 - 2x_3,$$

where x_3 can be chosen arbitrarily.

If the matrix A is nonsingular and the inverse matrix A^{-1} is known, then the solution of the linear system $Ax = b$ can be written as $x = A^{-1}b$. However, the techniques that we will describe for solving linear equations will not form A^{-1}. There are a number of reasons for not forming A^{-1}. The most notable is that it is both more expensive and unnecessary to compute A^{-1} than to use the elimination techniques described below. On small problems where most of the coefficients in the matrix are nonzero (a *dense* matrix), forming the inverse is three times as expensive as applying elimination. On large problems, the savings can be even more dramatic.

If the matrix A is large (if the number of variables is large), it is common in practical applications for most of the entries in the matrix to be zero (the matrix will be *sparse*). The entry in the matrix at row i and column j might represent the connection between two components in a machine (components i and j), and if the two components are not connected, then this coefficient will be zero. Since a large machine will have many components, most of which are only connected to a few nearby components, most of the entries in the matrix will be zero. For a sparse matrix it is not unusual for the number of nonzero entries in the matrix to be approximately αn, where n is the number of rows in the matrix. This says that there are about α nonzero entries in each row of the matrix. A dense matrix would have n^2 nonzero entries. For a problem with $n = 10,000$ and $\alpha = 5$ (5 nonzero entries per row), the sparse matrix would have 50,000 nonzero entries, while the corresponding dense matrix would have 100,000,000 entries, or 2,000 times as many.

When the elimination methods described below are applied to sparse matrices, the costs of solving a linear system are usually proportional to the number of nonzero entries, αn. (Here the term "costs" refers to the number of arithmetic operations required.) Special techniques are used that avoid the calculations associated with zero entries, such as adding zero to a number. If on the other hand the inverse A^{-1} is computed, the costs will normally be proportional to n^3, a dramatically higher cost. (See also the discussion in Appendix A.6.) If $n = 10,000$ and $\alpha = 5$, then the sparse matrix could be factored using about 50,000 arithmetic operations, while the corresponding dense matrix would require about 1,000,000,000,000 operations, or 20,000,000 times as many. If algorithms did not take advantage of sparsity, it would not be practical to solve large optimization problems.

A.5 Solving Systems of Linear Equations by Elimination

The most commonly used method for solving systems of linear equations is Gaussian elimination. Carl Friedrich Gauss described this technique in the early 1800s. The name of the method is derived from this reference, but the method is much older than this.

We describe the technique using an example. Consider the system

$$2.0x_1 - 4.0x_2 - 1.0x_3 = 2.0$$
$$0.4x_1 + 2.2x_2 + 1.8x_3 = 2.4$$
$$0.8x_1 - 0.1x_2 - 1.4x_3 = 5.8.$$

This system will be transformed into an equivalent but simpler system. The new system will be equivalent in the sense that the two systems of equations will have the same solution. The new system will be simpler in the sense that its solution will be easy to compute.

To perform the transformation, we subtract multiples of one equation from another in order to eliminate variables. We use the first equation to eliminate x_1 from the other two equations. If we subtract 0.2 times the first equation from the second equation, and subtract 0.4 times the first equation from the third equation, we obtain the reduced system

$$3.0x_2 + 2.0x_3 = 2.0$$
$$1.5x_2 - 1.0x_3 = 5.0.$$

We now apply elimination to this smaller system, subtracting 0.5 times the first equation from the second to obtain

$$-2.0x_3 = 4.0.$$

Elimination transforms the original system into the simpler system

$$2.0x_1 - 4.0x_2 - 1.0x_3 = 2.0$$
$$3.0x_2 + 2.0x_3 = 2.0$$
$$-2.0x_3 = 4.0.$$

It is called *upper triangular* because of the pattern of nonzero coefficients.

The third equation gives $x_3 = -2.0$. Then we can substitute this value into the second equation to obtain $x_2 = 2.0$. Finally, we substitute the values of x_2 and x_3 into the first equation and solve for $x_1 = 4.0$.

The technique for solving the triangular system is called *backsubstitution* (solving for x_3 in the third equation, and then substituting its value into the second equation, etc.).

The formulas for both elimination and backsubstitution are not difficult to derive. Suppose that the ith equation of the final triangular system is written as

$$\sum_{j=i}^{n} \hat{a}_{ij}x_j = \hat{b}_i,$$

where $\hat{a}_{ij}$ and $\hat{b}_i$ represent the coefficients and right-hand side of the transformed equation respectively. Then the formulas for backsubstitution are

$$x_i = \frac{1}{\hat{a}_{ii}} \left(\hat{b}_i - \sum_{j=i+1}^{n} \hat{a}_{ij}x_j \right) \qquad \text{for } i = n, n-1, \ldots, 1.$$

From this formula it is clear that the backsubstitution step can be carried out if $\hat{a}_{ii} \neq 0$ for all i, that is, if the triangular system is nonsingular. Hence the algorithm can be applied to any nonsingular triangular system. The cost of backsubstitution can also be determined. There are n steps (one for each x_i), each of which requires $2(n-i)+1$ arithmetic operations ($(n-i)$ multiplications, $(n-i-1)$ additions, 1 subtraction, and 1 division), so the total number of arithmetic operations is

$$\sum_{i=1}^{n}[2(n-i)+1] = n + 2\sum_{i=1}^{n}(n-i) = n + 2\sum_{i=0}^{n-1}i$$
$$= n + 2 \times (n-1)(n-1+1)/2 = n^2.$$

Since the number of coefficients in the simpler system is approximately $n^2/2$, and since each of these coefficients must be examined in order to determine the solution, *any* algorithm for solving a triangular system must have costs proportional to n^2—hence backsubstitution is an efficient algorithm for determining the solution.

The formulas for elimination are slightly more complicated. Let us assume that the triangular system is overwritten on the original system of equations. Denote the original linear system by $Ax = b$ with the ith equation having the form

$$\sum_{j=1}^{n} a_{ij}x_j = b_i.$$

The central step in elimination is the subtraction of a multiple of the ith equation from the kth equation:

$$a_{kj} \leftarrow a_{kj} - \frac{a_{ki}}{a_{ii}} a_{ij} \qquad \text{for } j = i, \ldots, n \text{ and } k = i+1, \ldots, n.$$

This is carried out for $i = 1, \ldots, n-1$. Counting operations shows that the cost of elimination is approximately $\frac{2}{3}n^3$ operations if n is large. Notice that elimination is about n times more expensive than backsubstitution.

In this and many other algorithms the number of additions and subtractions is almost the same as the number of multiplications and divisions. It is common to count only the latter, and also to treat divisions in the same way as multiplications. We will follow this practice. Hence we will say that backsubstitution requires $n^2/2$ multiplications, and that elimination requires $n^3/3$ multiplications.

The formulas for elimination will break down if $a_{ii} = 0$. This can happen even if the system of equations is nonsingular. However, with one simple modification, the above technique can be transformed into a general algorithm. As illustrated above, the technique will break down if it is applied to the linear system

$$0.0x_1 - 4.0x_2 - 1.0x_3 = 2.0$$
$$0.4x_1 + 2.2x_2 + 1.8x_3 = 2.4$$
$$0.8x_1 - 0.1x_2 - 1.4x_3 = 5.8.$$

If we try to apply elimination, we are attempting to subtract $0.4/0.0$ times the first equation from the second. This ratio is infinite, and so the step cannot be carried out. It is easy to see how to fix the method—just interchange the first equation with one of the others. It is common to choose the equation with the largest coefficient (in absolute value) of x_1. If we do this at every stage of elimination, we obtain the upper triangular system

$$0.8x_1 - 0.1x_2 - 1.4x_3 = 5.8$$
$$-4.0x_2 - 1.0x_3 = 2.0$$
$$1.9375x_3 = 0.625.$$

Backsubstitution can now be used to obtain the solution ($x_1 \approx 7.741936$, $x_2 \approx -0.580645$, $x_3 \approx 0.322581$).

Although it is only necessary to interchange equations when a division by zero will result, it is advisable to perform interchanges at every iteration to select the equation with the largest leading coefficient in absolute value. This is done to control the rounding errors that arise when Gaussian elimination is implemented on a computer. (See Appendix A.8.) This interchange technique is called *partial pivoting*. There is also a technique called *complete pivoting* that not only switches rows but also reorders the variables. It is more expensive than partial pivoting, and on most problems does not produce better answers, so it is rarely used.

If partial pivoting is used, then Gaussian elimination will not break down (i.e., encounter a zero pivot element) if the system of equations has a unique solution. In this case it is also possible to derive bounds on the error in the solution. (See the Notes.)

A.6 Gaussian Elimination as a Matrix Factorization

There is an alternative point of view of Gaussian elimination that will be useful to us. Consider a linear system written in matrix form $Ax = b$. Our new point of view will attempt to isolate the influence of the matrix A (often representing a general model of our situation) from the right-hand side b (often representing the data in our specific case). The basis for this point of view will be a representation of the matrix A as a factorization of A into "simpler" matrices.

For Gaussian elimination, A is represented in terms of its LU factors

$$A = PLU.$$

The matrix U is upper triangular and corresponds to the upper triangular system of equations that is obtained as a result of Gaussian elimination. The matrix L is lower triangular and is a record of the elimination calculations that were used to produce U. The matrix P (not always present) is a *permutation matrix* that records the interchanges of equations used during the elimination; the columns of P are a permutation of the columns of the identity matrix. Although the interchanges are done one at a time during the elimination algorithm, they can be collected together as if they were all done at once.

Before illustrating the entire factorization, we first would like to look at the matrices P and L. The permutation matrix P serves to reorder the equations. For a four-variable problem, for example,

$$\begin{pmatrix} 0 & 0 & 1 & 0 \\ 1 & 0 & 0 & 0 \\ 0 & 0 & 0 & 1 \\ 0 & 1 & 0 & 0 \end{pmatrix} \begin{pmatrix} \text{row}_1 \\ \text{row}_2 \\ \text{row}_3 \\ \text{row}_4 \end{pmatrix} = \begin{pmatrix} \text{row}_3 \\ \text{row}_1 \\ \text{row}_4 \\ \text{row}_2 \end{pmatrix}.$$

A permutation matrix satisfies $P^T P = I$, so that $P^{-1} = P^T$.

The matrix L records the steps in the elimination process. It is built up from simpler pieces, each of which corresponds to the addition of a multiple of one equation to another. For example,

$$\begin{pmatrix} 1 & 0 & 0 & 0 \\ 0 & 1 & 0 & 0 \\ 0 & 5 & 1 & 0 \\ 0 & -3 & 0 & 1 \end{pmatrix} \begin{pmatrix} \text{row}_1 \\ \text{row}_2 \\ \text{row}_3 \\ \text{row}_4 \end{pmatrix} = \begin{pmatrix} \text{row}_1 \\ \text{row}_2 \\ \text{row}_3 + 5\,\text{row}_2 \\ \text{row}_4 - 3\,\text{row}_2 \end{pmatrix}.$$

Each entry below the main diagonal in L corresponds to an elimination step.

We will illustrate this factorization using the 3×3 example

$$A = \begin{pmatrix} 1.6 & -4.2 & -0.8 \\ 4.0 & 1.5 & 3.0 \\ 8.0 & -1.0 & 1.0 \end{pmatrix}$$

$$= \begin{pmatrix} 0 & 1 & 0 \\ 0 & 0 & 1 \\ 1 & 0 & 0 \end{pmatrix} \begin{pmatrix} 1 & 0 & 0 \\ 0.2 & 1 & 0 \\ 0.5 & -0.5 & 1 \end{pmatrix} \begin{pmatrix} 8 & -1 & 1 \\ 0 & -4 & -1 \\ 0 & 0 & 2 \end{pmatrix} = PLU.$$

The coefficients in U are the coefficients from the final upper triangular system obtained after elimination. The coefficients in L are the multipliers from the elimination.

Before discussing the usage of the factorization, let us look in more detail at how it is obtained. If partial pivoting is used, then the first and third rows are interchanged via

$$P_1 A = \begin{pmatrix} 8.0 & -1.0 & 1.0 \\ 4.0 & 1.5 & 3.0 \\ 1.6 & -4.2 & -0.8 \end{pmatrix}, \quad \text{where} \quad P_1 = \begin{pmatrix} 0 & 0 & 1 \\ 0 & 1 & 0 \\ 1 & 0 & 0 \end{pmatrix}.$$

At the first stage of elimination we add $-4.0/8.0 = -0.5$ times the first equation to the second equation, and add $-1.6/8.0 = -0.2$ times the first equation to the third equation. These operations can be represented in matrix form as

$$L_1(P_1 A) = \begin{pmatrix} 1 & 0 & 0 \\ -0.5 & 1 & 0 \\ -0.2 & 0 & 1 \end{pmatrix} \begin{pmatrix} 8.0 & -1.0 & 1.0 \\ 4.0 & 1.5 & 3.0 \\ 1.6 & -4.2 & -0.8 \end{pmatrix}$$

$$= \begin{pmatrix} 8.0 & -1.0 & 1.0 \\ 0 & 2.0 & 2.5 \\ 0 & -4.0 & -1.0 \end{pmatrix}.$$

At the second and final step, the second and third rows are interchanged:

$$P_2(L_1 P_1 A) = \begin{pmatrix} 8.0 & -1.0 & 1.0 \\ 0 & -4.0 & -1.0 \\ 0 & 2.0 & 2.5 \end{pmatrix}, \quad \text{where} \quad P_2 = \begin{pmatrix} 1 & 0 & 0 \\ 0 & 0 & 1 \\ 0 & 1 & 0 \end{pmatrix}.$$

Then elimination is performed using

$$L_2(P_2 L_1 P_1 A) = \begin{pmatrix} 1 & 0 & 0 \\ 0 & 1 & 0 \\ 0 & 0.5 & 1 \end{pmatrix} \begin{pmatrix} 8.0 & -1.0 & 1.0 \\ 0 & -4.0 & -1.0 \\ 0 & 2.0 & 2.5 \end{pmatrix} = \begin{pmatrix} 8 & -1 & 1 \\ 0 & -4 & -1 \\ 0 & 0 & 2 \end{pmatrix}.$$

Combining all of these steps together gives

$$L_2 P_2 L_1 P_1 A = U$$

or, since the matrices $\{ P_i \}$ are symmetric,

$$A = (P_1^T L_1^{-1} P_2^T L_2^{-1})U = (P_1 L_1^{-1} P_2 L_2^{-1})U$$

with

$$L_1^{-1} = \begin{pmatrix} 1 & 0 & 0 \\ 0.5 & 1 & 0 \\ 0.2 & 0 & 1 \end{pmatrix} \quad \text{and} \quad L_2^{-1} = \begin{pmatrix} 1 & 0 & 0 \\ 0 & 1 & 0 \\ 0 & -0.5 & 1 \end{pmatrix}.$$

If desired, these transformations can be combined into the more compressed form $A = PLU$, but this is not necessary for purposes of calculation. For this example, it is straightforward to check that $P_1 L_1^{-1} P_2 L_2^{-1} = PL$, for the matrices P and L defined earlier.

We now return to our discussion of the factorization and its usage. It is possible to solve the system of equations $Ax = b$ directly from this factorization by computing temporary vectors y and z as follows:

$$Ax = PL\underbrace{\underbrace{Ux}_{y} = b}_{z}.$$

First we solve for z in

$$Pz = b.$$

This just corresponds to interchanging the elements in b (the vector z is just a temporary result). Then we solve for y in

$$Ly = z.$$

This system is lower triangular and so it can be solved via "forward" substitution, analogous to backsubstitution (again, y is just a temporary result). Finally, we solve for x in

$$Ux = y$$

via backsubstitution to obtain the solution. Combining the steps gives

$$x = U^{-1}y = U^{-1}L^{-1}z = U^{-1}L^{-1}P^{-1}b$$
$$= (PLU)^{-1}b = A^{-1}b,$$

as expected. If the factorization has already been computed, the cost of these three steps is $O(n^2)$.

It might seem that the storage costs for the factorization would be expensive, since now three matrices must be stored, but this is not necessary. The diagonal entries in L are all equal to 1, and so need not be stored. The remaining entries in L can be stored in the lower triangle of U, which has only zero entries. The matrix P records only the interchanges of rows, and so can be represented by an integer work array that stores the permutation required. Hence the factorization can be overwritten on the original matrix, if an auxiliary integer work array is provided.

There are a number of advantages to this approach. First, if several systems of equations must be solved, all having the same matrix A (the same general model) but different right-hand sides $b^{(j)}$ (different data), then the matrix A need only be factored once, and then this three-step procedure can be applied to each of the right-hand sides in turn. Since the factorization of A is much more expensive ($O(n^3)$) than the three-step procedure ($O(n^2)$) this can result in significant savings. Second, the factorization can be used to obtain error bounds on the solution (see the discussion of sensitivity below). Third, the factorization can sometimes be used to check if the matrix A has desired properties such as positive-definiteness (see the discussion of other matrix factorizations below).

Since it is not too difficult to compute the inverse of a triangular matrix, it would be possible to use the factorization to compute A^{-1} via $A^{-1} = U^{-1}L^{-1}P^{-1}$. This is not normally recommended, for the reasons discussed in Appendix A.4. We repeat this here because it is now possible to give an example that makes clear the disadvantage of forming the inverse. Consider the matrix

$$A = \begin{pmatrix} 4 & 1 & 0 & 0 & 0 & 0 \\ 1 & 4 & 1 & 0 & 0 & 0 \\ 0 & 1 & 4 & 1 & 0 & 0 \\ 0 & 0 & 1 & 4 & 1 & 0 \\ 0 & 0 & 0 & 1 & 4 & 1 \\ 0 & 0 & 0 & 0 & 1 & 4 \end{pmatrix}.$$

If Gaussian elimination is applied to A, then we obtain the factors $P = I$,

$$L = \begin{pmatrix} 1.0000 & 0 & 0 & 0 & 0 & 0 \\ 0.2500 & 1.0000 & 0 & 0 & 0 & 0 \\ 0 & 0.2667 & 1.0000 & 0 & 0 & 0 \\ 0 & 0 & 0.2679 & 1.0000 & 0 & 0 \\ 0 & 0 & 0 & 0.2679 & 1.0000 & 0 \\ 0 & 0 & 0 & 0 & 0.2679 & 1.0000 \end{pmatrix}$$

and

$$U = \begin{pmatrix} 4.0000 & 1.0000 & 0 & 0 & 0 & 0 \\ 0 & 3.7500 & 1.0000 & 0 & 0 & 0 \\ 0 & 0 & 3.7333 & 1.0000 & 0 & 0 \\ 0 & 0 & 0 & 3.7321 & 1.0000 & 0 \\ 0 & 0 & 0 & 0 & 3.7321 & 1.0000 \\ 0 & 0 & 0 & 0 & 0 & 3.7321 \end{pmatrix}.$$

Notice that L and U have the same pattern of nonzero entries as A. However, the inverse of A has the form

$$A^{-1} = \begin{pmatrix} 0.2679 & -0.0718 & 0.0192 & -0.0052 & 0.0014 & -0.0003 \\ -0.0718 & 0.2872 & -0.0769 & 0.0206 & -0.0055 & 0.0014 \\ 0.0192 & -0.0769 & 0.2886 & -0.0773 & 0.0206 & -0.0052 \\ -0.0052 & 0.0206 & -0.0773 & 0.2886 & -0.0769 & 0.0192 \\ 0.0014 & -0.0055 & 0.0206 & -0.0769 & 0.2872 & -0.0718 \\ -0.0003 & 0.0014 & -0.0052 & 0.0192 & -0.0718 & 0.2679 \end{pmatrix},$$

and the pattern of nonzero entries is destroyed. Hence, in terms of both storage and computation, A^{-1} is more expensive to work with than the factors L and U.

Care must be taken when factoring sparse matrices. For

$$A = \begin{pmatrix} 4 & 1 & 1 & 1 & 1 \\ 1 & 4 & 0 & 0 & 0 \\ 1 & 0 & 4 & 0 & 0 \\ 1 & 0 & 0 & 4 & 0 \\ 1 & 0 & 0 & 0 & 4 \end{pmatrix}$$

the LU factors are

$$L = \begin{pmatrix} 1 & 0 & 0 & 0 & 0 \\ 0.25 & 1 & 0 & 0 & 0 \\ 0.25 & -0.0667 & 1 & 0 & 0 \\ 0.25 & -0.0667 & -0.0714 & 1 & 0 \\ 0.25 & -0.0667 & -0.0714 & -0.0769 & 1 \end{pmatrix}$$

and

$$U = \begin{pmatrix} 4 & 1 & 1 & 1 & 1 \\ 0 & 3.7500 & -0.2500 & -0.2500 & -0.2500 \\ 0 & 0 & 3.7333 & -0.2667 & -0.2667 \\ 0 & 0 & 0 & 3.7143 & -0.2857 \\ 0 & 0 & 0 & 0 & 3.6923 \end{pmatrix}.$$

(No row interchanges are required in this example.) Notice that the triangular factors L and U are dense even though the original matrix is sparse. If A is reordered by interchanging the first and last rows and columns, then we obtain the matrix

$$\hat{A} = \begin{pmatrix} 4 & 0 & 0 & 0 & 1 \\ 0 & 4 & 0 & 0 & 1 \\ 0 & 0 & 4 & 0 & 1 \\ 0 & 0 & 0 & 4 & 1 \\ 1 & 1 & 1 & 1 & 4 \end{pmatrix}$$

whose factors are

$$\hat{L} = \begin{pmatrix} 1 & 0 & 0 & 0 & 0 \\ 0 & 1 & 0 & 0 & 0 \\ 0 & 0 & 1 & 0 & 0 \\ 0 & 0 & 0 & 1 & 0 \\ 0.25 & 0.25 & 0.25 & 0.25 & 1 \end{pmatrix}, \quad \hat{U} = \begin{pmatrix} 4 & 0 & 0 & 0 & 1 \\ 0 & 4 & 0 & 0 & 1 \\ 0 & 0 & 4 & 0 & 1 \\ 0 & 0 & 0 & 4 & 1 \\ 0 & 0 & 0 & 0 & 3 \end{pmatrix}.$$

The permuted matrix has sparse factors.

Sparse matrix software attempts to reorder the rows and columns of a matrix so that the factors are as sparse as possible. The techniques used to do this are heuristic, but they are successful on a wide range of practical problems. These reordering schemes often must make a trade-off between the sparsity of the factors and the accuracy of the computed solution to the corresponding linear system, accepting smaller pivot entries than would be tolerated for dense matrices. Some examples of reordering schemes for sparse matrices are described in the papers by Markowitz (1957) and Suhl and Suhl (1990). We will describe here briefly the technique due to Markowitz.

For the sparse matrix A above, the factors L and U had nonzero entries in positions that were zero in the matrix A. The number of such entries is referred to as the amount of *fill-in* that occurs during elimination. The Markowitz technique attempts to reduce the amount of fill-in during the entire elimination process by reducing the amount of fill-in at every step of the process. This is only a heuristic technique since decisions are only made one step at a time without regard to what might happen several steps later.

At the first step of elimination every entry in the matrix is considered as a candidate for the pivot entry. We calculate r_i, the number of nonzero entries in row i, and c_j, the number of nonzero entries in column j. The pivot entry is chosen as the entry $a_{i,j} \neq 0$ in A for which

$$(r_i - 1)(c_j - 1)$$

is minimal. Then row i is interchanged with row 1, and column j is interchanged with column 1, to bring $a_{i,j}$ into the $(1, 1)$ position of A. Elimination is performed, and the process repeated and applied to the remaining $n - 1$ columns and rows of the transformed matrix A.

The quantity $(r_i - 1)(c_j - 1)$ is an upper bound on the amount of fill-in that will occur if $a_{i,j}$ is chosen as a pivot entry. If $a_{i,j}$ is ignored, there are $r_i - 1$ remaining nonzero entries in row i, and $c_j - 1$ nonzero entries in column j. Each of these $c_j - 1$ entries would have to be eliminated, and each of these eliminations could introduce up to $r_i - 1$ new nonzero entries into the corresponding row.

There is one other detail that is commonly incorporated into the choice of the pivot entry. There is a danger that the pivot entry $a_{i,j}$ will be close to zero, leading to numerical difficulties. It is useful to insist that

$$|a_{i,j}| \geq \epsilon,$$

where ϵ is some tolerance for zero that may depend on the other entries in A. If this test is violated, then $a_{i,j}$ is rejected as a pivot entry, and the "next best" candidate pivot entry is considered. ("Next best" is measured in terms of $(r_i - 1)(c_j - 1)$.) If the tolerance ϵ is small, then it is less likely that candidate pivot entries will be rejected.

If no pivot entries are rejected, then the choice of pivot entries depends only on the sparsity pattern of the matrix (the pattern of zero and nonzero entries). This means that the elimination scheme can be analyzed prior to performing elimination, and a storage scheme set up to handle all the interchanges and fill-in that will occur. This is referred to as a *symbolic* factorization. The sparsity pattern of a matrix can be represented in a condensed form and can be manipulated more efficiently than the data for the matrix itself. With the symbolic factorization available, the actual *numerical* factorization can be performed to solve the linear system. Because of the possibility of small pivots, the pivot choices made during the symbolic factorization might have to be modified, with the potential for increases in the amount of computation and storage required.

Example A.4 (Sparse Matrix Factorization). Consider the sparse matrix

$$A = \begin{pmatrix} 3 & 1 & 0 & 0 & 0 \\ 0 & 0 & 1 & 2 & 1 \\ 4 & 0 & 0 & 0 & 2 \\ 0 & 2 & 0 & 0 & 0 \\ 0 & 0 & 5 & 1 & 3 \end{pmatrix}.$$

The number of nonzero entries in each row is given by the array

$$r = (2 \quad 3 \quad 2 \quad 1 \quad 3)$$

and the number of nonzero entries in each column is given by

$$c = (2 \quad 2 \quad 2 \quad 2 \quad 3).$$

We can calculate $(r_i - 1)(c_j - 1)$ for each nonzero entry in A:

$$\begin{pmatrix} 1 & 1 & - & - & - \\ - & - & 2 & 2 & 4 \\ 1 & - & - & - & 2 \\ - & 0 & - & - & - \\ - & - & 2 & 2 & 4 \end{pmatrix}.$$

The smallest of these values is 0 and occurs at position $(4, 2)$, so the pivot entry is $a_{4,2}$. This result predicts that no fill-in will occur if this pivot entry is used. If we interchange row 4 with row 1, and column 2 with column 1, then A is transformed into

$$A = \begin{pmatrix} 2 & 0 & 0 & 0 & 0 \\ 0 & 0 & 1 & 2 & 1 \\ 0 & 4 & 0 & 0 & 2 \\ 1 & 3 & 0 & 0 & 0 \\ 0 & 0 & 5 & 1 & 3 \end{pmatrix}.$$

After elimination we obtain

$$A = \begin{pmatrix} 2 & 0 & 0 & 0 & 0 \\ 0 & 0 & 1 & 2 & 1 \\ 0 & 4 & 0 & 0 & 2 \\ 0 & 3 & 0 & 0 & 0 \\ 0 & 0 & 5 & 1 & 3 \end{pmatrix}.$$

Now the process repeats. The number of nonzero entries in the remaining rows and columns are given by the arrays

$$r = (-\quad 3 \quad 2 \quad 1 \quad 3) \quad \text{and} \quad c = (-\quad 2 \quad 2 \quad 2 \quad 3).$$

Again, we calculate $(r_i - 1)(c_j - 1)$ for each nonzero entry in A:

$$\begin{pmatrix} - & - & - & - & - \\ - & - & 2 & 2 & 4 \\ - & 1 & - & - & 2 \\ - & 0 & - & - & - \\ - & - & 2 & 2 & 4 \end{pmatrix}.$$

The smallest value is 0 and occurs at position $(4, 2)$, so the pivot entry is $a_{4,2}$. And so on. Note that, if $|a_{4,2}|$ had been small, this entry would have been rejected as a pivot entry, and $a_{3,2}$ considered instead. ∎

A.6.1 Sparse Matrix Storage

The costs of storing and factoring a sparse matrix are somewhat higher than was suggested above, because there is overhead associated with sparse matrices. This overhead is a consequence of the special storage techniques used to represent sparse matrices on a computer.

Before discussing sparse matrix storage, let us look at a dense matrix

$$\begin{pmatrix} 1 & 2 & 4 \\ 1 & 3 & 9 \\ 1 & 5 & 25 \end{pmatrix}.$$

A dense matrix is stored on the computer one entry after another. In FORTRAN this is done one column at a time:

$$1, 1, 1, 2, 3, 5, 4, 9, 25.$$

Nine storage locations are needed to store the nine entries in the matrix.

Now consider the sparse matrix

$$A = \begin{pmatrix} 1 & 3 & 0 & 0 & 0 \\ 2 & 0 & 5 & 0 & 0 \\ 0 & 4 & 0 & 0 & 8 \\ 0 & 0 & 6 & 0 & 0 \\ 0 & 0 & 0 & 7 & 9 \end{pmatrix}.$$

If this matrix were written out as a dense matrix, 25 storage locations would be needed. If we just stored the nonzero entries in the array AVAL,

$$\text{AVAL} = \{\, 1, 2, 3, 4, 5, 6, 7, 8, 9 \,\},$$

only nine locations would be needed, but it would not be possible to reconstruct the matrix, since the locations of these entries have not been specified. Clearly a more elaborate storage scheme must be used.

We will describe one technique for storing a sparse matrix. As with the dense matrix, we will store it one column at a time. For each column we will indicate the number of nonzero entries in that column. For each entry we will record the row of the entry and its value. This example would be represented as

$$
\begin{aligned}
&\text{Column 1}: \quad 2, (1:1), (2:2) \\
&\text{Column 2}: \quad 2, (1:3), (3:4) \\
&\text{Column 3}: \quad 2, (2:5), (4:6) \\
&\text{Column 4}: \quad 1, (5:7) \\
&\text{Column 5}: \quad 2, (3:8), (5:9).
\end{aligned}
$$

In the computer, this information could be represented using three arrays: the array AVAL above, an array NCOL that records the number of nonzero entries in each column, and an array IROW that records the rows where these entries occur:

$$
\begin{aligned}
\text{AVAL} &= \{\, 1, 2, 3, 4, 5, 6, 7, 8, 9 \,\} \\
\text{NCOL} &= \{\, 2, 2, 2, 1, 2 \,\} \\
\text{IROW} &= \{\, 1, 2, 1, 3, 2, 4, 5, 3, 5 \,\}.
\end{aligned}
$$

Some effort is required to identify an entry in the matrix. To identify the entries in column 3 of the matrix, we first look at array NCOL to determine that these entries are in positions 5 and 6 or AVAL (NCOL indicates that there are two entries each in columns 1 and 2 of A, and hence the entries for column 3 start in positions 5; since there are two such entries, they are in positions 5 and 6). Then from positions 5 and 6 of IROW we know that these entries are in rows 2 and 4. Hence $A_{2,3} = 5$ and $A_{4,3} = 6$.

The sparse representation of this matrix requires 23 storage locations, 14 integer locations for the indexing information, and 9 real locations for the numbers themselves. (Integer storage can require less space than real storage on some computers.) Since the dense representation requires only 25 entries, this is not much of a savings. However, if this matrix were 100×100 with at most 2 nonzero entries per column, then the savings would be more dramatic. The dense representation of such a matrix would require $100^2 = 10,000$ locations, whereas the sparse representation would require at most 500, of which 300 would be integer locations.

A.7 Other Matrix Factorizations

We have previously discussed the LU factorization of a square matrix and the spectral decomposition of a symmetric matrix. There are three other matrix factorizations that are of use to us as we examine various optimization problems. Two are minor modifications to the

LU factorization discussed in the previous section and are used to factor positive definite matrices. (A brief discussion of positive-definite matrices is given below.) The third is an orthogonal factorization of a (possibly rectangular) matrix that is based on different techniques.

A.7.1   Positive-Definite Matrices

A symmetric matrix A is positive definite if and only if

$$x^T A x > 0$$

for all nonzero vectors x. This can be a difficult condition to verify, but there are equivalent definitions that are sometimes more practical. For example, A will be positive definite if all of its eigenvalues are positive. Also, if Gaussian elimination is applied to A *without pivoting* to transform A into upper triangular form

$$A \to U = \begin{pmatrix} u_{1,1} & u_{1,2} & \cdots & u_{1,n} \\ & u_{2,2} & \cdots & u_{2,n} \\ & & \ddots & \vdots \\ 0 & & & u_{n,n} \end{pmatrix}$$

and if $u_{i,i} > 0$ for all i, then A is positive definite.

Similarly a symmetric matrix A is

- *positive semidefinite* if $x^T A x \geq 0$ for all x (or equivalently, all the eigenvalues of A are nonnegative),
- *negative definite* if $x^T A x < 0$ for all $x \neq 0$ (all the eigenvalues of A are negative),
- *negative semidefinite* if $x^T A x \leq 0$ for all x (all the eigenvalues of A are nonpositive),
- *indefinite* if $x^T A x$ can take on both positive and negative values (A has both positive and negative eigenvalues).

A positive-definite matrix is automatically positive semidefinite, and likewise a negative-definite matrix is automatically negative semidefinite. Examples of these definitions are given below.

Example A.5 (Positive-Definite Matrices). Consider the matrix

$$A = \begin{pmatrix} 4 & 1 & 0 \\ 1 & 5 & 2 \\ 0 & 2 & 6 \end{pmatrix}.$$

Then

$$x^T A x = 4x_1^2 + 2x_1 x_2 + 5x_2^2 + 4x_2 x_3 + 6x_3^2$$
$$= 3x_1^2 + 3x_2^2 + 2x_3^2 + (x_1 + x_2)^2 + (x_2 + 2x_3)^2 > 0$$

if $x \neq 0$. So A is positive definite. If Gaussian elimination is applied to A, then

$$A \to U = \begin{pmatrix} 4 & 1 & 0 \\ 0 & 4.75 & 2 \\ 0 & 0 & 5.1579 \end{pmatrix}$$

and all the diagonal elements of U are positive. The eigenvalues of A,

$$\{\, 2.8549, 4.4760, 7.6691 \,\},$$

are also all positive.

The matrix

$$B = \begin{pmatrix} 4 & 2 & 0 \\ 2 & 6 & 2 \\ 0 & 2 & \frac{4}{5} \end{pmatrix}$$

is positive semidefinite but not positive definite. To see this, form

$$\begin{aligned} x^T B x &= 4x_1^2 + 4x_1 x_2 + 6x_2^2 + 4x_2 x_3 + \tfrac{4}{5}x_3^2 \\ &= (2x_1 + x_2)^2 + 5(x_2 + \tfrac{2}{5}x_3)^2 \ge 0. \end{aligned}$$

If $x = (\tfrac{1}{5}, -\tfrac{2}{5}, 1)^T$, then $x^T B x = 0$. When Gaussian elimination is applied to B,

$$B \rightarrow U = \begin{pmatrix} 4 & 2 & 0 \\ 0 & 5 & 2 \\ 0 & 0 & 0 \end{pmatrix}$$

and the diagonal entries in U are nonnegative, with $u_{3,3} = 0$. The eigenvalues of B are $\{\, 0, 3.1284, 7.6716 \,\}$.

The matrix

$$C = \begin{pmatrix} 3 & 5 & 0 \\ 5 & 4 & 7 \\ 0 & 7 & 2 \end{pmatrix}$$

is indefinite. For $x = (1, 0, 0)^T$, we have $x^T C x = 3 > 0$, but for $x = (1, -1, 0)^T$, we have $x^T C x = -3 < 0$. Gaussian elimination applied to C produces

$$C \rightarrow U = \begin{pmatrix} 3 & 5 & 0 \\ 0 & -4.3333 & 7 \\ 0 & 0 & 13.3077 \end{pmatrix}$$

and the matrix U has both positive and negative diagonal entries. The eigenvalues of C are also both positive and negative: $\{\, -5.4885, 2.6662, 11.8223 \,\}$.

The matrix $-A$ is negative definite, and $-B$ is negative semidefinite.   ∎

A.7.2   The LDL^T and Cholesky Factorizations

These two matrix factorizations are primarily of interest when factoring a positive-definite matrix A, although variants of the LDL^T factorization can be applied to more general symmetric matrices. They are used in nonlinear optimization problems to represent the Hessian matrix of the objective function. If A is symmetric and positive definite, it can be shown that Gaussian elimination can always be applied *without* partial pivoting, with no danger of the method trying to divide by zero, and with no danger of near-zero pivots that can lead to numerical difficulties.

If no row interchanges are used, the LU factorization takes the form

$$A = LU.$$

The first two factorizations are obtained by manipulating this formula.

Let D be the diagonal matrix whose entries are the diagonal entries of U: $d_{i,i} = u_{i,i}$. Then define $\hat{U} = D^{-1}U$ so that $D\hat{U} = U$. Hence $A = LD\hat{U}$. If A is positive definite, then it is also symmetric, so that

$$A^T = \hat{U}^T D^T L^T = \hat{U}^T D L^T = A = LD\hat{U}.$$

It is then easy to verify that $\hat{U} = L^T$, so that

$$A = LDL^T.$$

This is the first of the new factorizations, a factorization of A into the product of a lower triangular matrix, a diagonal matrix, and the transpose of the lower triangular matrix.

Slightly more can be deduced. If A is positive definite, then $x^T A x > 0$ for all $x \neq 0$. Using the factorization,

$$0 < x^T A x = x^T L D L^T x = (L^T x)^T D (L^T x) \equiv y^T D y,$$

where $y = L^T x$. Since L is nonsingular (it is triangular and all of its diagonal entries are equal to 1), $y \neq 0$ if and only if $x \neq 0$. Hence

$$0 < y^T D y = \sum_i d_{i,i} y_i^2$$

for all $y \neq 0$. This can happen only if $d_{i,i} > 0$ for all i. Hence D is a diagonal matrix with positive diagonal entries. It should be noted that the reverse is also true; i.e., if A can be represented as $A = LDL^T$, where D has positive diagonal entries, then A must be symmetric and positive definite. If we discover that $d_{i,i} \leq 0$ at some stage during the computation of the factorization, then A is not positive definite. This property will be useful when we apply Newton's method to multidimensional optimization problems (see Section 11.4).

Example A.6 ($A = LDL^T$). To illustrate this factorization, consider the positive-definite matrix

$$A = \begin{pmatrix} 4 & -2 & 2 \\ -2 & 2 & -3 \\ 2 & -3 & 14 \end{pmatrix}.$$

Then A can be represented as LDL^T, where

$$L = \begin{pmatrix} 1 & 0 & 0 \\ -0.5 & 1 & 0 \\ 0.5 & -2 & 1 \end{pmatrix} \quad \text{and} \quad D = \begin{pmatrix} 4 & 0 & 0 \\ 0 & 1 & 0 \\ 0 & 0 & 9 \end{pmatrix}. \quad \blacksquare$$

The second factorization is obtained easily from the first. Since D has positive diagonal entries we can write

$$D = \hat{D}\hat{D},$$

where $\hat{D}$ is a diagonal matrix with $\hat{d}_{i,i} = \sqrt{d_{i,i}}$. If we then define

$$\hat{L} = L\hat{D},$$

then $\hat{L}$ is also a lower triangular matrix, and

$$A = \hat{L}\hat{L}^T.$$

The $\hat{}$ is often omitted, and we simply write $A = LL^T$. This is the second of the factorizations, also referred to as a *Cholesky* factorization.

Example A.7 (LDL^T to Cholesky). We can modify the above factorization to obtain

$$\hat{D} = \begin{pmatrix} 2 & 0 & 0 \\ 0 & 1 & 0 \\ 0 & 0 & 3 \end{pmatrix}$$

and

$$\hat{L} = L\hat{D} = \begin{pmatrix} 2 & 0 & 0 \\ -1 & 1 & 0 \\ 1 & -2 & 3 \end{pmatrix},$$

so that

$$A = \hat{L}\hat{L}^T = \begin{pmatrix} 2 & 0 & 0 \\ -1 & 1 & 0 \\ 1 & -2 & 3 \end{pmatrix} \begin{pmatrix} 2 & -1 & 1 \\ 0 & 1 & -2 \\ 0 & 0 & 3 \end{pmatrix}. \qquad \blacksquare$$

It is also possible to compute the coefficients of the Cholesky factorization directly, by comparing coefficients of A and LL^T.

Example A.8 (Cholesky Factorization).   We compare the coefficients of A with those of LL^T:

$$
\begin{aligned}
a_{11} &= l_{11}^2 & &\rightarrow & l_{11} &= \sqrt{a_{1,1}} = \sqrt{4} = 2 \\
a_{2,1} &= l_{2,1}l_{11} & &\rightarrow & l_{2,1} &= a_{2,1}/l_{1,1} = -2/2 = -1 \\
a_{2,2} &= l_{2,1}^2 + l_{2,2}^2 & &\rightarrow & l_{2,2} &= \sqrt{a_{2,2} - l_{2,1}^2} = \sqrt{2-1} = 1 \\
a_{3,1} &= l_{3,1}l_{11} & &\rightarrow & l_{3,1} &= a_{3,1}/l_{3,1} = 2/1 = 1 \\
a_{3,2} &= l_{3,1}l_{1,1} + l_{3,2}l_{2,2} & &\rightarrow & l_{3,2} &= (a_{3,2} - l_{3,1}l_{1,1})/l_{2,2} = -2 \\
a_{3,3} &= l_{3,1}^2 + l_{3,2}^2 + l_{3,3}^2 & &\rightarrow & l_{3,3} &= \sqrt{a_{3,3} - l_{3,1}^2 - l_{3,2}^2} = \sqrt{14 - 1 - 4} = 3. \qquad \blacksquare
\end{aligned}
$$

Because of the symmetry in the matrix A, the computational costs of the LDL^T and Cholesky factorizations are about half those of the PLU factorization of a general matrix.

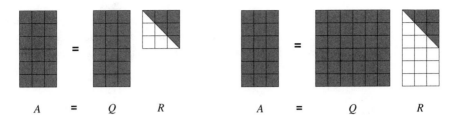

$$A \quad = \quad Q \qquad R \qquad\qquad A \quad = \quad Q \qquad R$$

Figure A.1. *Forms of orthogonal factorizations.*

A.7.3 An Orthogonal Matrix Factorization

The third matrix factorization is based on entirely different principles and applies in a quite different setting. It uses orthogonal transformations and can be applied to any matrix. The matrix A need not have an inverse; in fact, the matrix need not even be square. It is typically used when A is rectangular, in particular, when A has more rows than columns. Its most common usage is in the context of least-squares problems, but it is frequently used to represent the linear constraints in an optimization problem. It is called a QR factorization because it represents A as

$$A = QR,$$

where Q is an orthogonal matrix (i.e., $Q^T Q = I$) and R is an upper triangular (or "right" triangular) matrix. There is a slight ambiguity in the definition of the QR factorization. If A is an $m \times n$ matrix and $m > n$, then there are two forms of this factorization. The first chooses Q to be the same size as A, and then R is $n \times n$; the second chooses R to be the same size as A, and then Q is $m \times m$. See Figure A.1. The first form is all that is normally required to solve least-squares problems; the second form is often more useful when solving constrained optimization problems.

There are three techniques that are commonly used to compute the QR factorization of a matrix A. In the absence of rounding errors, they all can be considered equivalent, but on a computer there are important differences in their properties. The oldest, and the one most familiar to mathematicians, is called the *Gram–Schmidt orthogonalization process*. Its emphasis is on the orthogonal matrix Q, and it is often described without the triangular matrix R being mentioned. It should be noted that the traditional formulas for Gram–Schmidt orthogonalization have undesirable computational properties, but that a slight modification of the formulas is safe to use. The other two techniques are closer in spirit to Gaussian elimination in that they introduce zeroes into the matrix A so as to transform it into an upper triangular matrix. The version based on Givens rotations introduces one zero at a time, while the version based on Householder reflections introduces a whole column of zeroes at a time. We will only describe in detail the Householder approach.

Orthogonal matrix factorizations are most often used when solving least-squares problems. To see why, consider a least-squares problem written in the form

$$\text{minimize } f(x) = \| Ax - b \|_2^2 \,,$$

where A is an $m \times n$ matrix with $m \geq n$. Least-squares problems cannot be solved by applying elimination to the matrix A. When linear systems of equations are solved, the techniques of elimination result in a sequence of *equivalent* linear systems. Elimination is not applied to least-squares problems because the techniques of elimination do not leave the least-squares problem unchanged. However, if P is an orthogonal matrix so that $P^T P = I$, then

$$\| Py \|_2^2 = (Py)^T(Py) = y^T P^T P y = y^T y = \| y \|_2^2 \,;$$

that is, an orthogonal transformation does not affect the 2-norm of a vector. Hence

$$\| Ax - b \|_2^2 = \| P(Ax - b) \|_2^2$$

and so orthogonal transformations can be used to generate a sequence of equivalent least-squares problems.

Just as with linear equations, it is easy to solve least-squares problems when A is upper triangular. This is illustrated in the following example.

Example A.9 (Triangular Least-Squares Problems). Consider the least-squares problem

$$\text{minimize } f(x) = \| Ax - b \|_2^2$$

with

$$A = \begin{pmatrix} 5 & 2 & 1 \\ 0 & 3 & 4 \\ 0 & 0 & 2 \\ 0 & 0 & 0 \end{pmatrix}, \quad b = \begin{pmatrix} 13 \\ 17 \\ 4 \\ 12 \end{pmatrix}.$$

This is equivalent to solving

$$\text{minimize}_x \ (5x_1 + 2x_2 + 1x_3 - 13)^2 + (3x_2 + 4x_3 - 17)^2 + (2x_3 - 4)^2 + (12)^2.$$

The solution can be obtained via backsubstitution. This is a sum of squared terms, and the smallest value each term can achieve is zero. The term $(12)^2$ remains unchanged no matter what the values of the x_i are. The other three terms can be made equal to zero by solving the triangular system of equations

$$\begin{aligned} 5x_1 + 2x_2 + 1x_3 &= 13 \\ 3x_2 + 4x_3 &= 17 \\ 2x_3 &= 4. \end{aligned}$$

The solution is $x = (x_1, x_2, x_3)^T = (1, 3, 2)^T$. ∎

If the matrix A has been factored as $A = QR$, where $Q^T Q = I$ and R is upper triangular, then

$$\| Ax - b \|_2^2 = \| QRx - b \|_2^2 = \left\| Rx - (Q^T b) \right\|_2^2.$$

Hence the QR factorization allows us to transform a general least-squares problem into a triangular least-squares problem that can be solved via backsubstitution. Techniques for forming the QR factorization are discussed in the book by Golub and Van Loan (1996). An example of a QR factorization is given below.

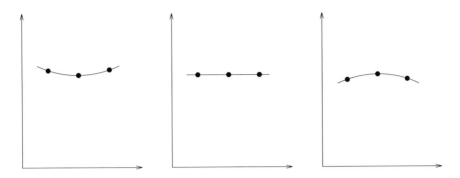

Figure A.2. *Fitting three quadratic functions.*

Example A.10 (The QR Factorization). Let

$$A = \begin{pmatrix} 3 & 4 & 2 \\ 2 & 7 & 1 \\ 5 & 3 & 6 \\ 2 & 9 & 1 \end{pmatrix}.$$

Then $A = QR$, where

$$Q = P_{(1)} P_{(2)} P_{(3)} = \begin{pmatrix} -0.4629 & 0.0252 & 0.8526 & 0.2412 \\ -0.3086 & -0.4934 & 0.0761 & -0.8096 \\ -0.7715 & 0.4738 & -0.4232 & -0.0345 \\ -0.3086 & -0.7290 & -0.2970 & 0.5340 \end{pmatrix}$$

and

$$R = A_{(3)} = P_{(3)} A_{(2)} = \begin{pmatrix} -6.4807 & -9.1039 & -6.1721 \\ 0 & -8.4923 & 1.6710 \\ 0 & 0 & -1.0548 \\ 0 & 0 & 0 \end{pmatrix}.$$

It can be verified that $Q^T Q = I$ (except for rounding errors).   ∎

A.8   Sensitivity (Conditioning)

If we had the ability to compute exactly, with no rounding errors, then systems of linear equations could be divided cleanly into those that were singular and those that were nonsingular. When data and calculations are inaccurate, however, there is not such a clear distinction between singular and nonsingular systems.

Before giving a precise discussion of conditioning let us look at a set of examples. They are variants of the data-fitting problem in Chapter 1. Consider fitting a quadratic function to the three data sets illustrated in Figure A.2. The only difference between the three data sets is the position of the middle data point.

The coefficient matrix for this problem is

$$A = \begin{pmatrix} 1 & 0.01 & 0.0001 \\ 1 & 0.02 & 0.0004 \\ 1 & 0.03 & 0.0009 \end{pmatrix}$$

and the three sets of right-hand sides are

$$b_{(1)} = \begin{pmatrix} 1 \\ 0.99 \\ 1 \end{pmatrix}, \quad b_{(2)} = \begin{pmatrix} 1 \\ 1 \\ 1 \end{pmatrix}, \quad b_{(3)} = \begin{pmatrix} 1 \\ 1.01 \\ 1 \end{pmatrix}.$$

The three sets of coefficients are

$$x_{(1)} = \begin{pmatrix} 1.03 \\ -4.00 \\ 100.00 \end{pmatrix}, \quad x_{(2)} = \begin{pmatrix} 1 \\ 0 \\ 0 \end{pmatrix}, \quad x_{(3)} = \begin{pmatrix} 0.97 \\ 4.00 \\ -100.00 \end{pmatrix};$$

the fits are illustrated in Figure A.2.

These examples represent *ill-conditioned* problems; that is, a small change in the data leads to a large change in the solution. In this case, the change in the data is 0.01 but the change in the third component of the solution is 100. The change in the data has been magnified by 10,000. Hence the solution is "sensitive" to changes in the data.

In this case the change in the data was deliberate, and it represented the solution of a different problem. The reason that ill-conditioning is worrisome is that often the data in a problem will be inaccurate. The numbers may be subject to measurement error, or they may be the results of earlier calculations and be subject to computer errors. This means that the "true" data will be unknown and that the numbers used in the calculations will be inaccurate. If the problem is ill conditioned, then the errors in the solution (to the "true" problem) can be large, *even if no errors are made when solving the linear system of equations.*

Going back to the example, let us now suppose that $b_{(2)}$ represents the exact data, and that $b_{(1)}$ and $b_{(3)}$ are in error. The exact data are on the line $y = 1$. The other two fits are dramatically different, both in their coefficients and in their graphs. However, all three fits go close to the data point $(1, 1)$, with "errors" of at most 0.01. The size of these "residuals" is proportional to the errors in the data, even though this problem is ill conditioned. This is typical for data-fitting problems.

We can quantify the sensitivity of a linear system with respect to changes in the data using the condition number of its matrix:

$$\text{cond}(A) = \|A\| \cdot \|A^{-1}\|.$$

Consider, for example, a diagonal matrix A where $A_{i,i} = 1$ except for $A_{n,n} = 10^{-n}$. If we use the 2-norm, then $\text{cond}(A) = 10^n$, and the condition number tends to infinity as A tends towards a singular matrix. This is true in general—the larger the condition number, the closer A is to being singular. Here "closeness to singularity" is measured relative to the norm of A. For the example at the beginning of this section, $\text{cond}(A) = 2.1234 \times 10^4$.

We would now like to show how the condition number is related to the errors in the solution of linear systems. Consider the case where all the errors are in the matrix A:

$$Ax = b \qquad \text{and} \qquad (A + E)\hat{x} = b,$$

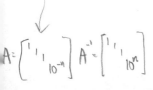

with $\|E\| = \epsilon \|A\|$. (Here the first system represents the "true" data, and the second is corrupted by errors E.) The lemma below shows that we can bound the change in the solution by

$$\frac{\|\hat{x} - x\|}{\|\hat{x}\|} \le \mathrm{cond}(A)\epsilon.$$

ie $\|\hat{x} - x\| \le \|\hat{x}\| \, \mathrm{Cond}(A)\epsilon$

Thus perturbations in the data matrix are magnified by $\mathrm{cond}(A)$ when the solution of the linear system is computed. A similar result holds for perturbations in the right-hand side.

Lemma A.11. *Let A be a square nonsingular matrix. Let x and $\hat{x}$ be the solutions to the linear systems*

$$Ax = b \quad \text{and} \quad (A + E)\hat{x} = b,$$

with $\|E\| = \epsilon \|A\|$. Then

$$\frac{\|\hat{x} - x\|}{\|\hat{x}\|} \le \mathrm{cond}(A)\epsilon.$$

Proof. From

$$E\hat{x} = b - A\hat{x} = Ax - A\hat{x} = A(x - \hat{x})$$

we obtain

$$x - \hat{x} = A^{-1}E\hat{x}.$$

Taking norms gives

$$\|x - \hat{x}\| = \|A^{-1}E\hat{x}\| \le \|A^{-1}\| \, \|E\| \, \|\hat{x}\| = \|A^{-1}\| \, (\epsilon \|A\|) \, \|\hat{x}\|.$$

Since $\mathrm{cond}(A) = \|A\| \cdot \|A^{-1}\|$, the result follows. $\square$

The above result has even greater significance. If Gaussian elimination with partial pivoting is used to solve the linear system $Ax = b$, then it can be shown that the computed solution $\hat{x}$ is the *exact* solution to a perturbed linear system of the form $(A + E)x = b$, where $\|E\| = C\epsilon_{\mathrm{mach}} \|A\|$. (The constant C is usually near 1; the constant depends on the number of equations in the system, and the size of the intermediate calculations relative to the size of the original matrix A.)

This is a most impressive result. Note that errors of size ϵ_{mach} are made when we store the matrix A in the computer, so that just entering the problem can cause the coefficient matrix A to be perturbed. Hence even if we could compute the *exact* solution to the *stored* system we could not guarantee a better error bound than that given by Gaussian elimination. Other methods, in particular computing the solution via the computation of A^{-1}, do not satisfy a result of this type.

This result also shows that Gaussian elimination will always produce a solution with small residuals. Since $(A + E)\hat{x} = b$ with $\|E\| = C\epsilon_{\mathrm{mach}} \|A\|$ (and C is the constant mentioned above) we can easily deduce that

$$\frac{\|A\hat{x} - b\|}{\|A\| \cdot \|\hat{x}\|} \le C\epsilon_{\mathrm{mach}}.$$

Hence the scaled residuals will always be proportional to ϵ_{mach}, even if the matrix is nearly singular.

Let us examine these bounds for the data-fitting example above. We will consider the true solution to be $x_{(2)} = (1, 0, 0)^T$, and the computed solution to be $x_{(1)} = (1.03, -4.00, 100.00)^T$. The errors are proportional to $\epsilon = 0.01$, which is much larger than $\epsilon_{\text{mach}} \approx 10^{-7}$. For the errors in the parameters x we obtain

$$100.0800 = \frac{\|\hat{x} - x\|}{\|x\|} \leq \text{cond}(A)\epsilon = 212.3442,$$

and for the residuals we obtain

$$5.7674 \times 10^{-5} = \frac{\|A\hat{x} - b\|}{\|A\| \cdot \|\hat{x}\|} \leq \epsilon = 0.01.$$

The bounds on the errors are worst-case bounds, and hence give only upper bounds on the errors.

There is a subtle point underlying the above comments. All of these error bounds are general error bounds made without regard to any special properties of the matrix A or the right-hand side b. They are worst-case bounds made before the system is solved. As a result, they can be pessimistic for particular linear systems. They are useful in understanding the general properties of linear systems and algorithms to solve them, but they are not always accurate error bounds. If accurate error bounds are desired, then they can be computed based on the particular linear system that is being solved.

Although we have already indicated that the condition number is a useful theoretical tool for analyzing linear systems, it is not so clear that it is a practical tool. Its value depends on knowing A^{-1}, and computing A^{-1} is three times as expensive as solving the linear system—a lot to pay for an error bound. However, it is possible to estimate the condition number at low cost as a by-product of solving a system of linear equations, and software packages such as LAPACK provide this capability.

In summary, if a system of linear equations is solved using Gaussian elimination with partial pivoting, then a computed solution with near-optimal accuracy can be obtained. This does not mean that the computed solution will be near the true solution, but rather that the errors in the computed solution will be proportional to the condition number of the coefficient matrix. (This result also applies to cases where the data in the coefficient matrix or the right-hand side are either inaccurate or subject to perturbations.) If the matrix is ill conditioned (i.e., the condition number is large), then the errors will be large, and if the matrix is well conditioned, the errors will be small. Regardless of the conditioning of the matrix, however, the residuals will be small. Software for Gaussian elimination can provide an estimate of the condition number so that the accuracy of the solution can be estimated.

A.9 The Sherman–Morrison Formula

If the inverse of a matrix is known, but then the matrix is perturbed, it is useful to be able to compute the inverse of the perturbed matrix efficiently. The Sherman–Morrison formula can be used to do this, for perturbations of a particular form.

Suppose that the original matrix is A and the perturbed matrix is

$$\hat{A} = A - uv^T,$$

where u and v are vectors. Then

$$\hat{A}^{-1} = A^{-1} + \alpha(A^{-1}u)(v^TA^{-1}),$$

where

$$\alpha = \frac{1}{1 - v^TA^{-1}u}.$$

This is the Sherman–Morrison formula.

It is easy to verify that this formula is correct by checking that

$$(A^{-1} + \alpha(A^{-1}u)(v^TA^{-1})) \times (A - uv^T) = I.$$

It is also easy to verify that A^{-1} exists if and only if $v^TA^{-1}u \neq 1$, that is, if and only if the scalar α is finite.

Example A.12 (Sherman–Morrison Formula). If

$$A = \begin{pmatrix} 1 & 0 & 0 \\ 0 & 2 & 0 \\ 0 & 0 & 3 \end{pmatrix}, \quad u = \begin{pmatrix} 1 \\ 1 \\ 1 \end{pmatrix}, \quad \text{and} \quad v = \begin{pmatrix} -1 \\ 4 \\ -1 \end{pmatrix},$$

then it is easy to calculate A^{-1} but not as easy to compute the inverse of

$$\hat{A} = A - uv^T = \begin{pmatrix} 2 & -4 & 1 \\ 1 & -2 & 1 \\ 1 & -4 & 4 \end{pmatrix}.$$

In this case

$$A^{-1} = \begin{pmatrix} 1 & 0 & 0 \\ 0 & \frac{1}{2} & 0 \\ 0 & 0 & \frac{1}{3} \end{pmatrix}, \quad A^{-1}u = \begin{pmatrix} 1 \\ \frac{1}{2} \\ \frac{1}{3} \end{pmatrix}, \quad v^TA^{-1} = (-1 \quad 2 \quad -\frac{1}{3}),$$

and $\alpha = 1/(1 - v^TA^{-1}u) = 3$, so the Sherman–Morrison formula gives

$$\hat{A}^{-1} = A^{-1} + \alpha(A^{-1}u)(v^TA^{-1}) = \begin{pmatrix} -2 & 6 & -1 \\ -\frac{3}{2} & \frac{7}{2} & -\frac{1}{2} \\ -1 & 2 & 0 \end{pmatrix}. \quad \blacksquare$$

A perturbation of the form

$$\hat{A} = A - uv^T$$

is called a "rank-one" perturbation because uv^T is a matrix whose rank is equal to one. The change to the basis matrix at an iteration of the simplex method corresponds to a change of this type. For example, suppose that the current basis matrix is

$$A = \begin{pmatrix} -1 & -4 & -5 & 4 \\ 1 & 8 & 5 & -1 \\ -1 & 7 & 4 & 4 \\ 9 & 8 & 7 & 1 \end{pmatrix}.$$

If at the current iteration the third column is replaced by

$$(6 \quad -3 \quad 4 \quad -9)^T,$$

then

$$\hat{A} = \begin{pmatrix} -1 & -4 & 6 & 4 \\ 1 & 8 & -3 & -1 \\ -1 & 7 & 4 & 4 \\ 9 & 8 & -9 & 1 \end{pmatrix} = A - uv^T$$

with

$$u = (-11 \quad 8 \quad 0 \quad 16)^T \quad \text{and} \quad v = (0 \quad 0 \quad 1 \quad 0).$$

The Sherman–Morrison formula is of great value when implementing algorithms, such as the simplex method, that are based on updating of matrices.

There is a generalization of the formula to the case where

$$\hat{A} = A - UV^T,$$

where U and V are matrices with $k > 1$ columns. (This will be a rank-k perturbation to the matrix if the columns of U and V are linearly independent.) In this case,

$$\hat{A}^{-1} = A^{-1} + (A^{-1}U)\alpha(V^T A^{-1}),$$

where $\alpha = (I - V^T A^{-1} U)^{-1}$. Here, α is a $k \times k$ matrix. This generalization is often referred to as the *Sherman–Morrison–Woodbury formula*.

A.10   Notes

Gaussian Elimination—We mentioned that it is possible to derive bounds on the error in the computed solution. In fact it can be shown that Gaussian elimination is as good an algorithm as you could hope for solving this problem. If Gaussian elimination with partial pivoting is applied to the system $Ax = b$, then the computed solution $\hat{x}$ can be shown to be the *exact* solution to a "nearby" linear system

$$(A + E)\hat{x} = b,$$

where E is some (unknown) perturbation to the matrix A. The size of the perturbation $\|E\|$ is not predictable in advance, but for most problems $\|E\|$ is comparable to the rounding errors made when storing the matrix A in the computer. Hence the simple act of storing the problem on the computer can introduce errors comparable to the errors incurred by using Gaussian elimination. It would be hard to think of a more favorable result. Return to Appendix A.8 for a detailed discussion of these ideas.

The most obvious reasons for using elimination techniques instead of computing the matrix inverse are based on the costs of the computations. However, there are other reasons. If elimination is used, then the computed solution is the exact solution to a nearby problem, but this is not guaranteed if the solution is obtained using a computed inverse. Elimination is also more flexible: As a side effect of elimination it is possible to compute Av, A^Tv,

$A^{-1}v$, and $A^{-T}v$ for any vector v. If the inverse is computed, then both A and A^{-1} must be stored to do this, doubling the storage costs for the algorithm.

The method given here for solving linear equations is the most traditional as well as the most widely used, but it is not the most efficient. There are methods for solving linear equations that have lower asymptotic costs than elimination (costs of $O(n^\omega)$ with $\omega < 3$), although they are only more efficient in practice for larger values of n ($n \geq 100$ or more, depending on the method). They are discussed in the paper by Higham (1990).

Sparse Matrices—Several other schemes for storing sparse matrices have been proposed. The scheme given above has deficiencies; for example, to modify an entry in column 72 of the matrix, the algorithm would have to step through the storage locations of the first 71 columns of the matrix. More elaborate schemes allow more direct access to entries in the matrix. For example, it might be useful to have an auxiliary integer array JCOL that indicated the location of the first entry in each column within the arrays AVAL and IROW. For the above example JCOL = { 1, 3, 5, 7, 8 }. The storage scheme is often related to the operations that must be performed on the matrix. For more information on sparse matrices, see the books by Davis (2006); George and Liu (1981); and Saad (2003).

Sherman–Morrison Formula—A detailed discussion of the Sherman–Morrison formula and related formulas can be found in the survey paper by Hager (1989). Although it is routine to refer to the Sherman–Morrison formula and the Sherman–Morrison–Woodbury formula, these formulas were first discovered by others. The survey paper gives detailed citations of this earlier work.

Appendix B
Other Fundamentals

B.1 Introduction

This appendix gives a brief introduction to some background ideas that are used in various places throughout the book. In many cases, this material may be familiar, but a limited discussion is provided here in an attempt to make the book self-contained. Further information on these topics can be found in a wide variety of sources. Some sample references are the following:

- Computer Arithmetic—T. Sauer, *Numerical Analysis*, Addison-Wesley (Boston), 2006.
- Big-O Notation—J. M. Ortega and W. C. Rheinboldt, *Iterative Solution of Nonlinear Equations in Several Variables*, Academic Press (New York), 1970. Reprinted by SIAM (Philadelphia), 2000.
- The Gradient, Hessian, and Jacobian—R. Courant and D. Hilbert, *Methods of Mathematical Physics, Volume* I, Interscience Press (New York), 1953. Reprinted by Wiley-Interscience (New York), 1989.
- Continuous Functions and the Implicit Function Theorem—R. Creighton Buck, *Advanced Calculus (Third Edition)*, McGraw–Hill (New York), 1978. Reprinted by Waveland Press (Long Grove, IL), 2003.

B.2 Computer Arithmetic

The arithmetic used by computers is not the same as the "pencil and paper" arithmetic that most of us learned as children. The differences are not great but their effects can sometimes be dramatic, particularly if the problem being solved is sensitive to small changes in its data.

On most computers, arithmetic is performed according to the IEEE standard, and our discussion here will reflect that standard. It will not be necessary for us to have a detailed understanding of computer arithmetic, and the description here will be brief. A more complete discussion can be found in the reference given above.

There are two types of numbers used in computer calculations, *integer* and *real*. ("Real" arithmetic is often called *floating-point* arithmetic for historical reasons.) Integer numbers behave much as we are used to, that is, if two integers are added together, then the result is another integer, and no error is made in the calculation. The only unusual property of computer integers is that there are only finitely many of them. On many computers an integer is stored using 32 bits (32 binary digits). This means that at most

$$2^{32} = 4,294,967,296$$

distinct integers can be represented. As a result there is a biggest integer,

$$2,147,483,647$$

and a smallest integer, $-2,147,483,648$. (The particular values given here are based on the IEEE standard for computer arithmetic.) If an attempt is made to compute an integer outside of this range, an *overflow* occurs.

The situation for "real" arithmetic is more complicated. Real numbers are represented in a form of scientific notation, for example,

$$0.5167 \times 10^{-6}.$$

Each number consists of a *mantissa* (0.5167) and an *exponent* (-6). We will represent numbers here using decimal digits, but the computer will normally use binary digits. Just as with integers, the computer can represent only finitely many real numbers. It is common to use 64 bits to store each real number (with some of the bits being used to store the mantissa, and the rest used to store the exponent); then the largest real number is approximately 10^{308}. Numbers larger than this (or smaller than about -10^{308}) result in an overflow. Overflows are considered to be serious errors and will often cause a program to stop executing.

There is also a smallest positive real number, the number closest to zero that can be represented on the computer. Numbers smaller than this are often reset to zero, and an *underflow* occurs. This is often considered harmless, and no warning message is generated when it occurs on most computer systems. The smallest positive number is approximately 10^{-324}.

When discussing the results of computer calculations we will often refer to the machine precision or the *machine epsilon*, denoted by ϵ_{mach}. It represents the accuracy of the computer calculations. It is defined to be the smallest positive number ϵ such that

$$1 + \epsilon > 1$$

when calculated on the computer. For example, if four decimal digits are available and rounding is used, then

$$1.000 + 0.0004 = 1.0004 \rightarrow 1.000,$$

whereas

$$1.000 + 0.0005 = 1.0005 \rightarrow 1.001 > 1.000.$$

In this simple case $\epsilon_{mach} = 0.0005$. When real numbers are stored using 64 bits, $\epsilon_{mach} \approx 10^{-16}$.

The machine epsilon represents how accurately calculations on the computer can be carried out. If a number is stored in the computer, then a *relative* error of up to ϵ_{mach} can occur. For example, if we try to store 12.256 using only four decimal digits, then it would be represented as 12.26 with a relative error of

$$\frac{|12.256 - 12.26|}{|12.256|} \approx 0.0003,$$

which is less than the machine epsilon of 0.0005. Similarly, if two numbers are added, the result can have a relative error of as much as ϵ_{mach}. These errors are called *rounding errors*. In general it is unreasonable to expect computed results to have a relative accuracy better than ϵ_{mach}.

In the discussion here, real numbers are stored using 64 bits. We will refer to these as *double precision* real numbers. This is the precision typically used for computations involving real numbers. Most computers are also capable of working with real numbers based on 32 bits, often referred to as *single precision*. If single precision is used, then the thresholds for overflow and underflow change, as does the value of ϵ_{mach}. On some computers it is also possible to compute in extended precisions that use additional bits to store real numbers.

B.3   Big-O Notation, $O(\cdot)$

"Big-O" notation is used to provide an order-of-magnitude estimate of a function. We will say that $g(n) = O(f(n))$ if for sufficiently large n,

$$g(n) \le Cf(n),$$

where C is some positive constant. For example,

$$3n^3 + 2n^2 + 5 = O(n^3)$$

because

$$3n^3 + 2n^2 + 5 \le 4n^3$$

for $n \ge 3$.

This notation is often used when measuring the cost of an algorithm (see Chapter 9). If the cost of an algorithm is $O(n^3)$, then, for large n, doubling n increases the cost of the algorithm by about 8. In many cases, this information is sufficient to determine if an algorithm is practical or competitive for a given problem.

This notation is also used when analyzing the accuracy of approximation formulas, such as the Taylor series. In this case it is more common to write that $g(h) = O(f(h))$ if

$$g(h) \le Cf(h)$$

as $h \to 0$. (This is equivalent to the other definition for $n \equiv 1/h$.) For example, if $|h| < 1$, then

$$\frac{1}{1-h} = 1 + h + h^2 + h^3 + \cdots .$$

From this we can conclude that

$$\frac{1}{1-h} = 1 + h + O(h^2)$$

since for $|h| < \frac{1}{2}$,

$$h^2 + h^3 + h^4 + \cdots = h^2(1 + h + h^2 + \cdots)$$

$$= h^2 \left(\frac{1}{1-h} \right) < h^2 \left(\frac{1}{1 - \frac{1}{2}} \right) = 2h^2.$$

The big-O notation gives an order-of-magnitude indication of the accuracy of the approximation.

B.4 The Gradient, Hessian, and Jacobian

We will first discuss the gradient and Hessian. Let f be a real-valued function of n variables

$$f(x) = f(x_1, x_2, \ldots, x_n).$$

The vector of first derivatives of f is called the *gradient* of the function and is notated as

$$\nabla f(x) \equiv \left(\frac{\partial f(x)}{\partial x_1}, \frac{\partial f(x)}{\partial x_2}, \ldots, \frac{\partial f(x)}{\partial x_n} \right)^T.$$

The matrix of second derivatives of f is called the *Hessian matrix* or simply the *Hessian* and is notated as $\nabla^2 f$. It is the matrix with entries

$$[\nabla^2 f(x)]_{ij} \equiv \frac{\partial^2 f(x)}{\partial x_i \partial x_j}.$$

For functions with continuous second derivatives, it will always be a symmetric matrix:

$$\frac{\partial^2 f(x)}{\partial x_i \partial x_j} = \frac{\partial^2 f(x)}{\partial x_j \partial x_i}.$$

These concepts are illustrated in the example below.

Example B.1 (The Gradient and Hessian). Consider the function

$$f(x_1, x_2) = 2x_1^4 + 3x_1^2 x_2 + 2x_1 x_2^3 + 4x_2^2.$$

The gradient of this function is

$$\nabla f(x) = \begin{pmatrix} 8x_1^3 + 6x_1 x_2 + 2x_2^3 \\ 3x_1^2 + 6x_1 x_2^2 + 8x_2 \end{pmatrix}$$

and the Hessian matrix is

$$\nabla^2 f(x) = \begin{pmatrix} 24x_1^2 + 6x_2 & 6x_1 + 6x_2^2 \\ 6x_1 + 6x_2^2 & 12x_1 x_2 + 8 \end{pmatrix}.$$

At the point $x_0 = (-2, 3)$ these become

$$\nabla f(x_0) = \begin{pmatrix} -46 \\ -72 \end{pmatrix}, \qquad \nabla^2 f(x_0) = \begin{pmatrix} 114 & 42 \\ 42 & -64 \end{pmatrix}. \qquad \blacksquare$$

To define the Jacobian we will use a vector-valued function:

$$f(x) = f(x_1, x_2, \ldots, x_n) = \begin{pmatrix} f_1(x_1, x_2, \ldots, x_n) \\ f_2(x_1, x_2, \ldots, x_n) \\ \vdots \\ f_m(x_1, x_2, \ldots, x_n) \end{pmatrix},$$

where each of the functions f_i is a real-valued function. Then ∇f is the matrix with entries

$$(\nabla f(x))_{i,j} \equiv \frac{\partial f_j(x)}{\partial x_i}.$$

The *Jacobian* of f at the point x is defined as $\nabla f(x)^T$. Note that the jth column of ∇f is the gradient of f_j.

Example B.2 (The Jacobian). Consider the vector-valued function

$$f(x) = f(x_1, x_2) = \begin{pmatrix} \sin x_1 + \cos x_2 \\ e^{3x_1 + x_2^2} \\ 4x_1^3 + 7x_1 x_2^2 \end{pmatrix}.$$

Then

$$\nabla f(x) = \begin{pmatrix} \cos x_1 & 3e^{3x_1+x_2^2} & 12x_1^2 + 7x_2^2 \\ -\sin x_2 & 2x_2 e^{3x_1+x_2^2} & 14x_1 x_2 \end{pmatrix}.$$

The Jacobian is equal to

$$\nabla f(x)^T = \begin{pmatrix} \cos x_1 & -\sin x_2 \\ 3e^{3x_1+x_2^2} & 2x_2 e^{3x_1+x_2^2} \\ 12x_1^2 + 7x_2^2 & 14x_1 x_2 \end{pmatrix}.$$

At the point $x = (1, 2)^T$ these matrices are equal to

$$\begin{pmatrix} \cos 1 & 3e^7 & 40 \\ -\sin 2 & 4e^7 & 28 \end{pmatrix} \quad \text{and} \quad \begin{pmatrix} \cos 1 & -\sin 2 \\ 3e^7 & 4e^7 \\ 40 & 28 \end{pmatrix},$$

respectively. ∎

If an optimization problem includes a set of linear constraints

$$Ax = b,$$

or equivalently

$$f(x) \equiv Ax - b = 0,$$

then the Jacobian has a simple form. In this case

$$\nabla f(x) = A^T$$

and the Jacobian is equal to A.

B.5   Gradient and Hessian of a Quadratic Function

For a quadratic function, the formulas for the gradient and Hessian are especially simple. Consider the quadratic function

$$f(x) = \tfrac{1}{2}x^T Q x - b^T x,$$

where $x = (x_1, \ldots, x_n)^T$, b is a vector of length n, and Q is an $n \times n$ symmetric matrix. The function can also be written as

$$f(x_1, \ldots, x_n) = \frac{1}{2}\sum_{j=1}^{n}\sum_{k=1}^{n} Q_{j,k}x_j x_k - \sum_{j=1}^{n} b_j x_j.$$

The only terms involving the variable x_i are

$$\frac{1}{2}Q_{i,i}x_i^2 + \frac{1}{2}\sum_{j \neq i} Q_{j,i}x_j x_i + \frac{1}{2}\sum_{k \neq i} Q_{i,k}x_i x_k - b_i x_i.$$

Taking the partial derivative of these terms with respect to x_i gives

$$\frac{\partial f}{\partial x_i} = Q_{i,i}x_i + \frac{1}{2}\sum_{j \neq i} Q_{j,i}x_j + \frac{1}{2}\sum_{k \neq i} Q_{i,k}x_k - b_i$$

$$= \sum_{j=1}^{n} Q_{i,j}x_j - b_i = (Qx - b)_i.$$

(We have used the fact that Q is symmetric.) Hence

$$\boxed{\nabla f(x) = Qx - b.}\qquad \left(\text{since } Q^T = Q\right)$$

The formula for $\partial f / \partial x_i$ can be used to show that

$$\frac{\partial^2 f}{\partial x_i \partial x_j} = Q_{i,j}$$

so that

$$\nabla^2 f(x) = Q.$$

Example B.3 (Derivatives of a Quadratic Function). Consider the quadratic function

$$f(x_1, x_2, x_3) = \tfrac{1}{2}x^T Q x - b^T x$$

with

$$Q = \begin{pmatrix} 4 & 2 & 1 \\ 2 & 5 & 3 \\ 1 & 3 & 7 \end{pmatrix} \quad \text{and} \quad b = (-2 \quad 8 \quad -9)^T.$$

The matrix-vector formulas for the gradient and Hessian give

$$\nabla f(x) = Qx - b = \begin{pmatrix} 4x_1 + 2x_2 + x_3 + 2 \\ 2x_1 + 5x_2 + 3x_3 - 8 \\ x_1 + 3x_2 + 7x_3 + 9 \end{pmatrix}$$

and

$$\nabla^2 f(x) = Q = \begin{pmatrix} 4 & 2 & 1 \\ 2 & 5 & 3 \\ 1 & 3 & 7 \end{pmatrix}.$$

The function f can also be written as

$$f(x_1, x_2, x_3) = \tfrac{1}{2}(4x_1^2 + 2x_1x_2 + x_1x_3 + 2x_1x_2 + 5x_2^2 + 3x_2x_3$$
$$+ x_1x_3 + 3x_2x_3 + 7x_3^2) - (-2x_1 + 8x_2 - 9x_3).$$

Taking the partial derivatives of this formula gives the same result as the matrix-vector formulas. ∎

B.6 Derivatives of a Product

Suppose that

$$f(x) = g(x)h(x),$$

where f and g are both continuously differentiable functions of the n-dimensional vector x. Then

$$\nabla f(x) = \nabla g(x)h(x) + \nabla h(x)g(x).$$

To obtain the Hessian, we first differentiate the term

$$\nabla g(x)h(x)$$

with respect to x. It can be verified that this gives the matrix

$$\nabla^2 g(x)h(x) + \nabla g(x)\nabla h(x)^T.$$

Similarly, differentiating

$$\nabla h(x)g(x)$$

with respect to x gives the matrix

$$\nabla^2 h(x)g(x) + \nabla h(x)\nabla g(x)^T.$$

Combining the two expressions, we obtain that

$$\nabla^2 f(x) = \nabla^2 g(x)h(x) + \nabla^2 h(x)g(x) + \nabla g(x)\nabla h(x)^T + \nabla h(x)\nabla g(x)^T.$$

Example B.4 (Derivative of a Product). Consider the function

$$f(x) = (a^T x)(b^T x),$$

where a, b, and x are n-dimensional vectors. Then

$$\nabla f(x) = a(b^T x) + b(a^T x)$$

and

$$\nabla^2 f(x) = ab^T + ba^T. \quad ∎$$

B.7   The Chain Rule

The rule for obtaining the derivative of a function of a function is called the *chain rule*. Consider a function $g(x) = g(x_1, \ldots, x_n)$, and suppose that each x_i is in turn a function of the variables $t_1, \ldots, t_m$; that is, $x_i = x_i(t_1, \ldots, t_m)$ for $i = 1, \ldots, n$. We examine the composite function

$$h(t) = g(x(t)).$$

The chain rule states that if g is continuously differentiable in $\Re^n$, and $x_1, \ldots, x_m$ are continuously differentiable in $\Re^m$, then h is continuously differentiable in $\Re^m$ and

$$\nabla h(t) = \nabla x(t) \nabla g(x(t)),$$

where

$$\nabla x(t) = (\nabla x_1(t) \quad \cdots \quad \nabla x_n(t)) .$$

The chain rule can be generalized to the case where g is a k-dimensional vector of functions g_i. In this case h will also be a k-dimensional vector of functions. If ∇h denotes the $n \times k$ matrix whose jth column is ∇h_i, and ∇g denotes the $n \times k$ matrix whose jth column is ∇g_i, then the above formula remains valid.

Example B.5 (Chain Rule). Suppose that

$$g_1(x) = x_1^2 - x_1 x_2$$
$$g_2(x) = -x_1^4 + 2x_2^2,$$

where

$$x_1 = x_1(t_1, t_2, t_3) = t_1 + 2t_2 - 3t_3$$
$$x_2 = x_2(t_1, t_2, t_3) = t_1^2 + t_2$$

and let $h(t) = g(x(t))$. Then

$$\nabla x(t) = \begin{pmatrix} 1 & 2t_1 \\ 2 & 1 \\ -3 & 0 \end{pmatrix}$$

and

$$\nabla g(x(t)) = \begin{pmatrix} 2x_1(t) - x_2(t) & -4x_1^3(t) \\ -x_1(t) & 4x_2(t) \end{pmatrix}$$
$$= \begin{pmatrix} 2(t_1 + 2t_2 - 3t_3) - (t_1^2 + t_2) & -4(t_1 + 2t_2 - 3t_3)^3 \\ -(t_1 + 2t_2 - 3t_3) & 4(t_1^2 + t_2) \end{pmatrix};$$

hence

$$\nabla h(t) = \begin{pmatrix} 1 & 2t_1 \\ 2 & 1 \\ -3 & 0 \end{pmatrix} \begin{pmatrix} 2(t_1 + 2t_2 - 3t_3) - (t_1^2 + t_2) & -4(t_1 + 2t_2 - 3t_3)^3 \\ -(t_1 + 2t_2 - 3t_3) & 4(t_1^2 + t_2) \end{pmatrix}. \qquad \blacksquare$$

A particular application of the chain rule is if

$$x = \begin{pmatrix} y(t) \\ t \end{pmatrix}.$$

If $h(t) = g(x(t))$, then

$$\nabla h(t) = \nabla x(t)\nabla g(x(t)) = (\nabla y(t) \quad I) \begin{pmatrix} \nabla_y g(x(t)) \\ \nabla_t g(x(t)) \end{pmatrix}$$

$$= \nabla y(t)\nabla_y g(y(t), t) + \nabla_t g(y(t), t).$$

Note that ∇_t refers to the gradient of a function with respect to the vector of variables t.

The chain rule can also be used to obtain second derivatives. We will assume here that g is a scalar function. If g and x_i are twice continuously differentiable, then h is twice continuously differentiable in $\Re^m$ and

$$\nabla^2 h(t) = \nabla^2 x(t)\nabla g(x(t)) + \nabla x(t)\nabla^2 g(x(t))\nabla x(t)^T,$$

where a product of the form $(\nabla^2 x)v$ is interpreted as

$$(\nabla^2 x)v = \sum_{i=1}^{n}(\nabla^2 x_i(t))v_i.$$

B.8 Continuous Functions; Closed and Bounded Sets

A set S is said to be *bounded* if every point $x \in S$ satisfies $\|x\| < M$ for some number M. S is *closed* if for any sequence of points $x_1, x_2, \ldots$ with $x_i \in S$ for all i and

$$\lim_{i \to \infty} x_i = x,$$

we have $x \in S$ also. A point $x \in S$ is *interior* to the set S if

$$\{ y : \|y - x\| < \epsilon \} \subset S$$

for some $\epsilon > 0$. A set S is *open* if every point $x \in S$ is interior to S.

For example, consider the sets

$$S_1 = \{ x : 0 < x \}, \quad S_3 = \{ x : 0 \leq x < 2 \},$$
$$S_2 = \{ x : 0 \leq x \}, \quad S_4 = \{ x : 0 \leq x \leq 2 \}.$$

Then S_1 is open but not bounded, S_2 is closed but not bounded, S_3 is bounded but neither closed nor open, and S_4 is both closed and bounded.

If f is a continuous function defined on a closed and bounded set S, then f takes on its minimum and maximum values somewhere on the set S. That is, there exist points $x_{\min} \in S$ and $x_{\max} \in S$ such that

$$f(x_{\min}) = \min_{x \in S} f(x)$$
$$f(x_{\max}) = \max_{x \in S} f(x).$$

This is not guaranteed unless S is both closed and bounded. To see this, consider the function $f(x) = x^2$ on the sets S_i defined above.

This theorem is due to Weierstrass. For a proof of this result see the book by Buck (1978, reprinted 2003).

Let $F(x)$ be an m-vector of functions of $x = (x_1, \ldots, x_n)^T$. Then F is *Lipschitz continuous* on an open set S if for some constant L,

$$\|F(x) - F(y)\| \le L \|x - y\|$$

for all $x, y \in S$.

The following technical result is used in Chapter 11.

Theorem B.6. *Let $F(x)$ be an m-vector of functions of $x = (x_1, \ldots, x_n)^T$. Assume that the Jacobian of F is Lipschitz continuous on an open convex set S with constant L. Then for any $x, y \in S$,*

$$\left\| F(y) - F(x) - \nabla F(x)^T (y - x) \right\| \le \frac{L}{2} \|y - x\|^2 .$$

Proof. See the book by Dennis and Schnabel (1983, reprinted 1996). □

B.9   The Implicit Function Theorem

The concept of an "implicit" function arises in connection with constraints. For example, the constraint $x_1 + x_2 - 1 = 0$ could be rewritten as $x_1 = 1 - x_2$, so the constraint implicitly determines x_1 as a function of x_2. Once x_2 is known, x_1 is determined and can be eliminated from the problem.

For simple constraints this is a straightforward procedure, but for systems of nonlinear constraints, such as

$$x_1^5 x_2 x_3^4 - 5 x_2^3 x_3^3 - 7 x_1^3 + 5 = 0$$
$$x_1^7 x_2^3 x_3 + 8 x_1^6 x_2^2 + 8 x_3^2 - 2 = 0,$$

it is less obvious how to express, say, x_1 and x_2 in terms of x_3, or even if the constraints implicitly define such functions. It is possible that one of the constraints is redundant, or reduces to a formula such as

$$x_1 = \pm\sqrt{x_3}$$

which would not correspond to a single-valued function.

The implicit function theorem states conditions under which such an implicit function is defined. The theorem does not give formulas for the implicit function, but just determines when such a function exists. The main requirement for the existence of such a function is that the Jacobian of the constraints be nonsingular. A version of the theorem is stated below; for a proof see the book by Ortega and Rheinboldt (1970, reprinted 2000).

Theorem B.7. *Consider the system of m equations in n variables*

$$g_i(x_1, \ldots, x_n) = 0 \quad \text{for } i = 1, \ldots, m,$$

and let $\hat{x}$ be a point that satisfies these equations. Assume that the gradients of the functions $\{g_i(x)\}$ *are continuous, and that the Jacobian matrix of the equations with respect to the variables* $x_1, \ldots, x_m$,

$$\begin{pmatrix} \dfrac{\partial g_1(\hat{x})}{\partial x_1} & \cdots & \dfrac{\partial g_1(\hat{x})}{\partial x_m} \\ \vdots & \ddots & \vdots \\ \dfrac{\partial g_m(\hat{x})}{\partial x_1} & \cdots & \dfrac{\partial g_m(\hat{x})}{\partial x_m} \end{pmatrix}$$

is nonsingular. Define $\hat{y} = (\hat{x}_{m+1}, \ldots, \hat{x}_n)^T$ *and* $y = (x_{m+1}, \ldots, x_n)^T$. *Then for some* $\epsilon > 0$, *if* $\|\hat{y} - y\| < \epsilon$ *there exist functions* $h_1(y), \ldots, h_m(y)$ *such that* (i) *the functions* $\{h_i(y)\}$ *have continuous gradients,* (ii) $\hat{x}_i = h_i(\hat{y})$ *for* $i = 1, \ldots, m$, *and* (iii) $g_i(h_1(y), \ldots, h_m(y), y) = 0$, *for* $i = 1, \ldots, m$.

To illustrate the theorem, we consider the equations

$$g_1(x) = x_1^5 x_2 x_3^4 - 5x_2^3 x_3^3 - 7x_1^3 + 5 = 0$$
$$g_2(x) = x_1^7 x_2^3 x_3 + 8x_1^6 x_2^2 + 8x_3^2 - 2 = 0.$$

These equations are satisfied at the point $\hat{x} = (0, 2, \frac{1}{2})^T$. At this point the Jacobian matrix with respect to the first two variables is

$$\begin{pmatrix} 5x_1^4 x_2 x_3^4 - 21x_1^2 & x_1^5 x_3^4 - 15x_2^2 x_3^3 \\ 7x_1^6 x_2^3 x_3 + 48x_1^5 x_2^2 & 3x_1^7 x_2^2 x_3 + 16x_1^6 x_2 \end{pmatrix}_{x=\hat{x}} = \begin{pmatrix} 0 & 0 \\ -15/2 & 0 \end{pmatrix}.$$

This matrix is singular, the conditions of the implicit function theorem are not satisfied, and so at this point the constraints cannot be rewritten to express the first two variables in terms of the third variable.

For the equation

$$g_1(x_1, x_2) = x_1 + x_2 - 1 = 0$$

the Jacobian matrix with respect to the first variable is (1). This matrix is nonsingular for all values of x_1 and x_2, so in this case an implicit function exists at all points where the constraint is satisfied.

The theorem attempts to define the first m variables in terms of the remaining $n - m$ variables. There is nothing special about the first m variables, since the variables could be relabeled so they could be put in any order.

A slightly more general version of the implicit function theorem is used in Chapter 16; it can be found in the book by Bertsekas (1982, reprinted 1996).

Appendix C

Software

C.1   Software

In the last two decades, optimization software has improved dramatically, and this has made it possible to solve important problems in many areas of science and engineering. Advances in modern optimization algorithms and in computing technology enable us to solve problems with hundreds of thousands and even millions of variables and constraints. Even on a laptop computer using general-purpose optimization software, we can solve problems with thousands of variables and constraints.

It is not feasible here to give a detailed discussion of optimization software. Instead we will discuss five categories of optimization software defined by their functionality.

The first category is general-purpose software environments for scientific computing and data analysis. Examples include Excel, *Mathematica*, MATLAB®, and SAS. These software systems include optimization modules along with many other computational and graphical tools. They offer a broad range of capabilities, but may not offer the advanced capabilities necessary for solving large or especially challenging optimization problems. Also, because these systems are not specialized for optimization, they may lack modeling tools that make it easier to develop and implement an optimization model.

The second software category is modeling software specifically oriented for optimization. This software makes it easy for a user to formulate an optimization problem and apply the whole arsenal of available optimization tools. The modeling software is usually based on some modeling language for describing optimization problems. A modeling language is a convenient way to convey an optimization problem to a solver. It makes it simple for a user to define objective functions and constraints of an optimization problem. The modeling language can then transform the model (represented in a form convenient to the user) into the mathematical form required by the optimization software. In the case of nonlinear models, the software can also derive formulas for derivative calculations. Also, by conveniently parameterizing a model, modeling languages provide a simple way to create optimization models based on data sets. This is particularly convenient in applications where the general model is fixed, but in which the data change regularly. Finally, once the model has been developed, it is straightforward to apply a variety of optimization algorithms in which cases where a small change in the model may require a different solver. This can occur, for example, when a nonlinear term is added to what had been a linear model. All these useful features allow a simple and fast way to model and solve optimization problems.

The third software category is linear programming solvers. In this category we also include software for solving integer programming problems. There are several reasons why we separate linear programming software from more general optimization algorithms. One reason is that it is simpler to describe the models, since it is only necessary to specify vectors and matrices and not general nonlinear functions. Another is that software for linear programming includes many features beyond that typically provided by nonlinear optimization algorithms; for example, it will be common to have modules for sensitivity analysis. In addition, software for linear programming often includes capabilities for solving problems with integer variables, while nonlinear optimization software typically does not (with the exception of some software for quadratic programming).

The fourth software category is quadratic programming solvers. Quadratic models are nonlinear, but they (like linear programs) can be described by specifying vectors and matrices rather than general nonlinear functions. Convex quadratic problems with linear constraints can be solved efficiently using an interior-point method or a modification of the simplex method. Therefore solvers for quadratic programming and linear programming are often combined in one software package that contains useful features such as the ability to handle integer variables or to perform sensitivity analysis. If these features are not required, then a convex quadratic program could be solved using software for nonlinear optimization. Since nonlinear optimization methods are often based on quadratic models, and often treat linear constraints separately from nonlinear constraints, nonlinear optimization software should often work well for convex quadratic programs as well. However, if it is necessary to solve a nonconvex quadratic problem or a problem with quadratic constraints—problems that may have local solutions—then it may be necessary to use general nonlinear optimization solvers.

The last software category is general nonlinear optimization solvers. There are many varieties of nonlinear optimization models, and a corresponding array of nonlinear optimization algorithms. If the optimization problem is especially challenging to solve, the user may have to provide additional guidance in choosing an appropriate solver, or even in reformulating the model. Another issue to keep in mind is that it can be difficult to express a nonlinear optimization model in software. In some cases it may be easy to write the formulas for the model, for example, if the model includes a quadratic function. In other cases the model may be defined only in terms of other software. This can happen, for example, if the constraints include a differential equation that is solved using auxiliary algorithms. A guide to optimization algorithms and modeling languages is available through NEOS, a software system at the Argonne National Laboratory.

In conclusion, we highlight several points.

First, optimization software continues to be developed and improved at a rapid pace. Any recommendations that we might make in a book such as this would quickly go out of date. For this reason, it may be helpful to consult with online sources of information. For example, the Web site for the magazine *OR/MS Today* includes surveys of various types of software. Another source of guidance is the NEOS Server for Optimization Software. Links to these and other resources can be found on the Web site for this book (see http://www.siam.org/books/ot108).

Second, modern linear programming software is able to solve problems with millions of variables and constraints. Moreover, the solver always finds the global solution to a linear programming problem. However, if the model includes integer-valued variables, it will typically be harder to solve, and this will limit the size of the problem that can be solved. In this case the default version of the solver may not compute the optimal solution, but instead may return an estimate that is only guaranteed to be within some tolerance of the optimal solution. The software is designed in this way because verifying optimality can be a time-consuming calculation when integer variables are included in the optimization model. In contrast, it is often relatively easy to compute a bound on the accuracy of the objective value.

Third, nonlinear optimization problems may have multiple local solutions, and most optimization software will only guarantee that a local solution will be found. If the model includes nonlinear constraints, the software may not be able even to find a feasible point. Therefore a good initial guess in the neighborhood of the global solution can be helpful. In some cases it may be necessary to try several different initial guesses of the solution.

Fourth, using a flexible optimization modeling environment with several different solvers is, in general, rewarding. Some problems can be easily solved by certain optimization algorithms but not others. As software improves, more and more optimization problems can be solved by general-purpose algorithms using the default settings of the algorithm parameters. But as this happens, users may become more ambitious and attempt to solve larger and more challenging optimization problems.

Finally, besides choosing an appropriate solver or a modeling language, it may also be important to select an appropriate form of the optimization model. For example, subsection 1.7.6 described the catenary problem, which had several formulations. The formulations all had the same solution, but they had different computational properties. One of these formulations was easier to solve than the others. For challenging optimization problems, it may be necessary to investigate whether another formulation may be worthwhile.

Bibliography

[1] J. ABADIE, *On the Kuhn-Tucker theorem*, in Nonlinear Programming, J. Abadie, ed., North–Holland, Amsterdam, 1967, pp. 19–36.

[2] ———, *The GRG method for nonlinear programming*, in Design and Implementation of Optimization Software, H. Greenberg, ed., Sijthoff and Noordhoff, The Netherlands, 1978, pp. 335–362.

[3] J. ABADIE AND J. CARPENTIER, *Généralisation de la méthode du gradient réduit de Wolfe au cas de contraintes non-linéaires*, Note HRR 6678, Électricité de France, Paris, 1965.

[4] C. ABLOW AND G. BRIGHAM, *An analog solution of programming problems*, Operations Research, 3 (1955), pp. 388–394.

[5] I. ADLER, *The expected number of pivots needed to solve parametric linear programs and the efficiency of the self-dual simplex method*, manuscript, Department of Industrial Engineering and Operations Research, University of California, Berkeley, CA, 1983.

[6] R. K. AHUJA, T. L. MAGNANTI, AND J. B. ORLIN, *Network Flows: Theory, Algorithms, and Applications*, Prentice–Hall, Englewood Cliffs, NJ, 1993.

[7] M. AIZERMAN, E. BRAVERMAN, AND L. ROZONOER, *Theoretical foundation of the potential function method in pattern recognition learning*, Automation and Remote Control, 25 (1964), pp. 821–837.

[8] M. AL-BAALI, *Descent property and global convergence of the Fletcher-Reeves method with inexact line search*, IMA Journal on Numerical Analysis, 5 (1985), pp. 121–124.

[9] E. D. ANDERSEN AND K. D. ANDERSEN, *Presolving in linear programming*, Mathematical Programming, 71 (1995), pp. 221–245.

[10] E. D. ANDERSEN, J. GONDZIO, C. MESZAROS, AND X. XU, *Implementation of interior point methods for large scale linear programming*, in Interior Point Methods in Mathematical Programming, T. Terlaky, ed., Kluwer Academic Publishers, Dordrecht, The Netherlands, 1996, pp. 189–252.

[11] K. M. ANSTREICHER, *A monotonic projective algorithm for fractional linear programming*, Algorithmica, 1 (1986), pp. 483–498.

[12] K. M. ANSTREICHER AND R. A. BOSCH, *A new infinity-norm path following algorithm for linear programming*, SIAM Journal on Optimization, 5 (1995), pp. 236–246.

[13] M. AVRIEL, *Nonlinear Programming: Analysis and Methods*, Prentice–Hall, Englewood Cliffs, NJ, 1976. Reprinted by Dover Publications, Mineola, New York, 2003.

[14] E. R. BARNES, *A variation on Karmarkar's algorithm for solving linear programming problems*, Mathematical Programming, 36 (1986), pp. 174–182.

[15] C. BARNHART, E. L. JOHNSON, G. L. NEMHAUSER, AND P. H. VANCE, *Crew scheduling*, in Handbook of Transportation Science, R. W. Hall, ed., Kluwer Academic Publishers, Dordrecht, The Netherlands, 1999, pp. 493–521.

[16] R. H. BARTELS AND G. H. GOLUB, *The simplex method of linear programming using LU decomposition*, Communications of the ACM, 12 (1969), pp. 266–268.

[17] J. BARUTT AND T. HULL, *Airline crew scheduling: Supercomputers and algorithms*, SIAM News, 23 (1990), p. 1.

[18] M. S. BAZARAA, J. J. JARVIS, AND H. D. SHERALI, *Linear Programming and Network Flows*, Wiley, New York, 1990.

[19] E. BEALE, *An alternative method for linear programming*, Proceedings of the Cambridge Philosophical Society, 50 (1954), pp. 513–523.

[20] ———, *Cycling in the dual simplex algorithm*, Naval Research Logistics Quarterly, 2 (1955), pp. 269–275.

[21] D. P. BERTSEKAS, *Constrained Optimization and Lagrange Multiplier Methods*, Academic Press, New York, 1982. Reprinted by Athena Scientific, Belmont, MA, 1996.

[22] D. P. BERTSEKAS AND P. TSENG, *The relax codes for linear minimum cost network flow problems*, Annals of Operations Research, 13 (1988), pp. 125–190.

[23] N. BIĆANIĆ AND K. JOHNSON, *Who was "Raphson"?*, International Journal for Numerical Methods in Engineering, 14 (1979), pp. 148–152.

[24] R. E. BIXBY, *Implementing the simplex method: The initial basis*, ORSA Journal on Computing, 4 (1993), pp. 267–284.

[25] R. E. BIXBY, J. W. GREGORY, I. J. LUSTIG, R. E. MARSTEN, AND D. F. SHANNO, *Very large-scale linear programming: A case study in combining interior point and simplex methods*, Operations Research, 40 (1992), pp. 885–897.

[26] R. G. BLAND, *New finite pivoting rules for the simplex method*, Mathematics of Operations Research, 2 (1977), pp. 103–107.

[27] R. G. BLAND, D. GOLDFARB, AND M. J. TODD, *The ellipsoid method: A survey*, Operations Research, 29 (1981), pp. 1039–1091.

[28] G. A. BLISS, *Calculus of Variations*, Open Court, Chicago, 1925.

[29] P. T. BOGGS AND J. W. TOLLE, *Sequential quadratic programming*, Acta Numerica, 4 (1995), pp. 1–52.

[30] O. BOLZA, *Lectures on the Calculus of Variations*, University of Chicago Press, Chicago, 1904. Reprinted by Scholarly Publishing Office, University of Michigan Library, Ann Arbor, MI, 2005.

[31] K.-H. BORGWARDT, *The average number of pivot steps required by the simplex method is polynomial*, Zeitschrift für Operations Research, 26 (1982), pp. 157–177.

[32] ———, *Some distribution-independent results about the asymptotic order of the average number of pivot steps of the simplex method*, Mathematics of Operations Research, 7 (1982), pp. 441–462.

[33] ———, *The Simplex Method*, Springer-Verlag, New York, 1987.

[34] B. BOSER, I. M. GUYON, AND V. VAPNIK, *A training algorithm for optimal margin classifiers*, in Proceedings of the Fifth Annual Workshop on Computational Learning Theory, ACM, New York, 1992, pp. 144–152.

[35] A. BREARLY, G. MITRA, AND H. WILLIAMS, *Analysis of mathematical programming problems prior to applying the simplex method*, Mathematical Programming, 8 (1975), pp. 54–83.

[36] R. C. BUCK, *Advanced Calculus, (third edition)*, McGraw–Hill, New York, 1978. Reprinted by Waveland Press, Long Grove, IL, 2003.

[37] C. J. BURGES, *A tutorial on support vector machines for pattern recognition*, Data Mining and Knowledge Discovery, 2 (1998), pp. 121–167.

[38] R. H. BYRD, M. E. HRIBAR, AND J. NOCEDAL, *An interior point algorithm for large-scale nonlinear programming*, SIAM Journal on Optimization, 9 (1999), pp. 877–900.

[39] R. H. BYRD, P. LU, J. NOCEDAL, AND C. ZHU, *A limited memory algorithm for bound constrained optimization*, SIAM Journal on Scientific Computing, 16 (1995), pp. 1190–1208.

[40] R. H. BYRD, J. NOCEDAL, AND Y.-X. YUAN, *Global convergence of a class of quasi-Newton methods on convex problems*, SIAM Journal on Numerical Analysis, 24 (1987), pp. 1171–1190.

[41] A. CAUCHY, *Mémoire sur la détermination des orbites des planètes et des comètes*, Compte Rendu des Séances de L'Académie des Sciences, XXV (1847), pp. 401–413.

[42] ———, *Mémoire sur les maxima et minima conditionnels*, Compte Rendu des Séances de L'Académie des Sciences, XXIV (1847), pp. 757–763.

[43] A. Cauchy, *Méthode générale pour la résolution des systèmes d'équations simultanées*, Compte Rendu des Séances de L'Académie des Sciences, XXV (1847), pp. 536–538.

[44] A. Charnes, *Optimality and degeneracy in linear programming*, Econometrica, 20 (1952), pp. 160–170.

[45] K. Chen, *Matrix Preconditioning Techniques and Applications*, Cambridge University Press, Cambridge, UK, 2005.

[46] J. Cheriyan and S. N. Maheshwari, *Analysis of preflow push algorithms for maximum network flow*, SIAM Journal on Computing, 18 (1989), pp. 1057–1086.

[47] V. Chvátal, *Linear Programming*, W. H. Freeman and Company, New York, 1983.

[48] T. F. Coleman, *A superlinear penalty function method to solve the nonlinear programming problem*, Ph.D. thesis, University of Waterloo, Waterloo, Ontario, Canada, 1979.

[49] P. Concus, G. H. Golub, and D. P. O'Leary, *A generalized conjugate gradient method for the numerical solution of elliptic partial differential equations*, in Sparse Matrix Computations, J. Bunch and D. Rose, eds., Academic Press, New York, 1976, pp. 309–332.

[50] A. R. Conn, *Constrained optimization using a nondifferentiable penalty function*, SIAM Journal on Numerical Analysis, 10 (1973), pp. 760–784.

[51] A. R. Conn, N. I. M. Gould, and P. L. Toint, *Trust-Region Methods*, SIAM, Philadelphia, 2000.

[52] A. R. Conn, N. M. Gould, and P. L. Toint, *LANCELOT: A Fortran Package for Large-Scale Nonlinear Optimization*, Springer-Verlag, Berlin, 1992.

[53] S. D. Conte and C. W. de Boor, *Elementary Numerical Analysis: An Algorithmic Approach*, McGraw–Hill, New York, 1980.

[54] R. Courant and D. Hilbert, *Methods of Mathematical Physics*, vol. I, Interscience, New York, 1953. Reprinted by Wiley-Interscience, New York, 1989.

[55] R. Courant, *Variational methods for the solution of problems of equilibrium and vibrations*, Bulletin of the American Mathematical Society, 49 (1943), pp. 1–23.

[56] N. Cristianini and J. Shawe-Taylor, *An Introduction to Support Vector Machines and Other Kernel-Based Learning Methods*, Cambridge University Press, Cambridge, UK, 2000.

[57] J. K. Cullum and R. A. Willoughby, *Lanczos Algorithms for Large Symmetric Eigenvalue Computations*. Vol. 1: Theory, Birkhäuser, Boston, 1985. Reprinted by SIAM, Philadelphia, 2002.

[58] A. Curtis, M. J. Powell, and J. K. Reid, *On the estimation of sparse Jacobian matrices*, Journal of the Institute of Mathematics and Its Applications, 13 (1974), pp. 117–119.

[59] G. B. DANTZIG, *Computational algorithm of the revised simplex method*, Report RM 1266, The Rand Corporation, Santa Monica, CA, 1953.

[60] ——, *Linear Programming and Extensions*, Princeton University Press, Princeton NJ, 1963. Reprinted by Princeton University Press, 1998.

[61] ——, *Making progress during a stall in the simplex algorithm*, Linear Algebra and Its Applications, 114/115 (1989), pp. 251–259.

[62] G. B. DANTZIG AND W. ORCHARD-HAYS, *The product form for the inverse in the simplex method*, Mathematical Tables and Other Aids to Computation, 8 (1954), pp. 64–67.

[63] G. B. DANTZIG, A. ORDEN, AND P. WOLFE, *The generalized simplex method for minimizing a linear form under linear inequality restraints*, Pacific Journal of Mathematics, 5 (1955), pp. 183–195.

[64] G. B. DANTZIG AND P. WOLFE, *The decomposition principle for linear programs*, Operations Research, 8 (1960), pp. 101–111.

[65] W. C. DAVIDON, *Variable metric method for minimization*, SIAM Journal on Optimization, 1 (1991), pp. 1–17.

[66] T. A. DAVIS, *Direct Methods for Sparse Linear Systems*, SIAM, Philadelphia, 2006.

[67] R. S. DEMBO, S. C. EISENSTAT, AND T. STEIHAUG, *Inexact Newton methods*, SIAM Journal on Numerical Analysis, 19 (1982), pp. 400–408.

[68] D. DEN HERTOG, *Interior point approach to linear, quadratic and convex programming*, Kluwer Academic Publishers, Dordrecht, The Netherlands, 1994.

[69] J. E. DENNIS, JR. AND J. J. MORÉ, *A characterization of superlinear convergence and its application to quasi-Newton methods*, Mathematics of Computation, 28 (1974), pp. 549–560.

[70] ——, *Quasi-Newton methods, motivation and theory*, SIAM Review, 19 (1977), pp. 46–89.

[71] J. E. DENNIS, JR. AND R. B. SCHNABEL, *Numerical Methods for Unconstrained Optimization and Nonlinear Equations*, Prentice–Hall, Englewood Cliffs, NJ, 1983. Reprinted by SIAM, Philadelphia, 1996.

[72] ——, *A view of unconstrained optimization*, in Optimization, G. Nemhauser, A. Rinnooy Kan, and M. J. Todd, eds., Elsevier, Amsterdam, 1989, pp. 1–72.

[73] E. W. DIJKSTRA, *A note on two problems in connexion with graphs*, Numerische Mathematik, 1 (1959), pp. 269–271.

[74] I. DIKIN, *Iterative solution of problems of linear and quadratic programming*, Doklady Akademiia Nauk SSSR, 174 (1967), pp. 747–748.

[75] A. S. DRUD, *CONOPT—a large-scale GRG code*, ORSA Journal on Computing, 6 (1992), pp. 207–216.

[76] J. EDMONDS AND R. M. KARP, *Theoretical improvements in algorithmic efficiency of network flow problems*, Journal of the ACM, 19 (1972), pp. 248–264.

[77] K. EISEMANN, *The trim problem*, Management Science, 3 (1957), pp. 279–284.

[78] J. FARKAS, *Über die Theorie der einfachen Ungleichungen*, Journal für die reine und angewandte Mathematik, 124 (1901), pp. 1–27.

[79] A. V. FIACCO, *Historical Survey of Sequential Unconstrained Methods for Solving Constrained Minimization Problems*, Technical Paper RAC–TP–267, Research Analysis Corporation, 1967.

[80] A. V. FIACCO AND G. P. MCCORMICK, *Nonlinear Programming: Sequential Unconstrained Minimization Techniques*, John Wiley and Sons, New York, 1968. Reprinted by SIAM, Philadelphia, 1990.

[81] R. FLETCHER, *Practical Methods of Optimization*, 2nd ed., Wiley, New York, 2000.

[82] R. FLETCHER, N. I. M. GOULD, S. LEYFFER, P. L. TOINT, AND A. WÄCHTER, *Global convergence of a trust-region SQP-filter algorithm for general nonlinear programming*, SIAM Journal on Optimization, 13 (2002), pp. 635–659.

[83] R. FLETCHER AND S. LEYFFER, *Nonlinear programming without a penalty function*, Mathematical Programming, 91 (2002), pp. 239–270.

[84] R. FLETCHER, S. LEYFFER, AND P. L. TOINT, *A brief history of filter methods*, SIAG/OPT Views and News, 18 (2007), pp. 2–12.

[85] R. FLETCHER AND M. J. POWELL, *A rapidly convergent descent method for minimization*, Computer Journal, 6 (1963), pp. 163–168.

[86] R. FLETCHER AND C. REEVES, *Function minimization by conjugate gradients*, Computer Journal, 6 (1964), pp. 149–154.

[87] C. A. FLOUDAS AND P. M. PARDALOS, eds., *Recent Advances in Global Optimization*, Princeton University Press, Princeton, NJ, 1992. Reprinted by Princeton University Press, 2007.

[88] L. FORD, JR. AND D. FULKERSON, *A suggested computation for maximal multi-commodity network flows*, Management Science, 5 (1958), pp. 97–101.

[89] ———, *Flows in Networks*, Princeton University Press, Princeton, NJ, 1962.

[90] J. J. FORREST AND D. GOLDFARB, *Steepest-edge simplex algorithms for linear programming*, Mathematical Programming, 57 (1992), pp. 341–374.

[91] J. J. FORREST AND J. TOMLIN, *Updating triangular factors of the basis to maintain sparsity in the product-form simplex method*, Mathematical Programming, 2 (1972), pp. 263–278.

[92] R. FOURER, D. M. GAY, AND B. W. KERNIGHAN, *AMPL: A Modeling Language for Mathematical Programming (second edition)*, Thomson/Brooks/Cole, Pacific Grove, CA, 2003.

[93] M. FREDMAN AND R. E. TARJAN, *Fibonacci heaps and their uses in improved network optimization algorithms*, Journal of the ACM, 34 (1987), pp. 596–615.

[94] K. FRISCH, *The Logarithmic Potential Method of Convex Programming*, Memorandum 13, University Institute of Economics, Oslo, Norway, 1955.

[95] P. GÁCS AND L. LOVÁSZ, *Khachiyan's algorithm for linear programming*, Mathematical Programming Study, 14 (1981), pp. 61–68.

[96] T. GAL, *Postoptimal Analysis, Parametric Programming and Related Topics*, McGraw–Hill, New York, 1979.

[97] D. GALE, H. W. KUHN, AND A. W. TUCKER, *Linear programming and the theory of games*, in Activity Analysis of Production and Allocation, T. Koopmans, ed., Wiley, New York, 1951, pp. 317–329.

[98] S. I. GASS AND T. SAATY, *The computational algorithm for the parametric objective function*, Naval Research Logistics Quarterly, 2 (1955), pp. 39–45.

[99] C. F. GAUSS, *Theoria Motus Corporum Cælestium in Sectionibus Conicus Solem Ambientum*, Dover Press, New York, 1809.

[100] ——, *Bestimmung des kleinsten Werthes der Summe $x_1^2 + x_2^2 + \cdots + x_n^2 = R^2$ für m gegebene Ungleichungen $u \geq 0$*, in Werke, vol. X (part II), Gedruckt in der Dieterichschen Universitätsdruckerei (W.F. Kaestner), Göttingen, 1850–51, pp. 473–482.

[101] D. M. GAY, *Computing optimal locally constrained steps*, SIAM Journal on Scientific and Statistical Computing, 2 (1981), pp. 186–197.

[102] ——, *A variant of Karmarkar's algorithm for problems in standard form*, Mathematical Programming, 37 (1987), pp. 81–90.

[103] I. GELFAND AND S. FOMIN, *Calculus of Variations*, Prentice–Hall, Englewood Cliffs, NJ, 1963. Reprinted by Dover Publications, New York, 2000.

[104] A. GEORGE AND J. LIU, *Computer Solution of Large Sparse Positive Definite Systems*, Prentice–Hall, Englewood Cliffs, NJ, 1981.

[105] J. C. GILBERT AND C. LEMARÉCHAL, *Some numerical experiments with variable storage quasi-Newton algorithms*, Mathematical Programming, 45 (1989), pp. 407–436.

[106] J. C. GILBERT AND J. NOCEDAL, *Global convergence properties of conjugate gradient methods for optimization*, SIAM Journal on Optimization, 2 (1992), pp. 21–42.

[107] P. E. GILL AND W. MURRAY, *Quasi-Newton methods for unconstrained optimization*, Journal of the Institute for Mathematics and Its Applications, 9 (1972), pp. 91–108.

[108] ———, *Newton-type methods for unconstrained and linearly constrained optimization*, Mathematical Programming, 28 (1974), pp. 311–350.

[109] ———, *Safeguarded Steplength Algorithms for Optimization Using Descent Methods*, Report NAC 37, National Physical Laboratory, Teddington, England, 1974.

[110] P. E. GILL, W. MURRAY, M. A. SAUNDERS, J. A. TOMLIN, AND M. H. WRIGHT, *On projected Newton barrier methods for linear programming and an equivalence to Karmarkar's projective method*, Mathematical Programming, 36 (1986), pp. 183–209.

[111] P. E. GILL, W. MURRAY, AND M. H. WRIGHT, *Practical Optimization*, Academic Press, New York, 1981.

[112] ———, *Numerical Linear Algebra and Optimization*, vol. 1, Addison–Wesley, Redwood City, CA, 1991.

[113] P. C. GILMORE AND R. E. GOMORY, *A linear programming approach to the cutting stock problem*, Operations Research, 9 (1961), pp. 849–859.

[114] ———, *A linear programming approach to the cutting stock problem—Part* II, Operations Research, 11 (1963), pp. 863–888.

[115] ———, *Multistage cutting stock problems of two and more dimensions*, Operations Research, 13 (1965), pp. 94–120.

[116] A. GOLDBERG, *A New Max-Flow Algorithm*, Technical Report MIT LCS TM–291, Laboratory for Computer Science, MIT, Cambridge, MA, 1985.

[117] A. GOLDBERG AND R. E. TARJAN, *A new approach to the maximum flow problem*, Journal of the ACM, 35 (1988), pp. 921–940.

[118] D. GOLDFARB AND J. K. REID, *A practical steepest-edge simplex algorithm*, Mathematical Programming, 12 (1977), pp. 361–371.

[119] A. J. GOLDMAN AND A. W. TUCKER, *Theory of linear programming*, in Linear Inequalities and Related Systems, H. W. Kuhn and A. W. Tucker, eds., Princeton University Press, Princeton, NJ, 1956, pp. 53–97.

[120] H. GOLDSTINE, *A History of the Calculus of Variations from the* 17th *through the* 19th *Century*, Springer-Verlag, New York, 1980.

[121] G. H. GOLUB AND C. VAN LOAN, *Matrix Computations (third edition)*, The Johns Hopkins University Press, Baltimore, 1996.

[122] J. GONDZIO, *Multiple centrality corrections in a primal-dual method for linear programming*, Computational Optimization and Applications, 6 (1996), pp. 137–156.

[123] C. C. GONZAGA, *Path-following methods for linear programming*, SIAM Review, 34 (1992), pp. 167–224.

[124] R. GOPALAN AND K. T. TALLURI, *Mathematical models in airline schedule planning: A survey*, Annals of Operations Research, 76 (1998), pp. 155–185.

[125] F. J. GOULD AND J. W. TOLLE, *A necessary and sufficient qualification for constrained optimization*, SIAM Journal on Applied Mathematics, 20 (1971), pp. 164–172.

[126] J. GREGORY AND C. LIN, *Constrained Optimization in the Calculus of Variations and Optimal Control Theory*, Van Nostrand Reinhold, New York, 1992. Reprinted by Springer-Verlag, New York, 2007.

[127] A. GRIEWANK, *Evaluating Derivatives: Principles and Techniques of Algorithmic Differentiation*, SIAM, Philadelphia, 2000.

[128] A. GRIEWANK AND G. F. CORLISS, eds., *Automatic Differentiation of Algorithms: Theory, Implementation, and Application*, Proceedings of the First SIAM Workshop (Breckenridge, CO, January 6–8, 1991), SIAM, Philadelphia, 1991.

[129] A. GRIEWANK AND P. L. TOINT, *Partitioned variable metric updates for large structured optimization problems*, Numerische Mathematik, 39 (1982), pp. 119–137.

[130] I. GRIVA AND R. J. VANDERBEI, *Case studies in optimization: Catenary problem*, Optimization and Engineering, 6 (2005), pp. 463–482.

[131] M. GUIGNARD, *Generalized Kuhn–Tucker conditions for mathematical programming problems in a Banach space*, SIAM Journal on Control, 7 (1969), pp. 232–241.

[132] O. GÜLER AND Y. YE, *Convergence behavior of interior-point algorithms*, Mathematical Programming, 60 (1993), pp. 215–228.

[133] W. W. HAGER, *Updating the inverse of a matrix*, SIAM Review, 31 (1989), pp. 221–239.

[134] M. HAIMOVICH, *The Simplex Method Is Very Good!—On the Expected Number of Pivot Steps and Related Properties of Random Linear Programs*, preprint, Columbia University, New York, 1983.

[135] E. HANSEN AND G. W. WALSTER, *Global Optimization Using Interval Analysis (second edition)*, CRC Press, Boca Raton, FL, 2003.

[136] P. M. HARRIS, *Pivot selection methods of the Devex LP code*, Mathematical Programming, 5 (1973), pp. 1–28.

[137] M. R. HESTENES, *Calculus of Variations and Optimal Control Theory*, John Wiley & Sons, New York, 1966.

[138] ——, *Multiplier and gradient methods*, Journal of Optimization Theory and Applications, 4 (1969), pp. 303–320.

[139] M. R. HESTENES AND E. STIEFEL, *Methods of conjugate gradients for solving linear systems*, Journal of Research of the National Bureau of Standards, 49 (1952), pp. 409–436.

[140] N. HIGHAM, *Is fast matrix multiplication of practical use?*, SIAM News, 23 (1990), pp. 12–14.

[141] J. HO, T. LEE, AND R. SUNDARRAJ, *Decomposition of linear programs using parallel computation*, Mathematical Programming, 42 (1988), pp. 391–405.

[142] A. HOFFMAN, M. MANNOS, D. SOKOLOWSKY, AND N. WIEGMANN, *Computational experience in solving linear programs*, J. Soc. Indust. Appl. Math., 1 (1953), pp. 17–33.

[143] A. J. HOFFMAN, *Cycling in the Simplex Algorithm*, Report 2974, National Bureau of Standards, Gaithersburg, MD, 1953.

[144] R. A. HORN AND C. R. JOHNSON, *Topics in Matrix Analysis*, Cambridge University Press, Cambridge, UK, 1991. Reprinted by Cambridge University Press, 1994.

[145] R. HORST, P. M. PARDALOS, AND N. V. THOAI, *Introduction to Global Optimization (second edition)*, Kluwer, Dordrecht, The Netherlands, 2000.

[146] F. JOHN, *Extremum problems with inequalities as subsidiary conditions*, in Studies and Essays Presented to R. Courant on his 60th Birthday, K. Friedricks, O. E. Neugebauer, and J. Stoker, eds., Wiley-Interscience, New York, 1948, pp. 187–204.

[147] C. A. JOHNSON AND A. SOFER, *A primal-dual method for large-scale image reconstruction in emission tomography*, SIAM Journal on Optimization, 11 (2001), pp. 691–715.

[148] K. L. JONES, I. J. LUSTIG, J. M. FARVOLDEN, AND W. B. POWELL, *Multicommodity network flows: The impact of formulation on decomposition*, Mathematical Programming, 62 (1993), pp. 95–117.

[149] A. R. KAN AND G. TIMMER, *Global optimization*, in Optimization, G. Nemhauser, A. R. Kan, and M. J. Todd, eds., Elsevier, Amsterdam, 1989, pp. 631–659.

[150] L. KANTOROVICH, *Mathematical methods of organizing and planning production*, Leningrad, 1939 (in Russian). English translation in *Management Science*, 6 (1959/1960), pp. 366–422.

[151] N. KARMARKAR, *A new polynomial-time algorithm for linear programming*, Combinatorica, 4 (1984), pp. 373–395.

[152] L. G. KHACHIYAN, *A polynomial algorithm in linear programming*, Doklady Akademii Nauk SSSR, 244 (1979), pp. 1093–1096.

[153] V. KLEE AND P. KLEINSCHMIDT, *Geometry of the Gass-Saaty parametric cost LP algorithm*, Discrete and Computational Geometry, 5 (1990), pp. 13–26.

[154] V. KLEE AND G. J. MINTY, *How good is the simplex algorithm?*, in Inequalities, III, O. Shisha, ed., Academic Press, New York, 1972, pp. 159–175.

[155] M. KOJIMA, S. MIZUNO, AND A. YOSHISE, *A primal-dual interior point algorithm for linear programming*, in Progress in Mathematical Programming: Interior Point and Related Methods, N. Megiddo, ed., Springer-Verlag, New York, 1989, pp. 29–47.

[156] T. G. KOLDA, R. M. LEWIS, AND V. TORCZON, *Optimization by direct search: New perspectives on some classical and modern methods*, SIAM Review, 45 (2003), pp. 385–482.

[157] H. W. KUHN, *Nonlinear programming: A historical note*, in History of Mathematical Programming, J. Lenstra, A. Rinnooy Kan, and A. Schrijver, eds., North–Holland, Amsterdam, 1991, pp. 82–96.

[158] H. W. KUHN AND A. W. TUCKER, *Nonlinear programming*, in Proceedings of the Second Berkeley Symposium on Mathematical Statistics and Probability, J. Neyman, ed., University of California Press, Berkeley, CA, 1951, pp. 481–492.

[159] J. L. LAGRANGE, *Oeuvres de Lagrange*, vols. XI and XII, Gauthier-Villars, Paris, 1888–1889.

[160] C. LANCZOS, *An iteration method for the solution of the eigenvalue problem of linear differential and integral operators*, Journal of Research of the National Bureau of Standards, 45 (1950), pp. 255–282.

[161] K. LANGE AND R. CARSON, *Reconstruction algorithms for emission and transmission tomography*, Journal of Computer Assisted Tomography, 8 (1984), pp. 306–316.

[162] L. LASDON, A. WAREN, A. JAIN, AND M. RATNER, *Design and testing of a GRG code for nonlinear optimization*, ACM Transactions on Mathematical Software, 4 (1978), pp. 34–50.

[163] E. K. LEE AND J. O. DEASY, *Optimization in intensity modulated radiation therapy*, SIAG/OPT Views and News, 17 (2006), pp. 20–32.

[164] C. LEMARÉCHAL, *An extension of Davidon methods to non-differentiable problems*, Mathematical Programming Study, 3 (1975), pp. 95–109.

[165] C. LEMKE, *The dual method of solving the linear programming problem*, Naval Research Logistics Quarterly, 1 (1954), pp. 36–47.

[166] K. LEVENBERG, *A method for the solution of certain problems in least squares*, Quarterly of Applied Mathematics, 2 (1944), pp. 164–168.

[167] L. LIBERTI AND N. MACULAN, eds., *Global Optimization: From Theory to Implementation (Nonconvex Optimization and Its Applications)*, Springer-Verlag, New York, 2006.

[168] D. LIU AND J. NOCEDAL, *On the limited memory BFGS method for large scale optimization*, Mathematical Programming, 45 (1989), pp. 503–528.

[169] D. G. LUENBERGER, *Linear and Nonlinear Programming (second edition)*, Springer-Verlag, New York, 2003.

[170] I. J. LUSTIG, R. E. MARSTEN, AND D. F. SHANNO, *The interaction of algorithms and architectures for interior point methods*, in Advances in Optimization and Parallel Computing, P. Pardalos, ed., North–Holland, Amsterdam, 1992, pp. 190–205.

[171] ———, *Computational experience with a globally convergent primal-dual predictor-corrector algorithm for linear programming*, Mathematical Programming, 66 (1994), pp. 123–135.

[172] ———, *Interior point methods for linear programming: Computational state of the art*, ORSA Journal on Computing, 6 (1994), pp. 1–14.

[173] J. N. LYNESS AND C. B. MOLER, *Numerical differentiation of analytic functions*, SIAM Journal on Numerical Analysis, 4 (1967), pp. 202–210.

[174] T. L. MAGNANTI AND J. B. ORLIN, *Parametric linear programming and anti-cycling pivoting rules*, Mathematical Programming, 41 (1988), pp. 317–325.

[175] O. L. MANGASARIAN, *Nonlinear Programming*, McGraw–Hill, New York, 1969. Reprinted by SIAM, Philadelphia, 1994.

[176] A. MANNE, *Programming of economic lot sizes*, Management Science, 4 (1958), pp. 115–135.

[177] H. M. MARKOWITZ, *The elimination form of the inverse and its application to linear programming*, Management Science, 3 (1957), pp. 255–269.

[178] H. M. MARKOWITZ AND G. P. TODD, *Mean-Variance Analysis in Portfolio Choice and Capital Markets (revised reissue of 1987 edition)*, John Wiley and Sons, New York, 2000.

[179] D. W. MARQUARDT, *An algorithm for least-squares estimation of nonlinear parameters*, SIAM Journal on Applied Mathematics, 11 (1963), pp. 431–441.

[180] A. MAYER, *Begründung der Lagrange'schen Multiplicatorenmethode in der Variationsrechnung*, Mathematische Annalen, 26 (1886), pp. 74–82.

[181] D. MAYNE AND N. MARATOS, *A first-order, exact penalty function algorithm for equality constrained optimization problems*, Mathematical Programming, 16 (1979), pp. 303–324.

[182] N. MEGIDDO, *Pathways to the optimal set in linear programming*, in Progress in Mathematical Programming: Interior Point and Related Methods, N. Megiddo, ed., Springer-Verlag, New York, 1989, pp. 131–158.

[183] S. MEHROTRA, *On the implementation of a primal-dual interior point method*, SIAM Journal on Optimization, 2 (1992), pp. 575–601.

[184] J. MEIJERINK AND H. V. D. VORST, *An iterative solution method for linear equation systems of which the coefficient matrix is a symmetric M-matrix*, Mathematics of Computation, 31 (1977), pp. 148–162.

[185] S. MIZUNO, M. J. TODD, AND Y. YE, *An adaptive-step primal-dual interior point algorithms for linear programming*, Mathematics of Operations Research, 18 (1993), pp. 964–981.

[186] R. MONTEIRO AND I. ADLER, *Interior path following primal-dual algorithms: Part* I: *Linear programming*, Mathematical Programming, 44 (1989), pp. 27–41.

[187] J. J. MORÉ, *Recent developments in algorithms and software for trust region methods*, in Mathematical Programming: The State of the Art (Bonn, 1982), A. Bachem, M. Grötschel, and B. Korte, eds., Springer, Berlin, 1983, pp. 258–287.

[188] J. J. MORÉ AND D. C. SORENSEN, *Computing a trust region step*, SIAM Journal on Scientific and Statistical Computing, 4 (1983), pp. 553–572.

[189] J. J. MORÉ AND D. THUENTE, *Line search algorithms with guaranteed sufficient decrease*, ACM Transactions on Mathematical Software, 20 (1994), pp. 286–307.

[190] W. MURRAY, *Analytical expressions for the eigenvalues and eigenvectors of the Hessian matrices of barrier and penalty functions*, Journal of Optimization Theory and Applications, 7 (1971), pp. 189–196.

[191] K. G. MURTY, *Linear Programming*, Wiley, New York, 1983.

[192] ——, *Network Programming*, Prentice–Hall, Englewood Cliffs, NJ, 1992. Reprinted by Prentice–Hall, 1998.

[193] S. G. NASH, *Newton-type minimization via the Lanczos method*, SIAM Journal on Numerical Analysis, 21 (1984), pp. 770–788.

[194] ——, *Preconditioning of truncated-Newton methods*, SIAM Journal on Scientific and Statistical Computing, 6 (1985), pp. 599–616.

[195] ——, *A survey of truncated-Newton methods*, Journal of Computational and Applied Mathematics, 124 (2000), pp. 45–59.

[196] S. G. NASH AND J. NOCEDAL, *A numerical study of the limited memory BFGS method and the truncated-Newton method for large scale optimization*, SIAM Journal on Optimization, 1 (1991), pp. 358–372.

[197] S. G. NASH AND A. SOFER, *Assessing a search direction within a truncated-Newton method*, Operations Research Letters, 9 (1990), pp. 219–221.

[198] ——, *A general-purpose parallel algorithm for unconstrained optimization*, SIAM Journal on Optimization, 4 (1991), pp. 530–547.

[199] ——, *Algorithm 711: BTN: Software for parallel unconstrained optimization*, ACM Transactions on Mathematical Software, 18 (1992), pp. 414–448.

[200] ——, *A barrier method for large-scale constrained optimization*, ORSA Journal on Computing, 5 (1993), pp. 40–53.

[201] J. NAZARETH, *Computer Solution of Linear Programs*, Oxford University Press, New York, 1987.

[202] J. NELDER AND R. MEAD, *A simplex method for function minimization*, Computing Journal, 7 (1965), pp. 308–313.

[203] G. L. NEMHAUSER AND L. A. WOLSEY, *Integer and Combinatorial Optimization*, Wiley, New York, 1988. Reprinted by Wiley-Interscience, 1999.

[204] A. S. NEMIROVSKII, *Interior point polynomial time methods in convex programming*, Lecture Notes, Faculty of Industrial Engineering and Management, Technion—The Israel Institute of Technology, Haifa, Israel, 1994.

[205] Y. NESTEROV AND A. NEMIROVSKII, *Interior-Point Polynomial Algorithms in Convex Programming*, SIAM, Philadelphia, 1994.

[206] Y. E. NESTEROV AND M. J. TODD, *Primal-dual interior-point methods for self-scaled cones*, SIAM Journal on Optimization, 8 (1998), pp. 324–364.

[207] J. NOCEDAL, *Theory of algorithms for unconstrained optimization*, Acta Numerica, 1 (1992), pp. 199–242.

[208] W. ORCHARD-HAYS, *Background, Development and Extensions of the Revised Simplex Method*, Report RM 1433, The Rand Corporation, Santa Monica, CA, 1954.

[209] J. B. ORLIN, *Genuinely Polynomial Simplex and Non-Simplex Algorithms for the Minimum Cost Flow Problem*, Technical Report 1615–84, Sloan School of Management, MIT, Cambridge, MA, 1984.

[210] J. B. ORLIN, S. A. PLOTKIN, AND É. TARDOS, *Polynomial dual network simplex algorithms*, Mathematical Programming, 60 (1993), pp. 255–276.

[211] J. M. ORTEGA AND W. C. RHEINBOLDT, *Iterative Solution of Nonlinear Equations in Several Variables*, Academic Press, New York, 1970. Reprinted by SIAM, Philadelphia, 2000.

[212] G. OSTROVSKII, Y. WOLIN, AND W. BORISOV, *Über die Berechnung von Ableitungen*, Wissenschaftliche Zeitschrift der Technischen Hochschule für Chemie, Leuna-Merseburg, 13 (1971), pp. 382–384.

[213] C. C. PAIGE AND M. A. SAUNDERS, *Solution of sparse indefinite systems of linear equations*, SIAM Journal on Numerical Analysis, 12 (1975), pp. 617–629.

[214] E. R. PANIER AND A. L. TITS, *On combining feasibility, descent and superlinear convergence in inequality constrained optimization*, Mathematical Programming, 59 (1993), pp. 261–276.

[215] C. H. PAPADIMITRIOU AND K. STEIGLITZ, *Combinatorial Optimization: Algorithms and Complexity*, Prentice–Hall, Englewood Cliffs, NJ, 1982. Reprinted by Dover Publications, Mineola, NY, 1998.

[216] B. N. PARLETT, *The Symmetric Eigenvalue Problem*, Prentice–Hall, Englewood Cliffs, NJ, 1980. Reprinted by SIAM, Philadelphia, 1998.

[217] A. PERRY, *A Class of Conjugate-Gradient Algorithms with a Two-Step Variable-Metric Memory*, Discussion paper 269, Center for Mathematical Studies in Economics and Management Science, Northwestern University, Evanston, IL, 1977.

[218] T. PIETRZYKOWSKI, *An exact potential method for constrained maxima*, SIAM Journal on Numerical Analysis, 6 (1969), pp. 299–304.

[219] R. POLYAK, *Modified barrier functions (theory and methods)*, Mathematical Programming, 54 (1992), pp. 177–222.

[220] M. J. D. POWELL, *A method for nonlinear constraints in minimization problems*, in Optimization, R. Fletcher, ed., Academic Press, New York, 1969, pp. 283–298.

[221] ——, *A new algorithm for unconstrained optimization*, in Nonlinear Programming, J. Rosen, O. Mangasarian, and K. Ritter, eds., Academic Press, New York, 1970, pp. 31–65.

[222] ——, *Convergence properties of a class of minimization algorithms*, in Nonlinear Programming 2, O. Mangasarian, R. Meyer, and S. Robinson, eds., Academic Press, New York, 1975, pp. 1–27.

[223] ——, *Restart procedures for the conjugate gradient method*, Mathematical Programming, 12 (1977), pp. 241–254.

[224] ——, *Problems related to unconstrained optimization*, in Numerical Methods for Unconstrained Optimization, W. Murray, ed., Academic Press, London, New York, 1972, pp. 29–55.

[225] M. J. D. POWELL AND P. L. TOINT, *On the estimation of sparse Hessian matrices*, SIAM Journal on Numerical Analysis, 16 (1979), pp. 1060–1074.

[226] J. RENEGAR, *A polynomial-time algorithm based on Newton's method for linear programming*, Mathematical Programming, 40 (1988), pp. 59–93.

[227] T. R. ROCKAFELLAR, *Conjugate Duality and Optimization*, SIAM, Philadelphia, 1974.

[228] C. ROOS AND J.-P. VIAL, *A polynomial method of approximate centers for linear programming*, Mathematical Programming, 54 (1992), pp. 295–306.

[229] C. ROOS, J.-P. VIAL, AND T. TERLAKY, *Interior Point Methods for Linear Optimization (second edition)*, Springer-Verlag, New York, 2005.

[230] J. B. ROSEN, *The gradient projection method for nonlinear programming, Part I. Linear constraints*, J. Soc. Indust. Appl. Math., 8 (1960), pp. 181–217.

[231] A. RUSZCZYNSKI AND R. J. VANDERBEI, *Frontiers of stochastically nondominated portfolios*, Econometrica, 71 (2003), pp. 1287–1297.

[232] Y. SAAD, *Krylov subspace methods for solving large unsymmetric linear systems*, Mathematics of Computation, 37 (1981), pp. 105–126.

[233] ——, *Iterative Methods for Sparse Linear Systems (second edition)*, SIAM, Philadelphia, 2003.

[234] T. SAUER, *Numerical Analysis*, Addison–Wesley, Boston, 2006.

[235] R. B. SCHNABEL AND E. ESKOW, *A new modified Cholesky factorization*, SIAM Journal on Scientific and Statistical Computing, 11 (1990), pp. 1136–1158.

[236] B. SCHÖLKOPF, C. J. BURGES, AND A. J. SMOLA, eds., *Advances in Kernel Methods: Support Vector Learning*, MIT Press, Cambridge, MA, 1999.

[237] R. SCHRADER, *Ellipsoid methods*, in Modern Applied Mathematics—Optimization and Operations Research, B. Korte, ed., North–Holland, Amsterdam, 1982, pp. 265–311.

[238] ——, *The ellipsoid method and its implications*, OR Spektrum, 5 (1983), pp. 1–13.

[239] A. SCHRIJVER, *Theory of Linear and Integer Programming*, John Wiley & Sons, New York, 1986. Reprinted by John Wiley & Sons, 1998.

[240] D. F. SHANNO, *Conjugate-gradient methods with inexact searches*, Mathematics of Operations Research, 3 (1978), pp. 244–256.

[241] D. M. SHEPARD, M. C. FERRIS, G. H. OLIVERA, AND T. R. MACKIE, *Optimizing the delivery of radiation therapy to cancer patients*, SIAM Review, 41 (1999), pp. 721–744.

[242] L. A. SHEPP AND Y. VARDI, *Maximum likelihood reconstruction for emission tomography*, IEEE Transaction on Medical Imaging, 1 (1982), pp. 113–122.

[243] N. SHOR, *On the structure of algorithms for the numerical solution of optimal planning and design problems*, Ph.D. thesis, Cybernetics Institute, Academy of Sciences of the Ukrainian SSR, Kiev, 1964.

[244] T. SIMPSON, *Essays on Several Curious and Useful Subjects in Speculative and Mix'd Mathematicks, Illustrated by a Variety of Examples*, London, 1740.

[245] R. SKEEL, *Scaling for numerical stability in Gaussian elimination*, Journal of the ACM, 26 (1979), pp. 494–526.

[246] D. C. SORENSEN, *Newton's method with a model trust region modification*, SIAM Journal on Numerical Analysis, 19 (1982), pp. 409–426.

[247] D. A. SPIELMAN AND S.-H. TENG, *Smoothed analysis of algorithms: Why the simplex algorithm usually takes polynomial time*, Journal of the ACM, 51 (2004), pp. 385–463.

[248] W. SQUIRE AND G. TRAPP, *Using complex variables to estimate derivatives of real functions*, SIAM Review, 40 (1998), pp. 110–112.

[249] T. STEIHAUG, *The conjugate gradient method and trust regions in large scale optimization*, SIAM Journal on Numerical Analysis, 20 (1983), pp. 626–637.

[250] U. H. SUHL AND L. SUHL, *Computing sparse LU factorizations for large-scale linear programming bases*, ORSA Journal on Computing, 2 (1990), pp. 325–335.

[251] A. ŚWIETANOWSKI, *A new steepest edge approximation for the simplex method for linear programming*, Computational Optimization and Applications, 10 (1998), pp. 271–281.

[252] É. TARDOS, *A strongly polynomial minimum cost circulation algorithm*, Combinatorica, 5 (1985), pp. 247–255.

[253] S. THOMAS, *Sequential Estimation Techniques for Quasi-Newton Algorithms*, Ph.D. thesis, Cornell University, Ithaca, New York, 1975.

[254] M. J. TODD, *Semidefinite optimization*, Acta Numerica, 10 (2001), pp. 515–560.

[255] M. J. TODD AND B. BURRELL, *An extension of Karmarkar's algorithm for linear programming using dual variables*, Algorithmica, 1 (1986), pp. 409–424.

[256] P. L. TOINT, *Towards an efficient sparsity exploiting Newton method for minimization*, in Sparse Matrices and Their Uses, I. Duff, ed., Academic Press, New York, 1981, pp. 57–87.

[257] L. N. TREFETHEN AND D. BAU, III, *Numerical Linear Algebra*, SIAM, Philadelphia, 1997.

[258] A. W. TUCKER, *Dual systems of homogeneous linear equations*, Annals of Mathematics Studies, 38 (1956), pp. 3–18.

[259] A. M. TURING, *On computable numbers, with an application to the Entscheidungsproblem*, Proceedings of the London Mathematical Society, 42 (1936), pp. 230–265.

[260] R. J. VANDERBEI, *Linear Programming: Foundations and Extensions (third edition)*, Springer, Berlin, 2007.

[261] R. J. VANDERBEI, M. MEKETON, AND B. FREEDMAN, *A modification of Karmarkar's linear programming algorithm*, Algorithmica, 1 (1986), pp. 395–407.

[262] R. J. VANDERBEI AND D. F. SHANNO, *An interior-point algorithm for nonconvex nonlinear programming*, Computational Optimization and Applications, 13 (1999), pp. 231–252.

[263] V. VAPNIK, ed., *The Nature of Statistical Learning Theory (second edition)*, Springer-Verlag, New York, 1998.

[264] J. VON NEUMANN, *Über ein ökonomisches Gleichungssystem und eine Verallgemeinerung des Brouwerschen Fixpunktsatzes*, 1937. English translation in *The Review of Economic Studies* 13 (1945/1946), pp. 1–9.

[265] J. VON NEUMANN, *Discussion of a maximum problem*, in John von Neumann (Collected Works), A. Taub, ed., vol. VI, Pergamon Press, Oxford, 1963, pp. 89–95.

[266] A. WÄCHTER AND L. T. BIEGLER, *On the implementation of an interior-point filter line-search algorithm for large-scale nonlinear programming*, Mathematical Programming, 106 (2006), pp. 25–57.

[267] R. WENGERT, *A simple automatic derivative evaluation program*, Communications of the ACM, 7 (1964), pp. 463–464.

[268] D. WHITESIDE, ed., *The Mathematical Papers of Isaac Newton*, vols. 1–7, Cambridge University Press, Cambridge, UK, 1967–1976.

[269] R. WILSON, *A Simplicial Algorithm for Concave Programming*, Ph.D. thesis, Harvard University, Cambridge, MA, 1963.

[270] P. WOLFE, *Convergence conditions for ascent methods*, SIAM Review, 11 (1969), pp. 226–235.

[271] H. WOLKOWICZ, R. SAIGAL, AND L. VANDENBERGHE, eds., *Handbook of Semidefinite Programming—Theory, Algorithms, and Applications*, Kluwer Academic Publishers, Dordrecht, The Netherlands, 2000.

[272] L. A. WOLSEY, *Integer Programming*, Wiley-Interscience, New York, 1998.

[273] M. H. WRIGHT, *Interior methods for constrained optimization*, Acta Numerica, 1 (1992), pp. 341–407.

[274] ——, *Some properties of the Hessian of the logarithmic barrier function*, Mathematical Programming, 67 (1994), pp. 265–295.

[275] S. J. WRIGHT, *Primal-Dual Interior-Point Methods*, SIAM, Philadelphia, 1997.

[276] X. XU, P. HUNG, AND Y. YE, *A simplified homogeneous self-dual linear programming algorithm and its implementation*, Annals of Operations Research, 62 (1996), pp. 151–171.

[277] H. YAMASHITA AND H. YABE, *Superlinear and quadratic convergence of some primal-dual interior point methods for constrained optimization*, Mathematical Programming, 75 (1996), pp. 377–397.

[278] ——, *Quadratic convergence of a primal-dual interior point method for degenerate nonlinear optimization problems*, Computational Optimization and Applications, 31 (2005), pp. 123–143.

[279] Y. YE, *Interior-Point Algorithms: Theory and Analysis*, John Wiley and Sons, New York, 1997.

[280] Y. YE AND M. KOJIMA, *Recovering optimal dual solutions in Karmarkar's polynomial algorithm for linear programming*, Mathematical Programming, 39 (1987), pp. 305–317.

[281] Y. YE, M. J. TODD, AND S. MIZUNO, *An o($\sqrt{n}L$) iteration homogeneous and self-dual linear programming algorithm*, Mathematics of Operations Research, 19 (1994), pp. 53–67.

[282] T. J. YPMA, *Historical development of the Newton–Raphson method*, SIAM Review, 37 (1995), pp. 531–551.

[283] D. YUDIN AND A. S. NEMIROVSKII, *Informational complexity and efficient methods for the solution of convex extremal problems*, Ekonomika i Matematicheskie Metody, 12 (1976), pp. 357–369.

[284] N. ZADEH, *A bad network problem for the simplex method and other minimum cost flow algorithms*, Mathematical Programming, 5 (1973), pp. 255–266.

[285] ———, *Near Equivalence of Network Flow Algorithms*, Technical Report, Department of Operations Research, Stanford University, Stanford, CA, 1979.

[286] W. I. ZANGWILL, *Algorithm for the Chebyshev problem*, Management Science, 14 (1967), pp. 58–78.

[287] Y. ZHANG AND R. A. TAPIA, *A superlinearly convergent polynomial primal-dual interior-point algorithm for linear programming*, SIAM Journal on Optimization, 3 (1993), pp. 118–133.

Index

An italic *e* or *f* following a page number indicates that an example or a figure appears on that page.